Horses:

American Association of Equine Practitioners	www.aaep.org
American Horse Council	www.horsecouncil.org
American Horse Publications	www.americanhorsepubs.org
Blood Horse	www.bloodhorse.com
Chronicle of the Horse	www.chronofhorse.com
Cyber Steed	www.cybersteed.com
Horse Breeds	www.ansi.okstate.edu/BREEDS/index.htm
Horse Country	www.horse-country.com
United States Pony Club	www.ponyclub.org
Western Horseman	www.westernhorseman.com

Poultry:

American Egg Board	www.aeb.org
Egg Nutrition Center	www.enc-online.org
Goldkist	www.goldkist.com
Pilgrim's Pride	www.pilgrimspride.com
Poultry Breeds	www.ansi.okstate.edu/BREEDS/index.htm
Poultry Information Network	www.wattnet.com
Poultry Internet Information Resources	www.oneglobe.com/agrifood/aginform/poultry
Ross Breeders	www.rossbreeders.com
Tyson Foods, Inc.	www.tyson.com
USA Poultry and Egg Association	www.poultryegg.org

Sheep:

American Sheep Industry Association	www.sheepusa.org
Australian Wool Exchange	www.awex.com.au
Lands' End, Inc.	www.landsend.com
Livestock Marketing Information Center	www.lmic.info
National Sheep Improvement Program	www.nsip.org
Pendleton Woolen Mills	www.pendleton-usa.com
Sheep Breeds	www.ansi.okstate.edu/BREEDS/index.htm
Woolmark Company	www.wool.com.au

Swine:

Iowa Pork Industry Center	www.extension.iastate.edu/ipic
Livestock Marketing Information Center	www.lmic.info
National Pork Producers Council	www.nppc.org
National Swine Registry	www.nationalswine.com
Purdue Pork Page	www.anr.ces.purdue.edu/anr/anr/swine/porkpage.htm
Swine Breeds	www.ansi.okstate.edu/BREEDS/index.htm
Swine Information Network	www.swine.net
Swine Testing and Genetic Evaluation System	www.ansi.purdue.edu/stages/index.htm

Scientific Farm
Animal Production

NINTH EDITION

Scientific Farm Animal Production

AN INTRODUCTION TO ANIMAL SCIENCE

THOMAS G. FIELD
Colorado State University

ROBERT E. TAYLOR
late, Colorado State University

PEARSON

Prentice
Hall

Upper Saddle River, New Jersey
Columbus, Ohio

Library of Congress Cataloging-in-Publication Data

Taylor, Robert E. (Robert Ellis)
  Scientific farm animal production: an introduction to animal science
  / Robert E. Taylor, Thomas G. Field.—9th ed.
    p. cm.
  ISBN 0-13-244736-3
  1. Livestock. I. Field, Thomas G. (Thomas Gordon) II. Title.
    SF61.T39 2008
    636—dc22

                                                                    2007017002

Editor in Chief: Vernon R. Anthony
Editorial Assistant: Sonya Kottcamp
Production Editor: Kevin Happell
Design Coordinator: Diane Ernsberger
Cover Designer: Aaron Dixon
Cover Art: Superstock
Production Manager: Deidra Schwartz
Director of Marketing: David Gesell
Marketing Manager: Jimmy Stephens
Marketing Assistant: Les Roberts

This book was set in Melior by Carlisle Publishing Services. It was printed and bound by Edwards Brothers. The cover was printed by Coral Graphics Services.

Credits and acknowledgments borrowed from other sources and reproduced, with permission, in this textbook appear on appropriate pages within text.

Pearson Prentice Hall™ is a trademark of Pearson Education, Inc.
Pearson® is a registered trademark of Pearson plc
Prentice Hall® is a registered trademark of Pearson Education, Inc.

Pearson Education Ltd.
Pearson Education Singapore, Pty. Ltd.
Pearson Education, Canada, Ltd.
Pearson Education—Japan

Pearson Education Australia Pty. Limited
Pearson Education North Asia Ltd.
Pearson Educación de Mexico, S.A. de C.V.
Pearson Education Malaysia, Pte. Ltd.

10 9 8 7 6 5 4 3 2 1
ISBN-13: 978-0-13-244736-2
ISBN-10:   0-13-244736-3

This book is dedicated to the many teachers and students who have invested themselves in the process of improving animal agriculture so that humanity might one day be free of hunger. Since the first edition, this book has been inspired by the marvelous relationships that exist when teachers and students are motivated to learn and discover.

Robert Taylor was my teacher and mentor. He was the embodiment of good stewardship. He was a great stockman who took care of the land, the herds, the people, and the abundant natural resources God has bestowed upon us. He nurtured his students while challenging them to always strive for excellence. He was a source of wisdom, a builder of community, and a visionary leader.

Finally, this work is dedicated to my sons, Justin, Sean, and Trae, who have been my greatest joy.

Contents

Preface

Scientific Farm Animal Production is distinguished by an appropriate coverage of both breadth and depth of livestock and poultry production and their respective industries. The book gives an overview of the biological principles applicable to the animal sciences with chapters on reproduction, genetics, nutrition, lactation, consumer products, and other subjects. The book also covers the breeding, feeding, and management of beef cattle, dairy cattle, horses, sheep, swine, poultry, goats, and aquaculture. Although books have been written on each of these separate topics, the authors have highlighted the significant biological principles, scientific relationships, and management practices in a condensed but informative manner.

TARGET AUDIENCE

This book is designed as a text for the introductory animal science course typically taught at universities and junior or community colleges. It is also a valuable reference book for livestock producers, vocational agriculture instructors, and others desiring an overview of livestock production principles and management. The book is basic and sufficiently simple for the urban student with limited livestock experience, yet challenging for the student who has a livestock production background.

KEY FEATURES

Chapters 1 through 9 cover animal products; Chapters 10 through 22 discuss the biological principles; and livestock, poultry, and aquaculture management practices and issues are presented in Chapters 23 through 38.

The glossary of terms used throughout the book has been expanded so that students can readily become familiar with animal science terminology. Many of the **Key Terms** in the text are included in the glossary. Additionally, key words are provided at the end of each chapter as an aid to student learning.

Photographs and line drawings are used throughout the book to communicate key points and major relationships. If "a picture is worth a thousand words," the numerous photographs and drawings expand the usefulness of the book beyond its pages.

Selected references are provided for each chapter to direct students into greater depth and breadth as they become intrigued with certain topics. Instructors can also use the references to expand their knowledge in current background material.

ACKNOWLEDGMENTS

Appreciation is expressed to the reviewers of the eighth edition, who offered suggestions to strengthen the book. Mike Amstutz, The Ohio State University—ATI; W.E. Beal, Virginia Polytechnic Institute and State University; Keith Cummins; Lee Denzer, Black Hawk College; Nathaniel Shelton; Elizabeth L. Walker, Missouri State University; and Weston Walker, Missouri State University.

About the Authors

Dr. Field was raised on a Colorado cow-calf and seedstock enterprise. He managed a seedstock herd of cattle after completing his B.S. degree. A competitive horseman as a youth, he has had practical experience with seedstock cattle, commercial cow-calf production, stockers, and horses. He has a B.S., M.S., and Ph.D. in animal science from Colorado State University.

Dr. Field has received teaching awards from the USDA National Excellence in Teaching program, the National Association of Colleges and Teachers of Agriculture, the American Society of Animal Sciences, and Colorado State University.

Dr. Taylor was raised on an Idaho livestock operation where several livestock species were produced. He received a B.S. degree in animal husbandry and a Master's degree in animal production from Utah State University. This background, combined with his Ph.D. work in animal breeding and physiology from Oklahoma State University, provided much depth to his knowledge of livestock production. He had practical production experience with beef cattle, dairy cattle, horses, poultry, sheep, and swine.

Dr. Taylor received teaching awards at Iowa State University (where he also managed a swine herd) and at Colorado State University. He also received the Distinguished Teaching Award from the American Society of Animal Science in recognition of his ability to organize and present materials to students. Many of his concepts for effective teaching are used in this book. Dr. Taylor passed away in 1998.

CHAPTER 1

Animal Contributions to Human Needs

LEARNING OBJECTIVES

- Describe the global distribution of livestock
- Quantify the role of animal products in the global food supply
- Evaluate differences in food production and agricultural productivity between developed and developing nations
- Compare food expenditures for at-home and away-from-home consumption in the United States
- Compare food consumption across diverse nations and cultures
- Describe changes in the U.S. agricultural productivity
- Describe the nonfood contributions of livestock

Humans domesticated animals some 6,000 to 10,000 years ago and, thus, began a fascinating relationship that has evolved over the centuries to provide food, shelter, power, clothing, fuel, and companionship. In many ways the history of civilization is told in the application of human creativity to the task of feeding, clothing, and raising the standard of living for the world's various societies via animal agriculture.

Wide differences have developed among people in various regions of the world in using agricultural technology, management practices, and production and distribution systems to improve their standard of living. But, in all countries, domestic livestock are a source of food, clothing, by-products used for consumer goods and animal feeds, draft power, manure for fuel (Fig. 1.1) and fertilizer, information from research using experimental animals, sport and recreation, and pleasure for those who keep animals. Table 1.1 outlines the major domesticated livestock species, their approximate numbers, and their primary uses. Chickens are the most numerous (16.7 billion), followed by cattle (1.35 billion), sheep (1.08 billion), and swine (960 million).

FIGURE 1.1   A load of cow dung en route to a market in India. Courtesy of R. E. McDowell, North Carolina State University.

TABLE 1.1   Major Domesticated Animal Species—Their Numbers and Uses in the World

Animal Species	World Numbers (mil)	Leading Countries or Areas with Numbers[a] (mil)	Primary Uses
Ruminants			
Cattle	1,355	India (332), Brazil (174), China (141), United States (97), Argentina (50)	Meat, milk, hides
Sheep	1,081	China (171), Australia (106), India (62), Iran (54), United States (6)	Wool, meat, milk, hides
Goats	808	China (196), India (120), Pakistan (57), Sudan (39), United States (3)	Milk, meat, hair, hides
Buffalo	174	India (98), China (23), Pakistan (26)	Draft, milk, meat, hides
Camels	19	Sudan (3), India (1), Mauritania (1), Pakistan (1)	Packing, transport, draft, meat, milk, hides
Nonruminants			
Chickens	16,740	China (4,360), United States (1,950), Indonesia (1,249), Brazil (1,100)	Meat, eggs, feathers
Swine	960	China (438), United States (60), EU-25 (149), Russian Fed. (36)	Meat
Turkeys	280	United States (88), France (35), Italy (26), Chile (26)	Meat, eggs, feathers
Ducks	1,046	China (725), Vietnam (50), Indonesia (34), United States (7)	Meat, eggs, feathers
Horses	55	China (8), Mexico (6), Brazil (6), United States (5), Argentina (4)	Draft, riding, sport, occasionally meat
Donkeys	41	China (9), Ethiopia (5), Pakistan (4), Mexico (3), Egypt (3)	Draft, transport
Mules	12	China (4), Mexico (3), Brazil (1)	Draft, transport

[a] U.S. numbers are given for comparison; may not always be among the leading countries.
Source: Adapted from USDA, FAO.

CONTRIBUTIONS TO FOOD NEEDS

When opportunity exists, most humans consume both plant and animal products (Fig. 1.2). Meat and dairy products are nearly always consumed in quantity when available. Animal protein availability in most countries is closely related to the economic status of the people and their agricultural technology. Vegetarianism in

FIGURE 1.2 Animal food products such as meat, milk, and eggs are preferred foods in countries with high standards of living. Photo courtesy of Certified Angus Beef LLC. *Certified Angus Beef*® is a registered trademark of Certified Angus Beef LLC and is used with permission.

countries such as India may be the long-term result of intense population pressures and scarcity of feed for animals because of competition between humans and animals for food. Rising population pressures, particularly in developing regions (Southeast Asia, Africa, and Latin America), force people to consume foods primarily of plant origin. Some major groups in human society practice vegetarianism for ethical reasons. In the Buddhist philosophy and some religions of India, for example, all animal life is considered sacred.

The contribution of animal products to the **per-capita calorie and protein supply** in food is shown in Table 1.2. Animal products constitute approximately 16% of the calories, 37% of the protein, and 45% of the fat in the total world food supply. Large differences exist between developed countries and developing countries in total daily supply of calories, protein, and fat.

For example, consumers in developed nations derive 26% of their calories from animal products with just over one-half of their total protein and fat supply from animal products. Consumers in developing nations derive 13% of their calorie supply, 29% of their protein, and 41% of their fat from animal products. The United States ranks higher than the world average for percent of calories and protein from animal sources but about average for percent of fat from animal products.

Changes in per-capita calorie supply and protein supply during the past 40 years are shown in Figure 1.3. Per-capita caloric supplies of both calories and protein have increased in most areas of the world. The contribution of animal products to the per-capita protein supply has increased in most of the world. In China, for example, the percent of calories has increased from animal products from 13% in 1994 to 22% in 2005, and the percent of protein from animal sources has increased from 24% in 1994 to 40% in 2005. These rises have resulted from a significant effort to increase livestock and poultry production. In the five-year period from 2000 to 2005, China added 35 million head of cattle, 38 million head of sheep, 39 million goats, 589 million chickens, and 39 million ducks to its national livestock inventory.

TABLE 1.2   Animal Product Contribution to Per-Capita Calorie, Protein, and Fat Supply

Country	Total Kilo Calories	Animal Products Kilo Cal	%	Total Protein (g/day)	Animal Products g/Day	%	Total Fat	Animal Products g/Day	%
Australia	3,176	1,049	33	107	71	66	138	72	52
Bangladesh	2,103	67	3	45	6	13	22	4	18
Brazil	2,985	615	21	80	41	51	89	42	47
China	3,029	583	19	85	30	35	84	49	58
Egypt	3,346	256	8	93	18	19	60	18	30
Germany	3,451	1,035	30	95	57	60	152	82	54
India	2,428	194	8	57	10	17	48	13	27
Japan	2,762	569	21	92	51	55	83	35	42
Kenya	1,965	234	12	50	15	30	47	15	32
Mexico	3,165	583	18	88	38	43	89	4	46
Nigeria	2,850	87	3	65	8	12	68	6	7
United Kingdom	3,334	1,002	30	98	55	56	145	77	53
United States	3,772	1,043	27	114	73	64	151	71	47
Developed	3,260	857	26	99	55	55	119	62	52
Developing	2,679	348	13	69	22	29	63	26	41
World Average	2,805	459	16	76	28	37	75	34	45

Source: Adapted from USDA, FAO.

The large differences among countries in the importance of animal products in their food supply can be partially explained by available resources and development of those resources. Most countries with only a small percentage of their population involved in agriculture have higher standards of living and a higher per-capita consumption of animal products. Comparing Table 1.3 with Table 1.2, note that the countries in Table 1.3 are listed by percentage of population involved in agriculture.

Agriculture **mechanization** (note tractor numbers in Table 1.3) has been largely responsible for increased food production and allowing people to turn their attention to professions other than production agriculture. This facilitates the provision of many goods and services, raises standards of living, and allows for the creation of more diverse economies. Note that 50% of the people in developing nations once engaged in agriculture while only 7% of the citizens in developed countries are active in the agricultural sector.

The tremendous increase in the productivity of U.S. agriculture (Table 1.4) has lowered the relative cost of food as vividly demonstrated in Table 1.5. Historical data show that agricultural productivity doubled in the 100-year span of 1820–1920. For example, at the turn of the century a team of horses, one handler, and a moldboard plow could plow 2 acres per day. Today, one tractor pulling three plows, each with five moldboards, plows 110 acres per day, accomplishing the work that once required 110 horses and 55 workers.

Livestock productivity since 1925 has progressively increased to extraordinary levels. The mix of animal enterprises on U.S. farms has shifted from a typical situation involving a vast number of species being raised on an average farm in the 1920s to contemporary scenarios where animal agriculture is considerably more specialized. These improvements in productivity have occurred primarily because people had an incentive to progress under a free-enterprise system.

In the United States, releasing people from producing their own food has given them the opportunity to improve their per-capita incomes. Increased

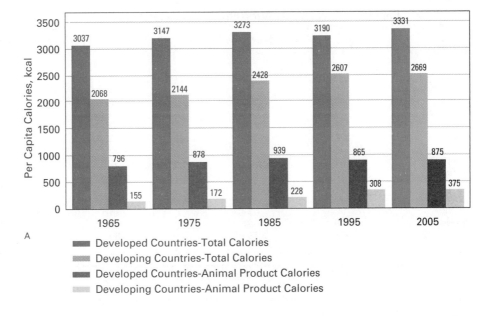

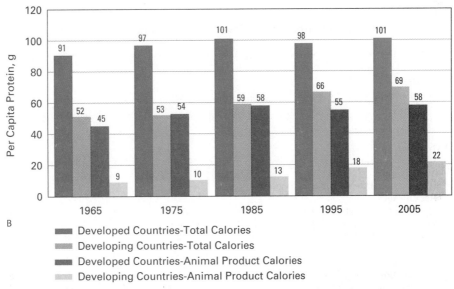

FIGURE 1.3 Caloric and protein intake from animal products. Adapted from USDA.

per-capita income associated with an abundance of animal products has re-sulted in reduced relative costs of many animal products with time. United States consumers allocate a smaller share of their disposable income for food than do people in many other countries. For example, per-capita expenditures for food as a percent of household expenses in Canada, France, Mexico, South Africa, and India are 9.9, 14.8, 24.5, 26.6, and 39.7%, respectively.

Table 1.6 shows that cereal grains are the most important source of energy in world diets. The energy derived from cereal grains, however, is twice as impor-tant in developing countries (as a group; there are exceptions) as in developed

TABLE 1.3   Population Involved in Agriculture in Selected Countries

Country	Population (mil)	Population in Agriculture[a] (mil)	Percent of Economically Active Population in Agriculture[b]	Number of Tractors (thou)
United States	294	5.9	2	4,760
United Kingdom	59	1.0	2	500
Germany	82	21.8	2	944
Australia	19	0.8	4	315
Japan	128	4.1	3	2,028
Brazil	178	26.4	15	806
Mexico	103	22.4	22	325
Nigeria	124	37.9	31	30
China	1,312	851.0	65	995
India	1,065	556.6	52	2,528
Bangladesh	147	77.4	53	5
Kenya	32	23.7	74	13
Developed	1,329	90.9	7	19,216
Developing	4,973	2,504.0	50	8,409
World Total	6,056	2,567	41	27,625

[a] *Agricultural population* is defined as all persons depending for their livelihood on agriculture. This comprises all persons actively engaged in agriculture and their nonworking dependents.
[b] Includes all economically active persons engaged principally in agriculture, forestry, hunting, or fishing.
Source: Adapted from USDA and FAO.

TABLE 1.4   Productivity Changes in Several Farm Animal Species in the United States

Species and Measure of Productivity	1925	1950	1975	1990	1995
Beef cattle					
Carcass weight (per year) marketed per breeding female (lb)	220	310	482	524	540
Sheep					
Liveweight marketed per breeding female (lb)	60	90	130	145	145
Dairy cattle					
Milk marketed per breeding female (lb)	4,189	5,313	10,500	14,000	16,400
Swine					
Liveweight marketed per breeding female (lb)	1,600	2,430	2,850	3,500	4,590
Broiler chickens					
Age to market weight (weeks)	15.0	12.0	7.5	7.4	7.3
Feed per pound of gain (lb)	4.0	3.3	2.1	1.9	1.8
Liveweight at marketing (lb)	2.8	3.1	3.8	4.5	5.09
Turkeys					
Age to market weight (weeks)	34	24	19	16	16
Feed per pound of gain (lb)	5.5	4.5	3.1	2.6	2.5
Liveweight at marketing (lb)	13.0	18.6	18.4	21.1	23.1
Laying hens					
Eggs per hen per year (no.)	112	174	232	250	254
Feed per dozen eggs (lb)	8.0	5.8	4.2	4.0	3.2

Source: Adapted from *Food from Animals*, CAST Report 82.

countries. Table 1.6 also illustrates that meat and milk are the major animal products contributing to the world supply of calories and protein.

Most of the world meat supply comes from cattle, swine, sheep, goats, chickens, and turkeys. There are, however, 20 or more additional species that collectively contribute about 6.5 billion pounds of edible protein per year or approximately 10% of the estimated total protein from all meats. These include the alpaca, llama, yak, horse, deer, elk, antelope, kangaroo, rabbit, guinea pig,

TABLE 1.5 Expenditures for Food in the United States (gross dollars and as percent of personal disposable income)

Year	At Home ($ bil)	At Home (%)	Away from Home ($ bil)	Away from Home %	Total ($ bil)	Total (%)
1930	15.3	19.3	2.7	3.4	18.0	22.7
1940	11.8	17.9	2.5	3.7	14.3	21.5
1950	32.2	18.1	7.5	4.2	39.7	22.3
1960	50.3	15.0	11.8	3.5	62.2	18.6
1970	73.7	11.4	25.1	3.9	98.9	15.3
1980	170.8	10.4	81.1	4.9	251.9	15.4
1985	221.1	8.9	123.5	5.0	344.5	13.9
1990	291.9	8.1	184.5	5.1	476.5	13.2
1995	320.0	7.1	221.4	4.9	541.7	12.1
2000	420.0	5.8	290.1	4.0	710.2	9.9
2005	524.3	5.8	369.8	4.1	894.1	9.9

Source: USDA.

TABLE 1.6 Contributions of Various Food Groups to the World Food Supply

Food Group	Calories (%)	Protein (%)
Cereals	50	45
Roots, tubers, pulses	8	7
Nuts, oils, vegetable fats	11	4
Sugar and sugar products	8	2
Vegetables and fruits	7	5
All animal products	16	37
Meat	*7*	*16*
Eggs	*1*	*3*
Fish	*1*	*7*
Milk and dairy	*5*	*10*
Other	*2*	*1*

Source: Adapted from USDA and FAO.

capybara, fowl other than chicken (duck, turkey, goose, guinea fowl, pigeon), and wild game exclusive of birds. For example, the Russian Federation cans more than 110 million pounds of reindeer meat per year, and in West Germany the annual per-capita consumption of venison exceeds 3 pounds. Peru derives more than 5% of its meat from the guinea pig.

Meat is important as a food for two scientifically based reasons. The first is that the assortment of amino acids in animal protein more closely matches the needs of the human body than does the assortment of amino acids in plant protein. The second is that vitamin B_{12}, which is required in human nutrition, may be obtained in adequate quantities from consumption of meat or other animal products but not from consumption of plants.

Milk is one of the largest single sources of food from animals. In the United States, 99% of the milk supply comes from cattle, but on a worldwide basis, milk from other species is important. Domestic buffalo, sheep, goat, alpaca, camel, reindeer, and yak supply significant amounts of milk in some countries. Milk and products made from milk contribute protein, energy, vitamins, and minerals for humans.

Besides the nutritional advantages, a major reason for human use of animals for food is that most countries have land areas unsuitable for growing cultivated crops. Approximately two-thirds of the world's agricultural land is permanent pasture, range, and meadow; of this, about 60% is unsuitable for producing cultivated crops that would be consumed directly by humans. This land, however, can produce feed in the form of grass and other vegetation that is digestible by grazing ruminant animals, the most important of which are cattle and sheep (Fig. 1.4).

FIGURE 1.4   Ruminant animals produce food for humans by utilizing grass, crop residues, and other forages from land that cannot produce crops to be consumed directly by humans. (A) Cattle grazing stubble in New South Wales, Australia. (B) Cattle grazing hillsides in Georgia. (C) Cattle grazing native range in Arizona. (D) Sheep utilizing corn stalk residue following grain harvest. Courtesy of Winrock International. (E) Sheep grazing native range. Courtesy of American Sheep Industry Association.

These animals can harvest and convert the vegetation, which is for the most part indigestible by humans, to high-quality protein food. In the United States, about 385 million acres of rangeland and forest, representing 44% of the total land area, are used for grazing. Although this acreage now supports only about 40% of the total cattle population, it could carry twice this amount if developed and managed intensively.

Ruminant **animal agriculture** therefore does not compete with human use for production of most land used as permanent pasture, range, and meadow. On the contrary, the use of animals as intermediaries provides a means by which land that is otherwise unproductive for humans can be made productive (Fig. 1.5).

People are concerned about energy, protein, population pressures (Fig. 1.6), and land resources as they relate to animal agriculture. Quantities of energy and protein present in foods from animals are smaller than quantities consumed by animals in their feed because animals are inefficient in the ratio of nutrients used to nutrients produced. More acres of cropland are required per person for diets high in foods from animals than for diets including only plant products. As a consequence, animal agriculture has been criticized for wasting food and land resources that could otherwise be used to provide persons with adequate diets. Consideration must be given to economic systems and consumer preferences to understand why agriculture perpetuates what critics perceive as resource-inefficient practices. These practices relate primarily to providing food-producing animals with feed that could be eaten by humans and using land resources to produce crops specifically for animals instead of producing crops that could be consumed by humans.

Hunger continues to be a challenge in some regions of the world. The factors that contribute to the hunger problem are varied and complex. Hunger takes two forms—chronic persistent hunger and famine. **Chronic persistent hunger (CPH)** results from a combination of poverty, climatic change, political instability, water shortages, loss of soil fertility, poor infrastructure (transportation, storage facilities, banking services, etc.), and illiteracy. Note that food scarcity is not a significant contributing factor to CPH. In fact, global food production has exceeded the population growth rate. **Famine**, unlike CPH, is typically a relatively short period of crisis resulting from the breakdown in food production and distribution infrastructure resulting from catastrophic events such as hurricanes, drought, or civil war. The international community is relatively adept at reacting to and minimizing the effects of famine.

The International Food Policy Research Institute suggests that while the number of malnourished children will decline from 1993 to 2020, there will still be 150 million babies and toddlers who are insufficiently fed in 2020. An additional 500 million people will also suffer from hunger. Africa, Latin America, the Caribbean, and West Asia are the regions most likely to bear the brunt of the problem in the future. Of the 11 countries with daily per-capita consumption of less than 2,000 calories, 10 of these are located in Africa.

Per-capita food availability is estimated to increase by nearly 7% by 2020, with China and East Asia experiencing the greatest increase. Evidence of a decline in global population increases is becoming apparent. However, slowing population increases is a gradual process, and for the next several decades approximately 80 million people will be added to the global population annually. Over 90% of this increase will occur in developing nations in or near urban areas. The 70 most susceptible countries to the effects of hunger are also the

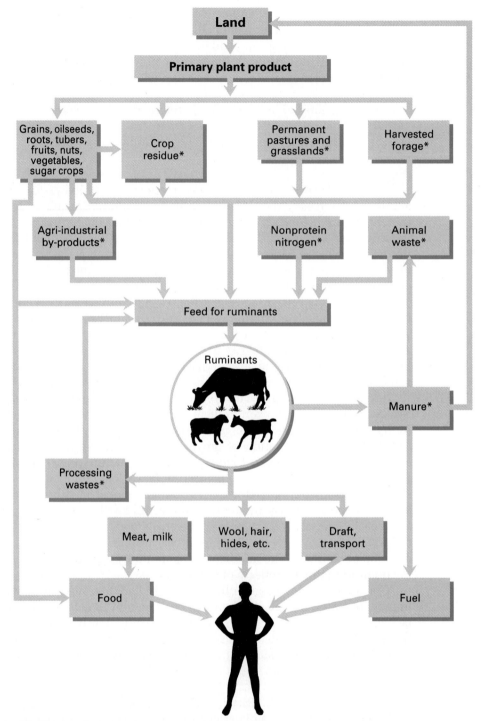

FIGURE 1.5 A graphic illustration of the land-plant-ruminant-animal-human relationship. Products marked with an asterisk (*) are not normally consumed by humans, but ruminants convert many of these products into useful products for humans. Courtesy of Winrock International.

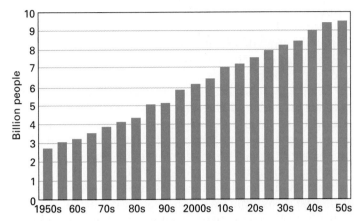

FIGURE 1.6 Past, present, and projected world population. *Source:* U.S. Census Bureau.

world's poorest nations. Sub-Saharan African nations have a per-capita income of approximately $380 per year.

Conquering hunger in developing nations involves a multifaceted strategy that includes increasing literacy rates, particularly in women; reducing poverty; improving health care; enhancing agricultural production; and improving the total food system infrastructure. As demand for food increases, some regions of the world will become more import dependent while others will become more export focused.

Even developed nations are not immune to the effects of hunger for part of the population. For example, 88% of U.S. households are **food secure,** where all members of the household have access to enough food for a healthy lifestyle and sufficient financial resources to acquire food, as it is needed. Households deemed **food insecure** (12%) are further divided into categories of without hunger (8%) or with hunger (4%). Food insecure households without hunger are able to gain access to groceries via food assistance programs or community outreach programs. Fortunately, only 0.7% of children in the United States live in food insecure environments coupled with periods of hunger. Those households considered food insecure typically have incomes below the poverty line of $19,000, and are comprised of single adults or single parents with children.

There will be annual demand increases of 1–1.5% for cereal grains, while worldwide demand for meat is expected to approach 2% per annum. Demand for meat will increase at the highest rate in developing countries (approximately 3%) with developed nations only accounting for less than 1% of annual demand growth. In the final analysis, the ability to pay will dictate food distribution. The need for economic growth in developing regions is of paramount importance.

Agriculture producers generate what consumers want to eat as reflected in the prices consumers are able and willing to pay. Eighty-five percent of the world's population desires food of animal origin in its diets, perhaps because foods of animal origin are considered more palatable than foods from plants. In most countries, as per-capita income rises, consumers tend to increase their consumption of meat and animal products.

If many consumers in countries where animal products are consumed at a high rate were to decide to eat only food of plant origin, consumption and price of foods from plants would increase, and consumption and price of foods from animals would decrease. Agriculture would then adjust to produce greater quantities of food from plants and lesser quantities of food from animals. Ruminant animals can produce large amounts of meat without grain feeding. The amount of grain feeding in the future will be determined by cost of grain and the price consumers are willing to pay for meat.

Some people advocate shifting from the consumption of foods from animals to foods from plants. They see this primarily as a moral issue, believing it is unethical to let people elsewhere in the world starve when our own food needs could be met by eating foods from plants rather than feeding plants to animals. The balance of plant-derived foods could then be sent abroad. These people believe that grain can be shipped with comparative ease because a surplus of grain exists in many developed nations and because any surplus should be provided at no cost. Providing free food to other countries has met with limited success in the past. In some situations, it upsets their own agricultural production, and in many cases the food cannot be adequately distributed in the recipient country because transportation and marketing systems are poorly developed.

There are strong feelings that the United States has a moral obligation to share its abundance with other people in the world, particularly those in developing countries. It appears that sharing our time and basic—but not necessarily advanced—technology can best do this. People need to have self-motivation to improve, access to knowledge and appropriate technology, and sufficient resources to develop agricultural productivity and infrastructure appropriate to their own culture.

Advances in agricultural production and related topics must be shared to minimize the effects of hunger on civilization. These achievements have been built on knowledge gained through experience and research, the extension of knowledge to producers, and the development of an industry to provide transportation, processing, and marketing in addition to production. Dwindling dollars currently being spent to support agricultural research and extension of knowledge may not provide the technology needed for future food demands. The next generation of agricultural leaders should view the decline in resources allocated to agricultural research, extension, and education as a potential crisis.

About 20% of the world human population and 32% of the ruminant animal population live in developed regions of the world, but ruminants of these same regions produce two-thirds of the world's meat and 80% of the world's milk. In developed regions, a higher percentage of animals are used as food producers, and these animals are more productive on a per-animal basis than animals in developing regions. This is the primary reason for the higher level of human nutrition in developed countries of the world.

Possibly many developing regions of the world could achieve levels of plant and animal food productivity similar to those of developed regions. Except perhaps in India, abundant world supplies of animal feed resources that do not compete with production of food for people are available to support expansion of animal populations and production. It has been estimated that through changes in resource allocation, an additional 8 billion acres of arable land (twice what is now being used) and 9.2 billion acres of permanent pasture and meadow (23% more than is now being used) could be put into production in the world. These estimates, plus the potential increase in productivity per acre and per animal in developed countries, demonstrate the magnitude of world food-production potential. This potential cannot be realized, however, without coordinated planning and increased incentive to individual producers.

Fortunately progress can be made in reducing hunger. For example, Asia has reduced hunger (percent of population consuming less than 2,100 calories per day) by 30% in the 10-year period from 1994 to 2004. During the same time, Bangladesh reduced the number of hungry people by 70% by making significant strides in food production and distribution. This is particularly impressive given that Bangladesh was once considered the epicenter of famine and chronic persistent hunger. Interestingly, significant changes in governmental policy focused not only on increasing food production but also on enhancing exports as a means to infuse foreign exchange into the economy. Furthermore, government policy focused on private-sector investment in irrigation systems, seeds, and fertilizer to stimulate food production. These policies increased irrigated acreage by 50% from 1994 to 2004.

In the long run, each nation must assume the responsibility of producing its own food supply by efficient production, barter, or purchase and by keeping future food-production technology ahead of population increases and demand. Extensive untapped resources that can greatly enhance food production exist throughout the world, including an ample supply of animal products. The greatest resource is the human being, who can, through self-motivation, become more productive and self-reliant.

CONTRIBUTIONS TO CLOTHING AND OTHER NONFOOD PRODUCTS

Products other than food from ruminants include wool, hair, hides, and pelts. Synthetic materials have made significant inroads into markets for these products. For example, the world's production of wool peaked in 1990 but since has declined to 40-year lows. It is important to note that in more than 100 countries, ruminant fibers are used in domestic production and cottage industries for clothing, bedding, housing, and carpets.

Annual production of animal wastes from ruminants contains millions of tons of nitrogen, phosphorus, and potassium. The annual value of these wastes for fertilizer is estimated at more than $1 billion.

Inedible tallow and greases are animal **by-products** used primarily in soaps and animal feeds and as sources of fatty acids for lubricants and industrial use. Additional tallow and grease by-products are used in the manufacture of pharmaceuticals, candles, cosmetics, leather goods, woolen fabrics, and tin plating. The individual fatty acids can be used to produce synthetic rubber, food emulsifiers, plasticizers, floor waxes, candles, paints, varnishes, printing inks, and pharmaceuticals.

Gelatin is obtained from hides, skins, and bones and can be used in foods, films, and glues. Collagen, obtained primarily from hides, is used to make sausage casings.

CONTRIBUTIONS TO WORK AND POWER NEEDS

The early history of the developed world abounds with examples of the importance of animals as a source of work energy through draft work, packing, and human transport. The horse made significant contributions to winning wars and exploration of the unknown regions of the world.

In the United States during the 1920s, approximately 25 million horses and mules were used, primarily for **draft purposes.** The tractor has replaced all but a few of these draft animals. In parts of the developing world, however, animals provide as much as 99% of the power for agriculture even today.

In more than half the countries of the world, animals—mostly buffalo and cattle, but also horses, mules, camels, and llamas—are kept primarily for work and draft purposes (Fig. 1.7). About 20% of the world's human population depends largely or entirely on animals for moving goods. According to the Food and Agriculture Organization of the United Nations, in developing countries animals provide 52% of the cultivation power, with an additional 26% derived from human labor. Developed countries, in contrast, use tractors for 82% of the cultivation, with animals and humans providing 11% and 7% of the power, respectively. There are more than 3 times as many tractors and harvesting

A

B

C

FIGURE 1.7 Animals provide significant contributions to the draft and transportation needs of countries lacking mechanization in their agricultural technology. In developed countries, the use of draft animals is more oriented to recreation than necessity. (A) Cattle used for draft purposes in Honduras. Courtesy of Winrock International. (B) Donkeys being used to thresh barley. Courtesy of FAO. Photo by J. Bravo. (C) Carriage horses being used for show purposes.

machines and twice as many milking machines in use in developed nations as compared to developing countries. It is estimated that India alone would have to spend more than $1 billion annually for gasoline to replace the animal energy it uses in agriculture.

ANIMALS FOR COMPANIONSHIP, RECREATION, AND CREATIVITY

Estimates of the number of **companion animals** in the world are unavailable. There are an estimated 59 million family-owned dogs and 75 million family-owned cats in the United States, in addition to the animals identified in Table 1.1. Approximately one-half of all U.S. households have at least one pet dog or cat. The U.S. pet food industry annually processes more than 3 million tons of cat and dog food valued at more than $1 billion. Many species of animals would qualify as companions where people derive pleasure from them. The contribution of animals as companions, especially to the young and elderly, is meaningful, even though it is difficult to quantify the emotional value.

Animals used in rodeos, bullfighting, and other sports provide income for thousands of people and recreational entertainment for millions (Fig. 1.8). Numerous people who have made money from nonagricultural businesses have invested in land and animals for recreational and emotional fulfillment away from the urban environment.

Our livestock heritage involves the interactions of humans with animals over the centuries. Historically, animals have been highly respected, revered, and even worshiped by humans. Early humans expressed the sacred, mysterious qualities of some animals through art on cave walls between 15,000 and 30,000 B.C. Thus, through these early paintings and sculptures, animals found their way into an expression of the things of humans—the humanities. Several historians have noted that an extremely high form of art is the intelligent manipulation of animal life—the modeling and molding of different types through the application of breeding principles.

ADDITIONAL ANIMAL CONTRIBUTIONS

The use of livestock species in human health research is also of significance. Most human health research involving livestock is focused on smaller animals such as miniature pigs, swine, and sheep. Biomedical research using these species has focused on such topics as human aging, diabetes, arteriosclerosis, and development of replacement joints. This research is conducted under strict federal guidelines governing the care and use of animals in laboratory settings.

Transgenic technologies also offer significant potential in terms of utilizing livestock to produce specialized proteins for use in the creation of therapeutic drugs and other medical applications. This approach is often referred to as "pharming" or the use of agricultural animals to produce pharmaceuticals. Transgenic dairy cows and goats have been utilized to produce hepatitis B antigens, tissue plasminogen activator for treatment of heart disease, and the clotting agent antithrombin III for example.

A

B

C

FIGURE 1.8 (A) Many people enjoy riding horses as part of a back-country experience. (B) Traditional use of the horse and dog in simulated hunts continues to be popular in some parts of the world. (C) Polo is an action-packed sport enjoyed by many.

D

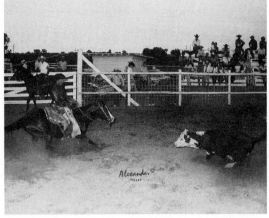

E

F

FIGURE 1.8, CONT'D   (D) The use of the stock dog is prevalent on many farms and ranches. Photo courtesy of Nancy Bard Duley. (E) The cutting horse event is founded on stock handling skills on the open range. Photo courtesy of Fred Field. (F) Team roping is a popular sport with increasing participation on the amateur and professional levels. Photo courtesy of Professional Rodeo Cowboys Association.

CHAPTER SUMMARY

■ Domesticated animals (>16 billion) contribute to the well-being of 5.6 billion humans throughout the world by providing food, clothing, shelter, power, recreation, and companionship.

■ Animal products contribute significantly to the world's human protein needs and energy supply.

■ As people increase their standard of living, the per-capita consumption of animal products also increases.

■ Human health is improved via the continuing research utilizing domestic animals.

KEY WORDS

per-capita calorie supply
per-capita protein supply
mechanization

animal agriculture
chronic persistent hunger (CPH)
famine
food secure

food insecure
by-products
gelatin
draft purposes
companion animals

REVIEW QUESTIONS

1. *True or False*: Developed nations have less of their population economically involved in agriculture than do developing nations.

2. *True or False*: People in developed nations consume more of their daily supply of protein and calories from animal products than do people from developing nations.

3. Compare the United States to other countries for percent of income spent on food.

4. What are the two scientifically based reasons why meat is important food for humans?

5. What was the single most important reason for increased food production in the mid 1900s?

6. What percentage of the world's agricultural land is unsuitable for cultivation of crops and is therefore used to pasture or graze livestock?

7. What are the basic human needs to which animals and animal products contribute?

8. Compare and contrast developed versus developing nations relative to agricultural productivity, management practices, and consumer diets.

SELECTED REFERENCES _____

Baldwin, R. L. 1980. *Animals, Feed, Food and People: An Analysis of the Role of Animals in Food Production.* Boulder, CO: Westview Press.

Devendra, C. 1980. Potential of sheep and goats in less developed countries. *J. Anim. Sci.* 51:461.

Economic Research Service: USDA. 2006. *Food Security in the United States: Conditions and Trends.* Washington, DC.

Meade, B., S. Rosen, and S. Shapouri. 2006. *Food Security Assessment, 2005.* Economic Research Service: USDA. Washington, DC.

National Research Council. 1983. *Little-Known Asian Animals with a Promising Economic Future.* Washington, DC: National Academy Press.

Pearson, R. A. 1994. Draft animal power. *Encyclopedia of Agriculture Science.* San Diego: Academic Press, Inc.

Pinstrup-Andersen, P., R. Pandya-Lorch, and M. W. Rosegrant. 1997. The world food situation: Recent developments, emerging issues, and long-term prospects. *Food Policy Report.* The International Food Policy Research Institute.

Reid, J. T., O. D. White, R. Anrique, and A. Fortin. 1980. Nutritional energetics of livestock: Some present boundaries of knowledge and future research needs. *J. Anim. Sci.* 51:1393.

Willham, R. L. 1985. *The Legacy of the Stockman.* Morrilton, AR: Winrock International.

CHAPTER 2

An Overview of the Livestock and Poultry Industries

LEARNING OBJECTIVES

- Quantify the economic impact of the U.S. livestock industry
- Describe the role of international trade in livestock and livestock products
- Overview the international and domestic beef, dairy, horse, poultry, sheep, goat, and swine industries
- Overview nontraditional livestock enterprises

It is important to see the broad picture of the livestock and poultry industries before studying the specific biological and economic principles that explain animal function and production. These industries are typically described with numbers of animals, pounds produced, production systems, prices, products, people (producers and consumers), and profitability. Products and consumers are covered in the next several chapters.

An understanding of the animal industries begins with basic terminology, especially the various species and sex classifications (see Table B at the end of this book). Refer to the glossary, which follows Chapter 38, for definitions of many commonly used animal terms.

Readers are advised to read Chapter 37 to gain a more in-depth understanding of the career opportunities associated with animal agriculture.

U.S. ANIMAL INDUSTRIES: AN OVERVIEW _____

Historically, farms in the United States were highly diversified in both crop and livestock enterprises. For example, during the depression most Iowa farms had cattle, chickens, hogs, and horses (Fig. 2.1). Over time, however, as fewer people chose to work in production agriculture, the industry had to develop more specialization in order to create the productivity required to meet consumer demand. The livestock and poultry industries in the United States must generate large volumes of output to meet the high animal product preference of more than 283 million U.S. consumers and also to supply the growing export market. Table 2.1 shows the number of livestock and poultry producers, the inventory number, and value of animals at one point in time during the year.

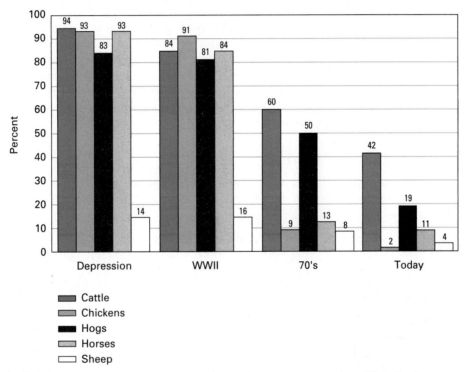

FIGURE 2.1   Percent of Iowa farms raising livestock species. *Source:* Adapted from Carolan, 2002.

TABLE 2.1   Numbers of Producers and Animals in the U.S. Livestock and Poultry Industries, 2006

Species	Number of Producers (1,000)	Number of Animals (mil head)	Cash Receipts ($ bil)
Beef cows	770.2	32.9	49.7
Swine	67.3	61.3	13.7
Dairy cattle	78.3	9.0	26.9
Broilers	NA	8.870	19.2
Sheep	68.2	4.6	0.5
Turkeys	NA	256.2	3.3
Goats—meat/milk/fiber	NA	2.8	NA

Source: Adapted from USDA.

Cash Receipts

An evaluation of farm **cash receipts** from the sale of animals and animal products provides another perspective of U.S. animal industries. Table 2.2 shows the cash receipts for animal commodities ranked against all agricultural commodities. The top five states for each commodity are also shown in this table. Note the cash receipts for all livestock products comprise 51% of all agricultural commodities in the United States.

Figure 2.2 shows the farm cash receipts from livestock and poultry products for each state. Note that 22 states each have annual cash receipts exceeding $2 billion.

World Trade

World trade influences profitability of U.S. animal industries. Table 2.3 shows the export and import markets for several animal commodities. While world trade for all U.S. products shows a deficit, agricultural and animal products show a positive trade balance. The economics of the U.S. export market for animal products is significantly influenced by cattle hides, meat (beef), fat/tallow, and dairy products. The highest import expenditures are for beef, dairy products, pork, and live cattle.

Commodity Prices

The profitability of U.S. animal industries is partly influenced by the prices paid to producers for animals and animal products. Prices can fluctuate

TABLE 2.2   Leading U.S. States for Farm Cash Receipts, 2005

Commodity	Rank[a]	Value ($ mil)	Five Leading States ($ mil)				
			1	2	3	4	5
All commodities	—	$241,241	CA $31,835	TX $16,498	IA $14,653	NE $11,780	MN $9,795
Livestock/poultry/products	—	123,481	TX 11,107	CA 8,623	NE 7,338	IA 7,284	KS 6,420
Cattle and calves	1	47,296	TX 7,580	NE 6,458	KS 6,089	CO 3,138	OK 2,697
Dairy products	2	27,368	CA 5,366	WI 3,688	NY 1,950	PA 1,769	MN 1,336
Broilers	4	20,446	GA 2,857	AR 2,731	AL 2,407	NC 2,042	MS 1,930
Hogs	7	14,348	IA 3,801	NC 2,079	MN 1,724	IL 1,028	NE 761
Chicken eggs	10	5,303	GA 394	AR 362	PA 340	OH 334	IN 292
Turkeys	14	2,996	MN 516	NC 449	MO 280	AR 287	VA 183
Horses	23	1,161	NA	NA	NA	NA	NA
Sheep/lambs	—[a]	521	CO 111	CA 58	TX 58	IA 35	SD 33

[a] Ranking is in comparison to all agricultural commodities. Ranking of sheep/lambs, wool, and mohair is below 25th.
Source: Adapted from USDA.

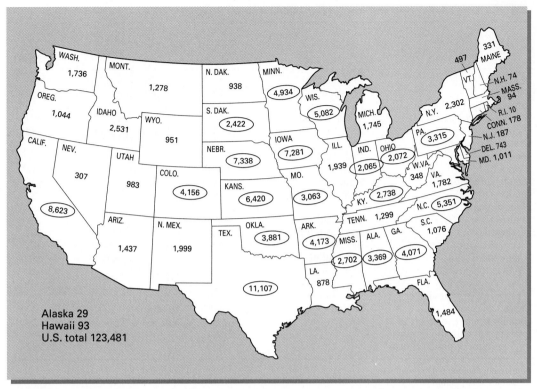

FIGURE 2.2 Cash receipts for livestock and products, 2004 ($ mil). *Source:* Adapted from USDA.

monthly, weekly, even daily. These price changes are influenced primarily by supply and demand.

Those individuals interested in animal profitability should know the average prices of animal products. In addition, an understanding of the prices for specific classes and grades of animals, of what causes prices to fluctuate, and of how high prices are obtained is also necessary. Figure 2.3 compares market prices of several livestock and poultry commodities. This overview shows average prices and variability in prices producers receive for the animals and products they sell. Details of how some factors influence prices are discussed in later chapters.

Biological Differences in Meeting Market Demand

Changes in consumer demand, feed prices, weather, and other factors dictate the need to increase or decrease animal numbers and amount of product produced. Figure 2.4 shows the large differences between some farm animal species and how quickly or slowly numbers can change. Broiler numbers can be increased or decreased in a couple of months, while several years are needed to make significant changes in cattle numbers.

TABLE 2.3   U.S. Exports and Imports of Major Animal Products, 2005

Commodity	Exports		Imports	
	Quantity	Value ($ mil)	Quantity	Value ($ mil)
Cattle and calves	21 (thou head)	7[a]	1,815 (thou head)	1,038
Hogs	149 (thou head)	26	8,190 (thou head)	598
Sheep, lambs, and goats	76 (thou head)	7	NA	NA
Horses, mules, burros	61 (thou head)	461	42 (thou head)	284
Poultry	49,221 (thou head)	105	17,931 (thou head)	35
Meats and meat products				
Beef and veal	1,043 (mil lb)	1,365	2,258 (mil lb)	3,658
Pork	1,995 (mil lb)	2,541	962 (mil lb)	1,313
Lamb and mutton	9 (mil lb)	10	184 (mil lb)	496
Poultry	5,857 (mil lb)	2,545	99 (mil lb)	137
Other products				
Tallow, grease, fat, and lard	2,286 (mil lb)	466	NA	NA
Variety meats	1,119 (mil lb)	765	NA	NA
Sausage casings	74 (mil lb)	107	NA	NA
Cattle hides	22.7 (mil pieces)	1,285	3.7 (mil pieces)[c]	79
Wool and mohair	21 (mil lb)	33	10 (mil lb)	26
Fluid dairy products	1,468 (mil lb)	16	NA	NA
Processed dairy products	2,177 (mil lb)	1,609	1,761 (mil lb)	2,592
Eggs (table)	70 (mil lb)	249	10 (mil lb)	22
Bovine semen and embryos	NA	64	NA	29
Total for Animals and Products		12,368		11,389
Total for Agricultural Products		62,398[b]		59,282

[a] Most major markets to U.S. beef closed.
[b] Grains are most important.
[c] All hides and skins.
Source: Adapted from USDA.

THE BEEF INDUSTRY

Global Perspective

Cattle were probably domesticated in Asia and Europe during the New Stone Age. Humped cattle *(Bos indicus)* were developed in tropical countries; the *Bos taurus* cattle were developed in more temperate zones.

Cattle, including the domestic water buffalo, contribute food, fiber, fuel, and draft animal power to the 5.8 billion people of the world. For most developed countries, beef (meat) is a primary product. For developing countries, beef is a secondary product as draft animal power and milk are the primary products. In some countries, cattle are still a mode of currency or a focus of religious beliefs and customs.

For 30 years cattle numbers have continued to increase due to (1) greater demand for beef in developing countries, and (2) increased export demand (the result of more liberal trade policies in some countries).

Table 2.4 shows the leading countries for cattle numbers, beef production, and beef consumption. India has the largest cattle population; however, its per-capita consumption is low because religious customs forbid cattle (considered sacred) from being slaughtered. The United States produces the most beef, but Uruguay and Argentina are higher in per-capita consumption.

Countries with a high cattle population relative to their human population typically have a high per-capita consumption of beef and high export tonnage.

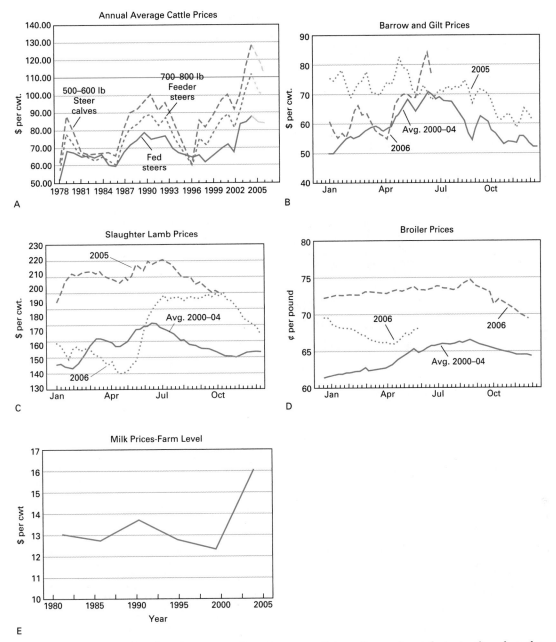

FIGURE 2.3   Livestock and product commodity prices are cyclic over time and contribute significantly to the economic challenges faced by agricultural producers. *Source:* Livestock Marketing Information Center, USDA.

For example, the cattle versus human population in Australia is 27.5 million versus 19 million, and in Argentina it is 50 million versus 37 million. Both countries rank high in per-capita consumption of beef. Japan, with its 4.5 million cattle, 127 million people, and rapidly expanding economy, has an increasing demand for beef (Table 2.5).

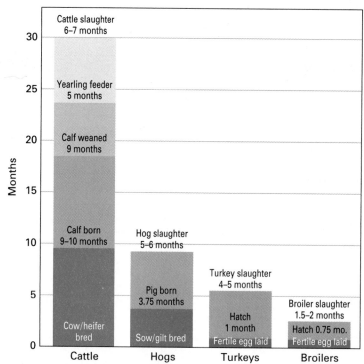

FIGURE 2.4   Biological time lag in changing numbers of several species of farm animals. *Source:* USDA.

TABLE 2.4   World Cattle Numbers, Production, and Consumption

Country	No. Cattle (mil head)	Country	Production (bil (lb))[a,b]	Country	Per-Capita Supply (lb)
1. India	332	1. United States	26.2	1. Argentina	124
2. Brazil	174	2. Brazil	16.6	2. Uruguay	120
3. China	141	3. China	13.7	3. United States	95
4. United States	97	4. Argentina	6.2	4. Australia	89
5. Argentina	50	5. Australia	4.6	5. Brazil	76
World Total	1,352	**World Total**	130	**World Average**[b]	23

[a] Does not include buffalo meat.

[b] Carcass weight.

Source: Adapted from USDA.

United States

The U.S. **beef industry** is made up of a series of producing, processing, and consuming segments that relate to each other but that operate independently. Table 2.6 identifies the various segments and the products they produce.

Figure 2.5 shows total cattle and amount of carcass beef produced in the United States from 1977 through 2002. Interestingly, 97 million head of cattle currently produce as much carcass beef as 120 million head produced in the 1970s. There are several reasons for this relatively high production of beef from fewer

TABLE 2.5 U.S. Beef Trade, 2003 (pre BSE)

Exports		Imports	
Country	($ mil)	Country	($ mil)
1. Japan	$1,391	1. Canada	$1,248
2. Mexico	876	2. Australia	917
3. South Korea	815	3. New Zealand	626
4. Canada	331	4. Brazil	203
5. Taiwan	76	5. Uruguay	430
World Total	3,031	**World Total**	3,651

Note: Due to loss of international markets at the end of 2003, the U.S. lost significant market share. By 2005 the U.S. exported only $8 million to Japan, $1 million to South Korea, and $42 million to Taiwan.
Source: Adapted from USDA and FAS.

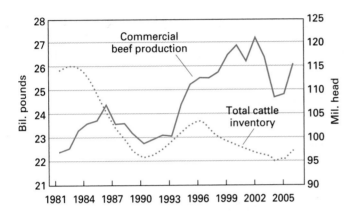

FIGURE 2.5 Total U.S. cattle inventory versus commercial beef production. *Source:* Livestock Marketing Information Center.

numbers of cattle: (1) the average carcass weight has increased from 579 lb in 1975 to 738 lb in 2001, (2) an increased number of cattle are fed per feedlot (2.4 times the feedlot capacity), (3) the market age of fed cattle has decreased, (4) more crossbreeding and faster-gaining European breeds (e.g., Simmental, Limousin, and Charolais) are being used in commercial breeding programs, and (5) more Canadian and Mexican calves are imported and fed in the United States.

Cattle Production

Most commercial beef cattle production occurs in three phases: the **cow-calf, stocker-yearling,** and **feedlot** operations. The cow-calf operator raises the young calf from birth to 6–10 months of age (400–650 lb). The stocker-yearling operator then grows the calf to 600–850 lb, primarily on roughage. Finally, the feedlot operator uses high-energy rations to finish the cattle to a desirable slaughter weight, approximately 900–1,300 lb. Most fed steers and slaughter heifers are between 15 and 24 months of age when marketed.

However, there are alternatives to the typical three-phase operation. In an integrated operation, for instance, the cattle may have a single owner from cow-calf to feedlot, or ownership may change several times before the cattle are ready for slaughter. Alternative production and marketing strategies are diagrammed in Figure 2.6.

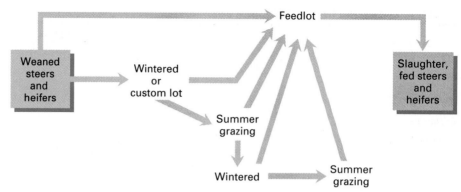

FIGURE 2.6   Alternative production and marketing strategies in the U.S. beef industry.

TABLE 2.6   **An Overview of the Beef Industry Segments in the United States**

Segment of the Beef Industry	Products Produced or Utilized	Approximate Number
Seedstock Producers ↓ ↑ ᵃ↓ ↑ᵇ	Breeding stock—primarily bulls (14–30 months old) and some heifers and cows. Some steers and heifers for feeding. Slaughter cows and bulls.	70,000 breeders 10 AI studs
Commercial Cow-Calf Producers ↓ ↑ ↓ ↑ ↓ ↑	Calves (6–10 months old), weighing 300–700 lb. Slaughter cows and bulls with majority over 5 years of age, weighing 800–1,500 lb (cows) and 1,000–2,500 lb (bulls).	774,630 producersᶜ
Yearling or Stocker Operator ↓ ↑ ↓ ↑	Feeder steers and heifers (most of them 12–20 months old, weighing 500–900 lb).	
Feeders ↓ ↑ ↓ ↑ ↓ ↑	Market steers, heifers, cows, and bulls (mostly steers and heifers 16–30 months old, weighing 900–1,400 lb).	1,781 feedlots with capacity >1,000 head
Packers ↓ ↑ ↓ ↑ ↓ ↑	Carcasses (approx. 600–800 lb). Boxed beef (carcasses into subprimal cuts).	706 packing plants
Retailers ↓ ↑ ↓ ↑ ↓ ↑	Retail cuts. By-products.	128,000 grocery stores
Consumers	Cooked products. By-products (leather, pharmaceuticals, variety meats, etc.).	300 million consumers (U.S.); 5.9 billion consumers (world)

ᵃ Animal and/or product flow.
ᵇ Demands and expectations.
ᶜ In addition, there are approximately 100,000 dairy farms that produce about 20% of U.S. beef.
Source: Field and Taylor, 2007.

Cow-Calf Production.   U.S. cow-calf production involves some 33 million head of beef cows that are distributed throughout the country. Most of the cows are concentrated in areas where forage is abundant. As Figure 2.7 shows, 16 states each have over 700,000 head of cows (75% of the U.S. total), most of them

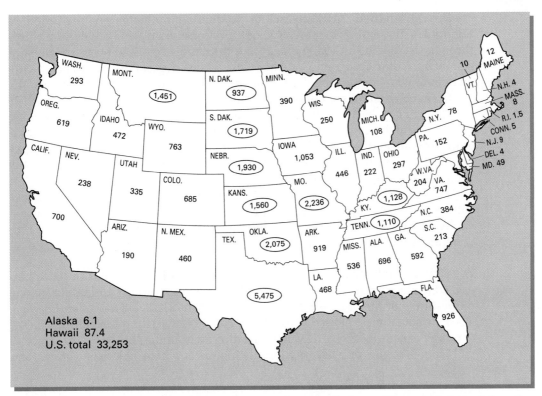

FIGURE 2.7 Beef cows that have calved January 1, 2006 (1,000 head). *Source:* Livestock Marketing Information Center.

TABLE 2.7 U.S. Beef Cow Operations and Inventory

Herd Size (No. Cows)	Operations Percent of Total	Inventory Percent of Total
1–49	78	28
50–99	12	19
100–499	9	38
500+	1	15

Source: Adapted from USDA.

located in the Plains, Corn Belt, and southeastern states. Approximately 60% of the 830,000 beef cow operations have less than 50 cows per operation. However, more than 70% of the beef cow inventory is in operations with more than 100 cows (Table 2.7). Cow numbers fluctuate over the years, depending on drought, beef prices, and land prices.

There are two kinds of cow-calf producers. Commercial cow-calf producers raise most of the potential slaughter steers and heifers. Seedstock breeders—specialized cow-calf producers—produce primarily breeding cattle and semen.

Stocker-Yearling Production.   Stocker-yearling producers feed cattle for growth prior to their going into a feedlot for finishing. Replacement heifers intended for the breeding herd are typically included in the stocker-yearling category. Our focus here, however, is on steers and heifers grown for later feedlot finishing.

Several alternate stocker-yearling production programs are identified in Figure 2.6. In some programs, a single producer owns the calves from birth through the feedlot-finishing phase, and the cattle are raised on the same farm or ranch. In other programs, one operator retains ownership, but the cattle are custom-fed during the growth and finishing phases. In still other programs, the cattle are bought and sold once or several times.

The primary basis of the stocker-yearling operation is to market available forage and high-roughage feeds, such as grass, crop residues (e.g., corn stalks, grain stubble, and beet tops), wheat pasture, and silage. Stocker-yearling operations also make use of summer-only grazing areas that are not suitable for the production of supplemental winter feed.

Stocker-yearling operations are desirable for early-maturing cattle. These cattle need slower gains to achieve heavier slaughter weights without being excessively finished. Larger-framed, later-maturing cattle usually are more efficient and profitable if they go directly to the feedlot after weaning.

Feedlot Cattle Production.   Feedlot cattle are fed in small pens or fenced areas, where harvested feed is brought to them. Some cattle are finished for market on pasture, but they represent only 10–15% of the slaughter steers and heifers. They are sometimes referred to as *nonfed cattle* because they are fed little, if any, grain or concentrate feeds. The cattle-feeding areas in the United States (Fig. 2.8) correspond to the primary feed-producing areas where cultivated grains and

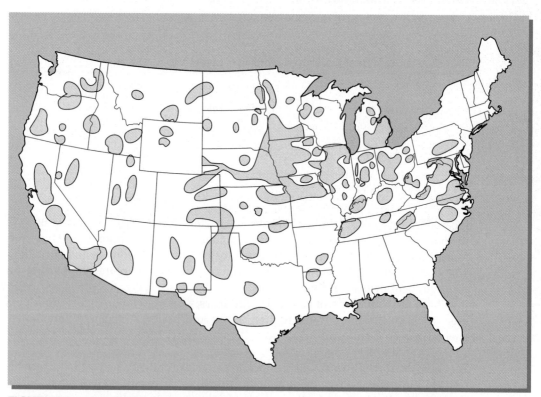

FIGURE 2.8   Cattle-feeding areas in the United States. The areas (in blue) represent location but not volume of cattle fed or backgrounded. *Source:* USDA.

roughage are grown. These locations are determined primarily by soil type, growing season, and amount of rainfall or irrigation water. Figure 2.9 shows where the approximately 23 million feedlot cattle are fed in the various states. The total number of fed cattle marketed in the leading states is shown in Figure 2.10. The 20.8 million head marketing by the leading states represent 97% of the 22.2 million head of fed cattle in all states. By contrasting Figures 2.9 and 2.10, the number of cattle marketed for each state is considerably higher than the number on feed. This is because most commercial feedlots feed more than 1.5 times their one-time capacity in a given year.

Cattle Feeding

The two basic types of cattle-feeding operations are (1) **commercial feeders** and (2) **farmer-feeders.** The two operations are distinguished by type of ownership and size of feedlot (Figs. 2.11 and 2.12).

The farmer-feeder operation is usually owned and operated by an individual or a family and has a feedlot capacity of under 1,000 head. An individual or partnership sometimes owns the commercial feedlot, but more often a corporation owns it, especially as feedlot size increases. It has a feedlot capacity of over 1,000 head. Approximately 95% of fed cattle are fed in feedlots with over 1,000-head capacity. A number of U.S. commercial feedlots have capacities of 40,000 head or higher, and a few have capacities of over 100,000 head. Some commercial feedlots custom-feed cattle; that is, the commercial feedlot provides the feed and feeding service to the cattle owner.

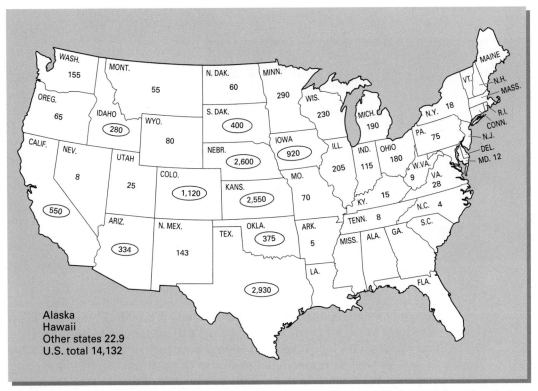

FIGURE 2.9   Cattle on feed January 1, 2006 (1,000 head). *Source:* Livestock Marketing Information Center.

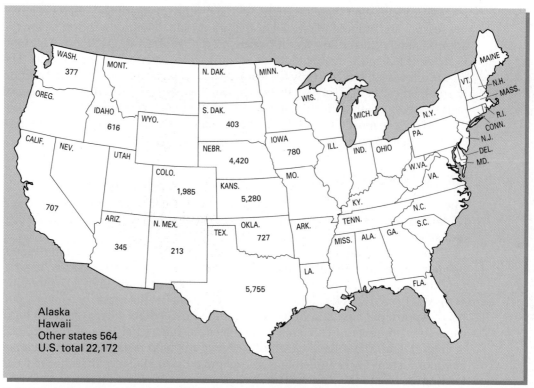

FIGURE 2.10   Fed cattle marketings 1,000+ head feedlots 2005 (1,000 head). *Source:* Livestock Marketing Information Center.

FIGURE 2.11   The feed mill associated with a large commercial cattle-feeding enterprise. Rations can be formulated and delivered to meet the needs of thousands of head of cattle.

Each type of feeding operation has its advantages and disadvantages. What is an advantage to one type of feedlot is usually a disadvantage to the other type. The large commercial feedlot usually enjoys economic advantages associated with its size as well as professional expertise in nutrition, health, marketing, and financing. The farmer-feeder has the advantages of distributing labor over several enterprises by using high-roughage feeds effectively, creating a market

FIGURE 2.12 Cattle on feed at a small farmer-feeder enterprise. The feed being consumed was raised on site and fed to cattle to add value to these crops.

TABLE 2.8 World Dairy Cattle Numbers, Milk Production, and Consumption

Country	No. Dairy Cattle (mil head)	Country	Fluid Milk Production (mil ton)	Country	Production per cow[a] (lb)
1. India	38	1. EU	119	1. Japan	20,105
2. EU	26	2. United States	73	2. United States	19,929
3. Brazil	15	3. India	35	3. Canada	16,138
4. Russian Fed.	10	4. China	31	4. EU	12,654
5. United States	9	5. Russian Fed.	28	5. Australia	11,596
World Total	225	**World Total**	481	**World Average**	4,275

[a] Whole and skim milk.
Source: Adapted from USDA.

of homegrown feeds through cattle, and more easily closing down the feeding operation during times of unprofitable returns.

Chapters 25 and 26 examine the breeding, feeding, and management of beef cattle in more detail.

THE DAIRY CATTLE INDUSTRY

Global Perspective

Milk and milk products are produced and consumed by most countries of the world. Although buffalo, goat, and other types of milk are important in some areas, our focus here is on cow's milk and its products.

Table 2.8 shows world dairy cattle numbers, fluid milk production, and per-capita consumption of milk. Of the 225 million dairy cows in the world, India leads all other countries with its 38 million. The United States has the highest total fluid milk production. The average annual world production per cow is 4,860 lb, with production in Israel the highest at nearly 22,000 lb. Countries with high milk-production levels per cow have excellent breeding, feeding, and health-management programs.

World butter production is 16.5 billion pounds. India, Pakistan, France, Germany, and the United States are the leading countries in total butter production, while New Zealand, France, Finland, and Germany have the highest per-capita consumption of butter.

The leading countries in cheese production are the United States, Germany, France, and Italy, with all countries producing 36 billion pounds of cheese. Greece, France, and Italy have the highest per-capita consumption of cheese.

Milk is produced worldwide, but manufactured dairy products dominate dairy trade. Dairy products traded internationally account for about 5% of the global milk production on a milk-equivalent basis. Trade restrictions keep trade volume and prices lower than if trade was more liberalized. Most developed countries, including the United States, extensively regulate their dairy industries by subsidizing production and often exportation.

Table 2.9 shows the U.S. import and export trade in dairy products. Cheese and nonfat dry milk are the most significant export items with cheese going to the Pacific Rim and NAFTA partners and dry milk to countries with minimal dairy production. Cheese and milk components are imported from countries in Europe and Oceania.

United States

The U.S. **dairy industry** has changed dramatically since the days of the family milk cow. Today it is a highly specialized industry that includes the production, processing, and distribution of milk. A large investment is required in cows, machinery, barns, and milking parlors where cows are milked. Dairy operators who produce their own feed need additional money for land on which to grow the feed. They also require machinery to produce, harvest, and process the crops.

Although the size of a dairy operation can vary from less than 30 milking cows to more than 5,000 milking cows (Fig. 2.13), the average U.S. dairy has approximately 100 milking cows, 30 dry cows, 30 heifers, and 25 calves. Average dairy producers farm 200–300 acres of land, raise much of the forage, and market the milk through cooperatives, of which they are members. The producers sell about 4,100 lb of milk daily, or about 1.5 million lb annually, valued at about $200,000. Their average total capital investment may exceed $500,000. The average dairy producer has a partnership (with a family member or another person) to make the management of time and resources easier.

Nearly 70% of the U.S. dairy herd was concentrated in large dairy farms (with 100 or more milk cows) in 2001. These large dairies represented just 20% of all U.S. farms with milk cows. New technologies have required extensive capital investment that is most feasible for large dairy operations. Since 1977, farms with fewer than 30 milk cows have declined continuously as a share of

TABLE 2.9 Value of U.S. Dairy Product Imports and Exports, 2001

Exports		Imports	
Commodity	($ mil)	Commodity	($ mil)
Nonfat dry milk	556	Nonfat dry milk	570
Cheese and curd	201	Cheese	201
Whey	203	Whey	223
Total	1,625	**Total**	1,681

Source: Adapted from USDA.

FIGURE 2.13 Dairies in the United States continue to increase in herd size. Many dairy cow herds are intensively managed in excellent facilities. Courtesy of Colorado State University.

all farms with milk cows. The share of farms having 100 or more milk cows is increasing in both number and share of all farms with milk cows. The largest farms are typically found in the West and Southwest. The traditional milk-producing states of the Northeast and Great Lakes states have seen their share of milk production become stable and then decline in recent years.

One way to stress the importance of dairying in the United States is to examine the number of cows in the states (Figure 2.14). The five leading states in thousands of dairy cows are, respectively, California (1,770), Wisconsin (1,240), New York (652), Pennsylvania (558), and Idaho (473). The five leading states in milk production per cow are, respectively, California (21,815 lb), Arizona (22,855 lb), Washington (23,055 lb), Colorado (23,155 lb), and New Mexico (21,515 lb). And the five leading states in total pounds of fluid milk produced (given in million pounds) are, respectively, California (38.3), Wisconsin (23.4), New York (12.0), Pennsylvania (10.7), and Minnesota (8.4).

As Figure 2.15 shows, today's 9 million dairy cows are approximately one-third the number of cows 50 years ago, yet total milk production continues to increase due to increased production per cow which is currently 19,951 pounds of milk per cow. This marked improvement is the result of effective breeding, feeding, health, and management programs.

Factors affecting dairy cattle productivity and profitability are covered in Chapters 27 and 28.

THE HORSE INDUSTRY

Global Perspective

The horse was domesticated about 5,000 years ago. It was one of the last farm animals to be domesticated. Horses were first used as food, then for war and sports, and also for draft purposes. They were used for transporting people

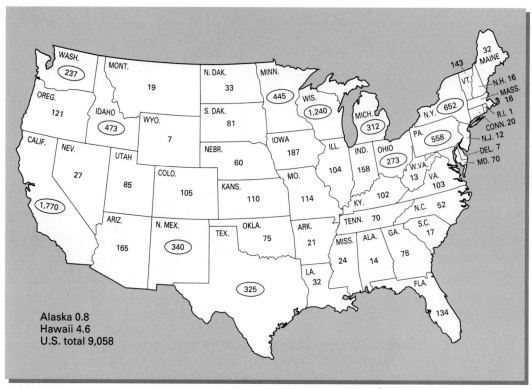

FIGURE 2.14   U.S. milk cow numbers, 2006 (1,000 head). The top 10 states are circled. *Source:* USDA.

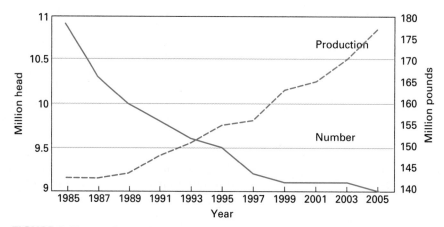

FIGURE 2.15   Trends in milk cow numbers and milk production. *Source:* USDA.

swiftly and for moving heavy loads. In addition, horses became important in farming, mining, and forestry.

The donkey, descended from the wild ass of Africa, was domesticated in Egypt prior to the domestication of the horse. Donkeys and zebras are in the same genus *(Equus)*, but are different species from horses (see Fig. 14.9, Chapter 14). Horses mate with donkeys and zebras but the offspring produced are sterile.

TABLE 2.10   World Horse, Donkey, and Mule Numbers

Horses		Donkeys		Mules	
Country	(mil head)	Country	(mil head)	Country	(mil head)
1. China	7.6	1. China	7.9	1. China	3.7
2. Mexico	6.3	2. Pakistan	4.1	2. Mexico	3.3
3. Brazil	5.9	3. Ethiopia	3.8	3. Brazil	1.3
4. United States	5.3	4. Mexico	3.2	4. Colombia	0.6
5. Argentina	3.6	5. Egypt	3.0	5. Morocco	0.5
World Total	55.0	**World Total**	40.7	**World Total**	12.4

Source: Adapted from FAO.

Table 2.10 shows the world numbers of horses (55 million), donkeys (41 million), and mules (13 million). The five leading countries are given for each of the three species, with China having the most of all three. Mexico and Brazil rank second and third in horse numbers, while the United States has approximately 5 million head.

Horses have been companions for people since their domestication. Once important in wars, mail delivery, farming, forest harvesting, and mining, the horse today is used in shows, in racing, in the handling of livestock, and for companionship, recreation, and exercise.

United States

There were no horses on the North American continent when Columbus arrived. However, there is fossil evidence that the early ancestor of the horse was here some 50–60 million years ago. The **Eohippus** (or dawn horse), a four-toed animal less than a foot high, is believed to be the oldest relative of the horse. Through evolutionary changes, the **Mesohippus** (about the size of a collie dog) was believed to have foraged on the prairies of the Great Plains.

The early ancestors of the horse disappeared in pre-Columbian times, supposedly by crossing from Alaska into Siberia. It is from these animals that horses may have evolved in Asia and Europe. The draft horses and Shetland ponies developed in Europe, whereas the lighter, more agile horses developed in Asia and the Middle East. The Spaniards and colonists introduced modern horses to the Americas.

In the early 1900s, there were approximately 25 million horses and mules in the United States. Shortly after World War I, however, horse numbers began a rapid decline. The war stimulated the development and use of motor-powered equipment, such as automobiles, trucks, tractors, and bulldozers. Railroads were heavily used for transporting people and for moving freight long distances. By the early 1960s, horses and mules had declined to a mere 3 million in the United States.

With the shorter work week and greater affluence of the working class, there has been more time and money available for recreation. An estimated 2 million American horse owners and an additional estimated 5 million people acting as service providers, employees, and volunteers are involved in the industry. Approximately 3.5% of consumer expenditures for recreation are spent in the **horse industry.** The U.S. horse business generates more than $25 billion in goods and services annually.

The American Horse Council estimates that there were 6.9 million horses in the United States in 1997, with about 43% utilized in recreational activities,

TABLE 2.11   Horse Population in Selected States by Primary Use

State	Racing	Showing (Percent)	Recreation	Other[a]
Texas	11	19	27	44
California	11	30	43	17
Florida	12	52	18	14
Oklahoma	20	34	10	36
Illinois	24	30	39	8
Colorado	10	26	29	35
Ohio	21	34	35	9
Kentucky	45	21	25	10
New York	18	27	37	18
U.S. Total	10	28	43	18

[a] Includes agricultural, other sports, nonspecific breeding, etc.
Source: Adapted from American Horse Council.

29% as show animals, 10% involved in racing, and the remainder utilized in rodeo, polo, ranch work, or other activities. The leading states for horse numbers are identified in Table 2.11. The export of horses generated more than $270 million in 1997 while imports were valued at $134 million.

Horse owners tend to be well-educated, upper-middle income (median income of $60,000), and middle-aged. However, approximately one-third of horse owners have an annual income of less than $50,000. The annual cost of horse maintenance ranges widely from $1,000 to $15,000 per horse depending on region and use. Monthly boarding costs (stall space, bedding and feed) are typically $150 to $250 per head. Training and riding lessons would typically run from $300 to $600 per month and $20 to $50 per hour, respectively. Horse owners typically pay $48 to $75 per horse for farrier service (horseshoeing and hoof care). The average cost of participating in a horse show ranges from $100 to $540 per show.

Horse numbers in the United States increased rapidly from 1960 to 1976, with slower growth after that time period. Horse numbers declined in the mid-1980s when applicable tax laws (that removed some of the advantages of horse ownership) were passed by Congress. Horse numbers are slowly rebuilding at the present time. Interest in diverse horse activities has risen as evidenced by increases in membership of the American Driving Society, the U.S. Combined Training Association, and the U.S. Dressage Federation by 4%, 14%, and 23%, respectively, over the past half-decade.

Horse breeding, feeding, and management practices are presented in Chapters 33 and 34.

THE POULTRY INDUSTRY

Global Perspective

The term *poultry* applies to chickens, turkeys, geese, ducks, pigeons, peafowls, and guineas. Chickens, ducks, and turkeys dominate the world poultry industry. In parts of Asia, ducks are commercially more important than broilers (young chickens), and in areas of Europe, there are more geese than other poultry because they are more economically important.

Chickens originated in Southeast Asia and were kept in China as early as 1400 B.C. Charles Darwin concluded in 1868 that domestic chickens originated

TABLE 2.12   Poultry and Egg Production

Broiler	(mil ton)	Turkey	(mil ton)	Eggs	(mil ton)
U.S.	14.7	U.S.	2.7	China	22.1
China	9.2	EU	2.1	U.S.	1.8
Brazil	8.5	Brazil	0.3	India	2.3
EU	6.9	Canada	0.2	Japan	2.2
Mexico	2.3			Russian Fed.	1.9
World	55.2	World	5.4	World	53.7

Source: Adapted from USDA.

TABLE 2.13   Poultry and Egg Consumption

All Poultry	Per capita (lbs)	Eggs	Per capita (lbs)
Netherlands Antilles	122	Japan	42
Saint Vincent	121	China	40
Bahamas	118	Denmark	38
Antigua	117	Hungary	36
U.S.	111	Paraguay	36
World	26	World	19

Source: Adapted from USDA and FAO.

TABLE 2.14   Poultry and Egg Imports-Exports for the United States

Commodity	Exports (mil $)	Imports (mil $)
Chicken	1,973	137[a]
Turkey	338	—
Eggs	249	22

[a] All poultry meat combined.
Source: Adapted from USDA.

from the Red Junglefowl, although three other Junglefowl species were known to exist.

Tables 2.12 and 2.13 show the leading countries in world poultry production and consumption. China has the highest poultry numbers, while the United States leads in poultry production. World hen egg production is approximately 54 million tons with China, the United States, and India as the leading producers. Eggs are used for both hatching and human consumption.

Japan, Hong Kong, Canada, the Russian Federation, and Mexico are the primary export markets for the U.S. **poultry industry** (Table 2.14).

World governments influence world poultry trade by controlling production and pricing and by placing **tariffs** on incoming goods—all barriers to free international trade. Trade liberalization by countries with industrial market economies would likely increase the trade of and decrease the price of poultry meat. Countries with efficient producers (such as the United States, Brazil, and Thailand), combined with consumers from countries with considerable trade protection (such as Japan, Canada, and the European Union), would benefit most from liberalized trade.

Poultry is the fastest-growing source of meat for people. The industrialized countries produce and export more than 50% of the poultry meat in the world.

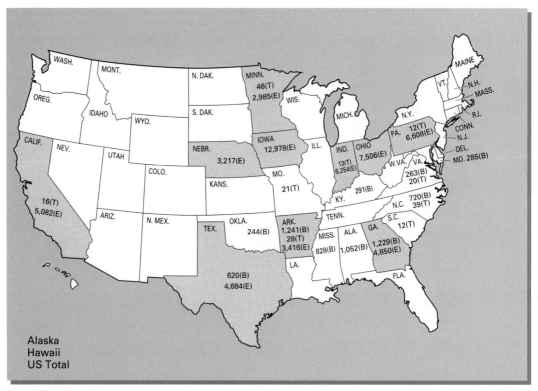

FIGURE 2.16   Top 10 states for broiler (B), turkey (T), and egg (E) production in the United States (million). *Source:* Adapted from USDA.

United States

The annual U.S. income from broilers, turkeys, and eggs exceeds $21 billion (broilers, $13.9 billion; eggs, $4.3 billion, and turkeys, $2.8 billion). Figure 2.16 shows the leading 10 states in broiler, turkey, and egg production. The U.S. poultry industry is concentrated primarily in the southeastern area of the country.

From 1900 to 1940, the primary concerns of the U.S. poultry industry were egg production by chickens and meat production by turkeys and waterfowl. Meat production by chickens was largely a by-product of the egg-producing enterprises. The **broiler** industry, as it is known today, was not yet established. Egg production was well established near large population centers, but the quality of eggs was often low because of seasonal production, poor storage, and the absence of laws to control grading standards.

Before 1940, large numbers of small farm flocks existed in the United States, but the management practices applied today were practically unknown. The modern mechanized poultry industry of the United States emerged during the late 1950s as the number of poultry farms and hatcheries decreased and the number of birds per installation dramatically increased. Larger cage-type layer operations appeared, and egg-production units grew, with production geared to provide consumers with eggs of uniform size and high quality. Huge broiler farms (Fig. 2.17) that provided consumers with fresh meat throughout the year were established, and large dressing plants capable of dressing 50,000 or more broilers daily were built. The U.S. poultry industry was thus revolutionized.

FIGURE 2.17   Modern broiler houses are ventilated facilities where broilers are raised from chicks to maturity. Each house holds thousands of broilers.

One of the most striking achievements of the poultry industry is its increased production of eggs and meat per hour of labor. Poultry operations with 1 million birds at one location are not uncommon. Automatic feeding, watering, egg collecting, egg packing, and manure removal have accomplished significant labor reduction.

Dramatic changes came between 1955 and 1975 with the introduction of integration. It began in the broiler industry and is applied to a lesser degree in egg and turkey operations. Integration brings all phases of an enterprise under the control of one head, frequently the corporate ownership of breeding flocks, hatcheries, feed mills, raising, dressing plants, services, and marketing and distribution of products (Fig. 2.18). After 1975, poultry entered a new period of consolidation, in which vertically integrated companies purchased, acquired, or merged with each other, creating a small number of superintegrated companies.

The actual raising of broilers is sometimes accomplished on a contract basis between the person who owns the houses and equipment and furnishes the necessary labor, and the corporation that furnishes the birds, feed, field service, dressing, and marketing. Payment for raising birds is generally based on a certain price for each bird reared to market age. Bonuses are usually paid to those who do a commendable job of raising birds. An important advantage to the integration system is that all phases are synchronized to ensure the utmost efficiency.

Broiler production is concentrated in the southern and southeastern states. California and Texas are the only states with large production located outside the "broiler belt." The "broiler belt" includes Alabama, Arkansas, Georgia, Mississippi, North Carolina, and Virginia. This region produces two-thirds of U.S. broilers.

Broiler production has increased tremendously over the past three decades, from 3.7 million pounds of ready-to-cook broilers in 1960 to 36.5 million

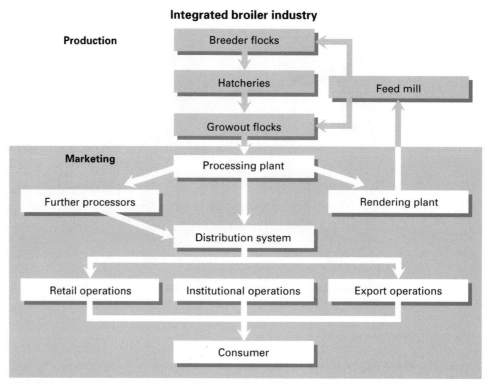

FIGURE 2.18   Structure of the integrated broiler industry.

pounds in 1996. The U.S. broiler industry is concentrating its growout operations into fewer but larger farms.

Although the number of laying hens in the United States has declined over the years, egg production is higher because of improved performance of the individual hen. In 1880, for example, the average laying hen produced 100 eggs per year; in 1950, the average was 175; in 1986, it increased to 250; and in 1997, it was 255 eggs.

The USDA estimates that 85% of the market eggs produced in the United States are from large commercial producers (those that maintain 1–11 million birds). The 45 largest egg-producing companies in the United States have more than 97 million layers, which is 35.5% of the nation's total.

Important changes have also occurred in U.S. turkey production. Fifty years ago, turkeys were raised in small numbers on many farms; today, they are raised in larger numbers on fewer farms. The average turkey farmer is producing well over 50,000 turkeys per year. The leading turkey-producing states are North Carolina, Minnesota, Virginia, and California.

Most of the turkeys produced today are the heavy or large, broad-breasted white type. At one time, the fryer-roaster-type turkey (5–9 lb) was popular; it bridged the gap between the turkey and the broiler chicken. Today, the broiler industry has taken over the fryer-roaster market because it can produce fryer-roaster chickens at a much lower price.

It takes 16 weeks for turkey hens and 19 weeks for turkey toms to reach market size. Because there is a heavy demand for turkey meat in November and December, turkey eggs are set in large numbers in April, May, and June. However, eggs are set year-round because there is a constant, though lower, demand for fresh turkey meat throughout the year. The normal incubation period for turkey eggs is 28 days.

In the recent past, most consumers considered turkey a seasonal product, with consumption occurring primarily at Thanksgiving. Increasingly, turkey is consumed throughout the year, and consumers have available to them a large variety of turkey products. The greater demand for turkey has meant that approximately 1 billion pounds more of ready-to-cook turkey is produced today than was produced in the early 1980s.

The export market has become increasingly important to the poultry industry with 20% of broiler production, 9% of turkey, and 2% of eggs sold to customers outside of the United States. However, disruptions in foreign markets can have significant impact on the industry. For example, when Russian buyers stopped purchasing U.S. chicken in 2001, not only did broiler prices go down but the competing meats also experienced price slippage as the supply of animal protein overwhelmed short-term demand.

Poultry breeding, feeding, and management practices are presented in Chapter 35.

THE SHEEP AND GOAT INDUSTRY

Global Perspective

Sheep and goats are closely related, with both originating in Europe and the cooler regions of Asia. Sheep are distinguished from goats by the absence of a beard, less odor (males only), and glands in all four feet. The horns are spiral in different directions; goat horns spiral to the left, while sheep horns spiral to the right (like a corkscrew).

Sheep and goats are important ruminants in temperate and tropical agriculture. They provide fibers, milk, hides, and meat, making them versatile and efficient, especially for developing countries. Sheep and goats are better adapted than cattle to arid tropics, probably because of their superior water and nitrogen economy. Cattle, sheep, and goats often are grazed together because they utilize different plants. Goats graze **browse** (shrubs) and some **forbs** (broad-leaved plants), cattle graze tall grasses and some forbs, and sheep graze short grasses and some forbs. Many of the forbs in grazing areas are broadleaf weeds.

More than 60% of all sheep are in temperate zones and fewer than 40% are in tropical zones. Goats, though, are mostly (80%) in the tropical or subtropical zones (0–40°N). Temperature and type of vegetation are the primary factors encouraging sheep production in temperate zones and goat production in tropical zones.

Sheep originated in the dry, alternately hot-and-cold climate of Southwest Asia. To succeed in tropical areas, then, sheep had to adapt the abilities to lose body heat, resist diseases, and survive in an adverse nutritional environment. For sheep to lose heat easily, they require a large body surface-to-mass ratio. Such a sheep is a small, long-legged animal. In addition, a hairy coat allows ventilation and protects the skin from the sun and abrasions. In temperate areas, sheep with large, compact bodies, a heavy fleece covering, and storage of subcutaneous fat have an advantage. Sheep can also constrict or relax blood vessels to the face, legs, and ears for control of heat loss.

The productivity of sheep is much greater in temperate areas than it is in tropical environments. This difference is the result not only of a more favorable environment (temperature and feed supply) but also of a greater selection emphasis on growth rate, milk production, lambing percentage, and fleece weight. Sheep from temperate environments do not adapt well to tropical environments; therefore, it may be more effective to select sheep in the production environment, rather than introduce sheep from different environments.

World sheep numbers of nearly 1.1 billion head in 2002 are the highest on record. China, Australia, India, Iran, and Sudan are the leading sheep countries (Table 2.15). New Zealand, Australia, and Saudi Arabia have the highest per-capita consumption of mutton, lamb, and goat meat.

The United States exports a relatively small amount of lamb and mutton with most of its export in the form of mutton from cull ewes (Table 2.16). However, Australia and New Zealand provide a significant amount of lamb to the U.S. market. The importation of lamb from Oceania is a point of contention in the U.S. **sheep industry.**

The 738 million head of goats in the world are concentrated primarily in India and China, with Pakistan, Sudan, and Bangladesh also having large goat populations (Table 2.17). In many countries, goats are important for both milk

TABLE 2.15 World Sheep Numbers, Production, and Consumption

Country	No. Sheep (mil head)	Country	Production (bil lb)[a]	Country	Per-Capita Supply[c] (lb)
1. China	171	1. China	2.7	1. Mongolia	116
2. Australia	106	2. Australia	0.6	2. New Zealand	55
3. India	62	3. New Zealand	0.5	3. Iceland	50
4. Iran	54	4. Turkey	0.4	4. Turkmenistan	41
5. Sudan	4	5. Iran	0.3	5. Kuwait	37
World Total	1,081	**World Total**	8.5	**World Average**[b]	4

[a] Carcass weight of lamb and mutton.
[b] Estimated by dividing world production by world population.
[c] Lamb and goat meat combined.
Sources: Adapted from USDA and FAO.

TABLE 2.16 U.S. Imports and Exports of Lamb and Wool

	Exports (mil $)	Imports (mil $)
Wool (raw, carpet grade, apparel grade)	20	70.1
Lamb/goat/mutton	10	196.0

Source: USDA.

TABLE 2.17 World Goat Numbers, Meat Production, and Goat Milk Production

Country	No. Goats (mil head)	Country	Goat Meat Production (mil ton)[a]	Country	Milk Production (mil ton)
1. China	196	1. China	2.1	1. India	2.5
2. India	120	2. India	0.5	2. Bangladesh	1.3
3. Pakistan	57	3. Pakistan	0.4	3. Sudan	1.2
4. Bangladesh	37	4. Nigeria	0.2	4. Pakistan	0.6
5. Nigeria	28	5. Bangladesh	0.1	5. Greece	0.5
World Total	808	**World Total**	5.0	**World Total**	11.3

[a] Carcass weight.
Sources: Adapted from USDA and FAO.

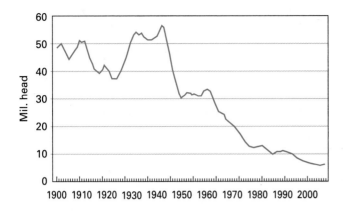

FIGURE 2.19   Total sheep and lamb population, January 1, U.S., annual. *Source:* Livestock Marketing Information Center.

and meat production. Table 2.17 shows the leading countries in goat meat and milk production.

United States

The number of sheep in the United States reached a high of 56 million in 1942, thereafter declining to approximately 6.5 million in 2006 (Fig. 2.19). United States sheep are located on about 65,000 different operations. Because of the decline in sheep numbers, the United States is increasingly dependent on imports to supply the relatively limited demand for lamb.

The number of ewes on U.S. farms in 2006 is shown in Figure 2.20. The five leading states in ewe numbers are, respectively, Texas (0.69 million), Wyoming (0.28 million), California (0.27 million), South Dakota (0.23 million), and Utah (0.21 million). The Midwest and West Coast areas are important in **farm-flock** sheep production, whereas other western states are primarily **range-flock** areas. The Midwestern states generally raise more lambs per ewe than the western states do, though Washington, Oregon, and Idaho resemble the Midwestern states in this regard.

Table 2.18 shows average ewe flock sizes in selected states. Most of the decline in U.S. sheep numbers during the past 50 years has occurred in the West. However, this region, with its extensive public and private rangelands, still produces 80% of the sheep raised in the country today. Most U.S. sheep growers have small flocks (50 or fewer sheep) and raise sheep as a secondary enterprise. About 40% of the West's sheep producers maintain flocks of more than 50 sheep; these flocks contain about 93% of the sheep in that region. Only about one-third of the operators in the West who have flocks of 50 or more sheep specialize in sheep; the other two-thirds have diversified livestock operations.

About 23% of all sheep born in the western United States and 13% of those born in the North-Central region are lost before they are marketed. In the West, predators, especially coyotes, and weather are the most frequent causes of death. In the North-Central region, the primary causes of death include weather and disease before docking, and disease and attacks by dogs after docking.

The range operator produces some slaughter lambs if good mountain range is available, but most range-produced lambs are feeders (e.g., lambs that must have additional feed before they are slaughtered). Feeder lambs are fed by producers, sold to feedlot operators, or fed on contract by feedlot operators. Some

TABLE 2.18 Average Ewe Flock Size in Selected States in the United States

Range Flocks		Farm Flocks	
State	Ewes (N)	State	Ewes (N)
Wyoming	573	Minnesota	48
Arizona	389	Nebraska	43
Colorado	362	Michigan	41
New Mexico	309	Pennsylvania	38
Nevada	259	Iowa	34

Source: USDA.

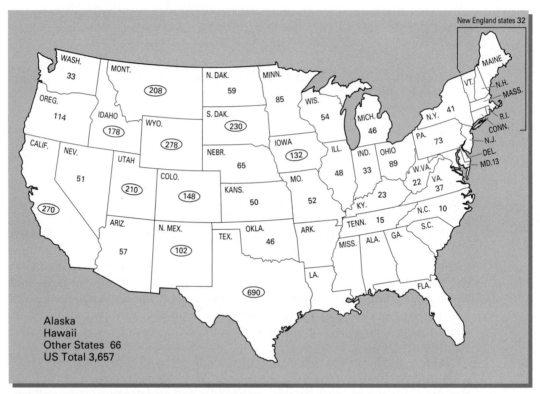

FIGURE 2.20 Breeding ewes, 1 year of age and older, 2006 (1,000 head). The top 10 states are circled. *Source:* Livestock Marketing Information Center.

feedlots have a capacity for 20,000 or more lambs on feed at one time, and many feedlots that have a capacity of 5,000 are in operation. The leading lamb-feeding states are Colorado, California, Texas, Wyoming, South Dakota, and Kansas.

Purebred Breeder

Some producers raise purebred sheep and sell rams for breeding. **Purebred breeders** have an important responsibility to the sheep industry: They determine the genetic productivity of commercial sheep. The purebred breeder should rigidly select breeding animals and then sell only those rams that will improve productivity and profitability for the commercial producer.

Records are essential to indicate which ram is bred to each of the ewes and which ewe is the mother of each lamb. All ram lambs are usually kept together to identify those that are desirable for sale. The purebred breeder often retains the ram lambs until they are 1 year of age, at which time they are offered for sale.

Commercial Market Lamb Producers

Commercial lamb producers whose pastures are productive can produce market wether and ewe lambs on pasture, whereas those whose pastures are less desirable produce feeder lambs. Lambs that are neither sufficiently fat nor large for market at weaning time usually go to feedlot operators, where they are fed to market weight and condition. However, most feedlot lambs are obtained from producers of range sheep. The producer of market lambs, whose feed and pasture conditions are favorable, strives to raise lambs that finish at 90–120 days of age, weighing approximately 120–130 lb each.

Commercial Feedlot Operator

Lambs are sent to the feedlot if they need additional weight prior to slaughter. The feedlot operator hopes to profit by increasing the value of the lambs per unit of weight and by increasing their weight. Lambs on feed that gain about 0.5–0.8 lb per head per day are considered satisfactory gainers. Feeder buyers prefer feeder lambs weighing 70–80 lb over larger lambs because of the need to put 20–30 lb of additional weight on the lambs to finish them. A feeding period of 40–60 days is usually sufficient for finishing healthy lambs.

Chapters 31 and 32 discuss in detail sheep breeding, feeding, and management practices. Chapter 36 covers goat production.

THE SWINE INDUSTRY

Global Perspective

Swine are widely distributed throughout the world, even though 50% of the world's pigs are located in China (Table 2.19). China far exceeds all other countries in number of pigs and total pork production. However, the swine herd in the United States is more productive on a per head basis. Other countries with large swine numbers and production include Brazil, Germany, and the Russian Federation (Table 2.19).

Table 2.20 shows the import and export of pork for the United States. Japan is the primary customer for the U.S. industry while the largest supplier to the United States is Canada.

TABLE 2.19 World Swine Numbers, Production, and Consumption

Country	No. Swine (mil head)	Country	Production (mil ton)[a]	Country	Per-Capita Supply (lb)
1. China	482	1. China	45	1. Austria	164
2. EU	151	2. EU	19	2. Spain	147
3. U.S.	61	3. U.S.	9	3. Denmark	146
4. Brazil	32	4. Brazil	2	4. Germany	119
5. Russian Fed.	16	5. Canada	1.7	5. Hungary	115
World Total	843	**World Total**	93	**World Average**	34

[a] Carcass weight.
Sources: Adapted from USDA and FAO.

TABLE 2.20   U.S. Imports and Exports of Pork and Lard

Commodity	Exports (1,000 $)	Imports (1,000 $)
Lard	23,930	—
Pork	2,280,000	1,280,817

Source: Adapted from USDA.

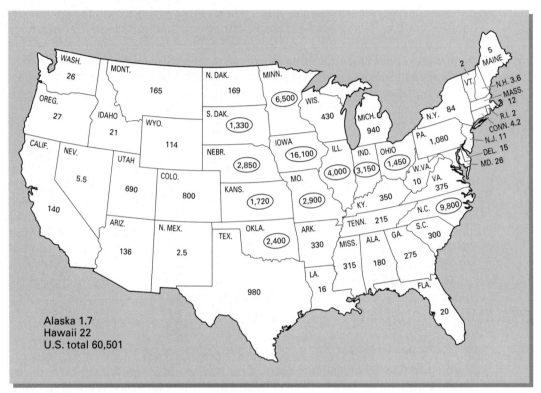

FIGURE 2.21   Total hogs and pigs December 1, 2004 (1,000 head). *Source:* Livestock Marketing Information Center.

United States

Cash receipts from the **swine industry** totaled nearly $12 billion in 2005 (Table 2.2). Swine production in the United States is concentrated heavily in the nation's midsection known as the **Corn Belt.** This area produces most of the nation's corn, the principal feed used for swine. Many swine producers farm several hundred acres of land and have other livestock enterprises. The Corn Belt states of Iowa, Missouri, Illinois, Indiana, Ohio, northwest Kentucky, southern Wisconsin, southern Minnesota, eastern South Dakota, and eastern Nebraska produce nearly 70% of the nation's swine, with Iowa alone accounting for approximately 25% (Fig. 2.21). Since 1990, North Carolina has increased swine numbers faster than most other states and currently there are more than 9 million pigs in the state. Oklahoma has also experienced dramatic growth in swine numbers since the early 1990s.

Figure 2.22 shows that during the past 20 years the number of hog farms has significantly decreased. In fact, fewer than half of the farms that raised pigs in

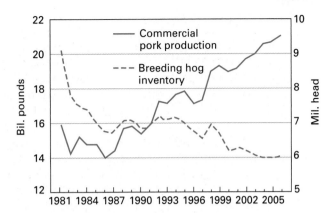

FIGURE 2.22   Pork production versus breeding hog inventory. *Source:* Livestock Marketing Information Center.

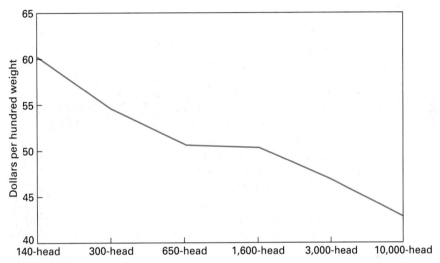

FIGURE 2.23   Average hog production costs in the United States. *Source:* USDA (*Cost of Production—Livestock and Dairy*, 1990).

TABLE 2.21   Trends in Market Share by Herd Size of Producer

	Percent Market Share			
Herd Size	1988	1997	2001	2005
1,000	32	25	13	11
1,001–1,999	40	14	12	10
2,000–4,999	9	15	22	26
5,000 +	19	40	53	53

Source: Adapted from USDA.

1987 still did so in 1997. However, during this same period of time, pork production has continued to increase. This implies that the number of hogs raised per farm has increased. Production costs, based on per hundredweight of hogs produced, decrease as number of hog production per farm increases (Fig. 2.23). However, as market share increases in the hands of large producers (Table 2.21), the industry has faced increasing pressure from environmental and animal welfare concerns.

Types of Swine Operations

There are four primary types of swine operations. In **feeder-pig production**, the producer maintains a breeding herd and produces feeder pigs for sale at an average weight of 40 lb. In **feeder-pig finishing**, the feeder pigs are purchased and then fed to slaughter weight. The **farrow-to-finish** operation maintains a breeding herd as pigs are produced and fed to market weight on the same farm. Finally, seedstock operations are similar to farrow-to-finish, except the salable product is primarily breeding boars and gilts. Due to biosecurity issues, economics of scale, and other factors, the feeder-pig producer and finisher are being absorbed into farrow-to-finish enterprises.

The various swine operations and the management factors affecting them are discussed in Chapters 29 and 30.

OTHER ANIMAL INDUSTRIES

Aquaculture

Aquaculture is the farming or rearing of aquatic plants and animals. In a sense, it is underwater agriculture. Farming implies some form of intervention in the rearing process to enhance production such as stocking, feeding, and protection from predators. The most common aquatic animals raised in the United States include finfish (catfish, trout, and salmon), crustaceans (shrimp and crawfish), and mollusks (oysters and clams).

Global fish production has expanded at such a large rate that the per-capita supply has nearly doubled in the last half century to a current level of nearly 38 pounds per person. Worldwide fish consumption accounts for one-sixth of all animal protein intake. Fish production also provides feed ingredients for livestock and aquaculture. Aquaculture produces 26% of the world's supply of food fish with approximately 60% originating from inland waters and 40% from marine environments.

Aquaculture sales in the United States exceed $1 billion with about two-thirds accounted for by food fish sales (catfish, salmon, trout, etc.). Production has increased significantly over the past several years with 365,566 acres involved in aquaculture. Catfish farming is the largest aquaculture activity in the United States with annual sales of $462 million involving approximately 600 million pounds. Mississippi produces about 60% of the U.S. total. Other major catfish producing states are Alabama, Arkansas, and Louisiana.

Trout production is the second largest sector in the U.S. aquaculture industry, with Idaho accounting for approximately one-half of the $79 million annual sales.

The third largest sector in the U.S. aquaculture industry is crawfish production. Concentrated mostly in Louisiana, the industry produces around 36 million pounds a year valued at $21 million.

There has been increased interest in producing other aquatic species, although most are in the experimental stages. Two species that have garnered publicity are hybrid striped bass and tilapia. Tilapia is a warm-water fish, native to Africa and the Middle East, but its production is expanding throughout the world. The Chinese have dominated the world's production of fish and they have claimed 26% of the world's production. In so doing, per-capita fish production has increased three times over the past decade. Imported fish, shrimp and crayfish tonnage is significant. For example, over $1 billion in shrimp is imported from Thailand and Indonesia. Crayfish imports grew so fast that a substantial tariff was implemented to protect domestic producers.

One of the most attractive aspects of aquatic farming is its low feed conversion ratio. Because fish are suspended in water, they don't expend many calories to move about, and because they are cold-blooded, they are more efficient to produce than any of the major land-raised meats. Today, fish farmers are achieving average feed conversion ratios as low as 1.3.

U.S. aquaculture producers have a number of competitive advantages. First, the United States is one of the world's largest seafood markets, and the domestic live market is one that pays premium prices. Second, although most aquaculture firms are in rural areas, they still are close to good transportation facilities. Third, U.S. producers have access to a growing network of researchers and companies working on aquaculture. Fourth, the United States is a major producer of grains and has large supplies of livestock by-products for use in fish feeds, which can account for up to 50% of variable costs. Fifth, the United States has a wide variety of climates and long coastlines for marine aquaculture, in addition to large freshwater resources.

On the other hand, U.S. growers also have a number of disadvantages. First, many of the biggest farm-raised species are tropical and, with the exception of Hawaii, the United States does not have any tropical areas. Second, land costs are generally higher in the United States than in less developed countries, especially the coastal properties needed for marine aquaculture. Third, in most cases, labor costs are considerably higher in the United States than in competing countries. Fourth, many countries have fewer regulations controlling aquaculture production and processing practices.

Bison

Over 230,000 **bison** are found on 4,000 private farms and ranches in the United States. An additional 20,000 wild bison can be found on public lands in the United States and Canada. In 2005, 30,000 head of bison were processed under federal inspection. The National Bison Association has focused its promotional efforts to increase consumption of buffalo meat via merchandising to chefs, food editors, and special events.

Development of a bison enterprise requires that owners/managers build a suitable set of facilities with fences and handling equipment well adapted to the behavioral patterns of the buffalo. Just as is the case with other livestock enterprises, a carefully managed nutritional program is required for high levels of reproductive and growth performance. Selection tends to be focused on fertility testing, weaning weight, growth rate, carcass performance, and maternal traits.

Elk

The farming of elk and deer is a relatively new enterprise to the U.S. (Fig. 2.24). In addition to the sale of breeding animals and meat, antlers or velvet from males are sold. A 2-year-old bull elk yields approximately 5–6 pounds of velvet while mature males may produce 35–40 pounds per year. A majority of the velvet is sold into the Asian market for use in traditional, holistic medicines.

FIGURE 2.24   Farming of elk is a relatively new enterprise that gains income from the sale of velvet from the horns, meat, and hides.

TABLE 2.22   Productivity of the Emu and Ostrich

Name	Adult Size (male)		No. Eggs Per Year	Egg Weight (lb)	Products
	Height (ft)	Weight (lb)			
Emu	5–6	150	30–40	1–1 1/2	Meat, leather, oil, feathers, carved eggs
Ostrich	up to 8	300	30–50	3–3 1/2	Meat, hides, feathers, egg shells

Source: Adapted from Kidwell, et al. 1954.

The breeding market is the dominant focus for most elk and deer breeders. As is often the case with new enterprises, speculation often creates unsustainable high markets. The use of domesticated elk and deer as food is typically found only in limited venues, such as specialty meat shops or in upscale restaurants that offer game meats as a menu item. Approximately 3 million pounds of venison from farmed deer is consumed in the United States.

Ostrich and Emu Farming and Ranching

Ostrich and **emu,** native to Africa and Australia, respectively, belong to the ratite family of flightless birds. The ostrich is the largest living bird. A general description of emu and ostrich is provided in Table 2.22.

Emu farming in the United States is relatively new. In 1989, Texas emu farmers started a national association. Only 1,200 ostriches were processed under federal inspection in 1995. However, 61,634 were inspected in 1997 with 100,000 birds processed in 1998.

INDUSTRY ISSUES AND CHALLENGES

There are several significant issues and challenges that the animal industries have encountered and will continue to face. These issues are discussed in Chapter 23.

CHAPTER SUMMARY

- The animal industries are described by numbers, cash receipts, per-capita product consumption, and their contribution to world trade.

- U.S. farm cash receipts in 2005 were $202.3 billion with livestock contributing $92.8 billion. Cash receipts for the major livestock species were cattle ($49.7 bil), dairy products ($26.9 bil), broilers ($19.2 bil), hogs ($13.7 bil), and turkeys ($3.3 bil).

- The major animal industries are becoming more integrated and concentrated. The poultry industry is highly integrated and swine production units continue to become larger. In the beef cattle industry, four major packers slaughter most of the cattle; however, most of the cattle are produced in herd sizes of less than 100 cows.

- Sheep numbers have shown a significant decline in the past 50 years. Dairy cow numbers have decreased; however, milk production has remained about the same. Milk production per cow continues to increase.

- Aquaculture and other types of animal farming (e.g., ostrich and emu) are increasing in the United States.

KEY WORDS

cash receipts	Mesohippus	purebred breeder
world trade	horse industry	swine industry
beef industry	poultry industry	Corn Belt
cow-calf	tariff	feeder-pig producer
stocker-yearling	broiler	feeder-pig finisher
feedlot	browse	farrow-to-finish
commercial feeders	forbs	aquaculture
farmer-feeders	sheep industry	bison
dairy industry	farm flock	ostrich
Eohippus	range flock	emu

REVIEW QUESTIONS

1. *True or False*: Agricultural products contribute to the trade deficit in the United States.

2. *True or False*: In some countries, the most important product of cattle is draft power and milk rather than meat.

3. What are the two reasons world cattle numbers have continued to increase during the last 30 years?

4. What country has the greatest number of cattle?

5. What country has the greatest production of beef?

6. What are the three phases of beef cattle production?

7. From what segment of cow-calf production do most of the potential slaughter steers and heifers come from?

8. Seedstock cow-calf producers primarily produce what two products?

9. The primary basis of stocker-yearling operations is to market _____.

10. What two segments comprise the feedlot beef industry?

11. *True or False*: Commercial feeders have greater production capacity (head of cattle) than farmer-feeders.

12. What country produces the most fluid milk?

13. What are the three dairy products produced throughout the world in the greatest quantity after fluid milk?

14. What country has the greatest number of horses, donkeys, and mules?

15. Why did the numbers of horses, donkeys, and mules in the United States decline after World War I?

16. Today, horses in the United States are used primarily for what purposes?

17. What is the leading country in egg production?

18. What is the leading country in production of poultry meat?

19. *True or False*: Most sheep are located in temperate zones, whereas goats occur overwhelmingly in tropical and subtropical zones.

20. What is the leading country for number of sheep?

21. What is the leading country for number of goats?

22. What is the leading country for goat meat production?

23. What is the leading country for number of swine and production of pork?

24. What are the segments of the swine industry?

SELECTED REFERENCES

American Poultry History. 1996. Mount Morris, IL: Watt Publishing Co.

Carolan, M. 2002. *Summary of U.S. Census of Agriculture Data for Iowa Farms.* Dept. of Sociology. Ames, IA: Iowa State Univ.

Directions: A 21st Century Vision. National Cattlemen, July, 2006. Englewood, CO: National Cattlemen's Association.

Field, T. G. and R. E. Taylor. 2007. Beef Production and Management Decision. 5th Edition. Upper Saddle River, NJ: Pearson Prentice Hall.

Helming, B. 1990. Meat's decade of danger. *Meat and Poultry* 36:30.

Horse Industry Directory. 2004–05. Washington, DC: American Horse Council, Inc.

Rankings of States and Commodities by Cash Receipts. 2006. Washington, DC: USDA.

Stillman, R., T. Crawford, and L. Aldrich. 1990. *The U.S. Sheep Industry.* Washington, DC: USDA, ERS.

USDA. 2006. *Economic Indicators of the Farm Sector: Costs of Production—Livestock and Dairy.* Washington, DC.

USDA. 2006. *Foreign Agricultural Trade of the United States.* Washington, DC.

USDA. 2006. *Livestock and Poultry Situation and Outlook Report.* Washington, DC.

CHAPTER 3

Red Meat Products

LEARNING OBJECTIVES

- Describe trends in red meat production
- Define the major wholesale cuts of beef, veal, pork, and lamb
- Identify the physical structures of meat and muscle
- Compare the nutrient value of beef, pork, and lamb
- Quantify domestic consumption patterns of red meat
- Explain the marketing of red meat

Red meat products come primarily from cattle, swine, sheep, goats, and, to a lesser extent, horses and other animals. Poultry meat, sometimes called *white meat*, is discussed in Chapter 4. Pork while advertised as "the other white meat" is a red meat.

Red meats are named according to their source: **beef** is typically from cattle over a year of age; **veal** is from calves 3 months of age or younger (veal carcasses are distinguished from beef by the grayish-pink color of their lean); **pork** is from swine; **mutton** is from mature sheep; **lamb** is from young sheep; **chevon** is from goats, but it is more commonly called **goat meat**.

PRODUCTION

World **red meat** production is approximately 265 million tons, with China, the United States, and Brazil leading all other countries. The leading 15 countries for red meat production account for nearly 85% of all production.

Red meat production in the United States is shown in Figure 3.1. Beef and pork comprise most of the annual production of nearly 50 billion pounds. The total production has been relatively stable over the past several years. Some red meat has to be imported into the United States. (Table 2.3, Chapter 2) to meet consumer demand and global trade agreements. Most of the imported meat product is beef needed to fill the U.S. deficit in processing beef.

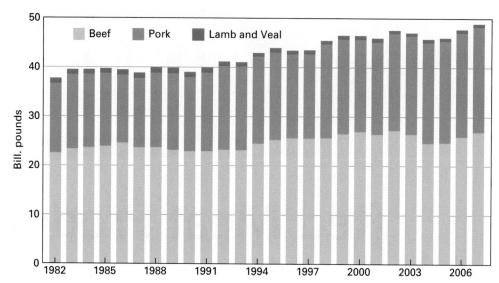

FIGURE 3.1   Annual commercial red meat production by type of meat. *Source:* Livestock Marketing Information Center.

Figures 3.2, 3.3, and 3.4 show where cattle, hogs, and sheep, respectively, are harvested in the United States. Kansas, Nebraska, Texas, and Colorado are the leading states in cattle harvest. Almost 30% of the nation's hogs were harvested in Iowa in 2005. The leading lamb-harvesting states are Colorado and Iowa.

Meat animals are processed in packing plants located near areas of live animal production. This reduces the transportation costs because carcasses, wholesale cuts, and retail cuts can be transported at less cost than live animals. The top 10 meat packers in the United States are identified in Table 3.1.

PRODUCTS

Animals are transported to packing plants where they are processed. During the initial processing stage, the animals are made unconscious by using carbon dioxide gas or by stunning (electrical or mechanical). The jugular vein and/or carotid artery is then cut to drain the blood from the animal. After bleeding, the hides are removed from cattle and sheep. Hogs are scaled to remove the hair, but the skin is usually left on the carcass. A few packers (especially small plants) skin hogs, as it is more energy efficient than leaving the skin on.

After the hide or hair is removed, certain parts, sometimes referred to as the drop, viscera, offal, or by-products, are separated from the carcass. Typically these are the head, hide, hair, shanks (lower parts of legs and feet), and internal organs.

Dressing percentage (sometimes referred to as yield) is the relation of hot or cold carcass weight to live weight. It is calculated as follows:

$$\text{Dressing percentage} = \frac{\text{Hot or cold carcass weight}}{\text{Live weight}} \times 100$$

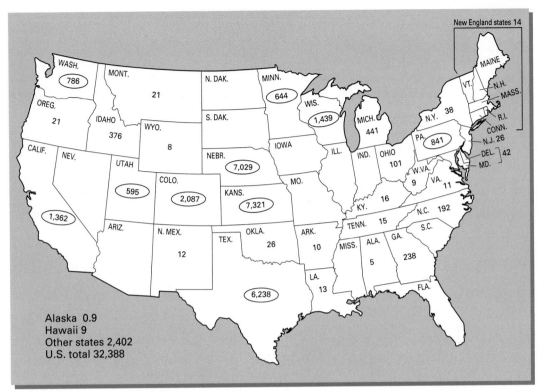

FIGURE 3.2 Commercial cattle harvest in the United States, 2005 (1,000 head). The top 10 states are circled. *Source:* Livestock Marketing Information Center.

Table 3.2 shows the average live slaughter weights, carcass weights, and dressing percentages for cattle, sheep, and hogs. For practical purposes, the average dressing percentage for hogs is 72%; for cattle, 60%; and for sheep, 50%; however, dressing percentage can vary several percentage points within each species. The primary factors affecting dressing percentage are fill (contents of digestive tract), fatness, muscling, weight of hide, and, in sheep, weight of wool.

Beef and pork carcasses are split down the center of the backbone (sheep carcasses are not split), giving two sides approximately equal in weight. After the carcasses are washed, they are put into a cooler to chill. The cooler temperature of approximately 28–32°F removes the heat from the carcass and keeps the meat from spoiling. Carcasses can be stored in coolers at 32–35°F for several weeks; however, this is uncommon in large packing plants as profitability is dependent on processing live animals one day and shipping carcasses or products out the next day or two. Smaller packers may keep carcasses in the coolers for as long as 2 weeks to allow the carcasses to age. This aging process increases meat tenderness through an enzyme activity that decreases the tensile strength of muscle fibers. Recent research has shown that injection of $CaCl_2$ (calcium chloride) into beef carcasses improves tenderness.

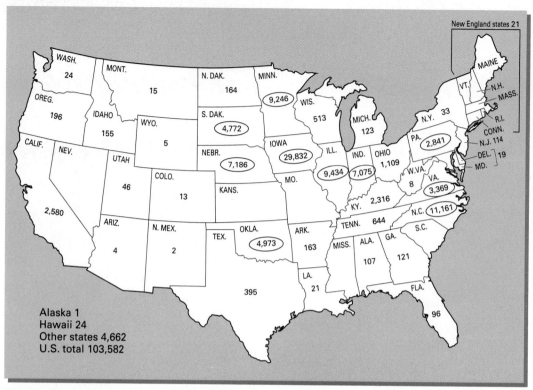

FIGURE 3.3   Commercial hog slaughter 2005 (1,000 head). *Source:* Livestock Marketing Information Center.

In the past, most red meat was shipped in carcass form, but today most packing plants process the carcasses into **wholesale** or primal **cuts** and even into retail cuts (Table 3.3; Figs. 3.5–3.7). The latter process involves shipping meat in boxes; thus the terms **boxed beef, boxed pork** and **boxed lamb.**

Most red meats are marketed as fresh products. Some are cured, smoked, processed, or canned. Plates A, B, C, and D identify the forms in which major meat products are prepared for sale.

KOSHER AND MUSLIM MEATS

Traditional Jewish and Muslim laws have specific requirements for the slaughter of religiously acceptable animals. The primary difference from the general practices in the United States is that the animals are not stunned prior to slaughter.

All **kosher meat** comes from animals with split hooves, that chew their cud, and that have been slaughtered in a manner prescribed by the Halakha (Orthodox Jewish law). Meat from undesirable animals or from animals not properly slaughtered or with imperfections is called nonkosher (trefah). In addition to the specification of which animals are considered kosher, the Talmud prohibits blood and the mixing of meat and milk. Foods that are kosher bear the endorsement symbol ∪.

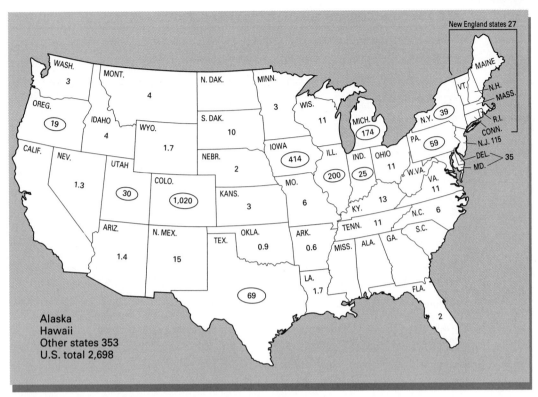

FIGURE 3.4   Commercial sheep slaughter during 2005 (1,000 head). *Source:* Livestock Marketing Information Center.

TABLE 3.1   Leading Red Meat Packing and Processing Companies

Rank	Company (location)	2005 Sales ($ bil)	No.[a] Plants	No. Employees	Products[b]
1	Tyson Foods, Inc. (Springdale, AR)	$26.0	120	114,000	B-P-PL
2	Cargill Meat Solutions (Wichita, KS)	15.0	21	33,000	B-P-PL
3	ConAgra Foods, Inc. (Wichita, KS)	14.5	NA	38,000	B-P
4	Smithfield Foods, Inc. (Smithfield, VA)	11.3	24	51,000	B-P-PL
5	Swift & Co. (Greeley, CO)	9.7	14	21,000	B-P-L
6	SYSCO Corporation (Houston, TX)	9.0	17	47,500	B-P-L
7	Hormel Foods Corp. (Austin, MN)	5.4	NA	17,600	P-PL
8	Oscar Mayer Kraft Foods, Inc. (Madison, WI)	4.5	175	94,500	P-PL
9	Sara Lee Corporation (Philadelphia, PA)	4.2	13	137,000	B-P-PL
10	National Beef Packing Co. Inc. (Kansas City, MO)	4.0	4	6,000	B

[a]Includes harvest, processing, rendering, etc.
[b]B = beef; P = pork; L = lamb; PL = poultry.
Source: Adapted from multiple sources.

Kosher slaughter is performed by a person specifically trained, known as a *Shochet.* The trachea and esophagus of the animal must be cut in swift continuous strokes with a perfectly smooth, sharp knife—the *chalef*—causing instant death and minimum pain to the animal. Various internal organs are inspected for defects and adhesions.

TABLE 3.2 U.S. Livestock Slaughter and Meat Production, 2005

Species/ Class	Meat Products	No. Head (mil)	Average Live weight (lb)	Total Carcass Weight (bil lb)	Average Carcass Weight (lb)	Average Dressing Percentage
Cattle	Beef	32.3	1,280	24.6	817	63.5
Calves[a]	Veal	0.7	352	0.16	216	61
Hogs	Pork	103.5	269	20.6	197	73
Sheep/lamb	Mutton/lamb	2.7	140	0.19	71	50

[a] Part of these are classified as calf carcasses, which are intermediate in age and size between veal and beef carcasses.
Source: Adapted from USDA.

TABLE 3.3 Wholesale Cuts of Beef, Veal, Pork, and Lamb

Beef	Veal	Pork	Lamb
Round	Leg/round	Leg/ham	Leg
Sirloin	Sirloin	Loin	Loin
Short loin	Loin	Blade shoulder	Rib
Rib	Rib	Jowl	Shoulder
Chuck	Shoulder	Arm shoulder	Neck
Foreshank	Foreshank	Spareribs	Foreshank
Brisket	Breast	Side	Breast
Short plate			
Flank			

FIGURE 3.5 A thick, tender, juicy steak is the most highly preferred beef product. *Source:* Photo courtesy of Certified Angus Beef LLC. *Certified Angus Beef*® is a registered trademark of Certified Angus Beef LLC and is used with permission.

FIGURE 3.6 A favorite pork dish is a smoked, boneless ham. *Source:* Courtesy of the National Pork Producer's Council.

FIGURE 3.7 Processed beef and pork are popular meat items for sandwich lovers. *Source:* Courtesy of the National Cattlemen's Beef Association and the National Pork Producer's Council.

The Torah forbids the eating of blood, including blood of arteries, veins, and muscle tissue. For this reason, certain arteries and veins, as well as the prohibited sciatic nerve must be removed from cattle; neck veins of poultry must be severed to allow the free flow of blood. Blood is extracted from meat by broiling and salting. There are specific steps required in proper broiling and salting to endorse the meat as kosher. Purchases of kosher foods in the United States exceed 100 million dollars annually.

The Muslim code is found in the Quran. Any Muslim may slaughter an animal while invoking the name of Allah. In cases where Muslims cannot kill their own animals, they may use meat killed by a "person of the book" (e.g., a Christian or a Jew). Usually this Halal slaughter is done by regular plant workers while Muslim religious leaders are present and reciting the appropriate prayers.

COMPOSITION

Meat consumption can be defined in either physical or chemical terms. Physical composition is observed visually and with objective measurements; chemical composition is determined by chemical analysis.

Physical Composition

The major physical components of meat are lean (muscle), fat, bone (Fig. 3.8), and connective tissue. The proportions of fat, lean, and bone change from birth to slaughter time. Chapter 18 discusses these compositional changes in more detail. Connective tissue, which to a large extent determines meat tenderness, exists in several different forms and locations. For example, tendons are composed of connective tissue (collagen) that attaches muscle to bone. Other collagenous connective tissues hold muscle bundles together and provide covering to each muscle fiber. Figure 3.8 shows the physical structure of muscle: myofibrils are component parts of muscle fibers, muscle fibers combined together make up a muscle bundle, and groups of muscle bundles are components of whole muscles or muscle systems. Within the myofibrils are two types of **myofilaments**—namely thick (myosin) and thin (actin) filaments.

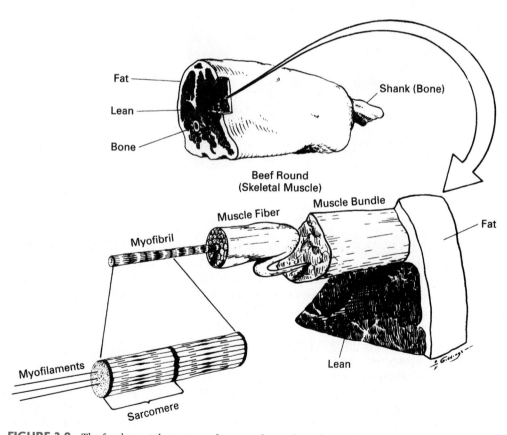

FIGURE 3.8 The fundamental structure of meat and muscle in the beef carcass.

Chemical Composition

Since muscle or lean meat is the primary carcass component consumed, only its chemical composition is discussed here. Chemical composition is important because it largely determines the nutritive value of meat.

Muscle consists of approximately 65–75% water, 15–20% protein, 2–12% fat, and 1% minerals (ash). As the animal increases in weight, water and protein percentages decrease and fat percentage increases.

Fat-soluble vitamins (A, D, E, and K) are contained in the fat component of meat. Most B vitamins (water-soluble) are abundant in muscle.

The major protein in muscle is **actomyosin,** a globulin consisting of the two proteins **actin** and **myosin.** Most of the other nitrogenous extracts in meats are relatively unimportant nutritionally. However, these other extracts provide aroma and flavor in meat, which stimulate the flow of gastric juices.

Simple carbohydrates in muscle are less than 1%. Glucose and glycogen are concentrated in the liver. They are not too important nutritionally, but they do have an important effect on meat quality, particularly muscle color and water-holding capacity.

NUTRITIONAL CONSIDERATIONS

Nutritive Value

Heightened health awareness in the United States has made today's consumers more astute food buyers than ever before (Fig. 3.9). Consumers look for foods that meet their requirements for a nutritious, balanced diet as well as for convenience, versatility, economy, and ease of preparation. Processed meat products have long been among Americans' favorite foods. Consumers today can choose from more than 200 styles or forms of meat, including over 50 cold cuts, dozens of sausages, and a wide variety of hams and bacon. Nutrient-rich red

FIGURE 3.9 Examples of Nutrition Facts labels for ground beef and ham.

TABLE 3.4 Nutrient Contents of Red Meats (3 oz of cooked, lean portions) and their Contribution to RDAs (Recommended Daily Allowances)[a]

Nutrient	Beef		Pork		Lamb	
	Amount	% RDA	Amount	% RDA	Amount	% RDA
Calories	192	10%	198	10%	176	9%
Protein	25 g	56	23 g	51	24 g	54
Fat[b]	9.4 g	14	11.1 g	17	8.1g	12
Cholesterol[b]	73 mg	24	79 mg	26	78 mg	26
Sodium	57 mg	2	59 mg	2	71 mg	2
Iron	2.6 mg	14	1.1 mg	6	1.7 mg	10
Zinc	5.9 mg	39	3.0 mg	20	4.5 mg	30
Thiamin	0.08 mg	6	0.59 mg	39	0.09 mg	6
Niacin	3.4 mg	17	4.3 mg	22	5.3 mg	27
B_{12}	2.4 mg	79	0.7 mg	24	2.2 mg	37

[a]Based on a 2,000-calorie diet and the RDA for women aged 23–51.
[b] The American Heart Association recommends that fat contribute not more than 30% of total daily calories and that cholesterol be limited to 300 mg/day.
Source: Adapted from USDA.

meat—beef, pork, lamb, and veal—are primary ingredients in these products. Meat products in all delicious and unique varieties contain the same nutrients as the fresh meats from which they are made (Table 3.4).

Meat is nutrient-dense and is among the most valuable sources of important vitamins and minerals in a balanced diet. **Nutrient density**, a measurement of food value, compares the amount of essential nutrients to the number of calories in food.

Red meats are abundant sources of iron, a mineral necessary to build and maintain blood hemoglobin, which carries oxygen to body cells. Heme iron, one form of iron found in red meat, is used by the body most effectively, and enhances the absorption of non-heme iron from meat and other food sources. Red meat is also a rich source of zinc, an essential trace element that contributes to tissue development, growth, and wound healing.

Red meats contain a rich supply of the B vitamins. Pork is a particularly rich source of thiamine, which is essential to convert carbohydrates into energy. Vitamin B_{12} occurs only in animal products, such as meat, fish, poultry, and milk, and is essential for the protection of nerve cells and the formation of blood cells in bone marrow. Niacin, riboflavin, and vitamin B_6 are also found in significant amounts in red meats.

Finally, red meats are an excellent source of protein, a fundamental component of all living cells. Fresh and processed meats contain high-quality or complete protein with all essential amino acids in quantity. The variety of retail cuts that can be generated from a beef or pork carcass offer opportunities for consumers to include red meat into their diet (Table 3.5).

New product development can have a significant impact both on consumer demand and profitability. For example, the Beef Value Cuts project, funded by beef checkoff dollars has resulted in the development of new products such as the flat iron steak, the western griller steak, western tip, ranch steak, and the petite tender. These products are the result of better muscle separation from wholesale cuts that are traditionally undervalued. The introduction of the flat iron steak contributed to a 60% increase in the value of beef chucks in the five-year period following market introduction.

Issues related to red meat consumption and human diet and health are presented in Chapter 23.

TABLE 3.5 Beef Cuts That Meet USDA Guidelines for Lean[a]

Beef Cut	Calories	Saturated Fat (g)	Total Fat (g)
Top round	157	1.6	4.6
Top sirloin steak	156	1.9	4.9
95% Lean ground beef	139	2.4	5.1
Round steak	154	1.9	5.3
Chuck pot roast	147	1.8	5.7
Western griller steak	155	2.2	6.0
Strip steak	161	2.3	6.0

[a] A representative sample of 29 cuts is provided based on 3-ounce serving.

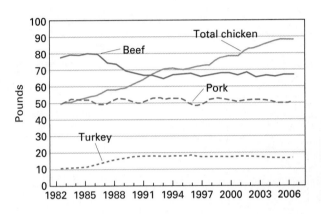

FIGURE 3.10 Annual U.S. red meat and poultry consumption (boneless weight). *Source:* Livestock Marketing Information Center.

CONSUMPTION

Meat is consumed for both nutrient content and eating satisfaction (Figs. 3.10–3.12). There are food products other than meat that cost less per unit of protein. However, the substantial palatability advantage of meat is a significant driver of consumer demand. These palatability attributes are enhanced by visual displays, like the one shown in Figure 3.9, that often appear at retail meat counters. They show that a cooked, trimmed, 3-oz serving of meat (lean portion only) is relatively low in calories, fat, and cholesterol, and that it supplies an abundance of other essential nutrients. Furthermore, the use of point-of-purchase recipes, in-store cooking demonstrations, product tasting opportunities, and other enticements can be utilized to attract consumers.

Consumer acceptance of red meat and poultry is shown in Figure 3.10. Poultry is included to demonstrate that over the past 20 years the total annual per-capita meat consumption has been increasing, primarily due to increased demand for poultry.

Tables 3.6 and 3.7 show the **per-capita consumption (disappearance)** of red meat, poultry, and fish on a carcass weight, retail weight, and boneless basis. Even the boneless data do not reflect the actual amount consumed, as there is fat trim, bone, cooking loss, and plate waste. Surveys have shown that most retail cuts of beef, pork, and lamb have 0.1 in. or less of external fat covering the lean muscle tissue.

Because actual consumption data is difficult to obtain, the use of production data coupled with population size has served as a proxy for per-capita intake of a given product. In essence much of the information provided as per-capita

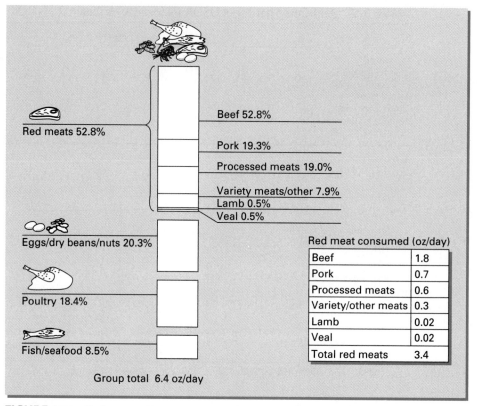

Red meat consumed (oz/day)	
Beef	1.8
Pork	0.7
Processed meats	0.6
Variety/other meats	0.3
Lamb	0.02
Veal	0.02
Total red meats	3.4

FIGURE 3.11 Daily per-capita meat consumption as part of meat, poultry, fish, dry beans, eggs, and nuts group (Food Guide Pyramid). *Source:* National Live Stock and Meat Board.

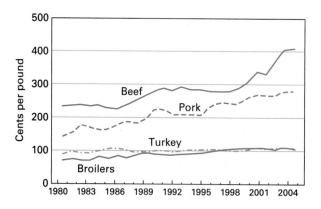

FIGURE 3.12 Annual retail meat and poultry prices (Nominal basis). *Source:* Livestock Marketing Information Center.

consumption is actually a better barometer of per-capita production because it is the net value of changes in cold storage, imports, and exports. Thus, data on the consumption and disappearance of meat and other food items can be misleading. What people say they eat and what they purchase does not determine the amount actually consumed. In one study, Dr. Bill Rathje (University of Arizona) evaluated garbage in the Tucson area in an attempt to determine what people eat by what they throw away. Over 10 years of his research data show that (1) people throw away approximately 15% of all the solid food they

TABLE 3.6 Per-Capita Consumption (disappearance) of Beef, Pork, and Lamb[a]

Year	Beef (lb)			Pork (lb)			Lamb (lb)		
	Carcass	Retail	Boneless	Carcass	Retail	Boneless	Carcass	Retail	Boneless
1970	114.4	84.6	79.6	73.1	56.0	48.0	3.2	2.9	2.1
1975	119.2	88.2	83.0	55.9	43.0	38.7	2.0	1.8	1.3
1980	103.4	76.6	72.1	74.0	57.3	52.1	1.5	1.4	1.0
1985	106.7	78.9	74.6	66.3	51.7	47.7	1.6	1.4	1.1
1990	96.1	67.8	63.9	64.1	49.8	46.4	1.6	1.4	1.0
1995	96.6	67.1	63.2	67.5	52.4	48.8	1.3	1.2	0.9
2000	96.5	67.5	64.5	65.5	50.8	47.8	1.3	1.1	0.8
2005	94.1	65.8	62.9	65.6	50.9	47.8	1.3	1.1	1.8

[a] Multipliers for converting carcass weight to retail and boneless equivalents are: beef (.70/.67); pork (.78/.73); lamb (.89/.66).
Source: Adapted from USDA.

TABLE 3.7 Per-Capita Consumption (disappearance) of Red Meat, Broilers, and Fish

Year	Red Meat (lb)		Poultry (lb)		Fish (lb)	Total Meat (lb)	
	Retail	Boneless	Retail	Boneless	Retail	Retail	Boneless
1970	145.9	131.7	48.7	34.1	11.8	204.2	177.3
1975	136.3	125.8	47.7	33.4	12.2	193.8	170.9
1980	136.8	126.5	59.1	41.4	12.5	204.7	179.8
1985	133.8	124.9	65.7	46.0	15.1	210.4	185.5
1990	120.1	112.3	80.5	56.7	15.0	209.6	183.5
1995	122.1	114.2	88.2	62.0	15.0	221.3	192.1
2000	120.2	113.7	77.4	67.9	15.2	212.8	196.8
2005	118.4	112.0	84.5	72.7	16.5	219.4	201.2

Source: Adapted from USDA.

TABLE 3.8 Disposable Per-Capita Income and Meat Expenditures

Year	Annual Per-Capita Food Expenditures ($)	Annual Per-Capita Meat, Poultry, Fish & Eggs Expenditures ($)	Expenditure for Food as a Percent of Disposable Income	
			At-home (%)	Food Service (%)
1980	NA	NA	9.0	4.2
1985	NA	NA	7.5	4.1
1990	4,296	668	7.0	4.2
1995	4,505	752	6.4	4.2
2000	5,158	795	5.8	4.0
2005	5,340	825	5.4	4.1

Source: Adapted from USDA.

purchase; and (2) when questioned, consumers in middle- and low-income brackets consistently overestimate their meat consumption, while those in higher-income brackets underestimate their meat consumption.

Per-capita disposable income does affect the total amount of money spent on meat and the amount of product consumed (Table 3.8). However, consumers are spending a smaller percentage of their disposable income on meat and food, indicating that these items are relatively inexpensive compared to other purchases.

TABLE 3.9   Consumption and Prices of Various Beef Products, 1970-2000

	Per-Capita Consumption (lb retail)				Price ($/lb)	
Year	Hamburger	Processed Beef	Beef Cuts	Total Beef[a]	Hamburger	Choice Beef
1970	21.9	10.8	51.9	84.6	0.66	1.00
1975	30.1	11.7	46.4	88.2	0.88	1.52
1980	24.7	7.4	44.5	76.6	1.36	2.34
1985	27.2	6.0	45.7	78.9	1.24	2.29
1990	28.5	5.9	31.3	67.8	1.59	2.81
1995	26.6	9.0	31.4	67.3	1.37	2.84
2000	28.3	9.4	31.7	69.4	1.57	3.06
2005	26.9	9.7	29.2	65.8	2.51	3.47

[a] Retail weight basis.
Source: Adapted from USDA.

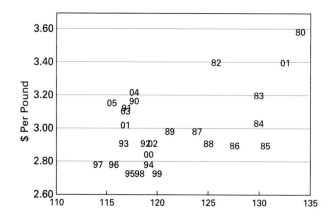

FIGURE 3.13   Price-quantity relationship for beef and pork—annual retail weight (deflated price basis). *Source:* Livestock Marketing Information Center.

Lower poultry prices relative to red meat prices have influenced some consumers to shift their preference from red meat to poultry (Fig. 3.12).

Table 3.9 shows consumption patterns for hamburger, processed beef, and beef cuts (steaks and roasts). Comparative prices for hamburger and average prices of retail cuts from choice beef are also shown.

DEMAND

Demand is a more important measure than is per-capita consumption (disappearance) because it deals not only with quantity but also the price at which consumers will buy a specific quantity all other factors being equal. Demand is a response curve of the quantities of product consumers will buy at various prices. Demand is based on a complex set of determinants such as prices of competing products, levels of per-capita disposable income, cultural preferences, consumer perceptions, and a host of other demographic factors.

Demand should not be confused with per-capita disappearance, share of expenditures, or share of consumption as none of these measures incorporates a schedule of prices into the equation. Demand can be more readily interpreted by evaluation of quantity and price relationships (Figs. 3.13 and 3.14). In both cases, there are periods of demand increase when consumers are willing to buy more products at higher prices (demand gain) and periods of time when consumption could only be increased by lowering prices (demand loss).

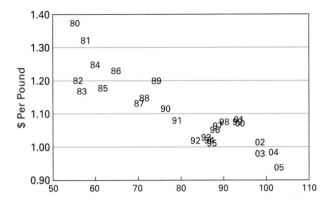

FIGURE 3.14 Price-quantity relationship for chicken and turkey—annual retail weight (deflated price basis). *Source:* Livestock Marketing Information Center.

TABLE 3.10 Market Outlets Used by U.S. Packers in Purchasing Red Meat Animals for Harvest

Animals	Total (mil head)	Direct Purchases [a] (%)	Public Markets [b] (%)
Cattle	32.4	88	12
Calves	0.7	86	14
Hogs	98.6	98	2
Sheep and lambs	2.5	92	8

[a] Direct purchases at packing plants, buying stations, country points, and feedlots.
[b] Public markets consist of terminal markets and auction markets.
Source: Adapted from USDA, *Packers and Stockyards' Statistical Resume.*

MARKETING

Meat production originates at the farm level and moves through feeder/finishers, marketing points, packers/processors, and food retailers or food service enterprises before reaching consumers. On hog farms, most pigs are raised to harvest weight. Young cattle and sheep are typically managed in off-farm feedlots to add weight under higher-energy diets prior to being harvested.

Packers and processors purchase the animals through several different marketing channels (Table 3.10). Direct purchases can occur at the packing plant, the feedlot, or a buying station near where the animals are produced.

Approximately 2,000 auctions, sometimes called **sale barns,** are located throughout the United States. The auction company has pens, a sale ring, and an auctioneer. Auctioneers sell the livestock to the highest bidder among the buyers attending the auction. Producers pay a marketing fee to the auction company for the sale service.

Most cattle, sheep, and swine have traditionally been purchased on a live weight basis, with the buyer estimating the value of the carcass and other products. Increasingly, animals are purchased on the basis of their carcass weight and the desirability of the carcass produced. For cattle, hogs, and sheep there has been an increased number of purchases on a carcass basis during the past several years (Table 3.11). In this method of marketing, sometimes called **grade and yield,** the seller is paid on a carcass-merit basis. The advantage of this trend is that price is established one step closer to the consumer. However, critics suggest that increased price determination at the packer undercuts the viability of the live market.

TABLE 3.11   Livestock Purchased by Packers on Carcass Grade and Weight Basis

Year	Cattle		Hogs		Sheep	
	No. Head (mil)	Percent of Total Purchases	No. Head (mil)	Percent of Total Purchases	No. Head (mil)	Percent of Total Purchases
1975	8.6	23%	6.0	9%	0.82	10%
1980	8.4	27	10.2	11	1.54	28
1985	10.2	30	13.0	16	2.19	37
1990	12.1	38	9.3	12	1.40	30
1995	15.7	46	39.3	43	2.01	46
2000	18.2	52	69.1	74	1.67	56
2004	17.3	53	75.5	77	1.13	46

Source: Adapted from USDA, *Packers and Stockyards' Statistical Resume.*

TABLE 3.12   Retail Food Chains in the United States

Company	Sales ($ bil)	Total Outlets (N)	Market Share (%)
Wal-Mart Stores, Inc.	155,500	1,829	18
Kroger Co.	57,200	3,302	10
Albertson's	41,300	2,476	5
Safeway	38,500	1,811	4
Costco Wholesale	31,800	412	4

Source: Adapted from multiple sources.

The USDA enforces the Packers and Stockyards Act, originally passed in 1921. The act provides for uniform and fair marketing practices. Federal supervision is given at market outlets, including packing plants, for fee and yardage charges, testing of scales, commission charges, and other marketing transactions. Carcass grade and weight sales are covered, as well as live animal sales. Marketing protection is provided to both the buyer and the seller.

The sale of food in both food service and retail formats has undergone dramatic shifts with food service sales capturing an increasing share of the consumer dollar. Growth of the food service industry has put additional demands on livestock industries to improve the consistency and value-added attributes of their product offering.

Simultaneously, national chains and conglomerates have captured a relative high market share of food sales (Table 3.12). These marketing entities will be increasingly focused on developing source-verified supply chain approaches. As such, the livestock industry will have to develop quality assurance and total quality management formats to assure access to markets.

The retail grocery business has undergone a dramatic shift from the 1950s when the independently owned corner grocery store dominated the market. Increasingly the supercenters, dollar stores, and club store formats have taken sales away from more traditional grocery and mass merchandising stores. The retail grocery segment is comprised of the following categories:

Conventional supermarkets—full offering of grocery, meat, and produce as well as an in-store deli and bakery, generating at least $2 million in sales, inventory of 15,000 items

Superstore—larger version of conventional supermarket, inventory of 25,000 items, and more than 40,000 ft² of floor space

Combination food/drug store—superstore format with more than 85% of sales generated from food products

Warehouse store—low-margin grocery with reduced variety, lower service levels, and lower prices

Super warehouse—high volume, more upscale combination of a superstore and warehouse format

Limited-assortment foodstore—low priced, limited service format with less than 2,000 product inventory

Specialty/gourmet retailer—single theme focus such as organic/natural, ethnic, or locally grown

Supercenter—large food-drug combination store and mass merchandiser under a single roof, 40% of shelf space devoted to grocery, more than 170,000 ft^2 of floor space

Wholesale club—bulk sales format in combination membership/wholesale store with 120,000 ft^2 or more of floor space, approximately one-third of revenues from food sales

Mass merchandiser—predominately household items, electronic goods, and apparel retailers offering limited packaged food items

Dollar store—limited assortment format offering general merchandise and food at industry lowest prices

Regardless of store format, consumers want grocery outlets to provide stores that are clean and conveniently formatted with high-quality meat and produce offerings. Consumers are increasingly time constrained and the food chain must acknowledge the following trends:

- Just over half of all food prepared away from home is carried out for at-home consumption and nearly 50% of consumers consider meals prepared away from home as essential to their lifestyle.
- One-half of white-collar workers eat one meal daily at their desk.
- One-quarter of calories are consumed in the form of snack foods.
- Ten percent of all food is consumed in a vehicle.
- The meat case confuses consumers in terms of what to buy and how to best prepare their selections.
- Three-quarters of consumers have not made an evening meal plan by the end of the workday.

Growth in the restaurant and food service has fueled increased demand for red meat and poultry. With $511 billion in annual sales, the food service sector has doubled its share of consumer expenditures over the last 50 years. This sector of the food industry accounts for approximately 70 billion meal or snack servings annually to 130 million daily customers.

Understanding the consumer market is not an easy task, as human attitudes and perceptions are diverse and in a constant state of flux. **Consumer** preferences are subject to change and can be affected by factors such as income, size of household, gender, age, geographic region, and ethnicity. Table 3.13 outlines some of the factors affecting consumption of beef. Marketing, product development, and

TABLE 3.13 Beef Consumption (annual lbs per capita) by Ethnicity, Region, Gender, and Residence

	lbs		lbs
Ethnicity		**Region**	
Black	77.2	Midwest	72.9
Hispanic	68.5	South	65.2
White	64.5	West	65.2
Other	62.4	Northeast	62.5
Residence		**Gender**	
Rural	75.4	Males	85.7
Urban	66.2	Females	48.1
Suburban	62.3		

Source: Adapted from USDA.

TABLE 3.14 Beef Purchase Motivators vs. Demotivators as Perceived by Consumer

Motivators	%	Demotivators	%
Family favorites	48%	Too expensive	27%
Easy to prepare	44	Don't know how to prepare	25
Know how to prepare	40	Takes too long to prepare	15
Good value	33	Too difficult to prepare	15
Versatility—variety	31	Kids don't like product	14
Convenience	21	Lack of recipes	11

Source: Adapted from NCBA, 1998.

business strategy decisions require an awareness of the dimensions of demand. For example, marketing strategies targeted to low users (women in this case) are likely to be different than those targeted at maintaining or increasing purchases by high demand consumers such as men, blacks, and rural residents (Table 3.13).

In the early 1980s, the livestock industries made a concentrated effort to produce leaner products. The beef industry, for example, responded to the market demand for leaner products with genetic changes, processing innovations, and a shift to closely trimmed subprimals. However, consumer demand actually slipped despite these efforts. As products became leaner, the taste of meat changed. Only when the industry optimized leanness with flavor did consumption begin to rise, especially in the case of the beef industry. Consumers also wanted better convenience, packaging, menu versatility, and value-added attributes. A list of factors that contributed to or inhibited beef purchases is provided in Table 3.14. As food producers work to evolve into consumer-driven industries, they must act on the consumer trends that have emerged in the past decade (Table 3.15).

Figure 3.15 shows consumer expenditures for red meat and poultry. Currently, consumers spend $250 more compared to dollars spent in 1980, with much of the long-term increase accounted for by demand for chicken. Percent of total expenditures for pork has remained relatively steady, while beef's market share losses incurred into the mid-1990s have reversed with recapture of lost share occurring over the past decade. However, beef continues to retain the largest percentage of dollar sales at the retail meat case (Table 3.16). The movement from commodity orientation to a value-added product focus is considered to be a key to gaining market share.

TABLE 3.15   Consumer Preferences for Food are Based on Multidimensional Expectations

Source of comfort and well-being
Simplicity in preparation—45 min. of preparation time/meal on weekends, 30 min. on weekdays
Important element of family and social gatherings
Ethnic cuisine in high demand
24-hour per day availability, take-out from traditional sit-down restaurants growing in demand
Diet is upgraded as incomes rise
Quality is multifaceted (style, taste, image, freshness, packaging, etc.)
Value-added foods in highest demand, desire high-quality recipes
Consumers losing culinary skills thus demand is high for meal solution kits or food service
Quick-fix diets
Expects safe food supply

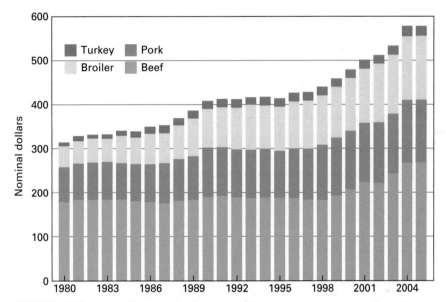

FIGURE 3.15   Annual per-capita U.S. expenditures for meat and poultry. *Source:* Livestock Marketing Information Center.

TABLE 3.16   Market Share of Meat Spending (%)

	Beef	Pork	Chicken	Turkey
1980	54	27	16	3
1985	50	26	19	4
1990	45	27	23	4
1995	40	29	26	4
2000	43	28	25	4
2005	47	25	26	3

Source: Adapted from USDA.

CHAPTER SUMMARY

■ Most of the 45 billion pounds of red meat production in the United States is beef (25 bil lb) and pork (20 bil lb).

■ The major contributions of red meats to human nutrition are the abundant sources of protein, iron, and vitamin B_{12}.

■ Per-capita red meat consumption in the United States is 119 pounds of retail weight with beef (67.9 lb) and pork (50.9 lb) being the primary contributors.

KEY WORDS

beef	dressing percentage	myosin
veal	wholesale cuts	nutrient density
pork	boxed beef, lamb, and	per-capita consumption
mutton	pork	(disappearance)
lamb	kosher meat	sale barns
chevon	myofilaments	grade and yield
goat meat	actomyosin	consumer
red meat	actin	

REVIEW QUESTIONS

1. What does dressing percentage represent?
2. How is dressing percentage calculated?
3. What is the rule of thumb or approximate dressing percentages of lamb, beef, and hogs?
4. What are the primary factors affecting dressing percentage?
5. What are the four major physical components of meat?
6. _____ are components of muscle fibers and are comprised of two types of myofilaments, myosin and actin.
7. What is the amount of essential nutrients to the number of calories in food referred to as?
8. *True or False*: Meat is nutrient-dense, meaning that it provides a large proportion of certain nutrients such as essential amino acids, B vitamins, and iron.
9. *True or False*: Most retail cuts of beef, pork, and lamb have greater than 0.1 inch of external fat covering lean muscle.
10. Describe the motivators and demotivators relative to the purchase of beef.

SELECTED REFERENCES

Boggs, D. L., and R. A. Merkel. 1993. *Live Animal, Carcass Evaluation and Selection Manual*. Dubuque, IA: Kendall-Hunt.

Cassens, R. G. 1994. Meat processing. *Encyclopedia of Agricultural Science*. San Diego, CA: Academic Press, Inc.

Eating in America today. 1994. Chicago, IL: National Live Stock and Meat Board.

Facts About Beef; Facts About Pork; Facts About Lamb. Chicago, IL: The National Live Stock and Meat Board.

Hedrick, H. B., E. D. Aberle, J. C. Forrest, M. D. Judge, and R. A. Merkel. 1993. *Principles of Meat Science*. Dubuque, IA: Kendall/Hunt.

Kashruth. Handbook for Home and School. 1972. New York, NY: Union of Orthodox Jewish Congregations of America.

Meat and Poultry Facts. 2005. Washington, DC: American Meat Institute.

Regenstein, J. M., and T. Grandin. 1992. Religious slaughter and animal welfare—An introduction for animal scientists. *Reciprocal Meat Conference Proceedings* 45:155.

Romans, J. R., K. W. Jones, W. J. Costello, C. W. Carlson, and P. T. Ziegler. 1985. *The Meat We Eat*. Danville, IL: Interstate Printers and Publishers.

CHAPTER 4

Poultry and Egg Products

LEARNING OBJECTIVES

- Describe the composition of poultry meat and eggs
- Describe the processing of poultry and eggs
- Explain the nutritive value of poultry meat and eggs
- Quantify poultry and egg consumption trends
- Overview the marketing of poultry and eggs

Poultry meat and eggs are nutritious and relatively inexpensive animal products used by humans throughout the world. Feathers, down, livers, and other **offal** are additional useful products and by-products obtained from poultry.

Application of genetics, nutrition, and disease control, along with sound business practices, has advanced the commercial poultry industry to the point where eggs and poultry meat can be produced very efficiently. The industry has developed two specialized and different types of chickens—one for meat production and one for egg production. Broiler lines are superior in efficient, economical meat production but have a lower egg-producing ability than the egg-production lines. Egg-type birds have been selected to produce large numbers of eggs profitably. They have been bred to mature at lightweights and they are therefore slower growing and less efficient meat producers.

POULTRY MEAT AND EGG PRODUCTION

Broiler chickens provide most of the world's production and consumption of poultry meat. Turkeys and roaster chickens, mature laying hens (fowl), ducks, geese, pigeons, and guinea hens are consumed in smaller quantities, although some of these are important food sources in some areas of the world. World poultry and egg production is presented in Chapter 2. The more common kinds of poultry used for meat production in the United States are shown in Table 4.1. The top 10 broiler producers are listed in Table 4.2.

Meat-type chickens are classified by size, weight, and age at the time of processing. As the demands of various market channels have differentiated, the industry has responded by producing birds specific to the needs of each market outlet. The common classes of chickens used for meat are as follows:

Broiler—a chicken raised specifically for meat as opposed to a *layer* that is utilized for egg production.

Poussin—a 1-pound or less bird that is less than 24 days of age.

Cornish game hens—birds less than 30 days of age and weighing 2 pounds.

Fast-food oriented broiler—birds weighing between 2 pounds 6 ounces and 2 pounds 14 ounces, less than 42 days of age, delivered cut-up without necks, giblets, tail fat, and leaf fat.

3's and up—birds weighing 3 to 4.75 pounds and between 40 to 45 days of age, delivered in whole, parts, and cut-up forms. This is the standard bird sold at retail.

TABLE 4.1   Ages and Weights of Various Kinds of Poultry Used for Meat Production

Classification	Typical Age (mo)	Average Slaughter Wt (lb)	Average Dressed Wt (lb)	Dressing (%)
Young chickens				
Broiler (fryer)	1.5–2	4–5	3–4	78
Rock Cornish	1.0	2	1.5	75
Roaster	3–5	6–8	5	75
Fowl (mature for stewing)	19	5	3.8	75
Turkeys	4–6	10–25	8–20	83
Ducks	<2	6–7	4–5	70
Geese	6	11	8–9	75

Sources: USDA and various others.

TABLE 4.2   Top 10 Integrated Broiler Producing Companies—2006

Rank	Company	Average Weekly Production[a] (mil lbs—ready to cook)	Market Share (%)
1	Tyson Foods, Inc.	150	22
2	Pilgrim's Pride Corp.	113	16
3	Gold Kist, Inc.	62	9
4	Perdue Farms	40	6
5	Wayne Farms	30	4
6	Sanderson Farms	30	4
7	Mountaire Farms, Inc.	25	3
8	Foster Farms	17	2
9	Peco Foods, Inc.	17	2
10	House of Raeford Farms, Inc.	16	2

[a] Ready-to-cook weight.
Source: Adapted from *National Provisioner* and National Broiler Council.

Roaster—birds weighing 5 to 8 pounds and between 55 and 60 days of age.

Broilers for deboning—males weighing 7 to 9 pounds, 47 to 56 days of age, processed to make nuggets, patties, and other value-added products.

Capon—once the standard, these are surgically castrated males approximately 14–15 weeks of age weighing 7 to 9 pounds. This product is very desirable in plumpness and tenderness but is considered a specialty item.

Heavy hens—these 15-month-old birds are spent breeders (culled from the laying flock) and typically weigh 5–5.5 pounds. These birds are used to produce cooked, diced, or pulled meat for inclusion in precooked and canned items such as soups and stews.

Poultry are slaughtered and processed in large plants owned by integrated poultry companies. Two such companies are Tyson Foods and Pilgrim's Pride, with 2005 sales of $26 billion and $6 billion, respectively.

COMPOSITION

Meat

Table 4.1 shows that broilers yield approximately 78% of their live weight as carcass weight, whereas the dressing percent for turkeys is 83%. The proportion of edible raw meat from carcasses of chickens and turkeys is approximately 67% and 78%, respectively.

The gross chemical composition of chickens and turkeys is shown in Table 4.3. **Dark meat** is higher in calories, lower in protein, and higher in cholesterol than **light meat** for both chicken and turkey. If the skin remains on poultry, it usually adds cholesterol and total fat.

Eggs

An egg has a spherical shape, with one end being rather blunt and larger than the other smaller, more pointed end. Figure 4.1 shows a longitudinal section of a hen's egg. The mineralized shell surrounds the contents of the egg. Immediately inside the shell are two membranes; one is attached to the shell itself, and

TABLE 4.3   Composition of Chicken and Turkey per 100 g, Edible Portion

	Water (g)	Food Energy (calories)	Protein (g)	Total Fat (g)	Cholesterol (mg)	Ash (g)
Young chicken						
Light meat (no skin)	74.9	114	23.2	1.6	58	0.98
Dark meat (no skin)	76.0	125	20.1	4.3	80	0.94
Turkey (hen)						
Light meat (no skin)	73.6	116	23.6	1.7	58	1.0
Dark meat (no skin)	74.0	130	20.1	4.9	62	0.95

Source: Adapted from USDA.

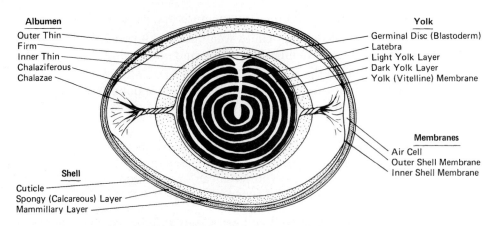

FIGURE 4.1   Longitudinal section of a hen's egg. *Source:* USDA.

TABLE 4.4   Chemical Composition of the Egg, Including the Shell

	Percent of Total	Water (%)	Protein (%)	Fat (%)	Ash (%)
Whole egg	100	65.5	11.8	11.0	11.7
White	58	88.0	11.0	0.2	0.8
Yolk	31	48.0	17.5	32.5	2.0

	Percent of Total	Calcium Carbonate (%)	Magnesium Carbonate (%)	Calcium Phosphate (%)	Organic Matter (%)
Shell and shell membranes	12	94.0	1.0	1.0	4.0

Source: Adapted from USDA.

the other tightly encloses the content of the egg. The air cell usually forms between the membranes in the blunt end of an egg shortly after it is laid, as the contents of the egg cool and contract.

Seven thousand to 17,000 tiny pores are distributed over the shell surface, with a greater number at the larger end. As the egg ages, these tiny holes permit moisture and carbon dioxide to move out and air to move in to form the air cell. The shell is covered with a protective covering called the **cuticle** or **bloom.** By blocking the pores, the cuticle helps to preserve freshness and prevent microbial contamination of the contents.

The egg white, or **albumen,** exists in four distinct layers. One of these layers surrounds the yolk and keeps the yolk in the center of the egg. The spiral movement of the developing egg in the magnum causes the mucin fibers of the albumen to draw together into strands that form the chalaziferous layer and chalazae. The twisting and drawing together of these mucin strands tend to squeeze out the thin albumen to form an inner thin albumen layer.

The yellowish-colored yolk is in the center of the egg, its contents surrounded by a thin, transparent membrane called the **vitelline membrane.**

Egg weight varies from 1.5 to 2.5 oz, with the average egg weighing approximately 2 oz. The component parts of an egg are shell and membranes (11%), albumen (58%), and yolk (31%). The gross composition of an egg is shown in

Table 4.4. The mineral content of the shell is approximately 94% calcium carbonate. The yolk has a relatively high percentage of fat and protein, although there is more total protein in the albumen.

The weight classes and grades of eggs are discussed in Chapter 8. Egg formation is presented in Chapter 10.

POULTRY PRODUCTS

Meat

More than 60% of all broilers leave the processing plant as cut-up chickens (Table 4.5) or selected parts rather than whole birds (compared to less than 30% in 1970).

Approximately 60% of the broilers, fowl (mature chickens), and turkeys are further processed beyond the ready-to-cook carcass or parts stage. The meat is separated from the bone and formed into products (e.g., turkey roll fingers and nuggets), cut and diced (e.g., for preparation of salads and casseroles), canned, or ground. Some of the cured and smoked ground products are frankfurters, bologna, pastrami, turkey ham, and salami. An even higher percentage of the broilers is expected to be further processed in future years. The poultry industry has led the meat business in terms of understanding and responding to the demands of consumers for further processed, value-added products. A review of a leading retail food store found 70 prepackaged, consumer-ready nonpoultry meat products—58 for pork, 6 for lamb, and 5 for beef. In contrast, several hundred poultry products of the same type were available.

Eggs

Eggs are unique, prepackaged food products that are ready to cook in their natural state. Eggs find their way into the human diet in numerous ways. They are most popular as a breakfast entrée; however, they are commonly used in sandwiches, salads, beverages, desserts, and several main dishes.

Shell eggs can be further processed into pasteurized liquid eggs and dried eggs and other egg products. Pasteurized liquid egg can be blast-frozen and sold to bakeries and other food-processing plants as an ingredient in other fabricated food products, or transported to users as a liquid product in refrigerated tank trucks. The dried egg products are sold to food manufacturers to be used in the production of products such as cake mixes, candy, and pasta.

TABLE 4.5   The Retail Parts and Cooked Edible Meat of a Broiler-Fryer Chicken

Part	% of Carcass	Edible[a]
Breast	28	63
Thighs	18	60
Drumsticks	17	53
Wings	14	30
Back and neck	18	27
Giblets	5	100
Total	100	53

[a] Net average yield of meat is 67% without neck and giblets.
Sources: Adapted from USDA; National Broiler Council.

A

B

C

FIGURE 4.2 Automated egg-handling equipment. (A) Overall view. (B) Automatic egg washer. (C) Eggs of different sizes are weighed on inline scales and then ejected at different points on the line. Courtesy of the USDA.

The initial step in making egg products is breaking the shells and separating the yolks, whites, and shells. This is done in most egg-breaking plants by equipment completely automated to remove eggs from egg-filler flats, wash and sanitize the shells, break the eggs for individual inspection of the contents by an operator, and separate the yolks from the whites. Processing equipment can also be used to produce various white and yolk products or mixtures of them. Present automatic systems are capable of handling 18,000–50,000 eggs per hour (Fig. 4.2).

Some breeds of chickens lay white-shelled eggs and other breeds produce brown-shelled eggs. The brown shell color comes from a reddish-brown pigment (ooporphyrin) derived from hemoglobin in the blood. Although there are shell color preferences by consumers in certain parts of the United States, there are no significant nutritional differences related to eggshell color.

Feathers and Down

In some areas of the world, ducks and geese are raised primarily for feathers and down. **Down** is a layer of small, soft **feathers** found beneath the outer feathers of ducks and geese. Feathers and down provide stuffing for pillows, quilts, and upholstery. They are also used in the manufacture of hats, clothing (outdoor wear), sleeping bags, fishing flies, and brushes.

The duck and goose industry in the United States is too small to meet the manufacturing demand for down. Thus, approximately 90% of raw feathers and down are imported, primarily from China, France, Switzerland, Poland, and several other European and Asian countries.

Breeder geese are often used for down production, as the older birds produce the best down. Four ducks or three geese will produce 1 lb of feather-and-down mixture, of which 15–25% is down. Down and feathers are separated by machine, then washed and dried.

Other Products and By-Products

Goose livers are considered a gourmet product in European countries. Force-feeding corn to geese three times a day for 4–8 weeks produces these enlarged fatty livers, weighing as much as 2 lb each. The livers are used to make a flavored paste called *pâté de foie gras* (for hors d'oeuvres or sandwiches).

Poultry by-products include hydrolyzed feather meal and poultry meat and bone scraps. These by-products are occasionally included in livestock and poultry rations as protein sources (refer to Chapter 7 for more discussion on these by-products).

NUTRITION CONSIDERATIONS

Nutritive Value of Poultry Meat

The white meat from fowl is approximately 33% protein, and the dark meat is about 28% protein. This protein is easily digested and is of high quality, containing all the essential amino acids. The fat content of poultry meat is lower than that found in many other meats. Poultry meat is also an excellent source of vitamin A, thiamin, riboflavin, and niacin.

Nutritive Value of Eggs

Since the egg (Fig. 4.3) contains many essential nutrients, it is recognized as one of the important foods in the major food groups recommended in *Dietary Guidelines for Americans*. These guidelines were developed by the U.S. Department of Agriculture to use in planning diets contributing to good health.

Eggs have a high nutrient density in that they provide excellent protein and a wide range of vitamins and minerals and have a low calorie count. The abundance of readily digestible protein, containing large amounts of essential amino acids, makes eggs a least-cost source of high-quality protein.

Table 4.6 lists the amounts of selected nutrients found in eggs. When the values of these nutrients are compared with the dietary allowances recommended by the Food and Nutrition Board of the National Research Council (1980), it is found that one large egg supplies a relatively high percentage of essential nutrients (Table 4.7). Eggs are an especially rich source of high-quality protein, vitamins (A, D, E, folic acid, riboflavin, B_{12}, and pantothenic acid), and minerals (phosphorus, iodine, iron, and zinc). Some consumers believe that dark

FIGURE 4.3 "The incredible edible egg." Photo courtesy of the American Egg Board.

TABLE 4.6 Estimated Nutrient Values for a Large Egg (based on 60.9-g shell weight with 55.1-g total liquid whole egg, 38.4-g white, and 16.7-g yolk)

Nutrients and Units (proximate)	Whole	White	Yolk
Solids (g)	13.47	4.6	8.81
Calories (kcal)[a]	84	19	64
	6.60	3.88	2.74
Total lipids (g)	6.00	—	5.80
Cholesterol (mg)	213	—	2.13
Ash (g)	0.55	0.26	0.29

[a] A medium-sized egg has 66 calories, while a jumbo egg has 94 calories.
Source: Adapted from USDA.

yellow yolks are of higher nutrient value than paler yolks, but there is no evidence to support this belief.

Eggs are relatively high in cholesterol, with most of the cholesterol and fat concentrated in the yolk (Table 4.6). Individuals who desire to reduce fat and cholesterol in their diet and still enjoy eggs can (1) limit egg yolks to one serving for scrambled eggs and omelets (use more whites for larger servings) and (2) substitute two egg whites for one whole egg when baking.

Table 4.7 lists the U.S. recommended daily allowances (RDAs) supplied by one large egg.

TABLE 4.7 U.S. Recommended Daily Allowance (RDA) in Relation to Nutrient Content of One Large Egg

Nutrient	Quantity in One Large Egg	Percent RDA One Egg (%)
Calories	75	4
Protein	6.3g	13
Total fat	5 g	8
Vitamins		
A	243 IU	5
D	17 IU	4
E	0.75 IU	3
Folate	24 mcg	6
Thiamin	0.031 mg	2
Riboflavin	0.24 mg	14
B_6	0.07 mg	4
B_{12}	0.6 mcg	8
Minerals		
Phosphorus	95 mg	10
Iron	0.55 mg	3
Zinc	0.91 mg	5

Source: Adapted from USDA and American Egg Board.

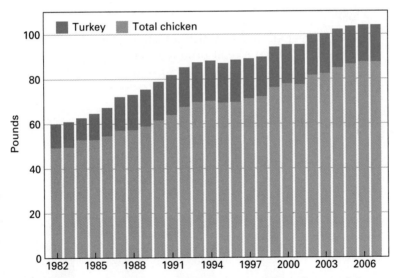

FIGURE 4.4 U.S. poultry consumption per capita, retail weight, annual. *Source:* Livestock Marketing Information Center.

CONSUMPTION

Meat

Worldwide annual poultry and egg consumption averages 24 pounds and 18 hen eggs per capita. The annual per-capita consumption of chicken and turkey in the United States is approximately 100 lb (Fig. 4.4). Poultry consumption is approximately one-half of the total per-capita consumption of red meat and poultry. In recent years, poultry consumption has increased at the expense of red meat consumption on a weight basis. This is due in part to the price differential existing between the various meats.

An important reason for the marked increase in chicken consumption has been the speed of preparing and serving chicken. The Kentucky Fried Chicken Corporation (KFC) led the way in preparing and serving chicken in a short time, and now chicken has become a major item in fast-food service. The KFC restaurants are found in 100 countries. In recent years, other fast-food restaurants have featured chicken or added more chicken items to their menus. Nearly 90% of consumers surveyed stated that they didn't plan to change their chicken consumption, suggesting the market is maturing. Taste, health considerations, and convenience were the primary demand drivers. Price was relatively low on the list of reasons for consuming chicken.

Heavy hens, the type used to produce hatching eggs for broiler production, are sometimes available in retail stores as "stewing hens" or "baking hens." These account for about 10% of the poultry meat consumed by Americans each year. Consumption of turkey meat per person has increased from 1.7 lb in 1935 and 6.1 lb in 1960 to 8.0 lb in 1970 and 18 lb in 2001. Most of this increase in turkey consumption was stimulated by development of processed products such as turkey rolls, roasts, potpies, and frozen dinners. The selling of prepackaged turkey parts in small packages has also increased consumption.

Eggs

Per-capita egg production in the United States is approximately 255 eggs per year resulting from a total production of nearly 77 billion eggs. Most of these are consumed as shell eggs (70%) rather than processed eggs (Fig. 4.5). The

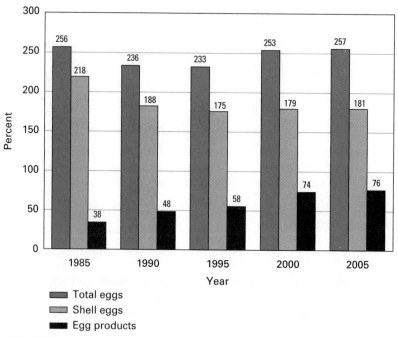

FIGURE 4.5   Per-capita production of eggs. *Source:* Adapted from USDA.

per-capita consumption is nearly the number of eggs (254) laid by the average hen per year. Egg consumption reached its low in 1991 at 234 eggs per capita. The growth in consumption over the past 15 years has been the result of an aggressive marketing and education campaign promoting the versatility, affordability, healthfulness, and value of eggs and egg products.

MARKETING

In the United States, large amounts of poultry meat are processed in commercial facilities and then transported under refrigeration to local and world markets. Chickens are increasingly marketed in further processed form as consumers demand more convenience, variety, and meal solution approaches (Table 4.8). The primary drivers of demand for chicken consumption include taste, convenience of preparation, and perceived health benefits (Table 4.9). The poultry industry has aggressively worked to meet the demands of consumers with a variety of value-added strategies that have resulted in slow and steady growth in consumption (Table 4.10). Consumer studies have shown that among heavy, medium, and light buyers of chicken, the leading products were salads using chicken as an ingredient, chicken strips, ready-to-eat rotisserie-roasted chicken

TABLE 4.8   Form in Which Broilers Are Marketed in Retail Groceries

Year	Whole (%)	Cut-up (%)	Further Processed (%)
1970	70	26	4
1980	50	40	10
1990	18	56	26
2000	10	44	46
2005	7	41	52

Source: Adapted from USDA and National Chicken Council.

TABLE 4.9   Consumer Perceptions about Chicken

Response	%
Primary reason for eating chicken	
Taste	38
Health	28
Ease of preparation	10
Like specific parts (white meat, legs, breast, etc.)	7
Method of preparation	7
Price	5
Other	5

Source: Adapted from Broiler Industry and National Broiler Council.

TABLE 4.10   Frequency of Chicken Consumption on a Weekly Basis

Point of Purchase	2001	2003	2005
Supermarket	1.2	1.45	1.55
Foodservice	0.85	1.0	1.0
Total	2.0	2.4	2.45

Source: Adapted from USDA and National Chicken Council.

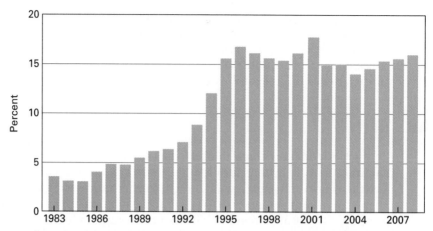

FIGURE 4.6   U.S. Broiler exports as a percentage of broiler production. *Source: Livestock Marketing Information Center.*

from supermarkets, spicy chicken wings, and chicken nuggets. In each case, the product represents value to consumers in versatility, flavor, convenience, and price. Another study found that boneless, skinless chicken breasts were clearly the preferred chicken product in retail groceries with two-thirds of chicken consumers usually buying chicken in this form.

Broilers are the primary poultry meat exported, valued at more than $2.1 billion. Approximately 62 million cases (dozen per case) of eggs valued at $38 million as well as processed egg products worth an additional $123 million are exported annually from the United States. The share of broiler meat exported is shown in Figure 4.6.

Chicken and turkey meat is marketed in a variety of forms. Most of the broilers are sold to consumers through retail stores. Fresh chicken can reach the retail market counter the day following slaughter. The "sell-by" date on each package is usually 7 days after processing and is the last date recommended for sale of chicken. However, with proper refrigeration (28–32°F), shelf life can be extended up to 3 days longer. Some processors guarantee 14 days of shelf life for their broilers.

Federal or state government employees inspect almost all commercially available chicken and turkey meat for wholesomeness. The USDA quality grades about two-thirds of the chicken harvested in the United States.

In an effort to differentiate product lines to meet the unique desires of niche markets, poultry companies have utilized labeling strategies that align production methods with the desire of some consumers for products touted as organic, natural, and free range. The use of labeling claims has in reality created confusion among some consumers. For clarification, the basic definitions for these "attribute"-based labeling categories are as follows:

Free range—birds have access to an enclosed pen outside of the poultry house. Because feed and water are provided in the house structure birds are often reluctant to utilize the pen for long periods of time. Less than 1% of chickens are raised under free-range conditions.

Natural—this is perhaps the most misunderstood term in food production. USDA defines "natural" meats as minimally processed and having no added ingredients. Almost all fresh poultry meets these standards.

Organic—this production system prohibits the use of antibiotics, pesticides, and many farm chemicals and specifies utilization of feedstuffs made from organic ingredients.

Produced without hormones—this label is misleading in that no artificial or added hormones are used by the poultry industry in accordance with FDA regulations.

Antibiotic-free—this label indicates that the production system did not incorporate any animal health care products classified as antibiotics in either a preventative or treatment form. Health care products not classified as antibiotics may still be incorporated.

Fresh—this term means that the product has not been chilled to temperatures below 26 degrees Fahrenheit.

In-store promotions are frequently utilized to enhance sales of food products in retail outlets. Reducing the price of the lead item in a menu (chicken breast for example) and buy one–get one free offers are the most effective strategies (Table 4.11).

Before and during the 1950s, the marketing of turkey was highly seasonal as 90% of the turkeys were consumed during the last quarter of the year for the Thanksgiving and Christmas holidays. Today less than 40% of turkey sales are made during that period. Current trends indicate that turkey is being consumed more on a year-round basis and has become a meat staple. However, turkey consumption has leveled off in the past decade.

The egg industry has invested significant resources into expanding the consumption of eggs beyond the traditional use of eggs in breakfast items (Table 4.12). Consumers rank freshness, product integrity (no broken/cracked

TABLE 4.11　Effectiveness of Various Retail Sales Promotions

Promotion	Effectiveness*
Price reduction of lead item	4.68
Buy one–get one free offer	4.02
Price reduction of secondary item	3.26
Multiple purchases (2-for-1, etc.)	3.02
Coupon	2.63
Free with purchase of another item	2.56
Extended price reduction	2.28

*5-point scale, with 5 being most effective.
Source: Adapted from American Egg Board, 2002.

TABLE 4.12　Egg Use by Consumers

Use	Consumers (% mentioning)
Entrees	31
Cooking	30
Breakfast and cooking	28
Breakfast	23
Recipes for meals	11
Ingredient	8
Salads	4
All ways	3

Source: Adapted from American Egg Board, 2002.

TABLE 4.13 Items Simultaneously Purchased Along With Eggs

Item	% of Consumers
Bacon, ham, sausage	41
Baking ingredients	26
Cheese	16
Breads, pastries	16
Butter	11
Milk	11
Produce	9
Condiments, spices	8

Source: Adapted from American Egg Board, 2002.

eggs), cleanliness, and price as leading factors affecting purchase of table eggs at retail. Consumers typically group purchases around a meal concept and, thus, by increasing demand for an item such as eggs, the sale of other products (many of animal origin) can be increased (Table 4.13). Breakfast meats, cheese, milk, and butter are frequently purchased in conjunction with egg sales.

Retail sales strategies in the future will include cooperative merchandising efforts by diverse food providers to enhance the overall sales of several items. For example, pork processors and egg wholesalers will find mutually beneficial approaches in selling their products. Furthermore, it will be increasingly important for the livestock industry to correctly interpret consumer signals, implement effective sales strategies, and develop profitable partnerships with both food retailers and food service companies as a means to assure market access.

CHAPTER SUMMARY

- Poultry meat and egg products are nutritious and relatively inexpensive animal products used by humans throughout the world.

- Poultry meat and eggs are excellent sources of protein, vitamin A, and several B vitamins for human nutrition.

- Annual per-capita consumption of broilers is 84 lb and has been increasing rapidly over the past several years. Per-capita turkey consumption is 18 lb and stabilizing. Egg consumption has been increasing and currently is 257 eggs per year.

- Broiler and turkey retail prices are much lower than prices of beef and pork.

KEY WORDS

offal	3's and UP	light (white) meat
Broiler	Roaster	cuticle (bloom)
Poussin	Broilers for deboning	albumen
Cornish game hens	Capon	vitelline membrane
Fast-food oriented broiler	Heavy hens	down
	dark meat	feathers

REVIEW QUESTIONS

1. What major products come from poultry?
2. What essential nutrient is poultry meat very high in?
3. What is the primary poultry meat exported by the United States?

SELECTED REFERENCES

Eggcyclopedia. 2006. American Egg Board, Park Ridge, IL 60068.

Martinez, S. W. 1999. *Vertical Coordination in the Pork and Broiler Industries: Implications for Pork and Chicken Products.* ERS: USDA.

Moreng, R. E., and J. S. Avens. 1985. *Poultry Science and Production.* Reston, VA: Reston Publishing.

Quality of U.S. Agricultural Products (poultry, p. 195). 1996. Ames, IA: Council for Agricultural Science and Technology.

Scanes, C. G., G. Brandt, and G. E. Ensminger. 2004. *Poultry Production.* Upper Saddle River, NJ: Pearson Prentice Hall.

Scanes, A. R. 1994. Poultry processing and products. *Encyclopedia of Agricultural Science.* San Diego, CA: Academic Press, Inc.

Stadelman, W. J. 1994. Egg production, processing, and products. *Encyclopedia of Agricultural Science.* San Diego, CA: Academic Press, Inc.

CHAPTER 5

Milk and Milk Products

LEARNING OBJECTIVES

- Describe the composition of milk
- Quantify the production of milk and dairy products
- Describe the processing of milk and dairy products
- Outline the nutritive value of milk
- Describe the consumption patterns of milk and dairy products
- Overview the marketing of milk products

The world's population obtains most of its milk and milk products from cows, water buffalo, goats, and sheep (Fig. 5.1). Horses, donkeys, reindeer, yaks, camels, and sows contribute a smaller amount to the total milk supply. Milk, with its well-balanced assortment of nutrients, is sometimes called "nature's most nearly perfect food." While milk is an excellent food product in many ways, it is not perfect, nor is any other food. Milk and numerous milk products (e.g., cheese, butter, ice cream, and cottage cheese) are major components of the human diet in many countries.

Changes in dietary preferences and milk marketing are some of the challenges facing the dairy industry. Health considerations and new products are changing the consumption patterns of milk and milk products. Milk surpluses in the United States and the world result in an abundance of consumer products, but the surpluses represent economic challenges to milk producers.

This chapter focuses primarily on milk as human food, but the importance of milk in nourishing suckling farm mammals should not be overlooked. When the term *milk* is used, reference is to milk from dairy cows unless otherwise specified.

MILK PRODUCTION

Table 5.1 shows the most important sources of milk for human consumption. Total world milk production has increased during the past 25 years, but not at the same rate as world human population. Dairy cows produce over 80% of the

FIGURE 5.1 Milk plays an important role in human nutrition throughout the world. (A) Goats being milked in Mexico. Courtesy of Winrock International. (B) Traditional milking in Ugandan cattle herd. Courtesy of FAO. (C) Herdsmen milking camels at a dairy plant in Mauritania. (D) Milking cows in a modern U.S. dairy. Courtesy of Colorado State University.

TABLE 5.1 Major Sources of Milk Production

Species	Total Production (bil lb)	Leading Regions (Milk from Cows) (bil lb)
Cows	1,088	North America—221
Buffalo	153	South America—77
Goats	27	European Union—292
Sheep	17	Former Soviet Union—101
All others	4	Asia—93
World Total	1,202	Oceania—55

Source: Adapted from FAO.

FIGURE 5.2 Dairy cows have the ability to convert feedstuffs into milk. Milk is an important source of nutrients to humans throughout the world.

FIGURE 5.3 Dairy goat milk can be processed into a variety of high-quality cheeses.

world fluid milk supply (Fig. 5.2). During the past 30 years, milk yield from buffalo and sheep has been increasing while goat milk production has been decreasing. However, most milk from goats is produced and consumed by individuals or families who raise a few goats. Goat's milk can be processed into numerous value-added cheeses (Fig. 5.3).

Leading sources of global milk supply are also shown in Table 5.1.

The national dairy herd in the United States has decreased from over 20 million cows in 1956 to approximately 9 million in 2006. Yet total milk production

FIGURE 5.4   Milk, cheese, and yogurt are among the most highly preferred milk products. Courtesy of American Dairy Farmers.

has increased. Reduction in cow numbers has been offset by an increased production per cow—from 5,800 lb in 1956 to more than 18,200 lb in 2002. The 18,200 lb per cow converts to 2,111 gal (1 gal = 8.62 lb), which would supply the annual per-capita fluid milk consumption for nearly 85 people in the United States. However, since part of the milk is made into other dairy products (Fig. 5.4), the milk from one cow provides milk and manufactured products (cheese, frozen dairy products, and others) for about 40 people.

Fluid milk sales are important to the U.S. economy. Consumer expenditures for dairy products in the United States are in excess of $90 billion annually.

MILK COMPOSITION

Milk is a colloidal suspension of solids in liquid. Fluid whole milk is approximately 88% water, 8.6% **solids-not-fat (SNF)**, and 3–4% milk fat. The SNF is the total solids minus the milk fat. It contains protein, lactose, and minerals.

Even though milk is a liquid, its 12% total solid is similar to the solids content of many solid foods. Differences in milk composition for several species of animals are given in Chapter 19.

The first milk a female produces after the young is born is called **colostrum.** True colostrum is milk obtained during the first milking. For the next several milkings, the milk is called *transitional milk,* and it is not legally salable until the 11th milking. Colostrum differs greatly in composition from milk. Colostrum is higher in protein, minerals, and milk fat, but it contains less lactose than milk. The most remarkable difference between colostrum and milk produced later is the extremely high immunoglobulin content of colostrum. These protein compounds accumulate in the mammary gland and they are the antibodies that are transferred through milk to the suckling young. People seldom use colostrum from animals because of its unpleasant appearance and odor, although a few may use it for pudding. Additionally, the high globulin content causes a precipitation of proteins when colostrum is pasteurized.

Milk contains a high nutrient density—more than 100 milk components have been identified. **Nutrient density** refers to the concentration of major nutrients in relation to the caloric value of the food. In other words, milk contains important amounts of several nutrients while being relatively low in calories. The major components of milk solids can be grouped into nutrient categories including proteins, fats, carbohydrates, minerals, and vitamins.

Milk Fat

In whole milk, the approximate 3–4% **milk fat** is a mixture of lipids existing as microscopic globules suspended in the milk. The fat contributes about 48% of the total calories in whole milk. Fat-soluble vitamins (A, D, E, and K) are normal components in milk fat. Milk fat contains most of the flavor components of milk, so when milk fat is decreased, there may be a concurrent reduction in flavor.

Approximately 500 different fatty acids and fatty-acid derivatives from 2 to 26 carbon atoms in length have been identified in milk. Milk fat from ruminant animals contains both short-chain and long-chain fatty acids.

Carbohydrates

Lactose, the predominant carbohydrate in milk, is synthesized in the mammary gland. Approximately 4.8% of cow's milk is lactose. It accounts for approximately 54% of the SNF content in milk. Lactose is only about one-sixth as sweet as sucrose, and it is less soluble in water than other sugars. It contributes about 30% of the total calories in milk. Milk is the only natural source of lactose.

Proteins

Milk contains approximately 3.3% protein. Protein accounts for about 38% of total SNF and about 22% of the calories of whole milk. The proteins of milk are of high quality. They contain varying amounts of all amino acids required by humans. A surplus of the amino acid lysine offsets the low lysine content of vegetable proteins and particularly cereals.

Casein, a protein found only in milk, is approximately 82% of the total milk protein. Whey proteins, primarily lactalbumin and lactoglobulin, constitute the remaining 18%. Immunoglobulins, the antibody proteins of colostral milk, were discussed earlier in the chapter.

Vitamins

All vitamins essential in human nutrition are found in milk. Fat-soluble vitamins are in the milk fat portion of milk, and water-soluble vitamins are in the nonfat portion. Milk is usually fortified with vitamin D during processing.

Milk fat from Jersey and Guernsey cows has a rich, yellow color due to carotene (a precursor of vitamin A). Milk fat from Holstein cows has a pale yellow color, and goat milk fat is white. Carotene can be split into two molecules of vitamin A. The milk fat of Holstein cows is pale yellow because most of the carotene has been split, and the milk fat of goats is white because all of the carotene has been changed to colorless vitamin A.

Water-soluble vitamins (C and B) are relatively constant in milk and are not greatly influenced by the vitamin content of the cow's ration. Rumen microorganisms produce the B vitamins, and vitamin C is formed by healthy epithelial tissue in most animals (excluding humans, other primates, and guinea pigs).

Minerals

Milk is a rich source of calcium for the human diet and a reasonably good source of phosphorus and zinc. Calcium and vitamin D are needed in combination to contribute to bone growth in young humans and to prevent **osteoporosis** in adults, particularly women. Milk is not a good source of iron, and the iodine content of milk varies with the iodine content of the animal's feed.

MILK PRODUCTS IN THE UNITED STATES

Table 5.2 and Figure 5.5 show how the milk supply is used to produce various milk products. More than 80% of the milk produced is marketed as fluid milk, cream, cheese, and butter.

The amount of milk needed to produce each product depends primarily on the milk fat content of the milk. Table 5.3 gives the approximate amount of milk used to produce each product.

TABLE 5.2   Milk and Milk Products Resulting from the 2004 Milk Supply

Product	Total Production (bil lb)
Fluid milk and cream	59.8
Cheese	8.9
Butter	1.3
Ice cream	1.0
Nonfat dry milk	1.4
Other uses	24.1

Source: Adapted from USDA.

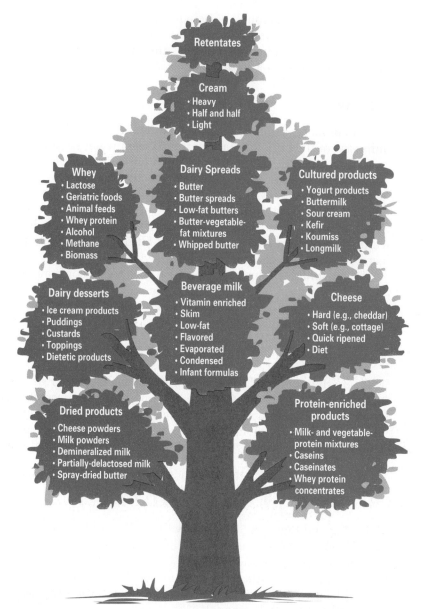

FIGURE 5.5   The modern dairy tree showing the many products and by-products of milk. *Source: J. Dairy Sci.* 64:1005.

Fluid Milk

Approximately 92% of the 156 billion pounds of milk produced in the United States is grade A quality milk. Grade A milk in excess of fluid milk demand is processed into other milk products. Figure 5.6 shows how fluid milk typically moves from farm to consumer.

Depending on the milk fat content and the processing at the plant, the fresh fluid product is labeled whole milk, low-fat milk, or skim milk. In further discussion in this chapter, *milk* refers to whole milk. Whole milk is defined as a

TABLE 5.3 Approximate Amount of Milk Used to Produce Several Selected Dairy Products

Dairy Product (lb)	Whole Milk (lb)
Butter	21.8
Whole-milk cheese	9.23
Evaporated milk	2.1
Condensed milk	7.4
Ice cream (1 gal)	12.0
Cottage cheese	7.2
Nonfat dry milk	2.15

Source: USDA.

A

B

C

D

FIGURE 5.6 Milk moves from the farm (A) via tanker trucks (B) to processing facilities where it is packaged into half-gallon (C) or gallon containers (D) for distribution to consumers. Photos A, C, and D courtesy of Dairy Management, Inc.

lacteal secretion, and when it is packaged for beverage use it must contain not less than 3.25% milk fat and not less than 8.25% milk SNF. Low-fat milk usually has had some or most milk fat removed to have one of the following milk fat levels: 0.5%, 1.0%, 1.5%, or 2.0%, with not less than 8.25% milk SNF. Skim milk has had most of the milk fat removed (it contains less than 0.5% milk fat). It must contain at least 8.25% SNF and may be fortified with nonfat solids to 10.25%.

Most fluid milk consumed in the United States is **homogenized** to prevent milk fat from separating from the liquid portion and rising to the top. Homogenization is a physical process resulting in a stable emulsion of milk fat.

A "cream line" does not appear in homogenized milk, nor does it form butter if churned.

Fat globules in raw milk average about 6 μm in diameter. To visualize how small the fat globules in homogenized milk are, note that 25,000 μm equal approximately 1 in. Homogenized milk will likely deteriorate by becoming rancid more rapidly than nonhomogenized milk. Rancidity of homogenized milk is forestalled by pasteurization of milk prior to or immediately following homogenization, thus destroying the action of lipolytic enzymes.

Evaporated and Condensed Milk

The process of preheating to stabilize proteins and then removing about 60% of the water produces **evaporated milk.** It is sealed in the container and then heat-treated to sterilize its contents. Milk fat and SNF of evaporated milk must be at least 7.5% and 25%, respectively. Evaporated skim milk must have at least 20% SNF and not more than 0.5% milk fat. Evaporated milk requires no refrigeration until opened. Once the can is opened, refrigeration is necessary to avoid spoilage.

Concentrated, or **condensed, milk** has milk fat (7.5%) and SNF (25.5%) requirements similar to those for evaporated milk. Concentrated milk also has water removed but, unlike evaporated milk, it is not subjected to further heat treatment to prevent spoilage. Most concentrated milk is sold bulk for industry use. A common form of concentrated milk on grocery shelves is sweetened (sugar added) condensed milk; it is used for candy and other confections.

Dry Milk

Dry milk is prepared by removing water from milk, low-fat milk, or skim milk. All of these products are to contain no more than 5% moisture by weight. Dry milk can be stored for long periods of time if it is sealed in an atmosphere of nonoxidizing gas, such as nitrogen or carbon dioxide. The spray-dried or foam-dried product can be reconstituted easily in warm or cold water with agitation.

Fermented Dairy Products

Buttermilk, yogurt, sour cream, and other **cultured** milk products are produced under rigidly controlled conditions of sanitation, inoculation, incubation, acidification, and/or temperature. The word *cultured* appearing in a product name indicates the addition of appropriate bacteria cultures to the fluid dairy product and subsequent fermentation. Selected cultures of bacteria convert lactose into lactic acid, which produces a tart flavor. The specific bacterial cultures used and other controls in the fermentation process determine whether buttermilk, sour cream, yogurt, or cheese is the end product.

Today buttermilk is a cultured product rather than the by-product of churning cream into butter as was the procedure in the past. At the correct stage of acid and flavor development, the buttermilk is stirred gently to break the curd that has formed. It is then cooled to stop the fermentation process.

The word **acidified** in the product name or production process indicates that the food was produced by souring milk or cream with or without the addition of microbial organisms.

Cottage cheese dry curd is soft unripened cheese; it is produced by culturing or direct acidification. Finished dry curd contains less than 0.5% milk fat and not more than 80% moisture. Cottage cheese is prepared for market by mixing cottage cheese dry curd with a pasteurized creaming mixture called *dressing.* The finished product contains not less than 4% milk fat and not over 80% moisture. The same process makes low-fat cottage cheese; however, the milk fat range is 0.5–2%.

Sour cream is generally made from lactic acid fermentation, although rennet extract is often added in small quantities to produce a thicker-bodied product. Federal standards provide that cultured sour cream contain not less than 18% milk fat. When stored for more than 3–4 weeks, sour cream may develop a bitter flavor as a result of continued bacterial proteolytic enzyme activity.

Yogurt, as a liquid or gel, can be manufactured from fresh whole, low-fat, or skim milk that is heated before fermentation. Federal standards specify that yogurt contain not less than 3.25% milk fat, low-fat yogurt between 0.5–2% milk fat, and nonfat yogurt not more than 0.5% milk fat before any bulky flavors are added.

Today three main types of yogurt are produced: (1) flavored, containing no fruit; (2) flavored, containing fruit (fruit may be at the bottom or blended in); and (3) unflavored.

Cream

Cream is a liquid milk product, high in fat that has been separated from milk. Federal standards require that cream contain not less than 18% milk fat.

Several cream products are marketed (Fig. 5.7). Half-and-half is a mixture of milk and cream containing not less than 10.5% but less than 18% milk fat. Light

FIGURE 5.7 A variety of cream products are produced for the consumer market. Courtesy of American Dairy Farmers.

cream (coffee or table cream) contains not less than 18% but less than 30% milk fat. Light whipping cream or whipping cream contains not less than 30% but less than 36% milk fat. Heavy cream or heavy whipping cream contains not less than 36% milk fat. Dry cream is produced by removal of water only from pasteurized milk and/or cream. It contains not less than 40% but less than 75% milk fat and not more than 5% moisture.

Butter

It is known from food remnants in vessels found in early tombs that the Egyptians cooked with butter and cheese. For many centuries, butter making and cheese making were the only means of preserving milk and cream.

Butter churns are the oldest dairy equipment, dating from prehistoric days when nomads carried milk in a type of pouch made from an animal's stomach. Slung on the back of a horse or camel, the pouch bounced as the animal moved, churning the cream or milk into butter.

Wooden or crockery dasher-type churns used at the turn of the century are today's attic treasures. Jouncing a long-handled wooden dasher up and down in the deep churn sloshed the cream around until butter particles separated from the remaining liquid, called *buttermilk*.

Butter is made exclusively from milk or cream, or both, and contains not less than 80% milk fat by weight. Federal standards establish U.S. grades for butter based on flavor, color, and salt characteristics.

In modern butter making, fresh sweet milk is weighed, tested for milk fat content, and checked for quality. The cream is then separated by centrifugation to contain 30–35% milk fat for batch-type churning or 40–45% milk fat for continuous churning.

Continuous butter-making operations, which can produce 1,800–11,000 lb an hour, are an industry trend. New continuous butter-making processes employ several different scientific principles to form butter. On a global basis, India is the leading butter-producing nation generating nearly 3 times more than all of North America.

Cheese

There are more than 400 different kinds of cheese that can be made (Fig. 5.8). The cheeses have more than 2,000 names, since the same cheese may have two or more different names. For example, cheddar cheese is one of the American-type cheeses in the United States, along with Colby, washed or stirred curd, and Monterey. Cheese types as a percent of production in the United States are cheddar (34%), Mozzarella (33%), American (8%), other Italian (8%), Swiss (3%), and all other (14%).

Most cheeses and cheese products are classified into one of four main groups: (1) soft, (2) semisoft, (3) hard, and (4) very hard. Classification is based on moisture content in the cheese. Thus the body and texture of cheeses range from soft unripened cheese (such as cottage cheese with 80% moisture) to very hard, grated, shaker cheeses such as Parmesan and Romano. These latter, ripened cheeses have 32–34% moisture.

Cheese flavor varies from bland cottage cheese to pungent Roquefort and Limburger. Modern microbiology makes it possible to add specific species and

FIGURE 5.8 A variety of cheeses have been produced to satisfy the varied preferences of consumers.

strains of microorganisms required to produce the desired product. For example, a fungus *(Penicillium roqueforti)* is used to produce Roquefort cheese.

The most dramatic increase in cheese production in the United States has occurred with Italian varieties. The phenomenal growth of the pizza industry has caused an increased production and importation of Mozzarella cheese.

Cheese making involves a biochemical process called **coagulation** or **curdling.** First the milk is heated, and then a liquid starter culture is added. Bacteria from the culture form acids and turn the milk sour. At a later time, rennet is added to thicken the milk. **Rennet,** obtained from the stomachs of young calves, contains the enzyme rennin. Other enzymes of microbial origin have been used successfully to replace a declining supply of rennin. After this mixture is stirred, a custard-like substance called **curd** is formed. Then the liquid part **(whey)** is removed. The cheese-making process reduces 100 lb of milk to 8–16 lb of cheese.

In cheese plants, disposal of whey is a problem. It is currently estimated that 35 billion pounds of whey containing more than 4 billion pounds of solids are produced. Slightly more than half the whey is used to produce human food or animal feeds. The remaining whey poses waste-disposal problems. There continues to be an increasing industrial use of whey for production of food, feed, fertilizer, alcohol, and insulation.

Table 5.4 identifies the leading countries in yearly cheese production and the most important cheese-producing states in the United States.

Ice Cream

Although they come in many variations, ice cream and a group of similarly frozen foods are made in a similar way and have many of the same ingredients.

TABLE 5.4   Leading Cheese-Producing Regions and States

Country	Total Production (bil lb)	State	Total Production[a] (mil lb)
European Union	14.0	Wisconsin	2,356
North America	9.8	California	1,996
South America	1.8	Idaho	718
Oceania	1.5	New York	700
Former Soviet Union	1.2	Minnesota	623
World Total	29.7	**U.S. total**	8,876

[a] Does not include cottage cheese.
Source: USDA.

Ice cream, frozen custard, French ice cream, and French custard ice cream are frozen dairy products high in milk fat and milk solids. Ice cream may contain egg yolk solids. If egg yolks are in excess of 1.4% by weight, the product is called frozen custard, French ice cream, or French custard ice cream. The leading states for ice cream production (% U.S. production) are California (14%), Indiana (10%), Illinois (5%), Pennsylvania (5%), and New York (3%).

Ice milk contains less milk fat, protein, and total solids than ice cream. Ice milk usually has more sugar than does ice cream. Soft ice milk and soft ice cream are soft and ready to eat when drawn from the freezer. About three-fourths of the soft-serve products are ice milks. Some of the soft-serve frozen dairy foods contain vegetable fats that have replaced all or most of the milk fat.

Frozen yogurt has become a popular dairy product. Frozen yogurt has less milk fat and higher acidity than ice cream and less sugar than sherbet.

Sherbet is low in both milk fat and milk solids. It has more sugar than ice cream. The tartness of fruit sherbet comes from the added fruit and fruit acid. Sherbet without fruit is flavored with such ingredients as spices, coffee, or chocolate.

Water ices are nondairy frozen foods. They contain neither milk ingredients nor egg yolk. Ices are made similarly to sherbet.

Mellorine differs from ice cream in that milk fat may be replaced partially by a vegetable fat, another animal fat, or both.

Eggnog

Eggnog contains milk products, egg yolk, egg white, and a nutritive carbohydrate sweetener. In addition, eggnog may contain salt, flavoring, and color additives. Federal standards specify that eggnog shall contain not less than 6% milk fat and 8.25% SNF.

Imitation Dairy Products

The FDA established regulations in 1973 to differentiate between imitation and substitute products. It defined an imitation product as one that looks like, tastes like, and is intended to replace the traditional counterpart but is nutritionally inferior to it. A substitute product resembles the traditional food and also meets the FDA's definition of nutritional equivalency.

Imitation milks usually contain ingredients such as water, corn syrup solids, sugar, vegetable fats, and protein from soybean, sodium caseinate, or other sources. Although imitation fluid milk may not contain dairy products per se, it may contain derivatives of milk such as whey, lactose, casein, salts of casein, and milk proteins, other than casein.

The dairy industry has developed a program for identification of real dairy foods. A "REAL" seal on a carton or package identifies milk or other dairy foods made from U.S.-produced milk that meets federal and/or state standards. This seal assures consumers that the food is not an imitation or substitute.

HEALTH CONSIDERATIONS

Nutritive Value of Milk

Milk and other dairy products make a significant contribution to the nation's supply of dietary nutrients. Particularly noteworthy are the relatively large percentages of calcium, phosphorus, protein, and B vitamins (Table 5.5).

Human milk is regarded as the best source of nourishment for infants. Cow's milk for infant feeding is modified to meet the nutrient and physical requirements of infants. Cow's milk is heated, homogenized, or acidified so infants can utilize the nutrients efficiently. Sugar is usually added to cow's milk for infant feeding to make the milk more nearly like human milk.

Milk is low in iron; therefore, young animals consuming nothing but milk may develop anemia. Baby pigs produced in confinement need a supplemental source of iron, usually given as an injection. Babies typically receive supplemental iron, and young children who consume large amounts of milk at the expense of meat should be given supplemental iron.

Milk and milk products are excellent sources of nutrients to meet the dietary requirements of young children, adolescents, and adults. With aging, there is an increased prevalence of osteoporosis, a problem of loss of bone

TABLE 5.5 Percentage Contribution of Dairy Foods (excluding butter) to Nutrients Consumed in the United States

Nutrients	1970	1980	1990	2000
Energy	10	10	9	9
Protein	20	20	20	19
Fat	12	12	12	13
Carbohydrate	6	6	5	5
Cholesterol	14	13	15	16
Minerals				
Calcium	74	75	75	72
Phosphorus	35	34	34	32
Magnesium	20	19	18	16
Iron	2	2	2	2
Zinc	18	18	19	16
Vitamins				
Ascorbic acid	4	3	3	2
Thiamin	9	7	7	5
Riboflavin	34	31	31	26
Vitamin B_6	11	11	10	9
Vitamin B_{12}	17	18	20	22
Vitamin A	17	17	18	15

Source: USDA.

mass. Since osteoporosis involves the loss of bone matrix and minerals, a diet generous in protein, calcium, vitamin D, and fluoride is recommended to overcome the problem. Thus, milk, owing to its content of calcium and other nutrients, is an important food for this segment of the population. Furthermore, for the elderly as well as for the infant and young child, milk is an efficient source of nutrients readily tolerated by a sometimes-weakened digestive system.

Wholesomeness

Milk is among the most perishable of all foods owing to its excellent nutritional composition and its fluid form. As it comes from the cow, milk provides an ideal medium for bacterial growth. Properly processed milk can be kept for 10–14 days under refrigeration. With **ultra-high-temperature (UHT) processing,** milk can be kept for several weeks at room temperature.

Protecting the quality of milk is a responsibility shared by public health officials, the dairy industry, and consumers. The U.S. Food and Drug Administration describes milk as "one of the best controlled, inspected and monitored of all food commodities."

The most important safeguards from a health standpoint are requirements that bacterial counts be low and that milk be pasteurized (Fig. 5.9). Milk sold through commercial outlets is certified to be from herds that are tested and found to be free from brucellosis and tuberculosis. The consumer who buys milk and milk products can be assured of obtaining a safe, desirable, wholesome food. There are definite health risks in consuming unpasteurized cow's or goat's milk. This concern is important where families or individuals have purchased an animal with an unknown health background or have purchased milk that has not been processed.

FIGURE 5.9   Milk samples are evaluated for undesirable levels of bacteria. Courtesy of Dairy Management, Inc.

FIGURE 5.10   Large tanker trucks move milk from dairy farms to processing plants. The tanks are washed and sanitized after each delivery. Courtesy of American Dairy Farmers.

Milk Processing

Milk is taken from the cow by a sanitized milking machine, and then transported through sanitized pipes into holding tanks. When withdrawn from the cow, milk is at the cow's body temperature of about 100°F (38°C). It flows into a refrigerated tank, where it is rapidly cooled to 40–42°F. Cold temperatures maintain the high quality of the milk while it is held for delivery.

Tank truck drivers who pick up milk inspect it to see if it is cold and has an acceptable aroma. They take a sample for testing later at the processing plant. Then the cold milk is pumped from the refrigerated farm tank through a sanitized hose into the insulated tank on the truck (Fig. 5.10).

After the milk sample passes several tests at the processing plant, the milk is pumped through sanitized pipes into the processing plant's refrigerated or insulated holding tanks.

Milk is pasteurized at the processing plant. **Pasteurization** is a process of exposing milk to a temperature that destroys all pathogenic bacteria but neither reduces the nutritional value of milk nor causes it to curdle. Milk is most commonly pasteurized at 161°F (71.5°C) for 15 seconds.

Ultrapasteurized milk and UHT processed milk are heated to 280°F (138°C) for at least 2 seconds; this sterilizes milk and increases shelf life.

As it flows out of the pasteurizer, most milk is homogenized by being pumped through a series of valves under pressure. The milk fat is broken up into particles too small to coalesce. As a result, they remain suspended throughout the milk rather than separating to the top as a layer of cream.

The pasteurized, ultrapasteurized, or UHT processed milk is cooled rapidly to below 45°F (7°C). UHT milk is packaged into presterilized containers and aseptically sealed. Since bacteria cannot enter the UHT milk, it can be kept unrefrigerated for at least 3 months. However, once the container is opened, the UHT milk picks up organisms from the air. Then the UHT milk must be handled and stored like any other fluid milk product.

Milk Intolerance

Milk is the main dietary source of a carbohydrate called *lactose*. In the 1960s, the potential problem of lactose intolerance was emphasized when reports revealed low levels of the enzyme lactase, which aids in the digestion of lactose, in the digestive tracts of 70% of black and 10% of white persons in the United States. Symptoms of lactose intolerance include bloating, abdominal cramps, nausea, and diarrhea. Worldwide, lactose intolerance is relatively high among nonwhite populations. However, results of many studies have disclosed the difference between lactose intolerance and milk intolerance. While a large segment of certain populations may be diagnosed as lactose-intolerant, most of these individuals, once adapted, can tolerate the amount of lactose contained in typical servings of dairy foods.

For the few individuals who are truly milk-intolerant, suitable alternatives are available. These include consumption of recommended amounts of milk in smaller but more frequent servings throughout the day, most cheeses, many fermented and culture-containing dairy products (e.g., yogurt), lactose-hydrolyzed milk, and some dairy products containing up to 75% less lactose.

Milk protein allergy may be considered as another form of milk intolerance. An allergic reaction to the protein component of cow's milk may occur in a few infants who experience allergies early in life. The incidence is probably 1% or less in the infant and child population in industrialized countries, although a range of 0.3–7.5% has been reported. The condition is usually outgrown by 2 years of age, beyond which time true allergic reactions to milk in the general population are rare.

CONSUMPTION

Population growth will no doubt cause an increase in consumption of dairy products; however, per-capita consumption is expected to decrease.

Table 5.6 shows the per-capita sales of milk and milk products in the United States over the past 30 years. In 2005, consumers purchased 205 lbs of total dairy products per capita. There is a noticeable consumer preference for low-fat milk. However, sales of butter, cream, and cheese have risen in the past decade. Figure 5.11 reflects the changes in consumer preferences during the past 30 years, with skim milk, low-fat milk, and cheese showing the most significant changes.

TABLE 5.6   Yearly Per-Capita Sales of Dairy Products in the United States (lb)

Product	1975	1985	1995	2000	2005
Plain whole milk	165.0	116.7	72.1	71.4	61.4
Low-fat milk/skim milk	64.7	95.9	125.8	128.9	117.1
Buttermilk	4.7	4.4	2.8	2.2	1.8
Yogurt	2.1	4.1	4.8	5.4	9.2
Eggnog	0.4	0.5	0.4	0.4	0.4
Half-and-half, light cream, heavy cream	3.4	4.4	3.1	6.3	7.9
Sour cream and dips	1.6	2.3	2.7	3.3	4.2
Butter	4.4	3.9	4.2	4.8	4.6
American cheese (includes cheddar)	8.2	12.2	11.8	12.9	12.9
Other cheese	6.1	10.4	15.4	16.8	17.0
Cottage cheese	4.6	4.1	2.8	2.7	2.6
Ice cream	18.6	18.1	16.1	16.8	15.4
Ice milk	7.6	6.9	7.7	7.9	7.8

Source: Adapted from USDA.

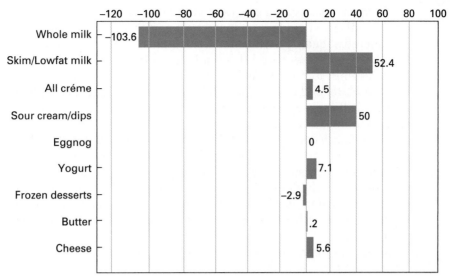

FIGURE 5.11 Percent changes in dairy product consumption 1975–2005. *Source:* USDA.

MARKETING

World

Only a small percentage of the world's production of milk products enters world trade: butter (4%), cheese (4%), skim milk powder (18%), and casein (80%). The European Union (EU) and New Zealand account for about 80% of total world exports of dairy products.

The EU, Canada, and the United States have surpluses of skim milk powder in part because of price stabilization programs and because production costs exceed world market prices. Exports of skim milk powder go principally to countries with a milk deficit and are used in a variety of recombined products. Much of the New Zealand and Australian production of casein is exported to the United States.

Nutritionally, the world may need milk, but there is no real market for its present surpluses of dairy products. Surpluses continue to be a major marketing challenge both in the United States and the world.

United States

Most milk produced on U.S. dairy farms goes to plants and dealers for processing. More than 50% of fluid milk is marketed by supermarkets, primarily in plastic gallon containers.

Prices

Cooperative milk marketing associations located near large population centers give producers more bargaining power, as they control approximately 90% of the milk produced. The cooperatives hire professionals to market the milk to prospective purchasers.

There are classes and grades of milk that determine price. Grade A (fluid or market milk) and grade B (manufacturing milk) are determined by the sanitary and microbial quality of the milk. Class I, the highest price class, is grade A milk for fluid use. Classes II and III are milk used in manufactured products. Class II includes milk for cottage cheese, cream, and frozen desserts; class III is milk used for butter and cheese. Class II is grade A milk, and class III is surplus grade A or grade B milk. Producers receive a "blend" price based on the proportion of milk used in each price class.

Forty-four federal "milk-marketing orders" and 14 states establish the prices that processors must pay dairy farmers for about 95% of the fluid milk and fluid milk products consumed in the United States. In some states, milk control commissions not only determine what farmers are to be paid but the prices stores can charge customers. Federal milk orders are intended to promote orderly marketing conditions for dairy producers and assure consumers an adequate supply of milk.

Milk prices in recent years are shown in Chapter 2, Figure 2.2. Total farm receipts from the sale of milk in 2005 were $27.4 billion.

CHAPTER SUMMARY

- Most of the 1,202 billion lb of world milk production is contributed by cows 1,088 bib lb, water buffalo (153 bib lb), goats (27 bib lb), and sheep (17 bib lb).

- In addition to fluid milk for drinking and cooking, cheese, butter, and dessert dairy products (e.g., ice cream, yogurt, and sherbet) are important consumer dairy products.

- Dairy products contribute significant amounts of protein, calcium, phosphorus, and riboflavin to the human diet.

- Sales of cheese, low-fat skim milk, and yogurt have increased in recent years, while whole milk and cottage cheese have experienced decreased per-capita sales.

KEY WORDS

solids-not-fat (SNF)	evaporated milk	curdling
colostrum	condensed milk	curd
nutrient density	dry milk	whey
milk fat	cultured	rennet
lactose	acidified	UHT processing
casein	yogurt	pasteurization
osteoporosis	cheese classifications	
homogenized	coagulation	

REVIEW QUESTIONS _____

1. What is the term used to describe all of the milk components exclusive of water and milk fat?

2. The solid nonfat component of milk consists of which three constituents?

3. What is the term for first milk produced by a female after giving birth?

4. What is the most remarkable difference between milk and colostrum?

5. Milk is a rich source of which mineral required in the diet of humans?

6. What is homogenization?

7. What is the process used to destroy all pathogens in milk during milk processing?

8. *True or False:* Pasteurization reduces the nutritional value of milk.

SELECTED REFERENCES _____

Dairy Producer Highlights. 2006. National Milk Producers Federation, Arlington, VA.

Hedrick, T. I., L. G. Harmon, R. C. Chandan, and D. Seiberling. 1981. Dairy products industry in 2006. *Journal of Dairy Science* 64:959.

Lowenstein, M., S. J. Speck, H. M. Barnhart, and J. F. Frank. 1980. Research on goat milk products: A review. *Journal of Dairy Science* 63:1629.

Milk Facts. 2005. Milk Industry Foundation, Washington, DC.

Miller, G. D., et al. 1995. *Handbook of Dairy Foods and Nutrition.* Rosemont, IL: National Dairy Council.

Quality of U.S. Agricultural Products. 1996. Ames, IA: Council for Agricultural Science and Technology.

Sellars, R. L. 1981. Fermented dairy foods. *Journal of Dairy Science* 64:1070.

Speckmann, E. W. L., M. F. Brink, and L. D. McBean. 1981. Dairy foods in nutrition and health. *Journal of Dairy Science* 64:1008.

Tobias, J., and G. A. Muck. 1981. Ice cream and frozen desserts. *Journal of Dairy Science* 64:1077.

Tong, P. 1994. Dairy processing and products. *Encyclopedia of Agricultural Science.* San Diego, CA: Academic Press, Inc.

CHAPTER 6

Wool and Mohair

LEARNING OBJECTIVES

- Describe the structures of wool
- Outline the factors affecting the value of wool
- Describe the classes and grades of wool
- Overview the production and processing of wool and mohair

Fibers that grow from the skin of animals give protection from abrasions to the skin and help keep animals warm. Skins of common mammals have a covering to which various terms are applied depending on the nature of the growth. For example, cattle, pigs, horses, and dairy goats have hair; sheep have wool; mink and non-Angora rabbits have fur; Angora rabbits have angora; and Angora goats have mohair.

Hair from most mammals has little commercial value (it is used mostly in padding and cushions). Fur of mink and non-Angora rabbits is either naturally beautiful or can be dyed to give attractive colors; therefore, it has considerable value as fur. Fur is composed of fine short fibers and relatively long, coarse guard hairs, in contrast to the skin covering of cattle, horses, and pigs, which is composed entirely of guard hairs.

Furs made from rabbits, mink, foxes, and bison (American buffalo) may be classed as status clothing because they are usually costly and attractive in appearance. White rabbit furs can be dyed any color (including pastels), but colored furs cannot. Because colored furs cannot be dyed, color mutations have been important in the mink business because they provide a wide array of fur colors. At present, the mink industry in the United States is suffering economically because of competition from furs made in other countries and high costs of producing mink.

Hides from young market lambs are often marketed with wool left on the pelt to be processed into heavy winter lambskin coats. Lambs from the Karakul breed of fat-tail sheep are used to produce Persian lambskins. The black, tight curls of fiber are more like fur than wool. Persian lambskins have been produced in the

United States, but larger numbers are currently produced in Afghanistan, Iran, and the former Soviet Union.

Hides from sheep with longer wool on the pelts also may be marketed intact for processing to produce ornamental rug pieces. The long wool may be dyed. The long wool may also be loosened and removed from the hides after slaughter and sold separately from the hides.

GROWTH OF HAIR, WOOL, AND MOHAIR

Hair or wool fiber grows from a **follicle** located in the outer layers of the skin. Growth occurs at the base of the follicle, where there is a supply of blood, and cells produced are pushed outward. The cells die after they are removed from the blood supply because they can no longer obtain nutrients or eliminate wastes. A schematic drawing of a wool follicle is presented in Figure 6.1.

The **cuticle** causes the fibers to cling together. The intermingling of wool fibers is known as felting. The felting of wool is advantageous in that wool fibers can be entangled to make woolens, but it is also responsible for the shrinkage that occurs when wool becomes wet.

All wool and hair fibers have a similar gross structure, consisting of an outer thin layer (cuticle) and a cortex that surrounds an inner core (**medulla**) in medullated fibers (Fig. 6.2). Only medium and coarse wools have a medulla. It is absent in fine wools.

Wool fibers have waves called **crimp** (Fig. 6.3). Crimp is caused by the presence of hard and soft cellular material in the cortex. The soft cortex is more elastic and is on the outer side of the crimp.

Hair does not exhibit any crimp because all the cells in the cortex are hard. Also, the inner core of hair is not solid, unlike the inner core of most wool fibers. Some fibers in wool of a low quality are large, lack crimp, and do not have a solid inner core. These fibers are called **kemp,** and they reduce the value of the fleece.

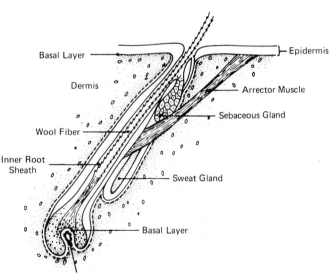

FIGURE 6.1 Schematic drawing of a wool follicle. *Source:* R. W. Henderson.

Cortex
Cuticle
Medulla

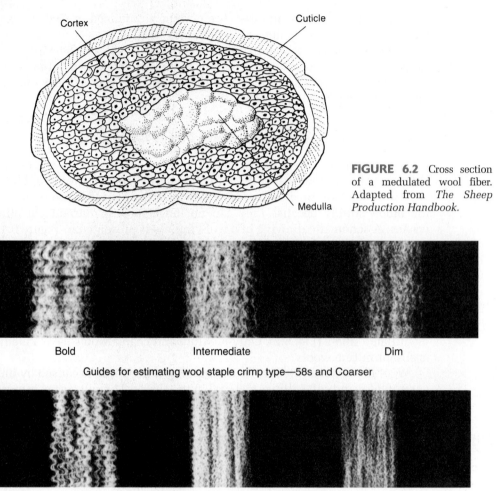

FIGURE 6.2 Cross section of a medulated wool fiber. Adapted from *The Sheep Production Handbook.*

Bold
Intermediate
Dim
Guides for estimating wool staple crimp type—58s and Coarser
Bold
Intermediate
Dim
Guides for estimating wool staple crimp type—60s and Finer

FIGURE 6.3 Several expressions of crimp in coarser and finer wool. Courtesy of USDA.

Mohair fiber follicles develop in groups consisting of three primary follicles each. Secondary follicles develop later. Mohair fibers have very little crimp (less than 1 crimp for each inch of length). The mohair cortex is composed largely of so-called *ortho* cells, in contrast to the cortex of wool, which is composed of both *ortho* and *para* cells. Mohair has a different scale structure from wool, but it does have scales and can be made into felt. Kemp and **medullated fibers** reduce the value of wool and mohair fleeces because such fibers do not dye well and because they show in apparel made from fleeces containing the fibers.

The skin of young lambs is made up of two layers: the surface skin (outer skin) is the **epidermis**, and the underlying skin is the **dermis** (Fig. 6.1). Between these two layers is the **basal layer.** Development of wool follicles occurs in skin of the fetus. The basal layer thickens and areas that are to become wool follicles push down into the dermis. Two glands develop: the **sebaceous gland,** which secretes sebum (a greasy substance) and the **sweat gland,** which produces and secretes sweat. The downward growth of the basal layer rests on the papilla, which has a supply of blood. The wool fiber grows toward the skin's surface, and as it becomes removed from the blood supply, the cells die.

Two major types of follicles are produced. The first type to develop is known as the **primary follicle.** These appear on the poll and face of the fetus by 35–50 days' gestation and over all other parts of the body by 60 days. They are arranged in groups of three. All primary follicles are fully developed and are producing fiber at birth.

A second type of follicle, associated with the primary follicles, develops later and is known as the **secondary follicle.** These follicles have an incomplete set of accessory structures—usually the sweat gland and erector muscle are absent. The primary and associated secondary follicles are grouped into follicle bundles.

Since all the primary follicles are formed prior to birth of the lamb, the density of primary follicles per square inch of skin declines rapidly during the first 4 months of postnatal life and then declines gradually until the lamb approaches maturity.

Secondary follicles show little activity during the first week of a lamb's life, after which they undergo a burst of activity. Follicular activity is at maximum between 1 and 3 weeks of age, with marked reduction after the third week.

Adverse prenatal environmental conditions can affect the number of secondary follicles initiated in development. Also, early postnatal influences could affect secondary follicle fiber production.

Growth of the wool fiber takes place in the **root bulb,** which is located in the follicle. Permanent dimensions of the fiber are determined in the area of the bulb, and there are no changes in growth characteristics of the fiber after it is formed. Defective portions of a fiber are caused by a reduction in the size of the fiber that creates a weakened area when sheep are under stressed conditions, such as poor nutrition and high body temperatures. Such fibers are likely to break under pressure.

Two types of undesirable fibers are *kemp* and *medullated* fibers. Both are hollow and brittle, and primary follicles rather than secondary follicles produce both, as is the case with true wool.

FACTORS AFFECTING THE VALUE OF WOOL

Ensuring wool quality is the result of best production practices both before and after shearing (Fig. 6.4). The two most important factors under the control of the producer that affect the amount of wool produced by sheep are nutrition and breeding. The amount of feed available to the sheep (energy intake) and the percentage of protein in the diet influence wool production. Wool production is decreased when sheep are fed diets having less than 8% protein. When the diet contains more than 8% protein, the amount of energy consumed is the determining factor in wool production.

Improving wool production per animal through breeding involves selection for increased clean **fleece weight, staple length, fineness** and uniformity of length, and fineness throughout the fleece. Performance testing rams for quantity and quality of wool production can make much progress, but for overall improvement of the flock, both lamb and wool production must be included (Fig. 6.5).

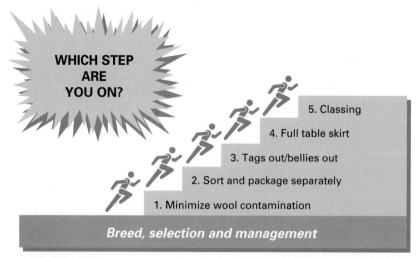

FIGURE 6.4   Preparation steps for wool quality improvement. *Source:* American Wool Council, division of the American Sheep Industry Association.

FIGURE 6.5   Depending on the goals of the production flock, lambs and wool are either seen as primary products or by-products. However, balanced selection can assist producers in optimizing production. Courtesy of American Sheep Industry Association.

Sheep and Angora goat producers are interested in maximizing net income from their sheep and goat operations. Increasing net income from wool and mohair can be accomplished by selecting to improve wool and mohair production and grade. Giving attention to the following items can enhance the values of wool and mohair that are produced:

1. The sheep or goats should be shorn when the wool or mohair is dry.
2. Inferior portions of the fleece should be removed at the skirting table and sacked separately. Offsorts (inferior portions) include parts of the clip that are matted or heavily contaminated with dung or vegetable matter, bellies, topknots, and floor sweepings (Fig. 6.6).

SKIRTING THE FLEECE

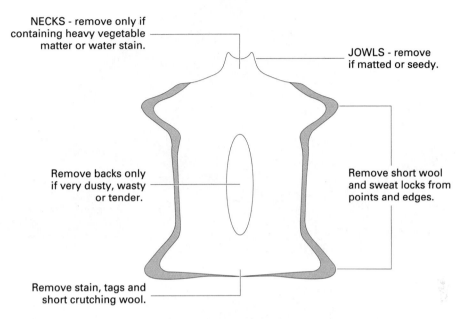

NECKS - remove only if containing heavy vegetable matter or water stain.

JOWLS - remove if matted or seedy.

Remove backs only if very dusty, wasty or tender.

Remove short wool and sweat locks from points and edges.

Remove stain, tags and short crutching wool.

NOTE - keep all strains separate
- make necessary cast lines
- remove skin pieces

FIGURE 6.6   Skirting the fleece. *Source:* American Wool Council, division of the American Sheep Industry Association.

3. Wool or mohair should be sacked by wool grades so that when **core samples** are taken from a sack of otherwise good wool, a few fleeces of low-grade wool (or mohair) will not cause the entire sack of wool or mohair to be placed in a low grade.

4. Wool should be shorn without making many double clips.

5. Fleeces should be tied with paper twine after they are properly folded with the clipped side out.

6. A lanolin-based paint should be used for branding animals. This type of paint is scourable.

7. Fleeces from black-faced sheep should be packed separately from other wool. Also, black fleeces should be packed separately. Black fibers do not take on light-colored dyes and consequently stand out in a garment made from black and white fibers dyed a light pastel color.

8. All tags, sweepings, and wool or mohair from dead animals should be packed separately.

9. Hay that has been baled with twine should be carefully fed to sheep; otherwise, pieces of twine will get into the wool or mohair and markedly lower its value.

10. Environmental stresses should be avoided. Being without feed or water for several days or having a high fever can cause a break or weak zone in the wool fibers.

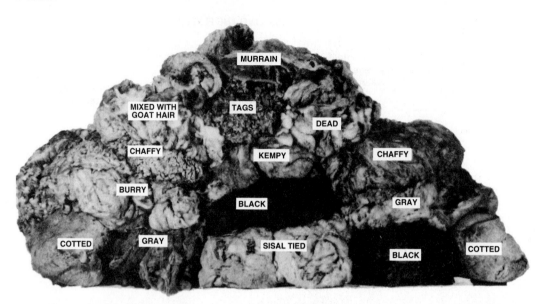

FIGURE 6.7   Several types of undesirable fleeces are identified. Each of these should be sorted and packed separately from the more desirable fleeces. Courtesy of USDA.

11. Wool containing coarse fibers, or kemp, should be avoided, or kempy fleeces should be packed separately. Cloth and wool containing these coarse fibers are highly objectionable and thus are low in price.

Figure 6.7 identifies several types of undesirable fleeces. These fleeces have a lower value than fleeces containing white fibers and minimal foreign material. Some of the fleeces in Figure 6.7 not previously discussed include the following:

Burry—wool that contains vegetable matter, such as grass seeds and prickly seeds, which adheres tenaciously to wool

Chaffy—wool that contains vegetable matter such as hay, straw, and other plant material

Cotted—wool fibers that are matted or entangled

Dead—wool pulled from sheep that have died but that have not been slaughtered

Murrain—wool obtained from decomposed sheep

CLASSES AND GRADES OF WOOL

The class and grade determine the value of wool. Staple length determines the class, and fineness of fibers determines the grade. The grades of wool are given by any one of three methods, all of which are concerned with a value that indicates the fineness (diameter) of the wool fibers. The three methods of reporting wool grades are these:

1. The American grade, or blood (outdated terminology), method is based on the theoretical amount of **fine-wool breeding** (Merino and Rambouillet)

represented in the sheep producing the wool (Fig. 6.8). The terms *fine,* ½-*blood*, ⅜, ¼-*blood*, and *low*-¼ *blood* are used to describe typical fiber diameter. Wool classified as *fine* or ⅝-*blood* has small fiber diameters, whereas ¼-*blood* wool is coarse.

2. The spinning count system refers to the number of *hanks* of yarn, each 560 yd long, that can be spun from 1 lb of **wool top.** These grades range from 80 (fine) to 36 (coarse). Thus, a grade of 50 by the spinning count method means that 28,000 yd (560 yd × 50 = 28,000 yd) of yarn can be spun from 1 lb of wool top.

3. The micron diameter method is based on the average of actual measurements of several wool fibers. This is the most accurate method of determining the grade of wool. A micron is $\frac{1}{25,400}$ of an inch.

Wool samples from each of the major grades are shown in Figures 6.8 and 6.9. The systems of grading, along with the breeds from which the grades of wool come, are presented in Table 6.1. The Merino and Rambouillet breeds are considered fine-wool breeds, with the Cotswold, Lincoln, and Romney identified as coarse-wool breeds. In Table 6.1, breeds listed between the fine- and coarse-wool breeds are called **medium-wool breeds.**

Classes of wool are staple, French combing, and clothing; however, fibers of finer grades that do not meet the length requirement may go into a higher class than they would if they were not fine. Much of the staple length wool is combed into **worsted**-type yarns for making garments of a hard finish that hold a press well.

The fineness of wool fibers from a sheep depends on the area of the body from which the fibers came. The body of a sheep can be divided into seven areas. Area 1 is the shoulder and area 2 is the neck, where wool grows longer and finer than average wool from the same sheep. Area 3 is the back. Its wool is long and has average fineness, though it has been exposed to the most weathering. Area 4 is the side. Wool from this area is long, has average fineness, and constitutes most of the wool. Area 5 is the tag, which yields long, coarse, dirty, and stained wool. Area 6 is the britch. Wool from this area is long and coarser than average and tends to be medullated. Area 7 is the belly. Its short, fine, matted wool may be very heavy in vegetable matter. Preference is given to a fleece that has uniformity of length and fineness in all areas of the sheep's body.

Qualitative factors of importance in wool marketing include strength, crimp, softness or handle, color, purity, character, and freedom from contaminants.

PRODUCTION OF WOOL AND MOHAIR

Production of wool can be expressed in terms of **greasy wool** or **scoured** (clean) **wool.** Greasy wool is wool or the fleece shorn once each year from sheep (Fig. 6.10). Scoured wool originates in the initial wool-processing stage. Scouring is the washing and rinsing of wool to remove grease, dirt, and other impurities.

Table 6.2 shows the production of greasy and scoured wool for several countries of the world. The world greasy wool production is approximately 4.8 billion pounds, with the United States producing approximately 37 million pounds. The U.S. wool production and total yearly supply is shown in Table 6.3. U.S. wool production by states is shown in Figure 6.11. Scoured or clean wool

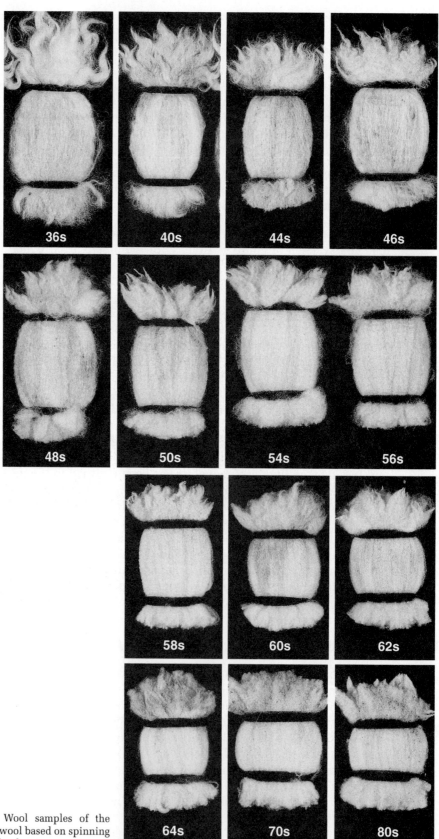

FIGURE 6.8 Wool samples of the major grades of wool based on spinning count and the blood system.

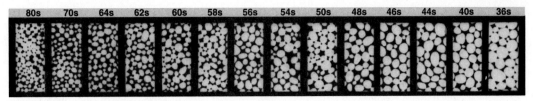

FIGURE 6.9 Cross section of magnified wool fibers demonstrating the wool grades based on fiber diameter measured in microns. Courtesy of USDA.

TABLE 6.1 Market Grades of Wool and Breeds of Sheep that Produce These Grades of Wool

USDA Grades Based on Spinning Count	USDA Grades Based on American (Blood) System	Grades Based on Micron Diameter	Grades According to Average Grades
80s	Fine	17.70–19.14	Merino
70s	Fine	19.15–20.59	Rambouillet
64s	Fine	20.60–22.04	Targhee, Southdown
62s	½ blood	22.05–23.49	Corriedale, Columbia
50s to 60s	¼ blood to ½ blood	23.50–30.99	Panama, Romeldale
48s to 56s	Low-¼ blood to ½ blood	26.40–32.69	Shropshire, Hampshire, Suffolk, Oxford, Dorset, Cheviot
46s to 48s	Common to low-¼ blood	32.69–38.09	Romney
36s to 40s	Common and braid	36.20–40.20	Cotswold
36s to 46s	Common and braid	32.70–40.20	Lincoln

Source: USDA.

FIGURE 6.10 Sheep being shorn for their yearly production of wool.

TABLE 6.2 World Production of Greasy Wool

Country	Greasy Wool (mil lb)
Australia	1,122
China	882
New Zealand	493
European Union	376
World total	4,850

Note that Australia, China, and New Zealand account for 50% of world production.
Source: Adapted from FAO.

TABLE 6.3   U.S. Shorn Wool Production

Year	No. Head Shorn (mil)	Ave. Fleece Wt. (lb)	Price (cents/lb)	Domestic Production (mil lb)
1985	11.2	7.9	$0.63	88.1
1990	11.2	7.8	0.80	88.0
1995	8.1	7.8	1.04	63.5
2000	6.1	7.6	0.33	46.4
2005	5.0	7.3	0.80	37.2

Source: Adapted from USDA.

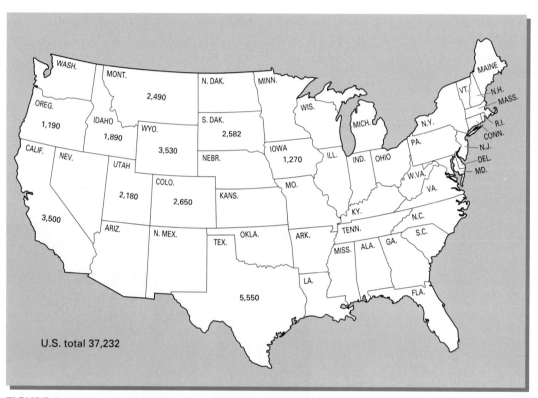

FIGURE 6.11   Top 10 states for wool production, 2005 (1,000 lb). *Source:* USDA.

production represents 50–60% of the greasy wool produced. Total U.S. wool production on a clean basis has been declining for the past two decades (Fig. 6.12).

Growth of wool varies greatly among sheep breeds. Fine-wool sheep grow 4–6 in. of wool per year, ¼- to ½-blood breeds of sheep grow 5–10 in. of wool per year, and common-to-braid-wool breeds grow 9–18 in. of wool per year. Mohair growth is approximately 12 in. per year.

Weights of **fleeces** also vary greatly among the breeds of sheep. Even though fine-wool breeds have fleeces of short length, weight of their fleeces varies from 9–12 lb; ¼- to ½-blood breeds of sheep produce fleeces weighing 7–15 lb; and coarse-wool (common and braid) breeds produce fleeces weighing 9–16 lb.

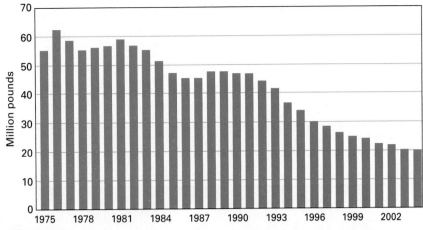

FIGURE 6.12   Annual U.S. wool production—clean basis.
Source: Livestock Marketing Information Center.

TABLE 6.4   Mohair Production

Year	No. Goats Clipped (mil)	Avg. Wt. Per Clip (lb)	Total Production (mil lb)	Price per lb	Total Value ($ mil)
1985[a]	1.7	7.7	13.3	$3.45	$45.9
1990[a]	1.9	7.8	14.5	0.95	13.8
1995	1.6	7.4	12.0	1.84	22.1
2000	0.3	6.6	1.7	2.20	3.8
2005	0.27	6.7	1.8	2.78	5.1

[a] Texas production alone.
Source: Adapted from USDA.

Annual world mohair production is approximately 59 million pounds, of which the United States produces between 29 and 30 million pounds or about half the world production. Texas produces about 97% of the mohair produced in the United States. The pounds and value of mohair production in Texas is shown in Table 6.4. Mohair growth is about 1–2 in. per month; therefore, goats are usually shorn twice per year, with length of clip averaging 6–12 in.

WOOL MARKETING

Wool sales are accomplished via several marketing alternatives including direct sale to warehouses, or via cooperative ventures where local producers pool their wool inventories to gain scale advantages in the market place. Worldwide improvement in descriptive technologies has refined the process of selling wool based on objective measurements of quality.

Manufacturers of wool products typically purchase from the central markets. Increasingly, wool is priced on a clean or scoured basis. Prices for raw wool are highly variable (Table 6.3). Clean wool prices for various grades and for U.S. versus Australian sources are provided in Table 6.5. Note that regardless of year, significant price advantages accrue to better grade wool with the fine wool from Australia commanding the highest prices.

TABLE 6.5 Prices (dollars per lb.) for Clean Wool Delivered to United States Mills

Year	U.S. Origin 64s Grade (20–22 micron)	U.S. Origin 58s Grade (25–28 micron)	Australian Origin 64s Grade (20 and less micron)
1995	2.49	1.70	2.81
2000	1.08	0.61	1.50
2001	1.21	0.72	1.66
2003	2.41	1.64	3.14
2005	2.35	1.62	2.75

Source: Adapted from USDA.

FIGURE 6.13 Angora goats with a full growth of mohair. Courtesy of Mohair Council of America.

USES OF WOOL AND MOHAIR

Fibers from sheep, Angora rabbits, and Angora goats (Fig. 6.13) are used in making cloth and carpets (Fig. 6.14). The consumption of wool in the United States is small when compared to the consumption of cotton and numerous manufactured fibers. Figure 6.15 shows the U.S. importation of wool which has declined along with domestic production of wool since the mid-1990s. These trends illustrate the decline in demand for wool. Wool trade from Australia accounts for approximately two-thirds of fine wool imports to the United States, while New Zealand provides in excess of 60% of imported medium grade wool.

Cloth made from wool has both highly desirable and undesirable qualities. Wool has a pleasant, warming quality and can absorb considerable quantities of

FIGURE 6.14 Mohair, kidmohair, and cashmere, strengthened by different proportions of synthetic fibers, are popular yarns for making attractive garments that "breathe" and reduce perspiration. Courtesy of George F. W. Haenlein, University of Delaware.

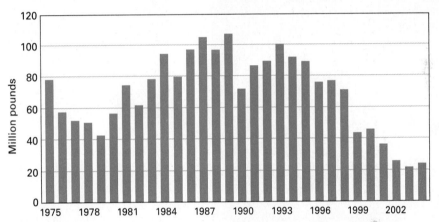

FIGURE 6.15 Annual U.S. raw wool imports—clean basis. *Source:* Livestock Marketing Information Center.

moisture while still providing warmth. It is also highly resistant to fire. However, wool has a tendency to shrink when it becomes wet, and cloth made from wool causes some people to itch.

Researchers have contributed greatly to the alteration of wool so that fabrics or garments will not shrink when washed. A process called the WURLAN treatment makes woolen garments machine-washable. In this treatment, the wool fibers are coated with a very thin layer of resin, which adds only 1% to the weight of the wool. The resin does not alter wool in any significant way except to prevent the fibers from absorbing water. There is an increasing trend toward blending wool with other fibers. This improves aesthetics and provides materials with more durability and comfort.

Only fabrics that have passed the quality tests demanded by the International Wool Secretariat and its U.S. branch, the Wool Bureau, Inc., are eligible to carry the Woolmark or Woolblend label.

Woolen garments are as old as the Stone Age, yet as new as today. Wool takes many forms—wovens, knits, piles, and felts—each the result of a different process. It is a long way from a bale of raw wool to a beautiful bolt of fabric and ultimately to beautiful, versatile garments (Fig. 6.16).

A

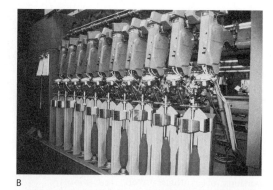

B

C

D

FIGURE 6.16 Natural fiber moves through a process of production, marketing, and processing to generate desirable wool and cashmere products for consumers. (A) Preparing wool for the skirting table. (B and C) Wool is dyed, carded, and spun before being woven into a desired fabric. (D) High-quality woolen garments are then made available to the public. Courtesy of Pendleton Woolen Mills, Portland, OR (B–D).

CHAPTER SUMMARY

- Wool, mohair, and hair from animals provide useful human products.
- Wool value is determined by fleece weight, cleanliness, staple length, and fineness (fiber diameter).
- Mohair production from Angora goats is concentrated primarily in Texas.
- Cotton, synthetic fibers, and wool imports have provided stiff competition for domestic wool production.

KEY WORDS

follicle
cuticle
medulla
crimp
kemp
medullated fibers
epidermis
dermis
basal layer
sebaceous gland

sweat gland
primary follicle
secondary follicle
root bulb
fleece weight
staple length
fineness
core sample
burry
chaffy

cotted
dead
murrain
fine-wool breeds
wool top
medium-wool breeds
worsted
greasy wool
scoured wool

REVIEW QUESTIONS

1. Wool and hair grow from what structure located in the outer layer of skin?
2. What is felting of wool? What is responsible for it and why is it important?
3. What are the three basic layers in a wool fiber?
4. What are waves in the wool fiber and what creates them?
5. What are hairlike fibers in wool called and why are they important?
6. What is staple length?
7. What are tags?
8. What three systems are used to grade wool?
9. *True or False:* The finer the wool fiber, the more valuable it is.
10. Which two breeds of sheep produce the finest grade of wool?
11. What is the process of cleaning wool called?

SELECTED REFERENCES

Botkin, M. P., R. A. Field, and C. L. Johnson. 1988. *Sheep and Wool: Science, Production, and Management.* Englewood Cliffs, NJ: Prentice Hall.

Code of Practice for Preparation of Wool Clips in the United States. American Sheep Industry Association, Inc., Englewood, CO.

Hyde, N. 1988. Fabric of history: Wool. *National Geographic* 173:552.

Lupton, C. J. 1994. Wool and mohair production and processing. *Encyclopedia of Agricultural Science.* San Diego, CA: Academic Press, Inc.

Quality of U.S. Agricultural Products. 1996. Ames, IA: Council for Agricultural Science and Technology.

Rogers, G. E., et al. 1989. *The Biology of Wool and Hair.* New York, NY: Chapman and Hall.

CHAPTER 7

By-Products of Meat Animals

LEARNING OBJECTIVES

- Identify the major edible and inedible by-products from livestock
- Describe the role of international by-product trade
- Explain the role of the rendering industry
- Discuss the options and challenges with dead livestock disposal

By-products are products of considerably less value than the major product. However, the value of by-products can account for as much as 8–10% of the total value of a fed steer. In the United States, meat animals produce meat as the major product; hides, fat, bones, and internal organs are considered by-products. In other countries, primary products may be draft (work), milk, hides, and skins, with meat considered a by-product when old, less useful animals are slaughtered.

There are numerous by-products resulting from animal slaughter and the processing of meat, milk, and eggs. Animals dying during their productive lives may also be processed into several by-products; however, these by-products are not used in human consumption.

By-products are commonly classified into two categories based on human consumption: **edible** and **inedible** (Fig. 7.1; Table 7.1). Some by-products that are not considered edible by humans are edible to other animals. By-products that are processed into animal feeds, such as feather meal or fishmeal, are good examples.

Figure 7.2 shows the major by-products obtained from a steer in addition to the retail beef products. The average total value of cattle by-products ranges between $0.05 and $0.09 per pound of live weight, with the hide comprising the largest part of the total value. Because hides' values have cyclic prices, as is often the case with commodity products, their value as a percent of the total by-product value has ranged from 45% to 65% during the past several years.

The use of by-products in a variety of applications allows the livestock industries the opportunity to assure that there is minimal wastage or lost value.

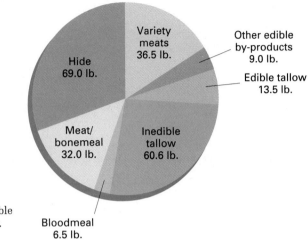

FIGURE 7.1 Edible and inedible by-products from a 1,100-lb steer. *Source:* USDA.

TABLE 7.1 Edible and Inedible By-Products from Cattle and Swine

By-Product	1,250-lb Steer		230-lb Hog	
	Expected Yield per Head (lb)	Market Value per lb	Expected Yield per Head (lb)	Market Value per lb
Edible				
Select liver	18.70	0.13	1.30	0.09
Heart	6.00	0.18	0.30	0.25
Cheek meat	5.60	0.95	0.30	0.65
Head meat	2.50	0.56	0.40	—
Salivary glands	1.25	0.06	0.10	0.25
Feet	—	—	0.80	0.31
Lips	1.21	1.49	0.25	—
Oxtail	3.25	1.90	—	—
Spleen	2.50	0.04	—	—
Tongue	4.88	1.35	0.20	0.74
Weasand meat[a]	3.75	0.68	0.12	—
Tripe (honeycomb)	2.50	0.52	—	—
Stomach	—	—	0.40	0.65
Sweetbread	1.14	0.49	—	—
Ears	—	—	0.26	0.81
Kidneys	2.90	0.15	0.20	—
Snouts	—	—	0.64	0.25
Tallow/lard	1.20	0.19	9.00	0.29
Inedible				
Lung	10.60	0.02	—	—
Gullet	1.50	0.04	—	—
Rendered fat	60.00	0.15	0.50	0.14
Meat and bone meal	32.00	0.06	1.50	1.22
Blood meal	6.50	0.22	1.95	4.54
Hide (native steer)	75.00	0.69	—	—

(*continued*)

TABLE 7.1   Edible and Inedible By-Products from Cattle and Swine (continued)

By-Product	1,250-lb Steer		230-lb Hog	
	Expected Yield per Head (lb)	Market Value per lb	Expected Yield per Head (lb)	Market Value per lb
Pharmaceutical				
Fetal blood[b]	—	48.00	—	—
Fresh bile	—	0.11	—	—
Pancreas	1.05	0.69	—	0.50
Adrenal glands	—	3.00	—	—
Pituitary glands	—	27.00	—	150.00
Thyroid glands	—	0.80	—	4.25
Pepsin linings	—	—	—	0.88
Ovaries	—	7.50	—	—
Spleens	—	—	—	—
Lungs (lobe only)	—	0.10	—	—
Trachea	1.50	0.34	—	—

[a] *Weasand* is the muscular lining that surrounds the esophagus from the larynx to the first stomach.
[b] Per liter; obtained from pregnant market females.
Sources: Adapted from USDA and *National Provisioner*.

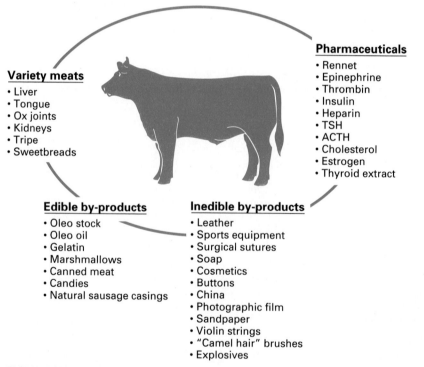

Variety meats
• Liver
• Tongue
• Ox joints
• Kidneys
• Tripe
• Sweetbreads

Pharmaceuticals
• Rennet
• Epinephrine
• Thrombin
• Insulin
• Heparin
• TSH
• ACTH
• Cholesterol
• Estrogen
• Thyroid extract

Edible by-products
• Oleo stock
• Oleo oil
• Gelatin
• Marshmallows
• Canned meat
• Candies
• Natural sausage casings

Inedible by-products
• Leather
• Sports equipment
• Surgical sutures
• Soap
• Cosmetics
• Buttons
• China
• Photographic film
• Sandpaper
• Violin strings
• "Camel hair" brushes
• Explosives

FIGURE 7.2   In addition to the retail product of beef are numerous by-products. Adapted from Field (1996).

EDIBLE BY-PRODUCTS

Variety meats are edible products originating from organs and body parts other than the carcass. Liver, heart, tongue, tripe, and sweetbread are among the more typical variety meats. **Tripe** comes from the lining of the stomach; **sweetbread** is the thymus gland.

TABLE 7.2   U.S. Consumption of Rendered Products

	1994 (mil lb)	1998 (mil lb)	2002 (mil lb)	2005 (mil lb)
Inedible tallow and greases	3,190	3,527	3,126	3,340
Soap	301	228	66	NA[a]
Feed	2,102	2,533	2,310	2,436
Lubricant	82	90	NA[a]	NA[a]
Fatty acid	634	608	NA[a]	NA[a]
Edible tallow	643	421	511	436
Edible use	411	265	249	212
Inedible use	232	174	262	225
Lard	452	403	300	234
Edible use	325	286	236	176
Inedible use	127	117	64	57

[a]Not reported to avoid disclosure of single company information.
Source: Adapted from USDA.

An average 1,100-lb slaughter steer produces approximately 36 lb of variety meats. Since the U.S. per-capita disappearance is 9 lb, surplus variety meats are exported into countries where consumer demand is more substantial.

Other edible products are fats used to produce **lard** and **tallow.** These products are eventually used in shortenings, margarines, pastries, candies, and other food items. Inedible tallow and greases are used in the manufacture of soap, lubricants, feed, and fatty acids (Table 7.2). Nearly 45% of the inedible tallow and nearly 20% of edible tallow and lard are exported. Rendered fats are sold on the basis of percent total fatty acids, unsaponifiable matter, insoluble impurities, free fatty acids, and moisture.

INEDIBLE BY-PRODUCTS

Tallow, hides (skins), and inedible organs are the higher-valued inedible by-products. Not all skins are inedible, as some pork skins are processed into consumable food items. Life-saving and life-supporting pharmaceuticals originate from animal by-products, even though many of the same pharmaceuticals are made synthetically. Table 7.3 shows many of these pharmaceuticals and their animal by-product sources.

Table 7.4 lists other inedible by-products and their uses. Animals contribute a large number of useful products. The many uses of inedible fats are discussed later in the chapter.

Hides and skins are valuable as by-products or as major products on a worldwide basis. Countries that are most important in hide and skin production are shown in Table 7.5.

Cattle and buffalo hides comprise 80% of the farm animal hides and skins produced in the world. The United States makes a major contribution in the world hide and skin market, with $1 billion worth of cattle hides produced being exported to other countries (Chapter 2). The United States produces very few goat skins but ranks 37th in sheepskins, with an annual production valued at $33 million. Pork skin usually remains on the carcass, although some pigs are skinned.

TABLE 7.3   Selected Pharmaceuticals from Red Meat Animals—Their Source and Utilization

Pharmaceutical	Source	Utilization
Amfetin	Amniotic fluid	Reducing postoperative pain and nausea; enhancing intestinal peristalis
Catalase	Liver	Preparation of contact lens care products
Cholesterol	Nervous system	Male sex hormone synthesis
Chondroitin sulfate	Trachea	Arthritis treatment
Chymotrypsin	Pancreas	Removing dead tissue; treatment of localized inflammation and swelling
Corticosteroids	Adrenal gland	Treatment for shock, Addison's disease
Corticotrophin (ACTH)	Pituitary gland	Diagnostic assessment of adrenal gland function; treatment of psoriasis, allergies, mononucleosis, and leukemia
Cortisone	Adrenal gland	Relief of hay fever, asthma, and other allergies; heart stimulation
Desoxycholic acid	Bile	Used in synthesis of cortisone for asthma and arthritis
Epinephrine	Adrenal gland	Relief of hay fever, asthma, and other allergies; heart stimulation
Fibrinolysin	Blood	Dead tissue removal; wound-cleansing agent; healing of skin from ulcers or burns
Glucagon	Pancreas	Counteracting insulin shock; treatment of some psychiatric disorders
Glycerin	Tallow	Cough syrups, lozenges, eyewashes, contraceptive creams, gel capsules
Heparin	Intestines, lungs	Anticoagulant used to thin blood; retarding clotting, especially during organ implants; preventing gangrene
Hyaluranidase	Testes	Enzyme that aids drug penetration into cells, cartilage and joint treatment
Insulin	Pancreas	Treatment of diabetes for 10 million Americans (primarily synthetic insulin is used)
Liver extracts	Liver	Treatment of anemia
Mucin	Stomach	Treating ulcers
Norepinephrine	Adrenal gland	Shrinking blood vessels, reducing blood flow and slowing heart rate
Ovarian hormone	Ovaries	Treating painful menstruation and preventing abortion
Ox bile extract	Liver	Treatment of indigestion, constipation, and bile tract disorders
Parathyroid hormone	Parathyroid gland	Treatment of human parathyroid deficiency
Plasmin	Blood	Digests fibrin in blood clots, used to treat patients with heart attacks
Rennet	Stomach	Assisting infants in digesting milk; cheese making
Serum	Blood	Variety of macromolecular proteins, buffering agents for cell culture mediums
Thrombin	Blood	Assisting in blood coagulation; treatment of wounds; skin grafting
Thyroid extract	Thyroid gland	Treatment of cretinism
Thyrotropin (TSH)	Pituitary gland	Stimulating functions of thyroid gland
Thyroxin	Thyroid	Treatment of thyroid deficiency
Vasopressin	Pituitary gland	Control of renal function

Source: Adapted from Field et al., 1996.

The U.S. hide export market yields approximately $1.3 billion on a yearly basis. Hides, skins, and pelts are made into useful leather products through the tanning process (Table 7.6). One cowhide can yield approximately 144 baseballs, 20 footballs, 18 volleyballs or soccer balls, 12 baseball gloves, or 12 basketballs. The surplus hides are exported primarily to Japan, Korea, Mexico, Taiwan, and Romania. Leather utilization in the United States is categorized as 40% for upholstery, 50% for shoes and shoe leather, and 10% for other uses.

The general term **hide** refers to a beef hide weighing more than 30 lb. Those weighing less than 30 lb are called **skins.** Skins come from smaller animals, such as pigs, sheep, goats, and small wild animals. Those skins from sheep with the wool left on are usually called **sheep pelts.** Hides, skins, and pelts are classified according to (1) species, (2) weight, (3) size and placement of brand, and (4) type of packer producing them.

TABLE 7.4   Other Inedible By-Products and Their Uses

By-Product	Use
Hog heart valves	Replacement of injured or weakened human heart valves; since the first operation in 1971, more than 35,000 hog heart valves have been implanted in humans.
Pig skins	Treating massive human burns; these skins help prepare the patient for permanent skin grafting.
Gelatin (from skin)	Coatings for pills and capsules.
Brains	Cholesterol for an emulsifier in cosmetics.
Blood	Sticking agent for insecticides; a leather finish; plywood adhesive; fabric painting and dyeing.
Hides and skins	Many leather goods from coats, handbags, and shoes to sporting goods.
Bones	Animal feed, glue, buttons, china, and novelties.
Gallstones	Shipped to Southeast Asia for use as ornaments in necklaces and pendants.
Hair	Paint and other brushes; insulation; padding in upholstery, carpet padding, filters.
Meat scraps and blood	Animal feed.
Inedible fats	Animal feed, fatty acids.
Wool (pulled from pelts)	Clothing, blankets, lanolin.
Poultry feathers	Animal feed, arrows, decorations, bedding, brushes.
Glycerin	Adhesives, antifreeze, cleaners, anticorrosive coatings, cosmetics, leather tanning and finishing, resins, and metal processing.

Source: Adapted from Field et al., 1996.

TABLE 7.5   Leading Countries in Fresh Hide and Skin Production

Cattle Hides		Sheep Skins		Goat Skins	
Country	(mil lb)	Country	(mil lb)	Country	(mil lb)
1. China	3,773	1. China	996	1. China	817
2. United States	2,305	2. EU	390	2. India	286
3. Brazil	1,746	3. New Zealand	309	3. Pakistan	274
4. Argentina	952	4. Australia	287	4. Bangladesh	92
5. India	891	5. Belarus	159	5. Nigeria	51
World total	17,597	**World total**	3,898	**World total**	2,171

Source: Adapted from FAO.

TABLE 7.6   Leather Uses Related to Types of Hides and Skins

Skin Origin	Use
Cow and steer	Shoe and boot uppers, soles, insoles, linings; patent leather; garments; work gloves; waist belts; luggage and cases; upholstery; transmission belting; sporting goods, packings
Calf	Shoe uppers; slippers; handbags and billfolds; hat sweatbands; book bindings
Sheep and lamb	Grain and suede garments; shoe linings; slippers; dress and work gloves; hat sweatbands; book bindings; novelties
Goat and kid	Shoe uppers, linings; dress gloves; garments; handbags
Pig	Shoe suede uppers; dress and work gloves; billfolds; fancy leather goods
Horse	Shoe uppers; straps; sporting goods

Source: Adapted from New England Tanners Club, *Leather Facts.*

The value of hides may be reduced by branding, nicking the hide while skinning, warbles (larvae of heel flies), mange, lice, biting and sucking insects, grubs, water, mud, and urine damage. Warbles emerge from the back region of cattle in the spring, making holes in the hide. Sheep hides may be damaged by outgrowths of grasses in the production of seeds (called beards), which can penetrate the skin, as well as by external parasites called **keds.**

A fed steer produces a 65–75-lb hide that is worth nearly $1 per pound. When the hide is converted into the "blue" stage (which means it is treated so it will not deteriorate during shipping), it loses about 15 lb. The value, however, has increased to $80–$90. A 60-lb blue hide will produce 40 ft^2 of shoe leather. Shoe leather sells for $2.50 a square foot, so the hide is now worth approximately $100.

Every year, U.S. tanneries convert millions of raw hides and skins into leather. The tanners' value added by manufacture constitutes over $500 million annually. Their product (leather) serves in turn as a raw material for the shoe and leather goods industries that provide jobs for over 200,000 people. Table 7.7 shows the quantity of exported hides and skins.

Table 7.7 shows that fewer than half of U.S. cattle hides are converted to leather domestically, most being exported for tanning. Some sheepskins are imported, as the domestic supply does not meet the demand. Pigskins are used primarily as food, but interest in leather production is increasing.

Just as meat is perishable, so too are hides and skins. If not cleaned and treated, they begin to decompose and lose leather-making substance within hours after removal from the carcass. Hides are commonly treated (cured) by adding salt as the principal curing agent. The salt solution penetrates the hides in about 12 hours, and then the hides are bundled for shipment. The hide is the single most valuable by-product of beef cattle. The 2000 Beef Industry Quality Audit found that hide damage due to management-related defects (parasite control, branding, etc.) reduced per-head value by $23.92. One of the largest causes of this loss is rib branding (Fig. 7.3).

Nearly 7 million tons of raw material by-products, at a value of $8 billion, are utilized by the pet food industry annually. Exports of U.S. pet foods have tripled since 1990. Dog food sales in the United States account for more than $9 million,

TABLE 7.7   U.S. Hide and Skin Exports

Item	1,000 Pieces			
	1992	1998	2000	2005
Cattle and buffalo hides	19,028	18,934	21,361	19,230
Sheep and lamb skins	4,825	18.7[a]	33.5[a]	20.3[a]
Calf hides	2,895	1,659	1,440	3,545

[a]Millions of dollars.
Source: USDA.

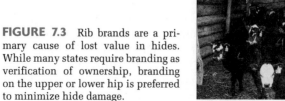

FIGURE 7.3 Rib brands are a primary cause of lost value in hides. While many states require branding as verification of ownership, branding on the upper or lower hip is preferred to minimize hide damage.

with an additional $1 billion spent on dog treats such as rawhide, cured bovine ears, and hooves. Rendered fats and greases have experienced significant export demand for use in livestock and poultry feed formulations. Asia, South America, and Mexico are the highest-volume markets for these by-products.

New products, such as biodiesel, have been developed as a means to extend demand for rendered fats and oils. Biodiesel is currently in use by some of the public transport system buses in several urban areas. The need for research and development is a critical component of assuring that the products of the livestock industry are not wasted.

THE RENDERING INDUSTRY

A focal point of by-products is the **rendering** industry, which recycles offal, fat, bone, meat scraps, and entire animal carcasses. The sources of these raw materials are packing and processing plants, butcher shops, restaurants, supermarkets, farmers, and ranchers. Some large packing and processing plants have their own rendering plants integrated with other operations.

Renderers have a regular pickup service that amounts to more than 70 million pounds of animal material daily. This amount has been declining in recent years because less fat is shipped out of packing plants. This pickup service is essential to public health, as it reduces a major garbage-disposal problem. Also, since animal by-products have value, consumers can buy meat at a lower price than would otherwise be possible.

Rendering of Red Meat Animal By-Products

Animal fat and animal protein are the primary products of the renderers' art. Originally, animal fats went almost entirely into soap and candles. Today, from the same basic material, renderers produce many grades of tallow and semiliquid fat. The major uses of rendered fat are in animal feeds, fatty-acid production, and soap manufacture.

Fatty acids are used in the manufacture of many products, such as plastic consumer items, cosmetics, lubricants, paints, deodorants, polishes, cleaners, caulking compounds, asphalt tile, printing inks, and others. The fatty-acid industry experienced tremendous growth since 1950.

The rendered animal proteins are processed into several high-protein (more than 50% protein) feed supplements, among which are meat and bone meals and blood meals. The supplements are more commonly fed to young monogastric animals—swine, poultry, and pet animals. These animals require a high-quality protein, particularly the amino acid lysine, which is typical of animal protein supplements.

Blood not processed into blood meal is used to produce products used in the pet food industry. Such products are blood protein (fresh, frozen whole blood) and blood cell protein (frozen, fresh, dewatered blood).

By-products enjoy a relatively strong export market, as indicated by Table 7.8. By-product sales grew in the 1990s, with fats, oils and greases, variety meats, and hides comprising the majority of sales. Gross sales have declined over the last several years. The major customers for U.S. exports of inedible tallow, edible tallow, lard, meat meal and tankage, and feather meal are listed in Table 7.9.

TABLE 7.8   Value of By-Product Exports and Rendered Items ($1,000)

	1990	1995	2000	2005
Fats, oils, greases	459,157	514,931	383,041	464,127
Lard, rendered pig fat	22,447	29,020	41,562	33,230
Tallow, inedible	344,079	335,375	225,292	273,293
Offals, edible, variety meat	370,674	524,715	766,816	766,710
Hides and skins	1,793,785	1,438,607	1,561,674	1,552,890

Source: Adapted from USDA, FAS.

TABLE 7.9   Leading Customers of U.S. Rendered Product Exports—2005

Inedible Tallow		Edible Tallow		Lard		Meat Meal and Tankage		Feather Meal	
Country	$1,000	Country	$1,000	Country	$1,000	Country	$1,000	Country	$1,000
Mexico	127,388	Mexico	40,300	Mexico	18,743	Indonesia	13,692	Indonesia	6,749
Turkey	53,589	Canada	10,130	Canada	2,697	Mexico	37,243	Canada	2,656
Nigeria	17,483	Rep. of Korea	6,682			Canada	4,313	Taiwan	954
Guatemala	11,440								
Colombia	7,258								

Source: USDA, FAS.

FIGURE 7.4   The export of value-added livestock by-products is economically significant and contributes to higher prices for producers. Courtesy of National Renderers Association.

Rendering of Poultry By-Products

Within a few hours after poultry is slaughtered, the offal is at the rendering plant. There it is cooked in steam-jacketed tanks at temperatures sufficient to destroy all pathogenic organisms. After being cooked for several hours, the material is passed through presses that remove most of the fat (poultry fat), while the remaining material becomes poultry by-product meal. While it is still hot, the poultry fat is piped to sterile tanks where appropriate antioxidants or "stabilizers" are added, and thence to tank cars or drums for shipment to feed manufacturers or for other uses as shown in Figure 7.4.

Poultry by-product meal (PBPM) is high in protein (approximately 58%) and contains 12–14% fat. PBPM consists of ground dry-rendered clean parts of the carcasses of slaughtered poultry, such as heads, feet, undeveloped eggs, and intestines, exclusive of feathers except in such trace amounts as might occur unavoidably in good factory practice. It should not contain more than 16% ash and not more than 4% acid ash.

DISPOSING OF DEAD LIVESTOCK

Unfortunately not all livestock survive to enter the market place (Table 7.10). As such, careful attention should be given to proper disposal of dead animals to avoid cross-contamination of disease to humans or other livestock, to assure environmental quality, and to avoid problems resulting from odor. The rendering industry plays a crucial role in this process by providing a safe and resource-efficient means to deal with the problem of dead animal disposal.

The generally approved protocols for disposal of dead livestock are:

1. Removal by a licensed rendering company;
2. Compost the carcass;
3. Burn the carcass in an incinerator (approved and permitted); or
4. Bury the carcass at least 4 feet deep.

Each of these options is suitable for certain circumstances, and all are associated with some level of cost (Table 7.11), with larger animals having higher costs of disposal. As state and federal policies are adopted in regards to this issue, arriving at workable solutions is critical from both the perspective of the

TABLE 7.10 Livestock Mortalities in the U.S., 2000

Species	Number (1,000 hd)	Weight (1,000 lb)
Dairy cattle	804	449,227
Beef cattle	3,328	1,482,952
Hogs	17,928	981,655
Sheep	281	21,957
Lambs	486	37,924
Goats	65	4,225
Chickens	50,507	154,951
Turkeys	31,946	191,679

Source: Adapted from National Renderers Association.

TABLE 7.11 Cost Estimates for Primary Protocols of Livestock Mortality Removal ($ per mortality)

| Species | Rendering[a] | | Burial | Incineration | Composting |
	MBM[a]:For Feed	MBM[b]: Not for Feed			
Cattle/calves	8.25	24.11	10.63	9.33	30.34
Weaned hogs	7.00	11.53	12.45	4.09	14.04
Pre-weaned hogs	0.50	0.70	2.01	0.30	1.02
Other	7.00	9.61	1.51	0.29	0.98

[a]Assumes all mortalities are rendered with a charge of $10.00 per mature cattle and $7.00 per calf.
[b]Meat and bone meal.
Source: Adapted from National Renderers Association.

profitable livestock enterprise owner and the maintenance of environmentally friendly approaches. Currently there are a myriad of state regulations, so producers are advised to contact their local health department or state department of agriculture for guidance in seeking the best solution.

Composting is becoming increasingly recognized as a viable solution in many situations. Composting is the process of accelerating decay processes. There are a variety of factors that affect the success of composting including moisture, type of co-composting material, carbon, oxygen, and nitrogen levels, and level of heat retention in the compost pile. Preferred moisture levels are 40–60% to assure rapid decay without the production of highly offensive odors. Wood chips, sawdust, corncobs, and poultry litter are popular co-composting materials that have proven to be effective. The preferred C:N ratio is 25 to 1 although a wider range is workable. Oxygen levels are most affordably maintained by avoiding overly wet compost, regular churning or turning of the compost pile, and the use of coarse co-composting ingredients. Compost temperatures need to reach the range of 120–150°F to assure optimal decay rates while assuring the destruction of pathogenic microorganisms.

CHAPTER SUMMARY

- Hides, fat, bones, and internal organs are the primary by-products, with hides usually having the highest value.
- Variety meats (liver, heart, tongue, tripe, etc.) are examples of edible by-products, while hides and other internal organs are inedible.
- Numerous inedible by-products produce useful human products—for example, pig heart valves (human heart valves), pig skin (human graft skin), hides and skins (leather, bone buttons, china, etc.), and many pharmaceuticals (e.g., corticosteroids, epinephrine, heparin, insulin, thyrotropin, etc.).
- Hides, fats, and variety meats are the by-products that comprise most of the total value of U.S. livestock exports.

KEY WORDS

edible by-products	lard	keds
inedible by-products	tallow	rendering
variety meats	hides	poultry by-product
tripe	skins	meal (PBPM)
sweetbread	pelts	composting

REVIEW QUESTIONS

1. What is the most valuable by-product obtained from the slaughter of livestock?
2. What important consumer product are hides used to produce?
3. What is the recycling of nonedible animal by-products and carcasses referred to as?

SELECTED REFERENCES

Field, T. G., J. Garcia, and J. Ahola. 1996. *Quantification of the Utilization of Edible and Inedible Beef By-products.* Colorado State University and The National Cattlemen's Beef Association, Englewood, CO.

Glanville, T. 1999. *Composting Dead Livestock: A New Solution to an Old Problem.* Ames, IA: Iowa State Univ.

Hog Is Man's Best Friend; Meat; By-products. Chicago, IL: The National Live Stock and Meat Board.

Kinsman, D. M. 1994. Animal by-products from slaughter. *Encyclopedia of Agricultural Science.* San Diego, CA: Academic Press, Inc.

Kirk-Othmer. 1993. *Encyclopedia of Chemical Technology.* New York: John Wiley & Sons.

Lawrence, J. D. 1994. An economic assessment of trends in pork by-product values. Ames, IA: Iowa State University Swine Research Report.

North American Renderin. 2002. National Renderers Association, Inc., Alexandria, VA.

Romans, J. R., K. W. Jones, W. S. Costello, C. W. Carlson, and P. T. Ziegler. 1985. Packing house by-products. Chap. 10 in *The Meat We Eat.* Danville, IL: Interstate Printers and Publishers.

CHAPTER 8

Market Classes and Grades of Livestock, Poultry, and Eggs

LEARNING OBJECTIVES

- Describe the market classes and grades of livestock
- Describe the USDA quality and yield grades of beef, pork, and lamb
- Describe the USDA grades for poultry and eggs

The annual production and movement of more than $99 billion worth of highly perishable livestock and poultry products to U.S. consumers is accomplished through a vast and complex marketing system. **Marketing** is the transformation and pricing of goods and services through which buyers and sellers move livestock and livestock products from the point of production to the point of consumption. Producers need to understand marketing if they are to produce products preferred by consumers, decide intelligently among various marketing alternatives, understand how animals and products are priced, and eventually raise productive animals profitably.

    Market classes and grades have been established to segregate animals, carcasses, and products into uniform groups based on preferences of buyers and sellers. The USDA Meat Grading and Certification Branch has established extensive classes and grades to make the marketing process simpler and more easily communicated. Use of USDA grades is voluntary. Some packers have their own private grades, though these are often used in combination with USDA grades. An understanding of market classes and grades helps producers recognize the quantity and quality of products they are supplying to consumers. **Meat inspection,** on the other hand, is administered by the Food Safety and Inspection Service of USDA and is a mandatory process.

MARKET CLASSES AND GRADES OF RED MEAT ANIMALS

Slaughter Cattle

Slaughter cattle are separated into classes based primarily on age and sex. Age of the animal has a significant effect on tenderness, with younger animals typically producing more tender meat than older animals. Age classifications for meat from cattle are veal, calf, and beef. **Veal** is from young calves, 1–3 months of age, with carcasses weighing less than 150 lb. **Calf** is from animals ranging in age from 3 to 10 months with carcass weights between 150–300 lb. **Beef** comes from mature cattle (over 12 months of age) having carcass weights higher than 300 lb. Classes and grades established by the USDA for cattle are based on sex, quality grade, and yield grade, all of which are used in the classification of both live cattle and their carcasses (Table 8.1).

The sex classes for cattle are **heifer, cow, steer, bull**, and **bullock.** Occasionally, the sex class of **stag** is used by the livestock industry to refer to males that have been castrated after their secondary sex characteristics have developed. Sex classes separate cattle and carcasses into more uniform carcass weights, tenderness groups, and processing methods. **Quality grades** are intended to measure certain consumer palatability characteristics, and **yield grades** measure amounts of fat, lean, and bone in the carcass. Slaughter steers representing some of the eight quality grades and five yield grades are shown in Figures 8.1 and 8.2.

Quality grades are based primarily on two factors: (1) **maturity** (physiological age) of the carcass and (2) amount of **marbling.** Maturity is determined primarily by observing bone and cartilage structures. For example, soft, porous, red bones with a maximum amount of pearly white cartilage characterize A maturity, whereas very little cartilage and hard, flinty bones characterize C, D, and E maturities. As maturity in carcasses increases from A to E, the meat becomes less tender.

Marbling is intramuscular fat or flecks of fat within the lean, and it is evaluated at the exposed ribeye muscle between the 12th and 13th ribs. Ten degrees of marbling, ranging from abundant to devoid, are designated in the USDA marbling standards. Various combinations of marbling and maturity that identify the carcass quality grades are shown in combinations of marbling and maturity in Figure 8.3. Figure 8.4 shows ribeye sections representing several different

TABLE 8.1   Official USDA Grade Standards for Live Slaughter Cattle and Their Carcasses

Class of Kind	Quality Grades (highest to lowest)	Yield Grades (highest to lowest)
Beef		
Steer and heifer	Prime, choice, select, standard, commercial, utility, cutter, canner	1, 2, 3, 4, 5
Cow	Choice, select, standard, commercial, utility, cutter, canner	1, 2, 3, 4, 5
Bullock	Prime, choice, select, standard, utility	1, 2, 3, 4, 5
Bull	(No designated quality grades)	1, 2, 3, 4, 5
Veal	Prime, choice, select, standard, utility	NA
Calf	Prime, choice, select, standard, utility	NA

Source: USDA.

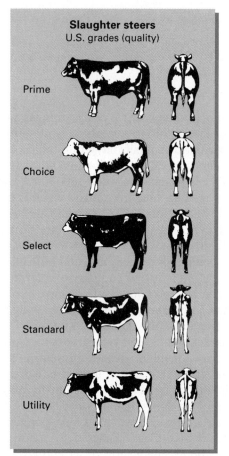

FIGURE 8.1 USDA quality grades (Commercial, Cutter, and canner omitted). Courtesy of USDA.

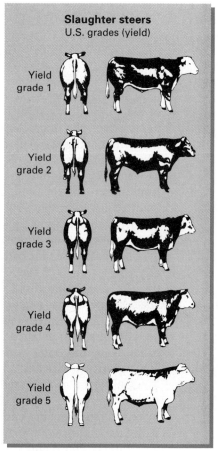

FIGURE 8.2 USDA yield grades for market cattle. Courtesy of USDA.

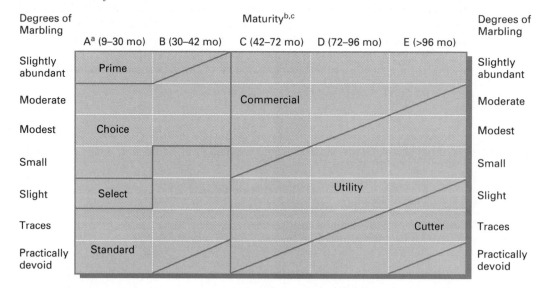

[a]Assumes that firmness of lean is comparably developed with the degree of marbling and that the carcass is not a "dark cutter."

[b]Maturity increases from left to right (A through E).

[c]The A maturity portion of the Figure is the only portion applicable to bullock carcasses.

FIGURE 8.3 Relationship between marbling, maturity, and carcass quality grade. *Source:* USDA.

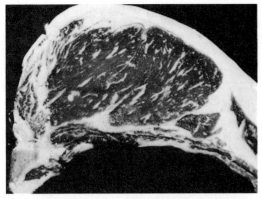

Prime (A and B maturity; moderately abundant marbling)

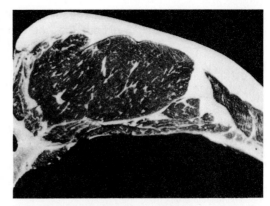

Choice (A and B maturity; modest marbling)

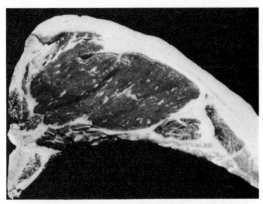

Select (A and B maturity; slight marbling)

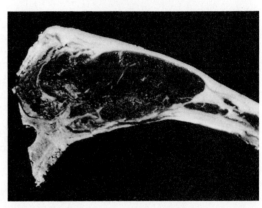

Standard (A and B maturity; traces of marbling)

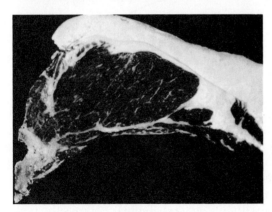

Commercial (C maturity; small marbling)

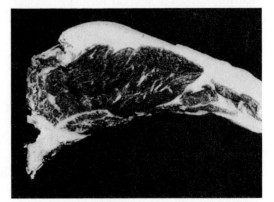

Utility (C maturity; slight marbling)

FIGURE 8.4 Exposed ribeye muscles (between the 12th and 13th ribs) showing various degrees of marbling associated with several beef carcass quality grades. Courtesy of the American Meat Science Association copyrighted 1997.

quality grades and several different degrees of marbling. Note that C maturity starts at 42 months of age. Slaughter cows more than 42 months of age will grade commercial, utility, cutter, or canner regardless of amount of marbling.

Yield grades, sometimes referred to as **cutability grades,** measure the quantity of **boneless, closely trimmed retail cuts (BCTRC)** from the major wholesale

TABLE 8.2 Beef Carcass Yield Grades and the Yield of BCTRC (boneless, closely trimmed retail cuts) from the Round, Loin, Rib, and Chuck

Yield Grade	BCTRC (%)	Yield Grade	BCTRC (%)	Yield Grade	BCTRC (%)
1.0	54.6	2.8	50.5	4.6	46.4
1.2	54.2	3.0	50.0	4.8	45.9
1.4	53.7	3.2	49.6	5.0	45.4
1.6	53.3	3.4	49.1	5.2	45.0
1.8	52.8	3.6	48.7	5.4	44.5
2.0	52.3	3.8	48.2	5.6	44.1
2.2	51.9	4.0	47.7	5.8	43.6
2.4	51.4	4.2	47.3	—	—
2.6	51.0	4.4	46.8	—	—

Source: Adapted from USDA.

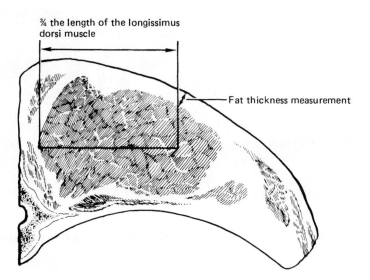

¾ the length of the longissimus dorsi muscle

Fat thickness measurement

FIGURE 8.5 Location of the fat measurement over the ribeye (longissimus dorsi) muscle. *Source:* Colorado State University.

cuts of beef (round, loin, rib, and chuck). BCTRC should not be confused with the total percentage of retail cuts (including hamburger) from a beef carcass. A numerical scale of 1–5 is used to rate yield grade, with 1 denoting the highest percentage of BCTRC. Although marketing communications generally quote yield grades in whole numbers, these yield grades are often shown in tenths in research data and information in carcass contests.

Packers can elect to have carcasses quality-graded. Table 8.2 shows the yield grades and their respective percentage of BCTRC. As an example, a carcass with a yield grade of 3.0 has a BCTRC of 50%; a 700-lb carcass with this yield grade would produce 350 lb of boneless, closely trimmed retail cuts from the round, loin, rib, and chuck.

Yield grades are determined from the following four carcass characteristics: (1) amount of fat measured in tenths of an inch over the **ribeye** muscle, also known as the **longissimus dorsi** (Fig. 8.5); (2) kidney, pelvic, and heart fat (usually estimated as percent of carcass weight); (3) area of the ribeye muscle, measured in square inches (Fig. 8.6); and (4) hot carcass weight. The last

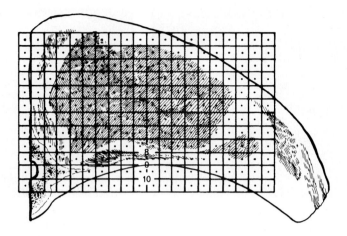

FIGURE 8.6 Plastic grid is placed over the ribeye muscle to measure the area. Each square represents 0.1 in. *Source:* Colorado State University.

measurement reflects amount of intermuscular fat. Generally, as the carcass increases in weight, the amount of fat between the muscles increases as well.

Measures of fatness in beef carcasses have the greatest effect in determining yield grade. Preliminary yield grade (PYG) is determined by estimating or measuring the outside fat of the ribeye muscle. Figure 8.7 shows the five yield grades with varying amounts of fat over the ribeye muscle and the area of the ribeye.

Quality grading and yield grading of beef carcasses by packers are voluntary. Approximately 90% of the 26 billion pounds of federally inspected carcass beef produced in 2004 was quality-graded and nearly 90% was yield-graded. Most of the graded beef originates from market steers and heifers fed in feedlots. The distribution of quality grades and yield grades is shown in Figure 8.8. Fifty-six percent of the beef that was graded in 2004 achieved USDA Choice or Prime. Of the beef quality graded, the Select grade has increased the most in recent years—from 4.6% in 1988 to 42% in 2001.

Feeder Cattle

The revised 2000 USDA **feeder grades** for cattle are intended to predict feedlot weight gain and the slaughter weight end point of cattle fed to a desirable fat-to-lean composition. The two criteria used to determine feeder grade are frame size and thickness. The three measures of frame size and four thickness descriptions are shown in Figures 8.9 and 8.10. Some examples of feeder cattle grade terminology are "large no. 1," "medium no. 2," and "small no. 2." Feeder cattle are given a USDA grade of "inferior" if the cattle are unhealthy or double-muscled. These cattle would not gain satisfactorily in the feedlot.

Although frame size and ability to gain weight in the feedlot are apparently related in the sense that large-framed cattle usually gain fastest, frame size appears to be a more accurate predictor of carcass composition or yield grade at different slaughter weights than of gaining ability. The USDA feeder grade specifications identify the different live weights from the three frame sizes when they reach the choice grade (Table 8.3).

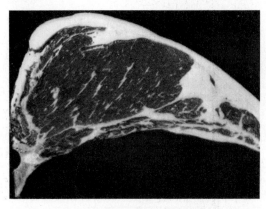

Yield Grade 1
(Fat 0.2 in., ribeye area 13.9 sq in.)

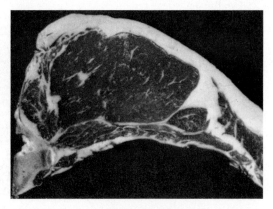

Yield Grade 2
(Fat 0.4 in., ribeye area 12.3 sq in.)

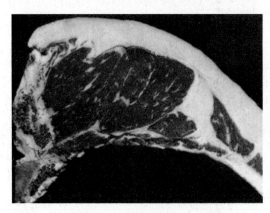

Yield Grade 3
(Fat 0.6 in., ribeye area 11.8 sq in.)

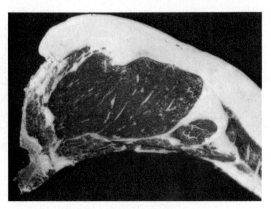

Yield Grade 4
(Fat 0.9 in., ribeye area 10.5 sq in.)

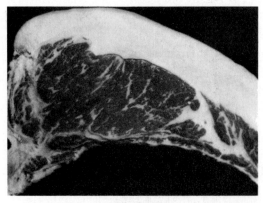

Yield Grade 5
(Fat 1.1 in., ribeye area 10.9 sq in.)

FIGURE 8.7 The five yield grades of beef shown at the 12th and 13th ribs. Courtesy of American Meat Science Association copyrighted 1997.

Color Plates

PLATE F

Courtesy of Iowa State University.

PLATE G

Courtesy of Iowa State University.

PLATE H

Courtesy of Iowa State University.

PLATE I

Courtesy of Iowa State University.

PLATE J

Courtesy of Iowa State University.

PLATE K

Courtesy of Iowa State University.

PLATE L

Courtesy of Iowa State University.

PLATE M

Some of the breeds of cattle important to beef production in the United States. Courtesy of the American Angus Association, Beefmaster Breeders Universal, American Brahman Breeders Association, International Brangus Breeders Association, American International Charolais Association, and American Hereford Association.

PLATE N

Some of the breeds of cattle important to beef production in the United States. Courtesy of American Gelbvieh Association, North America Limousin Foundation, Red Angus Association of America, Santa Gertrudis Breeders International, American Shorthorn Association, and Altenburg Simmentals.

PLATE O

Major breeds of dairy cows with origin, identifying characteristics, and production listed. Photos by Agri-Graphics.

PLATE P

Major breeds of dairy cows with origin, identifying characteristics, and production listed.

PLATE Q

Breeds of light horses. *Sources:* Appaloosa Horse Club (Appaloosa); International Arabian Horse Association (Arabian); American Morgan Horse Association (Morgan); American Paint Horse Association (Paint).

PLATE R

Breeds of light horses. *Sources:* American Quarter Horse Association (Quarter Horse); the U.S. Trotting Association (Standardbred); Tennessee Walking Horse Breeders and Exhibitors Association (Tennessee Walking Horse).

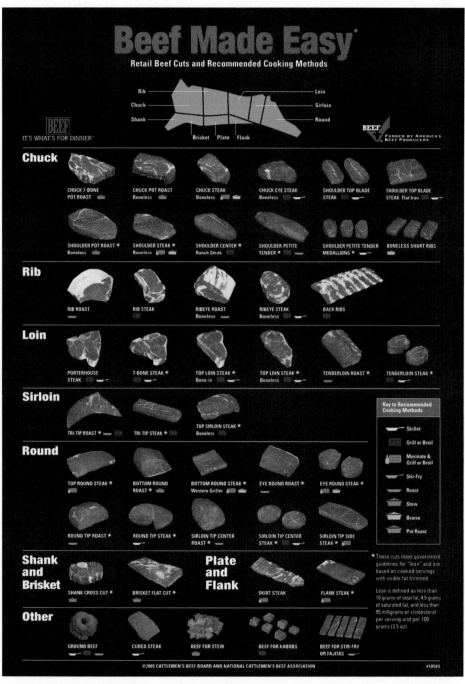

PLATE A Wholesale and retail cuts of beef with recommended types of cooking. Courtesy of the National Cattlemen's Beef Association © 1986.

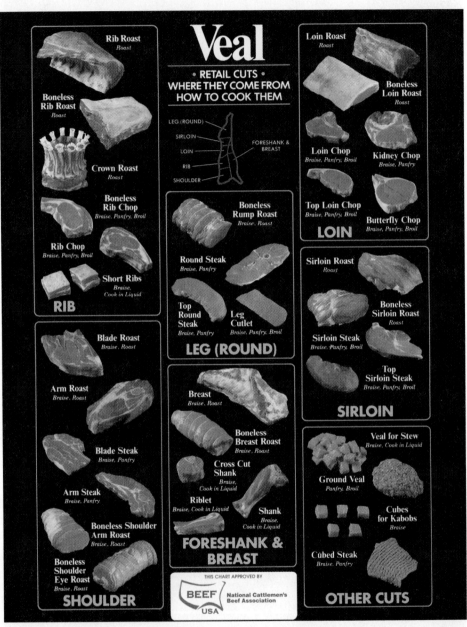

PLATE B Wholesale and retail cuts of veal with recommended types of cooking. Courtesy of the National Cattlemen's Beef Association © 1986.

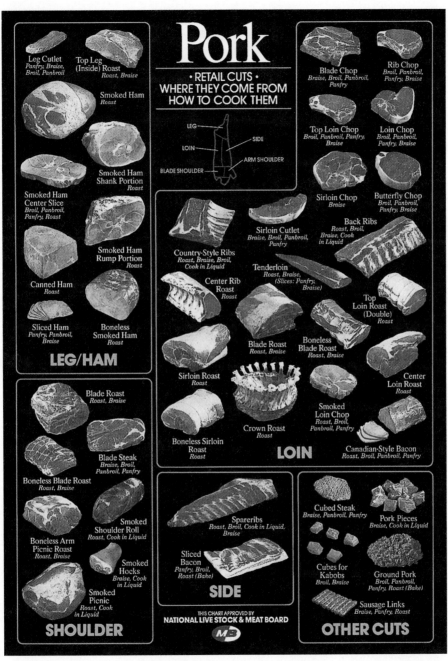

PLATE C Wholesale and retail cuts of pork with recommended types of cooking. Courtesy of the National Pork Producers Council © 1986.

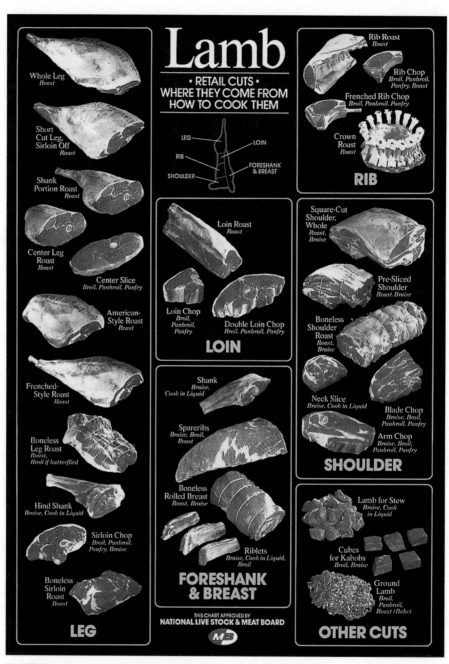

PLATE D Wholesale and retail cuts of lamb with recommended types of cooking. Courtesy of the National Livestock and Meat Board © 1986.

Color Plate Selection

CARCASS COMPOSITION DISPLAY

The proper ratio of fat to lean in an animal carcass has been a controversial is-sue among producers, feeders, and packers since the industry began. In the 1940s, fat-type animals, such as the lard-type hog, baby beef, and fat lamb, were the preferred type. Today, however, vegetable oils have largely replaced lard and other animal fats, lean beef has replaced the fatted calf, and the lamb has been made leaner through modern production techniques. Today's consumer demands lean meat, and today's grading system reflects those demands.

By visually appraising carcasses, a person can quickly identify the lean-to-fat compositional differences. Appraisal of similar differences in live animals, however, can be a more difficult and discriminating challenge. The following set of photographs graphically demonstrates a lean-fat ratio both on individual animals and as a comparison between different yield grades. The pictures are courtesy of Iowa State University and Colorado State University, where car-casses were frozen in a standing position. This process allowed technicians to remove different layers of tissue, such as the hide, fat, and muscling, and to cross-section entire carcasses for easy comparison. The photographs are orga-nized in color plates, with the caption for each photo presented here by plate and photo number:

Plate E compares body conformation of different yield grades in both live animals and their carcasses.

1. Thickness of muscling in two medium-framed feeder steers is shown using cutaway sections of the round. The steer on the left has more thickness of lean meat in the round than the steer on the right.

2.–3. Pounds of fat trimmed from one-half of the body of a yield grade 4 steer (92 lb) and a yield grade 2 steer (47 lb). Black stripes are hide and fat that remains at key locations on the body.

4. Rear view showing how the fat increases in thickness from the middle of the back to the edge of the loin.

5. Two yearling Hereford bulls of approximately the same weight. Breeding cattle can be visually appraised for fat and lean compositional differences. The bull on the left would sire slaughter steers having yield grade 4 or 5 carcasses, while the bull on the right would sire yield grade 1 or 2 steers. This assumes that the bulls would be bred to cows similar in frame-size to themselves, and that the steers would be slaughtered at approximately 1,150 pounds. Compare to plates B and C.

Plates F and G compare fat-to-lean composition on a yield grade 2 steer (plate F) and a yield grade 5 steer (plate G).

1. Live animals—side view.

2. Black ribbons identify where cross sections were made (3, 4, 5, and 6—cross sections from rump, hip, mid-carcass, and shoulder).

3. The percentages of fat, lean, and bone found in this carcass.

Plates H and I compare the fat-to-lean composition of lamb carcasses on yield grade 1 lamb (plate H) and yield grade 5 lamb (plate I).

1. Side view.

2. Rear view.

3. Black ribbons identify where cross sections were made (4, 5, and 6—cross sections from shoulder, mid-carcass, and rump).

Plates J and K compare fat-to-lean composition of a U.S. no. 1 pig (plate J) with a U.S. no. 4 pig (plate K).

1. Live animals—side view.

2. Rear view.

3. Black ribbons identify where cross sections were made (4, 5, and 6—cross sections from shoulder, mid-carcass, and rump).

Plate L shows the composition and appearance of a steer carcass at various stages of removal of hide, fat, and muscle.

1. Live steer with hair clipped from one side.

2. Frozen steer with hide removed and fat exposed.

3. Fat removed from one-half of steer's body.

4. Rear view with fat removed from the left side. Note cod and twist fat on the right side.

5. Skeleton of the beef animal after all the muscle has been removed.

1

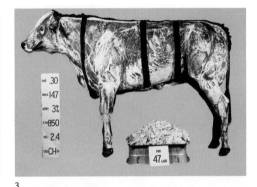

2

3

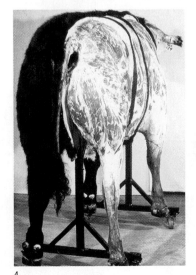

4

5

PLATE E   Courtesy of Colorado State University.

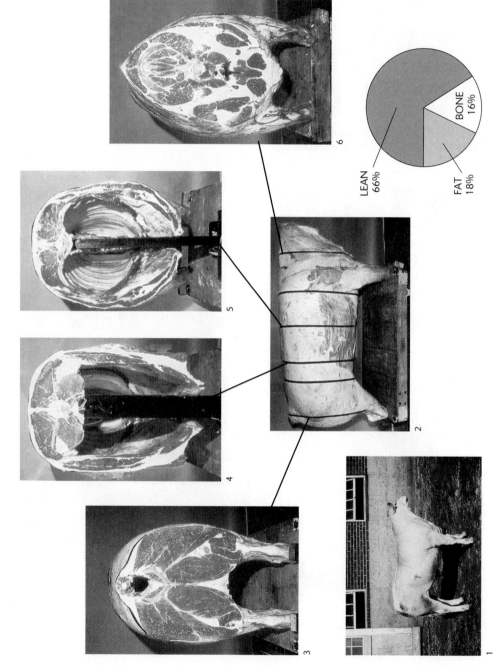

LEAN
66%

FAT
18%

BONE
16%

PLATE F   Courtesy of Iowa State University.

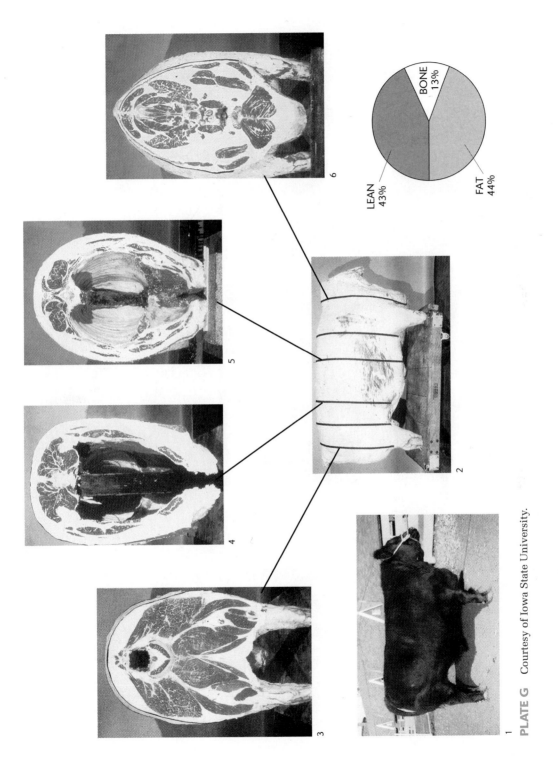

LEAN
43%

BONE
13%

FAT
44%

6

5

4

3

2

1

PLATE G   Courtesy of Iowa State University.

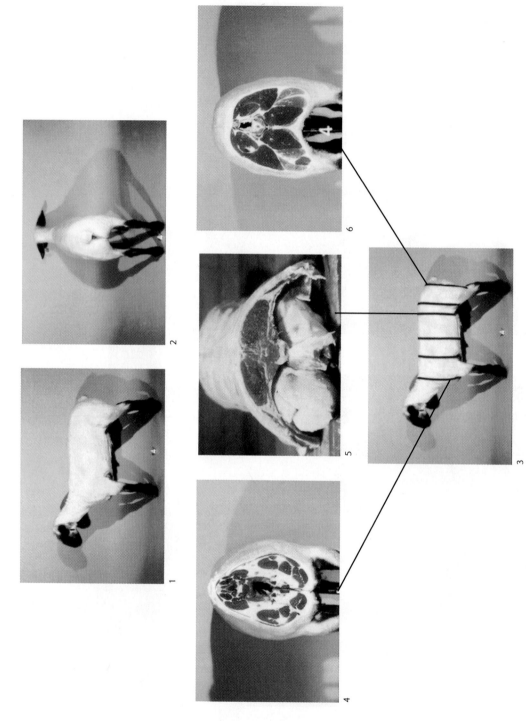

PLATE H Courtesy of Iowa State University.

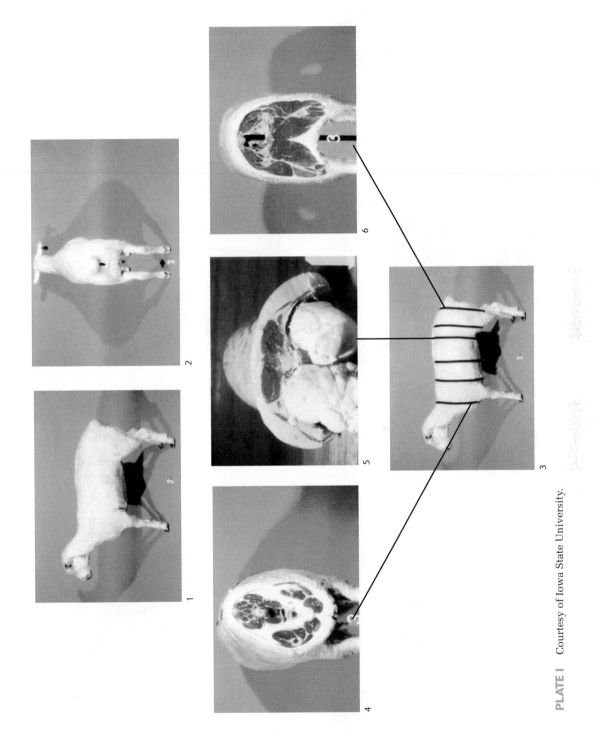

PLATE I Courtesy of Iowa State University.

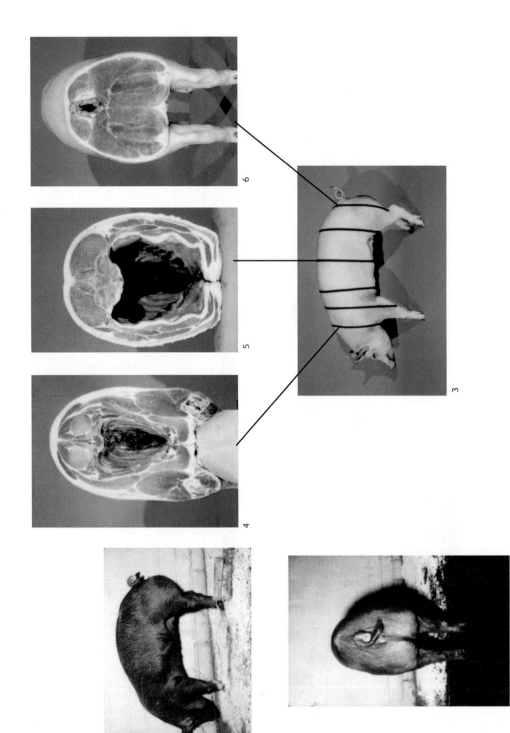

PLATE J   Courtesy of Iowa State University.

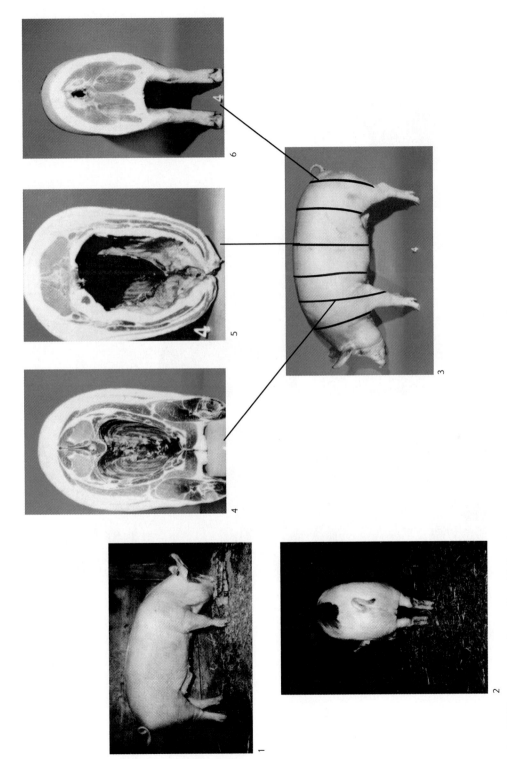

PLATE K   Courtesy of Iowa State University.

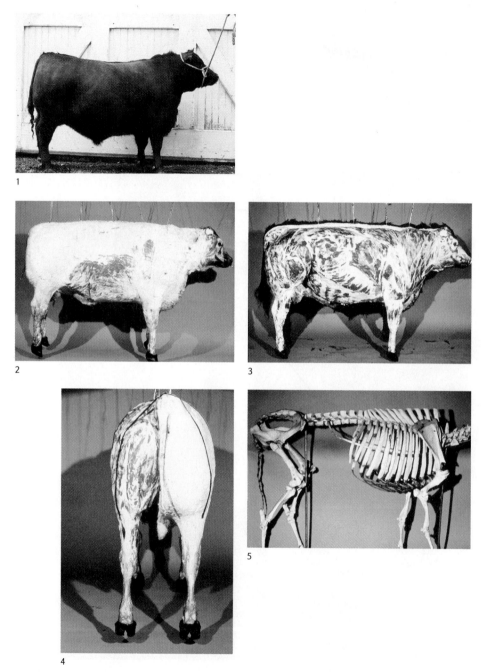

PLATE L Courtesy of Iowa State University.

Angus

Beefmaster

Brahman

Brangus

Charolais

Hereford

PLATE M Some of the breeds of cattle important to beef production in the United States. Courtesy of the American Angus Association, Beefmaster Breeders Universal, American Brahman Breeders Association, International Brangus Breeders Association, American International Charolais Association, and American Hereford Association.

Gelbvieh

Limousin

Red Angus

Santa Gertrudis

Shorthorn

Simmental

PLATE N Some of the breeds of cattle important to beef production in the United States. Courtesy of American Gelbvieh Association, North America Limousin Foundation, Red Angus Association of America, Santa Gertrudis Breeders International, American Shorthorn Association, and Altenburg Simmentals.

Origin: Scotland
Average weight:
    Bulls—1,850 lb
    Cows—1,200 lb
Color: Mahogany and white spotted,
    may have pigmented legs
Average milk yield: 15,094 lb
Percentage of fat: 3.89%

Ayrshire

Origin: Switzerland
Average weight:
    Bulls—2,000 lb
    Cows—1,400 lb
Color: Solid blackish, hairs dark with
    light tips
Average milk yield: 16,701 lb
Percentage of fat: 4.04%

Brown Swiss

Origin: Guernsey Island
Average weight:
    Bulls—1,600 lb
    Cows—1,100 lb
Color: Light red and white, yellow skin
Average milk yield: 14,070 lb
Percentage of fat: 4.49%

Guernsey

PLATE O Major breeds of dairy cows with origin, identifying characteristics, and production listed. Photos by Agri-Graphics.

Origin: Holland
Average weight:
  Bulls—2,200 lb
  Cows—1,500 lb
Color: Black and white
Average milk yield: 20,318 lb
Percentage of fat: 3.65%

Holstein

Origin: Jersey Island
Average weight:
  Bulls—1,500 lb
  Cows—1,000 lb
Color: Blackish hairs have white tips to give
  gray color or red tips to give fawn color;
  also can be solid black or white spotted.
Average milk yield: 14,275 lb
Percentage of fat: 4.65%

Jersey

Origin: Great Britain
Average weight:
  Bulls—2,000 lb
  Cows—1,250 lb
Color: Red and white, roan
Average milk yield: 13,930 lb
Percentage of fat: 3.6%

Milking Shorthorn

Origin: Holland
Average weight:
  Bulls—2,100 lb
  Cows—1,400 lb
Color: Red and white
Average milk yield: 19,967 lb
Percentage of fat: 3.69%

Red and White

PLATE P Major breeds of dairy cows with origin, identifying characteristics, and production listed.

Appaloosa

Arabian

Morgan

Paint

PLATE Q   Breeds of light horses. *Sources:* Appaloosa Horse Club (Appaloosa); International Arabian Horse Association (Arabian); American Morgan Horse Association (Morgan); American Paint Horse Association (Paint).

Quarter Horse

Standardbred

Tennessee Walking Horse

PLATE R Breeds of light horses. *Sources:* American Quarter Horse Association (Quarter Horse); the U.S. Trotting Association (Standardbred); Tennessee Walking Horse Breeders and Exhibitors Association (Tennessee Walking Horse).

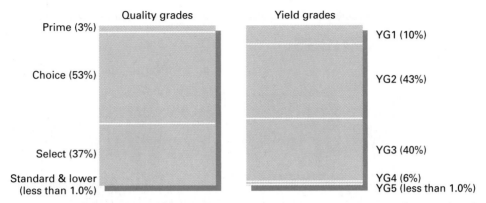

FIGURE 8.8 Quality grades and yield grades of beef, 2004. Note: Approximately 10% of total carcasses are not quality or yield graded, respectively. Quality and yield grade percentages do not equal 100% due to rounding error. *Source:* USDA.

Feeder Cattle
Official U.S. Grades

Frame Size

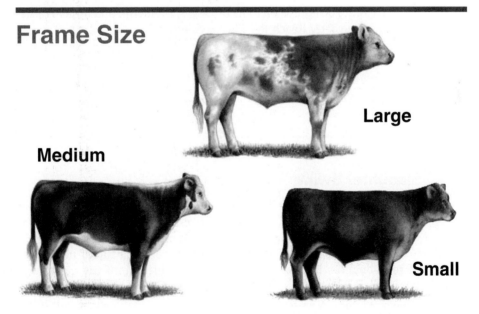

FIGURE 8.9 The three frame sizes of the USDA feeder cattle grade system. *Source:* USDA.

Thickness

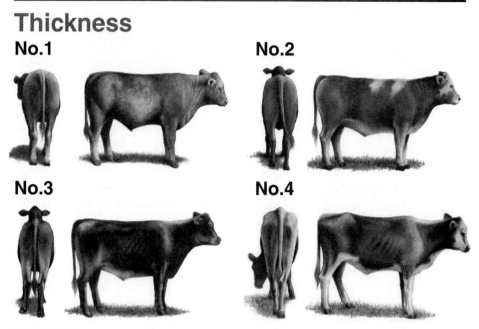

FIGURE 8.10   The four thickness standards of the USDA feeder cattle grade system. *Source:* USDA.

TABLE 8.3   Slaughter Weights of Large-, Medium-, and Small-Frame Slaughter Cattle When They Reach the Choice Grade

	Slaughter Weight	
Frame Size	Steers (lb)	Heifers (lb)
Large	>1,250	>1,150
Medium	1,100–1,250	1,000–1,150
Small	<1,100	<1,000

Slaughter Swine

Sex classes of swine are **barrow, gilt, sow, boar**, and **stag**. Boars and sows are older breeding animals, whereas gilts are younger females that have not produced any young. The barrow is the male pig castrated early in life, and the stag is the male pig castrated after it has developed boar flavor in the meat. Because of relationships between sex and sex condition and the acceptability of prepared meats to the consumer, separate grade standards have been developed for barrow and gilt carcasses and for sow carcasses. There are no official grade standards for boar and stag carcasses.

The traditional grades for barrow and gilt carcasses were based on two general criteria: (1) quality characteristics of the lean, and (2) expected combined yields of the four lean cuts (ham, loin, blade (Boston) shoulder, and picnic shoulder). There are only two quality grades for lean in a pork carcass: acceptable and unacceptable. Observing the exposed surface of a cut muscle, usually between the 10th and 11th ribs, assesses quality of the lean. Acceptable lean is gray-pink in color, has fine muscle fibers, and has fine marbling. Carcasses that have unacceptable lean quality (too dark or too pale, soft, or watery) or bellies too thin for suitable bacon production are graded U.S. utility, as they are soft and oily. Carcasses with acceptable lean quality are graded U.S. no. 1, U.S. no. 2, U.S. no. 3, or U.S. no. 4. These grades are based on expected yields of the four lean cuts, as shown in Table 8.4. Yield differences of the four wholesale cuts exist because of variation in the amount of muscling and fatness in relation to skeletal size. Measurements of backfat and estimates of muscling are the factors used to determine numerical yield grade (Table 8.5). Figure 8.11 illustrates the muscling scores used to determine grades.

Currently, pork packers are providing information that describes the pounds of lean produced per carcass (Table 8.6).

Assuming a 190-lb carcass with 0.8 in. of backfat measured by hand-held ruler, then the percent lean would be calculated as: $23.568 - (21.348 \times 0.8) + (0.503 \times 190) = 102.06$ lb of lean $\div$ 190 lb carcass = 53.72% lean.

The Fat-O-Meat'er is an electronic scanning technology that objectively measures fat thickness and loin eye depth with a line speed of 1,300 carcasses per hour. The development of electronic scanning technologies for application in the beef, pork, and lamb grading processes is a major focus and will likely replace the current yield estimation systems. Assuming the use of the Fat-O-Meat'er on a 190-lb carcass with an average fat depth of 0.9 and a loin depth of 2.4, the percent lean would be calculated as follows: $15.31 - (31.277 \times .9) + (3.813 \times 2.4) + (0.51 \times 190) = 93.21$ lb of lean $\div$ 190 lb carcass = 49.06% lean.

TABLE 8.4 Expected Yields of the Four Lean Cuts, Based on Percent of Chilled Carcass Weight

Grade	Four Lean Cuts (%)
U.S. no 1	>53.0
U.S. no 2	50.0–52.9
U.S. no 3	47.0–49.0
U.S. no. 4	<47.0

Source: Adapted from USDA.

TABLE 8.5 Preliminary Grade Based on Backfat Thickness over the Last Rib (assumes average muscle thickness)

Preliminary Grade[a]	Backfat Thickness (in.)
U.S. no. 1	<1.00
U.S. no 2	1.00–1.24
U.S. no 3	1.25–1.49
U.S. no 4	>1.50[b]

[a] Swine with thick muscling qualify for next higher grade; those with thin muscling are downgraded to next lower grade.
[b] Animals with an estimated last-rib backfat thickness of 1.75 in. or over have to be U.S. no. 4 and cannot be graded U.S. no. 3, even with thick muscling.
Source: USDA.

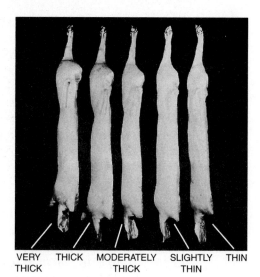

FIGURE 8.11 The five degrees of muscling associated with USDA pork grades. Courtesy of USDA.

VERY THICK   THICK   MODERATELY THICK   SLIGHTLY THIN   THIN

TABLE 8.6 Prediction Equations for Estimating Pounds of Fat-Free Lean from Pork Carcasses

Fat Measurement Protocol	Equation for lb of Fat-Free Lean
Stainless steel ruler to measure last-rib backfat	$23.568 - (21.348 \times \text{last-rib backfat}) + (0.503 \times \text{hot carcass weight})$[a]
Fat-O-Meter optical grading probe to measure fat and muscle depth between the 3rd and 4th from the last rib	$15.31 - (31.277 \times \text{fat depth}) + (3.813 \times \text{loin muscle}$
Ultrasound measurement of the average fat and muscle depth spanning the last rib to the 10th rib	$6.783 - (15.745 \times \text{avg. fat depth}) + (4.007 \times \text{avg. loin muscle depth}) + (0.47 \times \text{hot carcass weight})$[a]

[a] To convert to a percent fat-free lean, divide by hot carcass weight and multiply by 100.
Source: Adapted from USDA and American Meat Science Association.

TABLE 8.7 USDA Maturity Groups, Sex Classes, and Grades of Slaughter Sheep

Maturity Group	Sex Class	Quality Grade (highest to lowest)	Yield Grade (highest to lowest)
Lamb	Ewe, wether, or ram	Prime, choice, good, utility	1, 2, 3, 4, 5
Yearling mutton	Ewe, wether, or ram	Prime, choice, good, utility	1, 2, 3, 4, 5
Mutton	Ewe, wether, or ram	Choice, good, utility, cull	1, 2, 3, 4, 5

Source: USDA.

Slaughter Sheep

Slaughter sheep are classified by their sex and maturity. Live sheep and their carcasses are also graded for quality grades and yield grades (Table 8.7).

Lamb carcasses ranging from approximately 2 to 14 months of age always have the characteristic **break joint** on one of their shanks following the removal of their front legs (Fig. 8.12). **Mutton** carcasses are distinguished from lamb carcasses by the appearance of the spool joint instead of the break joint (Fig. 8.12). The break joint ossifies as the sheep matures. Yearling mutton carcasses ranging

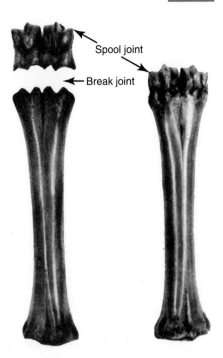

Spool joint

Break joint

FIGURE 8.12 Break and spool joints. The cannon bone on the left exhibits the typical break joint in which the foot and pastern are removed at the cartilaginous junction. In the cannon bone on the right, the cartilaginous junction has ossified, making it necessary for the foot and pastern to be removed at the spool joint. Courtesy of USDA.

from 12 to 15 months of age usually have the spool joint present but may occasionally have a break joint. Yearling mutton is also distinguished from lamb and mutton by color of the lean (intermediate between the pinkish red of lamb and the dark red of mutton) and shape of rib bones. Most U.S. consumers prefer lamb to mutton because it has a milder flavor and is more tender.

Quality grades are determined from a composite evaluation of conformation, maturity, and **flank streaking.** There are also minimum standards for **flank firmness and fullness;** however, most lambs meet these standards. Conformation is an assessment of overall thickness of muscling in the lamb carcass. Maturity of lamb carcasses is determined by bone color and shape and muscle color. Flank streaking (streaks of fat within the flank muscle) predicts marbling because lamb carcasses are not usually ribbed to expose marbling in the ribeye muscle. Flank firmness and fullness are determined by taking hold of the flank muscle with the hand. Most lamb carcasses grade either prime or choice, with few carcasses grading in the lower grades.

Lamb carcasses must be both quality-graded and yield-graded. Yield grades estimate boneless, closely trimmed retail cuts from the leg, loin, rack, and shoulder. The approximate percentage of retail cuts for selected yield grades are shown in Table 8.8.

Yield grades are determined by fat thickness over the loin eye muscle. Figure 8.13 shows cross sections of lamb carcasses representing the five yield grades and the amount of fat thickness over the loin eye for each yield grade. Multiplying the adjusted fat thickness by 10 and adding 0.4 estimates the yield grade. For example, a lamb with 0.25 in. of fat would have an estimated yield grade of 2.9 (e.g., [0.25 × 10] + 0.4 = 2.9).

TABLE 8.8 Lamb Carcass Yield Grades and Percentage of Retail Cuts

Yield Grade	Percent Retail Cuts	Yield Grade	Percent Retail Cuts
1.0	49.0	3.5	44.6
1.5	48.2	4.0	43.6
2.0	47.2	4.5	42.8
2.5	46.3	5.0	41.8
3.0	45.4	5.5	41.0

Source: USDA.

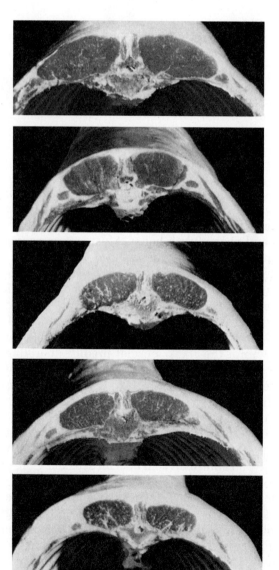

FIGURE 8.13 The five yield grades of lamb showing the progressive increases in the amount of fat over the loin eye at the 12th and 13th ribs. Courtesy of American Meat Science Association copyrighted 1997.

Feeder Lamb Grades

Choice and prime slaughter lambs are produced in relatively large numbers, fed on grass and their mothers' milk. Lambs weighing less than 100 lb at weaning are considered feeder lambs. They require additional feeding to produce a more desirable carcass.

There are no official USDA grades for feeder lambs. Some promising research has been done to evaluate some possible feeder lamb grades. These grades, based on frame size, project the slaughter weight of the lambs when they would reach 0.25 in. of fat thickness over the loin eye muscle. The frame sizes and slaughter weights are: *large* (over 120 lb), *medium* (100 lb–120 lb), and *small* (less than 100 lb). There are also body-thickness categories based on conformation of slaughter lambs: no. 1 (prime conformation), no. 2 (choice conformation), and no. 3 (less-than-choice conformation).

MARKET CLASSES AND GRADES OF POULTRY PRODUCTS

Poultry Meat

The class of poultry must be displayed on the package label or a tag on the wing of the bird. The age class indicates tenderness, as meat from younger birds is more tender than that from older birds. Table 8.9 shows the age-classification labels for poultry meat.

Broiling, frying, roasting, or barbecuing usually requires more tender meat from the young age classes. The mature, less tender poultry meat is best prepared by baking, stewing, or fabricating or by including it in other prepared dishes.

The grades are U.S. grade A, U.S. grade B, and U.S. grade C for each of the classes, with A being the highest grade (Fig. 8.14). Carcasses of A quality are free of deformities that detract from their appearance or that affect normal distribution of flesh. They have a well-developed covering of flesh and a well-developed

TABLE 8.9   Labels Used to Identify Poultry Age Classes

Type	Young	Mature
Chicken	Young chicken	Mature/old chicken
	Rock Cornish game hen	
	Broiler	Hen
	Fryer	Stewing or baking chicken
	Roaster	Fowl
	Capon	
Turkey	Young turkey	Mature/old turkey
	Fryer-roaster	Yearling turkey
	Young hen	
	Young tom	
Duck	Duckling	Mature/old
	Young duckling	
	Broiler duckling	
	Fryer duckling	
	Roaster duckling	
Goose	Young goose	Mature/old goose
Guinea	Young guinea	Mature/old guinea

Source: USDA.

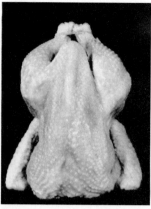
Broiler or fryer, A quality

Broiler or fryer, B quality

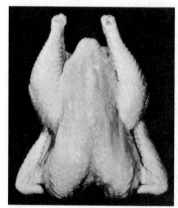
Hen or stewing chicken, A quality

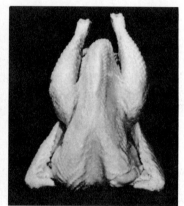
Hen or stewing chicken, B quality

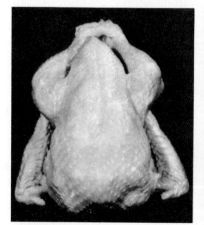
Young turkey, A quality

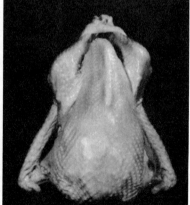
Young turkey, B quality

FIGURE 8.14 Classes and grades of ready-to-cook poultry. Courtesy of USDA.

layer of fat in the skin. They are free of pinfeathers and diminutive feathers, exposed flesh on the breast and legs, and broken bones. They have no more than one disjointed bone, and they are practically free of discolorations of the skin and flesh and defects resulting from handling, freezing, or storage. Carcasses of B quality may have moderate deformities. They have a moderate covering of flesh, sufficient fat in the skin to prevent a distinct appearance of the flesh through the skin, and no more than an occasional pinfeather or diminutive feather. They may have moderate areas of exposed flesh and discoloration of the skin and flesh. They may have disjointed parts but no broken bones, and they may have moderate defects resulting from handling, freezing, or storage.

Eggs

The grading of shell eggs involves classifying individual eggs according to established standards. Eggs are graded by sorting them into groups, each group having similar weight and quality characteristics. Table 8.10 shows how eggs are classified according to a size-and-weight relationship.

The USDA quality standards used to grade individual shell eggs are as follows:

Exterior Quality Factors	*Interior Quality Factors*
Cleanliness of shell	Albumen thickness
Soundness of shell	Condition of yolk
(cracks, and texture)	Size and condition of air cell
Shape	Abnormalities (e.g., blood spots, meat spots)

Exterior quality factors are apparent from external observation; **interior quality factors** involve an assessment of egg content. The latter is accomplished through a process called **candling** (visually appraising the eggs while light is shone through them).

Although shell color is not a factor in the U.S. standards and grades, eggs are usually sorted by color and sold as either "whites" or "browns." Eggs sell better when sorted by color and packed separately. Contrary to popular opinion, there are no differences other than color between similar-quality brown- and white-shelled eggs.

The U.S. standards for quality of individual shell eggs are applicable to eggs from domestic chickens only; these standards are summarized in Table 8.11. Consumer egg grades are U.S. grade AA, U.S. grade A, and U.S. grade B. Figure 8.15 shows selected shields for communicating the quality grade.

TABLE 8.10 Consumer Weight Classes of Eggs, Minimum Net Weight per Dozen

Size	Ounces (per doz)
Jumbo	30
Extra large	27
Large	24
Medium	21
Small	18
Peewee	15

Source: USDA.

TABLE 8.11 Summary of U.S. Standards for Quality of Individual Shell Eggs

Quality Factor	AA Quality	A Quality	B Quality	Dirty	Check
Shell	Clean, unbroken; practically normal	Clean, unbroken; practically normal	Clean to slightly stained,[a] unbroken; abnormal	Unbroken; adhering dirt or foreign material, prominent stains, moderate stained areas in excess of B quality	Broken or cracked shell but membranes intact, not leaking[c]
Air cell	$\frac{1}{8}$ in. or less in depth; unlimited movement and free or bubbly	$\frac{3}{16}$ in. or less in depth; unlimited movement and free or bubbly	Over $\frac{3}{16}$ in. in depth; unlimited movement and free or bubbly		
White	Clear, firm	Clear, reasonably firm	Weak and watery; small blood and meat spots present[b]		
Yolk	Outline slightly defined; practically free from defects	Outline fairly well defined; practically free from defects	Outline plainly visible; enlarged and flattened; clearly visible germ development, but no blood; other serious defects		

[a] Moderately stained areas permitted ($\frac{1}{32}$ of surface if localized, or $\frac{1}{16}$ if scattered).
[b] If they are small (aggregating not more than $\frac{1}{8}$ in. in diameter).
[c] Leaker has broken or cracked shell and membranes, and contents are leaking or free to leak.
Source: USDA, Egg Grading Standards.

FIGURE 8.15 USDA grade shields showing two egg quality grades. Courtesy of USDA.

CHAPTER SUMMARY

- Market classes and grades have been established to segregate animals, carcasses, and products into uniform groups based on preferences of buyers and sellers. However, movement toward branded value-added products reduces the need for commodity grading programs.

- The United States Department of Agriculture (USDA) classes and grades make the marketing process simpler and more easily communicated.

- Red meat animals are separated into classes based primarily on age and sex. Age of the animal significantly affects tenderness. Sex class separates animals and carcasses into more uniform carcass weights, tenderness groups, and processing methods.

- Feeder and slaughter grades exist for most red meat animals. For slaughter animals, quality grades are used to measure consumer palatability characteristics, while yield grades measure the amount of fat, lean, and bone in the carcass.

- Poultry meat classes are based on age (affects tenderness), and the grades evaluate thickness and fullness of meat in the breast and thigh.

- Eggs are graded for size (weight), shell condition, and interior quality (yolk, white, and air cell).

KEY WORDS

marketing	stag	barrow
market classes and grades	quality grades	gilt
	yield grades (cutability grades)	sow
meat inspection		boar
veal	maturity	break joint
calf	marbling	mutton
beef	boneless, closely trimmed retail cuts (BCTRC)	flank streaking
heifer		flank firmness and fullness
cow		
steer	ribeye	exterior quality factors
bull	longissimus dorsi	interior quality factors
bullock	feeder grades	candling

REVIEW QUESTIONS

1. Slaughter cattle are separated into market classes based upon what two characteristics?
2. Quality grades for carcasses, which estimate eating quality of meat, are based on what two characteristics?
3. What do beef carcass yield grades estimate?
4. What do U.S. numerical grades of pork carcasses estimate?
5. What do lamb carcass yield grades estimate?
6. What process is used to appraise the interior quality of table eggs from chickens?

SELECTED REFERENCES

Boggs, D. L., and R. A. Merkel. 1993. *Live Animal Carcass Evaluation and Selection Manual.* Dubuque, IA: Kendall/Hunt.

Meat Buyer's Guide. 2006. Chicago, IL: North American Meat Processor's Association.

Meat Evaluation Handbook. 2001. Chicago, IL: American Meat Science Association.

USDA. April 1995. *Egg Grading Manual.* USDA Agriculture Handbook no. 75.

USDA. April 1980. *Facts About U.S. Standards for Grades of Feeder Cattle.* USDA Agricultural Marketing Service AMS-586.

USDA. *Official United States Standards for Grades of Feeder Pigs; Grades of Slaughter Swine; Grades of Pork Carcasses; Grades of Veal and Calf Carcasses; Grades of Lamb, Yearling Mutton and Mutton Carcasses; Grades of Carcass Beef and Slaughter Cattle;* and *Grades of Poultry.* 2006. USDA: Agricultural Marketing Service.

CHAPTER 9

Visual Evaluation of Market Animals

LEARNING OBJECTIVES

- Identify the external parts of cattle, swine, and sheep
- Describe the conformational characteristics of market cattle
- Describe the location of wholesale cuts on live cattle, hogs, and sheep

Most redmeat animals ready for harvest, if purchased live, are evaluated based on visual appraisal of their apparent carcass merit. Combining meaningful performance records and effective visual appraisal best identifies productivity of breeding and slaughter meat animals. Opinions regarding the relationship of form and function in redmeat animals (cattle, sheep, and swine) differ widely among producers, who continually discuss the value of visual appraisal and objective measurements in breeding programs, in determining market grades, and in defining the so-called ideal types. It is important to separate true relationships from opinion in the area of animal form and function.

Type is defined as an ideal or standard of perfection combining all the characteristics that contribute to an animal's usefulness for a specific purpose. **Conformation** implies the same general meaning as *type* and refers to form and shape of the animal. Both *type* and *conformation* describe an animal according to its external form and shape that can be evaluated visually or measured more objectively with a tape, ultrasound machine, or other device.

Redmeat animals have three productive stages: (1) breeding (reproduction), (2) feeder (growth), and (3) slaughter (carcass or product). It has been well demonstrated that performance records are much more effective than visual appraisal in improving reproduction and growth stages. Visual appraisal does have importance in these stages, primarily in identifying reproductive soundness, skeletal soundness, and health status. Visual appraisal has some importance in complementing performance records (ultrasonic measurements for fatness), as visual appraisal is important in evaluating slaughter animals. Visual appraisal of slaughter redmeat animals can be used effectively to predict carcass composition (fat and muscling), primarily when relatively large differences exist.

EXTERNAL BODY PARTS

Effective communication in many phases of the livestock industry requires a knowledge of the external body parts of the animal. The locations of the major body parts for swine, cattle, and sheep are shown in Figures 9.1, 9.2, and 9.3, respectively.

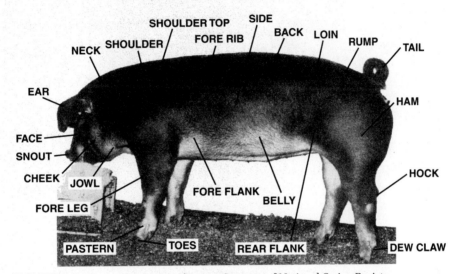

FIGURE 9.1   The external parts of swine. Courtesy of National Swine Registry.

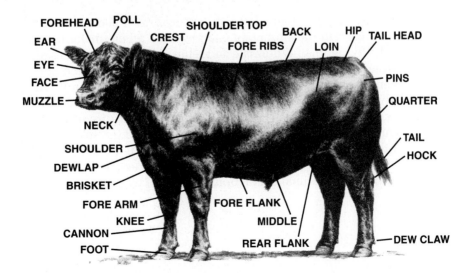

FIGURE 9.2   The external parts of cattle. Courtesy of the American Angus Association.

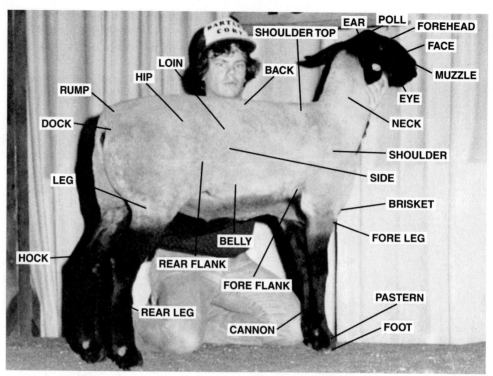

FIGURE 9.3   The external parts of sheep. Courtesy of *Sheep Breeder and Sheepman* magazine.

LOCATION OF THE WHOLESALE CUTS IN THE LIVE ANIMAL

The next step for effective visual appraisal, after becoming familiar with the external parts of the animal, is to understand where major meat cuts are located in the live animal. Wholesale and retail cuts of the carcasses of beef, sheep, and swine are identified in Chapter 3. A carcass is evaluated after the animal has been slaughtered, eviscerated, and, in the case of beef and swine, split into two halves. The lamb carcass remains as a whole carcass. Furthermore, when it is evaluated, the carcass is hanging by the hind leg from the rail in the packing plant. This makes it difficult to perceive how the carcass would appear as part of the live animal standing on all four legs. Figure 9.4, which shows the location of the wholesale cuts on the live animal, assists in correlating carcass evaluation to live-animal evaluation.

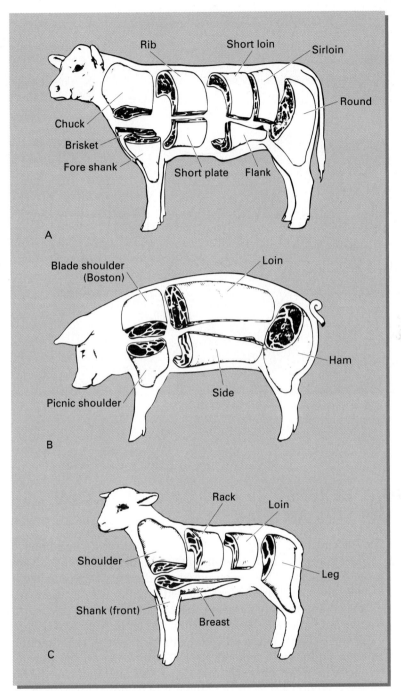

FIGURE 9.4 Location of the wholesale cuts on the live steer, pig, and lamb. *Source:* Colorado State University.

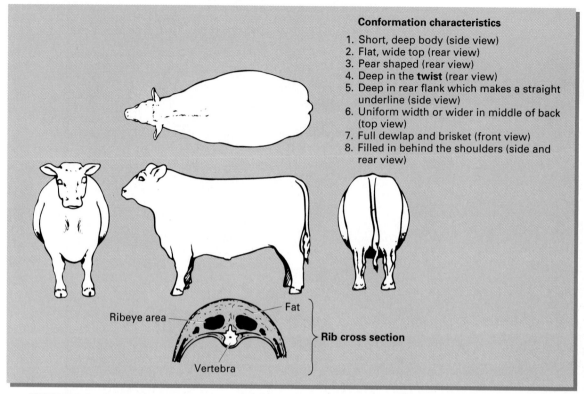

Conformation characteristics

1. Short, deep body (side view)
2. Flat, wide top (rear view)
3. Pear shaped (rear view)
4. Deep in the **twist** (rear view)
5. Deep in rear flank which makes a straight underline (side view)
6. Uniform width or wider in middle of back (top view)
7. Full dewlap and brisket (front view)
8. Filled in behind the shoulders (side and rear view)

Ribeye area — Fat

Vertebra — **Rib cross section**

FIGURE 9.5 Conformation characteristics of slaughter steer typical of yield grade 5. Compare with color plate G. *Source:* Colorado State University.

VISUAL PERSPECTIVE OF CARCASS COMPOSITION OF THE LIVE ANIMAL

The carcass is composed of fat, lean (red meat), and bone. The meat industry's goal is to produce large amounts of highly palatable lean and minimal amounts of fat and bone. These composition differences are reflected in the yield grades of beef and lamb and the lean percentage of swine discussed in Chapter 8. Effective visual appraisal of carcass composition requires knowing the body areas of the live animal where fat is deposited and muscle growth occurs. Color plates F and G show the fat and lean composition at several cross sections of a yield grade 2 steer and a yield grade 5 steer. The ribbons on the frozen carcasses show the location of the cross sections. Cross section 3 removes the bulge from the round, cross section 4 is in front of the hip bone down through the flank, cross section 5 is at the 12th and 13th rib, and cross section 6 is at the point of the shoulder down through the brisket. Note the contrasts in percentages of fat (18% versus 44%) and lean (66% versus 43%). Both steers graded choice and were slaughtered at approximately 1,100 lb. The conformation characteristics of the two steers, which are primarily influenced by fat deposits and some muscling differences, are contrasted in Figures 9.5 and 9.6. The conformation of these two steers is also contrasted to the conformation of a slaughter steer that is underfinished and thinly

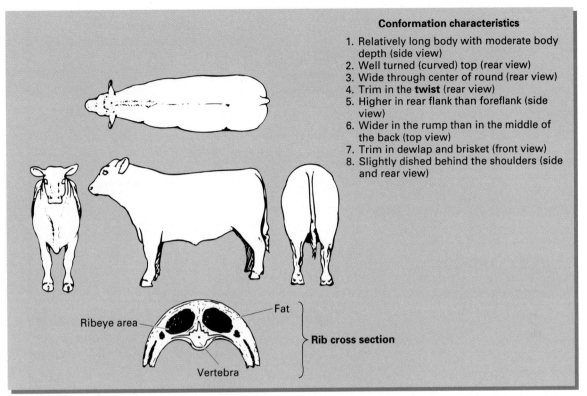

Conformation characteristics

1. Relatively long body with moderate body depth (side view)
2. Well turned (curved) top (rear view)
3. Wide through center of round (rear view)
4. Trim in the **twist** (rear view)
5. Higher in rear flank than foreflank (side view)
6. Wider in the rump than in the middle of the back (top view)
7. Trim in dewlap and brisket (front view)
8. Slightly dished behind the shoulders (side and rear view)

Ribeye area

Fat

Vertebra

Rib cross section

FIGURE 9.6 Conformation characteristics of slaughter steer typical of yield grade 2. Compare with color plate F. *Source:* Colorado State University.

muscled (Fig. 9.7). All body parts referred to in these figures can be identified in Figs. 9.2 and 9.4, with the exception of the twist—the distance from the top of the tail to where the hind legs separate as observed from a rear view.

Color plates H and I show the cross sections of a yield grade 1 lamb and a yield grade 5 lamb. Cross section 4 is cut through the shoulder area, cross section 5 is at the back, and cross section 6 is through the leg area. Note that many fat deposits and muscle areas are similar to those described in Figure 9.5.

Color plates J and K show the cross section of a U.S. no. 1 pig and a U.S. no. 4 pig. Percent lean cuts are used in pork carcass evaluation instead of yield grades. Although the terminology is different, both systems measure fat-to-lean composition. Percent lean cuts are measured by obtaining the total weight of the trimmed Boston shoulder, picnic shoulder, loin, and ham (Fig. 9.4), and then dividing the total by carcass weight. The lean cuts percentage for the U.S. no. 1 would be 53% or more, as contrasted with less than 47% for the U.S. no. 4. The ribbon on the frozen carcass marks the location of the cross sections. Cross section 4 is through the shoulder area, 5 is through the middle of the back, and 6 is through the ham area.

Even though the sizes and shapes of slaughter cattle, sheep, and swine are different, these three species are remarkably similar in muscle structure and fat-deposit areas. Therefore, what is learned from visual evaluation of one species can be applied to another species. Regardless of species, an animal that shows a square appearance over the top of its back and appears blocky and deep from

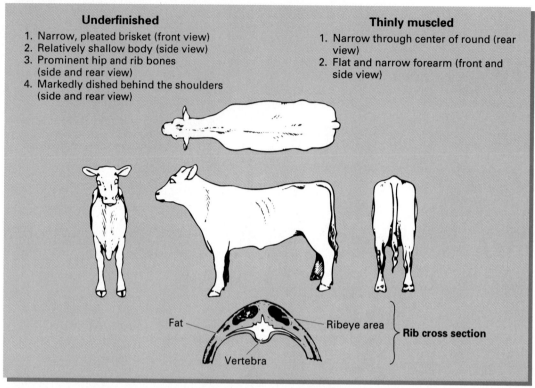

Underfinished	**Thinly muscled**
1. Narrow, pleated brisket (front view)	1. Narrow through center of round (rear view)
2. Relatively shallow body (side view)	2. Flat and narrow forearm (front and side view)
3. Prominent hip and rib bones (side and rear view)	
4. Markedly dished behind the shoulders (side and rear view)	

Fat — Ribeye area — Rib cross section
Vertebra

FIGURE 9.7 Conformation characteristics of slaughter steer that is underfinished (contrast with Fig. 9.5) and thinly muscled (contrast with Fig. 9.6). *Source:* Colorado State University.

a side view usually has a large accumulation of fat. **Fat** accumulates first in flank areas, **brisket, dewlap,** and throat (**jowl**); between the hind legs; and over the edge of the loin. Fat also fills in behind the shoulders and gives the animal a smooth appearance. Movement of shoulder blade can be observed when lean cattle and swine walk. A slaughter redmeat animal that has an oval turn to the top of its back and thickness through the center part of its hind legs (as viewed from the rear) has a high proportion of lean to fat. It is important that slaughter animals have an adequate amount of fat, because thin animals typically do not produce a highly palatable consumer product.

The wool covering of sheep can easily camouflage the amount of fat. The amount of fat in sheep can be determined by pressing the fingers of the closed hand lightly over the last two ribs and over the spinal processes of the vertebrae. Sheep that have a thick padding of fat in these areas will produce poor yield-grading carcasses.

Accuracy in visually appraising slaughter redmeat animals is obtained by making visual estimates of yield grades and percent lean cuts and their component parts, and then comparing the visual estimates with the carcass measurements. Accurate visual appraisal can be used as one tool in producing redmeat animals with a more desirable carcass composition of lean to fat.

Since the late 1980s and early 1990s the amount of fat on retail cuts has been significantly reduced—e.g., down to ¼ in., ⅛ in., or no fat on many cuts. This fat reduction has occurred primarily due to trimming of excess fat at the packing plant. Certain breeding and feeding practices could reduce the trimmable fat more economically; however, marketing procedures do not encourage the implementation of these practices (see the discussion on value-based marketing in Chapter 23).

CHAPTER SUMMARY

- Numerous red meat slaughter animals are evaluated and priced on the visual appraisal of the animal's carcass merit.
- Carcass composition (fat, lean, and bone) can be determined quite accurately by visually appraising live animals.
- An understanding of the anatomical structure, especially the location of major muscle and patterns of fat deposition, is needed for accurate visual appraisal.
- Thickness through the center of the round (cattle), ham (pig), and leg (lamb) gives a visual estimate of muscling. Squareness over the animal's top and rectangular appearance from a side view usually gives a visual assessment that the animal is excessively fat.

KEY WORDS

type	fat	dewlap
conformation	brisket	jowl

REVIEW QUESTIONS

1. What are the three productive stages of red meat-producing animals?
2. How do production records compare to visual appraisal for improving reproductive and growth stages of livestock?
3. What is being estimated during visual evaluation of livestock for carcass merit?
4. Carcasses are primarily composed of what three types of tissues?

SELECTED REFERENCE

Boggs, D. L., and R. A. Merkel. 1993. *Live Animal Carcass Evaluation and Selection Manual.* Dubuque, IA: Kendall/Hunt.

CHAPTER 10

Reproduction

LEARNING OBJECTIVES

- Identify the primary female reproductive structures and their associated functions
- Identify the primary male reproductive structures and their associated functions
- Describe the source, target, and functions of the primary hormones affecting reproduction
- Compare and contrast duration and timing of reproductive benchmarks (age at puberty, etc.) for the major livestock species
- Describe the process of normal and abnormal parturition

Reproductive efficiency in farm animals, as measured by number of calves or lambs per 100 breeding females or number of pigs per litter, is a trait of significant economic importance in farm animal production. Reproduction is typically considered at least twice as important to economic returns of livestock production as growth or carcass performance. It is essential to understand the reproductive process in the creation of new animal life because it is a focal point of overall animal productivity. Producers who manage animals for high reproductive rates must understand the production of viable sex cells, estrous cycles, mating, pregnancy, and birth. (Some aspects of reproductive behavior are presented in Chapter 22.)

FEMALE ORGANS OF REPRODUCTION AND THEIR FUNCTIONS

Figures 10.1 and 10.2 show the reproductive organs of the cow and sow. Figure 10.3 illustrates the reproductive tract of the mare. The female reproductive anatomy of the various farm animal species is similar, although there are a few obvious differences.

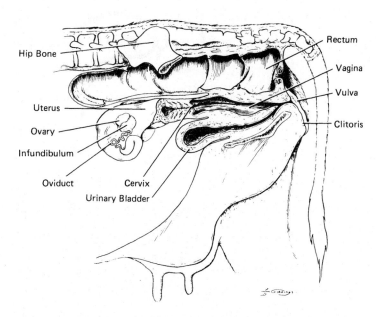

FIGURE 10.1 Reproductive organs of the cow. *Source:* Colorado State University.

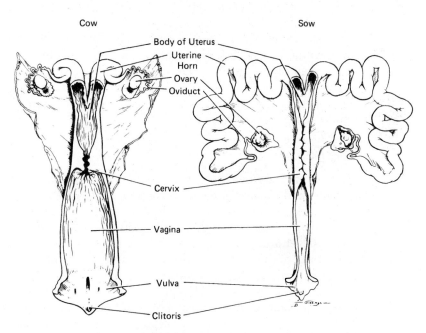

FIGURE 10.2 A dorsal view of the reproductive organs of the cow and the sow. The most noticeable difference is the longer uterine horns of the sow compared to the cow. *Source:* Colorado State University.

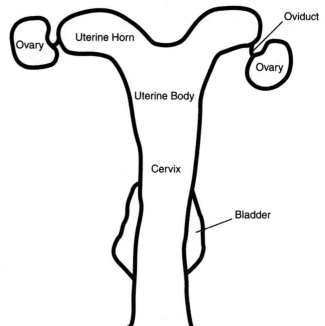

FIGURE 10.3 Reproductive tract of the mare. *Source:* Sean Field.

The organs of reproduction of the typical female farm mammal include a pair of ovaries, which are suspended by ligaments just back of the kidneys, and a pair of open-ended tubes, the **oviducts** (also called the **Fallopian tubes**), which lead directly into the uterus (womb). The **uterus** itself has two horns, or branches, that in farm mammals merge together at the lower part into the uterine body. The uterine body is protected by the **cervix,** which serves as a pathway to and from the uterus. Its surface is fairly smooth in the mare and the sow, but is folded in the cow and ewe. The cervix opens into the vagina, a relatively large canal or passageway that leads posterior to the external structures—the vulva and clitoris. The urinary bladder empties into the vagina through the urethral opening. Table 10.1 summarizes the primary functions of the organs of the female reproduction system.

Ovaries

Ovaries produce ova (female sex cells, also called *eggs*) and the hormones estrogen and progesterone. The ova (Fig. 10.4) are the largest single cells in the body; each ovum develops inside a recently formed **follicle** within the ovary (Fig. 10.5). Some tiny follicles develop and ultimately attain maximum size (about 0.8–1.5 in. in diameter) after having migrated from deep in the ovary to the surface of the ovary. These growing follicles produce estrogens. The mature (Graafian) follicles rupture, thus freeing the ovum in a process called ovulation.

Follicles can be classified as primordial (least mature), primary, secondary, tertiary, and Graafian (most mature). In addition to changes in size and complexity, follicles vary in their hormonal responsiveness as they mature. Follicles undergo a multistage process of development that may result in the formation of a Graafian follicle capable of ovulation. This developmental process can be defined as the steps of recruitment, selection, and dominance (Fig. 10.6). If the follicle fails to reach dominance, it will undergo a degenerative process known as atresia.

TABLE 10.1 Summary of the Primary Functions of the Organs of the Female Reproductive Tract

Organ	Function(s)
Oviduct	Transport of ova and sperm Site of fertilization and early cleavage
Ovary	Production of oocytes Production of estrogen (Graafian follicle) Production of progesterone (corpus luteum)
Uterus	Assisting in sperm transport Regulation of the corpus luteum Site of implantation and pregnancy Expulsion of fetus and fetal membranes
Cervix	Facilitating sperm transport (sow and mare) Prevention of uterine contamination
Vagina	Copulatory organ Birth canal

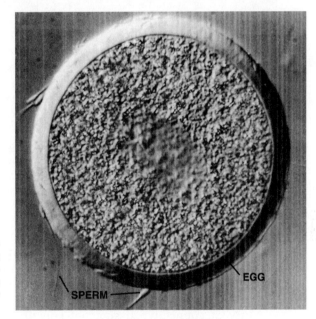

FIGURE 10.4 Bull sperm and cow egg magnified 300×. The ovum is about $\frac{1}{200}$ in. in diameter, while the sperm is $\frac{1}{6000}$ in. in diameter. Each is a single cell and contains half the chromosome number typical of other body cells. Courtesy of Colorado State University.

After the ovum is released from the mature follicle, cells of the follicle change into a **corpus luteum,** or "yellow body." The corpus luteum produces progesterone, which becomes a vitally important hormone for maintaining pregnancy.

The Oviducts

Immediately after ovulation, the ova are caught by the fingerlike projections of the **infundibulum,** which guide the ova into the tubular portion of the oviduct. The ova are tiny (approximately $\frac{1}{200}$ in. in diameter, about the size of a dot made by a sharp pencil). Sperm are transported through the uterus into the oviduct after the female is inseminated (naturally or artificially). Therefore, the oviducts are where ova and sperm meet to initiate fertilization. After fertilization, 3–5 days are required in cows and ewes (about the same amount of time in other farm animals) for the ova

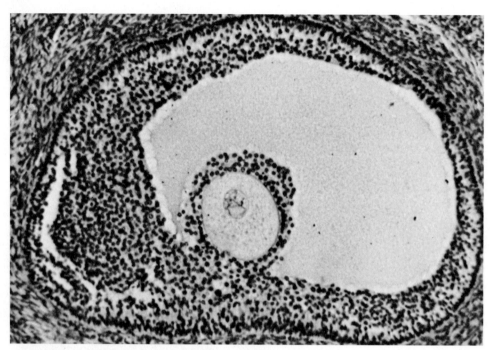

FIGURE 10.5   The large structure outlined with a circle of dark cells, is a follicle located on a cow's ovary (magnified 265×). The smaller circle, near the center, is the egg. The large light gray is the fluid that fills the follicle. When the follicle ruptures, the egg will move into the oviduct by anatomical action of the infundibulum. Courtesy of Colorado State University.

to travel down the remaining two-thirds of the oviduct. From the oviduct, the newly developing embryos pass to the uterus and soon attach to its lining.

The Uterus

The uterus varies in shape from the type that has long, slender left and right horns, as in the sow, to the type that is primarily a fused body with short horns, as in the mare. In the sow, the embryos develop in the uterine horn; in the mare, the embryo develops in the body of the uterus. Each surviving embryo develops into a fetus and remains in the uterus until **parturition** (birth).

The posterior outlet of the uterus is the cervix, an organ composed primarily of connective tissue that constitutes a formidable gateway between the uterus and the vagina. Like the rest of the reproductive tract, the cervix is lined with mucosal cells. These cells undergo significant changes as the animal experiences the various phases of the estrous cycle as well as the process of pregnancy. The cervical passage changes from being tightly closed or sealed in pregnancy to a relatively open, very moist canal at the height of estrus.

The Vagina

The vagina serves as the female organ of copulation at mating and as the birth canal at parturition. Its mucosal surface changes during the estrous cycle from very moist when the animal is ready for mating to almost dry, even sticky, between periods of heat. The tract from the urinary bladder joins the posterior ventral vagina; from this juncture to the exterior vulva, the vagina serves the dual role of a passageway for the reproductive and urinary systems.

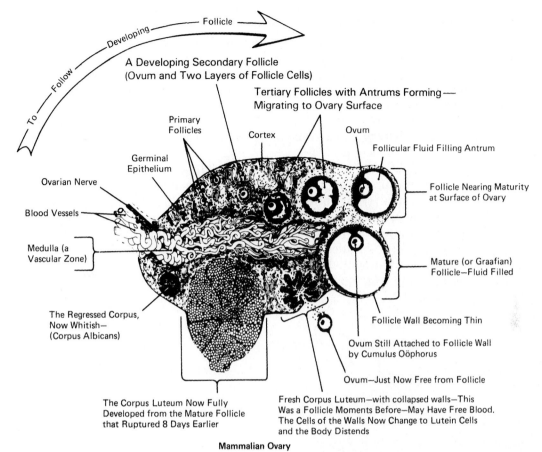

Mammalian Ovary

FIGURE 10.6 A cross section of the bovine ovary showing how a follicle develops of full size and than ruptures, allowing the egg to escape. The follicle then becomes a "yellow body" *(corpus luteum)*, which is actually orange in cattle. The corpus luteum degenerates in time and disappears. Of course, many follicles cease development, stop growing, and disappear without ever reaching the mature stage. *Source:* J. F. Bone, *Animal Anatomy and Physiology* 4th ed., (Corvallis: Oregon State University Book Stores © 1975).

The Clitoris

A highly sensitive organ, the clitoris is located ventrally and at the lower tip of the vagina. The clitoris is the homologue of the penis in the male (i.e., it came from the same embryonic source as the penis). Some research indicates that clitoral stimulation or massage following artificial insemination in cattle will increase the chance of conception.

Reproduction in Poultry Females

The hen differs from farm mammals in that the young are not suckled, the egg is laid outside the body, and there are no well-defined estrous cycles or pregnancy. However, variation in day length does affect poultry reproductive rates. Since eggs are an important source of human food, hens are selected and managed to lay eggs consistently throughout the year.

The anatomy of the reproductive tract of the hen is shown in Figures 10.7 and 10.8. At hatching time, the female chick has two ovaries and two oviducts. The right ovary and oviduct do not develop. Therefore, the sexually mature hen has a well-developed ovary and oviduct only on the left side. The ovary appears as a cluster of tiny gray eggs or yolks in front of the left kidney and attached to the back of the hen. The ovary is fully formed, although very small, when the chick is hatched. It contains approximately 3,600–4,000 miniature ova. As the hen reaches sexual maturity, some of the ova develop into mature yolks (yellow part of the laid egg). The remaining ova are variable in size from the nearly mature to microscopic.

The oviduct is a long, glandular tube leading from the ovary to the **cloaca** (common opening for reproductive and digestive tracts). The oviduct is divided into five parts: (1) the infundibulum (3–4 in. long), which receives the yolk; (2) the **magnum** (approximately 15 in. long), which secretes the thick albumen, or white, of the egg; (3) the **isthmus** (about 4 in. long), which adds the shell membranes; (4) the uterus (approximately 4 in. long), or shell gland, which secretes the thin white albumen, the shell, and the shell pigment; and (5) the vagina (about 2 in. long).

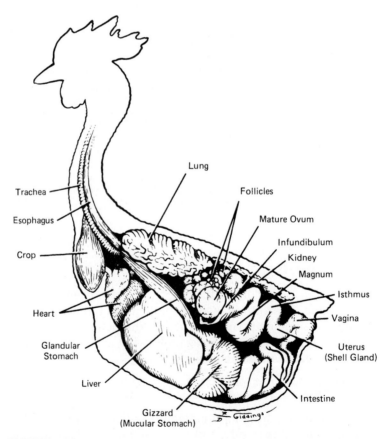

FIGURE 10.7 Reproductive organs of the hen in relation to other body organs. The single ovary and oviduct are on the hen's left side; an underdeveloped ovary and oviduct are sometimes found on the right side, having degenerated in the developing embryo.

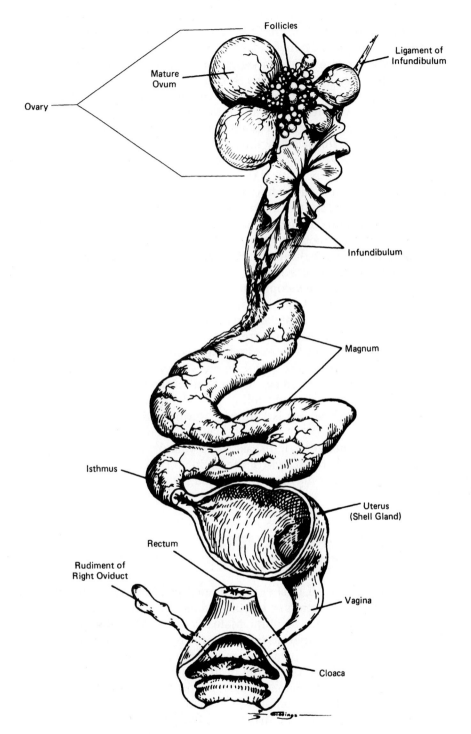

FIGURE 10.8 Reproductive organs of the hen. Sections of the uterus and cloaca are cut away to better view internal structure.

Ovulation is the release of a mature yolk (ovum) from the ovary. When ovulation occurs, the infundibulum engulfs the yolk and starts it on its way through the 25 to 27-in. oviduct. The yolk moves by peristaltic action through the infundibulum into the magnum area in about 15 minutes.

During the 3-hour passage through the magnum, more than 50% of the albumen is added to the yolk. The developing egg passes through the isthmus in about $1\frac{1}{4}$ hours. Here water and mineral salts and the two shell membranes are added. During the egg's 21-hour stay in the uterus, the remainder of the albumen is added, followed by the addition of shell and shell pigment. Moving finally into the vagina, the fully formed egg enters the cloaca and is laid. The entire time from ovulation to laying is usually slightly more than 24 hours. About 30 minutes after a hen has laid an egg, she releases another yolk into the infundibulum, and it will likewise travel the length of the oviduct.

After the fertilized egg is incubated for 21 days, the chick is hatched. The egg is biologically structured to support the growth and life processes of the developing chick embryo during incubation and for 3–4 days after the chick is hatched.

There are several egg abnormalities that occur because of factors affecting ovulation and the developmental process. Double-yolked eggs result when two yolks are released about the same time or when one yolk is lost into the body cavity for a day and is picked up by the infundibulum when the next day's yolk is released. Yolkless eggs are usually formed when a bit of tissue that is sloughed off the ovary or oviduct stimulates the secreting glands of the oviduct and a yolkless egg results. The abnormality of an egg within an egg is due to reversal of direction of an egg by the wall of the oviduct. One day's egg is added to the next day's egg, and shell is formed around both. Soft-shelled eggs generally occur when an egg is laid prematurely and insufficient time in the uterus prevents the deposit of the shell. Thin-shelled eggs may be caused by dietary deficiencies, heredity, or disease. Glassy- and chalky-shelled eggs are caused by malfunctions of the uterus of the laying bird. Glassy eggs are less porous and will not hatch but may retain their quality.

MALE ORGANS OF REPRODUCTION AND THEIR FUNCTIONS

Figures 10.9 and 10.10 show the reproductive organs of the bull and boar. The organs of reproduction of a typical male farm mammal include two **testicles,** which are held in the scrotum. Male sex cells (called **sperm** or *spermatozoa*) are formed in the **seminiferous tubules** of the testicles. The sperm from each testicle then pass through a ductal system into the **epididymis,** which is a highly coiled tube that is held in a covering on the exterior of the testicle. The epididymis functions to concentrate, store, transport, and facilitate maturation of sperm cells. The epididymis empties into a larger tube, the **vas deferens** (also called the **ductus deferens**). The two vasa deferentia converge at the upper end of the urethral canal, where the urinary bladder opens into the urethra. In some species, the wall of the upper end of the vas deferens is thickened and forms a secretory gland called the **ampulla.** The urethra is the large canal that leads through the penis to the outside of the body. The penis has a triple role: It serves as a passageway for semen and urine and it is the male organ of copulation.

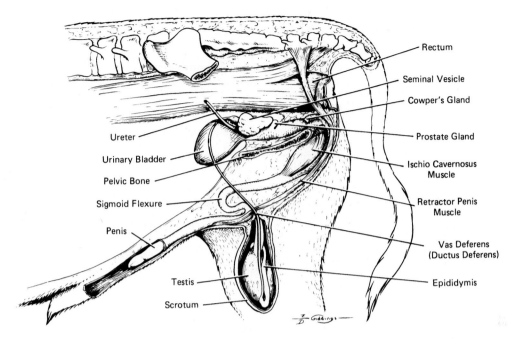

FIGURE 10.9   Reproductive organs of the bull.

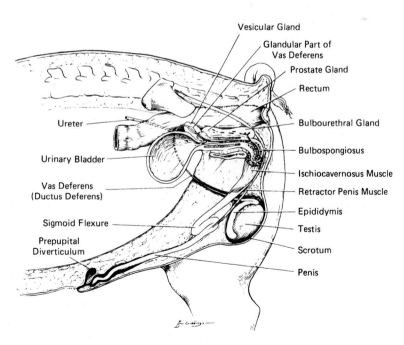

FIGURE 10.10   Reproductive organs of the boar.

TABLE 10.2   Summary of the Primary Functions of the Male Reproductive Tract

Organ	Primary Function(s)
Testicles	Testosterone production (interstitial cells) Spermatozoa production (seminiferous tubules)
Epididymis	Concentration, storage, maturation, and transport of spermatozoa
Scrotum	Support of testicles Temperature control (tunica dartos)[a]
Vas deferens	Sperm transport
Accessory glands	Addition of fluid volume, nutrients, and buffers to semen
Penis	Copulatory organ

[a] The cremaster and pampiniform plexus of the spermatic cord also assist with temperature regulation of the testes.

TABLE 10.3   Comparative Anatomy and Seminal Characteristics of Males

Animal	Weight of Paired Testes (g)	Volume per Ejaculate (ml)	Type of Ejaculate	Sperm Concentration (mil per ml)	Total Sperm per Ejaculate (bil)
Bull (cattle)	600	3–8	Rapid	800–2,000	5–15
Ram (sheep)	550	0.8–2.0	Rapid	2,000–3,000	2–4
Boar (swine)	750	150–350	Prolonged	200–300	30–60
Stallion (horse)	160	60–100	Prolonged	150–300	5–15
Buck (goat)	—	0.6–1.0	—	2,000–3,500	1–8
Dog (dog)	—	2–16	Rapid	10–200	0.6–3
Buck (rabbit)	6	0.5–6.5	—	300–1,000	1.5–6.5
Tom (turkey)	—	0.2–0.8	—	8,000–30,000	1–20
Cock (chicken)	25	0.2–1.5	—	3,000–7,000	0.6–3.5

Source: Adapted from Hafez, 1993; McDonald, 1989, and others.

The left and right parts of the seminal vesicles, which lie against the urinary bladder, consist of glandular tissue that secretes into the urethra a substance that supplies nutrients for the sperm. The prostate gland contains 12 or more tubes, each of which empties into the urethra. Another gland, the bulbourethral (Cowper's) gland, which also empties its secretions into the urethral canal, is posterior to (behind) the prostate. Table 10.2 outlines the primary activities of the male reproductive organs and tissues. Table 10.3 summarizes the comparative anatomical and seminal differences of males for several species.

Testicles

The testicles are suspended from the body cavity by the spermatic cord and produce (1) sperm cells that fertilize the ova of the female, and (2) a hormone called **testosterone** that conditions the male so that his appearance and behavior are masculine. Details of the structure of the spermatozoa of the bull are shown in Figures 10.11 and 10.12. The diameter of sperm is approximately ¹⁄₆₀₀₀ in.

If both testicles are removed (as is done in castration), the individual loses his sperm production capability and is sterile. Also, without testosterone his masculine appearance is not apparent, and he approaches the status of a neuter—an individual whose appearance is somewhere between that of a male and that of a female. A steer does not have the crest, or powerful neck, of the bull. The bull has heavier, more muscular shoulders and a deeper voice than a counterpart steer. If a bull calf is castrated, the reproductive organs, such as the vas deferens, seminal vesicles, and prostate and bulbourethral glands, all but

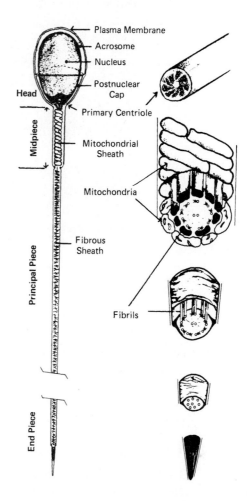

FIGURE 10.11 A diagrammatic sketch of the structure of bull sperm. *Source:* Dr. Arthur S. H. Wu, Oregon State University.

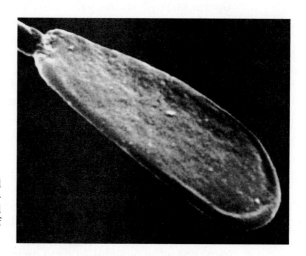

FIGURE 10.12 Bull spermatozoa (magnified 12,000×) viewed with the scanning electron microscope showing the depth of the sperm head covered by the raised acrosome. Courtesy of Oregon State University.

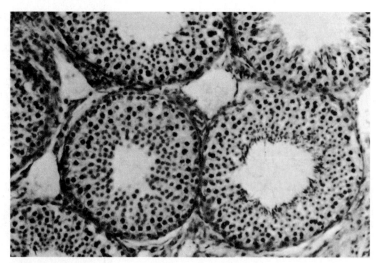

FIGURE 10.13 A cross section through the seminiferous tubules of the bovine testis (magnified 240×). The tubule in the lower right-hand corner demonstrates the more advanced stages of spermatogenesis as the spermatids are formed near the lumen (opening) of the tubule. Courtesy of Colorado State University.

cease further development. If castration is done in a mature bull, the remaining genital organs shrink in size and in function.

Within each testicle, sperm cells are generated in the seminiferous tubules, and testosterone is produced in the cells between the tubules, called **Leydig cells** or **interstitial cells** (Fig. 10.13).

The Epididymis

The epididymis is the storage site for sperm cells, which enter it from the testicle to mature. In passing through this very long tube (95–115 ft in the bull, longer in the boar and stallion), the sperm acquire the potential to fertilize ova. Sperm taken from the part of the epididymis nearest the testicle and inseminated into females are not likely to be able to fertilize ova, whereas sperm taken near the vas deferens have the potential to fertilize.

In the sexually mature male animal, sperm reside in the epididymis in large numbers. In time, the sperm mature, then degenerate, and are absorbed in the part of the epididymis farthest from the testicle, unless they have been moved on into the vas deferens to be ejaculated.

The Scrotum

The scrotum is a two-lobed sac that contains and protects the two testicles. It also regulates temperature of the testicles, maintaining them at a temperature lower than body temperature (3–7°F lower in the bull and 9–13°F lower in the ram and goat). When the environmental temperature is low, the tunica dartos muscle of the scrotum contracts, pulling the testicles toward the body and its warmth; when the environmental temperature is high, this muscle relaxes, permitting the testicles to drop away from the body and its warmth. The cremaster muscle that is aligned with the spermatic cord functions in a similar manner to the tunica dartos as an aid to temperature control. Furthermore, the pampiniform plexus acts as a heat exchanger to cool blood as it enters the testes. This heat-regulating mechanism of the scrotum begins at about the time of puberty. Hormone function precedes puberty by 40–60 days.

When the environmental temperature is elevated such that the testicles cannot cool sufficiently, the formation of sperm is impeded, and a temporary condition of lowered fertility results. Providing shade, keeping the males in the shade during the heat of the day, even providing air conditioning, are ways to manage and prevent this temporary sterility.

The Vas Deferens

The vas deferens is essentially a transportation tube that carries the sperm-containing fluid from each epididymis to the urethra. The vasa deferentia join the urethra near its origin as the urethra leaves the urinary bladder. In the mature bull, the vas deferens is about 0.1 in. in diameter except at its upper end, where it widens into a reservoir, or ampulla, about 4–7 in. long and 0.4 in. wide.

Under the excitement of anticipated mating, the secretion loaded with spermatozoa from each epididymis is propelled into each vas deferens and accumulates in the ampulla of the deferent duct. This brief accumulation of sperm in the ampulla is an essential part of sexual arousal. The sperm reside briefly in the ampulla until the moment of ejaculation, when the contents of each ampulla are pressed out into the urethra, and then through the urethra and the penis en route to their deposition in the female tract.

The ampulla is found in the bull, stallion, goat, and ram—species that ejaculate rapidly. It is not present in the boar or dog, animals in which ejaculation normally takes several minutes (8–12 minutes is typical in swine). In such animals, large numbers of sperm travel all the way from the epididymis through the entire length of the vas deferens and the urethra. On close observation of the boar at the time of mating, the muscles over the scrotum can be seen quivering rhythmically as some of the contents of each epididymis are propelled into the vasa deferentia and on into the urethra. This slow ejaculation of the boar contrasts to the sudden expulsion of the contents of the ampulla of the vas deferens at the height of the mating reaction, or orgasm, in the bull, stallion, ram, and goat.

The Urethra

The urethra is a large, muscular canal extending from the urinary bladder. The urethra runs posteriorly through the pelvic girdle and curves downward and forward through the full length of the penis. Very near the junction of the bladder and urethra, tubes from the seminal vesicles and tubes from the prostate gland join this large canal. The bulbourethral gland joins the urethra at the posterior floor of the penis.

Accessory Sex Glands

The ampullae, seminal vesicles, prostate, and bulbourethral glands are known as the **accessory sex glands.** Their primary functions are to add volume and nutrition to the sperm-rich fluid coming from the epididymis. Semen consists of two components: the sperm and the fluids secreted by the accessory sex glands. The semen characteristics of some farm animals are shown in Table 10.3.

The Penis

The penis is the organ of copulation. It provides a passageway for semen and urine. It is an organ characterized especially by its spongy, erectile tissue that fills with blood under considerable pressure during periods of sexual arousal, making the penis rigid and erect. The bull, ram, and boar have fibroelastic penile structure. The penis in the stallion, dog, and cat is vascular in structure.

The penis of the bull is about 3 ft in length and 1 in. in diameter, tapering to the free end, or glans penis. In the bull, boar, and ram, the penis is S-shaped when relaxed. This S curve, or sigmoid flexure, becomes straight when the penis is erect. The sigmoid flexure is restored after copulation, when the relaxing penis is drawn back into its sheath by a pair of retractor penis muscles. The stallion penis has no sigmoid flexure; it is enlarged by engorgement of blood in the erectile tissues.

The free end of the penis is termed the **glans penis.** The opening in the ram penis is at the end of a hairlike appendage that extends about 0.8–1.2 in. beyond the larger penis proper. This appendage also becomes erect and, during ejaculation, whirls in a circular fashion, depositing semen in the anterior vagina. It does not regularly penetrate the ewe's cervix, as some investigators claim it does. Only a small portion of the penis of the bull, boar, ram, and goat extends beyond its sheath during erection. The full extension awaits the thrust after entry into the vagina has been made. The stallion and ass usually extend the penis completely before entry into the vagina.

All the accessory male sex organs depend on testosterone for their tone and normal function. This dependence is especially apparent when the testicles are removed (as in castration); the usefulness of the accessory sex organs is then diminished or even terminated.

Reproduction in Male Poultry

The reproductive tract of male poultry is shown in Figure 10.14. There are several differences when compared to the reproductive tracts of the farm mammals

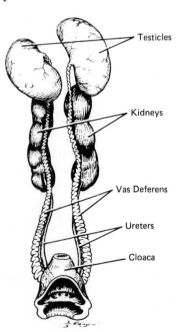

Testicles

Kidneys

Vas Deferens

Ureters

Cloaca

FIGURE 10.14 Male poultry reproductive tract (ventral view).

previously described. The testes of male poultry are contained in the body cavity. Each vas deferens opens into small papillae, which are located in the cloacal wall. The male fowl has no penis but does have a rudimentary organ of copulation. The sperm are transferred from the papillae to the rudimentary copulatory organ, which transfers the sperm to the oviduct of the hen during the mating process. The sperm are stored in primary sperm-host glands located in the oviduct. These sperm are then released on a daily basis and transported to secondary storage glands in the infundibulum. Fertilization occurs in the infundibulum. Sperm stored in the oviduct are capable of fertilizing the eggs for 30 days in turkeys and 10 days in chickens.

WHAT MAKES TESTICLES AND OVARIES FUNCTION

Testicular Function

Reproduction is controlled via a series of hormonal signals from the endocrine system (Table 10.4). Testicles produce their hormones under stimuli coming to them from the anterior pituitary (AP) gland situated at the base of the brain. The AP produces and secretes two hormones important to male reproductive performance. **Luteinizing hormone (LH)** and **follicle-stimulating hormone (FSH)** are known as **gonadotropic hormones** because they stimulate the gonads (ovary and testicle). LH produces its effect on the interstitial tissue (Leydig cells) of the testicle, causing the tissue to produce the male hormone, testosterone. FSH stimulates cells in the seminiferous tubules to nourish the developing spermatozoa.

Some species that respond to change in length of daylight exhibit more seasonal fluctuation than others in reproductive activities. These influences of daylight are exerted on the neurophysiological mechanism in the brain. The hypothalamus makes up the floor and part of the wall of the third ventricle of the brain and secretes through blood vessels releasing factors that affect the AP and its production of FSH and LH.

Ovarian Function

Ovarian hormones in the sexually mature female owe the cyclicity of their production to hormones that originate in the hypothalamus and AP (Fig. 10.15). The hypothalamus produces **gonadotropin-releasing hormone (GnRH)** that in turn stimulates the AP to release FSH and LH. GnRH is released in the presence of estrogen when progesterone levels are low.

FSH and LH work in concert to stimulate follicular development and then a surge of LH initiates ovulation. As the follicle matures, **estrogen** is released which stimulates not only the release of GnRH but also stimulates sexual behavior of the female.

Following ovulation, the luteal cells form into the corpus luteum that produces progesterone. Progesterone then acts to inhibit the release of LH and FSH. When the influence of progesterone is removed from the hypothalamus–AP axis then cyclicity is reinitiated.

Estrus or **heat** is the period of time when the female will accept the male for breeding purposes. The female of each species exhibits some behavior patterns that demonstrate she is in heat (see Chapter 22). For example, a mare in estrus,

TABLE 10.4　The Major Reproduction Hormones: Their Source, Target Organs, and Functions

Hormone	Source	Target	Function(s)
Gonadotropin-releasing hormone (GnRH)	Hypothalamus	Anterior pituitary	Release of LH & FSH
Luteinizing hormone (LH)	Anterior pituitary	Ovary (luteal cells)	Ovulation CL formation Progesterone production
		Testis (interstitial cells)	Testosterone production
Follicle-stimulating hormone (FSH)	Anterior pituitary	Ovary (granulosa cells) Testis (sertoli cells)	Development of follicle Estrogen synthesis Sperm production
Prolactin	Anterior pituitary	Mammary tissue	Milk synthesis Maternal behavior
Oxytocin	Posterior pituitary	Uterus Mammary tissue	Gamete transport Uterine contraction Milk let down
Estrogen	Follicle Placenta	Uterus Hypothalamus Mammary tissue	Mating behavior Uterine growth Secondary sex characteristics Promotes GnRH release
Progesterone	Corpus luteum Placenta	Uterus Mammary tissue Hypothalamus	Maintains pregnancy Mammary development Inhibits GnRH release
Testosterone	Interstitial cells of testis	Skeletal muscle Seminiferous tubules Male reproductive tract	Anabolic growth Sperm production Secondary sex characteristics
Prostaglandin $F_{2\alpha}$	Uterus (endometrium)	Corpus luteum Uterus (myometrium) Graafian follicles	Regression of CL Uterine contraction Ovulation
Placental lactogen	Placenta	Mammary gland	Promotes lactation
Relaxin	Ovary Placenta	Cervix, pelvis	Cervical dilation Pelvic expansion
Cortisol (fetal)	Adrenal gland (fetus)	Dam's uterus	Initiation of parturition
Melatonin	Pineal gland	Hypothalamus	Partial control of seasonal reproductive patterns

Source: Adapted from Hafez, 1993; McDonald, 1989; Senger, 2003, and others.

when "teased" with the presence of a stallion (Fig. 10.16), will not avoid or kick him. The mare in heat will stand solidly, sometimes squatting and urinating when approached by the stallion.

Synchronized with the **estrous cycle** is the important and essential phenomenon of ovulation. Ovulation occurs in the cow after estrus. It occurs in the sow, ewe, goat, and mare toward the latter part of, but nevertheless during, estrus. These species all ovulate spontaneously; that is, ovulation takes place whether copulation occurs or not. By contrast, copulation (or some such stimulation) is necessary to trigger ovulation in such animals as the rabbit, cat, ferret, and mink, which are considered "induced ovulators." Ovulation in these animals takes place at a fairly consistent time after mating. Hormonal action controls ovulation. The follicle of the ovary grows, matures, fills with fluid, and

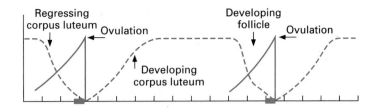

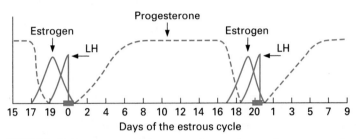

FIGURE 10.15 Events and hormone changes during the estrous cycle of the cow.

FIGURE 10.16 Estrus is determined in the mare by "teasing" her with the presence of a stallion. Courtesy of Colorado State University.

softens a few hours before rupturing. The follicle ruptures owing to a sudden release of LH rather than bursting as a result of pressure inside.

The FSH produced by the AP accounts for the increase in size of the ovarian follicle and for the increased amount of estradiol and estrone, which are products of the follicle cells. LH, another hormone from the AP, alters the follicle cells and granulosa cells of the ovary, changing them into luteal cells. LH from the AP in turn stimulates the luteal cells to produce progesterone.

In the course of 7–10 days, what was formerly an egg-containing follicle develops into the corpus luteum, a luteal body of about the same size and shape as the mature follicle. The luteal cells of the corpus luteum produce a sufficient quantity of progesterone to depress FSH secretion from the AP until the luteal cells reach maximum development (in nonpregnant animals), cease their development, and (in 3 weeks) lose their potency and disappear.

If pregnancy occurs, the corpus luteum continues to function, persisting in its progesterone production and preventing further estrous cycles. Thus, no more heat occurs until after pregnancy has terminated. If pregnancy does not occur, the uterus releases the hormone $PGF_{2\alpha}$ (**prostaglandin**). $PGF_{2\alpha}$ causes the regression of the corpus luteum and thus declining levels of progesterone. As previously stated, when the influence of progesterone is removed, GnRH secretion increases and another follicle reaches ovulatory status. The process of follicle development, ovulation, and corpus luteum development and regression is shown in Figure 10.6. Table 10.5 shows the length of estrus, estrous cycles, and time of ovulation for the different farm animals.

Seasonal Effects on Reproduction

The changing length of daylight is a potent factor that influences the estrous cycle, the onset of pregnancy, and the seasonal fluctuations in male fertility. The amount of daylight acts both directly and indirectly on the animal. It acts directly on the central nervous system by influencing the secretion of hormones and indirectly by affecting plant growth, thus altering the level and quality of nutrition available. When selection has resulted in improvements in the traits associated with reproduction, individuals exhibit higher levels of fertility (i.e., more intense expression of estrus, occurrence of estrus over more months of the year, or occurrence of spermatogenesis at a high level over more months of the year) than unselected individuals.

The two classic examples of the impact of seasonality on reproductive function are sheep and horses. Both species experience periods of the year with heightened fertility as well as periods of dormancy (anestrous). The sheep is considered a short-day breeder as maximum fertility is experienced in the late summer and fall months while the horse is a long-day breeder whereby the breeding season is aligned with increasing day length in the spring.

Seasonal species are responsive to melatonin, a hormone produced by the pineal gland in response to declining periods of light. Thus, the mare's estrous cycle is stimulated by decreasing periods of dark that results in declining levels

TABLE 10.5   **Age at Puberty, Length of Estrous Cycle, Duration of Heat, and Time of Ovulation**

Animal	Puberty (mo)	Estrous Cycle Length (days)		Duration of Estrus		Approx. Time from Onset of Ovulation
		Average	Range	Average	Range	
Heifer, cow (cattle)	6–18	21	14–29	18 hrs	12–30 hrs	18–48 hrs
Ewe (sheep)[a]	6–12	17	14–19	30 hrs	24–36 hrs	24–27 hrs
Mare (horse)[b]	20	22	19–24	7 days	4–8 days	24–48 hrs prior to end of heat
Gilt, sow (swine)	5–10	21	19–23	60 hrs	2–3 days	38–42 hrs
Doe (goat)[a]	4–8	21	18–22	39 hrs	1–3 days	24–36 hrs
Doe (rabbit)	5–8	—	—	Continuous	—	10 hr post-coitus
Bitch (dog)	6–12	—	—	8 days	7–9 days	First 1/3 of heat
Heifer, cow (bison)	24	21	18–22	2 days	—	18–45 hrs

[a]Typically fall breeders in Northern Hemisphere.
[b]Typically spring breeders in Northern Hemisphere.
Source: Adapted from Hafez, 1993; McDonald, 1989, and others.

of melatonin secretion resulting in increased GnRH secretion and thus initiation of the ovulatory cascade. Conversely, increased levels of melatonin resulting from the increasing levels of darkness in the fall months stimulate the ewe.

The use of artificial lighting systems and other day-length manipulation strategies can be used to influence the cyclicity of these species.

PREGNANCY

When the sperm and the egg unite (fertilization), conception occurs, which is the beginning of the gestation period. The fertilized egg begins a series of cell divisions (Fig. 10.17). About every 20 hours, embryonic cells duplicate their genes and divide, progressing through the 2-, 4-, 8-, and 16-cell stages, and so on. In farm mammals, the embryo migrates through the oviduct to the uterus in 3–4 days, by which time it has developed to the 16- or 32-cell stage. The chorionic and amniotic membranes develop around this new embryo, and the chorion attaches to the uterus. The embryo (and later the fetus) obtains nutrients and discharges wastes through these membranes. This period of attachment (20–30 days in cattle and 14–21 days in swine) is critical. Unless the environment is sufficiently favorable, the embryo dies. Embryonic mortality causes a significant economic loss in farm animals, especially swine, where multiple ovulations and embryos are typical of the species. Good management will help protect the female and embryos early in pregnancy.

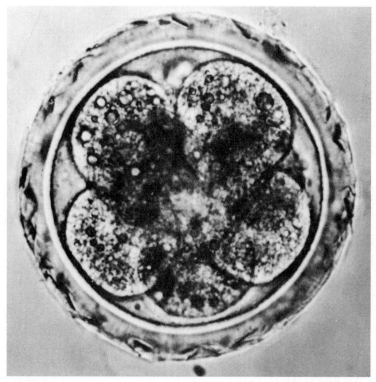

FIGURE 10.17 A bovine embryo in the six-cell stage of development (magnified 620×). Courtesy of Colorado State University.

High temperatures will cause embryonic death. The female should enter the breeding season in a thrifty, weight-gaining condition.

The **embryonic stage** is defined as that period when body parts differentiate and essential organs are formed. This period lasts 45 days in cattle.

When the embryonic stage is completed, the young organism is called a **fetus.** The **fetal period,** which lasts until birth, is mainly a time of growth. The duration of pregnancy in cattle is included in Table 10.6. Length of pregnancy varies chiefly with the breed and age of the mother.

TABLE 10.6 Gestation Length, Type of Placentation, and Number of Offspring Born

Animal	Gestation Length (days)	Type of Placentation	Usual Number of Offspring Born
Cow (cattle)	285	Cotyledonary	1
Ewe (sheep)	147	Cotyledonary	1–3
Mare (horse)	336	Diffuse	1
Sow (swine)	114	Diffuse	6–14
Doe (goat)	150	Cotyledonary	2–3
Doe (rabbit)	31	Discoid	4–8
Bitch (dog)	65	Zonary	7
Cow (bison)	270	Cotyledonary	1

Source: Adapted from multiple sources.

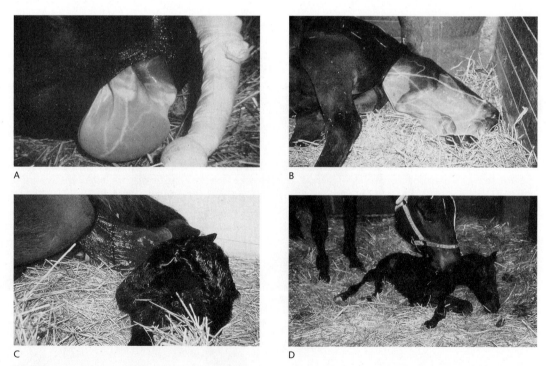

A

B

C

D

FIGURE 10.18 Parturition in the mare. (A) The water bag (a fluid-filled bag) becomes visible. (B) The head and front legs of the foal appear. (C) The foal is forced out by the uterine contractions of the mare. (D) In a few moments, the mare is licking the foal to increase circulation and stimulate the newborn. *Source:* Colorado State University.

PARTURITION

Parturition (birth) marks the termination of pregnancy (Fig. 10.18). Extraembryonic membranes that had been formed around the embryo in early pregnancy are shed at this time and are known as **afterbirth.** These membranes attach to the uterus during pregnancy and become known as the **placenta,** which is responsible for the transfer of nutrients and wastes between mother and fetus. The placenta produces hormones, especially estrogens and progesterone, in some farm animal species.

The **parturition** process is initiated by release of the hormone cortisol from the fetal adrenal cortex. Progesterone levels decline, whereas estrogen, prostaglandin $F_{2\alpha}$, and oxytocin levels rise, resulting in uterine contractions.

Parturition is a synchronized process. The cervix, until now tightly closed, relaxes. Relaxation of the cervix, along with pressure generated by uterine muscles on the contents of the uterus, permits the passage of the fetus into the vagina and on to the exterior. Another hormone, relaxin, is thought to aid in parturition. Relaxin, which originates in the corpus luteum or placenta, helps to relax cartilage and ligaments in the pelvic region.

At the beginning of parturition, the offspring typically assumes a position that will offer the least resistance as it passes into the pelvic area and through the birth canal. Fetuses of the cow, mare, and ewe assume similar positions in which the front feet are extended with the head between them (Fig. 10.19). Fetal piglets do not orient themselves in any one direction, which does not appear to affect the ease of birth. Calves, lambs, or foals may occasionally present themselves in a number of abnormal positions (Fig. 10.19). In many of these situations, assistance needs to be given at parturition. Otherwise, the offspring may die or be born dead, and in some instances, the mother may die as well. An abnormally small pelvic opening or an abnormally large fetus can cause some mild to severe parturition problems.

Producers can manage their herds and flocks for high reproductive rates. Management decisions are most critical that cause males and females selected for breeding to reach puberty at early ages, have high conception rates, and have minimum difficulty at parturition. Keeping animals healthy, providing adequate levels of nutrition, selecting genetically superior animals, and paying attention to parturition are some of the more critical management inputs to minimize reproductive loss. The ability of every female in the herd to give birth to live offspring is the key to successful farm animal reproduction (Fig. 10.20).

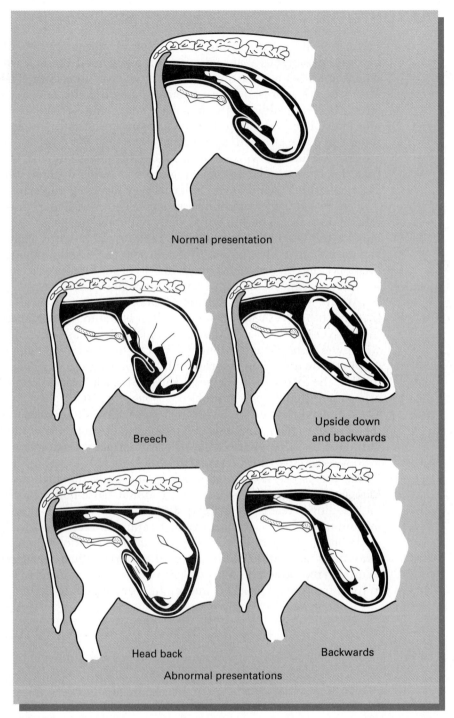

Normal presentation

Breech

Upside down and backwards

Head back

Backwards

Abnormal presentations

FIGURE 10.19 Normal and some abnormal presentations of the calf at parturition. *Source:* R. A. Battaglia and V. B. Mayrose, *Handbook of Livestock Management Techniques* (New York: Macmillan, 1981), pp. 131, 134, 135.

FIGURE 10.20 Live offspring are born to each breeding female in the herd or flock when the intricate mechanisms of reproduction function properly. (A) Beef cow and calf. Courtesy of the American Hereford Association. (B) Ewe and lambs. Courtesy of Colorado State University. (C) Nursing pigs. Courtesy of National Swine Registry. (D) Chicks starting on feed. Courtesy of Big Dutchman, Inc.

CHAPTER SUMMARY

- Reproductive efficiency, as measured by number of offspring born in a herd or flock of breeding females, is a trait of high economic importance.

- A knowledge of the anatomy and physiology of the male and female reproductive organs is essential in understanding the production of ova and sperm, estrous cycles, mating, pregnancy, and parturition.

- Follicle-stimulating hormone, luteinizing hormone, estrogen, progesterone, and testosterone are important hormones in the reproductive process.

- Reproduction in poultry is unique in that ova (eggs) are incubated outside the body, produced from only one ovary and there are no well-defined estrous cycles or pregnancy.

KEY WORDS

oviducts (Fallopian tubes)	seminiferous tubules	gonadotropic hormone
uterus	epididymis	gonadotropin-releasing
cervix	vas deferens	hormone (GnRH)
ovaries	ductus deferens	estrogen
follicle	ampulla	estrus (heat)
corpus luteum	interstitial cells (Leydig	estrous cycle
infundibulum	cells)	prostaglandin
parturition	testosterone	embryonic stage
cloaca	accessory gland	fetus
magnum	glans penis	fetal period
isthmus	leuteinizing hormone	afterbirth
testicles	(LH)	placenta
sperm	follicle-stimulating	parturition
	hormone (FSH)	

REVIEW QUESTIONS

1. What are the two functions of the female gonads or ovaries?
2. What structure on the ovary produces the ovum?
3. Where does fertilization take place?
4. In what structure does the fetus develop during pregnancy in mammals?
5. In what three features does reproduction in poultry differ from that in mammals?
6. How does the reproductive tract of female poultry differ from that of female mammals?
7. What are the two functions of the male gonads or testicles?
8. Where are the spermatozoa produced within the testicles?
9. What cells within the testicles produce testosterone?
10. What are two functions of the epididymis?

11. The scrotum of males helps to maintain the testicles at temperatures several degrees cooler than body temperature. Why is this knowledge important to livestock producers?

12. What are the functions of the accessory sex glands in male mammals?

13. What are the organs of copulation in male and female mammals? What is a second function of these organs?

14. What two hormones from the anterior pituitary gland are responsible for stimulating the gonads?

15. What is the term for the period when the female is sexually receptive to the male? Why is this knowledge important to livestock producers?

16. What hormone is responsible for the occurrence of estrus or heat?

17. What hormone is responsible for the male sex drive or libido?

18. What is the term for the termination of pregnancy, resulting in birth?

19. Why is the knowledge of the gestation length for individual species of livestock important to producers?

SELECTED REFERENCES

Battaglia, R. A. 2007. *Handbook of Livestock Management.* Upper Saddle River, NJ: Pearson-Prentice Hall.

Bearden, H. J., J. W. Fuquay, and S. T. Willard. 2004. *Applied Animal Reproduction.* Upper Saddle River, NJ: Prentice Hall.

Bone, J. F. 1988. *Animal Anatomy and Physiology.* Reston, VA: Reston Publishing.

Frandson, R. D. 1986. *Anatomy and Physiology of Farm Animals.* Philadelphia, PA: Lea & Febiger.

Hafez, E. S. E. (ed.). 1993. *Reproduction in Farm Animals.* Philadelphia, PA: Lea & Febiger.

King, G. J. (ed.). 1993. *Reproduction in Domesticated Animals.* Amsterdam, The Netherlands: Elsevier Science Publishers.

McDonald, L. E. 1989. *Veterinary Endocrinology and Reproduction.* Philadelphia, PA: Lea & Febiger.

Pickett, B. W., J. L. Voss, E. L. Squires, and R. P. Amann. 1981. *Management of the Stallion for Maximum Reproductive Efficiency.* Fort Collins: Colorado State University Press, Animal Reprod. Lab. Gen. Series 1005.

Senger, P. L. 2003. *Pathways to Pregnancy and Parturition.* Pullman, WA: Current Conceptions, Inc.

CHAPTER 11

Artificial Insemination, Estrous Synchronization, and Embryo Transfer

LEARNING OBJECTIVES

- Describe the process of semen collection and processing
- Describe the extent of AI utilization in the livestock industry
- Explain the process of artificial insemination
- Define the role of estrous synchronization in reproductive management
- Compare and contrast several synchronization protocols
- Describe the procedure to transfer embryos
- Explain the value of sexed semen to the livestock industry

Manipulation of the reproductive patterns of farm animals has long been of interest to livestock managers seeking to improve efficiency and profitability. The science of reproductive physiology has yielded tremendous technological breakthroughs over the past several centuries (Table 11.1).

In the process of **artificial insemination (AI),** semen is deposited in the female reproductive tract by artificial techniques rather than by natural mating. AI was first documented as successfully accomplished in the dog in 1780 and in horses and cattle in the early 1900s. AI techniques are also available for use in sheep, goats, swine, poultry, laboratory animals, and bees.

The primary advantage of AI is that it permits extensive use of outstanding sires to maximize genetic improvement. For example, a bull may sire 30–50 calves naturally per year over a productive lifetime of 3–8 years. In an AI program, a bull can produce 200–400 units of semen per ejaculate, with four ejaculates typically collected per week. If the semen is frozen and stored for later use, hundreds of thousands of calves can be produced by a single sire (one

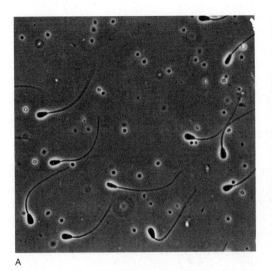 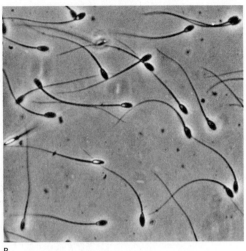

A   B

FIGURE 11.4   (A) Normal bull semen. (B) Normal stallion semen. Courtesy of Colorado State University.

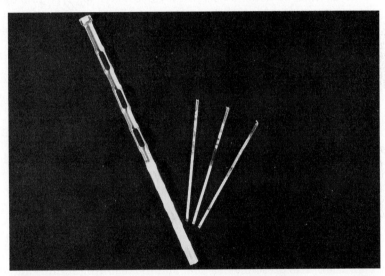

FIGURE 11.5   Bovine semen is typically packaged in plastic straws and then stored in liquid nitrogen. The bull's name, registration number, and inventory code are usually printed on each straw to ensure accurate identification of semen during breeding season.

is only modestly satisfactory; however, improvement has been noted in recent years. Turkey semen cannot be frozen satisfactorily. For maximum fertilization capacity, it should be used within 30 minutes of collection.

INSEMINATION OF THE FEMALE

Prior to insemination, the frozen semen is thawed. Thawed semen cannot be refrozen and used again because refreezing will kill the sperm cells.

High conception rates using AI depend on the female's cycling and ovulating; detecting estrus; using semen that has been properly collected, extended,

TABLE 11.2　Percent of AI Matings in Various Animal Industries

Animal	% AI
Beef cattle	5–10
Dairy cattle	80–90
Horses	40–50
Swine	50–60
Sheep	<5%
Turkeys	95%

and frozen; thawing and handling the semen satisfactorily at the time of insemination; insemination techniques; and avoiding extremes in stress and excitement to the animal being inseminated. Percent of total matings achieved by AI are summarized in Table 11.2. The turkey industry has the highest use (95%) with the sheep industry least likely to use the technology.

Detecting Estrus

Estrus must be detected accurately because it signals time of ovulation and determines proper timing of insemination. The best indication of estrus is the condition called **standing heat,** in which the female stands still when mounted by a male or another female.

Cows are typically checked for estrus twice daily, in the morning and evening. They are usually observed for 30 minutes to detect standing heat. Other observable signs are restlessness, attempting to mount other cows, and a clear, mucous discharge from the vagina. Some producers use sterilized bulls or hormone-treated cows as "heat checkers" in the herd. These animals are sometimes equipped with a chin marker that greases or paints marks on the back of the cow when she is mounted. Electronic systems are available but may be cost-prohibitive in many situations. More affordable alternatives include the use of chalk on the tail heads of cows or patches that are secured to the tail head and change color under the pressure of mounting behavior.

Estrus in sheep or goats is checked using sterilized males equipped with a brisket-marking harness. Gilts and sows in heat assume a rigid stance with ears erect when hands are placed firmly in their backs. The vulva is usually red and swollen. The presence of a boar and the resulting sounds and odors can help detect heat in swine, as the females in estrus will attempt to locate a boar. Signs of estrus in the mare are elevation of the tail, contractions of the vulva (winking), spreading of the legs, and frequent urination.

Proper Timing of Insemination

The duration of estrus and ovulation time are quite variable in farm animals. This variability poses difficulty in determining the best time for insemination. An additional challenge is that sperm are short-lived when put into the female reproductive tract. Also, estrus is sometimes expressed without ovulation occurring; sows, for example, typically show estrus 3–5 days after farrowing, but ovulation does not occur. Sows should not be bred at this time either artificially or naturally.

Insemination time should be as close to ovulation time as possible; otherwise sperm stay in the female reproductive tract too long and lose their fertilizing capacity. Cows found in estrus in the morning are usually inseminated the evening of the same day, and cows in heat in the evening are inseminated the following morning. Insemination, therefore, should occur toward the end or after estrus has been expressed in the cow. Ewes are usually inseminated in the second half of estrus, and goats are inseminated 10–12 hours after the beginning of estrus. Sows ovulate from 30 to 38 hours after the beginning of estrus, so insemination is recommended at the end of the first day and at the beginning of the second day of estrus. When sows are inseminated both days conception rate is improved compared to insemination on only one day.

Insemination of dairy cows occurs while the cow is standing in a stall or stanchion. Beef cows are penned and inseminated in a chute that restrains the animal. The most common insemination technique in cattle involves the inseminator having one arm in the rectum to manipulate the insemination tube through the cervix (Fig. 11.6). The insemination tube is passed just through the cervix, and the semen is deposited into the body of the uterus. The insemination procedure for sheep and goats is similar to that for cattle; however, a **speculum** (a tube approximately 1.5 in. in diameter and 6 in. long) allows the inseminator to observe the cervix in sheep and goats. The inseminating tube is passed through the speculum into the cervix, where the semen is deposited into the uterus or cervix.

The sow is usually inseminated without being restrained. The inseminating tube is easily directed into the cervix because the vagina tapers into the cervix. The semen is expelled into the body of the uterus. The mare is hobbled or adequately restrained prior to insemination. The vulva area is washed, and the tail is wrapped or put into a plastic bag. The plastic-covered arm of the inseminator is inserted into the vagina, and the index finger is inserted into the cervix. The insemination tube is passed through the cervix, and the semen is deposited into the uterus.

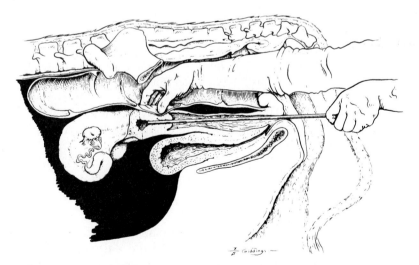

FIGURE 11.6 Artificial insemination of the cow. Note that the insemination tube has been manipulated through the cervix. The inseminator's forefinger is used to determine when the insemination rod has entered into the uterus. *Source:* Colorado State University.

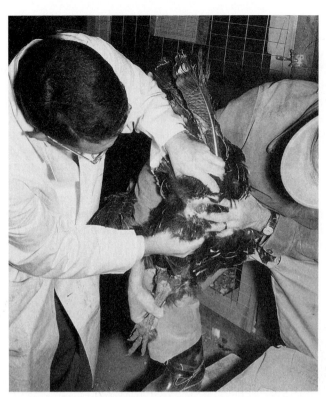

FIGURE 11.7 Insemination of a hen turkey is accomplished by everting the opening of the oviduct via the cloaca. The tube is inserted 1 to 2 in., pressure on the oviduct is relaxed, and the semen is deposited as the pipette is withdrawn. Courtesy of Colorado State University.

The turkey hen is inseminated by first applying pressure to the abdominal area to cause version of the oviduct (Fig. 11.7). The insemination tube is inserted into the oviduct approximately 2 in., and the semen is released. The first insemination is made when 5–10% of the flock has started laying eggs. A second insemination a week later ensures a high level of fertility. Thereafter, insemination is done at 2-week intervals. Fertility in the turkey usually persists at a high level for 2–3 weeks after insemination. This is possible because sperm are stored in special glands of the hen, where they are nourished and retain their fertilizing capacity. Most of these glands are located near the junction of the uterus and vagina.

EXTENT OF ARTIFICIAL INSEMINATION

The number of farm animals artificially inseminated each year is not well documented. AI is used primarily in the dairy industry, where 70% of the cows are bred artificially. Most of the inseminations are in the Holstein breed (Table 11.3). Less than 10% of the beef cows are inseminated, where semen from Angus and Red Angus bulls is used most extensively (Table 11.3). The greater use of AI in dairy compared to beef is reflected in Table 11.3, where in 1998

TABLE 11.3   Units of Semen Sold for Domestic Use, Custom Frozen, and Exported, 2005

Breed	Domestic Sales (1,000 units)	Custom Frozen (1,000 units)	Export Sales (1,000 units)
Holstein	15,990	1,599	8,608
Jersey	1,327	138	646
Brown Swiss	161	128	326
Total dairy	17,806	1,944	9,760
Angus	968	976	229
Simmental	90	111	46
Red Angus	82	105	135
Hereford (Polled and Horned)	40	155	35
Brahman	9	114	52
Charolais	12	100	25
Total beef	1,274	2,657	561

Source: Adapted from National Association of Animal Breeders.

13.1 million units of dairy semen was produced compared to 3.1 million units of beef semen.

More than one-half of all pigs born in the United States are conceived via artificial insemination. Use of AI techniques is highest on the largest farms. However, farms with fewer than 3,000 head of annual marketings are rapidly increasing their use of AI.

AI in horses is still limited because of difficulty in providing extended semen storage. AI in sheep and goats in the United States is limited because herds and flock are dispersed over wide areas and the cost per unit of semen is high. Little AI is done in chickens; however, AI is extensive in turkeys with approximately 95% being inseminated artificially. It is especially important in the broad-breasted turkey, which, because of the size of its breast, has difficulty mating naturally.

ESTROUS SYNCHRONIZATION

Estrous synchronization involves controlling or manipulating the estrous cycle so that females in a herd express estrus at approximately the same time. It is a useful part of an AI program because checking heat and breeding animals, particularly in large pasture areas, is time consuming and expensive. Estrous synchronization is a tool used in successful embryo transfer programs and it can be used with natural service where bulls are used intensively in breeding cows for a few days. Synchronization of estrus is used in approximately 12% of the U.S. cowherds (NAHMS 1998).

Success of a synchronization program depends on many factors, including facilities, available labor, the body condition, post calving interval, and fertility level of the cows and heifers. Also important are herd health, high-quality semen, qualified inseminators, females that are cycling, and accurate detection of estrus. But one of the most serious considerations affecting a herd's success with estrous synchronization is selecting the proper method. Several of the more common synchronization methods are presented based on a review by Geary (1997).

Prostaglandin

In 1979, prostaglandin was cleared for use in cattle. **Prostaglandins** are naturally occurring fatty acids that have important functions in several of the body systems. The prostaglandin that has a marked effect on the reproductive system is prostaglandin F_2 alpha. Lutalyse®, Estrumate®, and Prostamate® are commonly used prostaglandins.

Earlier in chapter 10, it is pointed out that the corpus luteum (CL) controls the estrous cycle in the cow by secreting the hormone progesterone. Progesterone prevents the expression of heat and ovulation. Prostaglandin destroys the CL, thus eliminating the source of progesterone. About 3 days after the injection of prostaglandin, the cow will be in heat. For prostaglandin to be effective, the cow must have a functional CL. It is ineffective in heifers that have not reached puberty or in noncycling mature cows. In addition, prostaglandin is ineffective if the CL is immature or has already started to regress. Prostaglandin is, then, only effective in heifers and cows that are in days 5–18 of their estrous cycle. Because of this relationship, prostaglandin is given in either a one- or two-injection system separately or in combination with other products that influence the estrous cycle.

One-Injection System with 10 Days of Breeding

Days 1–5:   The first 5 days are a conventional AI program of heat detection and insemination.

Day 6:   At this point, the herd owner must decide whether to proceed with the prostaglandin injection. If the percent cycling is satisfactory and drug costs are considered reasonable, the breeder injects the females not already detected in heat with prostaglandin.

Days 6–11:   The synchronized AI breeding season continues with 5 more days of heat detection and insemination (Fig. 11.8).

Days 27–33:   Cattle that synchronize but do not settle on the first service will return to heat. Repeat services can be performed at this time to produce additional AI calves.

This is the most popular protocol that uses only prostaglandin to synchronize estrus and can result in more than 90% of cyclic cows being bred the first 10 days of the breeding season. This system can be shortened by not inseminating during

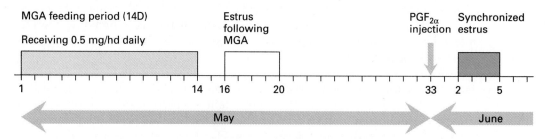

June 2 is the first day of the breeding season.

FIGURE 11.8 The MGA-prostaglandin ($PGF_{2\alpha}$) system for synchronizing estrus. *Source:* Colorado State University.

days 1–5. However, this will lower heat detection to 65–75%. In any event, if less than 20% of the females express heat in the first 5 days, success will be limited. Timed inseminating should not be used with this system.

One Injection of Prostaglandin with Breeding after Injection. Inject all cows and check heat and breed females 12 hours after standing heat. Approximately 75% of the cycling cows would be expected to display heat during the next 4–5 days following injection. For commercial herds, bulls can be turned in with the herd on day 6.

Two-Injection System. By using two injections of prostaglandin, it is possible to get all cycling cows and heifers in heat within a 5-day period. The following schedule describes how this is accomplished:

Day 1: All animals are injected with prostaglandin.

Days 11–14: All animals are again injected with prostaglandin. The second injection must be given on one of these days. While day 11 was traditionally recommended, more recent research suggests a better response if the second injection is administered on day 14.

Within 5 days following the second injection: All cycling cows and heifers should come into heat and can be heat-detected and inseminated. Most repeat services will occur approximately 20–25 days after the first heat is detected.

Days 32–38: Most repeat services occur between these days.

There are two different two-injection programs for synchronization with prostaglandin that allow breeding either after each injection or only after the second injection. Timed insemination is not recommended with the two-shot system.

MGA and Prostaglandin

The use of **MGA (melengestrol acetate)** in conjunction with prostaglandin results in an inexpensive, easily administered synchronization system. This system best fits the breeding of heifers being raised in semiconfinement. The MGA–prostaglandin system showing the most promise is the following 31- to 36-day system:

Days 1–14: Feed 0.5 mg MGA per head per day. The MGA prevents heifers from coming into heat during this period. MGA can be fed as topdressing or can be mixed in with the feed. It is important to provide enough bunk space for all heifers to have an opportunity to eat all the required amount of MGA.

Day 15: Stop feeding MGA. Heifers should come into heat within the next 7 days. The fertility at this heat is very low, so do not inseminate.

Day 31: Inject prostaglandin. Heifers and cows respond to the prostaglandin by coming into heat within the next 5 days.

Days 32–38: Inseminate according to estrus. While the synchronized estrus will be relatively tight, some females may not show heat until approximately 4 days after injection. Some heifers may not initiate a cycle, so semen is wasted if timed insemination is used (Fig. 11.8). For natural service, the prostaglandin shot can be waived with bull turnout occurring on that day.

Work at Virginia Tech suggests that tighter estrus synchronization is achieved if the prostaglandin is given on day 33, 19 days following the end of MGA feeding.

Prostaglandin is not a wonder drug; it will work only in well-managed herds where a high percentage of the females are cycling. Biologically, it has been well demonstrated that prostaglandin can synchronize estrus. Producers, however, must weigh the cost against the economic benefit. Caution should be exercised in administrating prostaglandin to pregnant cows as it may cause abortion. Estrous synchronization in cattle may not be advisable in areas where inadequate protection can be given to young calves during severe blizzards. Also, herd health programs must be excellent to prevent high losses from calf scours and other diseases that become more serious where large numbers of newborn calves are grouped together.

Select Synch (GnRH and Prostaglandin)

This system involves an injection of a gonadotropin-releasing hormone (GnRH) to each female followed 7 days later with an injection of prostaglandin. Estrous detection begins 24 hours prior to the prostaglandin injection and continues for the next 5 to 6 days. Most females will display standing estrus on day 2 through day 4 following the prostaglandin injection. Some producers administer a second GnRH injection to females that have not been detected in heat 72 hours past the prostaglandin injection. These females are then mass-inseminated at the time of the second injection.

Response to this estrus synchronization system, like the other systems, will vary from herd to herd. Data from seven different herds showed a 70% estrous response and a 60–65% average conception rate.

Co-Synch (GnRH and Prostaglandin)

This is a slight variation to the **Select Synch** system. It involves administering the GnRH injection on day 0, prostaglandin on day 7, and a second GnRH injection 48 hours after the prostaglandin injection coupled with mass or timed mating. The GnRH injection initiates a fertile ovulation in cows that have not yet exhibited estrus. Calf removal for 48 hours from the time of the prostaglandin injection until the timed insemination increases pregnancy rates by approximately 10%.

Ov-Synch (GnRH and Prostaglandin)

This is a variation of the **Co-Synch** system that involves an injection of GnRH on day 0, prostaglandin administration on day 7, and a second GnRH injection given 48 hours later. Sixteen to 18 hours after the second GnRH females are mass-mated. Pregnancy rates of about 50% can be expected. By using 48-hour calf removal, this rate can be increased to 60%. This system is labor intensive in that cows may go through the chute four times.

MGA Select

This system combines the MGA–PGF system with the Select Synch program as a means to tighten synchrony, initiate reproductive activity in anestrous cows, and improve pregnancy rates. MGA is fed at the rate of 0.5 mg/herd/day for 14 days. Twelve days later GnRH is given. Seven days following GnRH, prostaglandin is administered, and then heat detection and breeding (5 days) is initiated.

Controlled Internal Drug Release (CIDR)

The FDA approved the use of **CIDR** technology in synchronization of beef cows and heifers in 2002, thus providing the ability to use intravaginal release of progesterone. Marketed as the CIDR® Cattle Insert, the device is a T-shaped nylon structure coated with a layer of silicone containing 10% progesterone by weight. The device is inserted into the vagina with a lubricated applicator and it can be removed by grasping and pulling the flexible tail of the CIDR. Minimal levels of vaginitis may be observed but do not interfere with fertility. Retention rates typically exceed 95% when correct insertion protocols are followed.

The CIDR is inserted on day 0, females are administered an injection of prostaglandin on day 6, and the CIDR is removed on day 7. An alternate protocol calls for the injection on the same day as CIDR removal. Standing heat should be observed within 1 to 3 days following removal of the CIDR. This protocol has been effective in both noncycling and cycling heifers and cows in terms of percentage of animals expressing estrus and conceiving.

Synchronization with Natural Service

Some producers cannot economically use AI but would like the benefits of estrous synchronization. Research has shown that natural service can be used with estrous synchronization if managed properly. One bull per 15 to 20 females in a small pasture (or drylot), and rotated every 24 hours with a rested bull, is recommended during the synchronization period of 4–5 days. Observations have shown that one bull may service 5 to 20 females in a 24-hour period. Pregnancy rates during the synchronized period have ranged from 60–80% and 75–95% during a 30-day breeding season.

Bulls can be used in all synchronization programs. The most popular program is feed MGA for 14 days, wait 17 days, then place the bulls with the females. The advantages of this program are low drug costs, no heat detection, and less demand on the bulls in a short time period.

P.G. 600

P.G. 600 (pregnant mare serum gonadotropin) is utilized to enhance fertility efficiency in the swine industry. This product is used to induce fertile estrus in noncycling gilts over 5.5 months of age and weighing a minimum of 185 lb as well as weaned sows. The advantage of synchronizing gilts lies in the ability to more efficiently schedule breeding and farrowing facilities as well as lowering the days between weaning and return to estrus. However, as is the case with all synchronization schemes, the cost and benefit relationship must be carefully evaluated prior to adoption.

In a sense, there are some natural occurrences of estrous synchronization. The weaning process in swine is an example; the sow will typically show heat 3–8 days after the pigs are weaned. Estrus is suppressed through the suckling influence, so when the pigs are removed, the sow will show heat. Several beef cattle synchronization protocols call for 48-hour calf removal to enhance response rates.

EMBRYO TRANSFER

The primary function of **embryo transfer** is to increase the reproductive rate of valuable females by tenfold or more in a given year and fivefold or more in a cow's lifetime. Even greater increases in reproductive rates can be expected as new technologies are improved. Other uses of embryo transfer are to obtain offspring from infertile cows and to export or import breeding stock to reduce disease risk.

Embryo transfer is sometimes referred to as *ova transplant* or *embryo transplant*. In this procedure, an embryo in its early stage of development is removed from its own mother's (the donor's) reproductive tract and transferred to another female's (the recipient's) reproductive tract. The first successful embryo transfers were accomplished in rabbits in 1890 and in cattle in 1951. Commercial embryo transfer companies have been established in the United States and several foreign countries. More than 400,000 embryos are transferred annually throughout the world with about one-half of the embryos utilized in the United States. Approximately one-half of the embryos are fresh, with the remaining half frozen.

Superovulation is the production of a greater-than-normal number of eggs. Females that are donors for embryo transfer are injected with fertility drugs, which usually cause several follicles to mature and ovulate. The two most common methods of superovulation are using pregnant mare serum gonadotropin (PMSG) or follicle-stimulating hormone (FSH), with the latter usually producing more usable embryos. FSH injections are given over 3–4 days with prostaglandin $F_{2\alpha}$ (e.g., Lutalyse®) administered usually on the third day.

A nonsurgical embryo collection procedure occurs when a flexible rubber tube (Foley catheter) with three passageways is passed through the cervix and into the uterus. A rubber balloon, which is built into the anterior end of the tube, is then inflated to about half the size of a golf ball so that it expands to fill the uterine lumen and to prevent fluid from escaping around the edges. There are two holes in the tube anterior to the balloon that lead into separate passageways, one for fluid entering the uterus and the other for fluid draining from the uterus. A balanced salt solution (to which antibiotics and heat-treated serum are added) is placed in a container and held about 3 feet above the cow. The container is connected to the Foley catheter by an inflow tube. A second tube is connected to the other passageway of the catheter to drain off the medium that has washed (flushed) the ova and embryos out of the uterus. The solution is then collected in tall cylinders holding about 2 pints of fluid. Filtering the collected fluid through a cup-sized container with a fine stainless-steel filter in the bottom and then searching with a microscope can isolate the embryos.

At approximately 12 hours and 24 hours after first standing estrus, the donor is bred artificially. Usually frozen-thawed semen is used and 2–4 straws of high-quality semen is recommended for each insemination. The recipients

FIGURE 11.9   Ten Holstein embryo transfer calves resulting from one superovulation and transfer from the Holstein cow in the background. The 10 recipient cows are shown on the left-hand side of the fence. Courtesy of Colorado State University.

are usually synchronized with a prostaglandin. The recipient cow must be in estrus within 1.5 days of the donor's estrus for best results. Recipients are selected for calving ease, high milk production, and in excellent health status.

An ideal response is 5 to 12 usable embryos per donor per superovulation treatment; however, a range of 0 to more than 20 embryos can be expected (Fig. 11.9). On average, two to four calves will result per superovulated donor if fertile donors are utilized in a well-managed embryo transfer program. A potential donor cow usually produces more embryos if she is 3–10 years of age, has calved regularly each year of her productive life, usually conceives in two services or less, has exhibited regular estrous cycles, comes from fertile bloodlines, did not retain her placenta, and has no history of calving difficulties. The bull, to be bred to the donor cow, should have a history of excellent semen production and a high conception rate determined by his semen having been used in an AI program.

The key to the justification of embryo transfer is identifying genetically superior cows and bulls. Making repeat matings where genetically superior calves have previously been produced can eliminate much of the guesswork. Embryo transfer is usually confined to seedstock herds, where genetically superior females can be more easily identified and high costs can be justified. Embryo transfer calves have an estimated average break-even price of $1,500–$2,500 at 1 year of age (costs for embryo transfer are $500–$1,000 above the usual production costs). Seedstock producers must evaluate the marketability of the embryo transfer calves. Some genetically superior embryo transfer calves may not sell for a sufficient amount to cover costs. Some embryo transfer calves that are not genetically superior may be merchandised for high prices because of a demand that has been previously created.

Donor selection that combines expected high levels of financial return, genetic superiority, and high reproductive potential indicates that the numbers of donors worthy of an embryo transfer program are relatively few. This "relatively few" number could be justified into several thousand head, which is less than 1% of the total United States cowherd.

Although embryo transfer was done surgically in the past, today new nonsurgical techniques are used. The embryos are transferred by way of an artificial insemination gun shortly after being collected. The recipient females need to be in the same stage (within 36 hours) of the estrous cycle for highly successful transfers to occur. Large numbers of females must be kept for this purpose, or estrous synchronization of a smaller number of females will be necessary. The embryos can be frozen in liquid nitrogen and remain dormant for years or decades.

Although the conception rate is lower for frozen embryos and higher for fresh embryos, ongoing research is narrowing the difference in conception rate. In cattle, approximately 85% of frozen embryos are normal after thawing; however, 40–55% of those normal embryos result in confirmed pregnancies at 60–90 days. In contrast, fresh embryos transferred the same day of collection have a pregnancy rate of 55–65%.

While there is no decline in the embryo recovery rate the first three times a donor is superovulated, additional superovulation treatments result in a reduced number of embryos for some donors. Most donors can be superovulated three or four times a year. Season, breed, and lactating versus dry donors appear to have little effect on the success of embryo transfer. Surveys have shown that 2% of the recipients abort between 3 and 9 months of gestation, 4% of the embryo transfer calves die at birth, and another 4% die between birth and weaning—representing a total loss of 10% between 3 months gestation and weaning. These losses may be similar to those experienced in a typical natural service program.

Recent advances in reproduction technology include the birth of the first test-tube calf (fertilization occurred outside the cow's body) in 1981 and the first identical twin calves resulting from embryo splitting (Fig. 11.10). Splitting one embryo to produce two identical offspring is one of the initial steps in the cloning of livestock (Fig. 11.11) and other intriguing aspects of genetic manipulation.

Sexed Semen

The commercialization of sexed semen technology has only recently been achieved. Led by a collaborative effort between XY, Inc. and Colorado State University, the use of sexed semen offers a multitude of opportunities to livestock producers:

1. Initiating female pregnancies in first-calf heifers to help control rates of dystocia.
2. Creation of target-specific progeny. For example, production of only male calves from terminal sire AI matings.
3. Creating only female progeny from high-performance maternal genetics.

One of the challenges in using sexed semen as a genetic tool is the reality that because sorting protocols using flow cytometry are relatively slow, inseminations must be made with low-dose sperm concentrations. The normal sperm concentration for AI in sheep and cattle is 60 million and 20 million, respectively. Thus, the use of highly fertile sires coupled with superior female management protocols, superior estrus detection, and insemination is required to achieve favorable results.

The use of sexed semen results in predicting fetal sex rates of greater than 85–90% for cattle and sheep. The technology is being applied to cattle, sheep, swine, horses, elk, yaks, gorillas, sea mammals, dogs, and cats.

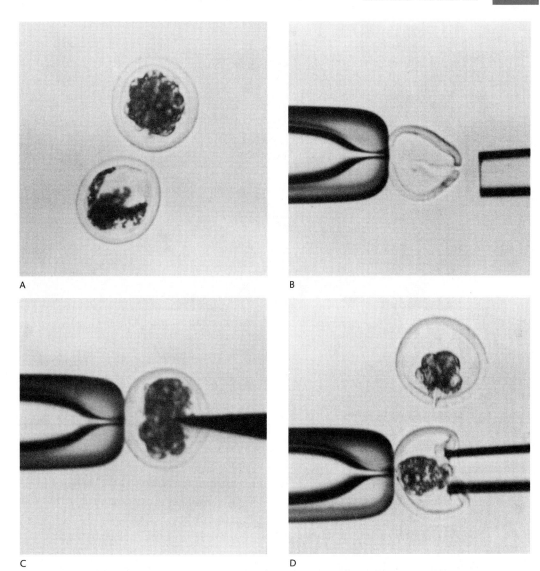

A

B

C

D

FIGURE 11.10 The process of embryo splitting. (a) A morula (40–70 cells of the fertilized egg) and an unfertilized egg. (B) The unfertilized egg with the cellular contents aspirated from it. (C) The morula is divided into two groups of cells with the microsurgical blade. (D) One-half of the cells are left in the morula while the other one-half of the cells are placed inside the other unfertilized egg. The result is two genetically identical embryos ready for transfer. Courtesy of Williams et al., *1983 NAAB Conference, Beef AI and Embryo Transfer.*

FIGURE 11.11 "Question" and "Answer" were the world's first split-embryo foals resulting from nonsurgical embryo transfer. Courtesy of Colorado State University.

CHAPTER SUMMARY

- Artificial insemination (AI) is the deposition of semen in the female reproductive tract by methods other than natural mating.
- The advantages of AI are the extended use of sires, controlling reproductive diseases, and using sires that have been injured or are dangerous when used naturally.
- AI is widely used in dairy cattle and turkeys and to a lesser extent in other farm animal species.
- Estrous synchronization is controlling or manipulating the estrous cycle so that females express estrus at approximately the same time.
- Prostaglandin F_2 alpha, melengestrol acetate (MGA), and GnRH are commonly used products to synchronize estrus in cattle.
- Embryo transfer involves removing embryos from females (donors) and transferring them into the reproductive tracts of other females (recipients).

KEY WORDS

artificial insemination AI	prostaglandin	CIDR
artificial vagina	melengestrol acetate	P.G. 600
standing heat	(MGA)	embryo transfer
speculum	Select Synch	superovulation
estrous synchronization	Co-Synch	

REVIEW QUESTIONS

1. What is artificial insemination?
2. What is the primary advantage of artificial insemination?
3. In relation to the time of ovulation, when should animals be inseminated?
4. When does ovulation usually occur?
5. In relation to estrus, when should insemination occur?
6. Why does fertility in turkey hens persist for 2 to 3 weeks after insemination?
7. Why is the use of artificial insemination especially important for turkeys.
8. What is estrous synchronization?
9. What are three methods of estrous synchronization in cattle?
10. How can estrus be synchronized in sows?
11. What is embryo transfer?
12. What is superovulation and why is it important?

SELECTED REFERENCES

American Breeder's Service. 1991. *A.I. Management Manual.* DeForest, WI.

Beal, W. E. 1998. Making Embryo Transfer Easier. Proc. Virginia Beef Genetics Management Conference. Roanoke, VA.

Beal, W. E., and J. M. DeJarnette. 2002. Understanding the Estrous Cycle of the Cow. Proc. Cattlemen's College, NCBA Convention. Denver, CO.

Ernst, R. A., F. X. Ogasawara, W. F. Rooney, J. P. Schroeder, and D. C. Ferebee. 1970. *Artificial Insemination of Turkeys.* University of California Agric. Ext. Pub. AXT-338.

Estienne, M. J., and A. F. Harper. 2001. *Uses of P.G. 600 in Swine Breeding Herd Management.* Virginia Cooperative Extension Livestock Update.

Foote, R. H. 1994. Embryo transfer in domestic animals. *Encyclopedia of Agricultural Science.* San Diego, CA: Academic Press, Inc.

Geary, T. 1997. Synchronization programs update. Proc. of the Range Beef Cow Symposium XV. Rapid City, SD.

Hafez, S. E. 1987. *Reproduction in Farm Animals.* Philadelphia, PA: Lea & Febiger.

Herman, H. A. 1981. *Improving Cattle by the Millions. NAAB and the Development and Worldwide Application of Artificial Insemination.* Columbia, MO: University of Missouri Press.

Kiessling, A. A., W. H. Hughes, and M. R. Blankevoort. 1986. Superovulation and embryo transfer in the dairy goat. *J. Am. Vet. Med. Assoc.* 188:829.

Lake, P. E, and J. M. Steward. 1978. *Artificial Insemination in Poultry.* Scotland Ministry of Agriculture, Fisheries, and Food - N. 213.

National Animal Health Monitoring System (NAHMS). 1998. Part 1: reference of 1997 beef cow-calf management practices. Washington, DC: USDA, APHIS, V.S.

Pickett, B. W., and D. G. Back. 1973. *Procedures, Collection, Evaluation and Insemination of Stallion Semen.* Fort Collins: Colorado State University Exp. Sta., Animal Reprod. Lab., Gen. Series 935.

Seidel, G. E., Jr. 1981. Superovulation and embryo transfer in cattle. *Science* 211:351.

Seidel, G. E. Jr., 1995. Reproductive Technologies for Profitable Beef Production. Proc. Natl. Assn. of Animal Breeders Conference. Sheridan, WY.

Seidel, G. E., Jr., S. M. Seidel, and R. A. Bowen 1978. *Bovine Embryo Transfer Procedures*. Fort Collion: Colorado State University Exp. Sta., Animal Reprod. Lab., Gen. Series 975.

CHAPTER **12**

Genetics

LEARNING OBJECTIVES

- Describe the formation of gametes
- Define the role of DNA and RNA
- Describe the six fundamental mating types
- Calculate genotypic and phenotypic outcomes
- Illustrate various forms of dominance
- Discuss the role of biotechnology in the livestock industry

Body tissues of animals and plants are composed of cells, whose structure can be observed microscopically. These cells, with certain exceptions, have an outer membrane, an internal cytoplasm, and a nucleus (see Chapter 18, Fig. 18.3). This nucleus contains rod-like bodies called *chromosomes* (Fig. 12.1). Body cells contain these chromosomes in pairs, and each chromosome contains genes, which are the functional units of inheritance. When cells divide to produce more body cells, the chromosomes replicate by a process called *mitosis* (Fig. 12.2). During mitosis, each member of each chromosome pair divides so that the two new cells *(daughter cells)* are identical to the original cell that divided.

Each species has a characteristic number of chromosomes (Table 12.1), which is maintained through meiosis and fertilization. Poultry have more chromosomes than livestock, whereas swine have the smallest number. For comparative purposes, humans have 23 pairs of chromosomes.

PRODUCTION OF GAMETES

The testicles of the male and the ovaries of the female produce sex cells (gametes) by a process called **gametogenesis.** The gametes produced by the testicles are called *sperm;* the gametes produced by the ovaries are called *eggs* or *ova.* Specifically, the production of sex cells that become sperm is called **spermatogenesis;** the production of ova is known as **oogenesis.** The unique type of cell division in which gametes (sperm or ova) are formed is called **meiosis.**

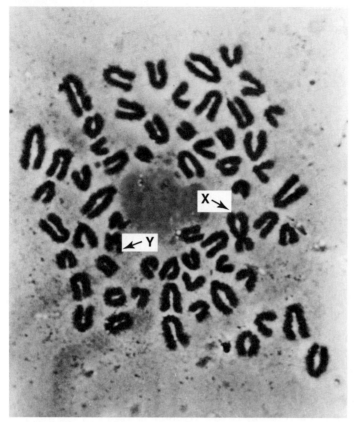

FIGURE 12.1 The 30 pairs of chromosomes of a bull magnified several hundred times. Note the X and Y chromosomes. Courtesy of Texas A&M University.

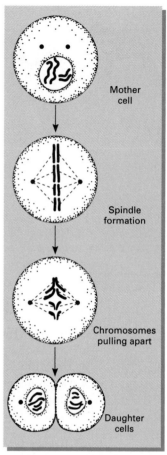

Mother cell

Spindle formation

Chromosomes pulling apart

Daughter cells

FIGURE 12.2 Mitosis. *Source:* Colorado State University.

TABLE 12.1   Pairs of Chromosomes in Livestock and Poultry

Species	Number of Pairs
Turkeys	41
Chickens	39
Horses	32
Cattle	30
Goats	30
Sheep	27
Humans[a]	23
Swine	19

[a]Shown for comparison.

Each newly formed gamete contains only one member of each of the chromosome pairs present in the body cells.

Let us examine gametogenesis in a theoretical species in which there are only two pairs of chromosomes in each body cell. Meiosis occurs in the primordial germ cells (cells capable of undergoing meiosis) located near the outer wall of the **seminiferous tubules** of each testicle and near the surface of each ovary. The initial steps in meiosis are similar for the male and female. The chromosomes replicate themselves so that each chromosome is doubled. Then each pair of chromosomes comes together in extremely accurate pairing called **synapsis.** After chromosome replication and synapsis, the cell is called a **primary spermatocyte** in the male and a **primary oocyte** in the female. Because subsequent differences exist between spermatogenesis and oogenesis, the two are described here separately.

SPERMATOGENESIS

The process of spermatogenesis in our theoretical species is shown in Figure 12.3. The primary spermatocyte contains two pairs of chromosomes in synapsis. Each chromosome is doubled following replication. Thus, the primary spermatocyte contains two bodies or structures formed by four parts. By two rapid cell divisions, in which no further replication of the chromosomes occurs, four cells, each of which contains two chromosomes, are produced. The spermatid cells each lose much of their cytoplasm and develop a tail. This process of spermatogenesis results in formation of sperm. Four sperm are produced from each primary spermatocyte. Whereas four chromosomes in two pairs are present in the primordial germ cell, only two chromosomes (one-half of each chromosome pair) are present in each sperm. Thus, the number of chromosomes in the sperm has been reduced to half the number as compared to the primordial germ cell.

OOGENESIS

The process of oogenesis in our theoretical species is shown in Figure 12.3. Like the primary spermatocyte of the male, the primary oocyte of the female contains a tetrad. The first maturation division produces one relatively large nutrient-containing cell (the secondary oocyte) and a smaller cell (the first polar body). Each of these two cells contains a dyad. The unequal distribution of nutrients that results from the first maturation division in the female serves to maximize the quantity of nutrients in one of the cells (the secondary oocyte) at the expense of the other (the first polar body). The second maturation division produces the egg (ovum) and the second polar body, each of which contains two chromosomes. The first polar body may also divide, but all polar bodies soon die and are reabsorbed. Note that the egg, like the sperm, contains only one chromosome of each pair that was present in the primordial germ cell. The gametes each contain one member of each pair of chromosomes that existed in each primordial germ cell.

FERTILIZATION

When a sperm and an egg of our theoretical species unite to start a new life, each contributes one chromosome to each pair of chromosomes in the fertilized egg, now called a **zygote** (Fig. 12.4). **Fertilization** is defined as the union of the sperm and the egg along with the establishment of the paired condition of the

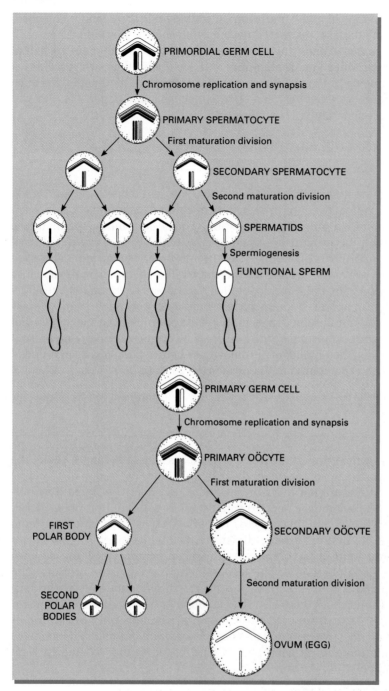

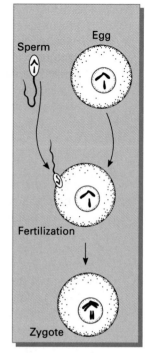

Figure 12.4 Combining of chromosomes through fertilization (two pairs of genes used for simplification of example). *Source:* Colorado State University.

FIGURE 12.3 Meiosis or reduction cell division in the testicle and ovary (example with two pairs of chromosomes). *Source:* Colorado State University.

chromosomes. The zygote is termed **diploid** (*diplo* means "double") because it has chromosomes in pairs; one member of each pair comes from the sire and one member comes from the dam. Gametes have only one member of each pair of chromosomes; therefore, gametes are termed **haploid** (*haplo* means "half"). Gametogenesis thus reduces the number of chromosomes in a cell to half the diploid number. Fertilization reestablishes the normal diploid number. It is important to remember that gametogenesis produces gametes that carry a random sample half of each parent's genetic code. Which sperm fertilizes the egg is also a random event. As such, genetic diversity in the population is assured.

DNA AND RNA

The two members of each typical pair of chromosomes in a cell are alike in size and shape and carry **genes** that affect the same hereditary characteristics (Fig. 12.5). Such chromosomes are said to be **homologous.** The genes are points of activity found in each of the chromosomes that govern the way in which traits develop. The genes form the coding system that directs enzyme and protein production. Thus, they control the development of traits.

Chromosomes in advanced organisms like poultry and livestock are composed of **deoxyribonucleic acid (DNA)** that is structured as two strands of nucleotides wrapped around each other and connected at the base. The structural configuration is termed "double-helix." Segments of DNA are the genes. Each gene is a specific DNA sequence that codes for a particular protein. The

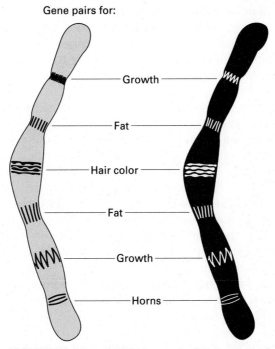

Gene pairs for:

Growth

Fat

Hair color

Fat

Growth

Horns

FIGURE 12.5   A simplified example showing a pair of chromosomes containing several pairs of genes. *Source:* Colorado State University.

location of a gene on the chromosome is called the **locus.** DNA is composed of deoxyribose sugar, phosphate, and four nitrogenous bases. The combination of deoxyribose, phosphate, and one of the four bases is called a **nucleotide;** when many nucleotides are chemically bonded to one another, they form a strand that composes one-half of the DNA molecules. (A molecule formed by many repeating sections is called a **polymer.**)

The bases of DNA are the parts that hold the key to inheritance. The four bases are adenine (A), thymine (T), guanine (G), and cytosine (C). In the two strands of DNA, A is always complementary to (pairs with) T, and G is always complementary to C (Fig. 12.6). During meiosis and mitosis, the unwinding and pulling apart of the DNA strands replicate the chromosomes, and a new strand is formed alongside the old. The old strand serves as a template, so that wherever an A occurs on the old strand, a T will be directly opposite it on the new, and wherever a C occurs on the old strand, a G will be placed on the new. Complementary bases pair with each other until two entire double-stranded molecules are formed where originally there was one.

Nearly all genes contain code for specific proteins. It is important to realize that DNA and protein are both polymers. For each of the 26 amino acids of which proteins are made, there is at least one "triplet" sequence of three nucleotides. For example, two DNA triplets, TTC and TTT, code for the amino acid lysine; four triplets, CGT, CGA, CGG, and CGC, all code for the amino acid alanine. If we think of a protein molecule as a word, and amino acids as the letters of the word, each triplet sequence of DNA can be said to code for a letter of the word, and the entire encoded message, the series of base triplets, is the gene.

The processes by which the code is "read" and protein is synthesized are called **transcription** and **translation.** To understand these processes, another group of molecules, the **ribonucleic acids (RNAs),** must be introduced. There are three types of RNA; **transfer RNA** (tRNA), which identifies both an amino acid and a base triplet in mRNA; **messenger RNA** (mRNA), which carries the information codes for a particular protein, and **ribosomal RNA** (rRNA), which is essential for ribosome structure and function. The DNA template codes all three RNAs.

The first step in protein synthesis is transcription. Just as the DNA molecule serves as a template for self-replication using the pairing of specific bases, it can also serve as a template for the mRNA molecule. Messenger RNA is similar to DNA but is single-stranded, coding for only one or a few proteins. The triplet sequence that codes for one amino acid in mRNA is called a **codon.** Through transcription, the encoded message held by the DNA molecule becomes transcribed onto the mRNA molecule. The mRNA then leaves the nucleus and travels to an organelle called a **ribosome,** where protein synthesis will actually take place. The ribosome is composed of rRNA and protein.

The second step in protein synthesis is the union of amino acids with their respective tRNA molecules. The DNA codes the tRNA molecules. They have a structure and contain an anticodon that is complementary to an mRNA codon. Each RNA unites with one amino acid. This union is very specific such that, for example, lysine never links with the tRNA for alanine but only to the tRNA for lysine.

The mRNA attaches to a ribosome for translation of its message into protein. Each triplet codon on the mRNA (which is complementary to one on DNA) associates with a specific tRNA bearing its amimo acid, using a base-pairing

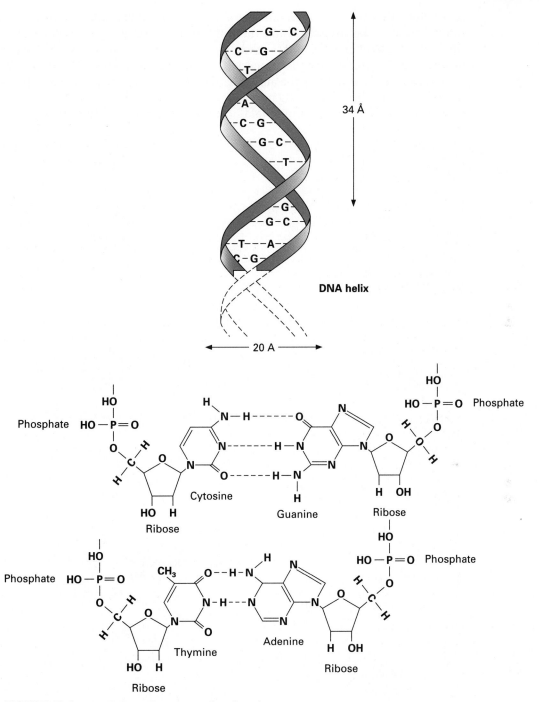

FIGURE 12.6 DNA helix and structure of nucleotides.

mechanism similar to that found in DNA replication and mRNA transcription. This matching of each tRNA with its specific mRNA triplets begins at one end of the mRNA and continues down its length until all the codons for the protein-forming amino acids are aligned in the proper order. The amino acids are chemically bonded to each other by so-called peptide bonding as the mRNA moves through the ribosome, and the fully formed protein disassociates from the tRNA-mRNA complex and is ready to fulfill its role as a part of a cell or as an enzyme to direct metabolic processes. Figure 12.7 summarizes the steps of protein synthesis. It depicts the amino acids arginine (arg), leucine (leu), threonine (thr), and valine (val) being moved into position to join a chain that already includes the amino acids alanine (ala), proline (pro), and leucine (leu).

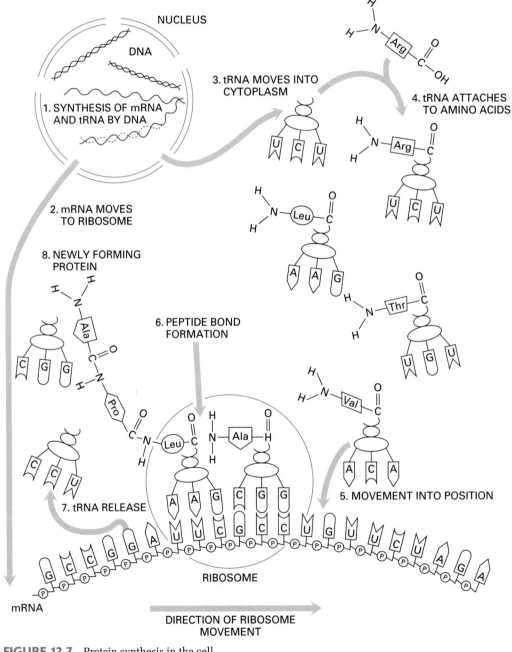

FIGURE 12.7   Protein synthesis in the cell.

GENES AND CHROMOSOMES

Because chromosomes are in pairs, genes are also in pairs. The location of a gene in a chromosome is called a *locus* (plural, *loci*). For each locus in one of the members of a pair of homologous chromosomes, a corresponding locus occurs in the other member of that chromosome pair. The transmission of genes from parents to offspring depends entirely on the transmission of chromosomes from parents to offspring.

A special pair of chromosomes, the so-called sex chromosomes (*X* and *Y*), exist as a pair in which one of the chromosomes does not correspond entirely to the other in terms of what loci are present. The *Y* chromosome is much shorter in length than the *X* chromosome.

The *X* and *Y* chromosomes determine the sex of mammals. A female has two *X* chromosomes, and a male has an *X* and a *Y* chromosome. The female, being *XX*, can contribute only an *X* chromosome to the offspring. The male contributes either an *X* or a *Y* (Fig. 12.1), thus determining the sex of the newborn.

In all bird species, including chickens and turkeys, the female determines the sex of the offspring. The sex chromosomes are identical in the sperm *(XX)*, but they are different in the egg *(XY)*.

The genes located at corresponding loci in *homologous chromosomes* may correspond to each other or contrast with each other in the way that they control a trait. If they correspond, the individual is said to be **homozygous** at the locus (*homo* means "alike"; *zygous* refers to the individual); if they differ, the individual is said to be **heterozygous** (*hetero* means "different"). Those genes that occupy corresponding loci in homologous chromosomes but that affect the same character in different ways (e.g., black or red color) are called **alleles.** Genes that are alike and that affect the character in the same way are called *identical alleles*.

The geneticist usually illustrates the chromosomes as lines and indicates the genes by alphabetical letters. When the genes at corresponding loci on homologous chromosomes differ, one of the genes often overpowers, or *dominates,* the expression of the other. This allele is called **dominant.** The allele whose expression is suppressed is called **recessive.** The dominant allele is symbolized by a capital letter; the recessive allele is symbolized by a lowercase letter. For example, in cattle, black hair color is dominant to red hair color, so we let *B* = black and *b* = red. Three combinations of genes are possible:

Both *BB* animals and *bb* animals are homozygous for the genes that determine hair color, but one is homozygous-dominant *(BB)* and the other is homozygous-recessive *(bb)*. The animal that is *Bb* is heterozygous; it has allelic genes. Keep in mind that dominant genes are not necessarily "good" nor are recessive genes always "bad."

SIX FUNDAMENTAL TYPES OF MATING

With three kinds of individuals (homozygous-dominant, heterozygous, and homozygous-recessive) and one pair of genes considered, six types of mating are possible. Using the genes designated as *B* = black and *b* = red in cattle, the six mating possibilities are *BB* × *BB*, *BB* × *Bb*, *BB* × *bb*, *Bb* × *Bb*, *Bb* × *bb*, and *bb* × *bb*. Keep in mind that our discussion of these genes is applicable to any other pair of genes in any species.

Homozygous-Dominant × Homozygous-Dominant *(BB × BB)*. When two homozygous-dominant individuals are mated, each can produce only one kind of gamete—the gamete carrying the dominant gene. In our particular example, this gamete carries gene *B*. The union of gametes from two homozygous-dominant parents results in a zygote that is homozygous-dominant *(B × B = BB)*. Thus, homozygous-dominant parents produce only homozygous-dominant offspring (Fig. 12.8).

Homozygous-Dominant × Heterozygous *(BB × Bb)*. The mating of a homozygous-dominant with a heterozygous individual results in an expected ratio of 1 homozygote-dominant:1 heterozygote. The homozygous-dominant parent produces only one kind of gamete *(B)*. The heterozygous parent produces in equal proportion two kinds of gametes, one carrying the dominant gene *(B)* and one carrying the recessive gene *(b)*. The chances are equal that a gamete from the parent producing only the one kind of gamete (the one having the dominant gene) will unite with each of the two kinds of gametes produced by the heterozygous parent; therefore, the number of homozygous-dominant offspring *(B × B = BB)* and heterozygous offspring *(B × b = Bb)* produced should be approximately equal (Fig. 12.9).

Homozygous-Dominant × Homozygous-Recessive *(BB × bb)*. The homozygous-dominant individual can produce only gametes carrying the dominant gene *(B)*. The recessive individual must be homozygous-recessive to express the recessive trait; therefore, the individual produces only gametes carrying the recessive gene *(b)*. When the two types of gametes unite, all offspring produced receive both the dominant and the recessive gene *(B × b = Bb)* and are thus all heterozygous (Fig. 12.10).

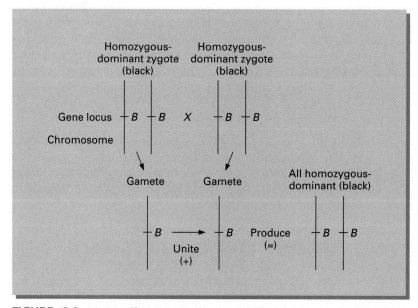

FIGURE 12.8   Mating of homozygous-dominant *(BB)* × homozygous-dominant *(BB)*.

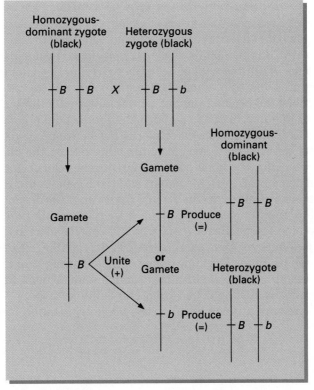

FIGURE 12.9 Mating of homozygous-dominant *(BB)* × heterozygote *(Bb)*.

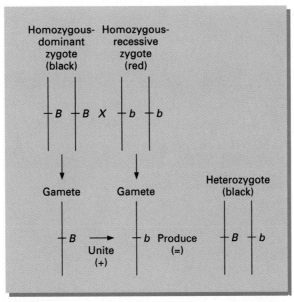

FIGURE 12.10 Mating of homozygous-dominant *(BB)* × homozygous-recessive *(bb)*.

Heterozygous × Heterozygous *(Bb × Bb)*.   Each of the two heterozygous parents produces two kinds of gametes in approximately equal ratios: one kind of gamete carries the dominant gene *(B)*; the other kind carries the recessive gene *(b)*. The two kinds of gametes produced by one parent each have an equal chance of uniting with each of the two types of gametes produced by the other parent. Thus, four equal-chance unions of gametes are possible. If the gamete carrying the *B* gene from one parent unites with the gamete carrying the *B* gene from the other parent, the offspring produced are homozygous-dominant *(B × B = BB)*. If the gamete carrying the *B* gene from one parent unites with the gamete carrying the *b* gene from the other parent, the offspring are heterozygous *(B × b = Bb)*. The latter combination of genes can occur in two ways: the *B* gene can come from the male parent and the *b* gene from the female parent, or the *B* gene can come from the female parent and the *b* gene from the male parent. When the gametes carrying the *b* gene from both parents unite, the offspring produced are recessive *(b × b = bb)*. The total expected ratio among the offspring of two heterozygous parents is 1*BB*:2*Bb*:1*bb* (Fig. 12.11). This 1:2:1 ratio is the *genotypic* ratio. The appearance of the offspring is in the ratio of 3 dominant:1 recessive because the *BB* and *Bb* animals are all black (dominant) and cannot be genetically distinguished from one another visually. This 3:1 ratio, based on external appearance, is called the *phenotypic* ratio.

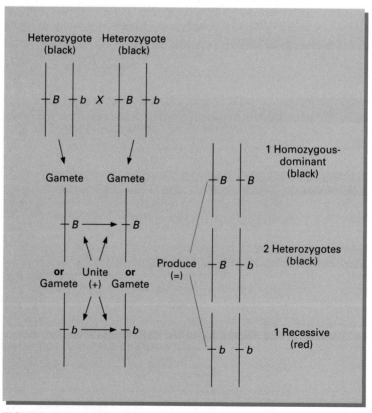

FIGURE 12.11   Mating of heterozygote *(Bb)* × heterozygote *(Bb)*.

Heterozygous × Homozygous-Recessive *(Bb × bb)*.   The heterozygous individual produces two kinds of gametes, one carrying the dominant gene *(B)* and the other carrying the recessive gene *(b),* in approximately equal numbers. The recessive individual produces only the gametes carrying the recessive gene *(b).* There is an equal chance that the two kinds of gametes produced by the heterozygous parent will unite with the one kind of gamete produced by the recessive parent *(B × b = Bb; b × b = bb).* The offspring produced when these gametes unite thus occur in the expected ratio of 1 heterozygous:1 homozygous-recessive (Fig. 12.12).

Homozygous-Recessive × Homozygous-Recessive *(bb × bb)*.   The recessive individuals are homozygous; therefore, they can produce gametes carrying only the recessive gene *(b).* When these gametes unite *(b × b = bb),* all offspring produced will be recessive (Fig. 12.13). This example illustrates the principle that recessives, when mated together, breed true.

Knowledge of the combinations resulting from each of the six fundamental types of matings provides a background for understanding complex crosses. When more than one pair of genes is considered in a mating, one can understand the expected results; that is, one can combine each of the combinations of one pair of genes with each of the other combinations of one pair of genes to obtain the expected ratios.

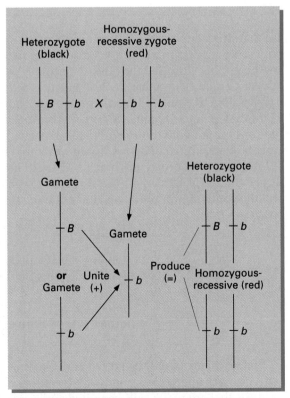

FIGURE 12.12   Mating of heterozygote *(Bb)* × homozygous-recessive *(bb).*

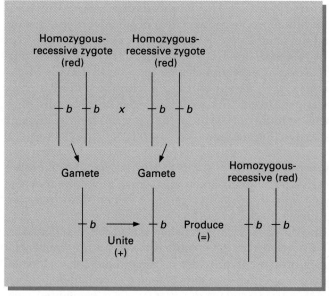

FIGURE 12.13   Mating of homozygous-recessive *(bb)* × homozygous-recessive *(bb)*.

MULTIPLE GENE PAIRS

Suppose that there are two pairs of genes to be considered, each pair independently affecting a particular trait. Let us consider two pairs of genes; one determines coat color in cattle; and the other determines whether the animal is polled (hornless) or horned. The genes are designated as follows: B = black (dominant), b = red (recessive), P = polled (dominant), and p = horned (recessive).

If a bull that is heterozygous for both traits *(BbPp)* is mated to cows that are also heterozygous for both traits *(BbPp)*, the expected phenotypic and genotypic ratios can be determined. The results of crossing $Bb \times Bb$ (giving a ratio of 3 black:1 red in the offspring, or ¾ black and ¼ red) have previously been shown (Fig. 12.11). Similarly, a cross of $Pp \times Pp$ gives a ratio of 3 polled:1 horned. The $BbPp \times BbPp$ combination produces the following phenotypes:

$$BbPp \times BbPp$$

$$3 \text{ black} \times \begin{cases} 3 \text{ polled} = 9 \text{ black polled} \\ 1 \text{ horned} = 3 \text{ black horned} \end{cases}$$

$$1 \text{ red} \times \begin{cases} 3 \text{ polled} = 3 \text{ red polled} \\ 1 \text{ horned} = 1 \text{ red horned} \end{cases}$$

Of the ¾ that are black, ¾ will be polled and ¼ will be horned. Therefore, ¾ of ¾ (or ⁹⁄₁₆) will be black and polled; and ¼ of ¾ (or ³⁄₁₆) will be black and horned. Similarly, of the ¼ that are red, ¾ will be polled and ¼ will be horned, and ¾ of ¼ (or ³⁄₁₆) will be red and polled and ¼ of ¼ (or ¹⁄₁₆) will be red and horned. This

TABLE 12.2   Genotypes and Phenotypes with Two Heterozygous Gene Pairs

Sperm	Eggs			
	BP	*Bp*	*bP*	*bp*
Bp	*BBPP*[a] black, polled[b]	*BBPp* black, polled	*BbPP* black, polled	*BbPp* black, polled
Bp	*BBPp* black, polled	*BBpp* black, horned	*BbPp* black, polled	*Bbpp* black, horned
BP	*BbPP* black, polled	*BbPp* black, polled	*bbPP* red, polled	*bbPp* red, polled
Bp	*BbPp* black, polled	*Bbpp* black, horned	*bbPp* red, polled	*bbpp* red, horned

[a] Genotype. Number of different genotypes: 1 *(BBPP)*, 2 *(BBPp)*, 2 *(BbPp)*, 4 *(BbPp)*, 1 *(BBpp)*, 1 *(bbPP)*, 2 *(Bbpp)*, 2 *(bbPp)*, 1 *(bbpp)*.

[b] Phenotype. Number of different phenotypes: 9 black, polled; 3 black, horned; 3 red, polled; 1 red, horned.

distribution of phenotypes is more often expressed as a 9:3:3:1 ratio instead of a set of fractions. The following demonstrates the different types of genotypes and phenotypes resulting from the $BbPp \times BbPp$ mating:

		Genotypes (9)	**Phenotypes (3)**
1 *BB* ×	1 *PP* = 1 *BBPP*	black, polled
	2 *Pp* = 2 *BBPp*	black, polled
	1 *pp* = 1 *BBpp*	black, horned
2 *Bb* ×	1 *PP* = 2 *BbPP*	black, polled
	2 *Pp* = 4 *BbPp*	black, polled
	1 *pp* = 2 *Bbpp*	black, horned
1 *bb* ×	1 *PP* = 1 *bbPP*	red, polled
	2 *Pp* = 2 *bbPp*	red, polled
	1 *pp* = 1 *bbpp*	red, horned

Table 12.2 shows another method of demonstrating the different phenotypes and genotypes that can be produced from a $BbBp$ (bull) $\times BbPp$ (cow) mating. The four different sperm have the opportunity to combine with each of the four different eggs. Nine different genotypes and four different phenotypes are possible.

A related condition to horns is the presence of a loose bony growth called a *scur*. However, two different gene pairs control the scurred condition and the horned/polled condition.

GENE INTERACTIONS

A gene may interact with other genes in the same chromosome **(linear interaction)**, with its corresponding gene in a homologous chromosome **(allelic interaction)**, or with genes in nonhomologous chromosomes **(epistatic interaction)**. In addition, genes interact with the cytoplasm and with the environment. The environmental factors with which genes interact are internal, such as hormones, and external, such as nutrition, temperature, and amount of light.

Linear interactions are known to exist in lower animals *(Drosophila)*, but have not been demonstrated in farm animals.

Allelic Interactions

When contrasting genes occupy corresponding loci in the same (homologous) pair of chromosomes, each gene exerts its influence on the trait, but the effects of each of the genes depend on the relationship of dominance and recessiveness. Allelic interactions may also be called *dominance interactions*. When unlike genes occupy corresponding loci, complete dominance might exist. In this situation, only the effect of the dominant gene is expressed. A good example, as previously discussed, is provided by cattle, in which the hornless (polled) condition is dominant to the horned condition. The heterozygous animal is polled and is phenotypically indistinguishable from the homozygous polled animal.

There may be lack of dominance, in which heterozygous animals show a phenotype that is different from either homozygous phenotype and is usually intermediate between them. A good example is observed in sheep, in which two alleles for ear length lack dominance to each other. Sheep that are *LL* have long ears, *Ll* have short ears, and *ll* are earless. Thus the heterozygous *Ll* sheep are different from both homozygotes and are intermediate between them in phenotype.

Lack of dominance can also be considered additive gene action. Additive gene action occurs when each gene has an expressed phenotypic effect. For example, if *D* gene and *d* gene influence rate of gain (where $D = 0.10$ lb/day and $d = 0.05$ lb/day), then *DD* = increase in gain of 0.20 lb, *Dd* = increase in gain of 0.15 lb, and *dd* = increase of gain of 0.10 lb.

There is evidence that many pairs of genes affect production traits (like rate of gain) in an additive manner. For example, consider two pair of genes that influence daily gain, where $D = 0.10$ lb, $d = 0.05$ lb, $N = 0.10$ lb, and $n = 0.05$ lb:

$$DDNN \quad = 0.40 \text{ lb}$$

$$\left.\begin{array}{l} DDNn \\ DdNN \end{array}\right\} = 0.35 \text{ lb}$$

$$\left.\begin{array}{l} DdNn \\ DDnn \\ ddNN \end{array}\right\} = 0.30 \text{ lb}$$

$$\left.\begin{array}{l} Ddnn \\ ddNn \end{array}\right\} = 0.25 \text{ lb}$$

$$ddnn \quad = 0.20 \text{ lb}$$

Sometimes heterozygous individuals are superior to either of the homozygotes. This condition, in which the heterozygotes show overdominance, is an example of a selective advantage for the heterozygous condition. Overdominance means that heterozygotes possess greater vigor or are more desirable in other ways (such as producing more milk or being more fertile) than either of the two homozygotes that produced the heterozygote. Because the heterozygotes of breed crosses are more vigorous than the straightbred parents, they are said to possess **heterosis.** This greater vigor or productivity of crossbreds is also said to be an expression of **hybrid vigor.**

The effects of dominance on the expression of traits that are important in livestock production (production traits like fertility, milk production, growth

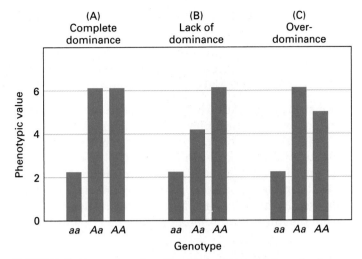

FIGURE 12.14 Bar graphs illustrating: (A) complete dominance; (B) lack of dominance; (C) overdominance.

rate, feed conversion efficiency, carcass merit, and freedom from inherited defects) can be illustrated by bar graphs (Fig. 12.14). Figure 12.14A, which illustrates complete dominance, shows equal performance of the homozygous-dominant and heterozygous individuals, both of which are quite superior to the performance of the recessive individual.

With lack of dominance, one homozygote is superior, the heterozygote is intermediate, and the other homozygote is inferior (Fig. 12.14B). When genes that show the effects of overdominance control traits, the heterozygote is superior to either of the homozygous types (Fig. 12.14C).

Another form of gene dominance is referred to as **partial dominance.** The heterozygote expresses a phenotype that is intermediate to either homozygote but more closely resembles the expression of the homozygous domiant pairing. Hyperkalemic periodic paralysis (HYPP) in horses is representative of partial gene dominance. HYPP causes erratic episodes of muscular tremor that ranges from trembling to acute paralysis.

HYPP was spread broadly in halter horses and stockseat horses via selection for increased muscularity and favoring a specific bloodline that corrects the deficit. The gene has been difficult to remove from the population because it is not completely lethal and because carriers of the gene tend to be heavily muscled—a trait that is favored by breeders of halter and stock horses.

A gene or a pair of genes in one pair of chromosomes may alter or mask the expression of genes in other chromosomes. A gene that interacts with other genes that are not allelic to it is said to be epistatic to them. In other cases, a gene or one pair of genes in one pair of chromosomes may influence many other genes in many pairs of chromosomes.

Several coat colors in horses are due to epistatic interactions. For example, the two basic coat colors in horses are black and chestnut. The recessive gene *e* produces the chestnut color; thus, *EE* and *Ee* are black, while *ee* is chestnut. Some horses possess a gene for white color that masks all other genes for color that exist in the genotype. This gene is called the dominant white gene *W*.

The recessive gene *w* allows other color genes in the genotype to express themselves. Thus, the genotypes and phenotypes are:

EEWW or *EEWw*: white	*EEww*: black
EeWW or *EeWw*: white	*Eeww*: black
eeWW or *eeWw*: white	*eeww*: chestnut

Epistasis can result from dominant or recessive genes or a combination of both. For example, in White Rock chickens, *C* = color, *c* = albino, *W* = color, and *w* = white. Homozygous *cc* prevents *W* from showing, and *ww* prevents *CC* from showing. In this example, the mating *CcWw* × *CcWw* would result in only two genotypes (colored or white). This contrasts with four phenotypes if *C* and *W* showed complete dominance, or nine phenotypes if there were no dominance expressed in the two pairs of genes. Thus, epistasis changes the phenotypic ratio, but the number of different genotypes remains the same.

INTERACTIONS BETWEEN GENES AND ENVIRONMENT

Genes interact with both external and internal environments. The external environment includes temperature, light, altitude, humidity, disease, and feed supply. Some breeds of cattle (Brahman) can withstand high temperatures and humidity better than others. Some breeds (Scottish Highland) can withstand the rigors of extreme cold better than others.

Perhaps the most important external environmental factor is feed supply. Some breeds or types of cattle can survive when feed is in short supply for considerable periods of time, and they may consume almost anything that can be eaten. Other breeds of cattle select only highly palatable feeds, and these animals have poor production when good feed is not available.

Allelic, epistatic, and environmental interactions all influence the degree to which genetic improvement can be made through selection. When the external environment has a large effect on production traits, genetic improvement is quite low. For example, if animals in a population are maintained at different nutritional levels, those that are best fed obviously grow faster. In such a case, much of the difference in growth may be due to the nutritional status of the animals rather than to differences in their genetic composition. The producer has two alternatives: (1) standardize the environment so that it causes less variation among the animals, or (2) maximize the expression of the production trait by improving the environment. The first approach is geared to increasing genetic improvement so that selection is more effective. Improvements made via this approach are permanent. The latter approach does not improve the genetic qualities of animals, but it allows animals to express their genetic potentials. The most sensible approach is to expose the breeding animals to an environment similar to one in which commercial animals are expected to perform economically.

BIOTECHNOLOGY

Biotechnology is defined as the use of living organisms to improve, modify, or produce industrially important products or processes. Microorganisms, for instance, have been used for centuries in the production of food and fermented substances, such as some dairy products. **Genetic engineering** of animals by selection and hybridization was first implemented at the turn of the century. Superovulation, sexing of semen, embryo splitting (cloning), embryo transfer (discussed in Chapter 11), identification of genetic markers, and genetic therapy are some of the more recent advances in the field of biotechnology.

During the past several years, research has provided the tools to identify and manipulate DNA genes at the molecular and cellular levels. These genetic-engineering methods, which are used to alter hereditary traits, have created a renewed excitement in biotechnology. The American Association for the Advancement of Science ranks genetic engineering as among the four major scientific advancements of the twentieth century, similar in significance to unlocking the atom, escaping the earth's gravity, and the computer revolution.

Applications

Enzymes are used as "genetic scissors" to cut the DNA at designated places. These genetic components can then be reconstructed into unique combinations not easily achieved with natural selection. In addition to altering existing genes, synthetic genes can be constructed.

Genes can be injected from other animals of the same or different species (Fig. 12.15). The technology mimics nature in that genes can be moved around within and occasionally between species by some types of virus infections. The manipulation of the genetic material does not occur at the whole animal level, but is performed **in vitro.** The genes are then either inserted into embryos that are transferred to recipient animals or inserted into cells that are grafted into living animals.

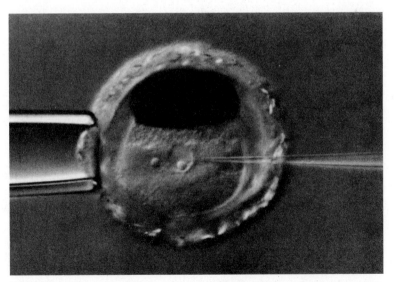

FIGURE 12.15 A fertilized swine egg photographed at the moment it is microinjected with new genetic material. The vacuum in the large pipette at the bottom anchors the cell while a mixture containing the genetic material is forced through the smaller pipette into one of the egg's pronuclei. Courtesy of R. E. Hammer and R. L. Brinster, University of Pennsylvania School of Veterinary Medicine.

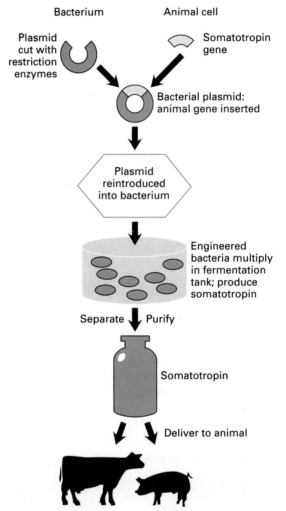

FIGURE 12.16   Somatotropin production for use in cows and pigs.

Bovine somatotropin (BST), a bovine growth hormone, is a natural protein produced in the pituitary glands of cattle. It helps direct the energy in feed to meet the animal's needs for growth and milk production. BST can be produced outside the animal's body by the genetic-engineering process illustrated in Figure 12.16. Providing supplemental BST to the dairy cow increases feed consumption and milk production. BST directs more of the additional feed toward increasing milk production than to body maintenance.

Porcine somatotropin (PST) has also been produced by a process similar to that used for BST (Fig. 12.16). Research has shown that PST given as a supplement increases litter size, reduces the time needed for pigs to reach market weight, improves feed efficiency, and reduces carcass fat production.

Table 12.3 illustrates the advancement of cloning technologies applied to cattle. The application of cloning and other biotechniques raises practical, legal, and ethical questions. In the short run, cloning is not likely to gain widespread utilization due to the low efficiency of gaining a pregnancy (< 25%), increased birth weights of clones, and a cost of $15,000 or more per clone.

TABLE 12.3 Progression of Events Involving the Cloning of Cattle

Date	Event
1986	First bovine produced via nuclear transfer using embryo blastomeres.
1993	First bovine cloned by nuclear transfer using cells from inner cell mass.
1993	Bovine fetuses produced by nuclear transfer using embryonic stem cells.
1997	First bovine clone born by nuclear transfer using nonembryo derived cells.
2000	A clone of the Holstein cow—Lauduc Broker Mandy—sold at auction for $82,000.
2001	Cloned Holstein 2-year-old heifers successfully deliver offspring.
2002	First successful clone of the Guernsey cow—Westlynn Tom Dee 'EX-96'—the highest scored living female in the breed.

FIGURE 12.17 The normal-appearing boar is a transgenic pig. He received a growth gene (from both mouse and cattle origin) by the process illustrated in Figure 12.15. Courtesy of R. E. Hammer and R. L. Brinster, University of Pennsylvania School of Veterinary Medicine.

Figure 12.17 shows a **transgenic** pig. While normal in appearance, the boar contains a new gene composed of the mouse metallothionien promotor/regulator fused to the bovine growth hormone structural gene. This gene, transmitted by the boar in Figure 12.17, causes elevated levels of growth hormone in the blood and results in a 10–15% increase in growth rate. The transgenic pigs are also more efficient in feed conversion to body weight and have significantly less body fat.

Nuclear fusion is the union of nuclei from two sex cells (sperm and eggs). In terms of biotechnology, this means not only can nuclei from a sperm and egg be united but also that the union of nuclei from two eggs or two sperm can occur. The union of nuclei from two females results in all female offspring, while the union of two sperm nuclei produces ½ females *(XX)*, ¼ surviving males *(XY)*, and ¼ lethal males *(YY)*. Thus, nuclei from two outstanding males (or females) can be united and then transferred to a recipient female. This has been accomplished in laboratory animals but not in larger, domestic farm animals. Even self-fertilization, once considered possible only in plants, may become a choice in genetically engineered farm animals.

Gene therapy involves inserting genes into a patient's cells to treat or cure certain diseases. In September 1990, the first federally approved gene therapy for humans took place. A girl with a hereditary disorder of the immune system had some of her white blood cells removed, genetically altered, then transfused back into her bloodstream. The genetically altered cells were to produce a needed enzyme. Early results of the therapy appear promising. Similar approaches have been used to treat advanced melanoma (skin cancer), immune deficiency disorders, and other diseases.

Future Expectations and Concerns

Future advances in biotechnology through genetic manipulation appear to be mind-boggling and ethically challenging. Many of these changes will take years to accomplish because genetic engineering tasks are complicated and expensive. The present state of the art has a high number of failures. Research in the following areas gives insight into future possibilities.

1. Identification of the **genome** (all the genes) of domestic animals creates unique opportunities in terms of better selection of breeding stock, improvement of animal health, and improving the nutritional value of animal products to name a few. This process allows the identification of genetic markers or regions of the chromosome that are responsible for significant levels of genetic variation within a trait.

2. Animal and plant productivity (i.e., growth, changing the fat-to-lean ratio, milk production) could be significantly increased and the cost of animal production could be reduced. The National Center for Food and Agriculture Policy evaluated six biotech crops in the United States and found that they yielded an additional 4 billion pounds from the same acreage, increased farm income by $1.5 billion, and cut pesticide use by 46 million pounds as compared to conventional plants.

3. Animals with resistance to specific disease could be genetically engineered. Feedstuffs can also be developed with the capability to provide vaccines against disease.

4. Parasites could be controlled by genetic interference with their immune systems.

5. Domestic and international markets could be expanded through new genetically engineered products. Furthermore, biotechnology may hold the key to preventing widespread hunger in the world.

6. Development of rapid, accurate tests for identifying food contaminants could potentially allow food inspection to move from visual assessment to a sensitive system able to detect specific pathogens or substances. The use of DNA probes and biosensors such as enzymes, antibodies, or microbial cells is currently under evaluation.

7. Diseases may be controlled via the ability to turn associated genes off and on. "Smart" drugs may be developed that can distinguish between healthy and cancerous cells and target only those that are diseased.

Future advances in biotechnology will likely bring many significant genetic changes and challenges that will benefit both humans and domestic animals.

However, genetic engineering of animals raises questions of ethics, food-product safety, and environmental safety. These issues are discussed in more detail in Chapter 23.

CHAPTER SUMMARY

- Cells contain several pairs of chromosomes that are composed of deoxyribonucleic acid (DNA). Genes are segments of DNA that are the basic units of inheritance.

- Two chromosomes (*X* and *Y*) determine the sex of animals. In livestock, *XX* is female while *XY* is male. In poultry, *XX* is male and *XY* is female. The parent carrying the *XY* chromosome determines the sex of the offspring.

- Genetic variation and genetic change occur through chromosome (gene) segregation and chromosome (gene) recombination resulting from sex cell formation and fertilization.

- Genes produce their effects through dominance, overdominance (heterosis), lack of dominance (additive), or epistasis (interactions between pairs of genes).

- Biotechnology and genetic engineering are bringing changes and challenges to both humans and domestic animals.

KEY WORDS

gametogenesis
spermatogenesis
oogenesis
meiosis
seminiferous tubules
synapsis
primary spermatocyte
primary oocyte
zygote
fertilization
diploid
haploid
genes
homologous
deoxyribonucleic
  acid (DNA)

locus
nucleotide
polymer
transcription
translation
ribonucleic acid (RNA)
transfer RNA
messenger RNA
ribosomal RNA
codon
ribosome
homozygous
heterozygous
alleles
dominant

recessive
interaction (linear,
  allelic, epistatic)
heterosis
hybrid vigor
partial dominance
biotechnology
genetic engineering
in vitro
transgenic
nuclear fusion
genome

REVIEW QUESTIONS

1. The functional units of inheritance called _____, are located on rod-like bodies called _____ located within the nucleus of each cell.

2. Cell division that yields two daughter cells that have identical chromosomes as compared to the parent cell is termed _____.

3. What is the term for production of sex cells or gametes? Specifically, what is this process called for the production of ova? For the production of sperm?

4. During gametogenesis, what is the unique type of cell division called that reduces the chromosome number of gametes to one-half of that possessed by all other body cells?

5. Where does meiosis occur in the testicle? In the ovary?

6. What is the union of the sperm and the ovum called?

7. What is a newly fertilized ovum called?

8. *True or False:* A zygote has the haploid number of chromosomes.

9. *True or False:* All cells in the body except gametes have the diploid number of chromosomes.

10. Within all cells except gametes, chromosomes are paired, being alike in size and shape and carrying genes that affect the same hereditary traits. Paired chromosomes are said to be _____.

11. What is the chemical component of genes?

12. DNA carries the code or instructions for control of enzymes and _____.

13. The DNA code of a gene is "read" by what process?

14. Genes are transcribed to produce what type of chemical compound?

15. Through what process does the information in RNA transcripts result in protein synthesis?

16. What is the location of a gene on a chromosome called?

17. What is the term for a pair of genes that occupy corresponding loci on homologous chromosomes?

18. An individual possessing alleles for a given trait that are identical is said to be _____. When the alleles for the trait are different, the individual is said to be _____.

19. When a pair of genes or alleles for a trait differ, and the individual is heterozygous for that trait, one of the genes in the pair is expressed and often overpowers the expression of the other gene. What is the term used to describe the gene that is expressed? To describe the gene whose expression is prevented?

20. What is the actual gene makeup for a trait called? What is the expression of the gene makeup or appearance of the individual called?

21. Crossbreeding animals of different breeds results in offspring with _____ or hybrid vigor, which means that the offspring are more vigorous than the parents.

SELECTED REFERENCES

Acquaah, G. 2004. *Understanding Biotechnology—An Integrated and Cyber-based Approach.* Upper Saddle River, NJ: Pearson Prentice Hall.

American Health Institute. 1988. *Bovine Somatotropin (BST).* Alexandria, VA: AHI.

Bailey, E. 2001. The Equine Genome. *Equus* 260:51.

Berkowitz, D. B. 1994. Transgenic animals. *Encyclopedia of Agricultural Science*. San Diego, CA: Academic Press, Inc.

Bourdon, R. M. 2000. *Understanding Animal Breeding*. Upper Saddle River, NJ: Prentice Hall, Inc.

Cundiff, L. V., L. D. Van Vleck, L. D. Young, and G. D. Dickerson. 1994. Animal breeding and genetics. *Encyclopedia of Agricultural Science*. San Diego, CA: Academic Press, Inc.

Genetic Engineering: A Natural Science. St. Louis, MO: Monsanto Company.

Genetically Modified Livestock: Progress, prospects, and issues. 1993. *J. Anim. Sci.* 71 (Supplement 3).

Green, R. D. 1999. DNA + EPDs = Market Assisted Selection. Proceedings of Cattlemen's College. Charlotte, NC.

Hartl, D. L., and E. W. Jones. 2001. *Genetics—Analysis of Genes and Genomes*. Sudbury, MA: Jones and Bartlett Publishers.

Legates, J. E. 1990. *Breeding and Improvement of Farm Animals*. New York: McGraw-Hill.

McGarvey, R. 2002. Biotech breakthroughs. *Harvard Business Review*, August: 51–59.

Petters, R. M. 1986. Recombinant DNA, gene transfer and the future of animal agriculture. *J. Anim. Sci.* 62:1759.

CHAPTER 13

Genetic Change Through Selection

LEARNING OBJECTIVES

- Explain the concept of genetic variation
- Compare and contrast qualitative and quantitative traits
- Describe the genetic model and its components
- Calculate adjusted 205-day weights
- Calculate the rate of genetic change
- Outline evidence of genetic change in the livestock industry
- Describe the various approaches to selection and the role of selection tools in genetic improvement

CONTINUOUS VARIATION AND MANY PAIRS OF GENES

Most economically important traits in farm animals, such as milk production, egg production, growth rate, and carcass composition, are controlled by hundreds of pairs of genes; therefore, it is necessary to expand one's thinking beyond inheritance involving one and two pairs of genes. The exact number of total genes in a species is not known, although estimates of more than 100,000 genes in livestock have been made.

Consider even a simplified example of 20 pairs of heterozygous genes (1 gene pair on each pair of 20 chromosomes) affecting yearling weight in sheep, cattle, or horses. The estimated numbers of genetically different gametes (sperm or eggs) and genetic combinations are shown in Table 13.1. Remember that for one pair of heterozygous genes, there are three different genetic combinations (i.e., *AA, Aa,* and *aa*), and for two pairs of heterozygous genes, there are nine different genetic combinations (Chapter 12, Table 12.2).

TABLE 13.1 Number of Gametes and Genetic Combinations with Varying Numbers of Heterozygous Gene Pairs

No. of Pairs of Heterozygous Genes	No. of Genetically Different Sperm or Eggs	No. of Different Genetic Combinations (genotypes)
1	2	3
2	4	9
n	2^n	3^n
20	$2^n = 2^{20} = $ ~1 million	$3^n = 3^{20} = $ ~3.5 billion

TABLE 13.2 Number of Gametes and Genetic Combinations with Eight Pairs of Genes with Varying Amounts of Heterozygosity and Homozygosity

Genes in Sire:	Aa	Bb	Cc	Dd	Ee	FF	GG	Hh	
Genes in Dam:	aa	Bb	CC	Dd	Ee	FF	gg	Hh	Total
No. different sperm for sire	2 ×	2 ×	2 ×	2 ×	2 ×	1 ×	1 ×	2 =	64
No. different eggs for dam	1 ×	2 ×	1 ×	2 ×	2 ×	1 ×	1 ×	2 =	16
No. different genetic combinations possible in offspring	2 ×	3 ×	2 ×	3 ×	3 ×	1 ×	1 ×	3 =	324

Most farm animals are likely to have some heterozygous and some homozygous gene pairs, depending on the mating system being utilized. Table 13.2 shows the number of gametes and genotypes where eight pairs of genes are either heterozygous or homozygous and each gene pair is located on a different pair of chromosomes.

Many economically important traits in farm animals show continuous variation primarily because many pairs of genes control them. As these many genes express them, and the environment also influences these traits, producers usually observe and measure large differences in the performance of animals for any one trait. For example, if a large number of calves were weighed at weaning (~205 days of age) in a single herd, there would be considerable variation in the calves' weights. Distribution of weaning weights of the calves would be similar to the examples shown in Figures 13.1 or 13.2. The bell-shaped curve distribution demonstrates that most of the calves are near the average for all calves, with relatively few calves having extremely high or low weaning weights when compared at the same age.

Figure 13.2 shows how the statistical measurement of standard deviation (SD) is used to describe the variation of differences in a herd where the average weaning weight is 440 lb and the calculated standard deviation is 40 lb. Using herd average and standard deviation, the variation in weaning weight is shown in Figure 13.2 and can be described as follows:

1 SD:	440 lb ± 1 SD (40 lb) = 400 – 480 lb (68% of the calves are in this range)
2 SD:	440 lb ± 2 SD (80 lb) = 360 – 520 lb (95% of the calves are in this range)
3 SD:	440 lb ± 3 SD (120 lb) = 320 – 560 lb (99% of the calves are in this range)

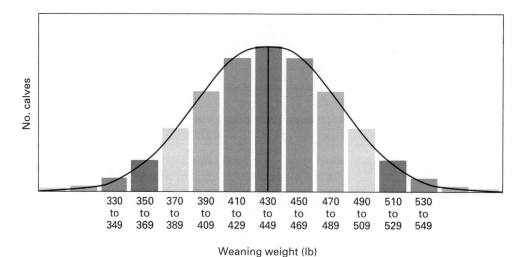

FIGURE 13.1 Variation or difference in weaning weight in beef cattle. The variation shown by the bell-shaped curve could be representative of a breed or a large herd. The dark vertical line in the center is the average or the mean—in this example, 440 lb.

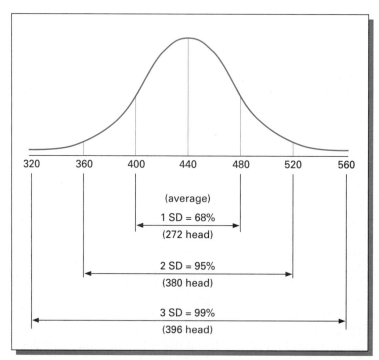

FIGURE 13.2 A normal bell-shaped curve for weaning weight showing the number of calves in the area under the curve (400 calves in the herd).

One percent of the calves (4 calves in a herd of 400) would be on either side of the 320–560-lb range. Most likely, two calves would be below 320 lb and two calves would weigh more than 560 lb.

Animals have multiple traits that can be measured or described. **Quantitative traits** are those that can be objectively measured, and the observations typically exist along a continuum. Examples include growth traits, skeletal size, speed, and

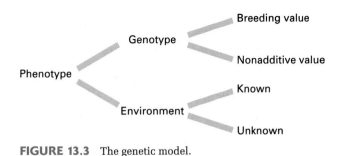

FIGURE 13.3 The genetic model.

TABLE 13.3 **Typical Phenotypic Means and Standard Deviations for a Sample of Traits from Farm Animals**

Species	Trait	Mean	Standard Deviation
Cattle (beef)	Birth weight	80 lb	10 lb
	Yearling weight (bulls)	950 lb	60 lb
	Mature weight	1,100 lb	85 lb
	Backfat thickness (steers)	0.4 in	0.1 in
Cattle (dairy)	Calving interval	404 days	75 days
	Milk yield	13,000 lb	560 lb
Horses	Wither height (mature)	60 in	1.8 in
	Time to run ¼ mile	20 seconds	0.6 seconds
	Time to run 1 mile	96 seconds	1.3 seconds
	Cutting score	209 points	10.3 points
Swine	Litter size (# born alive)	9.8 pigs	2.8 pigs
	Days to 230 lb	175 days	12 days
	Loineye area	4.3 in²	0.25 in²
Poultry	Hatchability (chickens)	90%	2.2%
	Egg weight (layers)	0.13 lb	0.01 lb
	Feed conversion ratio (broilers)	5.40 lb/lb	0.88 lb
	Breast weight (broilers)	0.64 lb	0.007 lb
Sheep	60-day weaning weight	45 lb	8 lb
	Grease fleece weight	8 lb	1.1 lb
	Staple length	2.5 in	0.5 lb

Source: Adapted from Bourdon, R. M. 2000.

others. **Qualitative traits** are descriptively or subjectively measured and would include hair color, horned versus polled, and so forth. Many gene pairs control quantitative traits, while few, if not just one, gene pairs control qualitative traits.

The observation or measurement of each trait is referred to as the **phenotype.** Phenotypic variation exists within the trait due to two primary sources of influence—genotype and environment (Fig. 13.3). Table 13.3 shows the typical phenotypic means and standard deviations for a sample of traits from the primary livestock species.

Weaning weight is a phenotype, since the expression of this characteristic is determined by the genotype (genes received from the sire and the dam) and the environment to which the calf is exposed. The genetic part of the expression of weaning weight is obviously not simply inherited. There are many pairs of genes involved, and, at the present time, the individual pairs of genes cannot be identified similarly to traits controlled by one to two pair of genes.

Genotype is the result of both the cumulative effects of the animal's individual genes for the trait and the effect of the gene combinations. **Environmental effects** can be thought of as the summation of all nongenetic influences.

Breeding value or parental worth for a trait can be defined as that portion of genotype that can be transferred from parent to offspring. **Breeding value** in mathematical terms is the total of all the independent genetic effects in an animal on a given trait. **Nonadditive value** is that portion of genotype that is attributed to the gene combinations unique to a particular animal. Because genetic combinations are reestablished in each successive graduation, nonadditive value does not pass from generation to generation. Therefore, breeding value responds to selection while nonadditive value is accessed via choice of mating system (i.e., cross breeding).

There are two basic types of environmental effects—known and unknown. **Known effects** have an average effect on individuals in a specific category. Examples include age, age of dam, and gender. Calves born earlier in the calving season and those older at time of weaning weigh more than their younger counterparts. **Unknown effects** are random in nature and are specific to an individual phenotype. Known effects can be quantified and used to adjust phenotypic measures to allow more accurate selection. Unknown effects are more difficult to account for, but breeders can use management to minimize their impacts.

One attempt to remove the effects of environmental influence is the use of adjusted records. For example, weaning weight can be adjusted to account for age of calf and age of dam. The formula used to compute an adjusted weaning weight or 205-day weight for beef cattle is:

$$\left[\left(\frac{\text{Actual Wean Weight} - \text{Birth Weight}}{\text{Age in Days at Weaning}}\right) \times 205\right] + \text{Birth Weight} + \text{Age-of-Dam Adjustment}$$

Assume that an adjusted weaning weight needs to be calculated for the following pair of bull calves:

	Bull A	Bull B
Actual weaning weight	550 lb	520 lb
Age at weaning	230 days	190 days
Birth weight	82 lb	75 lb
Age of dam	3 years	6 years

Use adjustments listed below.

Age of Dam (years)	Bull Calf (lb)	Heifer Calf (lb)
2	+60	+54
3	+40	+36
4	+20	+18
5–10	+0	+0
>10	+20	+18

Adjusted weaning weight for Bull A:

$$\left[\left(\frac{550 - 82}{230}\right) \times 205\right] + 82 + 40 = 539 \text{ lb}$$

Adjusted weaning weight for Bull B:

$$\left[\left(\frac{550 - 75}{190}\right) \times 205\right] + 75 + 0 = 555 \text{ lb}$$

The interpretation of these solutions is that if bulls A and B had been weaned at 205 days of age, and if their dams had been mature cows between the ages of 5 and 10, then their weights would be estimated at 539 and 555 pounds, respectively.

The phenotype will more closely predict the genotype if producers expose their animals to a similar environment; however, the latter must be within economic reason. This resemblance between phenotype and genotype, where many pairs of genes are involved, is predicted with the estimate of heritability. Use of **heritability,** along with selecting animals with superior phenotypes, is the primary method of making genetic improvement in traits controlled by many pairs of genes. The application of this method is discussed in more detail later in the chapter.

Traits influenced little by the environment can also show considerable variation. An example is the white belt, a breed-identifying characteristic in Hampshire swine (Fig. 13.4). Note that the white belt can be nonexistent in some Hampshires, whereas others can be almost completely white. To be eligible for pedigree registration, Hampshire pigs must be black with the white belt entirely circling the body, including both front legs and feet. Too much white can also limit registration. It is difficult to select a herd of purebred Hampshires that will breed true for desired belt pattern. Apparently, some of the genes for this trait exist in heterozygous or epistatic combinations.

FIGURE 13.4   Variation in belt pattern in Hampshire swine. Courtesy of National Swine Registry.

In a beef herd where the average weaning weight is 440 lb, or a dairy herd where annual milk production averages 12,500 lb per cow, the herd managers may want to increase the herd average. This assumes the increased production is economically feasible—that the economic benefits of improvement will cover the additional costs.

Selection

Selection is differential reproduction—preventing some animals from reproducing while allowing other animals to become parents of numerous offspring. In the latter situation, the selected parents should be genetically superior for the economically important traits. Factors affecting the rate of genetic improvement from selection include selection differential, heritability, and generation interval.

Selection Differential

Selection differential, sometimes called **reach,** is the superiority (or inferiority) of the selected animals compared to the herd average. To improve weaning weights in beef cattle, producers cull as many below-average-producing cows as economically feasible. Then replacement heifers and bulls are selected that are above the herd average for weaning weight. For example, if the average weaning weight of the selected replacement heifers is 480 lb in a herd averaging 440 lb, then the selection differential for the heifers would be 40 lb. Part of this 40-lb difference is due to genetic differences, and the remaining part is due to differences caused by the environment.

Heritability

A *heritability* estimate describes the percent of total phenotypic variation (phenotypic differences) that is due to breeding value. Heritability can also be defined as that portion of the selection differential that is passed from parent to offspring. If the parents' performance is a good estimate of progeny performance for that trait, then the heritability is said to be high.

Realized heritability is the portion actually obtained compared to what was attempted in selection. To illustrate realized heritability, let us suppose a farmer has a herd of pigs whose average postweaning gain is 1.80 lb/day. If the farmer selects from this original herd a breeding herd whose members have an average gain of 2.3 lb/day, the farmer is selecting for an increased daily gain of 0.5 lb/day.

If the offspring of the selected animals gain 1.95 lb, then an increase of 0.15 lb (1.95 − 1.80) instead of 0.5 lb (2.3 − 1.8) has been obtained. To find the portion obtained of what was reached for in the selection, 0.15 is divided by 0.5, which gives 0.3. This figure is the realized heritability. If 0.3 is multiplied by 100%, the result is 30%, which is the percentage obtained of what was selected for in this generation.

Table 13.4 shows heritability estimates for several species of livestock. Traits having heritability estimates of 40% and higher are considered highly heritable. Those with estimates 20–39% are classified as having medium heritability; and low-heritability traits have heritability estimates below 20%. Heritability is a measure determined by studying populations and, as such, is not specific to individuals. Furthermore, heritability may differ from breed to breed and environment to environment.

TABLE 13.4 Heritability Estimates for Selected Traits in Several Species of Farm Animals

Species Trait	Percent Heritability	Species Trait	Percent Heritability
Beef Cattle		Jumping (earnings)	20
Age at puberty	40%	Dressage (earnings)	20%
Weight at puberty	50	Cutting ability	5
Scrotal circumference	50	Thoroughbred racing	
Birth weight	40	Log earnings	50
Gestation length	40	Time	15
Body condition score	40	Pacer: best time	15
Calving interval	10	Trotter	
Percent calf crop	10	Log earnings	40
Weaning weight	30	Time	30
Postweaning gain	45	**Poultry**	
Yearling weight	40	Age at sexual maturity	35
Yearling hip (frame size)	40	Total egg production	25
Mature weight	50	Egg weight	50
Carcass quality grade	40	Broiler weight	40
Yield grade	30	Mature weight	50
Tenderness of meat	50	Egg hatchability	10
Longevity	20	Livability	10
Dairy Cattle		**Sheep**	
Services per conception	5	Multiple births	20
Birth weight	40	Birth weight	30
Milk production	25	Weaning weight	30
Fat production	25	Postweaning gain	40
Protein	25	Mature weight	50
Solids-not-fat	25	Fleece weight	40
Type score	30	Face covering	50
Feet and leg score	10	Loineye area	40
Teat placement	30	Carcass fat thickness	50
Udder score	20	Weight of retail cuts	40
Mastitis (susceptibility)	10	**Swine**	
Milking speed	20	Litter size	10
Mature weight	50	Birth weight	10
Excitability	25	Litter weaning weight	15
Goats		Postweaning gain	30
Milk production	30	Feed efficiency	30
Mohair production	20	Backfat (live animal)	40
Horses		Carcass fat thickness	50
Withers height	45	Loineye area	50
Pulling power	25	Percent lean cuts	45
Riding performance			

PREDICTING GENETIC CHANGE

The rate of **genetic change** that can be made through selection can be estimated using the following equation:

$$\text{Genetic change per year} = \frac{\text{heritability} \times \text{selection differential}}{\text{generation interval}}$$

Consider the following example of selecting for weaning weight in beef cattle and predicting the genetic change. The herd average is 440 lb and bulls and heifers are selected from within the herd. The selection differential measures the phenotypic superiority for inferiority of the selected animals compared to the herd average—e.g., bulls (535 lb − 440 lb = 95 lb); heifers (480 lb − 440 lb = 40 lb).

Heritability times selection differential measures the genetic superiority of the selected animals compared to the herd average.

Bulls	Weight
Average of selected bulls	535 lb
Average of all bulls in herd	440 lb
Selection differential	95 lb
Heritability	0.30
Total genetic superiority	28 lb
Only half passed on (28 ÷ 2)	14 lb

Heifers	Weight
Average of selected heifers	480 lb
Average of all heifers in herd	440 lb
Selection differential	40 lb
Heritability	0.30
Total genetic superiority	12 lb
Only half passed on (12 ÷ 2)	6 lb

Fertilization combines the genetic superiority of both parents, which in this example is 14 lb + 6 lb = 20 lb. The calves obtain half of their genes from each parent; therefore $\frac{1}{2}(28) + \frac{1}{2}(12) = 20$ lb.

The selected heifers represent only approximately 20% of the total cowherd, so the 20 lb is for one generation. Since the heifer replacement rate in the cowherd is 20%, it would take 5 years to replace the cowherd with selected heifers. Because of this generation interval, the genetic change per year from selection would be 20 lb ÷ 5 = 4 lb/year.

Generation Interval

This trait is defined as the average age of the parents when the offspring are born. **Generation interval** is calculated by adding the average age of all breeding females to the average age of all breeding males and dividing by 2. The generation interval is approximately 2 years in swine, 3–4 years in dairy cattle, and 5–6 years in beef cattle. When the rapid speed of genetic change in the poultry industry is evaluated, the rapid generation turnover is a major factor.

Genetic Change for Multiple Trait Selection

The previous example of 4 lb/year shows genetic change if selection was for only one trait. If selection is practiced for more than one trait, genetic change is $1\sqrt{n}$, where n is the number of traits in the selection program. If four traits were in the selection program, the genetic change per trait would be $1/\sqrt{4} = \frac{1}{2}$. This means that only half the progress would be made for any one trait compared to giving all the selection to one trait. This reduction in genetic change per trait should not discourage producers from multiple-trait selection, as herd income is dependent on several traits. Maximizing genetic progress in a single trait may not be economically feasible, and it may lower productivity in other economically important traits.

Sometimes traits are genetically correlated, meaning that some of the same genes affect both traits. In swine, the genetic correlation between rate of gain and feed per pound of gain (from similar beginning weights to similar end weights) is negative. This is desirable from a genetic improvement standpoint because animals that gain faster require less feed (primarily less feed for maintenance). This relationship is also desirable because rate of gain is easily measured, whereas feed efficiency is expensive to measure.

Yearling weight or mature weight in cattle is positively correlated with birth weight, which means that as yearling or mature weight increases, birth weight also increases. This weight increase may pose a potential problem, since birth weight to a large extent reflects calving difficulty and calf death loss.

EVIDENCE OF GENETIC CHANGE

The previously shown examples of selection are theoretical. Does selection really work, or is the observed improvement in farm animals a result of improving only the environment? Let's evaluate several examples.

There have been marked, visual meat-to-bone ratio changes in the thick-breasted modern turkey selected from the narrow-breasted wild turkey, which is the only animal native to the United States. Estimates for genetic change for meat-to-bone ratio have been 0.5% per year for the modern turkey. Genetic selection and improving the environment, particularly through improved nutrition and health, have produced the modern turkey. Tom turkeys can produce 27 lb of live weight (22 lb dressed weight) in 5 months; the wild turkey weighs 10 lb in 6 months.

The modern turkey needs an environment with intensive management and is not likely to survive in the same environment as the wild turkey. For example, turkeys are mated artificially because the heavy muscling in the breast prevents them from mating naturally. Selection has been effective in changing rate of growth and meatiness in turkeys, but it has necessitated a change in the environment for these birds to be productive. Therefore, the improvement in turkeys has resulted from improvement in both the genetics and the environment.

There are tremendous size differences in horses. Draft horses have been selected for large body size and heavy muscling to perform work. The light horse is more moderate in size, and ponies are small. Miniature horses have genetic combinations that result in a very small size. The miniature horse is considered a novelty and a pet. Table 13.5 shows the height and weight variations in different

TABLE 13.5   **Height and Weight Differences in Mature Draft Horses, Light Horses, Ponies, and Miniature Horses**

| Horse Type | Approximate Height at Withers | | Approximate Mature Weight |
	Hands	Inches	(lb)
Draft	17	68	1,600
Light	15	60	1,100
Pony	13	52	700
Miniature	8	32	250

TABLE 13.6   Trends in U.S. Pork Production on a Per Animal Basis

Year	Pork Production (lb per breeding animal)	Live (lb per hog)	Retail Meat Yield (lb per hog)
1955	NA	237	123
1960	NA	236	124
1965	1,315	238	127
1970	1,442	240	129
1975	1,531	240	130
1980	1,704	242	133
1985	2,120	245	136
1990	2,230	249	141
1995	2,523	256	144
1997	2,622	260	145
2001	2,971	264	151
2005	3,020	269	154

Source: National Pork Producers Council.

types of horses. Apparently these different horse types have been produced from horses that originally were quite similar in size and weight. The extreme size differences in horses result primarily from genetic differences, as the size differences are apparent when the horses are given similar environmental opportunities.

The effectiveness of selection for increased growth and retail yield in swine is shown in Table 13.6. The reduction in lard production per hog has decreased dramatically, even though the average slaughter weight has increased approximately 24 lb since 1960.

In past years, the backfat probe was used to measure differences in backfat thickness in live pigs. Use of the backfat probe improved the effectiveness of selection by measuring backfat thickness in breeding animals without having to slaughter them. Today, ultrasound has replaced the backfat probe in measuring fat thickness in live animals.

Selection for milk production in dairy cattle poses an additional challenge in a genetic improvement program because the bull does not express the trait. This is an example of a **sex-limited trait.**

Genetic evaluation of a bull is based primarily on how his daughters' milk production compares to that of their **contemporaries.** Milk production is a moderately heritable trait (25%), and an increase of approximately 11,000 pounds per cow over the past 50 years is most noteworthy, even though the trait is not expressed in the bull (Table 13.7). Even when multiple traits receive the attention of selection pressure, progress can be made. Figure 13.5 documents the genetic trends for the Red Angus cattle breed over a period of nearly 50 years. In the poultry industry, disciplined selection from 1947 to 1988 enabled producers to reduce the age of marketing from 12 weeks to less than 6 weeks. Over the same time period, the feed required to raise a broiler was cut in half.

However, even in traits of moderate heritability, improvement may not be achieved. In the All-American Quarter Horse Futurity (440 yd), 5-year average winning times during the 20-year period from 1986 to 2005 did not show appreciable improvement (Table 13.8). In this case, biomechanical limitations make it very difficult to make horses faster.

A project conducted at Colorado State University involved mating a group of commercial Hereford cows to Hereford bulls representative of the type favored in the 1950s, 1970s, and 1990s. The performance comparison of the

TABLE 13.7 Changes in Milk Production in the United States, 1940–2004

Year	No. Cows (mil)	Average Milk Per Cow (lb)	Total Milk (bil lb)
1940	23.7	4,622	109.4
1950	21.9	5,314	116.6
1960	17.5	7,029	123.1
1970	12.0	9,751	117.0
1980	10.8	11,875	128.4
1990	10.1	14,645	148.3
1995	9.5	16,443	155.6
1997	9.3	16,916	156.6
2000	9.2	18,204	167.6
2004	9.0	18,608	170.1

Source: USDA.

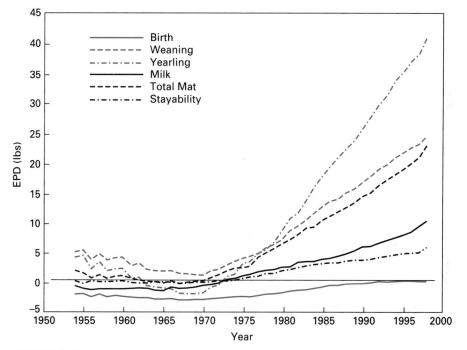

FIGURE 13.5 Genetic trends since 1954 for the six traits presented in this sire evaluation.

TABLE 13.8 Average Winning Times in the All-American Quarter Horse Futurity (440 yards)

Years	Average 5-year Winning Times
1986–1990	21.16
1991–1995	21.47
1996–2000	21.45
2001–2005	21.29

resulting progeny illustrated that breeders had been very successful in changing the growth rates of their cattle (Table 13.9). Note that birth weights increased along with growth rate and frame size. Intense selection for increased growth may lead to undesirable levels of birth weight and mature size.

TABLE 13.9 Growth of Hereford Cattle Sired by Different Generations of Bulls

Sire Generation	Birth Wt. (lb)	On-Test Wt. (lb)	Off-Test Wt. (lb)	Frame Size
1950	82.5	665	1,083	3.7
1970	85.9	717	1,158	4.9
1990	91.4	791	1,261	5.5

Source: Adapted from Tatum and Field, 1996.

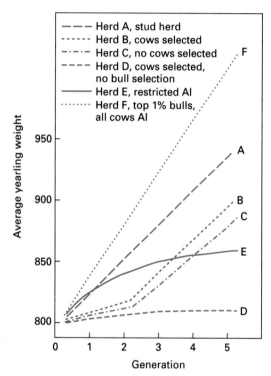

FIGURE 13.6 Expected genetic change in different types of beef herds. Adapted from Magee, *Mich. State Univ. Qtr. Bul.* 48:4.

GENETIC IMPROVEMENT THROUGH ARTIFICIAL INSEMINATION

The primary reason for using artificial insemination (AI) is that genetically superior sires can be used more extensively. AI is used more in dairy cattle than any other species of farm animals. Using AI dairy bulls has resulted in higher milk production than using natural service bulls.

Figure 13.6 shows the rate of genetic improvement for yearling weight in beef cattle under different selection schemes. The most significant points are these:

1. Marked difference between herd A (no AI, males and females selected for yearling weight) and herd D (bulls obtained from a herd where no selection is practiced for yearling weight).

2. Small difference between herd B (bulls selected from herd A on yearling weight and females selected for yearling weight) and herd C (bulls selected from herd A but no selection for yearling weight of females). Effective sire selection accounts for 80–90% of the genetic change in a herd.

3. Greatest change in herd F, where the top 1% of the sons of the top 1% of the bulls were used to build generation upon generation of AI selection. The rate of improvement using AI in herd F is 1.5 times that of herd A with natural service. The primary reasons for the greater improvement in herd F is a larger selection differential.

SELECTION METHODS

The three methods of selection are (1) tandem, (2) independent culling level, and (3) selection index. **Tandem** is selection for one trait at a time. This method can be effective if the situation calls for rapid change in a single, highly heritable trait. When the desired level is achieved in one trait, then selection is practiced for the second trait. However, the tandem method is rather ineffective if selection is for more than two traits or if the desirable aspect of one trait is associated with the undesirable aspect of another trait.

For example, in the case of dairy cattle, milk production and milk fat percentage are negatively correlated. If a producer were interested in increasing milk yield, selection pressure solely focused on that trait would eventually result in decreases in percent milk fat. If the producer then decided to practice single trait selection to increase percent milk fat, progress could be made in that trait. However, such an approach would lead to a decline in milk yield over time due to the negative correlation between the traits. Tandem selection makes it difficult to sustain progress when more than one trait is of importance. Thus, this approach is typically not recommended.

Independent culling level establishes minimum culling levels for each trait in the selection program. Even though it is the second most effective type of selection, it is the most prevalent. Independent culling level is most useful when the number of traits being considered is small and when only a small percentage of offspring is needed to replace the parents. For example, with poultry, 100 or more chicks might be produced from each hen and 1,500 or more chicks from each rooster. Here only 1 female out of 50 or 1 male out of 750 is needed to replace the parents. It is simple using this method to select from the upper 10% of the females and the upper 1% of the males.

Table 13.10 shows an example of using independent culling levels in selecting yearling bulls. Birth weight is correlated with calving ease. Yearling weight indicates rate of growth. Puberty and semen production are estimated with scrotal circumference. Any bull that does not meet the minimum or maximum level is culled. Birth weights are evaluated on a maximum level (upper limit) since high birth weights result in increased calving difficulty.

The boxed records in Table 13.10 indicate bulls that did not meet the independent culling levels. Bulls A, B, and D would be culled. A disadvantage of

TABLE 13.10 Independent Culling Level Selection in Yearling Bulls

Trait	Culling Level	Bull				
		A	B	C	D	E
Birth weight (lb)	85 (max)	105	82	85	93	76
Yearling weight (lb)	1,000 (min)	1,142	980	1,001	1,098	1,160
Scrotal circumference (cm)	30 (min)	34	37	31	29	35

TABLE 13.11 Selection Index Method of Ranking Yearling Bulls

Bull ID	Birth Weight (lb)	Yearling Weight (lb)	Index Value = YW − 5.8 (BW)	Ranking
A	105	1,142	533	3
B	82	980	504	5
C	85	1,001	508	4
D	93	1,098	559	2
E	76	1,160	719	1

this method is that it may cull a relatively superior animal for marginal performance in a single trait.

The **selection index** method recognizes the value of multiple traits and places an economic weighting on the traits of importance. Such a calculation allows an overall ranking of the animals from best to worst utilizing a highly objective approach. The selection index is the most effective system but the most difficult to develop. The disadvantages of this system include the possibility of shifts in economic value of traits over time and potential failure to identify functional defects or weaknesses.

A comparison of Tables 13.10 and 13.11 reveals several differences resulting from the application of independent culling level versus selection index. In both cases, bull E would be identified as the most desirable. Bull D is culled using independent culling level due to inadequate scrotal circumference. However, in the selection index, bull D is ranked second because scrotal measurement wasn't included. On the other hand, bull C barely escapes culling under the independent culling level system with just acceptable performance in all three traits. The selection index ranks bull C next to last.

An advantage of independent culling levels is that selection can occur during different productive stages during the animal's lifetime (e.g., at weaning). This is more cost-effective than the index method, where no culling would occur until records are recorded for all traits. For example, bull A would have been quickly culled by the independent culling level method as opposed to waiting until yearling weight and thus incurring additional cost, as would be the case with the selection index. A combination of the two methods may be most useful and cost-effective.

BASIS FOR SELECTION

Effective selection requires that the traits in question be heritable, relatively easy to measure, and associated with economic value; that genetic estimates or predictions be accurate; and that genetic variation be available. The notion of making sustained genetic progress in a herd is the basis for development of breed associations and utilization of performance data.

In the not-too-distant past, breeders depended almost entirely on visual appraisal as the basis for selection. Modern breeders have access to an amazing array of selection tools. The basis for modern selection is the availability of breeding value estimates often referred to as predicted differences or expected progeny differences.

Predicted Differences or Expected Progeny Differences (EPDs)

EPDs are calculated for a variety of traits by utilizing information on the individual, on siblings (half and full), on ancestors, and, best of all, on progeny. As more information is utilized in the calculation, the accuracy of the estimate improves.

The dairy industry has been the leader of the modern performance movement. By 1929, all 48 states had dairy cow record associations and in the mid-1930s a progeny test program for dairy sires had been initiated. Breed associations have maintained accurate pedigree records outlining the parentage of seedstock animals for several hundred years. However, only in the past 30–50 years have breed associations focused on developing performance databases for a multitude of traits.

The earliest efforts at objective across-herd comparisons utilized central tests. In these tests, young bulls, rams, or boars were brought together in a common environment to be evaluated, primarily for growth traits. These tests yield information on only a few traits and all the information is obtained from the individual. Because of these limitations, designed progeny tests were initiated to compare sires via information collected on their respective progeny. While an improvement, these designed progeny tests were relatively limited in scope and expensive to conduct.

The advent of **best linear unbiased prediction (BLUP)** techniques has allowed field data to be utilized in computation of valid breeding values that could be used to compare animals across herds. Breed associations or large seedstock companies sponsor most of these national sire or animal evaluation systems.

The poultry and dairy industries have made the best use of sophisticated genetic information. Genetic information is also widely available in the beef cattle industry. The swine industry initiated a national genetic evaluation system to evaluate maternal sow lines in 1997—a follow-up to the 1995 terminal sire line genetic evaluation program. The sheep industry genetic evaluation programs have been most extensive in Australia and New Zealand, with a focus on wool traits such as fiber diameter and fleece weight. Genetic prediction estimates for equines have received more attention in Europe than in the United States.

The utilization of performance data is one of the most profit-oriented decisions a livestock producer can make. Table 13.12 illustrates the progress of the

TABLE 13.12 Comparison of Average Per Cow Milk Production from DHIA vs. Non-DHIA Herds

Year	DHIA Herds	Non-DHIA Herds
1906	5,034	3,600
1950	9,000	5,300
1970	13,000	9,747
1990	18,031	14,782
1995	19,005	16,405
2000	20,462	17,771
2005	21,854	19,443

TABLE 13.13 Importance of Factors in Purchasing Breeding Bulls

Factor	Percent of Respondents by Level of Importance			
	Not	Moderate	Very	Extreme
Birth weight	20.3	20.0	38.0	21.7
Weaning weight/yearling weight	20.2	15.7	42.9	21.2
Hip height/frame score	14.2	27.0	42.6	16.2
Expected progeny differences	30.5	25.3	31.5	12.7
Appearance/structural soundness	2.5	3.0	43.3	51.2
Price	8.1	23.7	37.9	30.3

Source: National Animal Health Monitoring System, USDA, 1994.

dairy herds utilizing the Dairy Herd Improvement Association (DHIA) system versus those who did not. DHIA herds have had, on average, a clear productivity advantage.

Unfortunately, producers do not always accept or utilize the newest generation of genetic prediction tools. Table 13.13 illustrates that producers often tend to rely on visual appraisal or raw data rather than the more accurate EPDs that are available.

CHAPTER SUMMARY

■ Phenotype (what is seen or measured) is determined by genotype (genetic makeup and the environment to which the animal is exposed).

■ Heritability measures the proportion (0–100%) of the total phenotypic variation that is due to genetics. Traits high in heritability are ≥40%, while low-heritability traits are <20%.

■ Selection differential is the superiority (or inferiority) of the selected animals compared to the average of the group from which they came. Generation interval is the average age of the parents when the offspring are born.

■ Genetic change per year = heritability $\times \dfrac{\text{selection differential}}{\text{generation interval}}$

■ Independent culling level is the most common method of selection.

■ Expected progeny differences ought to be the basis for an effective selection program.

KEY WORDS

quantitative traits
qualitative traits
phenotype
genotype
environmental effects
breeding value
nonadditive value
known effects

unknown effects
heritability
selection
selection differential
reach
realized heritability
genetic change
generation interval

sex-limited traits
contemporaries
tandem
independent culling
  level
selection index
best linear unbiased
  prediction (BLUP)

REVIEW QUESTIONS

1. Phenotype = genotype + _____.
2. Phenotypic variation that is due to differences in an animal's genotype is called _____.
3. What is the term for superiority (or inferiority) of animals selected to be parents compared to the average of the herd?
4. The amount of genetic change made in a herd of animals during a year can be predicted by knowing what three variables?
5. What is the term for the average age of parents when their offspring are born?
6. What is a sex-limited trait?
7. Why are EPDs considered to be the most useful selection tools?
8. Which selection procedure involves selection for only one trait at a time and is least effective?
9. Which selection procedure involves ranking each individual animal against all others in the herd for the traits being selected and then obtaining a total score of the ranks for each animal, and is most effective?
10. What are the limitations to making selection decisions based on visual appraisal?

SELECTED REFERENCES

Bourdon, R. M. 2000. *Understanding Animal Breeding.* Upper Saddle River, NJ: Prentice Hall, Inc.

Bowling, A. T. 1996. *Horse Genetics.* Center for Agriculture and Biosciences International. Cambridge, UK: Cambridge University Press.

Cundiff, L. V., L. D. Van Vleck, L. D. Young, and G. D. Dickerson. 1994. Animal breeding and genetics. *Encyclopedia of Agricultural Science.* San Diego, CA: Academic Press, Inc.

Freeman, A. E., and G. L. Lindberg. 1993. Challenges to dairy management: Genetic considerations. *J. Dairy Sci.* 76:3143.

Genetics and Goat Breeding. Proceedings of the Third International Conference on Goat Production and Disease. 1982. *Dairy Goat Journal.* Scottsdale, AZ.

Hetzer, H. O., and W. R. Harvey. 1967. Selection for high and low fatness in swine. *J. Anim. Sci.* 26:1244.

Hintz, R. L. 1980. Genetics of performance in the horse. *J. Anim. Sci.* 51:582.

Legates, J. E. 1990. *Breeding and Improvement of Farm Animals.* New York: McGraw-Hill.

National Animal Health Monitoring System. 1994. Beef Cow/Calf Reproductive and Nutritional Management Practices. USDA: APHIS.

Van Vleck, L. D., E. Oltenacu, and J. Pollack. 1986. *Genetics for the Animal Sciences.* New York: W. H. Freeman.

CHAPTER 14

Mating Systems

LEARNING OBJECTIVES

- Describe the roles of the animal breeders
- Describe the advantages and disadvantages of inbreeding, linebreeding, outcrossing, and crossbreeding
- Describe heritability and heterosis
- Calculate percent heterosis
- Discuss the role of composite and hybrid breed formation

Seedstock breeders (sometimes called **purebred producers**) and commercial breeders (producers) are the two general classifications of animal breeders. **Seedstock** livestock historically come from the pure breeds for which ancestry is recorded on a pedigree by a breed association (Fig. 14.1). Most commercial slaughter livestock are crossbreds, resulting from crossing two or more breeds or lines of breeding.

Animal breeders make three critical decisions—choosing the individuals that become parents, determining the rate of reproduction from each individual (especially sires), and deciding which mating system is most beneficial. Genetic improvement can be optimized in most herds and flocks by utilizing a combination of selection and mating systems.

Mating systems are identified primarily by genetic relationship of the animals being mated. Two major systems of mating are **inbreeding** and **outbreeding**. Inbreeding is the mating of animals more closely related than the average of the breed or population. Outbreeding is the mating of animals not as closely related as the average of the population.

Since mating systems are based on the **relationship** of the animals being mated, it is important to understand more detail about genetic relationship. Proper pedigree evaluation also involves understanding the genetic relationships between animals. Relationship is best described as knowing which genes two animals have in common and whether the genes in an animal or animals exist primarily in a heterozygous or homozygous condition. Figure 14.2 shows the mating systems and their relationship to homozygosity and heterozygosity.

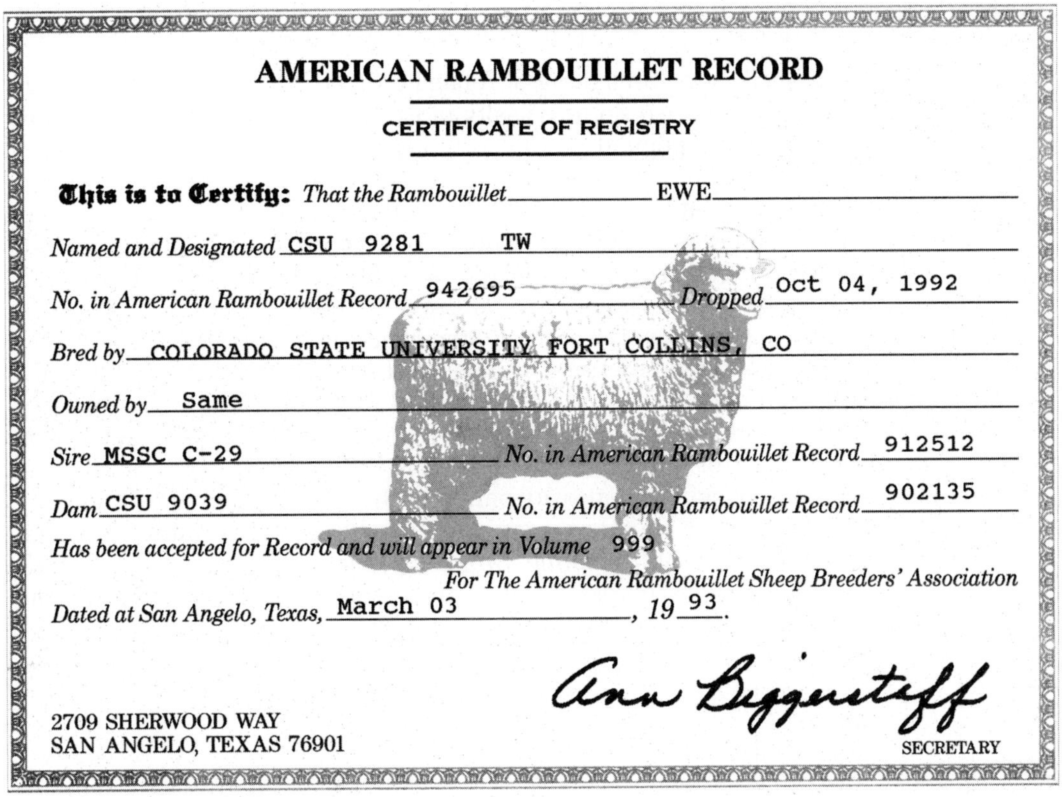

AMERICAN RAMBOUILLET RECORD

CERTIFICATE OF REGISTRY

This is to Certify: *That the Rambouillet*_____ EWE_____

Named and Designated CSU 9281   TW_____

No. in American Rambouillet Record 942695_____ *Dropped* Oct 04, 1992_____

Bred by COLORADO STATE UNIVERSITY FORT COLLINS, CO_____

Owned by Same_____

Sire MSSC C-29_____ *No. in American Rambouillet Record* 912512

Dam CSU 9039_____ *No. in American Rambouillet Record* 902135

Has been accepted for Record and will appear in Volume 999

For The American Rambouillet Sheep Breeders' Association

Dated at San Angelo, Texas, March 03_____, 19 93 .

Ann Biggerstaff

2709 SHERWOOD WAY
SAN ANGELO, TEXAS 76901
SECRETARY

FIGURE 14.1   Pedigree of a Rambouillet ewe (CSU #942695). Note that the sire and dam are listed as part of the pedigree. *Source:* Colorado State University.

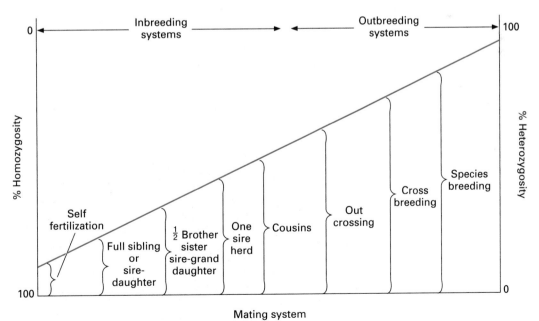

FIGURE 14.2   Relationship of the mating system to the amount of heterozygosity or homozygosity. Self-fertilization is currently not an available mating system in animals.

INBREEDING

Because inbreeding is the mating of related animals, the resulting inbred offspring have an increased homozygosity of gene pairs compared to noninbred animals in the same population (breed or herd). An example of inbreeding is shown in Figure 14.3, where animal A has resulted from mating B (the sire) to C (the dam). Note that B and C have the same parents (D and E). B and C are genetically related because they have a brother-sister relationship. B and C do not have the same identical genetic makeup because each has received a sample half of genes from each parent. In the arrow pedigree the arrow represents a sample half of genes from the parent to the offspring.

Animal A is inbred because it has resulted from the mating of related animals. Although the calculations are not shown in the figure, animal A has 25% more homozygous gene pairs as compared with a noninbred animal in the same population. Keep in mind that breeders have little control over assuring that only desired or beneficial genes are paired in the homozygous condition.

Figure 14.4 shows another example of inbreeding, where animal X has resulted from a sire-daughter mating. Sire A was mated to dam D, with D being sired by A. The increased homozygosity of animal X is 25%, which is the same as for animal A in Figure 14.3.

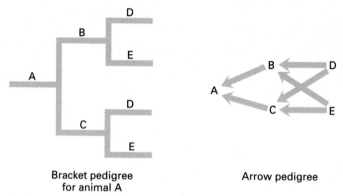

Bracket pedigree
for animal A

Arrow pedigree

FIGURE 14.3   Bracket pedigree and arrow pedigree showing animal A resulting from a full brother-sister mating.

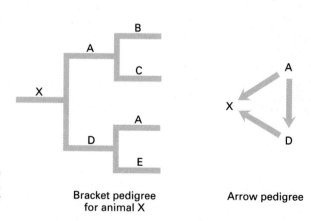

FIGURE 14.4   Bracket pedigree and arrow pedigree showing animal X resulting from a sire-daughter mating.

Bracket pedigree
for animal X

Arrow pedigree

The two different forms of inbreeding are these:

1. *Intensive inbreeding*—mating of closely related animals whose ancestors have been inbred for several generations.
2. *Linebreeding*—a mild form of inbreeding where inbreeding is kept relatively low while maintaining a high genetic relationship to an ancestor or line of ancestors.

Intensive Inbreeding

Intensive inbreeding occurs when closely related animals are mated for several generations. Figure 14.5 shows an example of intensive inbreeding. The increase in the homozygosity of animal H's genes is higher than 25% (compare with Fig. 14.4) because both the sire and grandsire of animal H were also inbred.

There are numerous genetically different inbred lines that can be produced in a given population such as a breed. The number of different, completely homozygous inbred lines is 2^n, where n is the number of heterozygous gene pairs. Thus, with two heterozygous gene pairs, 2^2 or 4 different inbred lines are possible. For example, with *BbPp* genes in a completely heterozygous herd (see Chapter 12), the different completely homozygous lines that can result from inbreeding are *BBPP, BBpp, bbPP,* and *bbpp.* Each of these four inbred lines is genetically pure (homozygous) for these two pair of genes and each will breed true when animals are mated within the same homozygous line.

There have been research projects with swine and beef cattle in which inbred lines have been produced; anticipating that the crossing of these inbred lines would produce results similar to those for hybrid corn (Fig. 14.6). One research project is at the San Juan Basin Research Center (SJBRC) in Hesperus, Colorado, where cattle have been inbred for more than 40 years. Figure 14.7 shows the arrow pedigree of an inbred bull (Royal 4160) produced in the Royal line of Hereford cattle. Note that College Royal Domino 3 is the sire of 10 animals in Royal 4160's pedigree, while Royal 3016 has sired 6 animals in the same

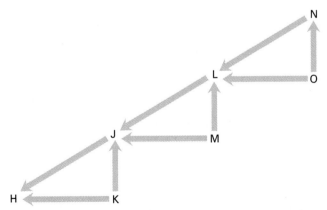

FIGURE 14.5 An arrow pedigree showing animal H resulting from three generations of sire-daughter matings. Note that J is the sire of H and also the sire of K, the latter being the dam of H. Animal J has resulted from two previous, successive generations of sire-daughter matings.

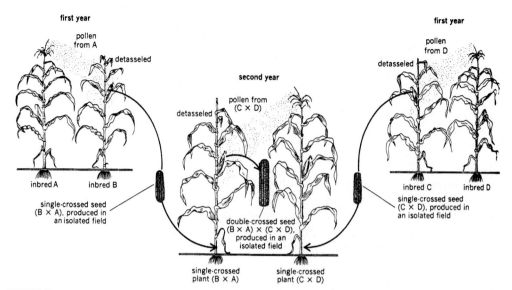

FIGURE 14.6 Crossing of inbred lines of corn to produce the double-crossing hybrid corn seed utilized widely today. Note the relative size of the corn ears from the inbred lines compared with the resulting hybrid cross. *Source:* USDA

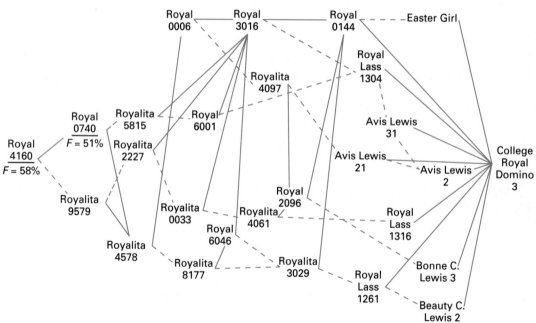

FIGURE 14.7 Arrow pedigree of Royal 4160, which has an inbreeding coefficient of 58%. Solid lines represent the genetic contribution of the bulls, whereas the cow's contribution is represented by a broken line. *Source:* CSU Expt. Sta. (San Juan Basin Research Center) General Series 982.

pedigree. The increased homozygosity of this bull is 58%, which represents some of the most highly inbred cattle in the world.

Information obtained from inbreeding studies with farm animals demonstrates the following results and observations:

1. Increased inbreeding is usually detrimental to reproductive performance and preweaning and postweaning growth. Also, inbred animals are more susceptible to environmental stresses. Whereas 60–70% of the inbred lines show the detrimental effects of increased inbreeding, 30–40% of the lines show no detrimental effect, with some lines demonstrating improved productivity.

2. In a Colorado research beef herd, the inbred lines showed a yearly genetic increase of 2.6 lb in weaning weight over a 26-year period, while the crosses of inbred lines made a 4.6 lb increase over the same time period. Heterosis is demonstrated in the line crosses, and the 4.6 lb increase is typical of what breeders might expect from using intense selection in an outbred herd.

3. Inbreeding quickly identifies some desirable genes and also undesirable genes, particularly the serious recessive genes that are hidden when left in the heterozygous state.

4. Inbred animals with superior performance are the most likely to have superior breeding values, which result in more uniform progeny with high levels of genetically influenced productivity.

5. Crossing of inbred lines results in heterosis; however, in most cases heterosis compensates for inbreeding depression.

6. Crossing of inbred lines of animals has not yielded the same results as crossing inbred lines of corn. The reasons appear to be: (1) inbreeding animals is slower (they cannot self-fertilize as corn does); (2) it is easier and less costly to produce more inbred lines of corn; (3) inbred lines of animals are eliminated because of extremely poor reproductive performance and being less adapted to environmental stress.

7. There is merit in using some inbreeding in developing new lines of poultry and swine. (This is discussed in more detail later in the chapter.)

It is not logical for breeders to develop their own lines of highly inbred beef cattle. Inbreeding depression usually affects the economics of the operation. Both seedstock and commercial producers can take advantage of highly productive inbred lines by crossing these inbred bulls with unrelated cows.

Inbreeding processes such as sire-daughter matings, are logical ways to test for undesirable recessive genes. Also, seedstock producers use inbreeding in well-planned linebreeding programs. Consideration should be given to implementing linebreeding when breeders have difficulty introducing sires from other herds that are genetically superior to those they are producing.

Linebreeding

Linebreeding is a mild form of inbreeding used to maintain a high genetic relationship to an outstanding ancestor, usually a sire. Seedstock producers who have high levels of genetic superiority in their herds and find it difficult to

Bracket pedigree for the quarter horse Impressive

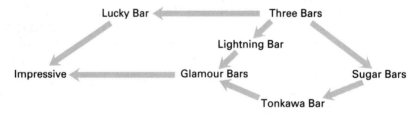

**Arrow pedigree showing the genetic pathways by which
Three Bars contributes to the inbreeding and linebreeding of Impressive**

FIGURE 14.8   Horse pedigree showing linebreeding.

locate sires that are superior to the ones they are raising in their herds best use this mating system.

Occasionally a breeder may produce a sire with a superior combination of genes that consistently produces high-producing offspring. Younger sires may not outproduce some of these sires. This is observed in some dairy bulls that remain competitively superior as long as they produce semen. These sires warrant use in a linebreeding program.

Figure 14.8 gives an example of linebreeding. Impressive, an outstanding quarter horse stallion is linebred to his ancestor, Three Bars, by three separate pathways. The inbreeding of Impressive is approximately 9%, whereas the genetic relationship of Impressive to Three Bars is approximately 44%. Inbreeding below 20% is considered low, whereas a genetic relationship is high when it approaches 50%.

Impressive has nearly the same genetic relationship as if Three Bars had been his sire (44 versus 50%). Progeny of Impressive have produced outstanding records, particularly as halter-point and working-point winners in the show ring. Unfortunately, Impressive has also been identified as a carrier of the gene that leads to hyperkalemic periodic paralysis, or HYPP. HYPP is the result of a genetic mutation, but it has not been confirmed if this mutation originated with Impressive.

OUTBREEDING

The four types of outbreeding are as follows:

1. *Species cross*—crossing of animals of different species (e.g., horse to donkey or cattle to bison).
2. *Crossbreeding*—mating of animals of different established breeds.
3. *Outcrossing*—mating of unrelated animals within the same breed.
4. *Grading up*—mating of purebred sires to commercial-grade females and their female offspring for several generations. Grading up can involve some crossbreeding or it can be a type of outcrossing system.

Species Cross

A species designation is part of the zoological classification (Fig. 14.9) used in taxonomy (a branch of zoology) to classify animals on the basis of similarities in body structure. These differences and similarities in body structure translate to genetic differences and similarities. For example, some animals of different species but the same genus can be crossed to produce viable offspring. Animals of different genus cannot be successfully crossed because chromosome number and other genes are different. Therefore, a **species cross** is the widest possible kind of outbreeding that can be achieved.

One of the most common species cross is the **mule,** resulting from crossing the jack of the ass species and the mare of the horse species *(Equus asinus × Equus caballus)*. Mules existed in large numbers as work animals before the advent of the tractor. The **hinny** is the reciprocal cross of the mule *(Equus caballus* stallion × *Equus asinus* jennet). The hinny never became as popular as the mule.

Mare mules are usually sterile, which gives verification to genetic differences between the ass and horse. There have been a few reports of fertile mare mules.

Crossing of the zebu or humped cattle with European-type cattle *(Bos indicus × Bos taurus)* is common in the southeastern part of the United States. These crosses are more adaptable and productive in hot, humid environments than either of the straight species. Some authorities raise questions on *Bos indicus* and *Bos taurus* being separate species, and their crosses are usually referred to as *crossbreds* rather than species crosses.

Numerous crosses of American bison and cattle have been made. Some of these crosses have been designated as separate breeds called **Cattalo** or **Beefalo.** These crosses are intended to be more adaptable to harsh environments (cold temperatures and limited forage). Fertility problems have existed in some of these crosses, and their numbers are limited.

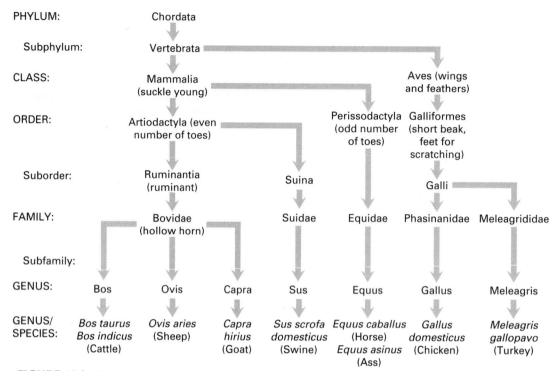

FIGURE 14.9 Zoological classification that identifies the major species of farm animals. Adapted form R. F. Plimpton and J. F. Stephens, *Animal and Science for Man: Study Guide* (Minneapolis: Burgess, 1979).

Sheep and goats have been crossed even though they have different genus classifications. Fertilization occurs but embryos die in early gestation.

There are other species crosses that have occurred. Most species crosses have little commercial value, although insight into the evolutionary process is interesting. Recent advances in genetic engineering might make some genetic combinations between species more feasible. Gene-splicing (inserting a gene or genes from one animal to another) has occurred between species. In the future, opportunity to combine desirable genes both within a species and between species could become a reality.

Crossbreeding

There are two primary reasons for using **crossbreeding:** To take advantage of (1) breed complementation and (2) heterosis (hybrid vigor). Breed complementation implies crossing breeds so their strengths and weaknesses complement one another. There is no one breed that is superior in all desired production characteristics; therefore, planned crossbreeding programs that use breed complementation can significantly increase herd productivity.

Crossbreeding, if properly managed, allows for the effective use of heterosis, which has a marked effect on productivity in swine, poultry, and beef cattle. **Heterosis** is defined as the increase in productivity in the crossbred progeny above the average of breeds or lines that are crossed. An example of calculating heterosis is shown in Table 14.1. The calculated heterosis for calf-crop percent in this example is 5%, whereas heterosis for weaning weight is 4%.

Crossbreeding is sometimes questioned when the performance of crossbred animals is less than that of the parent breed. In Table 14.1, for example, the weaning weight of the crossbreds is 520 lb, while one parent (Breed B) is 540 lb. However, for calf-crop percent, the crossbreds are 84%, which is higher than either Breed A or Breed B. The value of crossbreeding in this example is best demonstrated by combining calf-crop percent and weaning weight (calf-crop percent × weaning weight = lb calf weaned per cow exposed). Note in Table 14.1 that the value for lb of calf weaned per cow exposed is 447 lb for the crossbreds, whereas it is 377 lb for Breed A and 421 lb for Breed B.

The amount of heterosis expressed is related to the heritability of the trait. Table 14.2 shows that heterosis is highest for low-heritability traits and lowest for high-heritability traits. These relationships are helpful to commercial producers in selecting and crossbreeding to enhance genetic improvement. Figure 14.10 shows

TABLE 14.1 Computation of Heterosis for Percent Calf Crop and Weaning Weights

	Calf Crop (%)	Weaning Weight (lb)	Lb Calf Weaned per Cow Exposed
Breed A	82	460	377
Breed B	78	540	421
Average of the two breeds (without heterosis)	80	500	399
Average of crossbreds (with heterosis)	84	520	447
Superiority of crossbreds over average of two breeds	4	20	48
Percent heterosis	5% (4 ÷ 80)	4% (20 ÷ 500)	12% (48 ÷ 399)

TABLE 14.2 Relationship of Heritability and Heterosis for Most Traits

Traits	Heritability	Heterosis
Reproduction	Low	High
Growth	Medium	Medium
Carcass	High	Low

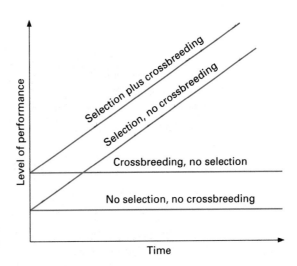

FIGURE 14.10 Improvement in performance with various combinations of selection and crossbreeding.

the relative importance of selection and crossbreeding in an improvement program. This figure demonstrates that selecting genetically superior animals is more important than crossbreeding. However, using the two methods in combination gives the highest level of performance.

Crossbreeding is most commonly used in swine, beef cattle, and sheep. Little crossbreeding is done in dairy cattle because of the primary emphasis on one trait (milk production) and the superiority of the Holstein breed for that trait. Poultry breeders utilize heterosis primarily through crossing lines that have been developed from crossing breeds and inbreeding separate and distinct lines.

Figures 14.11, 14.12, and 14.13 show the crossbreeding systems most frequently used in swine, beef cattle, and sheep. More specific detail in using these crossbreeding systems for these species is given in Chapters 25, 29, and 31.

Outcrossing

The most widely used breeding system for most species is **outcrossing.** As unrelated animals within the same breed are mated, the gene pairs are primarily heterozygous, although there is a slight increase in homozygosity over time. Homozygosity for several breeds has been estimated at between 10 and 20%. This slight increase in homozygosity occurs because the animals mated are somewhat related in that they are members of the same breed.

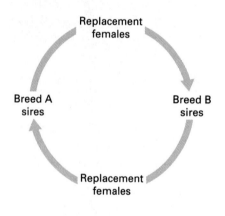

FIGURE 14.11 Two-breed rotation cross. Females sired by breed A are mated to breed B sires, and females sired by breed B are mated to breed A sires.

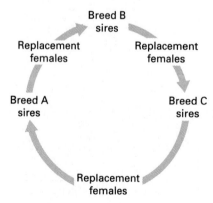

FIGURE 14.12 Three-breed rotation cross. Females sired by a specific breed are bred to the breed of sire next in rotation.

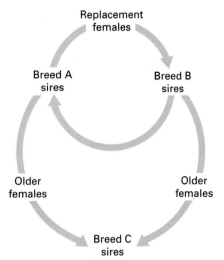

FIGURE 14.13 Terminal (Static) or modified-terminal crossbreeding system. It is terminal or static if all females in herd (A × B) are then crossed to breed C Sires. All male and female offspring are sold. It is a modified-terminal system if part of females are bred to A and B sires to produce replacement females. The remainder of the females are terminally crossed to breed C sires.

The usefulness of outcrossing is primarily dependent on the effectiveness of selection (selection differential × heritability). The gene pairs stay primarily in a heterozygous condition, as there is no attempt to maximize homozygosity or heterozygosity.

Grading Up

The continuous use of purebred sires of the same breed in a grade herd or flock is called **grading up.** In this situation, grading up is similar to outcrossing. The accumulated percentage of inheritance of the desired purebred is 50% (½), 75% (¾), 84.5% (⅞), and 94% (¹⁵⁄₁₆) for four generations when grading up is practiced. The fourth generation resembles the purebred sires so closely in genetic composition that it approximates the purebred level.

The grading-up system is useful in the breeding of cattle and horses, but it has little value in breeding sheep, swine, or poultry. High-producing purebred sheep, swine, and poultry breeding stock are available at reasonable prices; therefore, the breeder can buy them for less than he or she can produce them by grading up. The use of production-tested males that are above average in performance in a commercial herd can grade up the herd not only to a general purebred level but also to a high level of production.

A use of grading up on a large scale occurred with the introduction of many European beef cattle breeds to the United States. Most of the introduction was accomplished with males (bulls or semen), as females were less available and more expensive because of the numbers needed. Grading up allowed relatively rapid propagation of these imported breeds.

Imported bulls (or their semen) have been used on commercial cows or purebred cows of other breeds. Grading up, as used here, is a type of crossbreeding, although the intent is not to maintain heterosis but to increase the frequency of genes from the introduced breed.

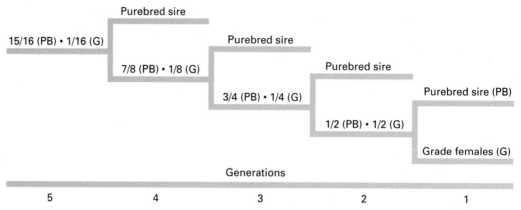

FIGURE 14.14 Utilizing grading up to produce purebred offspring from a grade herd.

After several successive generations of mating the new breed to cows carrying a certain percentage of the new breed, the resulting offspring have been designated purebreds. In most breeds, this designation has been given when the calves had ⅞ or ¹⁵⁄₁₆ of the genetic composition of the new breed. Figure 14.14 shows how these matings are made. It would require a minimum of 7 years to produce the first ¹⁵⁄₁₆ calves of the new breed.

FORMING NEW LINES OR BREEDS

New breeds have been formed and are currently being formed by crossing several breeds. These are sometimes given a general classification of **synthetic breeds** and **composite breeds.** In beef cattle, the Brangus, Barzona, Beefmaster, and Santa Gertrudis breeds are composite (synthetic) breeds formed several years ago, whereas MARC I (crosses of Charolais, Brown Swiss, Limousin, Hereford, and Angus breeds) and RX3 (crosses of Red Angus, Hereford, and Red Holstein breeds) are examples of composites more recently developed. Columbia, Targhee, and Polypay are examples of synthetic breeds of sheep.

Crossing several swine breeds, a practice associated with some inbreeding, has been used to develop the hybrid boars now being merchandised by several companies. Hybrid boars are used extensively in the swine industry.

In poultry, breeding for egg production differs from breeding for broiler production. Different traits are emphasized in the production of these two products. Both inbreeding and heterosis are utilized in the production of specific lines and strains of birds that are highly productive in the production of either eggs or broilers.

Figure 14.15 shows the change from poultry breeds to crossmated lines and strains of birds. The incrossmated strains represent many of the synthetic egg-type strains or lines. These are mainly two- and three-way crosses of primarily Mediterranean breeds (e.g., White Leghorns). The crossmated chickens are the commercial broiler chickens that are primarily cornish-type males or White Plymouth females. The area of the chart depicting crossmated strains shows the tremendous changes in breeding methods utilized by the broiler industry.

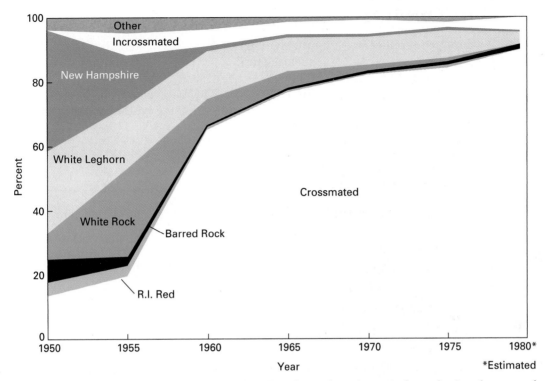

FIGURE 14.15   In poultry breeding the specific breeds are losing their identity in the production of crossmated and incrossmated lines. Data from over 30 million birds recorded in the National Poultry Improvement Plan. *Source:* USDA.

CHAPTER SUMMARY

- Genetic relationship estimates the genes two animals have in common because the same ancestors appear in the first six generations of their pedigrees.

- Mating systems are identified by the genetic relationship of the animals being mated.

- Inbreeding is the mating of animals more closely related than the average of the population, while outbreeding is the mating of animals not as closely related as the average of the population.

- Inbreeding increases genetic homozygosity, while outbreeding increases genetic heterozygosity.

- Linebreeding is a mild form of inbreeding while maintaining a high genetic relationship to an outstanding ancestor.

- Outcrossing is the mating of unrelated animals within the same breed.

- Crossbreeding is the mating of animals from different breeds, resulting in heterosis (hybrid vigor).

KEY WORDS

purebred producers	linebreeding	heterosis
seedstock	species cross	outcrossing
inbreeding	mule	grading up
outbreeding	hinny	composite breeds
relationship	Cattalo (Beefalo)	synthetic breeds
intensive inbreeding	crossbreeding	

REVIEW QUESTIONS

1. Animals derived from matings within a single breed are called _____.

2. What is inbreeding?

3. How does inbreeding affect homozygosity of traits?

4. What is a major disadvantage of inbreeding?

5. What is a major advantage of inbreeding?

6. What is line breeding?

7. What is crossbreeding?

8. What are the two primary reasons for crossbreeding?

9. What is outcrossing?

10. What is the term for newly developed breeds created by crossing several established breeds?

11. What is grading up?

SELECTED REFERENCES

Bourdon, R. M. 2000. *Understanding Animal Breeding.* Upper Saddle River, NJ: Prentice Hall, Inc.

Cundiff, L. V., L. D. Van Vleck, L. D. Young, and C. D. Dickerson. 1994. Animal breeding and genetics. *Encyclopedia of Agricultural Science.* San Diego, CA: Academic Press, Inc.

Legates, J. E. 1990. *Breeding and Improvement of Farm Animals.* New York: McGraw-Hill.

CHAPTER 15

Nutrients and Their Functions

LEARNING OBJECTIVES

- Discuss the role of water, carbohydrates, fats, proteins, minerals, and vitamins in nutrition
- Illustrate proximate analysis
- Calculate percent crude protein, percent digestibility, and percent dry matter
- Discuss the process of energy utilization
- Classify the basic feedstuffs

A **nutrient** is any feed constituent that functions in the support of life. There are many different feeds available to animals to provide nutrients.

Most animal feeds are classified as **concentrates** or **roughages.** Concentrates include cereal grains (e.g., corn, wheat, barley, oats, and milo), oil meals (e.g., soybean meal, linseed meal, and cottonseed meal), molasses, and dried milk products. Concentrates are high in energy, low in fiber, and highly (80–90%) digestible. Roughages include legume hays, grass hays, and straws, the latter being by-products from the production of grass, seed, and grain. Additional roughages are silage, stovers (dried corn, cane, or milo stalks and leaves with the grain portion removed), soilage (cut green feeds), and grazed forages. Roughages are less digestible than concentrates. Roughages are typically 50–65% digestible, but the digestibility of some straws is significantly lower.

NUTRIENTS

The six basic classes of nutrients—water, carbohydrates, fats, proteins, vitamins, and minerals—are found in varying amounts in animal feeds. Nutrients are composed of at least 20 of the more than 100 known chemical elements. These 20 elements and their chemical symbols are: calcium (Ca), carbon (C), chlorine (Cl), cobalt (Co), copper (Cu), fluorine (F), hydrogen (H), iodine (I), iron (Fe), magnesium (Mg), manganese (Mn), molybdenum (Mo), nitrogen (N), oxygen (O), phosphorus (P), potassium (K), selenium (Se), sodium (Na), sulphur (S), and zinc (Zn).

Water

Water contains hydrogen and oxygen. The terms **water** and **moisture** are used interchangeably. Typically, water refers to drinking water, whereas moisture is used in reference to the amount of water in a given feed or ration. The remainder of the feed, after accounting for moisture, is referred to as **dry matter.** Moisture is found in all feeds, ranging from 10% in air-dry feeds to more than 80% in fresh green forage. Livestock and poultry consume several times more water than dry matter each day and will die from lack of water more quickly than from lack of any other nutrient. Water in feed is no more valuable than water from any other source. This knowledge is important in assessing feeds that vary in their moisture content.

Water has important body functions. It enters into most of the metabolic reactions, assists in transporting other nutrients, helps maintain normal body temperature, and gives the body its physical shape (water is the major component within cells).

Carbohydrates

Carbohydrates are found in both plant and animal tissue. Carbohydrates contain carbon, hydrogen, and oxygen in either simple or complex forms. The more simple carbohydrates, such as starch, supply the major energy source for swine and poultry diets. The more complex carbohydrates, such as cellulose, are the major components of the cell walls of plants. These complex carbohydrates are not as easily digested as simple carbohydrates and they require host or microbial interaction for effective utilization. Thus, grazing ruminants, such as cattle or sheep, make the most effective use of forages, with horses intermediate to the monogastric species. Glucose and glycogen are carbohydrates stored in animal tissue that provide readily available sources of energy.

Fats

Fats and oils, also referred to as **lipids,** contain carbon, hydrogen, and oxygen, although there is more carbon and hydrogen in proportion to oxygen than with carbohydrates. Fats are solid and oils are liquid at room temperature due generally to relative saturation. Fats contain 2.25 times more energy per pound than carbohydrates.

Most fats are composed of three fatty acids attached to a glycerol backbone. For example:

$$
\begin{array}{l}
\text{H} \\
\text{HC—OH} \quad \text{HO—C—(CH}_2)_n\text{CH}_3 \\
\quad\qquad\qquad \| \\
\quad\qquad\qquad \text{O}
\end{array}
$$

Glycerol + three fatty acids ⟶ a triglyceride + water

Certain fatty acids are saturated or unsaturated, depending on their particular chemical composition. Saturated fatty acids have single bonds tying the carbon atoms together (e.g., —C—C—C—C—), whereas, unsaturated fatty acids have one or more double bonds (e.g., =C=C—C=C—). The term **polyunsaturated fatty acids** is applied to those having more than one double bond. Although more than 100 fatty acids have been identified, linoleic and α-linolenic have been determined to be a dietary essential for livestock. The two apparent functions of the essential fatty acids are (1) as precursors of prostaglandins and (2) as structural components of cells.

Proteins

Proteins are polypeptides of high molecular weight that contain hydrogen, oxygen, carbon, and nitrogen. If a protein contains only amino acids, it is said to be a **simple protein**. A **complex protein** contains additional non-amino acid substances, such as heme (hemoglobin), carbohydrate (glyco-proteins), or lipid (lipoproteins). Protein is the only nutrient class that contains nitrogen. Proteins in feeds contain 16% nitrogen on average. This is why feeds are analyzed for the percent nitrogen in the feed, with the percent multiplied by 6.25 (100% ÷ 16% = 6.25) to convert it to percent protein. If, for example, a feed is 3% nitrogen, 100 g of the feed contains 3 g nitrogen. Multiplying 6.25×3 gives 18.75%, meaning that 100 g of this feed contains 18.75 g protein.

Proteins are composed of various combinations of some 25 amino acids. Amino acids are called the *building blocks* of the animal's body. The building blocks for growth (including growth of muscle, bone, and connective tissue), milk production, and cellular and tissue repair are amino acids that come from proteins in feed. The interstitial (between cells) fluid, blood, and lymph require amino acids to regulate body water and to transport oxygen and carbon dioxide. All enzymes are proteins, so amino acids are also required for enzyme production. Amino acids have an amino group (NH_2) in each of their chemical structures. There are many different combinations of amino acids that can be structured together. The chemical, or peptide, bonding of amino acids is illustrated using alanine and serine, which results in the formation of a dipeptide:

It can be seen that amino acids have a basic portion, NH_2, and an acid portion, and it is because of these that they can combine into long chains to make proteins. When digestion occurs, the action is at the peptide linkage to free amino acids from one another.

Amino acids may be classified as either essential or nonessential (Table 15.1). Essential amino acids must be provided via dietary means, or in the case of ruminants, via microbial action. Nonessential amino acids are required by the animal for normal growth but can be synthesized by the animal.

Protein can also be evaluated for quality by evaluating the amount and ratio of essential amino acids present. Proteins vary in terms of absorption rate or biological value. Egg protein is considered the highest in value (>90%). Proteins from animal sources (60–80%) are greater than plant proteins (40–60%).

Minerals

Chemical elements other than carbon, hydrogen, oxygen, and nitrogen are called **minerals.** They are inorganic because they contain no carbon; organic nutrients do contain carbon. Some minerals are referred to as **macro** (required in larger amounts) and others are **micro** or trace minerals (required in smaller amounts) (Table 15.2).

Calcium and phosphorus are required in specified amounts and in a specified ratio to each other for bone growth and repair and for other body functions. The blood plasma contains sodium chloride; the red blood cells contain potassium chloride. The osmotic relations between the plasma and the red blood cells are maintained by proper concentrations of sodium chloride and potassium chloride. Excessive sweating that results from heavy physical work in hot

TABLE 15.1   Essential and Nonessential Amino Acids

Essential	Nonessential
Arginine	Alanine
Histidine	Asparganine
Isoleucine	Aspartic acid
Leucine	Cysteine
Lysine	Cystine
Methinonine	Glutamic acid
Phenylalanine	Glutamine
Threonine	Glycine[a]
Trytophan	Hydroxproline
Valine	Proline[a]
	Serine
	Tyrosine

[a] Under some scenarios, these may be essential in poultry diet.

TABLE 15.2 Micro and Macro Minerals

Macro	Micro
Calcium	Chromium
Chlorine	Cobalt
Magnesium	Copper
Phosphorus	Fluorine[a]
Potassium	Iodine
Sodium	Iron
Sulfur	Manganese
	Molybdenum[a]
	Selenium[a]
	Zinc

[a] Beneficial in some regions but toxic if fed in excess.

weather may deplete sodium chloride. It is essential that salt and plenty of water be available under such conditions. The acid-base balance of the body is maintained at the proper level by minerals.

Microminerals may become a part of the molecule of a vitamin (e.g., cobalt is a part of vitamin B_{12}) and they may become a part of a hormone (e.g., thyroxin, a hormone made by the thyroid gland, requires iodine for its synthesis).

Certain important metabolic reactions in the body require the presence of minerals. Selenium and vitamin E both appear to work together to help prevent white muscle disease, which is a calcification of the striated muscles, the smooth muscles, and the cardiac muscles. Both vitamin E and selenium are more effective if the other is present. Excesses of certain minerals may be quite harmful. For example, excess amounts of fluorine, molybdenum, and selenium are highly toxic.

Vitamins

Vitamins are organic nutrients needed in very small amounts to provide for specific body functions in the animal. There are 16 known vitamins that function in animal nutrition. Vitamins may be classed as either **fat-soluble** or **water-soluble.**

The fat-soluble vitamins are vitamins A, D, E, and K. Vitamin A helps maintain proper repair of internal and external body linings. Because the eyes have linings, lack of vitamin A adversely affects the eyes. Vitamin A is also a part of the visual pigments of the eyes. Vitamin D is required for proper use of calcium and phosphorus in bone growth and repair. A major function of vitamin D is to regulate the absorption of calcium and phosphorus from the intestine. Vitamin D is produced by the action of sunlight on sterols of the skin; therefore, animals that are exposed to sufficient sunlight make all the vitamin D they need. Vitamin K is important in blood clotting; hemorrhage might occur if the body is deficient in vitamin K. Vitamin E is an excellent antioxidant and has a role in preventing the breakdown of cell membranes by free radicals. Vitamin E exists in several forms classified as tocopherols.

The water-soluble vitamins are ascorbic acid (vitamin C), biotin, choline, cyanocobalamin (vitamin B_{12}), folic acid, niacin, pantothenic acid, pyridoxine (vitamin B_6), riboflavin (B_2), and thiamin (vitamin B_1). More diseases caused by inadequate nutrition have been described in the human than in any other animal, and among the best known are those caused by a lack of certain vitamins: beriberi (lack of thiamin); pellagra (lack of niacin);

pernicious anemia (lack of vitamin B_{12}); rickets (lack of vitamin D); and scurvy (lack of vitamin C).

In ruminant animals, microorganisms in the rumen make all of the water-soluble vitamins. Water-soluble vitamins also appear to be readily available to horses; perhaps some are made by fermentation in the cecum. Water-soluble vitamins cannot be synthesized by monogastric animals and must therefore be in their feed. Most fat-soluble vitamins are not synthesized by either ruminants or monogastrics and must be supplied in the diets of both groups (an exception is vitamin K, which is synthesized by rumen bacteria in ruminants). Many vitamins are supplied through feeds normally given to animals.

PROXIMATE ANALYSIS OF FEEDS

The nutrient composition of a feed cannot be determined accurately by visual inspection. A system has been devised by which the value of a feed can be approximated. **Proximate analysis** separates feed components into groups according to their feeding value. This analysis is based on a feed sample and analysis, and therefore is no more accurate than how representative the sample is of the entire feed source.

The inorganic and organic components resulting from a proximate analysis are water, crude protein, crude fat (sometimes referred to as *ether extract*), crude fiber, nitrogen-free extract, and ash (minerals). Figure 15.1 shows these components resulting from a feed that would have a laboratory analysis of 88% dry matter, 13% protein, 4% fat, 10% crude fiber, and 56% nitrogen-free extract (NFE) on a natural or air-dry basis. The analysis might be reported on a dry-matter basis (no

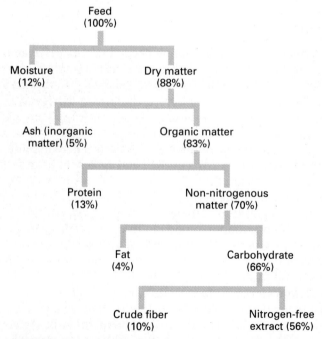

FIGURE 15.1 Proximate analysis showing the inorganic and organic components of a feed (similar to wheat) on a natural or air-dry basis.

moisture) as 14.8% protein, 4.5% fat, 11.4% crude fiber, and 63.6% nitrogen-free extract. Therefore, caution needs to be exercised in interpreting the proximate analysis results because different laboratories may report their analytical values on either an air-dry (or as-fed) basis or a dry-matter basis.

Values reported on a dry-matter basis can be converted to an as-is basis by multiplying the value, crude protein for example, by the dry-matter percentage and dividing by 100. For example, a hay sample measuring 15% crude protein (CP) and a dry matter of 90% would have a CP on an as-is basis of:

$$15 \times \frac{90}{100} = 13.5\% \text{ CP}$$

Conversely, values can be converted for an as-is or wet basis to a dry-matter basis. For example, an oat sample has a crude protein measurement of 13% on an as-is basis and contains 89% dry matter. The crude protein content on an as-is basis is determined by:

$$13 \times \frac{100}{8.9} = 14.6\% \text{ CP}$$

The proximate analysis for the six basic nutrients does not distinguish the various components of a nutrient. For example, ash content of a feed does not tell the amount of calcium, phosphorus, or other specific minerals. Figure 15.2 gives the chemical analysis for organic and inorganic nutrients. There are specific chemical analyses for each of these nutrients in a feed if such an analysis is needed.

DIGESTIBILITY OF FEEDS

Digestibility refers to the amount of various nutrients in a feed that are absorbed from the digestive tract. Different feeds and nutrients vary greatly in their digestibility. Many feeds have been subjected to digestion trials, in which feeds of known nutrient composition have been fed to livestock and poultry. Feces have been collected and the nutrients in the feces analyzed. The difference between nutrients fed and nutrients excreted in the feces is the apparent digestibility of the feed.

For example, the digestibility of protein is obtained by determining the digestibility of nitrogen in a feed. Digestibility is expressed as a percentage of nitrogen, for example, as follows:

$$\frac{\text{Nitrogen in feed} - \text{Nitrogen in feces}}{\text{Nitrogen in feed}} \times 100 = \text{Percentage digestibility}$$

As an example, if 25 lb of feed contains 3.2 g of nitrogen and 100 g of feces contains 0.8 g of nitrogen, the percent digestibility of nitrogen is:

$$\frac{3.2 - 0.8}{3.2} \times 100 = 75\%$$

Note that the determination of 3.2 g of nitrogen in 100 g of feed enables the percentage of protein in the feed to be estimated as 20% ($3.20 \times 6.25 = 20$).

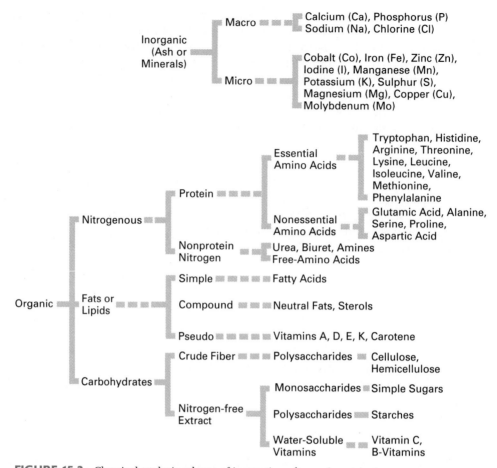

FIGURE 15.2   Chemical analysis scheme of inorganic and organic nutrients.

ENERGY EVALUATION OF FEEDS

Nutrients that contain carbon provide the energy for animals. Carbohydrates, fats, and proteins can all be used to provide energy; however, carbohydrates supply most of the energy, as they are a more economical energy source than proteins. Complete oxidation (burning or taking on oxygen) of carbon releases the energy. Energy is the force, or power, that is used to drive a variety of body systems.

Energy can be used to power movement of the animal, but most of it is used as chemical energy to drive reactions necessary to convert feed into animal products and to keep the body warm or cool.

Energy needs of animals generally account for the largest portion of feed consumed. Several systems have been devised to evaluate feedstuffs for their energy content. **Total digestible nutrient (TDN)** estimates of feeds were historically the most commonly used energy estimation system. TDN is typically expressed in pounds, kilograms, or percentages after obtaining the proximate analysis and digestibility figures for a feed. The formula for calculating TDN is TDN = (digestible crude protein) + (digestible crude fiber) + (digestible nitrogen-free extract) + (digestible crude fat × 2.25). The factor of 2.25 is used to equate fat to a carbohydrate basis, since fat has 2.25 times as much energy as an equivalent amount of carbohydrate.

TABLE 15.3   An Example of Calculating Total Digestible Nutrients (TDN)

Nutrient	Amount of Nutrient (g)	Digestibility (%)	Amount of Degestible Nutrient
Protein	20	75	15.00
Carbohydrates			46.75
Soluble (NFE)	55	85	46.75
Insoluble (fiber)	10	20	2.00
	11.15	85	9.50
			TDN = 73.25

An example of calculating TDN in 100 g of feed is shown in Table 15.3.

TDN is roughly comparable to **digestible energy (DE)** but it is expressed in different units. TDN and DE both tend to overvalue roughages.

Even though there are some apparent shortcomings in using TDN as an energy measurement of feeds, it works well in balancing rations for cows and growing cattle. However, TDN is being replaced by estimates of **net energy (NE)** in many ration formulation systems. The net energy system (NE) is a more precise energy measurement of feeds. This system usually measures energy values in megacalories per pound or kilogram of feed. The calorie basis, which measures the heat content of feed, is as follows:

Calorie (cal)—amount of energy or heat required to raise the temperature of 1 g of water 1°C.

Kilocalorie (kcal)—amount of energy or heat required to raise the temperature of 1 kg of water 1°C.

Megacalorie (Mcal)—equal to 1,000 kilocalories or 1 million calories.

Figure 15.3 shows various ways in which animals utilize the energy of feeds and the various energy measurements of feeds. Note that some energy in the feed is lost in feces (not digested), urine (digested but not used by the body cells), and gases from microbial fermentation of the feed, and heat resulting from digestion and metabolism of the feed.

Maintenance and Production

Feeds provide energy that the animal uses to supply two basic functions: (1) maintenance and (2) production. Maintenance energy is used to maintain basal metabolism, to provide for the voluntary activity of the animal, to generate heat to keep the body warm, and to provide energy to cool the body.

Production activities include fetal development, semen production, growth, fat deposition, and production of milk, eggs, and wool. These functions are only sustained if either stored or consumed energy is in excess of that required for maintenance.

Measurement of Energy

Gross energy (GE) is the quantity of heat (calories) released from the complete burning of the feed sample in an apparatus called a *bomb calorimeter*. GE has little practical value in evaluating feeds for animals because the animal does not metabolize feeds in the same manner as a bomb calorimeter. For example, oat straw has the same GE value as corn grain. Digestible energy (DE) is GE of feed

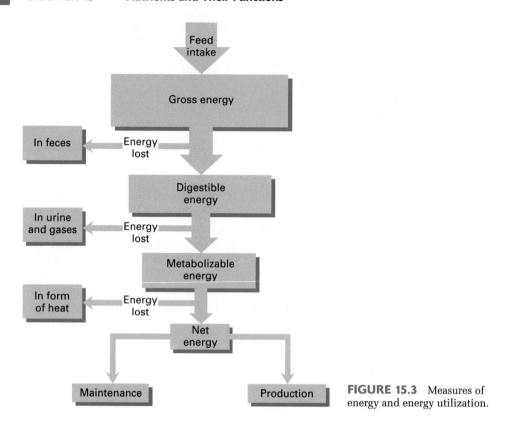

FIGURE 15.3   Measures of energy and energy utilization.

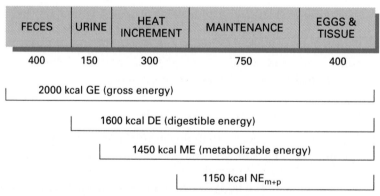

FECES	URINE	HEAT INCREMENT	MAINTENANCE	EGGS & TISSUE
400	150	300	750	400

2000 kcal GE (gross energy)

1600 kcal DE (digestible energy)

1450 kcal ME (metabolizable energy)

1150 kcal NE_{m+p}

FIGURE 15.4   Energy utilization by a laying hen. Data represent approximately 1 lb of feed containing 2,000 kcal of gross energy.

minus fecal energy. **Metabolizable energy (ME)** is DE of feed minus energy in urine and gaseous products of digestion. Net energy (NE) is the ME of feed minus the energy used in the consumption, digestion, and metabolism of the feed. This energy lost between ME and NE is called **heat increment.**

Another way to illustrate the several measures of feed energy and how they are utilized is shown in Figure 15.4. In this example, 2,000 kcal of GE (in approximately 1 lb of feed) is fed to a laying hen. The DE shows that 400 kcal were lost in the feces. The ME (1,450 kcal) is used for heat increment, maintenance, and production (eggs and tissue). There are 300 kcal lost in the heat increment, which leaves 1,150 kcal for maintenance and production. For maintenance, 750 kcal are needed, which leaves 400 kcal for egg production and tissue growth. Therefore only 20% (440/2,000) of the GE is used for production.

Net energy for maintenance (NE_m) and net energy for gain (NE_g) is more commonly used for formulating diets for feedlot cattle than any other energy system. Net energy for lactation (NE_l) is used in dairy cow ration formulation. NE_m in animals is the amount of energy needed to maintain a constant body weight. Animals of known weight, fed for zero energy gain, have a constant level of heat production.

NE_g measures the increased energy content of the carcass after feeding a known quantity of feed energy. All feed fed above maintenance is not utilized at a constant level of efficiency. Higher rates of gain require more feed per unit of gain as composition of gain varies with rate of gain.

FEEDS AND FEED COMPOSITION

Classification of Feeds

Feeds are naturally occurring ingredients in diets of farm animals used to sustain life. The terms *feed* and *feedstuffs* are generally used interchangeably; however, **feedstuffs** are a more inclusive term. Feedstuffs can include certain nonnutritive products such as additives to promote growth and reduce stress, to give flavor and palatability, to add bulk, or to preserve other feeds in the ration.

The National Research Council (NRC) classification of feedstuffs is as follows:

1. Dry roughages and forages

   Hay (legume and nonlegume)
   Straw
   Fodder
   Stover
   Other feeds with greater than 18% fiber (hulls and shells)

2. Range, pasture plants, and green forages

3. Silages (corn, legume, and grass)

4. Energy feeds (cereal grains, mill by-products, fruits, nuts, and roots)

5. Protein supplements (animal, marine, avian, and plant)

6. Mineral supplements

7. Vitamin supplements

8. Nonnutritive additives (antibiotics, coloring materials, flavors, hormones, preservatives, and medicants)

Roughages and forages are used interchangeably, although roughage usually implies a bulkier, coarser feed. In the dry state, roughages have more than 18% crude fiber. The crude fiber is primarily a component of cell walls that is not highly digestible. Roughages are also relatively low in TDN, although there are exceptions; for example, corn silage has over 18% crude fiber and approximately 70% TDN.

Feedstuffs that contain 20% or more protein, such as soybean meal and cottonseed meal, are classified as protein supplements. Feedstuffs with less than 18% crude fiber and less than 20% proteins are classified as energy feeds or concentrates. Cereal grains are typical energy feeds, which is reflected by their high TDN values.

Nutrient Composition of Feeds

Feeds are analyzed for their nutrient composition, as discussed earlier. The ultimate goal of nutrient analysis of feeds is to predict the productive response of animals when they are fed rations of a given composition.

The nutrient compositions of some of the more common feeds are shown in Table 15.4 (ruminants) and 15.5 (monogastric animals). The information in Table 15.4 represents averages of numerous feed samples. Feeds are not constant in composition, and an actual analysis should be obtained whenever economically feasible. The actual analysis is not always feasible or possible because of lack of available laboratories and insufficient time to obtain the analysis. Therefore, feed-analysis tables become the next-best source of reliable information on nutrient composition of feeds. It is not uncommon to expect the following deviations of actual feed analysis from the table values for several feed constituents: crude protein (±15%), energy values (±10%), and minerals (±30%).

Digestible protein is included in some feed-composition tables, but because of the large contribution of body protein to the apparent protein in the feces, digestible protein values are more misleading than crude protein values. For this reason, crude protein is more commonly found in feed-composition tables and used in formulating diets for ruminants.

Digestible protein (DP) can be calculated from crude protein (CP) content by using the following equation (%DP and %CP are on a dry-matter basis):

$$\%DP = 0.9 \, (\%DP) - 3$$

Five measures of energy values—TDN, ME, NE_m, NE_g, and NE_l)—are shown in Table 15.5. TDN is shown because there are more TDN values for feeds and because TDN has been a standard system of expressing the energy value of feeds. Some individuals seek ME (metabolizable energy) values for feed because these values are in calories rather than pounds. NE_m and NE_g values are used primarily to formulate feedlot diets and diets for growing replacement heifers, as these values offset the major problem associated with the TDN energy system. NE_l is used in formulating diets for dairy cows.

CHAPTER SUMMARY _____

- ■ Nutrients are feed constituents that support or sustain life.
- ■ The six classes of nutrients are water, carbohydrates, fats, proteins, vitamins, and minerals.
- ■ Carbohydrates and fats are the primary energy sources in feed.
- ■ Proteins are composed of some 25 amino acids; the latter are known as the building blocks of the animal's body.

TABLE 15.4 Nutrient Composition of Selected Feeds Commonly Used in Diets of Ruminants

Feed	Dry Matter (%)	TDN[1] (%)	NEm[2] (Mcal/lb)	NEg[3] (Mcal/lb)	NEl[4] (Mcal/lb)	Crude Protein (%)	Crude Fiber (%)	ADF[5] (%)	Calcium (%)	Phosphorus (%)
				On a Dry-Matter Basis (moisture-free)						
Alfalfa hay (early bloom)	90	59	0.59	0.28	0.59	19	28	5	1.41	0.26
Barley (grain)	89	84	0.92	0.61	0.87	12	5	7	0.06	0.38
Bermuda grass (hay)	89	53	0.53	0.18	0.53	10	30	39	0.46	0.20
Bluegrass (grazed)	36	69	0.71	0.43	0.70	15	27	32	0.37	0.30
Bone meal (steamed)	95	16	0.27	—	0.11	13	—	—	27.0	12.74
Brome (grazed early)	30	64	0.65	0.36	0.65	15	28	33	0.45	0.34
Corn (whole)	88	87	0.96	0.64	0.90	9	2	3	0.02	0.30
Corn (flaked)	85	93	1.04	0.71	0.97	9	2	3	0.02	0.28
Corn silage (mature)	34	72	0.75	0.47	0.74	8	21	27	0.28	0.23
Corn stover (no ears, husks)	80	0.59	0.59	0.28	0.59	5	35	44	0.35	0.19
Cottonseed meal	92	80	0.86	0.56	0.83	46	13	18	0.21	1.19
Dicalcium phosphate	96	0	0	0	0	0	0	0	22.0	18.65
Grass (hay)	88	58	0.58	0.26	0.58	10	33	41	0.60	0.21
Limestone (ground)	98	0	0	0	0	0	0	0	34.0	0.02
Meadow hay (native)	90	50	0.50	0.12	0.49	7	33	44	0.61	0.18
Milo (sorghum) (ground)	89	82	0.89	0.59	0.85	11	3	6	0.04	0.32
Molasses (cane)	76	75	0.79	0.50	0.77	5	0	0	1.00	0.10
Oats (grain)	89	76	0.81	0.52	0.78	13	11	15	0.05	0.41
Poultry litter (dried)	87	64	0.65	0.36	0.65	25	18	—	3.00	2.05
Prairie hay (midbloom)	91	50	0.50	0.12	0.49	7	34	47	0.40	0.15
Soybean meal (solvent)	91	84	0.92	0.61	0.87	49	6	10	0.38	0.71
Sudan grass (grazed)	18	70	0.73	0.44	0.71	17	23	29	0.46	0.36
Timothy (grazed)	26	64	0.65	0.36	0.65	11	31	36	0.40	0.28
Wheat (grazed early)	21	71	0.74	0.46	0.73	20	18	30	0.35	0.36
Wheat (straw)	91	42	0.43	0	0.40	3	43	58	0.16	0.05
Wheatgrass, crested (early)	37	60	0.60	0.30	0.60	11	26	28	0.46	0.32

Source: Adapted from Preston, 2006.
[1]Total digestible nutrients.
[2]Net energy-maintenance.
[3]Net energy-gain.
[4]Net energy-lactation.
[5]Acid detergent fiber.

TABLE 15.5 Nutrient Composition of Selected Feeds Commonly Used in Rations of Monogastric Animals (air-dry basis)

Feed	Poultry ME[1] (Kcal/lb)	Swine ME[1] (Kcal/lb)	Crude Protein (%)	Minerals Calcium (%)	Phos-phorus (%)	Iron (ppm)	Manga-nese (ppm)	Zinc ppm	Niacin (mg/lb)	Panto-thenic Acid (mg/lb)	Vitamins Ribo-flavin (mg/lb)	Choline (mg/lb)	B$_{12}$ (mg/lb)	Amino Acids Lysine (%)	Meth-ionine (%)	Tryp-tophan (%)
Alfalfa meal (dehydrated)	672	1,020	17.0	1.30	0.23	400	35	21	19	14	6	583	—	0.73	0.28	0.45
Barley (grain)	1,250	1,305	11.5	0.08	0.42	80	16	30	35	3.6	0.7	461	—	0.53	0.18	0.17
Blood meal	1,465	875	80.0	0.28	0.22	2,500	5	22	10	1.4	1.2	244	0.20	6.9	1.0	1.0
Corn (grain)	1,540	1,520	8.0	0.01	0.25	23	6	15	10	2.3	0.5	222	—	.24	0.18	0.7
Feather meal	1,310	1,030	85.0	0.20	0.70	70	9	55	8.9	3.8	0.9	379	0.27	.66	0.55	0.4
Fish meal (whole)	1,180	1,120	61.0	7.00	3.50	80	10	80	23.5	3.7	2.1	1,338	0.07	4.3	1.65	0.7
Meat and bone meal (45%)	1,080	1,090	45.0	11.0	5.9	500	10.1	9.0	22	2.3	7.9	864	0.03	2.2	0.53	0.18
Milo (sorghum)	1,505	1,470	16.0	0.04	0.29	52	13	14	16	5	0.5	288	—	0.27	0.1	0.1
Oats	1,160	1,215	11.0	0.01	0.35	70	38	38	6.2	4.3	0.6	427	—	0.4	0.2	0.18
Soybean meal (solvent)	1,020	1,405	44.0	0.25	0.60	120	27	60	12.1	7.2	1.3	1,151	—	2.9	0.65	0.6
Wheat, hard (red winter)	1,440	1,465	13.5	0.05	0.41	50	40	34	23.5	4.5	0.6	446	—	0.40	0.25	0.18
Whey (dried)	860	1,450	12.0	0.87	0.79	160	7	3	4.5	19.5	11.6	756	—	1.1	0.2	0.2

Source: Adapted from *Feedstuffs,* 2006.

[1]Metabolizable energy.

- Proteins contain 16% nitrogen and compose most of the muscle mass.

- Vitamins and minerals are required in smaller amounts than the other nutrient classes; however, they are necessary for certain metabolic reactions to occur.

- The proximate analysis of a feed sample identifies components that reflect the feeding value—that is, water, crude protein, fat (ether extract), crude fiber, nitrogen-free extract, and ash (minerals).

- Energy evaluation of feeds is measured by total digestible nutrients (TDN), digestible energy (DE), metabolizable energy (ME), and net energy (NE).

KEY WORDS

nutrient
concentrates
roughages
water
moisture
dry matter
carbohydrates
lipids (fats)
ployunsaturated fatty
  acids

proteins (simple and
  complex)
minerals (macro and
  micro)
vitamins (water- and
  fat-soluble)
proximate analysis
digestibility
total digestible nutrient
  (TDN)

digestible energy (DE)
net energy (NE)
calorie
metabolizable energy
  (ME)
heat increment
feedstuffs
digestible protein

REVIEW QUESTIONS

1. Any feed constituent that functions in the support of life is a _____.

2. What are the two classes into which most animal feeds fall?

3. What are the six basic classes of nutrients?

4. What is dry matter of feed?

5. Fats and oils are referred to as _____.

6. What are the three types of carbohydrates?

7. What are the building blocks of protein?

8. What are the two classes of amino acids?

9. What is an essential amino acid?

10. What are the two general classes of mineral nutrients?

11. What are the two groups of vitamins?

12. What are the fat-soluble vitamins?

13. What are the water-soluble vitamins?

14. What is the proximate analysis of feeds?

15. The amount of various nutrients in a feed that can be absorbed from the digestive tract is referred to as _____.

16. How is energy obtained from feed nutrients?

SELECTED REFERENCES

Cheeke, P. J. 2005. *Applied Animal Nutrition.* Upper Saddle River, NJ: Pearson Prentice Hall.

Church, D. C. 1991. *Livestock Feeds and Feeding.* Englewood Cliffs, NJ: Prentice Hall.

Ensminger, M. E., J. E. Oldfield, and W. W. Heinemann. 1990. *Feeds and Nutrition.* Clovis, CA: Ensminger Publishing Co.

Jurgens, M. H. 1993. *Animal Feeding and Nutrition.* 5th ed. Dubuque, IA: Kendall-Hunt.

Martin, D. W., P. A. Mayes, V. W. Rodwell, and D. K. Granner. 1985. *Harper's Review of Biochemistry.* Los Altos, CA: Lange Medical Publications.

National Research Council. *Nutrient Requirements of Beef Cattle* (2000); *Dairy Cattle* (2001); *Goats* (1981); *Horses* (1989); *Poultry* (1994); *Sheep* (1985); and *Swine* (1998). Washington, DC: National Academy Press.

Preston, R. L. 2006. Feed Composition Guide. *Beef.* St. Paul, MN.

CHAPTER 16

Digestion and Absorption of Feed

LEARNING OBJECTIVES

- Define carnivores, omnivores, and herbivores
- Describe the structures and functions of monogastric digestive systems
- Describe the structures and functions of ruminant digestive systems
- Describe the structures and functions of poultry digestive systems
- Compare and contrast digestion in monogastrics, ruminants, and poultry

Animals obtain substances needed for all body functions from the feeds they eat and the liquids they drink. Before the body can absorb and use them, feeds must undergo a process called **digestion.** Digestion includes mechanical action, such as chewing and contractions of the intestinal tract; chemical action, such as the secretion of hydrochloric acid (HCl) in the stomach and bile in the small intestine; and the action of enzymes, such as maltase, lactase, and sucrase (which act on disaccharides), lipase (which acts on lipids), and peptidases (which act on proteins). Enzymes are produced either by the various parts of the digestive tract or by microorganisms. The role of digestion is to reduce feed particles to molecules so they can be absorbed into the blood and eventually support body functions.

CARNIVOROUS, OMNIVOROUS, AND HERBIVOROUS ANIMALS

Animals are classed as carnivores, omnivores, or herbivores according to the types of feed they normally eat. **Carnivores,** such as dogs and cats, normally consume animal tissues as their source of nutrients; **herbivores,** such as cattle, horses, sheep, and goats, primarily consume plant tissues. Humans and pigs are examples of **omnivores,** which eat both plant and animal products.

Carnivores and omnivores are monogastric animals, meaning that their stomachs are simple in structure, having only one compartment. Some herbivores, such as horses and rabbits, are also monogastric. Other herbivores, such as cattle, sheep, and goats, are ruminant animals, meaning that their stomachs are complex in structure, containing four compartments.

The digestive tracts of pigs and humans are similar in anatomy and physiology; therefore, much of the information gained from studies on pig nutrition and digestive physiology can be applied to humans. Both the pig and the human are omnivores and both are monogastric animals. Neither can synthesize the B-complex vitamins or amino acids to a significant extent. Both pigs and humans tend to eat large quantities, which can result in obesity. Humans can control obesity by regulating food intake and exercising as a means of using, rather than storing, excess energy. Obesity in swine can be controlled by limiting the amount of feed available to them and through genetic selection of leaner animals. The latter has received the greater emphasis as pigs are typically fed ad libitum.

DIGESTIVE TRACT OF MONOGASTRIC ANIMALS

The anatomy of the digestive tract varies greatly from one species of animal to another. The basic parts of the digestive tract are the mouth, stomach, small intestine, and large intestine (or colon). The primary function of the parts preceding the intestines is to reduce the sizes of feed particles. The **small intestine** functions in splitting food molecules and in absorbing nutrients, whereas the **large intestine** absorbs water and forms indigestible wastes into a solid called **feces.** In a mammal having a simple stomach (such as the pig), the mouth has teeth and lips for grasping and holding feed that is masticated (chewed), and salivary glands that secrete saliva for moistening feed so it can be swallowed.

Feed passes from the mouth to the **stomach** through the **esophagus.** There is a sphincter (valve) at the junction of the stomach and esophagus, which can prevent feed from coming up the esophagus when stomach contractions occur. The stomach empties its contents into that portion of the small intestine known as the **duodenum.** The pyloric sphincter, located at the junction of the stomach and the duodenum, can be closed to prevent feed from moving into or out of the stomach. Feed goes from the duodenum to portions of the small intestine known as the jejunum and the ileum. It then passes from the small intestine to the large intestine, or colon. The ileocecal valve, located at the junction of the small intestine and the colon, prevents material in the large intestine from moving back into the small intestine.

The small intestine actually empties into the side of the colon near, but not at, the anterior end of the colon. The blind anterior end of the colon is the cecum, or, in some animals, the vermiform appendix. The large intestine empties into the rectum. The anus has a sphincter, which is under voluntary control so that the animal can prevent defecation until it actively engages in the process. The structures of the digestive system of the pig are shown in Figure 16.1

Animals such as pigs, horses, and poultry are classed as monogastric animals, but they differ markedly in certain ways. For example, the horse (Fig. 16.2) has a large structure called the **cecum,** where feed is fermented. Because the cecum is posterior to the area where most feed is absorbed, horses do not obtain all of the nutrients made by microorganisms in the cecum. Digestive tract sizes and capacities of monogastric animals are contrasted in Table 16.1.

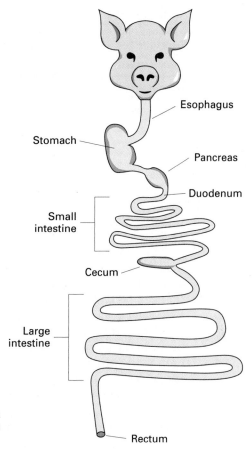

FIGURE 16.1 Digestive tract of the pig as an example of the digestive tract of a monogastric animal.

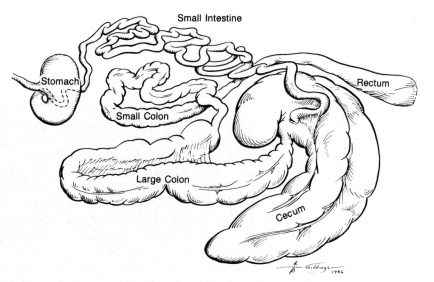

FIGURE 16.2 Digestive system of the horse. The posterior view shows the colon or large intestine proportionally larger than the rest of the digestive tract. Note particularly the location of the cecum at the anterior end of the colon.

TABLE 16.1   Digestive Tract Sizes and Capacities of Selected Monogastric Animals

Part of Digestive Tract	Species			
	Human	Pig	Horse	Chicken
Esophagus	—	—	4 ft	Total length of digestive tract in mature chickens is approximately 7 ft (beak to crop, 7 in; beak to proventriculus, 14 in; duodenum, 8 in; ileum and jejunum, 48 in; and cecum, 7 in)
Stomach	1 qt	2 gal	4 gal	
Small intestine	1 gal	2 gal	12 gal	
		60 ft	70 ft	
Large intestine	1 qt	3 gal	11 gal	
		12 ft	20 ft	
Cecum			8 gal	
			4 ft	

The digestive tracts of most poultry species differ from that of the pig in several respects. Because they have no teeth, poultry break their feed into a size that can be swallowed by pecking with their beaks or by scratching with their feet. Feed goes from the mouth through the esophagus to the **crop,** which is an enlargement of the esophagus where feed can be stored. Some fermentation may occur in the crop, but it does not act as a fermentation vat. Feed passes from the crop to the **proventriculus,** which is a glandular stomach in that it secretes gastric juices and hydrochloric acid but does not grind feed. Feed then goes to the **gizzard,** where it is ground into finer particles by strong muscular contractions. The gizzard apparently has no function other than to reduce the size of feed particles. Feed moves from the gizzard into the small intestine. Material from the small intestine empties into the large intestine. At the junction of the small and large intestines are two ceca, which contribute little to digestion. Material passes from the large intestine into the **cloaca,** into which urine also empties. Material from the cloaca is voided through the vent (Fig. 16.3).

STOMACH COMPARTMENTS OF RUMINANT ANIMALS

In contrast to the single stomachs of monogastric animals, the stomachs of cattle, sheep, and goats have four compartments—**rumen, reticulum, omasum,** and **abomasum** (Fig. 16.4 and 16.5). The rumen is a large fermentation vat where bacteria and protozoa thrive and break down roughages to obtain nutrients for their use. It is lined with numerous papillae, which give it the appearance of being covered with a thick coat of short projections. The papillae increase the surface area of the rumen lining. The microorganisms in the rumen can digest cellulose and can synthesize amino acids from nonprotein nitrogen. The B-complex vitamins are also synthesized in the rumen. Later, these microorganisms are digested in the small intestine to provide these nutrients for the ruminant animal's use.

The reticulum has a lining with small compartments similar to a honeycomb; thus it is occasionally referred to as the **honeycomb.** Its function is to interact with the rumen in initiating the mixing activity of the rumen and providing an additional area for fermentation. The omasum has many folds, so it is often called the **manyplies.** The omasum may not have a major digestive

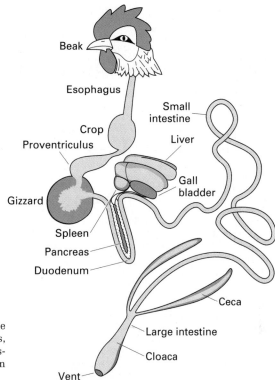

FIGURE 16.3 Digestive tract of the chicken showing crop, proventriculus, and gizzard, all of which are characteristic of poultry. *Source:* J. E. Parker, Oregon State University.

function, although some believe that the folds produce a grinding action on the feed. The abomasum, or true stomach, corresponds to the stomach of monogastric animals and performs a similar digestive function.

The size and capacity of the ruminant stomach and intestinal tract are given in Table 16.2. The data in Table 16.2 are for mature ruminants, as the relative proportions of the stomach compartments are considerably different in the young lamb and calf. At birth, the abomasum comprises 60% of the total stomach capacity, whereas the rumen is only 25% of the total.

Animals having the four-compartment stomach eat forage rapidly and later, while resting, regurgitate each bolus of feed (known as the **cud**). The regurgitated feed is chewed more thoroughly and swallowed, and then another bolus is regurgitated and chewed. This process continues until the feed is thoroughly masticated. The regurgitation and chewing of undigested feed is known as **rumination.** Animals that ruminate are known as **ruminants.** As microorganisms in the rumen ferment feeds, large amounts of gases (chiefly methane and carbon dioxide) are produced. The animal normally can eliminate the gases by controlled belching, also called **eructation.**

DIGESTION IN MONOGASTRIC ANIMALS

Feed that is ingested (taken into the mouth) stimulates the secretion of saliva. Chewing reduces the size of ingested particles and saliva moistens the feed. The enzyme amylase, present in the saliva of some species including pigs and humans, acts on starch. (Ruminants do not secrete salivary amylase.) However,

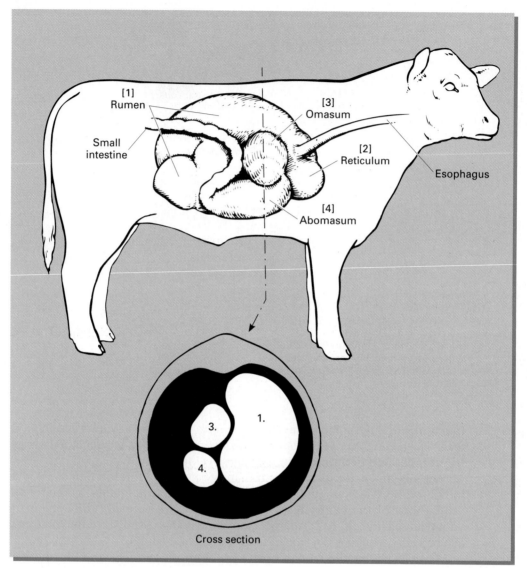

FIGURE 16.4 Beef cattle digestive tract.

little actual breakdown of starch into simpler compounds occurs in the mouth, primarily because feed is there for only a short time.

An **enzyme** is an organic catalyst that speeds a chemical reaction without being altered by the reaction. Enzymes are rather specific; that is, each type of enzyme acts on only one or a few types of substances. Therefore, it is customary to name enzymes by giving the name of the substance on which it acts and adding the suffix -*ase*, which, by convention, indicates that the molecules so named are enzymes (Table 16.3). For example, lipase is an enzyme that acts on lipids (fats); maltase is an enzyme that acts on maltose to convert it into two molecules of glucose; lactase is an enzyme that acts on lactose to convert it into one molecule of glucose and one molecule of galactose; and sucrase is an enzyme that acts on sucrose to convert it into one molecule of glucose and one

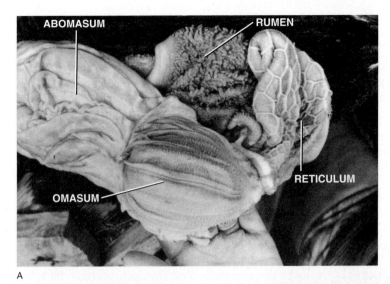

FIGURE 16.5 Lining of the four compartments of the ruminant stomach (goat). (A) Compartments intact. (B) Compartments separated. Courtesy of George F. W. Haenlein. University of Delaware.

TABLE 16.2   Digestive Tract Sizes and Capacities of Mature Ruminant Animals

	Species	
Part of Digestive Tract	Cow	Ewe
Stomach		
Rumen	40 gal	5 gal
Reticulum	2 gal	2 qt
Omasum	4 gal	1 qt
Abomasum	4 gal	3 qt
Small intestine	15 gal (130 ft)	2 gal (80 ft)
Large intestine	10 gal	6 qt

TABLE 16.3   Enzymes Important in the Digestion of Feed

Enzyme	Substrate	Substances Resulting from Enzyme Action
Amylase	Starch	Disaccharides, dextrin
Chymotrypsin	Peptides	Amino acids and peptides
Lactase	Lactose	Glucose and galactose
Lipase	Lipids	Fatty acids and glycerides
Maltase	Maltose	Glucose and glucose
Pepsin	Protein	Polypeptides
Peptidases	Peptides	Amino acids
Sucrase	Sucrose	Glucose and fructose
Trypsin	Protein	Polypeptides

molecule of fructose. Some lipase is present in saliva, but little hydrolysis of lipids into fatty acids and glycerides occurs in the mouth.

As soon as it is moistened by saliva and chewed, feed is swallowed and passes through the esophagus to the stomach. The stomach secretes HCl, mucus, and the digestive enzymes pepsin and gastrin. The strongly acidic environment in the stomach favors the action of pepsin which breaks proteins down into polypeptides. The HCl also assists in coagulation, or curdling, of milk. Little breakdown of proteins into amino acids occurs in the stomach. Mucous secretions help to protect the stomach lining from the action of strong acids.

In the stomach, feed is mixed well and some digestion occurs; the mixture that results is called **chyme.** The chyme passes next into the duodenum, where it is mixed with secretions from the pancreas, bile, and enzymes from the intestine.

Secretin, pancreozymin, and cholecystokinin, three hormones that are released from the duodenal cells stimulate secretion from the pancreas and discharge of bile from the gallbladder. The enzymes from the pancreas are lipase, which hydrolyzes fats into fatty acids and glycerides; trypsin, which acts on proteins and polypeptides to reduce them to small peptides; chymotrypsin, which acts on peptides to produce amino acids; and amylase, which breaks starch down to disaccharides, after which the disaccharides are broken down to monosaccharides. The liver produces bile that emulsifies fats; the bile is strongly alkaline and so helps to neutralize the acidic chyme coming from the stomach. Some minerals that are important in digestion also occur in bile.

By the time they reach the small intestine, amino acids, fatty acids, and monosaccharides (simple sugars or carbohydrates) are all available for absorption. Thus, the small intestine is the most important area for both digestion and absorption of feed. Absorption of feed molecules may be either passive or active. Passive passage results from diffusion, which is the movement of molecules from a region of high concentration of those molecules to a region of low concentration. Active transport of molecules across the intestinal wall may be accomplished through a process in which cells of the intestinal lining (**villi,** shown in Fig. 16.6) engulf the molecules and then actively transport these molecules to either the bloodstream or the lymph. Energy is expended in accomplishing the active transport of molecules across the gut wall.

When molecules of digested feed enter the capillaries of the blood system, they are carried directly to the liver. Molecules may enter the lymphatic system, after which they go to various parts of the body, including the liver. The liver is

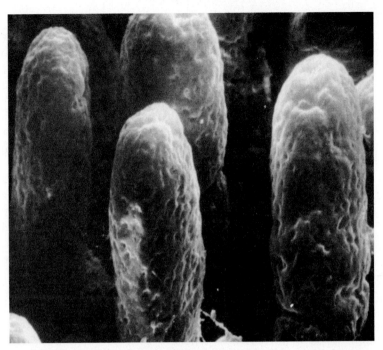

FIGURE 16.6 Electron micrograph of the lining of the small intestine. These projections (villi) increase the surface area and are covered with cells that digest and absorb nutrients from the feed (magnified 200×). Courtesy of Dr. G. L. Waxler, 1972, *Am. J. Vet. Res.* 33:1323.

an extremely important organ both for metabolizing useful substances and for detoxifying harmful ones.

In some monogastric animals, such as the horse, postgastric (cecal) fermentation of roughages occurs. In these animals, the feed that can be digested is digested and absorbed before the remainder reaches the cecum. These animals are perhaps more efficient than ruminants in their use of feeds such as concentrates. In the ruminant animals, the bacteria and protozoa use the concentrates given along with roughages. Because the microorganisms in ruminants use starches and sugars, little glucose is available to ruminants for absorption. The microorganisms do provide volatile fatty acids, which are absorbed by the ruminant and converted to glucose as an energy source. The postgastric fermentation that occurs in horses breaks down roughages, but this takes place posterior to the areas where nutrients are most actively absorbed; consequently, the animal does not obtain all nutrients resulting from postgastric fermentation.

DIGESTION IN RUMINANT ANIMALS

In mature ruminant animals (cattle, sheep, and goats), predigestive fermentation of feed occurs in the rumen and reticulum. The bacteria and protozoa in the rumen and reticulum use roughages consumed by the animal as feed for their growth and multiplication; consequently, billions of these microorganisms develop. The rumen environment is ideal for microorganisms because moisture, a warm temperature, and a constant supply of nutrients are present.

Excess microorganisms are continuously removed from the rumen and reticulum along with small feed particles that escape microbial fermentation and pass through the omasum into the abomasum. When feed passes into the abomasum, strong acids destroy the bacteria and protozoa. The ruminant animal then digests the microorganisms in the small intestine and uses them as a source of nutrients. The digested microbial cells provide the animal with most of its amino acid needs and some energy. Thus, ruminant animals and microorganisms mutually benefit each other. All digestive processes in ruminants are the same as those in monogastric animals after the feed reaches the abomasum, which corresponds to the stomach of monogastric animals.

The rumen fermentation process also produces **volatile fatty acids** (acetic, propionic, and butyric acids), which are waste products of microbial fermentation of carbohydrates. The animal then uses these volatile fatty acids (VFAs) as its major source of energy. In the process of fermenting feeds, methane gas is also produced by the microorganisms. The animal releases the gas primarily through belching. Occasionally, the gas-releasing mechanism does not function properly and gas accumulates in the rumen, causing **bloat** to occur. Death will occur owing to suffocation if gas pressure builds to a high level and interferes with adequate respiration.

A young, nursing ruminant consumes little or no roughage. Consequently, at this early stage of life, its digestive tract functions similarly to that of a monogastric animal. Milk is directed immediately into the abomasum in young ruminants by the **esophageal groove** (Fig. 16.7). The sides of the esophageal groove

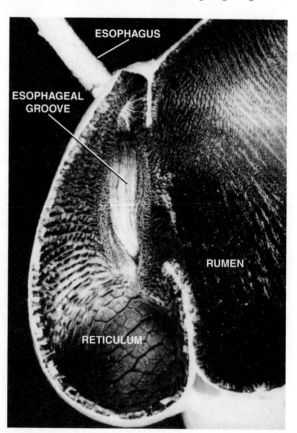

FIGURE 16.7   The esophageal groove, with its location relative to the esophagus, reticulum, and rumen. Courtesy of N. J. Benevenga et al., 1969. Preparation of the ruminant stomach for classroom demonstration. *J. Dairy Sci.* 52:1294.

extend upward by a reflex action and form a tube through which milk passes directly from the esophagus to the abomasum. This allows milk to bypass fermentation in the rumen. Rumen fermentation is an inefficient use of energy and protein in a high-quality feed such as milk.

When roughage is consumed it is directed into the rumen, where bacteria and protozoa break it down into simple forms for their use. The rumen starts to develop functionally as soon as roughage enters it, but some time is required before it is completely functional. Complete development of the rumen, reticulum, and abomasum requires about 2 months in sheep and about 3–4 months in cattle. The type of feed given to the animal can influence development of the rumen. If only milk and concentrated feeds are given, the rumen shows little development. If very young ruminants are forced to live on forage, the rumen develops much more rapidly.

Energy Pathways

Figure 16.8 shows the digestion and utilization of carbohydrates and fats contained in the ingested forages and grains. The primary energy end products of glucose and fatty acids supply energy in the body tissues and become milk fat and lactose in the lactating ruminant. Excess energy is stored as body fat in the body tissues.

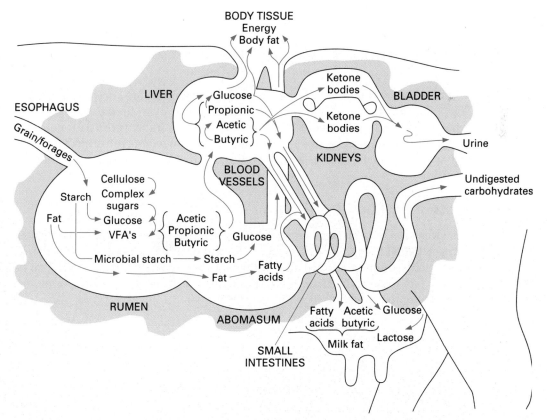

FIGURE 16.8 Energy pathways in the ruminant. *Source:* J. Bryant and B. R. Moss, Montana State University.

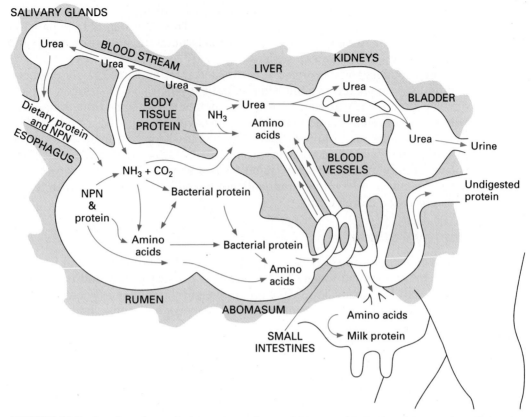

FIGURE 16.9　Protein pathways in the ruminant. *Source:* J. Bryant and B. R. Moss, Montana State University.

The primary organs and tissues in energy metabolism are shown in Fig. 16.8. These are the rumen, abomasum, small intestines, liver, blood vessels, mammary gland, and body tissues. Undigested carbohydrates (primarily complex carbohydrates such as lignin) are excreted through the large intestine. Other energy waste products, such as ketone bodies, are excreted through the kidneys in the urine.

Protein Pathways

The digestion, utilization, and excretion of dietary protein and **nonprotein nitrogen (NPN)** are shown in Fig. 16.9. The end products of protein and NPN are amino acids, ammonia (NH_3), and synthesized amino acids. Excess NH_3 can be formed into urea in the liver, then excreted through the urine, with some urea returning to the rumen as a component of saliva.

CHAPTER SUMMARY

- The digestive tract of a monogastric animal consists of the mouth, esophagus, stomach, small intestine, and large intestine (colon).
- The stomachs of ruminant animals have four compartments—rumen, reticulum, omasum, and abomasum. The latter functions like the stomach of a monogastric animal.

- Microorganisms in the rumen digest cellulose, synthesize amino acids, and synthesize B vitamins.
- Poultry differs from other monogastric animals in having a crop (for food storage), proventriculus (for chemical digestion), and gizzard (for physical grinding of the feed).
- The horse has a cecum (a large pouch between the small and large intestine) where feed is fermented as in the rumen of ruminant animals.
- Digestion breaks down the feed into smaller components that can be absorbed and utilized by the animal—carbohydrates to simple starches and sugars, proteins to amino acids, and fat to fatty acids.

KEY WORDS

digestion	crop	rumination
carnivore	proventriculus	ruminant
herbivore	gizzard	eructation
omnivore	cloaca	enzyme
small intestine	rumen	chyme
large intestine	reticulum	villi
feces	omasum	volatile fatty acids
stomach	abomasum	bloat
esophagus	honeycomb	esophageal groove
duodenum	manyplies	nonprotein nitrogen
cecum	cud	(NPN)

REVIEW QUESTIONS

1. What is the role of digestion?
2. How are animals classified?
3. What are the three classes of animals according to the types of feed they normally eat?
4. What are the basic parts of the digestive tract?
5. The large intestine absorbs water and forms indigestible wastes into a solid called _____.
6. What are the three regions of the small intestine?
7. *True or False:* The stomach of ruminants such as cattle, sheep, and goats has four compartments.
8. What is the function of the rumen?
9. What structure of the colon, in some species such as horses, performs the function of microbial fermentation of feed?
10. What feed component is most effectively digested by fermentation in the rumen and cecum?
11. What is the significance of fermentative digestion of cellulose by microorganisms in the rumen and cecum?
12. Why are ruminants more efficient at utilizing roughages than are animals, such as horses, that possess a large active cecum?

13. What are the three regions of the large intestine?

14. What is the glandular stomach in birds called?

15. What is the crop in birds?

16. *True or False:* Mechanical digestion in birds occurs in the mouth, just as it does in mammals.

17. What are the four stomach compartments in ruminants?

18. Ruminants regurgitate undigested roughage, which is referred to as _____, or chewing the _____.

19. _____ are organic catalysts that speed up a chemical reaction without being altered by the reaction and are important in digestion of specific nutrients.

20. What is the function of the esophageal groove?

21. What are the products of fermentative digestion in the rumen that ruminant animals utilize as energy?

SELECTED REFERENCES

Cheeke, P. J. 2005. *Applied Animal Nutrition.* Upper Saddle River, NJ: Pearson Prentice Hall.

Church, D. C. 1991. *Livestock Feeds and Feeding.* Englewood Cliffs, NJ: Prentice Hall.

Ensminger, M. E., J. E. Oldfield, and W. W. Heinemann. 1990. *Feeds and Nutrition.* Clovis, CA: Ensminger Publishing Co.

CHAPTER 17

Providing Nutrients for Body Functions

LEARNING OBJECTIVES

- Describe the high priority maintenance functions
- Discuss the most important minerals and vitamins involved in animal growth
- Explain the nutrient requirements for reproduction and lactation
- Describe the nutrient requirements for egg laying
- Describe the nutrient requirements for wool production
- Demonstrate the use of the Pearson square in ration balancing

Feeding animals is of fundamental importance to any livestock or poultry production program because animals must be healthy to function efficiently and yield maximum profits to the producer. The basic task of the producer, then, is to supply animals with feed that satisfies their body functions for maintenance, growth, fattening, reproduction, lactation, egg laying, wool production, and work. Each of these functions has a unique set of nutrient requirements, which are additive when more than one function is occurring.

Profits derived from any feeding program must be assessed against production costs. Knowledgeable producers can increase their profits by feeding their animals both adequately and economically.

NUTRIENT REQUIREMENTS FOR BODY MAINTENANCE

Body maintenance requires that nutrients be supplied to keep the body functioning in a state of well-being. There is no gain or loss of weight or production. Maintenance functions that have a high priority for nutrients are (1) body tissue repair, (2) control of body temperature, (3) energy to keep all vital organs (respiratory, digestive, and so on) functioning, and (4) water balance maintenance.

Nutrients first meet maintenance needs before supplying any of the other body functions. Approximately half of all feed fed to livestock and poultry is used to fill the maintenance requirement. Feedlot animals on full feed may only use 30–40% of their nutrients for maintenance, while some mature breeding animals may need 90% of their feed for maintenance. Highly efficient dairy cows, producing over 100 lb of milk per day, have a daily feed consumption four to five times their maintenance requirement.

Body Size and Maintenance

Maintenance needs are related to body size. A large animal obviously needs more feed than a small one, but maintenance requirements are not linearly related to body weight. Small animals require more feed per unit of body weight for maintenance than large ones. The approximate maintenance requirement in relation to weight is expressed as $Wt^{0.75}$ rather than $Wt^{1.00}$. Thus, if a 500-lb animal requires 15 lb of feed per day for maintenance, a 1,000-lb animal of the same type does not require twice as much feed even though the latter animal weighs twice as much as the first. The quantity of $1,000^{0.75}$ can be determined and applied to show that the 1,000-lb animal requires approximately 1.7 times as much feed for maintenance as the 500-lb animal. The 1,000-lb animal therefore requires approximately 25.5 lb daily (15 lb $\times$ 1.7 = 25.5 lb). Table 17.1 shows how the TDN requirement changes with increasing weight. Where corn is 91% TDN, on an air-dry basis, it would take approximately 12.5 lb (12.5 lb $\times$ 0.91 = 11.4 lb TDN) of corn to fill the maintenance requirement of a 1,000-lb animal.

NUTRIENT REQUIREMENTS FOR GROWTH

Growth occurs when protein synthesis is in excess of protein breakdown. When cells increase in number or size, or when a combination of both takes place. Growth at the tissue level is accomplished primarily through the building of muscle, bone, and connective tissue.

There are several important nutrient requirements for growth, including energy, protein, minerals, vitamins, and energy. The dry matter of muscle and connective tissue is composed largely of protein; therefore, young, growing animals that need feed to sustain growth in addition to maintenance have greater protein requirements. A young, growing animal is like a muscle-building factory, and protein in the feed is the raw material for the manufacturing process. If provided with only a maintenance amount of feed for an extended period, a young animal may be permanently stunted. The nutrient requirements for young growing beef calves are provided in Table 17.2. As weight and growth rate

TABLE 17.1   TDN Needed for Maintenance of Cattle in the Growing-Finishing Period

Weight of Cattle (lb)	TDN Needed Daily for Maintenance (lb)
400	5.7
600	7.7
800	9.7
1,000	11.4
1,100	12.3

TABLE 17.2 Nutrient Requirements of Growing Beef Calves

Weight (lb)	Average Gain (lb/day)	TDN (lb)	Metabolizable Protein (lb)	Ne$_m$ (Mcal/day)	NE$_g$ (Mcal/day)
440	1.0	7.5	.34	4.1	1.3
	2.0	7.9	.66	4.1	2.7
	3.0	8.1	.97	4.1	4.2
550	1.0	8.0	.34	4.8	1.5
	2.0	8.8	.66	4.8	3.2
	3.0	9.9	.97	4.8	5.0
660	1.0	8.4	.35	5.6	1.7
	2.0	9.4	.67	5.6	3.7
	3.0	11.9	.97	5.6	5.7

Source: National Research Council, *Nutrient Requirements of Beef Cattle*, 2000.

increase, total nutrient intake (TDN), protein intake, and net energy intake requirements also increase.

Monogastric animals not only need a certain quantity of protein, but they also must have certain amino acids for proper growth. The protein needs of hogs, for example, are usually supplied by feeding them soybean meal as a supplemental source of amino acids. Young ruminant animals cannot consume enough roughage to achieve maximum growth. If dams that produce adequate amounts of milk are nursing young ruminant animals, the young will do well on good pastures, good quality hay, or both together.

The mineral needs of a young, growing animal include calcium and phosphorus for proper bone growth, salt or a normal sodium level in the body, and any mineral that may be deficient in the area in which the animal lives. **Calcium** is usually plentiful in legume forages, and **phosphorus** is usually plentiful in grains, so a combination of hay and grain should provide all the calcium and phosphorus that young ruminant animals need. Animals fed on hay alone may need additional phosphorus, and those fed diets high in concentrates may need additional calcium. Some producers feed a mixture of steamed bone meal or dicalcium phosphate and salt at all times to ensure that their animals have the necessary calcium and phosphorus.

Two minerals, **iodine** and **selenium,** require special consideration. Some areas may be deficient in one or both of these elements. An insufficient amount of iodine in the ration of pregnant females might cause an iodine deficiency in the fetus, which prevents thyroxine from being produced and thus causes a goiter in the newborn. Young animals with goiters die shortly after they are born. Iodized salt can be easily provided to the pregnant females to avoid the iodine deficiency.

A lack of selenium might cause the young to be born with white muscle disease. Giving the pregnant female an injection of selenium can prevent this deficiency. The injectable selenium is distributed commercially. Directions for proper dosages that are supplied by the distributor should be followed closely because an overdose in natural feeds or supplements can kill the animal.

Young, growing animals need vitamins. Young ruminant animals are usually on pasture with their dams and thus are exposed to sunshine. The action of ultraviolet rays from the sun converts cholesterol in the skin into vitamin D, providing the animal with this vitamin. **Vitamin D** is needed for the proper use of calcium and phosphorus in bone growth. Because pigs, poultry, and rabbits are often raised inside, where sunshine is limited or lacking, they need a dietary source of vitamin D.

Most vitamins must be supplied to pigs and poultry through feeds. The only vitamin commonly fed to ruminant animals is **vitamin A,** and then only when they are on dry pasture or are being fed hay that is quite mature or that has been dried in the sun for several days due to being moistened in processing. Vitamin A is easily lost in sunlight and during extended dry storage. The activity of vitamin A in silage is often quite high because this vitamin is usually preserved by the acid fermentation that takes place when silage is made. Silage, however, is usually quite low in vitamin D because the plants used in making silage do not make this vitamin through the action of sunlight and they are not exposed to sunlight for long periods after they are cut.

Young animals need sufficient energy to sustain their growth, high metabolic rate, and activities. They obtain some energy from their mother's milk, and additional energy is supplied as carbohydrates (starch and sugar) and fats from grazed forage or supplemental feeds. Feed grains are high in carbohydrates and also contain some fats. Young ruminants on good pasture typically obtain sufficient energy from pasture and from milk.

The energy needs of young pigs and poultry are generally supplied by feeding them grains such as corn, barley, or wheat. Young horses can usually obtain their energy needs when they are on pasture with their dams since most mares produce much milk; however, for adequate growth after weaning, they need some grain in addition to good pasture or good quality hay.

Water availability and intake are required for normal function. Intake will vary based on animal age, stage of production (lactating versus dry), level of feed intake, and a variety of environmental factors. The range of expected intake of water for adult farm animals is provided in Table 17.3.

NUTRIENT REQUIREMENTS FOR FATTENING

Fattening is the storing of surplus feed energy as fat both within and around body tissues. Fattening is desirable to give meat some of its palatability characteristics and to provide energy reserves for postpartum reproductive performance.

Gain from growth is usually less costly than gain from fattening. It takes 2.25 times as much energy to produce a pound of fat as to produce a pound of protein tissue.

Fattening is the result of excess energy from carbohydrates, fats, or protein above that required for maintenance and growth. Usually fattening animals are

TABLE 17.3   **Water Intake of Adult Livestock[a]**

Species	Water (gal/day)
Cattle (beef)	7–20
Cattle (dairy)	10–29
Chickens	0.05–0.10
Horses	8–12
Sheep and goats	1–4
Swine	3–5
Turkeys	0.10–0.15

[a] Temperate climatic conditions.
Source: Adapted from Cheeke, 2005.

full-fed high-energy rations during the last phase of the growing-finishing feeding program.

NUTRIENT REQUIREMENTS FOR REPRODUCTION

The requirements for reproduction fall into two categories—those for (1) gamete production and (2) fetal growth in the uterus. In general, healthy males and females are capable of producing gametes. The energy needs for germ cell production are no greater than those needed to keep animals in a normal, healthy condition. For example, ruminant animals grazing on pastures of mixed grass and legumes are generally neither deficient in phosphorus nor lacking in fertility. A lack of phosphorus may cause irregular estrous cycles and impaired breeding in females.

Animals that are losing weight rapidly because of poor feed conditions and animals that are overly fat may be low in fertility. To attain optimum fertility from female animals, they should be in moderately low to moderate body fat conditions as breeding season approaches, but should, ideally, be increasing in condition (i.e., gaining weight) for 2–4 weeks before and during the breeding season.

The nutrients required by the **growing fetus** are much greater in the last trimester of pregnancy than earlier, as little fetal growth occurs during the first two trimesters of pregnancy. Because the fetus is growing, its requirements are the same as those for growth of a young animal after it is born. Healthy females can withdraw nutrients from their bodies to support the growing fetus temporarily while the amount or quality of their feed is low, but reproductive performance will be lower if nutrition is inadequate for 2–3 months in cattle and a few weeks in swine and sheep.

NUTRIENT REQUIREMENTS FOR LACTATION

Among common farm animals, dairy cows and dairy goats produce the most milk; however, most mammalian females are expected to produce milk for their young as part of normal enterprise management schedules. Milk production requires considerable protein, minerals, vitamins, and energy. The need for protein is greater because milk contains more than 3% protein. As an example, a cow that weighs 1,500 lb and produces 30 lb of milk per day needs at least 30 lb of feed per day, which contains 15% protein; this gives her 4.5 lb protein for her body and the milk she produces. If the protein she eats is 60% digestible, there is 2.7 lb of digestible protein, of which 1.5 lb is present in her milk. If a cow of this weight is to produce 100 lb of milk per day, she must now consume 50 lb of feed containing 15% protein to compensate for the 3 lb of protein in the 100 lb of milk that she gives. Generally, during peak milk production, feed consumption cannot compensate for nutrient output and the cow does mobilize some body protein. Actually, more body energy is mobilized than body protein to meet the nutrient deficit.

Calcium and phosphorus are the two most important minerals needed for lactation. Milk is rich in these minerals; their absence or imbalance may result in decreased lactation or may even cause death. The dairy cow may develop milk fever shortly after calving if there is an exceptionally heavy drain of calcium from her system. Cows afflicted by milk fever might become comatose and die if not treated. However, an intravenous injection of calcium gluconate usually helps the cow recover in less than a day. Milk fever rarely occurs in species or animals that produce relatively small quantities of milk.

Dairy cows produce milk that contains considerable quantities of vitamin A and most B-complex vitamins. Because cows are ruminants, it is unnecessary to feed them B-complex vitamins, and they require vitamin D supplementation only if confined indoors.

Exceptions occur with beef cows that give large amounts of milk. If they do not have adequate feed while they are nursing their calves, they may not conceive for the next calf crop. In those parts of the world where sheep and dairy goats are the principal dairy animals, the energy needs of these animals are quite similar to those of dairy cows that produce much milk.

The requirements for milk production in sows are usually provided in several ways—by increasing the percentage of protein in the ration, by increasing the amount of feed allowed, and by providing a mineral mix (a combination of minerals that usually contains calcium, phosphorus, salt, and some trace minerals).

Energy is perhaps the most vital requirement for the production of much milk. The energy need is based on the amount of milk being produced. A lactating cow needs energy for body maintenance while she also produces milk and provides the energy stored in it. She cannot eat enough hay to obtain the quantity of energy needed, so she must receive high-energy feeds such as concentrates; even then her production may be limited by the amount of feed she can eat. A high-producing dairy cow may need three to four times the energy of a nonlactating cow of the same size. Even when fed large amounts of concentrates, a cow that is producing much milk often loses weight and body condition because she cannot consume enough feed to produce at her maximum level; therefore, she draws on her body reserves to supply part of her energy needs.

In the dairy cow, the roughage-to-concentrate ratio should be approximately 40:60, as a certain amount of roughage is needed to maintain the desired fat content in milk. Therefore, simply feeding more concentrates is not the only factor in increased milk production.

NUTRIENT REQUIREMENTS FOR EGG LAYING

The nutrient requirements of poultry are dependent on the specific purposes of production. For example, broilers need nutrients primarily for growth, with less emphasis on egg production; Leghorn-type hens (layers) need nutrients with primary emphasis on eggs and less on growth. Nutrient requirements for growth are discussed earlier in this chapter.

Leghorn-type chickens are smaller in body size than broilers, so their maintenance requirements are less. They are prolific in egg production, so they are usually fed *ad libitum* during the growing and laying period. Because layers eat *ad libitum* to satisfy their energy needs, rations need to have adequate concentrations of energy, protein (amino acids), vitamins, and minerals.

NUTRIENT REQUIREMENTS FOR WOOL PRODUCTION

Nutrient requirements for wool production are in addition to nutrients needed for maintenance, growth, and reproduction. Insufficient energy owing to amount or quality of feed is usually the most limiting nutritional factor affecting wool production. As wool fibers are primarily protein in composition, the ration should be adequate in protein content.

Shearing removes the sheep's natural insulation and may cause an increase in energy requirements owing to heat loss. This is especially true when periods of cold weather occur shortly after shearing.

NUTRIENT REQUIREMENTS FOR WORK

Animals used for work, either for pulling heavy loads or for being ridden, require large amounts of energy in addition to their needs for maintenance. Horses are the primary work animals in the United States, but elsewhere, donkeys, cattle, and water buffalo are used.

Horses, mules, and donkeys rely partly on perspiration to remove nitrogenous wastes. If a horse is used for hard work 5 days of the week and is not allowed to exercise the next 2 days, a strain is placed on the kidneys and illness may result.

The primary requirement is energy above that needed for maintenance and growth. If energy in the ration is not sufficient to meet the work needs, then body fat stores will provide the additional energy needs.

RATION FORMULATION

This chapter is not intended to cover the details of ration formulation. Attempting to expose the reader to this area without providing the details can be misleading. Books are written and courses taught on feeds and feeding or animal nutrition that provide an in-depth coverage of this topic.

Typically producers or their consultants balance rations with computer-assisted formulation software systems. However, the **Pearson square method** is an approach to balancing simple rations such as in the case of two-ingredient formulations.

Assume that a producer wants to determine the appropriate amounts of soybean meal and corn to utilize in combination to meet a goal of providing 14% crude protein in the diet. Utilizing the scenario in which a feed test reveals that the crude protein of the available corn and soybean meal is 11 and 45%, respectively, the ingredient values are listed on the left side of the box with the desired CP% at the center. Values are subtracted across the diagonal (convert all negative values to positive) to determine the ratio of feed ingredients required to meet the 14% CP goal.

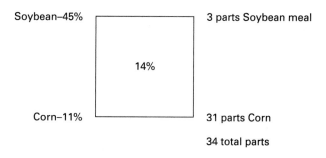

Corn represents 91% of the ration (31/34) while soybean meal comprises the other 9% (3/32) to yield a ration that meets the desired level of protein. The calculation can be verified as follows:

$$91 \text{ lb corn} \times 11\% \text{ CP} = 10.01 \text{ lbs CP}$$
$$9 \text{ lb SBM} \times 45\% \text{ CP} = \underline{4.05 \text{ lbs CP}}$$
$$14.06 \text{ lbs CP (just over } 14\%)$$

When using this method, it is important to remember that the number in the middle of the box must be intermediate in value to the numbers on the left side of the square and negative signs are ignored.

Chapters 15, 16, and 17 give a brief background on nutrition, which leads into ration formulation. The primary objective of ration formulation is economically matching the animal's nutrient requirements (Tables 17.4–17.11) with the available feeds, taking into consideration the nutrient content of the feeds. Additional considerations are the palatability of the ration, the physical form of the feed, and other factors that affect feed consumption.

Additional material on feeding farm animals can be found in the following chapters on individual species: beef cattle (Chapter 26), dairy cattle (Chapter 28), swine (Chapter 30), sheep (Chapter 32), horses (Chapter 34), poultry (Chapter 35), goats (Chapter 36), and aquaculture (Chapter 38).

NUTRIENT REQUIREMENTS OF RUMINANTS

The following are some major comparisons that demonstrate changes in nutrient requirements for maintenance, growth, lactation, and reproduction:

1. For ewes (maintenance), note the increased requirement in dry matter, energy (TDN or ME), protein, calcium, phosphorus, vitamin A, and

TABLE 17.4 Daily Nutrient Requirements of Sheep (dry-matter basis)

Weight (lb)	Gain (lb)	Dry Matter (lb)	TDN (lb)	ME (Mcal)	Crude Protein (lb)	Ca (g)	P (g)	Vitamin A (IU)	Vitamin E (IU)
Ewes (maintenance)									
110	0.02	2.2	1.2	2.0	0.21	3.0	2.8	2,350	15
176	0.02	2.9	1.6	2.6	0.27	3.3	3.1	3,760	20
Ewes (nonlactating and first 15 weeks of gestation)									
110	0.07	2.6	1.5	2.4	0.25	3.0	2.8	2,350	18
176	0.07	3.3	1.8	3.0	0.31	3.3	3.1	3,760	22
Ewes (last 4 weeks of gestation with 200% lambing rate)									
110	0.5	3.7	2.4	4.0	0.43	4.1	3.9	4,250	26
176	0.5	4.4	2.9	4.7	0.49	4.8	4.5	6,800	30
Ewes (first 8 weeks of lactation, suckling singles)									
110	−0.06	4.6	3.0	4.9	0.67	10.9	7.8	4,250	32
176	−0.06	5.7	3.7	6.1	0.76	12.6	9.0	6,800	39
Ewes (first 8 weeks of lactation, suckling twins)									
110	−0.13	5.3	3.4	5.6	0.86	12.5	8.9	5,000	36
176	−0.13	6.6	4.3	7.0	0.96	14.4	10.2	8,000	45
Lambs (finishing)									
66	0.65	2.9	2.1	3.4	0.42	4.8	3.0	1,410	20
88	0.06	3.5	2.7	4.4	0.41	5.0	3.1	1,880	24
110	0.45	3.5	2.7	4.4	0.35	5.0	3.1	2,350	24

Source: National Research Council, *Nutrient Requirements of Sheep*, 1985.

TABLE 17.5 Daily Nutrient Requirements (NRC) for Breeding Heifers and Cows

Weight (lb)	Daily Gain (lb)	Dry-Matter Consumption (lb)	Total Crude Protein (lb)	TDN (lb)	ME (Mcal)	Ca (g)	P (g)
Pregnant Heifers—Last Third of Pregnancy							
1,000[a]	0.73	29.7	1.8	11.7	20.9	29	22
1,200[a]	0.88	23.7	2.0	13.3	24.2	33	24
1,400[a]	1.02	26.6	2.2	14.8	26.9	37	27
Cows Nursing Calves—Average Milking Ability[b] (First 3 months postpartum)							
900	0	23.0	1.9	12.5	20.9	24	17
1,100	0	26.0	2.1	13.9	23.4	27	19
1,400	0	29.0	2.3	15.4	25.7	30	21
Cows Nursing Calves—Superior Milking Ability[c] (First 3 months postpartum)							
900	0	25.4	2.6	14.9	24.9	35	24
1,100	0	28.4	2.8	16.4	27.3	37	25
1,400	0	31.3	3.0	17.8	29.7	40	27
Dry Pregnant Mature Cows—Month 3 of Pregnancy							
900	0	16.7	1.2	8.2	13.4	14	14
1,100	0	19.5	1.4	9.5	15.6	17	17
1,400	0	23.3	1.6	11.4	18.7	21	21
Dry Pregnant Cows—Month 8 of Pregnancy							
900	0.9	21.0	1.6	10.9	18.3	23	14
1,100	0.9	24.1	1.8	12.6	20.9	27	17
1,400	0.9	27.0	2.2	14.2	23.9	32	21

[a] Mature weight potential.
[b] Ten pounds of milk per day (equivalent of approximately 450 lb of calf at weaning if there is adequate forage).
[c] Twenty pounds of milk per day (equivalent of approximately 650 lb of calf at weaning if there is adequate forage).
Source: National Research Council, *Nutrient Requirements of Beef Cattle*, 2000.

TABLE 17.6 Daily Nutrient Requirements for Growing-Finishing Heifers and Steers

Live Weight (lb)	TDN (%DM)	Daily Gain (lb)	Dry-Matter Intake (lb)	Protein Intake (lb)	Crude Protein (%)	NEm (Mcal/lb)	NEg (Mcal/lb)	Ca (%)	P (%)
1,100 lb at Finishing (28% body fat)									
600	50	0.68	17.5	1.2	7.1	0.45	0.20	0.21	0.13
	70	2.86	18.0	2.2	12.3	0.76	0.48	0.45	0.23
	90	4.00	15.7	2.7	17.1	1.04	0.72	0.66	0.32
770	50	0.68	19.6	1.3	6.8	0.45	0.20	0.19	0.12
	70	2.86	20.2	2.2	10.9	0.76	0.48	0.39	0.20
	90	4.00	17.6	2.6	14.8	1.04	0.72	0.56	0.28
880	50	0.68	21.7	1.4	6.5	0.45	0.20	0.19	0.12
	70	2.86	22.4	2.2	9.8	0.76	0.48	0.34	0.18
	90	4.00	19.5	2.5	13.1	1.04	0.72	0.48	0.25
1,300 lb at Finishing (28% body fat)									
780	50	0.76	19.8	1.4	7.1	0.45	0.20	0.21	0.13
	70	3.21	20.4	2.5	12.1	0.76	0.48	0.45	0.23
	90	4.48	17.8	3.0	16.9	1.04	0.72	0.66	0.32
910	50	0.76	22.2	1.5	6.7	0.45	0.20	0.20	0.13
	70	3.21	22.9	2.4	10.7	0.76	0.48	0.39	0.20
	90	4.48	22.0	2.9	14.6	1.04	0.72	0.56	0.28
1,040	50	0.76	24.5	1.6	6.5	0.45	0.20	0.19	0.13
	70	3.21	25.3	2.4	9.6	0.76	0.48	0.34	0.19
	90	4.48	22.1	2.8	12.9	1.04	0.72	0.98	0.25

Source: National Research Council, *Nutrient Requirements of Beef Cattle*, 2000.

TABLE 17.7   Daily Nutrient Requirements of Dairy Cows

Body Weight (lb)	NE$_l$ (Mcal)	TDN (lb)	Crude Protein (lb)	Ca (lb)	P (lb)	Vitamin A (IU)
Mature Lactating Cows (maintenance)						
800	7.16	6.9	0.70	0.029	0.024	30
1,100	8.46	8.1	0.80	0.044	0.031	38
1,300	9.70	9.3	0.89	0.053	0.037	46
1,550	10.89	10.5	0.99	0.062	0.044	53
Mature Dry Cows (last 2 months of gestation)						
800	9.30	9.1	1.96	0.057	0.035	30
1,100	11.00	10.8	2.32	0.073	0.044	38
1,300	12.61	12.4	2.66	0.086	0.053	46
1,550	14.15	13.9	2.98	0.101	0.062	53
Milk Production (meal or lb of nutrient per lb of milk for various fat percentages)						
Percentage fat						
3.0	0.291	0.282	0.077	0.0025	0.00170	
3.5	0.313	0.304	0.082	0.0026	0.00175	
4.0	0.336	0.326	0.087	0.0027	0.00180	
4.5	0.354	0.344	0.092	0.0028	0.00185	
5.0	0.377	0.365	0.098	0.0029	0.00190	

Source: National Research Council, *Nutrient Requirements of Dairy Cattle,* 2001.

TABLE 17.8   Suggested Dietary Amino Acid and Protein Allowances for Swine Fed Corn-Soybean Meal Diets

Amino Acids (% of Diet)	GrowingPigs[a]			Barrows			Gilts		
	20–45	45–80	80–120	120–170	170–220	220–280	120–170	170–220	220–280
Crude protein	20	18	16	14	13	12	16	14	13
Lysine	1.2	1.0	0.9	0.75	0.68	0.58	0.84	0.73	0.62
Methionine-cystine	0.72	0.62	0.56	0.48	0.43	0.37	0.53	0.47	0.39
Tryptophan	0.21	0.18	0.16	0.15	0.13	0.11	0.16	0.14	0.11
Threonine	0.78	0.67	0.60	0.52	0.48	0.40	0.58	0.51	0.43
Arginine	0.50	0.36	0.33	0.14	0.13	0.10	0.15	0.14	0.11
Histidine	0.39	0.32	0.29	0.24	0.21	0.19	0.27	0.23	0.20
Isoleucine	0.72	0.60	0.54	0.45	0.40	0.35	0.50	0.44	0.37
Valine	0.81	0.68	0.61	0.51	0.47	0.39	0.57	0.50	0.42
Leucine	1.20	1.00	0.90	0.75	0.68	0.58	0.84	0.73	0.62
Plenylalanine + Tyrosine	1.14	0.95	0.80	0.71	0.65	0.55	0.80	0.70	0.59

[a] Barrows and gilts.

Source: Adapted from Baker et al., 2002.

vitamin D as body weight changes from 110 to 176 lb (Table 17.4). More nutrients are needed to maintain a heavier body weight.

2. In Table 17.4, compare the requirements of ewes in the last 4 weeks of gestation with the maintenance requirements of ewes and the requirements of ewes that are nonlactating and in the first 15 weeks of gestation. During the latter part of gestation, there are greater nutrient demands owing to rapid fetal growth. During the latter part of gestation compared with maintenance only, energy and protein requirements almost double, with mineral and vitamin requirements showing significant increases.

3. During lactation, nutrient requirements are even higher than those during gestation. Even with larger amounts of dry matter being supplied, the ewes

TABLE 17.9 Protein, Mineral, and Vitamin Requirements for Gestating and Lactating Sows

	Gestation (% of diet)	Lactation (% of diet)
Crude protein	12	17
Calcium	0.75	0.75
Phosphorus, available	0.35	0.35
Salt	0.35	0.35
Iron (mg/lb)	36	36
Zinc (mg/lb)	23	23
Copper (mg/lb)	2.3	2.3
Manganese (mg/lb)	9.0	9.0
Iodine (mg/lb)	0.06	0.06
Selenium (mg/lb)	0.07	0.07
Vitamin A (IU/lb)	2,000	1,000
Vitamin D (IU/lb)	150	150
Vitamin E (IU/lb)	20	—
Vitamin K (IU/lb)	230	230
Riboflavin (mg/lb)	2	2
Niacin (mg/lb)	5	5
Pantothenic acid (mg/lb)	5	5
Vitamin B_{12} (mg/lb)	7	7
Choline (mg/lb)	600	450

Source: Adapted from Baker et al., 2002.

TABLE 17.10 Daily Nutrient Requirements for Horses

Status	Body Weight (lb)	DE[1] (mcal)	CP[2] (g)	Ca (g)	P (g)	Vit. A (1,000 IU)
Maintenance	900	13.4	536	16	11	12
	1,100	16.4	656	20	14	15
	1,300	19.4	776	24	17	18
Mares—11th month of gestation	900	16.1	708	31	23	24
	1,100	19.7	886	37	28	30
	1,300	23.3	1,024	44	34	36
Mares—early lactation	900	22.9	1,141	45	29	24
	1,100	28.3	1,427	56	36	30
	1,300	33.7	1,712	67	43	36
Working horses Light	900	16.7	670	20	15	18
	1,100	20.5	820	25	18	22.5
	1,300	24.2	970	30	21	27
Moderate	900	20.1	804	25	17	18
	1,100	24.6	984	30	21	22.5
	1,300	29.1	1,164	36	25	27
Intense	900	26.8	1,072	33	23	18
	1,100	32.8	1,312	40	29	22.5
	1,300	38.8	1,552	47	34	27

[1] Digestible energy.
[2] Crude protein.
Source: Adapted from NRC, 1989.

lose weight. Compare the requirements of ewes nursing single lambs versus twins and note the even higher requirements for ewes nursing twins.

4. The nutrient requirements for lambs being finished for slaughter show that the gains are approximately the same, with increased nutrient requirements due to increased body weight. The increased nutrient requirements are primarily due to an increased maintenance requirement.

TABLE 17.11   Suggested Amino Acid Recommendations for Broiler Diets[1]

	0–21 Days		22–42 Days		43–53 Days	
	Male	Female	Male	Female	Male	Female
Arginine	0.88	0.83	0.77	0.73	0.66	0.59
Lysine	0.81	0.76	0.70	0.67	0.53	0.50
Methionine + cystine	0.60	0.60	0.56	0.50	0.46	0.41
Tryptophan	0.16	0.15	0.12	0.11	0.11	0.10
Histidine	0.24	0.22	0.22	0.20	0.20	0.18
Leucine	0.84	0.64	0.81	0.76	0.76	0.72
Isoleucine	0.54	0.51	0.52	0.48	0.45	0.40
Phenylalanine + tyrosine	1.04	0.98	0.95	0.87	0.78	0.70
Threonine	0.53	0.49	0.43	0.41	0.42	0.38
Valine	0.66	0.63	0.62	0.57	0.51	0.46
Glycine + serine	1.00	0.94	0.80	0.74	0.70	0.63
Protein	15.25	15.25	13.75	13.75	12.25	12.25

[1]Expressed as a percent per metabolizable megacalorie per pound of feed. To calculate the percent of the nutrient required in the diet, multiply the tabular value by the metabolizable megacalorie per pound of feed.
Source: Adapted from Waldrop, 2002.

Diet Modification to Minimize Nitrogen and Phosphorus Pollution

One of the most critical challenges confronting livestock agriculture is the impact of animal waste on air and water quality. A key strategy to minimize these impacts is the alteration of rations to diminish nutrient output of nitrogen (N) and phosphorus (P).

Nitrogen is a structural component of the amino acids that form protein. Nitrogen is a by-product of protein digestion that has the potential to negatively affect surface water quality. Phosphorus is a mineral nutrient that also has the potential to be a pollutant when excreted in excess.

There are two environmental concerns from animal waste—the *volatilization* of N in the form of ammonia from manure that may enter the environment via rainfall, *dry precipitation*, or absorption directly from the waste product. As N content of manure increases the risk of ammonia loss rises. Therefore, control of N in the diet is a useful strategy to reduce contamination. The second issue is distribution of manure nutrients via the implementation of an integrated waste management plan. Manure is an excellent fertilizer and the environmental impact is negligible when manure nutrient application is matched to plant requirements. However, at excess application rates P can infiltrate the soil and contaminate surface water while N can pollute groundwater.

Specific strategies for minimizing the impact of N and P are discussed in Chapters 26, 28, 30, and 35. However, significant decreases in N and P excretion by livestock and poultry can be achieved via careful management of diet composition, better understanding of digestibility differences between feedstuffs, and assuring that appropriate diet changes are made in accordance with shifts in animal requirements for N and P. Historically, N and P have been overfed in the diets of many livestock. However, as pressure from environmental regulation increases, the need for more precise nutrient management will be assured.

CHAPTER SUMMARY

- Nutrients are used by animals for maintenance (maintaining body functions), growth, fattening, reproduction, egg laying, wool production, and work.
- Maintenance involves no change (gain or loss) in body weight. Maintenance requirements increase as body weight increases.
- Growth occurs when protein synthesis exceeds protein breakdown and loss.
- Fattening occurs when the energy in feeds (primarily from carbohydrates and fats) exceeds the body's need for maintenance and growth.
- Rations are formulated to match the animal's nutrient requirements with the nutrient content of available feeds.

KEY WORDS

body maintenance	selenium	growing fetus
growth	vitamin D	*ad libitum*
calcium	vitamin A	Shearing
phosphorus	fattening	Pearson square method
iodine	reproduction	

REVIEW QUESTIONS

1. What are nutrients needed for?
2. What body maintenance functions have a high priority for nutrients?
3. *True or False:* Small animals require less nutriton for body maintenance than large animals.
4. *True or False:* Diets of primarily roughage are adequate to meet nutrient requirements for growth in young ruminants.
5. What are the two categories of nutrient requirements for reproduction?
6. When is the nutrient requirement for fetal growth greatest?
7. What are the two most important minerals needed for lactation?
8. What is the most vital requirement for production of large quantities of milk during lactation?

SELECTED REFERENCES

Baker, D. H., R. A. Easter, G. R. Hollis, M. Ellis, and R. Zijlstra. 2002. Dietary nutrient allowances for swine. *Feedstuffs Reference Issue*, 74:28.

Cheeke, P. J. 2005. *Animal Nutrition*. Upper Saddle River, NJ: Pearson Prentice Hall.

Church, D. C. 1991. *Livestock Feeds and Feeding*. Englewood Cliffs, NJ: Prentice Hall.

Council for Agricultural Science and Technology. 2002. *Animal diet modification to decrease the potential for nitrogen and phosphorus pollution*. Issue paper no. 21. Ames, IA.

Kline, K. H. 2002. Horse feeds and feeding. *Feedstuffs Reference Issue*, 74:28.

National Research Council. *Nutrient Requirements of Beef Cattle* (2000); *Dairy Cattle* (2001); *Goats* (1981); *Horses* (1989); *Poultry* (1994); *Sheep* (1985); and *Swine* (1998). Washington, DC: National Academy Press.

Pond, W. G. 1995. *Basic Animal Nutrition and Feeding*. New York: John Wiley & Sons.

Waldrop, P. W. 2002. Dietary nutrient allowances for poultry and turkeys. *Feedstuffs Reference Issue*, 74:28.

CHAPTER 18

Growth and Development

LEARNING OBJECTIVES

- Describe prenatal growth in farm animals and in poultry
- Compare and contrast the skeletons of mammals and birds
- Define the three muscle types
- Define power, strength, and endurance
- Compare and contrast red and white muscle fibers
- Describe the primary hormonal influences on growth
- Describe the growth curve
- Explain the importance of the growth curve in animal management
- Describe the effects of frame size, gender, muscling, and age on growth
- Calculate average daily gain and weight per day of age
- Discuss the evaluation of teeth to estimate animal age

Profitable and efficient production of livestock and poultry involves understanding and effective management of growth and development. The use of integrated management protocols to affect genetic and environmental factors can alter growth patterns in farm animals.

Many aspects of growth and development are discussed in other chapters of this book. The material on reproduction, genetics, nutrition, and products should be integrated with this chapter for a deeper understanding of animal growth and development.

Generally, **growth** is an increase in body weight until mature size is reached. This growth is the result of an increase in cell size and cell numbers with protein deposition resulting. More specifically, growth is an increase in the mass of structural tissue (bone, muscle, and connective tissue) and organs accompanied by a change in body form and composition.

Development is defined as the directive coordination of all diverse processes until maturity is reached. It involves growth, cellular differentiation, and changes in body shape and form. In this chapter, growth and development are combined and discussed as one entity.

PRENATAL (LIVESTOCK)

The three phases of prenatal life—sex cells, the embryo, and the fetus—are briefly discussed in chapter 10. Embryological development is a fascinating process; a spherical mass of cells differentiates into specific cell types and eventually into recognizable organs (Figs. 18.1 and 18.2). The endoderm (Fig. 18.2D) differentiates into the digestive tract, lungs, and bladder; the mesoderm (Fig. 18.2D) into the skeleton, skeletal muscle, and connective tissue; the ectoderm (Fig. 18.2D) into the skin, hair, brain, and spinal cord. The growth, development, and differentiation processes, involving primarily protein synthesis, are directed by DNA chains of chromosomes and the organizers in the developing embryo (see chapter 12). Thus the nucleus is a center of activity for different types of cells, directing the growth and development process (Fig. 18.3).

The fetus undergoes marked changes in shape and form during prenatal growth and development. Early in the prenatal period, the head is much larger than the body. Later, the body and limbs grow more rapidly than other parts. The order of tissue growth follows a sequential trend determined by physiological importance, starting with the central nervous system and progressing to internal organs, bones, tendons, muscles, intermuscular fat, and subcutaneous fat.

During the first two-thirds of the prenatal period, most of the increase in muscle weight is due to **hyperplasia** (increase in number of fibers). During the last several months of pregnancy and postnatal, **hypertrophy** (increase in size of fibers) represents most of the muscle growth. Individual muscles vary in their rate of growth, with larger muscles (those of the legs and back) having the greatest rate of postnatal growth. Water content of fetal muscle declines with fetal age, and this decline continues through postnatal growth as well.

The relative size of the fetus changes during gestation, with the largest increase in weight occurring during the last trimester of pregnancy (Fig. 18.4).

BIRTH (LIVESTOCK)

Following birth, the number of muscle fibers does not appear to increase significantly; therefore, postnatal muscle growth occurs primarily via hypertrophy. In red-meat animals, all muscle fibers appear to be red at birth, but shortly thereafter some differentiate into white and intermediate muscle types.

At birth, the various body parts have considerably different proportions when compared with mature body size and shape. The head is relatively large, the legs are long, and the body is small; in the mature animal, the head is relatively small, the legs are relatively short, and the body is relatively large. Birth weight represents approximately 5–7% of the mature weight, while leg length at birth is approximately 60%; height at withers is approximately 50% of those same measurements at maturity. Hip width and chest width at birth are approximately one-third of the same measurements at maturity. This shows that the distal parts (leg and shoulder height) are developed earlier than proximal parts (hips and chest).

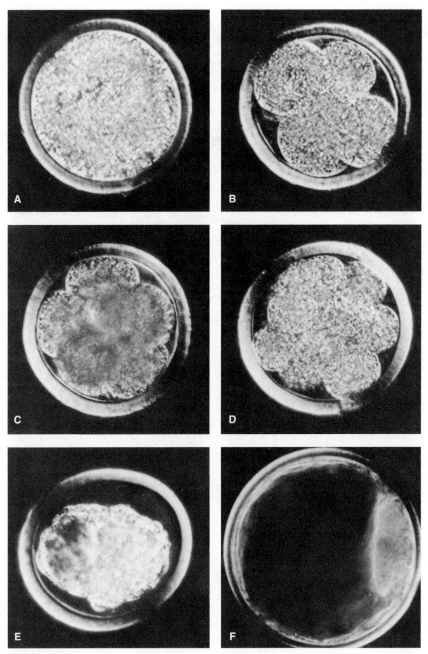

FIGURE 18.1   Embryonic development during the first 8 days of pregnancy in cattle. (A) Unfertilized egg. (B) Four-cell embryo on day 2 of pregnancy (estrus—day 0). (C) Eight-cell embryo on day 3 of pregnancy. (D) Eight- to 16-cell embryo on day 4 of pregnancy. (E) Very early blastocyst stage (approximately 60 cells). (F) Expanded blastocyst (>100 cells) recovered from the uterus on day 8 of pregnancy. Magnification approximately 300×. *Science* 211:351. Copyright 1981 by the American Association for the Advancement of Science.

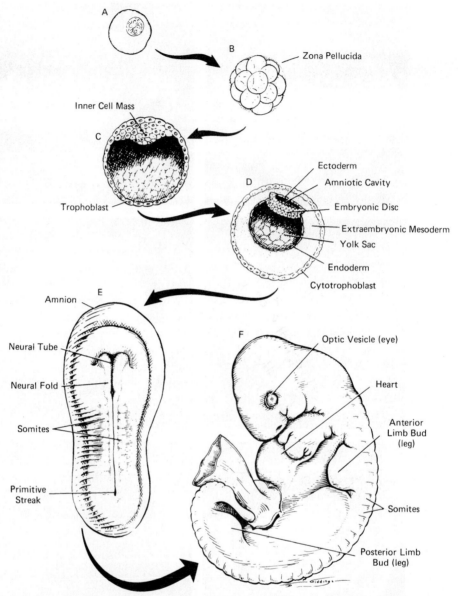

FIGURE 18.2   The morphogenesis of (A) a single egg cell into (B) a morula, then to (C) a blastocyst. (D) The stage at which the two cavities have formed in the inner cell mass; an upper (amniotic) cavity and a lower cavity yolk sac. The embryonic disc containing the ectoderm and endoderm germ layers is located between cavities. (E) A cattle embryo showing the neural tube and somites. (F) The development of the 14-day cattle embryo. *Source:* Colorado State University.

POULTRY

Embryonic Development

The development of a chick differs from that of mammals because there is no physiological connection with its mother. The chick develops in the egg, entirely outside the hen's body. Embryonic development is much more rapid in chicks than in farm mammals.

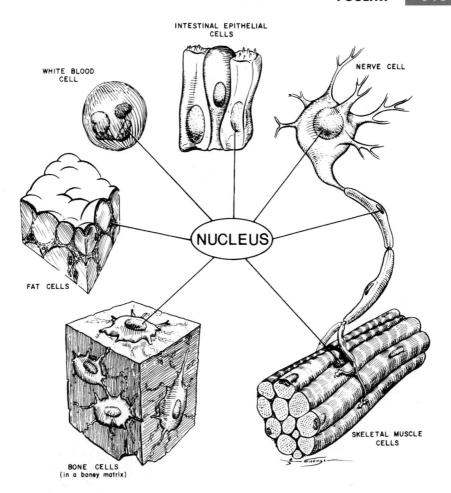

FIGURE 18.3 Cells of selected tissues showing similarity in cell structure but not cell shape. The nucleus gives direction to differentiation of cells and their function. *Source:* Colorado State University.

Every egg, whether fertile or nonfertile, has a germ spot called the **blastoderm** (see Fig. 4.1). This is where the chick embryo develops if the fertile egg is properly incubated.

A controlled environment must be maintained during incubation to produce live chicks from the fertile eggs. The major components of the controlled environment are (1) temperatures of 99.5–100°F; (2) 60–75% relative humidity; (3) turning of the egg every 1–8 hours; and (4) provision of adequate oxygen.

Three hours after fertilization, the blastoderm divides to form two cells. Cell division occurs until maturity except during the holding period before incubating the eggs.

There are four membranes that are essential to the growth of the chick embryo (Fig. 18.5). The **allantois** is the membrane that allows the embryo to breathe. It takes oxygen through the porous shell and oxygenates the blood of the embryo. The allantois removes the carbon dioxide, receives excretions from the kidneys, absorbs albumen used as food for the embryo, and absorbs calcium from the shell for use by the embryo. The **amnion** is a membrane filled with a

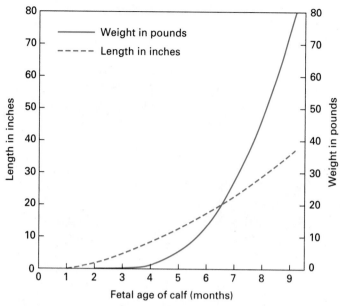

FIGURE 18.4 Growth of the fetal calf. *Source:* Colorado State University.

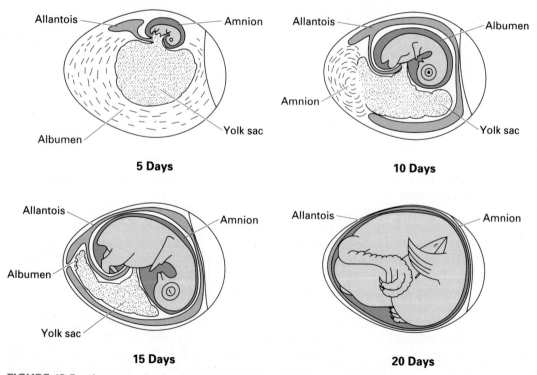

FIGURE 18.5 Changes in the development of the chick embryo with associated changes in membranes and other contents of the egg.

TABLE 18.1   Major Changes in Weight, Form, and Function of the Chick Embryo (White Leghorn) During Incubation

Day	Weight (g)	Developmental Changes
1	00.0002	Head and backbone are formed; central nervous system begins
2	00.0030	Heart forms and starts beating; eyes begin formation
3	00.0200	Limb buds form
4	00.0500	Allantois starts functioning
5	00.1300	Formation and reproductive organs
6	00.2900	Main division of legs and wings; first movements noted
7	00.5700	
8	01.1500	Feather germs appear
9	01.5300	Beak begins to form; embryo begins to look birdlike
10	02.2600	Beak starts to harden; digits completely separated
11	03.6800	
12	05.0700	Toes fully formed
13	07.3700	Down appears on body; scales and nails appear
14	09.7400	Embryo turns its head toward blunt end of egg
15	12.0000	Small intestines taken into body
16	15.9800	Scales and nails on legs and feet are hard; albumen is nearly gone; yolk is main food
17	18.5900	Amniotic fluid decreases
18	21.8300	
19	25.6200	Yolk sac enters body through umbilicus
20	30.2100	Embryo becomes a chick; it breaks amnion, then breathes air in air cell
21	36.3000	Chick breaks shell and hatches

Source: Adapted from Rollins, 1984.

colorless fluid that serves as a protection from mechanical shock. The **yolk sac** is a layer of tissue growth over the surface of the yolk. This tissue has special cells that digest and absorb the yolk material for the developing embryo. The **chorion** surrounds both the amnion and yolk sac.

Table 18.1 identifies some of the primary changes in the growth and development of the chick embryo. Many of these phenomenal changes occur rapidly, sometimes in only hours.

BASIC ANATOMY AND PHYSIOLOGY _____

In the developmental process, cells become grouped in appearance and function. Specialized groups of cells that function together are called **tissues.** The primary types of tissues are (1) muscle, (2) nerves, and (3) connective and epithelial; examples are shown in Figure 18.3.

Organs are groups of tissues that perform specific functions. For example, the uterus is an organ that functions in the reproductive process. A group of organs that function in concert to accomplish a larger, general function is known as a **system.** The reproductive system is discussed in chapter 10, the digestive system chapter 16, and the mammary system chapter 19.

It is not the intent of this chapter to discuss all the different systems even though they are important in growth and development. A few additional systems are briefly surveyed—those considered important in understanding farm animals and their productivity. Figures are provided, illustrating selected systems for one or two species. The systems are similar and generally comparable among the various species of farm animals, although some large differences between poultry and farm mammals exist.

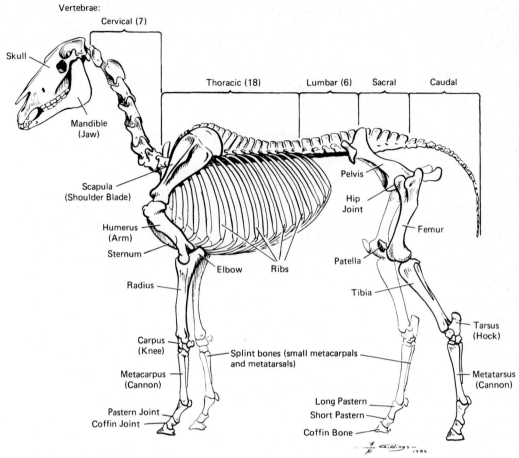

FIGURE 18.6   Skeletal system of the horse.

Skeletal System

Figure 18.6 shows the skeletal system of the horse and Figure 18.7 portrays the chicken's skeletal system. Even though only bones and some joints are shown in these figures, teeth and cartilage are also considered part of the skeletal system.

The skeleton protects other vital organs and gives a basic form and shape to the animal's body. Bones function as levers and in storing minerals, and the bone marrow is the site of blood cell formation.

Chicken bones are more pneumatic (bone cavities are filled with air spaces), harder, thinner, and more brittle than those of mammals, and have a different ossification process than occurs in mammals.

Muscle System

There are three types of muscle tissue—skeletal, smooth, and cardiac. Skeletal muscle is the largest component of meat animal products. Smooth muscle is located in the digestive, reproductive, and urinary organs. The heart is composed of cardiac muscle.

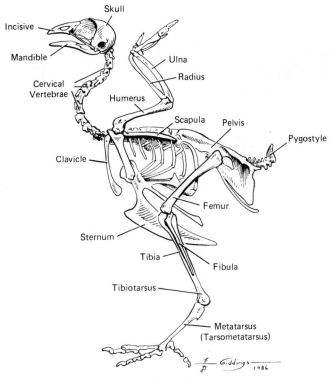

FIGURE 18.7 Skeletal system of the chicken.

Figure 18.8 identifies some of the major muscles similar in name and location in the meat animal species and horses. Of special note is the longissimus dorsi, which is discussed in Chapter 8. The size of this muscle and the marbling it contains are important factors in determining yield grades and quality grades of meat animals.

The primary muscles of the turkey are shown in Figure 18.9. The muscles of poultry are referred to as **dark meat** (legs and thighs) and **white meat** (breast and wings).

The role of muscle is to convert chemical energy to mechanical energy. Muscle performance depends on power, strength, and endurance. **Power** is determined by the amount of work that can be accomplished within a fixed time period. Thus, power is the result of the speed and strength of muscle contraction. **Strength** is the result of muscle size with larger muscles capable of greater contractible force. **Endurance** is defined as the length of time in which muscle can work without becoming fatigued. Endurance is significantly affected by the nutrient storage capacity of a muscle. Effective training programs for the working horse are designed to improve these three components of muscle performance.

Skeletal muscle fibers can be generally classified as fast-twitch or slow-twitch. Fast-twitch fibers are typically predominant in sprinters and jumpers. Fast-twitch fibers are designed for power and are efficient at anaerobic activity for short periods of time. Because they have less capillary supply than slow-twitch

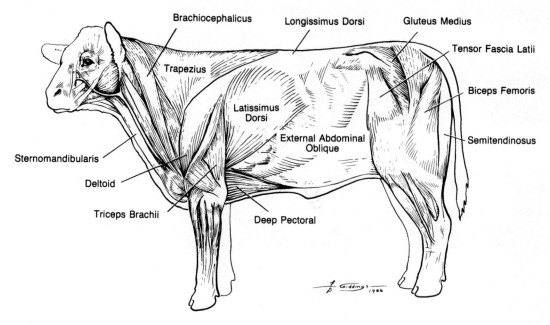

FIGURE 18.8   Primary muscles of the beef steer.

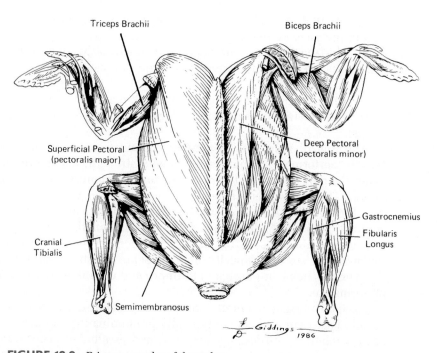

FIGURE 18.9   Primary muscles of the turkey.

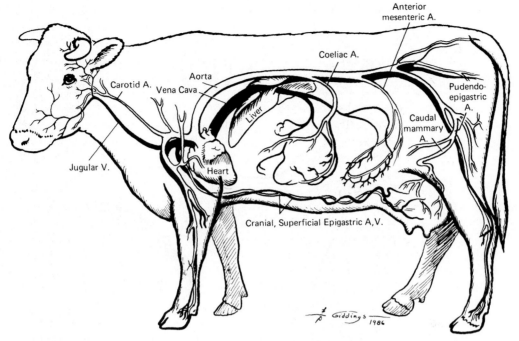

FIGURE 18.10 Circulatory system of the dairy cow (A = artery, V = vein).

fibers, fast-twitch fibers are often referred to as **white fibers.** Slow-twitch or **red fibers** have a more extensive capillary supply and are better suited to aerobic supply of nutrients and endurance activity. They are smaller in fiber diameter than fast-twitch fibers and contain more myoglobin as a means to enhance oxygen diffusion.

Circulatory System

Figure 18.10 shows the major components of the dairy cow's circulatory system. The circulatory systems of other farm animal species are similar to that of the dairy cow.

The heart, acting as a pump, and the accompanying vessels make up the circulatory system. **Arteries** are vessels that transport blood away from the heart, while the vessels that carry blood to the heart are called **veins.** The lymph vessels transport lymph (intercellular fluid) from tissues to the heart.

The circulatory system is important in growth and development; the blood transports oxygen, nutrients, cellular waste products, and hormones.

Milk production in all species is dependent on the nutrients in milk arriving through the circulatory system. Dairy cows producing 20,000–40,000 lb of milk per year (over 100 lb per day) must consume large amounts of feed, with these feed nutrients circulating in high concentration through the mammary blood supply. Approximately 400–500 lb of blood circulates through the dairy cow's mammary gland for each pound of milk produced.

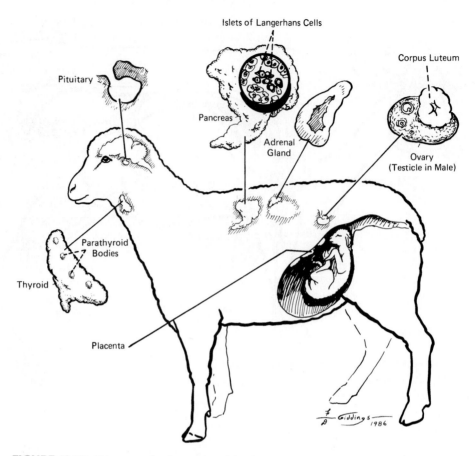

FIGURE 18.11 Primary endocrine organs of the sheep.

Endocrine System

Growth and development is highly dependent on the endocrine system. This system consists of several endocrine (ductless) glands that secrete **hormones** into the circulatory system. Hormones are chemical substances that affect a gland (or organ) or, in some cases, all body tissues.

The major endocrine glands are shown in Figure 18.11; the hormones produced and their major effects are identified in Table 18.2. Figure 19.3 (Chapter 19) shows the origin of most of the hormones noted in Table 18.2. The important roles of the pituitary and hypothalmus should also be observed in Figure 19.3.

GROWTH CURVES

Postweaning growth is a curved-line function regardless of how it is expressed mathematically. If growth is considered an increase in body weight, then most of the early growth follows a straight line or is linear in the age-weight relationship. As the animal increases in age and approaches puberty, rate of growth usually declines, and true growth ceases when the animal reaches maturity.

TABLE 18.2   Major Hormones Affecting Growth and Development in Farm Animals

Hormone	Source	Major Effect
Growth (Somatotrophic, STH)	Pituitary (anterior)	Body cell growth—especially muscle and bone cells
Adrenocorticotropic (ACTH)	Pituitary (anterior)	Stimulates adrenal cortex to produce adrenal cortical steroid hormones
Glucocorticoids	Adrenal (cortex)	Conversion of proteins to carbohydrates
Mineralocorticoids	Adrenal (cortex)	Regulates sodium and potassium balance; water balance
Thyroid-stimulating (TSH)	Pituitary (anterior)	Stimulates thyroid gland to produce thyroid hormones
Thyroid	Thyroid	Regulates metabolic rate
Testosterone	Testicles (interstitial cells)	Accessory sex gland development; male secondary sex characteristics and muscle growth
Estrogen	Ovary (follicle) Placenta	Female reproductive organ growth; female secondary sex characteristics; mammary gland duct growth
Progesterone	Corpus luteum; placenta	Uterine growth; alveoli growth in mammary gland growth
Antidiuretic (ADH) (Vasopressin)	Pituitary (posterior)	Controls water loss in kidney
Epinephrine	Adrenal (medulla)	Increases blood glucose concentration
Norepinephrine	Adrenal (medulla)	Maintains blood pressure
Insulin	Pancreas	Lowers blood sugar
Parathormone (PTH)	Parathyroid	Calcium and phosphorus metabolism
Glucagon	Pancreas	Raises blood sugar

Source: Compiled from several sources.

After an animal reaches maturity, it may initiate large fluctuations in body weight simply by increasing or decreasing the amount of fat or water that is stored. This increase in weight owing to fattening is not true growth because no net increase in body protein occurs. In fact, animals tend to lose body protein, as they grow older. The loss of protein is one of the phenomena in the aging process.

Figure 18.12 shows representative growth curves for most farm animals. The difference in curves for large and small breeds is primarily a function of differences in skeletal frame size. Maturity is reached at heavier weights in larger breeds, which have larger skeletal frame sizes than smaller-framed animals.

The relatively straight-lined growth shown for broilers does curve and level off as shown for the other species. This occurs when the birds reach approximately 11 weeks of age.

CARCASS COMPOSITION

Products from meat animals and poultry are composed primarily of fat, lean, and bone. A low proportion of bone, a high proportion of muscle, and an optimum amount of fat characterize a superior carcass. Understanding the growth and development of these animals is important in knowing when animals should be slaughtered to produce the desirable combinations of fat, lean, and bone. Figure 18.13 shows the variation in fat thickness, rib-eye area, and yield grade for two steers of comparable weight and age but of different sizes. The steer on the left was managed to a correct endpoint, while the steer on the right was fed too long and harvested at too heavy a weight.

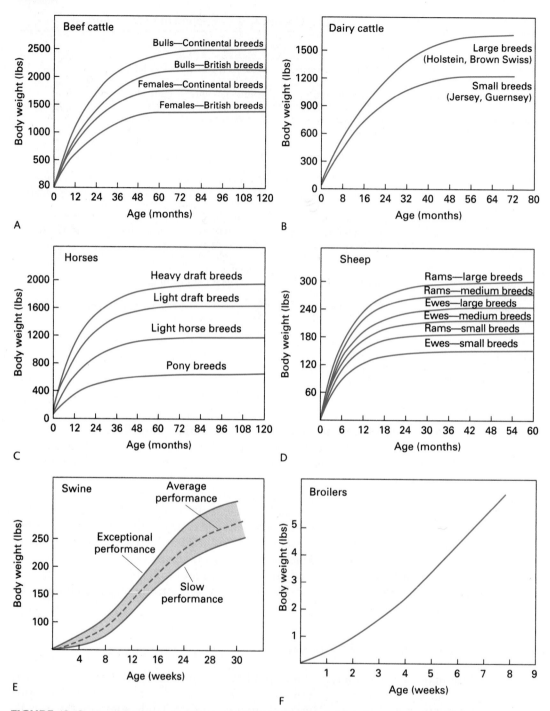

FIGURE 18.12 Representative growth curves for (A) beef cattle, (B) dairy cattle, (C) horses, (D) sheep, (E) swine, and (F) broilers. Adapted from Battaglia and Mayrose, *Handbook of Livestock Management Techniques* (New York: Macmillan, 1981). Copyright © Macmillan Publishing Co.

FIGURE 18.13 Two steers of varying composition at similar market weights. The steer on the left weighed 1,166 lb., was 52 in. tall, and had 0.30 in. of backfat, a 13.0-sq. in. ribeye, and a USDA yield grade of 2.4. The steer on the right weighed 1,160 lb., was 47 in. tall, and had 0.95 in. of backfat, a 9.9-sq. in. ribeye, and a USDA yield grade of 4.9.

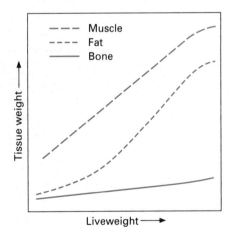

FIGURE 18.14 Tissue growth relative to increased live weight. Adapted from Berg and Walters (1983).

Figure 18.14 shows the expected changes in fat, muscle, and bone as animals increase in live weight during the linear-growth phase. As the animal moves from the linear-growth phase into maturity, the growth curves shown in Figure 18.14 are extended similarly to those shown in Figure 18.12.

At lightweights, fat deposition begins rather slowly, then increases geometrically when muscle growth slows as the animal approaches physiological maturity. Muscle makes up the greatest proportion, and its growth is linear during the production of young slaughter animals. Bone has a smaller relative growth rate than either fat or muscle. Because of the relative growth rates, the ratio of muscle to bone increases as the animal increases in weight.

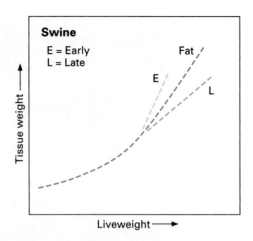

FIGURE 18.15 Effects of different maturity types on fat composition. Adapted from Berg and Walters (1983).

TABLE 18.3   Bull Hip Height (in.) and Associated Frame Scores

Age (mos.)	1	2	3	4	5	6	7	8
6	34.8	36.8	38.8	40.8	42.9	44.9	46.9	48.9
8	37.2	39.2	41.2	43.2	45.2	47.2	49.3	51.3
10	39.2	41.2	43.3	45.3	47.3	49.3	51.3	53.3
12	41.0	43.0	45.0	47.0	49.0	51.0	53.0	55.0
14	42.5	44.5	46.5	48.5	50.4	52.4	54.4	56.4
16	43.6	45.6	47.6	49.6	51.6	53.6	55.6	57.5
18	44.5	46.5	48.5	50.5	52.4	54.4	56.4	58.4

Source: Beef Improvement Federation.

Effects of Frame Size

Different maturity types of animals (sometimes referred to as **breed types** or **biological types**) have a marked influence on carcass composition at similar live weights (Fig. 18.15). The earlier-maturing types increase fat deposition at lighter weights than either the average or later-maturing types.

Skeletal **frame size,** measured as hip height, is a more specific way of defining maturity type. Examples of frame sizes for bulls at various ages and heights are provided in Table 18.3. Figure 18.16 shows the difference in fat, lean, and bone composition for small- (frame 3), medium- (frame 5), and large-framed (frame 7) beef steers at different live weights. Note that the three different frame sizes have a similar composition at <1,000 lb (small frame), 1,250 lb (medium frame), and >1,250 lb (large frame) (refer back to Fig. 18.13).

Effect of Gender

The effect of sex is primarily on the fat component, although there are differences among species. Heifers deposit fat earlier than steers or bulls, with bulls being leaner at the same slaughter weight (Fig. 18.17). Heifers are typically slaughtered at lighter weights (100–200 lb less) than steers in order to ensure a similar fat-to-lean composition.

Swine are different from cattle, as barrows are fatter than gilts or boars at similar slaughter weights (Fig. 18.17). The reason for this species difference is not known.

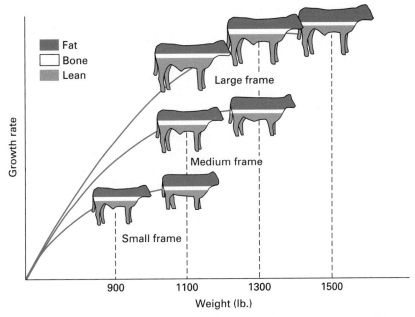

FIGURE 18.16 The relationship of frame size and weight to carcass composition in beef steers. *Source:* Colorado State University.

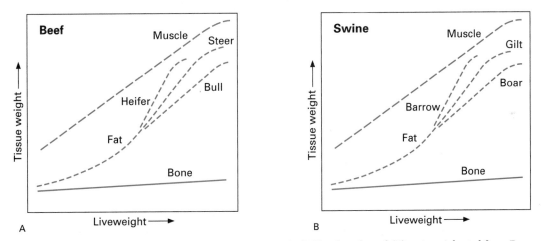

FIGURE 18.17 Influence of sex on carcass composition in (A) beef cattle and (B) swine. Adapted from Berg and Walters (1983). *Source:* American Society of Animal Science.

At typical slaughter weights, the sex differences in sheep do not have a marked effect on fat and lean composition. Apparently, sheep are harvested at an earlier stage of development (before puberty) than cattle and swine. Compositional differences in sheep are similar to those in cattle if sheep are fed beyond their usual slaughter weights.

Effect of Muscling

Relatively large differences exist in muscling among the various species of farm animals. The proportion of muscle in the carcass varies indirectly with fat, and a higher proportion of fat is associated with a lower proportion of muscle. Muscle has a much faster relative growth rate than bone. Muscle weight relative to live weight or muscle-to-bone ratio can be used as a valuable measurement of muscle yield.

Most of the current cattle population in the United States does not display large differences in muscling. There are, however, muscling differences in 10–15% of the cattle, which are best expressed by a muscle-to-bone ratio (pounds of muscle compared to pounds of bone). Muscle-to-bone ratios can range from 3 lb of muscle to 1 lb of bone in thinly muscled slaughter steers and heifers to 5 lb of muscle to 1 lb of bone in more thickly muscled cattle. Muscle-to-bone ratios higher than 5:1 are found in double-muscled cattle. These differences in muscle-to-bone ratio are economically important, particularly if more boneless cuts are merchandised.

Some of the muscling differences for cattle are shown in Figure 18.18: lighter-muscled, heavy-muscled, and double-muscled cattle are compared to

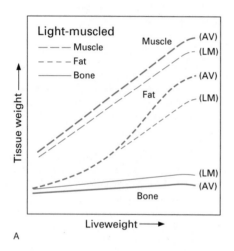

A

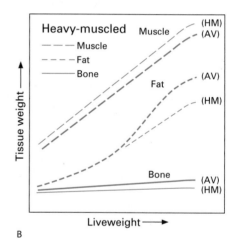

B

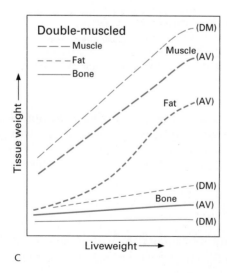

C

FIGURE 18.18 Effects of light, heavy, and double muscling on carcass composition compared to average muscling (bold lines). Adapted from Berg and Walters (1983).

cattle of average muscling. Double-muscled cattle do not have a duplicate set of muscles, but they do have an enlargement (hypertrophy) of the existing muscles.

Extreme muscling, as observed in double-muscled or heavy-muscled types of animals, can be associated with problems in total productivity. Some heavy-muscled swine might die when subjected to stressful conditions or produce undesirable meat that is pale, soft, and exudative (see porcine stress syndrome (PSS) and PSE pork in chapter 29). Reproductive efficiency is lower in some heavy-muscled animals. There is some latitude in increasing muscle-to-bone ratio in slaughter animals without encountering negative side effects.

Individual muscle dissection has shown a similarity of proportion of individual muscles to total body muscle weight. Small differences in muscle distribution have been shown, but the differences do not seem to be economically significant for the industry. Thus it appears that increasing muscle weight in areas of high-priced cuts at the expense of muscle weight in lower-priced cuts is not commercially feasible. Even though muscle distribution is not important, muscling as previously defined—as muscle-to-bone ratio—has high economic importance.

AGE AND TEETH RELATIONSHIP

For many years, producers have observed the teeth of farm animals to estimate their age or their ability to graze effectively with advancing age. Some animals are **mouthed** to classify them into appropriate age groups for show ring classification. The latter has not always proven exact because of variation in tooth condition and the ability of some exhibitors to manipulate the condition of the teeth.

Table 18.4 gives some of the guidelines used in estimating the age of cattle by observing their teeth. The **incisors** (front cutting teeth) are the teeth used. There are no upper incisors, and the molars (back teeth) are not commonly used to determine age.

Figure 18.19 shows the incisor teeth of cattle of different ages. If the grazing area is sandy and the grass is short, wearing of the teeth progresses faster than previously described. Also, some cows may be **broken-mouthed,** which means some of their permanent incisor teeth have been lost.

TABLE 18.4 Description of Cattle Teeth to Estimate Age

Approximate Age	Description of Teeth
Birth	Usually only one pair of middle incisors
1 month	All eight temporary incisors
1.5–2 years	First pair of permanent (middle) incisors
2.5–3 years	Second pair of permanent incisors
3.25–4 years	Third pair of permanent incisors
4–4.5 years	Fourth (corner) pair of permanent incisors
5–6 years	Middle pair of incisors begins to level off from wear; corner teeth might also show some wear
7–8 years	Both middle and second pair of incisors show wear
8–9 years	Middle, second, and third pair of incisors show wear
10 years and over	All eight incisors show wear

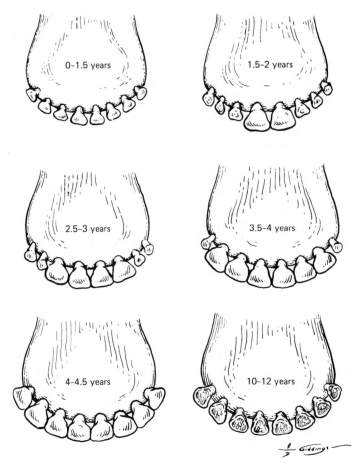

FIGURE 18.19 Incisor teeth of cattle of different ages.

MEASUREMENTS OF GROWTH _____

The sale of livestock has been founded on value determination based on typically one of two standards—the individual animal or weight. Because of the importance of weight on the market value of livestock, measurement of increases in weight is of interest to managers.

A common calculation is to determine average daily gains (ADG) by the formula:

$$\frac{W_2 - W_1}{T_2 - T_1}$$

Where $W_2 - W_1$ = change in weight, while $T_2 - T_1$ = change in time.

Assume a calf weighs 500 lb at the beginning of a weigh period and 650 lb 50 days later. The average daily gain would be

$$\frac{650 - 500}{50} = 3.0 \text{ lb per day}$$

Another measure of an animal's growth is weight per day of age (WDA). WDA is calculated by dividing an animal's weight by its age in days. For example, a heifer weighing 400 lb at 200 days of age would have a WDA of:

$$\frac{400}{200} = 2.0 \text{ lb per day of age}$$

Weight per day of age is a snapshot evaluation of an animal's growth rate at a fixed point in time. WDA is typically measured at ages up to and just beyond weaning.

While weight and weight changes are important traits to be monitored by livestock managers, it is important to keep in mind that the composition, quality, and consistency of the final product determines true value.

CHAPTER SUMMARY

- Growth is an increase in body weight until mature size is reached. Development is the directive coordination of all physiological processes until maturity is obtained.

- Prenatal growth is a unique series of events starting from the union of two sex cells, which then grow and differentiate into a fetus with different organs and tissues.

- Embryonic development of the chick differs from that of mammals, as the chick develops more rapidly in the egg outside the hen's body.

- Muscle growth occurs through an increase in fiber size (hypertrophy) and an increase in fiber number (hyperplasia).

- The study of the anatomy and physiology of farm animals involves understanding the structure and function of cells, tissues, organs, and systems. The most common systems are skeletal, nervous, muscular, circulatory, and endocrine.

- Growth curves are influenced by nutrition, sex, frame size, and muscling. They are important in determining when animals should be harvested to produce desirable combinations of fat, lean, and bone.

KEY WORDS

growth	organs	veins
development	system	hormones
hyperplasia	dark meat	breed types (biological
hypertrophy	white meat	types)
blastoderm	power	frame size
allantois	strength	mouthed
amnion	endurance	incisors
yolk sac	white fibers	broken-mouthed
chorion	red fibers	
tissues	arteries	

REVIEW QUESTIONS

1. What is the difference between growth and development?
2. What are the three phases of prenatal life?
3. When does the greatest increase in prenatal weight occur?
4. *True or False:* At birth the distal parts of the body (legs) are more developed than the proximal parts (hips and chest).
5. What is the germ spot on a fertilized or nonfertilized poultry egg called?
6. What are the four membranes essential for growth of the chick embryo?
7. Specialized groups of cells are called _____.
8. Groups of tissues that perform specific functions are _____.
9. What is the term for a group of organs that function together to accomplish a larger general function?
10. What are the two functions of the skeletal system?
11. What are the three types of muscle tissue?
12. _____ are blood vessels that transport blood away from the heart, whereas _____ are vessels that transport blood back to the heart.
13. Chemical substances secreted by a ductless gland into the circulatory system to affect a distant gland or organ is called _____.
14. When does the greatest postnatal growth occur?
15. What are the three primary tissues that compose a carcass?
16. When is fat deposition in the body greatest?
17. *True or False:* Females are typically leaner than intact or castrated males.
18. *True or False:* An animal's age can often be estimated by examining its teeth.

SELECTED REFERENCES

Berg, R. T., and L. E. Walters. 1983. The meat animal. Changes and challenges. *J. Anim. Sci.* 57:133 (Suppl. 2).

Berg, T. T., and R. M. Butterfield. 1976. *New Concepts of Cattle Growth.* Sydney, Australia: Sydney University Press.

Bone, J. F. 1988. *Animal Anatomy and Physiology.* Reston, VA: Reston Publishing Co.

Currie, W. B. 1988. *Structure and Function of Domestic Animals.* Boston, MA: Butterworth Publishers.

Frandson, R. D. 1986. *Anatomy and Physiology of Farm Animals.* Philadelphia, PA: Lea & Febiger.

Hammond, J., T. Robinson, and J. Bowman. 1983. *Hammond's Farm Animals.* Baltimore, MD: Edward Arnold University Park Press.

Prior, R. L., and D. B. Lasater. 1979. Development of the bovine fetus. *J. Anim. Sci.* 48:1546.

Rollins, F. D. 1984. Development of the embryo. *Arizona Coop. Ext. Serv. Publication* 8427.

Swatland, H. J. 1984. *Structure and Development of Meat Animals.* Englewood Cliffs, NJ: Prentice Hall.

Trenkle, A. H., and D. N. Marple. 1983. Growth and development of meat animals. *J. Anim. Sci.* 57:273 (Suppl. 2).

CHAPTER 19

Lactation

LEARNING OBJECTIVES

- Label the structures of the mammary gland
- Describe the development and function of the mammary gland
- Discuss the hormonal influences on milk production
- Describe nonhormonal influences on milk production
- Compare and contrast the milk composition of farm animals
- Discuss the role of colostrum

Lactation, the production of milk by the mammary gland, is a distinguishing characteristic of mammals, whose young at first feed solely on milk from their mothers. Even after they start to consume other feeds, the young continue to nurse until they are weaned (separated from their mothers so they cannot nurse).

The mammary gland serves two functions: (1) it provides nutrition to animal offspring, and (2) it is a source of passive immunity to the offspring. The importance of milk as a nutritional source to perpetuate each mammalian species has been known since the beginning of history. Only during the past few decades have the basic mechanisms of immunity through milk been determined.

Humans consume milk and recognize it as a palatable source of nutrients. Milk consumed in the United States comes primarily from dairy cows and, to a much lesser extent, from goats and sheep. In other countries, the milk supply also comes from water buffalo, yak, reindeer, camels, donkeys, and sows.

MAMMARY GLAND STRUCTURE

The mammary gland is an **exocrine gland** that produces the external secretion of milk transported through a series of ducts. The cow has four separate mammary glands that terminate into four teats; sheep and goats have two glands and two teats; and the mare has four mammary glands that terminate into two teats. The mammary glands of these species are located in the groin area (Fig. 19.1).

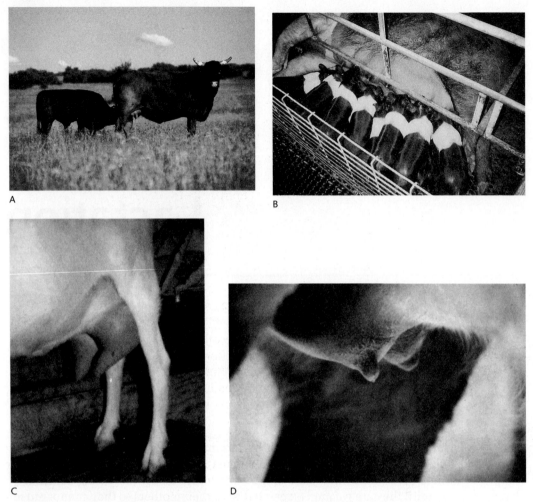

FIGURE 19.1   Mammary glands of the (A) cow, (B) sow, (C) doe, and (D) mare. Courtesy of (B) National Swine Registry, (C) George F. W. Haenlein, and (D) Colorado State University.

Sows have 6–20 mammary glands located in two rows along the abdomen, with each gland having a teat. Typically 10–14 of the sow's mammary glands are functional (Fig. 19.1). There is evidence that teat number in swine is not related to litter size or litter-weaning weight.

Figure 19.2 shows the basic structure of the cow's udder. The udder is supported horizontally and laterally by suspensory ligaments (Fig. 19.2A). The internal structure of the udder is similar for all farm animal species, except for the number of glands and teats.

The secretory tissue of the mammary gland (Fig. 19.2C) is composed of millions of grapelike structures called **alveoli.** Each **alveolus** has its own separate blood supply from which milk constituents are obtained by epithelial cells lining the alveolus. Milk collects in the alveolus lumen and, during milk letdown, travels through ducts to a larger collection area called the *gland cistern.* During milk letdown and the milking process, milk is forced into the teat cistern and through the streak canal to the outside of the teat.

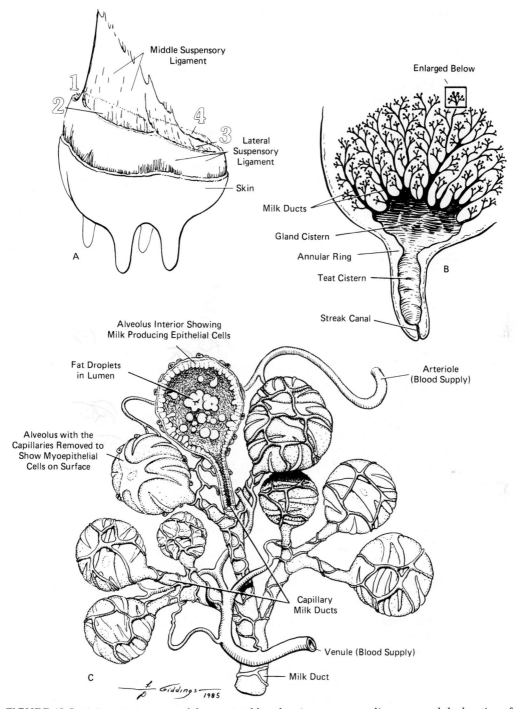

FIGURE 19.2 (A) Basic Structure of the cow's udder, showing suspensory ligaments and the location of the four separate quarters. (B) A section through one of the quarters, showing secretory tissue, ducts, and milk-collecting cisterns. (C) An enlarged lobe with several alveoli and their accompanying blood supply. *Source:* Colorado State University.

MAMMARY GLAND DEVELOPMENT AND FUNCTION

Development

Mammary gland growth and development occur rapidly as the female reaches puberty. The ovarian hormones (estrogen and progesterone) have a large effect on development. Estrogen is primarily responsible for duct and cistern growth, whereas progesterone stimulates growth of the alveoli.

Estrogen and progesterone are produced by the ovary under the stimulation of FSH and LH from the anterior pituitary (Fig. 19.3). The pituitary also has a direct influence on mammary growth through the production of growth hormone. Placental lactogen stimulates general cell growth of the mammary gland and may also influence fetal growth.

Figure 19.3 also identifies the additional indirect effect (other than FSH and LH) on mammary growth and development. Thyroid hormones are produced under the influence of TSH, and the adrenal gland produces the corticosteroids when stimulated by ACTH from the pituitary. All of these hormones work in concert to produce mammary growth and function.

Milk Secretion

Growth hormone, adrenal corticoids, and prolactin are primarily responsible for the initiation of lactation. These hormones become effective as parturition nears and when estrogen and progesterone hormone levels decrease.

Through milking or nursing, the milk in the gland cistern is soon removed. There remains a large amount of milk in the alveoli that is forced into the ducts by the contractions of the myoepithelial cells (Fig. 19.2). These cells contract under the influence of the hormone oxytocin, which is secreted by the posterior pituitary (Fig. 19.3).

Oxytocin, released by the suckling reflex, can also be released by other stimuli; a suckling calf will nudge the udder with its head to initiate the milk letdown response. The milk letdown can be associated with feeding the cows or washing the udder. (The latter is a typical management practice before milking cows.) Oxytocin release can be inhibited by pain, loud noises, and other stressful stimuli.

Maintenance of Lactation

Lactation is maintained primarily through hormonal influence. Prolactin, thyroid hormones, adrenal hormones, and growth hormone are all important in the maintenance of lactation.

Daily milk production typically increases during the first few weeks of lactation, peaks at approximately 4–6 weeks, and then decreases over the next several weeks of the lactation period. Persistency of lactation measures how milk production is maintained over time. For example, in dairy cattle, persistency is determined by calculating milk production of the current month as a percentage of the last month's production.

Figure 19.4 shows lactation curves for the various species of farm animals. The curve for dairy cows (Figure 19.4A) represents approximately 14,000 lb produced in 305 days.

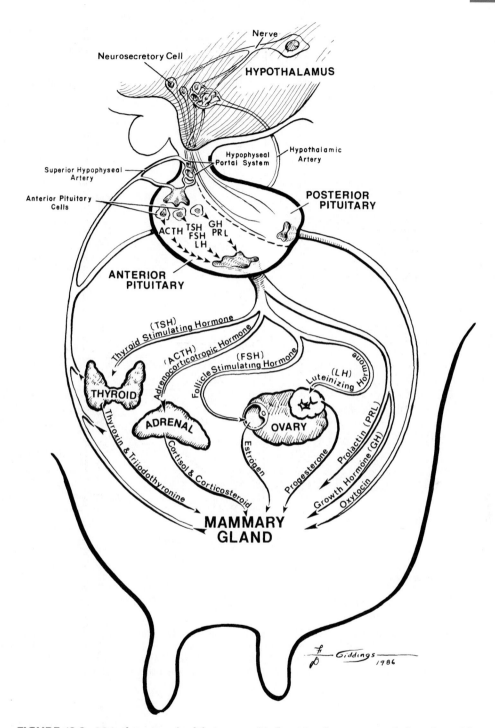

FIGURE 19.3   Major hormones (and their sources) influencing the anatomy and physiology of the mammary gland. Refer to Figure 18.11 for the location of the endocrine glands throughout the body. *Source:* Colorado State University.

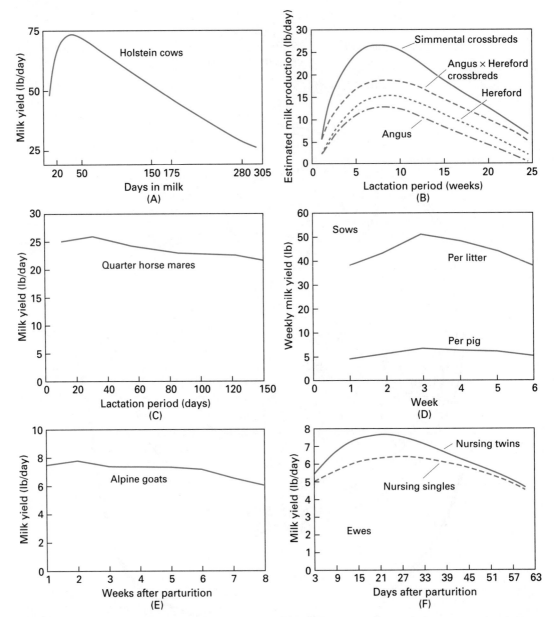

FIGURE 19.4 Lactation curves for several species of farm animals. (A) Dairy cows. *Source:* Colorado State University (CO, ID and UT DHIA records). (B) Beef cow. *Source:* Jenkins and Ferrell, 1984 Beef Cow Efficiency Forum. (C) Quarter horse mares. *Source: J. Anim. Sci.* 54:496. (D) Sows. *Source: Missouri Agric. Research Bull.* 712. (E) Alpine goats. *Source: J. Dairy Sci.* 63:1677. (F) Ewes. *Source: Michigan State Agric. Expt. Sta. Research Report* 491.

Decreased milk production during the lactation period is due primarily to a decreased number of active alveoli and less secretory tissue (epithelial cells) in the alveoli. These and other changes are associated with hormonal changes.

When milking or suckling is stopped, the alveoli are distended and the capillaries are filled with blood. After a few days the secretory tissue becomes involuted (reduced in size and activity) and the lobes of the mammary gland consist primarily of ducts and connective tissue. The female then becomes **dry,** as milk secretion is not occurring. After the dry period (approximately 2 months in the dairy cow) and when parturition approaches, hormones and other influences prepare the mammary gland to resume its secretion and production of milk.

FACTORS AFFECTING MILK PRODUCTION

Inheritance determines the potential for milk production, but feed and management determine whether or not this potential is attained. The best feed and care will not make a high-producing female out of one that is genetically a low producer. Likewise, a female with high genetic potential will not produce at a high level unless she receives proper feed and care. Production is also influenced by the health of the animal. For example, **mastitis** (an inflammation of the udder) in dairy cows can reduce production by 30% or more. Proper management of dairy cows, particularly adherence to routine milking and feeding schedules, contributes to a high level of production.

In lactating farm animals, the level of milk production is important because milk from these females provides much of the required nutrients for optimal growth in the young. If beef cows, for example, are inherently heavy milkers, their milk-producing levels will be established by the ability of the young to consume milk. Normally, if the cow is not milked, she will adjust her production to the consumption level of the calf. However, if the cow is milked early in lactation, her milk flow may be far greater than the calf can consume. This situation makes it necessary either to continue milking the cow until the calf becomes large enough to consume the milk or to put another calf onto the cow.

Anything that causes a female to reduce her production of milk is likely to cause some regression of the mammary gland, thereby preventing resumption of full production. If the young animal lacks vigor, it does not consume all the milk that its mother can supply, and her level of production is lowered accordingly. This is one reason inbred animals grow more slowly and crossbred animals more rapidly during the nursing period. Crossbreds are usually larger and more vigorous at birth and can stimulate a high level of milk production by their dams, whereas inbred animals are smaller and less vigorous and cause their dams to produce milk at a comparatively low rate.

Females with male offspring produce more milk than females with female offspring. Male offspring are usually heavier at birth and have a faster growth rate than females. This puts a greater demand on the milk-producing ability of the dam.

Females with multiple births usually produce more total milk than females with single births. Age of the female also affects milk yield, with younger and older females producing less milk compared to females that have had several lactations. For example, sows are usually at peak milk yields during third or fourth lactation, beef cows produce the most milk at 5–9 years of age, and dairy cows have the highest milk yields between 5–8 years of age.

Large amounts of nutrients are needed to supply the requirements of lactating females producing high levels of milk, because much energy is required for milk secretion and because milk contains large quantities of nutrients. Adequate nutrition is much greater for lactation than for gestation; for example, a cow producing 100 lb of milk daily could yield 4.0 lb of butterfat, 3.3 lb of protein, and almost 5.0 lb of lactose in that milk. If nutrition is inadequate in quality or quantity, a cow that has the inherent capacity to produce 100 lb of milk daily will draw nutrients from her own body as the body attempts to sustain a high level of milk production. Withdrawal of nutrients reduces the body stores. This withdrawal usually occurs to a limited extent in high producers even when they are well fed. Cows reduce their milk production in response to inadequate amounts or quality of feed, but they usually do so only after the loss of nutrients is sufficiently severe to cause a loss in body weight. In gestation, a cow becomes relatively efficient in digesting her feed, so feed restrictions in gestation are less harmful than they might be at other times.

MILK COMPOSITION

Species Differences

Milk composition is markedly different among the mammalian species (Table 19.1). Milk from reindeer and aquatic mammals is exceptionally high in **total solids,** whereas the mare's milk is low in this regard. Large differences in milk fat percentages exist between mammals; donkeys are low (1.4%) and fur seals are high (53.3%).

Milk from cows or goats is much higher in protein than milk from humans. This is not a problem when babies are bottle-fed, as the additional protein is not harmful.

In dairy cows, the amount or percentage of fat is easily changed through changing the diet fed. Carbohydrates **(lactose)** remain relatively constant even with dietary fluctuations. Varying the protein content of the ration has little change on the protein content of the milk.

COLOSTRUM

The fetus develops in a sterile environment. The microorganisms existing in the external environment have not yet challenged its immune system. Before or shortly after birth, the fetal immune system must be made functional or death is imminent.

Immunoglobulins (Ig) are involved in the passive immunity transfer from the mother to the offspring. In some animals, the Ig are transferred *in utero*

TABLE 19.1 Average Milk Composition of Several Species

Species	Total Solids (%)	Fat (%)	Protein (%)	Lactose (%)	Minerals (%)
Human	13.3	4.5	1.6	7.0	0.2
Cow *(Bos taurus)*	12.7	3.9	3.3	4.8	0.7
Cow *(Bos indicus)*	13.5	4.7	3.4	4.7	0.7
Goat	12.4	3.7	3.3	4.7	0.8
Water buffalo	19.0	7.4	6.0	4.8	0.8
Ewe	18.4	6.5	6.3	4.8	0.9
Sow	19.0	6.8	6.3	5.0	0.9
Mare	10.5	1.2	2.3	5.9	0.4
Reindeer	33.7	18.7	11.1	2.7	1.2

through the bloodstream, whereas in other animals immunoglobulins are transferred through **colostrums.** Certain animals utilize both methods of Ig transfer; however, the colostrum method is most common for larger, domestic animals.

These Ig antibodies give the newborn protection from harmful microorganisms that invade the body and cause illness. The intestinal wall of the newborn is quite porous, permitting colostrum antibodies to be absorbed and enter the bloodstream. Within a few hours (no more than 24), the gut wall becomes less porous, allowing little absorption of the antibodies to occur. Thus, passive immunity of the newborn is dependent on an adequate supply of antibodies in the colostrum and on consuming the colostrum within a few hours after parturition.

CHAPTER SUMMARY

- Lactation is the production of milk from the mammary gland.
- Milk from mammals provides nutrients and passive immunity to the young and is a nutrient source for humans.
- Colostrum, the first milk produced after the young are born, is the source of immunoglobulins that provide early immunity to the offspring.
- The mammary gland is an exocrine gland that produces an external secretion (milk) from structures called alveoli.
- Several hormones—namely, estrogen, progesterone, thyroid hormone, growth hormone, adrenal hormone, and prolactin—affect mammary gland development, milk secretion, and the maintenance of lactation.
- Lactation curves for most farm animals peak a few weeks after lactation is initiated, then decrease throughout the lactation period.
- Milk is composed primarily of water with lesser amounts of protein, fat, lactose, and minerals.

KEY WORDS

exocrine gland	dry	lactose
alveoli	mastitis	colostrum
alveolus	total solids	

REVIEW QUESTIONS

1. What are the two functions of the mammary gland?
2. What are the secretory units of the milk-secreting tissue within the mammary gland?
3. What two reproductive hormones directly influence growth and development of the mammary gland?
4. How is milk released from the alveoli during milking or suckling?
5. What hormone is responsible for milk letdown?
6. What determines an animal's potential for milk production?

7. What determines whether an animal's genetic potential for milk production is attained?

8. _____ is an inflammation of the udder which can reduce milk production by 30% or more.

9. *True or False:* Milk from cows and goats is similar to human milk in all components except protein.

SELECTED REFERENCES

Ackers, R. M. 1994. Lactation. *Encyclopedia of Agricultural Science.* San Diego, CA: Academic Press, Inc.

Allen, A. D., J. F. Lasley, and L. F. Tribble. Milk production and related factors in sows. *Missouri Agric. Research Bull.* 712.

Gibbs, P. G., G. D. Potter, R. W. Blake, and W. C. McMullan. 1982. Milk production of quarter horse mares during 150 days of lactation. *J. Anim. Sci.* 54:496.

Henry, M. S., and M. E. Benson. 1990. Milk production and efficiency of lactating ewe lambs. *Michigan State Agric. Expt. Sta. Report* 491.

Keown, J. F., R. W. Everett, N. B. Empet, and L. H. Wadell. 1986. Lactation curves. *J. Dairy Sci.* 69:769.

Larson, B. L. (ed.). 1985. *Lactation.* Ames, IA: Iowa State University Press.

Offedal, O. T., H. F. Hintz, and H. Schryver. 1983. Lactation in the horse: Milk composition and intake by foals. *J. Nutr.* 113:2196.

Sakul, H., and W. J. Boylan. 1992. Lactation curves for several U.S. sheep breeds. *Animal Production* 54:229.

CHAPTER 20

Adaptation to the Environment

LEARNING OBJECTIVES

- Describe the importance of animals being adapted to their environment
- Discuss the thermal neutral zone relative to effective ambient temperature
- Outline the factors influencing effective ambient temperature
- Discuss adjusting rations for changes in weather conditions
- Compare and contrast heat and cold stress
- Describe the role of hair coat and fat cover on critical temperature
- Discuss hypoxia, hypertension, and white muscle disease

Livestock throughout the world are expected to produce under an extremely wide range of environments. Variations in temperature, humidity, wind, light, altitude, feed, water, and exposure to parasites and disease organisms are some of the environmental conditions to which livestock are exposed (Fig. 20.1). This chapter will focus on responses of livestock to the climatic environment.

Under **intensive management,** such as integrated poultry enterprises, confinement swine operations, and large dairies, many environmental conditions are highly controlled. Feed and water are plentiful, rations are carefully balanced, excellent health programs are carefully monitored, and, in many cases, temperature, humidity, and other weather influences are controlled.

Extensive management involves less producer control over the environmental conditions in which animals are expected to produce (Fig. 20.2). Many ruminant animals are extensively managed while also exposed to numerous climatic conditions as they graze the forage, which is often sparse. Livestock that function under these less than perfect conditions are capable of doing so because of the ability to successfully **acclimate** or because they are **genetically adapted** to the existing environment.

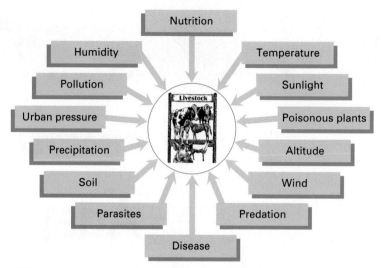

FIGURE 20.1 Environmental influences on livestock that must be managed by humans to assure profitability.

FIGURE 20.2 The ability of cattle to adapt to changing environmental conditions allows them to be managed under extensive conditions.

Economics dictate to a large extent whether animals will be managed intensively or extensively. Under intensive management, many environmental conditions are modified to better meet animal needs. Under extensive management, animals are changed (selected for adaptability) to fit the environment.

It is important to understand the impacts of the climatic environment on animal productivity in order to make sound information-based management decisions. These decisions may involve modifying the environment (provide shelter, feed, etc.) or it may be more cost-effective to select animals that are more genetically adapted to the existing climate. For example, a feedyard manager would evaluate whether it was more cost-effective to construct shades and/or sprinkler systems to enhance performance of *bos taurus* breeds of cattle exposed to heat or to utilize *bos indicus* cattle that are genetically more heat tolerant. In some cases, managers would choose a combination of environmental modifications and utilization of genetic selection to obtain optimal results.

RELATIONSHIP BETWEEN ANIMALS AND THE ENVIRONMENT

Adjusting to Environmental Changes

As the seasons change, two major kinds of changes occur in the environment: changes in temperature and changes in length of daylight. In the Western Hemisphere as summer approaches, temperature and length of daylight increase, thus increasing the amount of heat available to the animal. As autumn approaches, length of daylight and temperature decrease. Hormonal changes help the animal respond physiologically to these seasonal changes.

The thyroid gland, the lobes of which are located on both sides of the trachea, regulates the metabolic activity of the animal. When the environmental temperature is cool, the air inhaled tends to cool the thyroid gland. The thyroid responds by increasing its production of the hormone **thyroxine** that stimulates metabolic activity and thus heat is generated. The growing and shedding of insulating coats of fiber are slow changes that are also influenced by hormones. Sudden warm periods before shedding occur in spring, and sudden cool periods before a warm coat has been produced in autumn, can have severe effects on an animal. These sudden temperature changes are the greatest predisposers for the onset of respiratory diseases such as pneumonia, influenza, and shipping fever.

Temperature Zones of Comfort and Stress

Livestock are **homeotherms** and therefore must maintain constant body temperature. This requirement provides significant challenges for livestock that are exposed to wide changes in climate. The basis of climatic impact on livestock is, for the most part, a change in rate of energy flow. Farm animals must maintain constant body temperature in spite of the fact that the rate of energy gain and loss changes when animals are exposed to varying climatic conditions. One difficulty when discussing effect of climate on livestock is an accurate description of the climate. This is most conveniently done by use of an equivalent temperature referred to as **"effective ambient temperature."** The term effective ambient temperature (EAT) is a theoretical index of the heating or cooling power of the environment in terms of dry-bulb temperature and includes any environmental factor that alters environmental heat demand such as solar radiation, wind, humidity, or precipitation. Specific formulas for combinations of such variables as wind and temperature (windchill index) and temperature and humidity (heat humidity index) are available, although a comprehensive assessment that accounts for all climatic variables is not available. An example of effective temperature is the windchill index for cattle that is shown in Table 20.1. For example, a temperature of 20°F combined with a wind speed of 30 mph results in an effective ambient temperature of −16°F.

Evaluation of the relationship between animals and their thermal environment begins with the **thermal neutral zone (TNZ)** (Fig. 20.3). For purposes here,

TABLE 20.1 Windchill Factors for Cattle with Winter Coat

Wind Speed (mph)	Temperature (°F)												
	−10	−5	0	5	10	15	20	25	30	35	40	45	50
Calm	−10	−5	0	5	10	15	20	25	30	35	40	45	50
5	−16	−11	−6	−1	3	8	13	18	23	28	33	38	43
10	−21	−16	−11	−6	−1	3	8	13	18	23	28	33	38
15	−25	−20	−15	−10	−5	0	4	9	14	19	24	29	34
20	−30	−25	−20	−15	−10	−5	0	4	9	14	19	24	29
25	−37	−32	−27	−22	−17	−12	−7	−2	2	7	12	17	22
30	−40	−41	−36	−31	−26	−21	−16	−11	−6	−1	3	8	13
35	−60	−55	−50	−45	−40	−35	−30	−25	−20	−15	−10	−5	0
40	−78	−73	−68	−63	−58	−53	−48	−43	−38	−33	−28	−23	−18

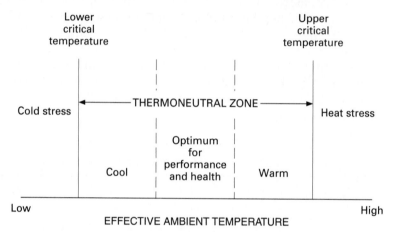

FIGURE 20.3 Range of stress related to changes in effective ambient temperature.

the thermal neutral zone is defined as the range in effective ambient temperature where rate and efficiency of performance is maximized and health is optimal. It must be emphasized that thermal comfort is a human term and that comfort for the stockman may be different from the TNZ of the animal; therefore, selection or assessment of animal environments must not be based on human comfort.

At temperatures immediately below optimum, but still within the TNZ, there is a cool zone (Fig. 20.3) where animals invoke mechanisms to conserve body heat. These are mainly behavioral responses such as postural adjustments and seeking protection from the wind but also may include reduced blood flow to the skin by a process termed vasoconstriction. Specific examples of behavioral responses include chickens fluffing their feathers to create a large space of dead air about themselves to provide added insulation and cattle congregating in an area that provides protection from the wind as they huddle together for protection.

The effectiveness of vasoconstriction and behavioral responses to cool conditions is maximal at the lower limit of the TNZ, a point called the **lower critical temperature (LCT)**. Below this point is the cold zone (Fig. 20.3) where the animal must increase its rate of metabolic heat production to maintain homeothermy (constant body temperature). This may be accomplished in the short term by increasing feed intake, voluntary physical activity, or shivering. Insulation in the form of hair, wool, feathers, or fat is a major factor in establishing LCT and rate of energy loss below LCT. As expected, increased insulation (hair, wool, feathers, fat) will raise the TNZ and LCT to a higher effective temperature. Clearly, thermal zones and critical temperatures change for a given animal depending on factors such as insulation, level of feed intake, and level of activity.

As effective temperature rises above optimum, the animal is in the warm zone (Fig. 20.3). Increasing blood flow to the skin by vasodilation, reducing feed intake and changing posture are typical mechanisms used to facilitate rate of heat loss and minimize heat production in the warm zone. When effective temperature exceeds the **upper critical temperature (UCT)**, animals are heat stressed and must employ evaporative heat loss mechanisms such as sweating and panting to maintain homeothermy. Higher producing animals have greater metabolic heat production and therefore tend to be more susceptible to hot climates. This is different from cold conditions where high-producing animals with their higher metabolic heat production are in a more advantageous position than poor-producing animals. Evaporation of moisture from the skin surface (sweating) or respiratory tract (panting) is the primary mechanism used by animals to lose excess body heat in a hot environment. Swine will use the behavior of getting wet and then drying off to cool themselves. Evaporation rate is limited by humidity but enhanced by air movement.

To avoid excessive heat loss, animals maintain the temperature of their extremities at a level below that of the rest of the body through a unique mechanism called **countercurrent blood flow action.** Blood in arteries coming from the core of the body is relatively warm and blood in veins in the extremities is relatively cool. The blood in veins in the extremities cools the warmer arterial blood so that the extremities are kept at a cool temperature. With the extremities thus kept at a relatively cool temperature, loss of heat to the environment is reduced.

Between 65 and 80°F, dilation of blood vessels near the skin and in the extremities occurs so that the surface of the animal becomes warmer, water consumption increases, respiration increases, and, in animals that can sweat, perspiration increases. Above 80°F, animals that have the capacity to sweat keep their body surfaces wet with sweat so that evaporation can help cool them. Producers often try to provide appropriate facilities to assist animals in adapting to climatic change. The use of snow fence suspended over a pen is one such example (Fig. 20.4).

When the environmental temperature exceeds 90°F, farm animals may die, have lower daily gains and poor feed conversion, and have lower reproductive performance due to higher numbers of embryonic deaths as well as diminished libido and reduced semen quality. During these periods of high temperature, animals become less active to reduce the amount of heat they generate, and lie down in the shade to reduce their exposure to the sun. Animals typically increase their consumption of water and excretion of urine. If the water consumed is cooler than the temperature of the animal, it can help considerably to cool the body. Cattle alter their water intake throughout the year in accordance with seasonal changes in temperature, doubling intake in the hottest months as compared to the winter months. Because the process of drying uses evaporation to

FIGURE 20.4 Snow fence is suspended over a feedlot pen to provide shade for cattle when heat stress is a factor.

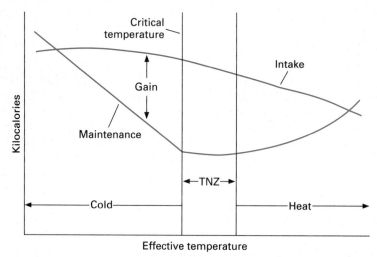

FIGURE 20.5 Effect of temperature on rate of feed intake, maintenance energy requirement, and gain. Adapted from Ames, 1980.

help cool the animal, managers may choose to use timed sprinkler systems in pens to help manage the effects of excessive heat.

CHANGING THE ENVIRONMENT

Adjusting Rations for Weather Changes

Optimal performance is reached in the TNZ or range of effective temperature where rate and efficiency of performance is high (Fig. 20.5).

Animal productivity is a function of feed intake and feed energy required for maintenance. Nutrient requirements and predicted gains of farm animals are published from research studies in which animals are protected from environmental extremes. The most common environmental factor that alters both performance and nutrient requirements is temperature; thus, livestock producers should be aware of critical temperatures that affect the performance of their animals and should consider making changes in their feeding and management programs if economics so dictate.

Feed intake and the feed energy required for maintenance are affected by temperature change. During periods of cold stress an animal's feed intake increases but at a slower rate than the rise in maintenance requirement (Fig. 20.5), resulting in an energy imbalance. Thus, as cold increases, managers must assure that sufficient additional feed is provided to avoid excessive weight loss.

In **heat stress** conditions, animals restrict their feed intake and experience rising maintenance costs. In both cases, managers must carefully evaluate cost of gain and other measures of productivity to determine whether or not environmental management strategies must be set in place.

The critical temperatures and optimal temperature ranges for some animals are shown in Figure 20.6. Effective ambient temperatures have not been calculated for all farm animals.

An example of temperature effect on rate of performance and efficiency of food animals is illustrated by data shown in Table 20.2 collected from swine grown in temperatures ranging from cold stress (32°F) to heat stress (86°F). Efficiency was reduced during both cold and heat stress, while being highest in the TNZ. Most dairy cows will experience a reduction in milk production when the environmental temperature exceeds 80°F. This occurs primarily because appetite is depressed and feed consumption declines. Milk yields will improve when thermal stress is alleviated by providing management such as shade, fans, sprinklers, or refrigerated air. While the temperature and efficiency values may differ for animals with different insulation, diet, etc., or for different species and products, the same general pattern of reduced efficiency is consistent among animals exposed to temperature stress.

Reproductive inefficiency is one of the most costly and production-limiting problems facing the livestock industry. Reproductive processes in both the male and female are sensitive to heat stress. As a general rule, increased temperature decreases ovulation rates, shortens intensity and duration of estrus, increases embryonic mortality, and decreases fertility of males. Seasonal variations in reproductive activity are well documented but may be attributed to either temperature or photoperiod. For example, seasonal activity in sheep is primarily controlled by day length, but heat (90°F) exposure of ewes during different phases of the reproductive cycle indicates that the embryo is most susceptible to heat stress shortly after conception. The impact of thermal stress on reproductive performance of the sow has been identified as a major problem in the swine industry. Reports indicate increased embryonic mortality when gilts are heat stressed immediately following conception and substantial increases in stillborn pigs when gilts are exposed to heat in late gestation. Obviously, production data for swine can be greatly affected by environment. Cattle are also susceptible to thermal stress. For example, reports indicate lowered (48% versus 0%) conception rates in cows exposed to 90°F compared with 70°F. An example of response of the male to thermal stress shows that only 59% of gilts mated with heat-stressed boars were pregnant 30 days postbreeding compared with 82% mated with control boars.

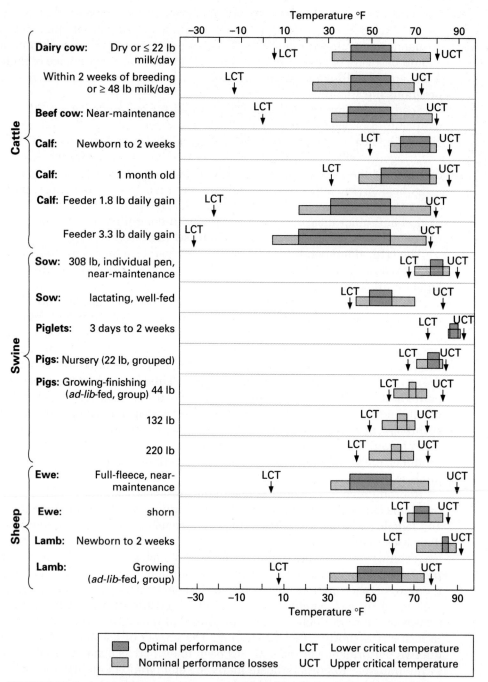

FIGURE 20.6 Critical ambient temperatures and temperature zones for optimal performance in cattle *(Bos taurus)*, swine, and sheep. Adapted from Yousef, 1985.

In summary, the thermal environment can have a drastic effect on animal performance. Fortunately, the impact of both cold and heat can be tempered by the ability of animals to adjust physiologically and behaviorally to temperature extremes. The cost of producing food and fiber of animal origin is increased by exposure to the climatic extremes.

TABLE 20.2 Effect of Temperature on Intake, Growth Rate, and Efficiency of Energy Conversion for Swine (70–100 kg)

Temperature (°C)	Caloric Intake (kcal DE/da)	Growth Rate (kg/da)	Product (kcal GE/da)	% Caloric Efficiency
0°	15,377	.54	2,991	19.4
5°	11,404	.53	2,936	25.7
10°	10,616	.80	4,432	41.7
15°	9,554	.79	4,376	45.8
20°	9,766	.85	4,709	48.2
25°	7,976	.72	3,988	50.1
30°	6,703	.45	2,493	37.1
35°	4,579	.31	1,717	37.4

Source: Ames, 1980.

MANAGING THE THERMAL ENVIRONMENT

Livestock producers are usually willing to incorporate management systems to improve energetic efficiency when it is economically advantageous. Initially, one thinks of modifying the existing environment to reduce the impact of thermal stress and improve energetic efficiency. Among the many possibilities for improving livestock environment to reduce cold stress are windbreaks (natural and man-made), sheds, confinement buildings without supplemental heat, and buildings environmentally controlled with supplemental heat. Modifications that provide optimum environments from an efficiency viewpoint require the most input (buildings with supplemental heat). Basically, the decision on the degree to modify animal environments depends on the cost of providing improved environment compared with the value of improved performance (cost-to-benefit ratio). Of course, such factors as cold tolerance of the animals, effect of diseases, and other determinants must be considered as well. Similar decisions regarding use of shades, sprinklers, and air-conditioned environments during heat stress are based on return for investment. Maximizing rate of performance of livestock is important but not necessarily the major goal. Economic considerations largely determine the level of efficiency selected for livestock systems.

A second management approach for dealing with both cold and heat is to consider the impacts of temperature extremes on performance and consider making changes in their feeding and management programs if economics so dictate.

Coldness of a specific environment is the value that must be considered when adjusting rations for cows. **Coldness** is simply the difference between effective temperature **(windchill)** and lower critical temperature. Using this definition for coldness, instead of using temperature on an ordinary thermometer, helps explain why wet, windy days in March might be colder for a cow than extremely cold but dry, calm days in January (Table 20.3).

TABLE 20.3 Estimated Lower Critical Temperatures for Beef Cattle

Coat Description	Critical Temperature (°F)
Summer coat or wet	59
Fall coat	45
Winter coat	32
Heavy winter coat	18

The major effect of cold on the nutrient requirements of cows is an increased need for energy, which usually means increasing the total amount of daily feed. Feeding tables recommend that a 1,200-lb cow receive 16.5 lb of good mixed hay to supply energy needs during the last one-third of pregnancy. How much feed should the cow receive if she is dry and has a winter hair coat but the temperature is 20°F with a 15-mph wind? The coldness is calculated by subtracting the windchill or effective temperature (4°F) from the cow's lower critical temperature (32°F). Thus the magnitude of coldness is 28°F. A rule of thumb (more detailed tables are available) is to increase the amount of feed 1% for each degree of coldness. A 28% increase in the original requirement of 16.5 lb would mean that 21.1 lb of feed must be fed to compensate for the coldness. However, if the cow is wet, the same increase in feed would be required at 31°F windchill which is also 28°F of coldness. This illustrates the importance of effective temperature when assessing animal environment.

Table 20.4 illustrates a partial list of adjustment factors for air temperature to account for the influence of floor type, air speed, and evaporative cooling systems for determining effective temperature in swine. Obviously, many factors influence the thermal environment of livestock.

Inability of Animals to Cope with Temperature Extremes

Animals are sometimes unable to control their body temperatures in conditions of extreme heat or cold. When the body temperature exceeds normal because the animal cannot dissipate its heat, a condition known as **fever** results. Fever is often caused by systemic infectious diseases and is often most severe when environmental temperatures are extremely high or low. Keeping the ill animal comfortable while medication is given will assist recovery.

When the temperature becomes excessively high, some animals (notably pigs) might lose normal control of their senses and do things that aggravate the situation. If nothing is done to prevent them from doing so, pigs that get too hot will often run up and down a fence line, squealing, until they collapse and die.

TABLE 20.4   Adjustment Factors to Air Temperature to Determine Effective Temperature for Swine

Factor	Temperature Adjustment (°F)
Type of flooring	
Mat	+5
Dry concrete	−9
Wet concrete	−18
Plastic-coated wire	−7
Uncoated wire	−9
Air speed	
10 yd/minute	−7
30 yd/minute	−13
100 yd/minute	−18
Evaporative cooling	
Fogger/mist system	−6
Drip coolers	−10
Sprinkler system	−10

Source: Adapted from Bell, 1996.

Pigs seen doing this should be cooled with water and encouraged to lie down on damp soil.

Most dairy cows will experience a reduction in milk production when the environmental temperature exceeds 80°F. This occurs primarily because appetite is depressed and feed consumption declines. Milk yields will improve when providing shade, fans, or refrigerated air to alleviate thermal stress.

ADAPTING ANIMALS TO THE ENVIRONMENT

Among the most impressive examples of adaptability under extensive management are the small Black Bedouin goats that graze the Negev desert of Israel and the eastern Sinai. Owing to the arid and desert conditions, they are watered only once every 2–4 days, which greatly increases their foraging range.

Ninety Bedouin goats (2,970 lb total body mass) can be herded successfully 2 days away from water points, where only 1 fat-tail Awassi sheep (110 lb) or 1.5 Black Mediterranean goats (165 lb) would be able to live because of more frequent watering needs.

During water deprivation for 4 days, the Bedouin goats lose 25–30% of their body weight from reduction of total body water and blood plasma volume. Nevertheless, these goats have relatively high milk yields, with 35-lb goats producing over 4.5 lb of milk per day. The Bedouin goats also have relatively low daily caloric demands for maintenance per unit of metabolic weight. Milk production efficiencies have exceeded 33% of energy consumed. The Bedouin goats have developed this adaptability over several centuries of natural selection as they have survived and produced in these arid, desert environments.

Inability of Animals to Cope with Other Stresses

Some animals are unable to adjust to environmental stresses. This inability may be genetically controlled.

When cattle are grazed at high altitudes, 1–5% develop high mountain (brisket) disease (Fig. 20.7). This disease usually occurs at altitudes of 7,000 feet or more. It is characterized as right-heart failure (deterioration of the right side of the heart). As the animal becomes more affected, marked subcutaneous edema (filling of tissue with fluid) develops in the brisket area. It was considered that this disease was caused by chronic **hypoxia** (lack of oxygen), which leads to pulmonary **hypertension** (high blood pressure). However, some recent studies at the U.S. Poisonous Plant Center at Logan, Utah, have shown that certain poisonous plants that grow only at high altitudes can cause brisket disease when fed to cattle maintained at low altitudes.

Some young cattle and sheep whose dams are fed plants grown in selenium-deficient areas develop **white muscle disease.** The disease is characterized by calcification of the heart and voluntary muscles, giving the muscles a white appearance. Administering selenium to the mother during pregnancy and to the young at birth can prevent white muscle disease. In the absence of treatment, some animals are affected and some are not; therefore, resistance and susceptibility appear to be inherited.

Equines may experience a variety of ailments due to winter conditions and failure of owners to make appropriate management shifts in accordance with climatic changes. Common problems of horses in winter include colic (intestinal

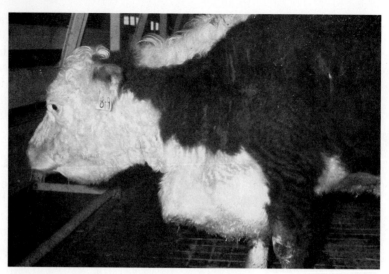

FIGURE 20.7   A classic case of high altitude or brisket disease. Note the swelling in the dewlap and brisket.

impaction), muscle injury, and respiratory ailments. Horse owners can minimize these problems by assuring water availability and appropriate intake, avoiding overfeeding of grain, maintaining a daily exercise program with thorough warm-up and cool-down periods, matching the ration to the activity level, and keeping barns and stables well ventilated, clean, and dry.

CHAPTER SUMMARY

- Extensively managed livestock are expected to be productive under a wide range of environmental conditions.
- Variations in environmental conditions are primarily due to changes in temperature, humidity, wind, daylight, altitude, feed, and water, and to exposure to parasites and disease-causing organisms.
- Producers may change animals genetically to adapt to different environmental conditions or may change the environment when economically feasible.
- The thermal neutral zone (TNZ) or comfort zone identifies a range of temperatures where heat production and heat loss from the animal's body are about the same. Temperatures above or below the TNZ significantly affect feed intake and the animal's performance.

KEY WORDS

intensive management	thermal neutral zone	windchill
extensive management	(TNZ)	fever
acclimate	critical temperature	hypoxia
genetically adapted	(upper and lower)	hypertension
thyroxine	countercurrent blood	white muscle disease
homeotherm	flow action	
effective ambient	heat stress	
temperature	coldness	

REVIEW QUESTIONS

1. Livestock operations, such as integrated poultry enterprises and confinement swine operations, which have highly controlled environmental conditions, are said to be under _____ management.

2. Livestock operations with less producer control over the environmental conditions in which animals are expected to produce are under _____ management.

3. What are the two major kinds of changes that occur in the environment during changes in season that influence productivity of livestock?

4. What is the term for the range of temperatures where heat production and heat loss from the body are about the same?

5. *True or False:* When temperatures are below the lower limit of an animal's thermoneutral zone, additional feed will be required to maintain body temperature.

6. *True or False:* Temperatures exceeding the upper limit of an animal's thermoneutral zone will have no effect on animals.

7. Describe animal performance, feed intake, and maintenance requirements during both heat and cold stress.

8. What steps can horse owners take to ensure that horses adapt to winter conditions?

SELECTED REFERENCES

Ames, D. R. 1980. Livestock nutrition in cold weather. *Anim. Nutr. and Health*, October.

Bell, A. 1996. Air temperature may be deceiving. *Pork 96*, October.

Bonsma, J. 1983. *Livestock Production*. Cody, WY: Ag Books.

Hahn, G. L. 1985. Weather and climate impacts on beef cattle. *Beef Research Progress Report* no. 2, MARC, USDA, ARS-42.

McDowell, L. R. (ed.). 1985. *Nutrition of Grazing Ruminants in Warm Climates.* New York: Academic Press.

National Research Council. 1981. *Effect of Environment on Nutrient Requirements of Domestic Animals.* Washington, DC: National Academy Press.

Yousef, M. K. 1985. *Stress Physiology in Livestock. Vol. II: Ungulates.* Boca Raton, FL: CRC Press.

CHAPTER **21**

Animal Health

LEARNING OBJECTIVES

- Describe the role of prevention in animal health
- Describe the approaches to assuring animal health that do not involve vaccines or treatment
- Compare and contrast the various product administration protocols
- List the physical signs a sick animal might display
- Discuss the role of quality assurance in the livestock industry

Productive animals are typically in excellent health. Death loss **(mortality rate)** is the most dramatic sign of health problems; however, lower production levels and higher costs of production due to sickness **(morbidity)** are economically more serious. For example, feedlot cattle that are healthy may return as much as $100 per head profit as compared to cattle that are sick. The $100 profit is realized from less death loss, improved cost of gain, increased medicine costs, and better carcass grading resulting from better health.

Disease is any deviation from normal health in which there are marked physiological, anatomical, or chemical changes in the animal's body. There are two major disease types: noninfectious and infectious. Noninfectious diseases result from injury, genetic abnormalities, ingestion of toxic materials, and poor nutrition. Microorganisms are not involved in noninfectious diseases. Examples of noninfectious diseases are plant poisoning, bloat, and mineral deficiencies.

Microorganisms such as bacteria, viruses, and protozoa cause infectious diseases. A **contagious disease** is an infectious disease that spreads rapidly from one animal to another. Brucellosis, ringworm, and transmissible gastroenteritis (TGE) are examples of infectious diseases.

PREVENTION

The old adage, "An ounce of prevention is worth a pound of cure," is an important part of herd health. Preventative herd health programs can eliminate or reduce most of the health-related livestock losses. Most major animal disease problems are associated with health management.

The components of a herd health-management program include (1) veterinarian-assisted planning, (2) sanitation, (3) proper nutrition, (4) record analysis, (5) physical facilities, (6) source of livestock, (7) proper use of biologics and pharmaceuticals, (8) minimization of stress, and (9) personnel training. All of these components require cooperative efforts between the producer and veterinarian.

Planning with the Veterinarian

Veterinarians are professionally educated and trained in animal health management. Practicing veterinarians are familiar with the disease and health problems common in their particular region. Producers who have the most successful herd health programs use veterinarians for planning and implementation of animal health-management systems. Important parts of the plan include regularly scheduled visits by the veterinarian throughout the year to assess preventative programs as well as to provide animal treatments. The veterinarian can also train farm and ranch personnel in simple herd health-management practices and serve as a reference for new products.

In most livestock and poultry operations, it is more economically feasible to have veterinarians assist in planning and implementing preventative health programs than to use their services in crisis situations only.

Producers should keep cost-effective health records as recommended by the veterinarian. These records are needed to assess problem areas and to develop a more effective herd health plan. The records can include such information as what was done and when in the preventative and treatment programs, descriptions of the conditions during health problems, and deaths. Possible causes of death also should be recorded, along with **necropsy** records when economically feasible.

Sanitation

The severity of some diseases is dependent on the number and virulence of microorganisms entering the animal's body. Many microorganisms live and multiply outside the animal, so implementing sanitation practices can reduce exposure to pathogens. These practices, in turn, reduce the incidence of disease outbreaks.

Manure and other organic waste materials are ideal environments for the proliferation of microorganisms. A good sanitation program includes cleaning of organic materials from buildings, pens, and lots. This allows the effective destruction of microorganisms via high temperatures and drying. Buildings, pens, and pastures should be well drained; preventing prolonged wet areas or mud holes. These sanitation practices help both in disease prevention and in controlling parasites.

Antiseptics and **disinfectants** are carefully selected and effectively utilized in a good sanitation program. Antiseptics are substances, usually applied to animal tissue, that kill or prevent the growth of microorganisms. Disinfectants are products that destroy pathogenic microorganisms. They are usually agents used on inanimate objects. In the absence of disinfectants, sanitizing with clean water may be helpful.

Other important sanitation measures include the prompt and proper disposal of dead animals, either to rendering plants or by burial.

Proper Nutrition

Well-nourished animals receive an adequate daily supply of essential nutrients. Undernourished animals usually have weak immune systems, thus making them more vulnerable to invading microorganisms.

Nutrition can be particularly critical during times of unusual stress, such as at weaning. Helping weaned calves, lambs, or pigs transition to the next production stage is of critical importance. Nutritional management of newly weaned animals can be improved with strategies such as supplementing with vitamins B and E, as well as trace minerals such as zinc.

Nutrition principles are presented in Chapters 15, 16, and 17. Feeding programs for the various species of farm animals are covered in Chapters 26, 28, 30, 32, 34, 35, 36, and 38.

Record Analysis

Herd health must be approached from an economic standpoint. Does a health problem exist, or is it just a perception? Proper records permit identification of health problems and determination of the cause, thereby allowing alternative methods of prevention and treatment to be assessed.

Physical Facilities

Physical facilities contribute to animal health problems by causing physical injury or stress, or by allowing dissemination of pathogens through a group of animals. They can contribute to the spread of disease by not preventing its transmission (e.g., venereal disease transmission owing to poor or inadequate fences). Even proper facilities can make transmission of disease easier if they are misused, for example feedlots that subject animals to crowding, stress, and poor sanitation.

Source of Livestock

Producers can reduce the spread of infectious diseases into their herds and flocks by (1) purchasing animals (entering their herds) from other producers who have effective herd health-management programs; (2) controlling exposure of their animals to people and vehicles; (3) providing clothing, boots, and disinfectant to people who must be exposed to the animals and facilities; (4) isolating animals to be introduced into their herds so that disease symptoms can be observed for several weeks; (5) controlling insects, birds, rodents, and other animals that can carry disease organisms; and (6) keeping their animals out of drainage areas that run through their farm from other farms.

Maintaining Biosecurity

The development of a detailed **biosecurity** plan is particularly critical where animals are intensively managed or where an enterprise is dependent on a marketing contract that specifies freedom from pathogens. The swine and poultry industries tend to have more sophisticated biosecurity measures than other livestock industries.

Biosecurity is of increasing importance as world trade in livestock and meat rises in both frequency and value. Furthermore, the threat of bioterrorism against the food supply must be considered a possibility. As a result, management systems designed to stop the spread of infectious diseases by minimizing the movement of biological organisms within and between groups of animals should be developed.

The components of an effective biosecurity system include verification of sources of animals in regards to herd health management, serological testing, treatment and vaccination protocols, monitoring incidence of disease, and record keeping. If outside sources of animals are utilized, a biosecurity plan also specifies design of isolation facilities, duration of isolation, protocols for employees working in the isolation units to prevent contamination of other farm sites, serological testing, and additional elements of a critical control point management system.

The biosecurity plan will also detail assuring integrity of farm perimeters, managing farm employees and visitors, pest management, quality assurance of feed and water supplies, managing outside trucks used to transport animals to and from the farm, disposal of dead and chronically sick animals, and manure disposal. A more detailed list of biosecurity strategies is outlined in Table 21.1.

TABLE 21.1   Components of an Effective Biosecurity Plan

Avoid introduction of diseased or infected animals
Select replacement stock from within the herd or flock
Introduce only young, virgin males
If replacements are purchased, buy only from known sources
Test for specific diseases of concern (recognize that tests may be limited in effectiveness)
Quarantine new animals for 30 days or more

Increase specific disease resistance
Implement an appropriate and effective vaccination program
Store and use the vaccine according to label guidelines
Follow booster schedules precisely

Increase overall disease resistance
Minimize dystocia
Provide balanced diet in adequate volume to meet animal requirements
Assure appropriate trace mineral supplementation
Control internal and external parasites
Minimize animal stress during handling and transport

Minimize exposure to disease agents
Isolate sick animals
Properly dispose of dead animals
Change needles appropriate to risk
Maintain clean facilities
Assure sanitation and proper waste management
Properly control rodents, birds, and insects
Assure fresh water supply
Store feed properly
Test and monitor feedstuffs
Maintain effective record-keeping systems
Avoid outside vehicle traffic through livestock areas
Provide protective clothing and foot covering for visitors

Source: Adapted from Smith, 2002.

More specific details can be obtained from the farm veterinarian, the extension service, and state and national livestock associations.

Use of Biologicals and Pharmaceuticals

Drugs are classified as biologicals or pharmaceuticals. **Biologicals** are used primarily to prevent diseases, whereas **pharmaceuticals** are used mainly to treat diseases. Both are needed in a successful herd health-management program.

Most biologicals are used to stimulate immunity against specific diseases. Vaccines are biological agents that stimulate active immunity in the animal.

Biologicals stimulate the body's immune system to produce antibodies that fight diseases. This is similar to the immunity attained through natural infection. An antibody is a protein molecule that circulates in the bloodstream and neutralizes disease-causing microorganisms. An animal is immunized when sufficient antibodies are produced to prevent the disease from developing. Specific antibodies must be produced for specific diseases; thus vaccination for several different diseases may be necessary. Also, periodic revaccination is often needed to maintain circulating antibodies at an adequate level.

Immunity via vaccination occurs only if the immune system responds properly. In some situations vaccination is not effective. In most animal species, for example, the immune systems of newborn young are not yet developed enough to respond. This is why their immunity is dependent on the maternal antibodies ingested from their mother's milk (colostrum). Also, undernourished or stressed animals, or those animals already exposed to a disease, rarely give a positive response to vaccination.

The amount of vaccine, frequency, route of administration, and duration of immunity vary with the specific vaccine. The manufacturer's directions should be followed carefully.

Vaccines (biologicals) and drugs (pharmaceuticals) can be administered as follows:

1. Topically—applied to the skin (Fig. 21.1).
2. Orally—through the mouth by feeding, drenching, or using balling guns. The latter are used to give pills such as capsules or boluses.
3. Injected directly into the animal's body by using a needle and syringe.
4. Intranasally—product delivered via inspired air.

Types of injections include (1) subcutaneous (under the skin but not into the muscle), (2) intramuscular (directly into the muscle), (3) intravenous (into a vein), (4) intramammary (through a cannula in the teat canal and into the milk cistern), (5) intraperitoneal (into the peritoneal cavity), and (6) intrauterine (using an infusion pipette or a tube through the cervix). Some injection sites are shown in Figures 21.2 and 21.3.

It is important to recognize that intramuscular injections in food animals should be minimized whenever possible as this route of administration may lead to injection-site blemishes (Fig. 21.4). These blemishes can affect the tenderness and visual appeal of meat not only at the injection site but throughout the surrounding tissue. Significant educational efforts by organizations

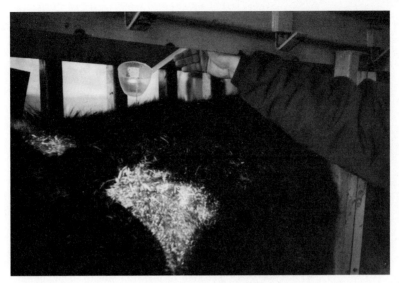

FIGURE 21.1   Topical administration of a dewormer.

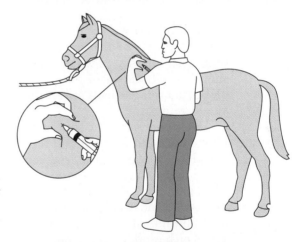

FIGURE 21.2   Subcutaneous injection in the horse. *Source:* R. A. Battaglia and V. B. Mayrose. *Handbook of Livestock Management Techniques* (New York: Macmillan, 1981). Copyright © Macmillan Publishing Co.

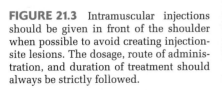

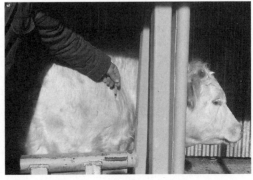

FIGURE 21.3   Intramuscular injections should be given in front of the shoulder when possible to avoid creating injection-site lesions. The dosage, route of administration, and duration of treatment should always be strictly followed.

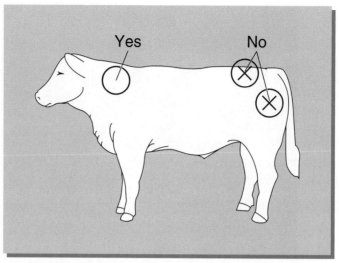

FIGURE 21.4 Selection of injection site is an important management decision in preventing blemishes to valuable cuts of meat. *Source:* NCBA.

such as the National Cattlemen's Beef Association and National Pork Producers' Council have limited the use of intramuscular injections in the livestock industry.

Pharmaceuticals are used to kill or reduce the growth of microorganisms in the treatment of diseases and infections. Pharmaceuticals come in a variety of forms: liquids, powders, boluses, drenches, and feed additives. They should be administered only after a specific diagnosis has been made. Examples of pharmaceuticals are antibiotics, steroids, sulfa compounds, hormones, and nitrofurans.

Other pharmaceuticals are used to control external and internal parasites. Losses owing to external and internal parasites are usually in the form of reduced weight gains, poor feed conversion, lower milk and egg production, reduced hide value, and excessive carcass trim, rather than high death losses.

Common external parasites are flies, lice, mites, and ticks, which live off the flesh and blood of animals. These parasites can also transmit infectious diseases from one animal to another. Insecticides are used to control external parasites. They can be applied as systemics or by spraying, fogging, dipping, or using back rubbers, ear tags, or dust bags.

Roundworms and tapeworms are the most common internal parasites, with flukes causing problems in certain environments. Internal parasites are controlled by interrupting their life cycle by (1) the presence of unfavorable climatic conditions (wet, warm weather favors the proliferation of many internal parasites); (2) destroying intermediate hosts; (3) managing animals so they will not ingest the parasites or their eggs; and (4) giving therapeutic chemical treatment. **Anthelmintics** are drugs that are given to kill the internal parasites.

Stress

Stress is any environmental factor that can cause a significant change in the animal's physiological processes. Physical sources of stress are temperature, wind velocity, nutritional deficiency or oversupply, mud, snow, dust, fatigue, weaning, ammonia buildup, transportation, castration, dehorning, and abusive handling. Social or behavior-related stress can result from aggression or over-crowding.

Prolonged stress can impair the body's immune system, causing a reduced resistance to disease. Stressful conditions in animals should be minimized within economic constraints.

Personnel Training

Maintaining good animal health requires continuous training of personnel. This is one of the most difficult areas to accomplish, particularly in a livestock operation that employs many people. While the owner understands what needs to be done, this may differ from what the employees actually accomplish. Cooperative training programs and informative sessions with the veterinarian can assist in the implementation of effective herd health-management programs. Training is the basis for the many quality assurance programs being implemented in the livestock and poultry industries.

DETECTING UNHEALTHY ANIMALS

Visual Observation

Detecting sick animals and separating them from the healthy animals is an important key to a successful treatment program. Animals treated in the early stages of sickness usually respond more favorably to treatment than do animals whose illness has progressed to advanced stages.

Early signs of sickness are not easily detected; however, the following are some of the observable signs, several of which are shown in Figure 21.5:

1. Loss of appetite is observed.
2. Animal appears listless and depressed.

FIGURE 21.5 This calf is sick with scours. Calves suffering from this condition typically appear listless and dehydrated.

TABLE 21.2 Vital Functions of Selected Livestock and Poultry Species

Animal	Rectal Temperature (± 1°F)	Respiration Rate[a] (per min)	Heart Rate[b] (per min)
Cattle	101.5	30[a]	60–70
Swine	102.0	16	60–85
Sheep	103.0	19	60–120
Goat	102.0	15	70–135
Horse	100.5	12	25–70
Poultry	107.1	12–36[b]	250–300

[a] Average.
[b] Normal range.

3. Ears may be droopy or not held in an alert position.
4. Animal has a hump in its back and holds its head in a lower position.
5. Animal stays separated from the rest of the herd or flock.
6. Coughing, wheezing, or labored breathing occurs.
7. Movement appears stiff and labored.

Vital Signs

Body temperature, respiration rate, and heart rate are the **vital signs** of the animal. Health problems are usually evident when one or more of the vital signs deviate from the normal range. The average and normal ranges for the vital functions are given in Table 21.2.

Body temperature is taken rectally. However, thermographic scanning technologies are being evaluated for possible use in the livestock industry. An elevated body temperature occurs in overheated animals or with most infectious diseases. A subnormal temperature indicates chilling or a critical condition of the animal.

Evaluating the animals' vital signs and visually observing their appearance can identify health problems identified in their early stages. This allows for early treatment, which usually prevents serious losses.

MAJOR DISEASES OF FARM ANIMALS

Table 21.3 identifies some of the major diseases and health problems of beef cattle, dairy cattle, swine, horses, poultry, and sheep.

QUALITY ASSURANCE

Quality assurance programs have been designed by a variety of agencies and agricultural organizations (Fig. 21.8) to help livestock producers improve their management practices, record keeping, and personnel training as a means to assure that the products they produce are wholesome and to offer consumers a high degree of confidence that food production practices contribute to food safety.

TABLE 21.3 Selected Major Diseases and Health Problems of Beef Cattle, Dairy Cattle, Swine, Horses, Poultry, and Sheep

Species/Disease/Cause	Signs	Prevention	Treatment
Beef Cattle			
Bovine viral diarrhea (BVD) Virus. Laboratory diagnosis imperative for accuracy	Feeder cattle: Ulcerations throughout digestive tract; diarrhea (often containing mucus or blood)	Vaccination prior to exposure; avoid contact with infected animals	Symptomatic treatment; antibiotics; sulfonamides; force feed
	Breeding cattle: abortions, repeat breeding	Vaccination annually 30 days prior to breeding	None; symptomatic treatment
Bovine respiratory syncytical virus (BRSV)	Labored breathing; pneumonia	Vaccination	Antihistamines, corticosteroids
Brucellosis Bacteria	Abortions	Calfhood vaccination (in some states) between the ages of 4–12 months	Test and slaughter; report reactors to state veterinarian
Campylobacteriosis (vibriosis) Bacteria	Repeat breeding; abortions (1–2%)	Vaccination of females and bulls prior to breeding; use of artificial insemination; virgin bulls on virgin heifers; avoid sexual contact with infected animals; cull open cows in infected herds	None (consult herd veterinarian)
Infectious bovine rhinotracheitis (IBR) Virus. Laboratory diagnosis imperative for accuracy	Pneumonia; fever, vaginitis, abortion and infertility in females; preputial infections in males	Vaccinate cows 40 days prior to breeding; vaccinate feeder cattle prior to exposure; semen from reputable bulls	Oxytetracycline, penicillin to minimize bacterial infections
Leptospirosis Leptospira spp. Several bacteria	Breeding cattle: Fever, off feed, abortions, icterus, discolored urine	Vaccination at least annually (in high-risk areas, more frequently); control of rodents, proper water management; avoid contact with wildlife and other infected animals	Streptomycin; penicillin; dihydrostreptomycin
Scours, calf colibacillosis "septicemia" E. coli K99 (pathogenic bacteria); several viruses plus stress factors	Diarrhea, weakness, and dehydration; rough hair and coat	Dry, clean calving areas; adequate colostrum	Fluid therapy; antibiotics to prevent secondary infections
Scours (viral) Virus (reovirus and corona virus)	Acute diarrhea, high mortality; affects calves shortly after birth (See Figs. 21.5 and 21.6)	Precalving dam vaccination with specific viral vaccines	Fluid therapy
Trichomoniasis Protozoa	Infertility and abortion (2–4 mos.); pyometra	Use of artificial insemination; virgin bulls on virgin heifers; maintain closed herd	Cull carrier animals
Dairy Cattle **Mastitis** Primarily infectious bacteria (Streptococci, Staphylococci, or E. coli)	Inflammation of the udder; decreased milk production	Use California mastitis test (CMT) or somatic cell counts (SCC)—checks for white blood cells in milk; avoid injury to udder; use proper milking techniques	Determine specific organism causing the problem, then select most effective antibiotic; frequent milking; anti-inflammatories

(continued)

TABLE 21.3   Selected Major Diseases and Health Problems of Beef Cattle, Dairy Cattle, Swine, Horses, Poultry, and Sheep *(continued)*

Species/Disease/Cause	Signs	Prevention	Treatment
Milk fever (parturient paresis) Metabolic disorder (low blood calcium level associated with a deficiency of vitamin D and phosphorus)	Occurs at onset of lactation; muscular weakness; drowsiness; cow lies down in curled position	Proper nutrition and management of cows during dry period	Intravenous injection of calcium borogluconate
Respiratory diseases (See *Infectious Bovine Rhinotracheitis*, "*Shipping Fever*" *parainfluenza*, under **Beef Cattle**)			
Uterine infections Usually results from bacterial contamination at the time of calving	Infections classified according to tissues involved; animal may be systemically or only locally affected; endometritis; inflammation of lining of uterus; metritis: infection of all tissues of uterus; pyometra: pus in the uterus	Sanitation; cleanliness when giving calving assistance	Early detection; antibiotics, hormone therapy, prostaglandins
Venereal diseases (See *BVD, IBR, leptospirosis, vibriosis and trichomoniasis*, under **Beef Cattle**)			
Swine **Atrophic rhinitis** *Bordetella bronchiseptica* (bacteria); *Pasteurella* (secondary invader)	Sneezing (most common); sniffling; snorting; coughing; twisting of snout; nasal infection (inflammation of membranes in nose). Diagnosis confirmed by observing turbinate bone atrophy (nasal) during postmortem examination (Fig. 21.7)	Monitor performance; monitor contact with animals from outside the herd; correct environmental deficiencies (sanitation, temperature, humidity, ventilation, dust, drafts, excessive ammonia, and overcrowding)	Vaccinate against *Bordetella* and *Pasteurella* organisms; improve environment and facilities; medicate sow feed with sulfamethazine or oxytetracycline
Colibacillosis *E. coli* Bacteria	Diarrhea (pale, watery feces); weakness; depression; most serious in pigs under 7 days of age	Good sanitation; good management practices (nutrition, pigs suckle soon after birth, prevent chilling); vaccinate sows to increase protective value of colostrum	Effective treatment is limited; identify strain of *E. coli*; consult with herd veterinarian
Mycoplasmal infections (Pneumonia and arthritis) Mycoplasma bacteria	Pneumonia: death loss low; dry cough; reduced growth rate; lesions in lungs Arthritis: inflammation in lining of chest and abdominal cavity; lameness; swollen joints; sudden death	Pneumonia: reduce animal contact; good nutrition, warm and dust-free environment, and parasite (ascarid and lung worm) control minimizes effects of the disease	Pneumonia: adequate treatment not available; sulfas and antibiotics prevent secondary pneumonia infections Arthritis: no satisfactory treatment; can depopulate and restock with disease-free animals

TABLE 21.3   *(continued)*

Species/Disease/Cause	Signs	Prevention	Treatment
Pseudorabies Virus of herpesvirus group	Baby pigs (less than 3 weeks old): sudden death; pigs (3 weeks to 5 months): fever, loss of appetite, labored breathing, trembling and incoordination of hind legs; mature pigs: less severe fever, loss of appetite; abortion and reproductive failure	Prevent direct contact with infected swine; humans should wear clean clothes and disinfect boots when entering and leaving the premises; recovered swine are immune but should be considered carriers	Drugs and feed additives are not effective; quarantine; blood test and slaughter those infected; repopulate
Transmissible gastroenteritis (TGE) Virus of coronavirus group	Vomiting; diarrhea; weakness; dehydration; high death rate in pigs under 32 weeks of age (see Fig. 21.6)	Sanitation is most cost effective; prevent transfer of virus from infected animals through exposure to other pigs, birds, equipment, and people	No drugs are effective; fresh water and a draft-free environment will reduce losses; antibiotics may prevent some secondary infections
Horses **Colic** Noninfectious: a general term indicating abdominal pain	Looking at flank; kicking at abdomen; pawing; getting up and down; rolling; sweating	Parasite control, avoiding moldy or spoiled feeds; prevent from overeating or drinking too much water when hot; have sharp points of teeth filed (floated) annually to assure proper chewing of feed	Quiet walking of horse; avoid undue stress; bran mash and aspirin; milk of magnesia; mild, soapy, warm water enema for impaction; always consult veterinarian as soon as possible
Lameness Most common are stone bruises, puncture wounds, sprains, and navicular disease	Departure from normal stance or gait; limping; head bobbing; dropping of the hip; alternate resting on front feet; pointing (extends one front foot in front of the other); stiffness; navicular disease: soft tissue bursitis in the foot	Bruise: avoid running horse on graveled roads or rocky terrain; puncture wounds: avoid areas where nails and other sharp objects may be present; sprains: avoid stress and strains on feet and legs; navicular disease: exact cause not known; avoid continual concussion on hard surfaces; provide proper shoeing and trimming	Varies with cause of lameness. Bruises: soaking foot in bucket of ice water or standing horse in mud; aspirin; puncture wound: tetanus protection; antibiotic; aspirin; soaking foot in hot epsom salts; navicular disease: no known cure; drug therapy and corrective shoes for temporary relief; surgery
Respiratory Three key viral diseases: viral rhinopneumonitis; viral arteritis; influenza; one key bacterial disease: strangles	Nasal discharge; coughing; lung congestion; fever; rhinopneumonitis may cause abortion	Isolate infected animals; avoid undue stress; draft-free shelter; parasite control; avoid chilling; rhinopneumonitis, influenza, and strangles vaccines are available	Follow prevention guidelines; strangles: drain abscess
Parasites 150 different internal parasites with strongyles (bloodworms), ascarids (roundworms), bots, pinworms, and stomach worms most common; external parasites (most common are lice, ticks, and flies)	Unthrifty appearance ("pot-bellied," rough hair coat); weakness; poor growth; lice, ticks, and flies can be visually observed on the horse	Good sanitation practices (clean feeding and watering facilities; regular removal of manure); periodic rest-periods for pastures; avoid overcrowding of horses; avoid spreading fresh manure on pastures	Internal parasites: worm horses twice a year (usually spring and fall); bot eggs can be shaved from the hairs of the legs; external parasites: application of insecticides

(continued)

TABLE 21.3 Selected Major Diseases and Health Problems of Beef Cattle, Dairy Cattle, Swine, Horses, Poultry, and Sheep *(continued)*

Species/Disease/Cause	Signs	Prevention	Treatment
Poultry			
Avian influenza Virus	Drop in egg production; sneezing; coughing; in the severe systemic form, deep drowsiness and high mortality are common	Vaccine (but has short immunity); select eggs and poults from clean flocks	No effective drug available
Coccidiosis Coccidia (depends on type of coccidiosis—there are nine or more types. Some signs, prevention, and treatment of three more common types are identified)	Weight loss; unthriftiness; palor; blood in droppings; lesions in intestinal wall; highly transmissible to other species	Use coccidiostat (kills coccidia organism)	Sulfa drugs in drinking water
Lymphoid leukosis Virus	Combs and wattles may be shriveled, pale, and scaly; enlarged, infected liver; lesions common in liver and kidneys	Sanitation; development of resistant strains through breeding methods	None
Marek's disease Herpesvirus	Can cause high mortality in pullet flocks; paralysis; death might occur without any clinical signs	Vaccination	None
Mycoplasma infections Mycoplasma organisms (there are several diseases caused by mycoplasma organisms; most important are chronic respiratory disease (CRD) and infectious synovitis)	CRD—difficult breathing; nasal discharge; rattling in windpipe; death loss may be high in turkey poults; swelling of face in turkeys; synovitis— swollen joints, tendons, tendon sheaths and footpads; ruffled feathers	For both CRD and synovitis: use chicks or poults from disease-free parent stock; sanitation (cleaning and disinfecting the premises)	Antibiotic in feed or drinking water
Newcastle disease	Gasping, coughing, hoarse chirping; paralysis; mortality may be high	Vaccination	None
Sheep			
Pseudotuberculosis Bacteria	Enlarged lymph nodes	Sanitation; reduce opportunity for wound infection at docking, castration, and shearing	Antibiotics
Enterotoxemia (overeating disease) Bacteria	Often sudden death occurs without warning; sick lambs might show nervous symptoms; e.g., head drawn back, convulsions	Vaccination (most effective in young lambs under 6 weeks of age nursing heavy-milking ewes and in weaned lambs on lush pasture or in feedlots)	None; antitoxins can be used but they are expensive and immunity is temporary (2–3 weeks)
Epididymitis Bacteria	Most common in western United States; swelling of epididymis; poor semen quality; low conception rates	Rigid culling and vaccination	Rigid culling

TABLE 21.3 *(continued)*

Species/Disease/Cause	Signs	Prevention	Treatment
Footrot Bacteria	Lameness; interdigital skin is usually red and swollen	Remove from wet pastures or stubble pastures; vaccines in some cases	Disinfectants such as 5% formalin or 10% copper sulfate; antibiotics
Pneumonia Many types; caused by viruses, stress, bacteria, and parasites	Sudden death; nasal discharge; depression; high temperature	Reduce stress factors	Antibiotics; sulfonamides
Pregnancy toxemia Undernutrition in late pregnancy; stress associated with poor body condition	Listlessness; loss of appetite; unusual postures; progressive loss of reflexes; hypoglycemia; coma; death	Prevent obesity in early pregnancy; provide good nutrition during last 6 weeks of pregnancy; reduce environmental stresses	In early stages the administration of some glucogenic materials may reduce mortality; in advanced stages no treatment improves survival

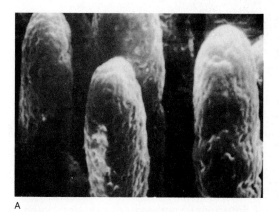
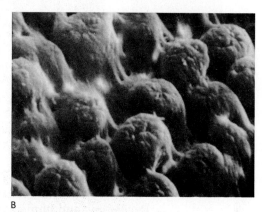

A   B

FIGURE 21.6 (A) Normal villi that line the intestinal wall. (B) Denuded villi after a viral infection. Several diarrhea results as the animals has lost the ability to digest and absorb nutrients. If the animal survives, the villi will regain their natural form in 1 to 2 weeks. Courtesy of Dr. G. L. Waxler, *Am. J. Vet. Res.* 33:1323.

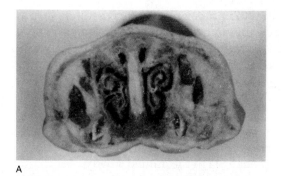
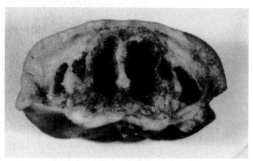

A   B

FIGURE 21.7 Cross sections through the hog's snout showing the effects of atrophic rhinitis. (A) Normal snout with turbinate bones. (B) Diseased snout showing severe atrophy or absence of turbinate bones. Courtesy of Elanco Products Company.

FIGURE 21.8 Quality assurance programs have been implemented in the livestock industry to ensure the production of wholesome and safe food.

The basic components of a quality assurance program include the following components:

1. Developing a list of critical control points where improved management practices are most likely to yield benefits.

2. Developing and implementing improved management practices such as treatment protocols, health care product inventory control, and record-keeping procedures.

3. Conducting an ongoing monitoring process of the critical control points while implementing improvements as needed.

4. Assuring a high level of employee training.

Improved record keeping is central to implementing a successful quality assurance program (Fig. 21.9). Records should be kept in regards to health product inventories, individual and group treatments, processing maps (Fig. 21.10) that indicate product and route of administration, pesticide use, and other appropriate documentation.

CHAPTER SUMMARY

■ Disease is any deviation from normal health in which there are marked physiological, anatomical, or chemical changes in the animal's body.

■ Death loss (mortality rate) and sickness (morbidity) rates are minimized in an excellent herd health-management program that includes (1) planning with a veterinarian, (2) sanitation, (3) proper nutrition, (4) record keeping and

Treatment Record for Individual Cattle

Animal ID: _____ Home Group/Pen: _____ Color: _____

Rx = medication name, WD = withdrawal time

Date:	Diagnosis	Temp	Severity (1-5)	Rx 1	Rx 2	Rx 3	Rx 4	Comments:	WD

FIGURE 21.9 Individual treatment records are important from both management and profitability perspectives.

Cattle Health Record

Name: _____, Address: _____

City: _____, State: _____, Zip: _____ Ph: _____

When possible select SubQ products, and never give injections in rear leg or top butt.

List "Treatment" Number on line connecting Injection Triangle & indicate ear implanted

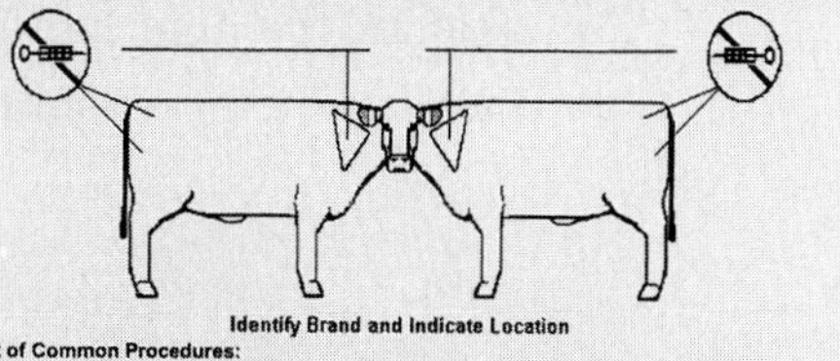

Identify Brand and Indicate Location

List of Common Procedures:

Respiratory virals,	Clostridials,	Pasteurella,	H.somnus,	Brucella,
Internal Parasites,	Coccidiostat,	External Parasites,	Implants,	Antibiotics,
Creep/Bunk Broke,	Micro-Nutrients,	Medicated Feed

Circle procedure preformed and list on numbered line in table below **AND** list number on line above that corresponds to the side of the cattle the injection was given.

NOTE: Use the Injection Triangle for all shots.

Procedure Procedure #	/Lot or Serial #	Company	Date Given	Date Withdrawal	Route Admin	Dose	Booster N/Y-when	Crew Initials
1.								
2.								
3.								
4.								
5.								
6.								
7.								
8.								
9.								
10.								

Number of Cattle _____ Date Weaned: _____, Dehorned (Yes / No)

Bulls ____, Steers _____ (method _____), Heifers ___ (Spayed: No/Yes = method _____),

ID: Right Ear or Left Ear/Group color and number: _____ /Individual (as appropriate): _____

Description/Comments:

Owner Signature: _____ Date: _____

Veterinarians Signature: _____ Phone: _____

FIGURE 21.10 Processing records facilitate maintenance of useful information about the health care of groups of animals.

analysis, (5) physical facilities, (6) choosing source of livestock, (7) proper use of biologics and pharmaceuticals, (8) minimizing stress, and (9) training personnel who are working with the animals.

■ Health problems are usually evident when one or more of the vital signs (temperature, respiration rate, and heart rate) deviate from the normal range.

■ The detection, prevention, and treatment protocols are given for the major diseases and health problems for cattle, swine, horses, poultry, and sheep.

■ Implementation of an effective quality assurance program is a key to increased profitability and improved consumer confidence in products from animal agriculture.

KEY WORDS

mortality rate	antiseptics	anthelmintics
morbidity	disinfectants	stress
disease	biosecurity	vital signs
contagious disease	biologicals	quality assurance
necropsy	pharmaceuticals	programs

REVIEW QUESTIONS

1. Death loss or _____ rate is the most dramatic sign of health problems among animals.

2. How is disease defined?

3. What is a contagious disease?

4. The best herd health-management programs involve _____ of disease.

5. What are the nine components of a herd health-management program?

6. What is a quality assurance program?

SELECTED REFERENCES

American Sheep Industry Association, USDA, and Colorado State University. 1995. *Producing High Quality Products from Sheep.* Denver, CO.

Battaglia, R. A., and V. B. Mayrose. 2002. *Handbook of Livestock Management Techniques.* Upper Saddle River, NJ: Prentice Hall. (esp. Chap. 10, "Animal Health Management").

Colorado State University. 2002. *Colorado Beef Quality Assurance Manual,* Colorado State University, Fort Collins, CO.

Gaafar, S. M., W. E. Howard, and R. E. Marsh (eds.). 1985. *Parasites, Pests and Predators.* New York: Elsevier Science.

Galyean, M. L., G. C. Duff, and L. J. Perino. 1999. Interaction of cattle health/ immunity and nutrition. *J. Anim. Sci.* 77:1120.

Jackson, N. S., W. J. Greer, and K. Baker. 2000. *Animal Health.* Danville, IL: Interstate.

Keeler, R. F., K. R. VanKampen, and L. F. James (eds.). 1978. *Effects of Poisonous Plants on Livestock.* New York: Academic Press.

Kirkbride, C. A. 1986. *Control of Livestock Diseases.* Springfield, IL: Charles C. Thomas.

Merck Veterinary Manual. 8th ed. 2002. Rahway, NJ: Merck & Co.

National Pork Producers' Council. 1997. *Pork Quality Assurance Program.* Des Moines, IA.

Naviaux, J. L. 1985. *Horses in Health and Disease.* Philadelphia, PA: Lea & Febiger.

Radostits, O. M., and D. C. Blood. 1985. *Herd Health.* Philadelphia, PA: W. B. Saunders.

Roeber, D. L., K. E. Belk, S. B. LeValley, J. S. Scanga, J. N. Sofos, and G. C. Smith. 2001. *Producing consumer products from sheep—the sheep safety and quality assurance program.* American Sheep Industry Association.

Smith, D. 2002. Biosecurity principles for livestock producers. Publication G1442. University of Nebraska-Lincoln, NC.

Stalheim, O. H. V. 1994. Animal diseases. *Encyclopedia of Agricultural Science.* San Diego, CA. Academic Press, Inc.

CHAPTER 22

Animal Behavior

LEARNING OBJECTIVES

- Describe the fields of animal behavior
- Discuss how an understanding of animal behavior is useful in managing:

  Reproduction
  Nutrition
  Handling and restraint
  Facility design and administration

- Describe the following behaviors:

  Caregiving
  Agonistic
  Eliminative
  Investigative
  Allelomimetic
  Maladaptive

Knowledge of animal behavior is essential to understanding the whole animal and its ability to adapt to various management systems utilized by livestock and poultry producers. For example, the value and performance of farm animals can be increased when managers apply their knowledge of animal behavior.

Animal behavior is a complex process involving the interaction of inherited abilities and learned experiences to which the animal is subjected. Behavioral changes enable animals to adjust to changing conditions, improve their chances of survival, and serve humans. Producers who understand patterns of behavior can manage and train animals more effectively and efficiently.

Basically, there are two major fields of animal behavior: one is psychology and the other is **ethology.** Historically, psychology has been directed toward studying learning in humans and applying insights gained from nonhuman animal studies to understanding human behavior. There are many different schools within the broad field of psychology. Ethology, however, originated with naturalists (going back to Aristotle) who originally emphasized instinctive behavior, but who also studied learned behavior.

Instinct (reflexes and behavioral patterns) is inherently present at birth. All mammals, at birth, have the instinct to nurse even though they must first learn the location of the teat. Shortly after hatching, chicks begin pecking to obtain feed.

Habituation is lack of response to a repeated stimulus such as a low-flying aircraft. When animals first see and hear the airplane, the novelty of it may frighten them. However, after they are repeatedly exposed to this experience, they become habituated and are no longer frightened.

Conditioning is the process whereby an animal makes an association between a previously neutral stimulus (e.g., a bell) or behavioral response (e.g., lifting its foot) and a previously significant stimulus, such as a shock or food. There are two types of conditioning:

1. **Classical conditioning**—e.g., Pavlov's noted study showed an association formed between an unconditioned stimulus (the sight of food, which caused salivation) and a neutral stimulus (the sound of a bell). The animal initially salivated at the sight of food; later, the mere sound of the bell produced salivation because of the previous association between the two stimuli.

2. **Operant conditioning** is learning to respond in a particular way to a stimulus as a result of **reinforcement** when the proper response is made. Reinforcement is a punishment or reward for making the proper response. Animals avoiding an electric fence and cattle coming to the feed bunk when they see a feed truck are examples of operant conditioning. In the first example, the animal is negatively reinforced by the shock. In the second example, cattle are positively reinforced with feed from the feed truck when they arrive at the feed bunk.

Trial and error is trying different responses to a stimulus until the correct response is performed, at which time the animal receives a reward. For example, newborn mammals soon become hungry and want to nurse. They search for someplace to nurse on any part of the mother's body until they find the teat. This is trial and error until the teat is located; then, when the young mammals nurse, they receive milk as a reward. Soon they learn where the teat is located and find it without having to go through trial and error. Thus, the young become conditioned in nursing behavior through reinforcement.

Reasoning is the ability to respond correctly to a stimulus the first time a new situation is presented. **Intelligence** is the ability to learn to adjust successfully to certain situations. Both short-term and long-term memory is part of intelligence.

Imprinting covers those processes where the helpless young bond to their caretaker—usually their dam. The way imprinting occurs varies between species. Odors and the dam licking the fluids from the newborn lead to bonding and rapid recognition in cattle and sheep. Creating a nest helps to assure that a sow bonds with her young pigs.

SYSTEMS OF ANIMAL BEHAVIOR

Farm animals exhibit several major systems or patterns of behavior: (1) sexual, (2) caregiving, (3) care-soliciting, (4) agonistic, (5) ingestive, (6) eliminative, (7) shelter-seeking, (8) investigative, and (9) allelomimetic. Some of these behavior systems are interrelated, though they are discussed separately in this chapter. It is not the intent here to describe in detail the different behavior patterns for all farm animals. The major focus is on identifying the behavioral activities that most significantly affect animal well-being, productivity, and profitability. By understanding animal behavior, producers can plan and implement more effective management systems for their animals.

SEXUAL BEHAVIOR

Observations on sexual behavior of female farm animals are useful in implementing breeding programs. Cows that are in heat, for example, allow themselves to be mounted by others. Producers observe this condition of "standing heat" or estrus to identify those cows to be **hand-mated** or bred artificially. Other ewes do not mount ewes in heat, but vasectomized rams can identify them.

Males and females of certain species produce **pheromones,** chemical substances that attract the opposite sex. Cows, ewes, and mares may have pheromones present in vaginal secretions and in their urine when they are in heat. Bulls, rams, and stallions will smell the vagina and the urine using a nasal organ that can detect pheromones. A common behavioral response in this process is called **flehmen,** during which the male animal lifts its head and curls its upper lip.

It appears that in a sexually active group of cows, the bull is attracted to a cow in heat most often by visual means (observing cow-to-cow mounting) rather than by olfactory clues. The bull follows a cow that is coming into heat, smells and licks her external genitalia, and puts his chin on her rump. When the cow is in standing heat, she stands still when the bull chins her rump. When she reaches full heat, she allows the bull or other females to mount (Fig 22.1).

When females are sexually receptive, they usually seek out a male if mating has not previously occurred. Females are receptive for varying lengths of time; cows are usually in heat for approximately 12 hours, whereas mares show heat for 5–7 days, with ovulation occurring during the last 24 hours of estrus.

FIGURE 22.1   A cow or heifer in estrus will allow either bulls or females to mount.

Vigorous bulls breed females several times a day. If more than one cow is in heat at the same time, bulls tend to mate with one cow once or a few times and then go to others. Other bulls may become attached to one female and ignore the others that are also in heat.

The ram chases a ewe that is coming into heat. The ram champs and licks, puts his head on the side of the ewe, and strikes with his foot. When the ewe reaches standing heat, she stands when approached by the ram.

The buck goat snorts when he detects a doe in heat. The doe shows unrest and may be fought by other does. Mating in goats is similar to that in sheep.

The boar does not seem to detect a sow that is in heat by smelling or seeing. If introduced into a group of sows, a boar chases any sow in the group. The sow that is in heat seeks out the boar for mating, and when the boar is located she stands still and flicks her ears. Boars produce pheromones in the saliva and preputial pouch, which attracts sows and gilts in estrus to the boars. Ejaculation requires several minutes for boars, in contrast to the instantaneous ejaculation of rams and bulls.

The sequence of events in estrus detection in horses appears to be similar to that in swine. The stallion approaches a mare from the front and a mare not in heat runs and kicks at the stallion. When the mare is in standing heat she stands, squats somewhat, and urinates as he approaches. Her vulva exhibits **winking** (opens and closes) when she is in heat.

In chickens and turkeys, a courtship sequence between the male and female usually takes place. If either individual does not respond to the other's previous signal, the courtship does not proceed further. After the courtship has developed properly, some females run from the rooster, which chases them until they stop and squat for mating. The male chicken or turkey stands on a squatting female and ejaculates semen as his rear descends toward the female's cloaca. Semen is ejaculated at the cloaca and the female draws it into her reproductive tract while the male is mounting her.

Male chickens and turkeys show a preference for certain females and may even refuse to mate with other females. Likewise, female chickens and turkeys may refuse to mate with certain males. This is a serious problem when pen matings of 1 male and 10–15 females are practiced, as the eggs of some females may be infertile. This preferential mating is a greater problem in chickens, as AI is the common breeding practice in turkeys.

Little relationship appears to exist between sex drive and fertility in male farm animals. In fact, some males that show extreme sex drive have reduced fertility because of frequent ejaculations that result in semen with reduced sperm numbers.

Research studies show that many individual bulls have sufficient sex drive and mating ability to fertilize more females than are commonly allotted to them. An excessive number of males, however, are used in multiple-sire herds to offset the few that are poor breeders and to cover for the social dominance that exists among several bulls running with the same herd of cows. The bull may guard a female that he has determined is approaching estrus. His success in guarding the female or actually mating with her is dependent on his rank of dominance in a multiple-sire herd. If low fertility exists in the dominant bull or bulls, then calf crop percentages will be seriously affected even in multiple-sire herds.

Tests have been developed to measure differences in libido and mating ability in young bulls. While behavioral differences are evident between different bulls, these tests, based on pregnancy rates, have not proven accurate for use by the beef industry. Performance may be improved in young bulls by exposing them to a female in heat prior to being placed in the breeding pasture.

Bulls being raised with other bulls commonly mount one another, have a penile erection, and occasionally ejaculate. Individual bulls can be observed arching their backs, thrusting their penises toward their front legs, and ejaculating.

Bulls can be easily trained to mount objects that provide the stimulus for them to experience ejaculation. AI studs commonly use restrained steers for collection of semen. Bulls soon respond to the artificial vagina when mounting steers, which provides them with a sensual reward.

Mating behavior has an apparent genetic base, as there is evidence of more frequent mountings in hybrid or crossbred animals.

Some profound behavior patterns are associated with sex of the animal and changes resulting from castration. This verifies the importance of hormonal-directed expression of behavior. Intact males display more aggressive behavior, whereas castrates are more docile after losing their source of male hormone.

CAREGIVING BEHAVIOR

Caregiving behavior can originate from the sire or dam; however, most caregiving is maternally oriented.

When the young of cattle, sheep, goats, and horses are born, the mothers clean the young by licking them. This stimulates blood circulation and encourages the young to stand and to nurse. Sows do not clean their newborns but encourage them to nurse by lying down and moving their feet as the young approach the udder region. They thus help the young to the teats.

Most animal mothers tend to fight intruders, especially if the young squeal or bawl. Often cows, sows, and mares become aggressive in protecting their young shortly after parturition. Serious injury can occur to producers who do not use caution with these animals.

Strong attachments exist between mother and newborn young, particularly between ewe and lamb and cow and calf. Beef cows diminish their output of milk about 100–120 days after birth of young, and ewes do the same after 60–75 days. This reduction in milk encourages the young to search for forage, the consumption of which stimulates rumen development. It is at this time that caregiving by the mother declines.

If young pigs have a high-energy feed available at all times, they nurse less frequently. Without a strong stimulus of nursing, sows reduce their output of milk. Some sows may wean their pigs early and show little concern for them a few days after they are weaned. Producers usually wean pigs at 21–35 days of age.

There is evidence that more cows calve during periods of darkness than during daylight hours. The calving pattern, however, can be changed by when cows are fed. Cows that are fed during late evening have a higher percentage of their calves during daylight hours.

CARE-SOLICITING BEHAVIOR

Care-soliciting behavior is manifested when young animals cry for help when disturbed, distressed, or hungry (Fig 22.2). Lambs bleat, calves bawl, pigs squeal, and chicks chirp. Even adult animals call for help when under stress. The female

FIGURE 22.2   A calf separated from its dam has a distinct bawl, which communicates distress of dissatisfaction. Courtesy of the American Hereford Association.

and her offspring may recognize each other's vocal sounds; however, it appears that the most effective way the dam recognizes her offspring is by smell. The offspring usually nurses with its rear end toward the female's head. This allows a dam to smell her offspring and decide to accept or reject it. The dam will bunt a rejected young animal with her head and kick it with her rear legs when it attempts to nurse. The young animals are less discriminate in their nursing behavior than are their dams.

AGONISTIC BEHAVIOR

Agonistic behavior includes behavior activities of fight and flight and those of aggressive and passive behavior when an animal is in contact (physically) with another animal or with livestock and poultry producers.

Interaction with Other Animals

Unless castrated when young, the males of all farm animals fight when they meet other unfamiliar males of the same species. This behavior has great practical implications for management of farm animals. Male farm animals are often run singly with a group of females in the breeding season, but it is often necessary to keep males together in a group at times other than the breeding season. The typical producer simply cannot afford to provide a separate lot for each male.

Bulls and other males may engage in prolonged physical activity when fighting, even to the point of exhaustion. Therefore, bulls and other potential fighting males should be put together either early in the morning or late in the evening when the environmental temperature is lower than at midday. If possible mix strange animals in a new pen to reduce the intensity of fighting. A resident animal may fight more intensely to defend its territory. Fighting to establish social dominance essentially moves through four stages: offense, defense, escape, and passivity. Mixing unfamiliar males often results in fighting behavior that is concluded once the "defeated" animal has escaped and assumed a passive posture. It is important, therefore, to assure that there is sufficient room to allow less dominant animals to escape from the more dominant individuals.

Cows, sows, and mares usually develop a pecking order, but fight less intensely than males. Sows that are strangers to each other sometimes fight. Ewes seldom, if ever, fight, so ewes that are strangers can be grouped together without harm. Young animals should be raised in a social group with their own species so they can learn proper social behavior with their own kind. Young intact males that are reared in isolation away from their own kind are more likely to attack people or be overly aggressive toward other animals.

Some cows withdraw from the group to find a secluded spot just before calving. Almost all animals withdraw from the group if they are sick.

Early and continuous association of calves is associated with greater social tolerance, delayed onset of aggressive behavior, and relatively slow formation of social hierarchies.

Status and social rank typically exist in a herd of cows, with certain individuals dominating the more submissive ones. The presence or absence of horns is important in determining social rank, especially when strange cows are mixed together. Also, horned cows usually outrank polled or dehorned cows where close contact is encountered, such as at feed bunks or on the feed ground.

Large differences in age, size, strength, genetic background, and previous experience have powerful effects in determining social rank. Once the rank is established in a herd, it tends to be consistent from one year to the next. There is evidence that genetic differences exist for social rank.

Animals fed together consume more feed than if they are fed individually. A competitive environment evidently is a stimulus for greater feed consumption. Dairy calves separated from their dams at birth appear to gain equally well whether fed milk in a group or kept separate. There is, however, evidence that they learn to eat grain earlier when group-fed compared with being individually fed. Cattle individually fed in metabolism stalls consume only 50–60% of the amount of feed they will eat if the animals are group-fed.

When fed in a group of older cows, 2-year-old heifers have difficulty getting their share of supplemental feed. Two-year-old heifers and 3-year-old cows can be fed together without any significant age effect in competition for supplemental feed. These behavior differences no doubt explain some of the age-related differences in nutrition, weight gains, and postpartum intervals that exist when cows of all ages compete for the same supplemental feed.

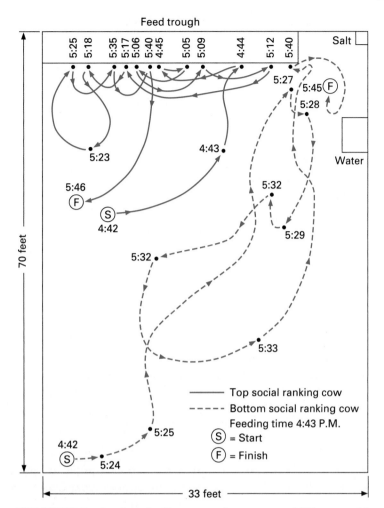

FIGURE 22.3 One-hour feeding pattern for two cows of different social rankings. Adapted from Schake and Riggs, 1972.

Dominant cows raised in a confinement operation usually consume more feed and wean heavier calves. More submissive cows wean lighter calves (25%), and fewer of them are pregnant compared with more aggressive cows. Figure 22.3 shows a 1-hour feeding pattern for two cows of different social rankings.

Interactions with Humans

Producers rank the disposition or temperament of animals from docile to wild or "high-strung." Evaluating animal posture can help handlers understand the mood and intent of an animal (Table 22.1). This evaluation is usually made when the animals are being handled through various types of corrals, pens, chutes, and other working facilities. The typical behavior exhibited by animals with poor dispositions is one of fear or of aggressive fighting or kicking.

There is evidence that farm animals develop good or poor dispositions from the way they have been treated or handled, though there is also evidence that

TABLE 22.1   Interpretation of Equine Mood via Assessment of Posture

Mood	Head-Neck	Ears	Eyes	Nostrils-Muzzle	Feet-Legs	Tail
		Physical Signs				
Anger	Neck outstretched	Pinned back	Narrowed	Lips pursed, nostrils flared	Stomping, stroking	Swishing
Challenge	Head and neck stretched upward and outward	Active movement	Clearly focused	Nostrils flared	Active pacing	Held high
Curiosity	Head and neck extended toward subject of interest	Pricked forward	Focused on subject	Sniffing	Firmly planted	Held up
Fear	Neck pulled in with head turned toward source	Fixed on source	Wide open	Nostrils flared	Rigid or fleeing	Clamped down
Relaxation	Head down	Drooped	Nearly closed	Lips drooped	Inactive	Low and relaxed
Sexual arousal	Arched and flexed	Forward and flicking	Dilated	Active	Pawing	Held high
Submission	Head low and averted	Held low	Averted	Low, lip smacking in young	—	—

Source: Adapted from multiple sources.

disposition has an inherited basis as well. A few heritability estimates for disposition are in the medium-to-high category, indicating that the trait would respond to selection. Some producers cull or eliminate animals with poor dispositions from their herds and flocks because of the potential for personal injury and economic loss (broken fences and facilities), as well as to reduce the excitability of other animals. In fact, at least one purebred cattle breed organization records temperament score as part of its national cattle evaluation program.

Recent research has shown that cattle with a nervous, excitable temperament (e.g., being highly agitated when restrained) experience lower weight gains in the feedlot. Cattle that ran out of a squeeze chute also had lower weight gains than those who exited more calmly.

Behavior during Handling and Restraint

Most animals are handled and restrained several times during their lifetimes. Ease of handling depends largely on the animal's temperament, size, and previous handling experience, and on the design of the handling facilities. Animals remember positive and negative experiences. Livestock with previous experience of calm, quiet handling will be less stressed and easier to handle in the future than animals that have had previous experiences with rough handling. Animals of all species will remain calmer and be easier to handle if they become acclimated to people quietly moving amongst them.

Understanding animal behavior can assist in preventing injury, undue stress, and physical exertion for both animals and producers. For example, knowing how to approach animals so they will respond to how the handler prefers the animals to move (Fig. 22.4) is critical to minimizing stress during movement of livestock. Most animals have a flight zone. When a person is outside this zone, the animal usually exhibits an inquisitive behavior. When a person moves inside the flight zone, the animal usually moves away. The size of the flight zone depends on the tameness or wildness of the animal.

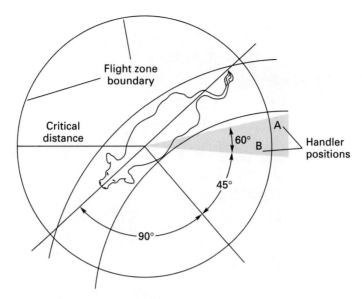

FIGURE 22.4 Handler positions for moving cattle. Positions A and B are the best places for the handler to stand. The flight zone is penetrated to cause the animal to move forward. Retreating outside the flight zone causes the animal to stop moving. Handlers should avoid standing directly behind the animal because they will be in the animal's blind spot. If the handler gets in front of the line extending from the animal's shoulder, the animal will back up. This is the point of balance. The solid curved lines indicate the location of the curved single-file chute. *Source:* Temple Grandin.

Blood odor appears to be offensive to some animals; therefore, the reduction or elimination of such odors may encourage animals to move through handling facilities with greater ease. Animals are easily disturbed by loud or unusual noises such as motors, pumps, and compressed air.

With their 310–360° vision, cattle are sensitive to shadows and unusual movements observed at the end of a chute or outside a chute (Fig. 22.5). For these reasons, cattle will move with more ease through curved chutes with solid sides (Fig. 22.6).

Round pens, having an absence of square corners, handle cattle that are more excitable with lower rates of injury.

Some breeders claim that AI facilities should permit beef cows to be handled quietly and carefully, and that using facilities where cows have previously felt pain should be avoided. They also state that pregnancy rates will thereby increase. This may sound logical, but research has not substantiated these claims.

INGESTIVE BEHAVIOR

Farm animals exhibit **ingestive behavior** when they eat and drink. Rather than initially chewing their feed thoroughly, ruminants swallow it as soon as it is well lubricated with saliva. After the animals have consumed a certain amount, they ruminate (regurgitate the feed for chewing). Cattle graze for 4–9 hours/day and sheep and goats graze for 9–11 hours/day. Grazing is usually done in periods, followed by rest and **rumination.** Sheep rest and ruminate more frequently

FIGURE 22.5 Shadows that fall across a chute can disrupt the flow of animals through the facilities. The lead animal often balks and refuses to cross the shadows. Courtesy of Temple Grandin.

FIGURE 22.6 Animals move more easily through curved chutes with solid sides because the solid sides prevent them from seeing people and other distractions through the fence. Courtesy of Temple Grandin.

when grazing than do cattle—cattle ruminate 4–9 hours/day; sheep 7–10 hours/day. Cows may regurgitate and chew between 300 and 400 **boluses** of feed per day; sheep between 400 and 600 boluses per day.

Under range conditions, cattle usually do not go more than 3 miles away from water, whereas sheep may travel as much as 8 miles a day. When cattle and sheep are on a large range, they tend to overgraze near the water area and to avoid grazing in areas far removed from water. Development of water areas, fencing, placing of salt away from water, and herding the animals are management practices intended to assure a more uniform utilization of the range forage.

Cattle, horses, and sheep have palatability preferences for certain plants and many have difficulty changing from one type of plants to other types. Most animals prefer to graze lower areas, especially if they are near water. These grazing behaviors tend to cause overgrazing in certain areas of the pasture and to reduce weight gains.

The behavior of cows grazing native range during winter is affected by cow age and weather conditions (Table 22.2). At the Range Research Station in Miles City, Montana, cows grazed less as temperatures dropped below 20°F, and at a −30°F, 3-year-olds grazed approximately 2 hours less than 6-year-olds. Also, with colder temperatures, cows waited longer before starting to graze in the morning. At 30°F, cows started grazing between 6:30 and 7:00 A.M., but at −30°F, they waited until about 10 A.M. to begin grazing.

ELIMINATIVE BEHAVIOR

Eliminative behavior involves fecal and urine deposition. Cattle, sheep, goats, and chickens void their feces and urine indiscriminately. Hogs, by contrast, defecate in definite areas of the pasture or pen. Knowing defecation patterns of the pigs can help to plan the ease of cleaning swine pens. Horses tend to void their feces on scent piles of other horses and genders.

Cattle, sheep, goats, and swine usually defecate while standing or walking. All these animals urinate while standing, but not usually when walking. Cattle defecate 12–18 times a day; horses, 5–12 times. Cattle and horses urinate 7–11 times per day. Animals on lush pasture drink less water than when they consume dry feeds; therefore, the amount of urine voided may not differ greatly under these two types of feed conditions.

TABLE 22.2 Activities of Cows Grazing on Winter Range

Activity	Hours
Grazing	9.45
Ruminating	
Standing	0.63
Lying	8.30
Idle	
Standing	1.11
Lying	3.93
Traveling	0.58
Total	24.00

All farm animals urinate and defecate more frequently and void more excreta than normal when stressed or excited. They often lose a minimum of 3% of their live weight when transported to and from marketing points. Much of the **shrink** in transit occurs in the first hour, so considerable weight loss occurs even when animals are transported only short distances. Weight loss can be reduced by handling animals carefully and quietly, and by avoiding any excessive stress or excitement of the animals.

SHELTER-SEEKING BEHAVIOR

Animal species vary greatly in the degree to which they demonstrate **shelter-seeking behavior.** Cattle and sheep seek a shady area for rest and rumination if the weather is hot, and pigs try to find a wet area. When the weather is cold, pigs crowd against one another when they are lying down to keep each other warm. In snow and cold winds, animals often crowd together. In extreme situations, they pile up to the extent that some of them smother. Unless the weather is cold and windy, cattle and horses often seek the shelter of trees when it is raining. This may be hazardous where strong electrical storms occur because animals under a tree are more likely to be killed by lightning than those in the open.

INVESTIGATIVE BEHAVIOR

Pigs, horses, and dairy goats are highly curious, investigating any nonthreatening strange object. They usually approach carefully and slowly, sniffing and looking as they approach. Cattle also do a certain amount of **investigating behavior** (Fig. 22.7). Sheep are less curious and more timid than some other farm animals. They may notice a strange object, become excited, and run away from it.

FIGURE 22.7 Cattle expressing investigative behavior as the result of interest in a novel item in the pen. Courtesy of Temple Grandin.

An object such as a paper cup on the ground can be either threatening or attractive to the animals. A novel object may attract the animals when they are on pasture. However, this same object may cause bolting and balking if the handler attempts to force the animals to walk over it. In one situation, the object triggers a fear reaction and in the other situation an investigative response.

ALLELOMIMETIC BEHAVIOR

Animals of a species tend to do the same thing at the same time. Cattle and sheep tend to graze at the same time and rest and ruminate at the same time. Range cattle gather at the watering place at about the same time each day because one follows another. This behavior is of practical importance because the producer can observe the herd or flock with little difficulty, notice anything that is wrong with a particular animal, and have that animal brought in for treatment. When artificially inseminating beef cattle, the best time to locate range cows in heat is when they gather at the watering place. **Allelomimetic behavior** is useful in driving groups of animals from one place to another.

OTHER BEHAVIORS

Communication and maladaptive behavior are two other behaviors that are common to the nine behavior systems previously presented. Some highlights of these two behaviors are given in the following text.

Communication

Communication exists when some type of information is exchanged between individual animals. This may occur with the transfer of information through any of the senses.

Females more easily adopt the young of others through transfer of the odor of one young animal to another. Cows may foster several calves if their own calves are removed at birth and the foster calves are smeared with amniotic fluid previously collected from the second "water bag." This is an example of imprinting.

Many farm animals learn to respond to the vocal calls or whistles of the producer who wants the animals to come to feed. The animals soon learn that the stimulus of the sound is related to being fed. This is an example of operant conditioning.

The bull vocally communicates his aggressive behavior to other bulls and intruders into his area through a deep bellow. This bellow and aggressive behavior is under the control of the male hormone (testosterone), as the castrated male seldom exhibits similar behavior. This communication behavior is part of the agonistic behavior system.

The bull also issues calls to cows and heifers, especially when he is separated from but still within sight of them. This type of communication could be included in the system of sexual behavior.

Horses have at least four vocal and three nonvocal sounds: (1) *squeals* (made during threats and encounters between individual animals—high pitched); (2) *nickers* (made by the stallion during mating and between the mare and foal prior to feeding—low pitched); (3) *whinnies* (begin as a squeal and end in a nicker—occurs between horses that are distressed or seek social contact); (4) *groans* (occur during discomfort or anguish); (5) *blows* (nonvocal sounds as air passes through the nostrils during prefeeding or when in alarm); (6) *snorts* (produced during nasal irritation, conflict, or relief); and (7) *snores* (produced by inhaling air—this has little part in communication).

Cattle are especially perceptive in their sight, as they have 310–360° vision. This affects their behavior in many ways—for example, when they are approached from different angles and when they are handled through various types of facilities.

Maladaptive or Abnormal Behavior

Animals that cannot adapt to their environment may exhibit inappropriate or unusual behavior. Some animals under extensive management systems, such as poultry and swine, are often kept in continuous housing to reduce costs of land and facilities. Frequently, both chickens and swine resort to cannibalism, which may lead to death if preventive measures are not taken. Some swine producers remove the tails of baby pigs to prevent tail chewing. Tail chewing can cause bleeding, and whenever bleeding occurs, the pigs are likely to become cannibalistic.

Some male uncastrated animals raised with other males masturbate and demonstrate homosexuality. In the latter situation, males mount other males, attempting to breed them. Some of the more submissive males may have to be physically separated from the more aggressive males to prevent injury or death.

The **buller-steer syndrome** is exhibited in steers that have been castrated before puberty. This demonstrates a masculine behavior of other than testosterone origin. Certain steers (bullers) are more sexually attractive for other steers to mount. As one steer mounts a buller, other steers are attracted to do the same. Thus, the activity associated with the buller-steer syndrome can cause physical injury, a reduction in feedlot gains, and additional labor and equipment expense as bullers are usually sorted into separate feedlot pens. On some feedlots 1–3% of the steers are bullers.

Growth implants and reduced pen space have been cited as reasons for an increased incidence of buller-steers. Some behaviorists cite evidence of similar homosexual behavior of males in free-ranging natural environments.

CHAPTER SUMMARY

- Animal behavior is a response to instincts and various stimuli to which the animals are exposed. It is a complex process involving the interaction of inherited abilities and learned experiences.

- Behavioral changes enable animals to adjust to changing conditions, improve their chance of survival, and serve humans by adapting to various management systems.

- Farm animals exhibit several major systems or patterns of behavior: (1) sexual, (2) caregiving, (3) care-soliciting, (4) agonistic, (5) ingestive, (6) eliminative, (7) shelter-seeking, (8) investigative, and (9) allelomimetic.

KEY WORDS

ethology	imprinting	rumination
instinct	hand-mated	boluses
habituation	pheromones	eliminative behavior
conditioning (classical	flehmen	shrink
and operant)	winking	shelter-seeking behavior
reinforcement	caregiving behavior	investigative behavior
trial and error	care-soliciting behavior	allelomimetic behavior
reasoning	agonistic behavior	buller-steer syndrome
intelligence	ingestive behavior	

REVIEW QUESTIONS

1. _____ is the scientific study of an animal's behavior in response to its natural environment.

2. _____, or reflexes and responses, is inherently present at birth, whereas _____ is a lack of response to a repeated stimuli.

3. What are the two types of conditioning?

4. What is imprinting?

5. What are the nine major systems or patterns of behavior exhibited by farm animals?

6. An example of sexual behavior displayed by female farm animals in which they allow a male, or sometimes other females, to mount them and which permits producers to detect their estrus is _____.

7. A mare in estrus will display standing heat and _____ of the vulva when a stallion approaches.

8. *True or False:* Caregiving behavior can originate from either the sire or the dam, but is most often maternally oriented.

9. Behavioral activities of fight and flight are examples of what type of behavior?

10. Interactions with other animals and interactions with humans are types of _____ behavior.

11. Producers can take advantage of the knowledge of agonistic animal behavior to handle animals in a low-stress manner by moving in and out of an animal's _____ zone.

12. What are reasons for culling animals with poor dispositions from the herd or flock?

13. Rumination or chewing the cud in ruminants is an example of what type of behavior?

14. Imitative behavior, when animals of the same species do the same thing at the same time—such as grazing or herding—is an example of what type of behavior?

15. What are two behaviors that are common to the nine systems of behavior?

16. Bellowing and aggressive behavior of bulls is under control of what hormone?

17. Animals under intensive management systems that cannot adapt to their environment and that exhibit inappropriate or unusual behavior are displaying what type of behavior?

18. What is an example of maladaptive behavior exhibited by chickens and swine under confinement housing?

SELECTED REFERENCES

Fraser, A. F., and D. M. Broom. 1990. *Farm Animal Behavior and Welfare.* London, UK: Bailliere Tindall.

Grandin, T. 1993. Teaching principles of behavior and equipment design for handling livestock. *J. Anim. Sci.* 71:1065.

Grandin, T. (ed.). 2000. *Livestock Handling and Transport.* 2nd ed. Wallingford, Oxon, UK: CAB International.

Hafez, E. S. E. (ed.). 1975. *The Behaviour of Domestic Animals.* 3rd ed. London: Bailliere Tindall.

Hart, B. L. 1985. *The Behavior of Domestic Animals.* New York: W. H. Freeman.

Houpt, K. A. 1991. *Domestic Animal Behavior for Veterinarians and Animal Scientists.* Ames, IA: Iowa State University Press.

Hurnik, J. F., A. B. Webster, and P. B. Siegel. 1995. *Dictionary of Farm Animal Behavior.* Guelph, Canada: Office of Educational Practice.

Lynch, J. J., G. N. Hinch, and D. B. Adams. 1992. *The Behavior of Sheep.* Sydney, Australia: CSIRO Publications.

McGlone, J. J. 1994. Animal behavior (ethology). *Encyclopedia of Agricultural Science.* San Diego, CA: Academic Press, Inc.

Monahan, P., and D. Wood-Gush. 1990. *Managing the Behavior of Animals.* New York: Chapman and Hall.

CHAPTER 23

Issues in Animal Agriculture

LEARNING OBJECTIVES

- Discuss the role of risk assessment in developing public policy
- Explain the perception of consumers relative to livestock production
- Discuss the concerns that some citizens have relative to animal welfare specific to each species
- Describe the role of a code of ethics in assuring proper animal care
- Discuss consumer perceptions about biotechnology
- Describe the primary environmental issues confronting the livestock industry
- Discuss the diet-health misconceptions about animal products
- Explain how the industry might change its products to better serve consumers
- Overview the food safety interventions used by the food industry
- Describe the primary foodborne pathogens of concern to the meat and poultry business
- Discuss the need for livestock industry participants to be active in the political process

The industries that make up animal agriculture face an increasingly diverse set of issues that left unresolved may have serious negative consequences. As the major political parties have lost their position as the sole method of participation in policy formation, the specialized interest group has become a significant force in determination of **public policy** at all levels of government.

The issues that currently face animal agriculture can be categorized as environmental, diet-health, animal rights, socioeconomic, and food safety (Table 23.1). However, before these issues can be adequately described it is critical to understand the underlying changes in policy formation that impact how

TABLE 23.1   Major Issues in the Animal Industries (ranked by species)

Issue	Species/Rank					
	Beef	Dairy	Horses	Poultry	Sheep	Swine
Animal welfare	12	3	1	2	7	9
Biotechnology	11	6	5	-	-	8
Consumer						
Diet and health	5	9	-	-	-	3
Food safety	1	7	-	1	-	2
Product perception	5	4	-	3	1	3
Quality assurance	6	-	-	-	5	4
Environment						
Air quality	10	2	-	-	-	5
Endangered species	4	-	-	-	-	-
Global warming	13	-	-	-	-	-
Water quantity and quality	9	2	3	-	-	5
Wildlife	13	-	-	-	-	-
Government policies						
Milk marketing	-	1	-	-	-	-
Predator control	-	-	-	-	4	-
Wool act	-	-	-	-	2	-
Marketing	2			-		
Integration & concentration	7	-	-	-	-	1
Global markets	3	5	4	-	6	6
Value-based marketing	8	-	-	-	-	7
Public lands		-	-		3	-

Source: Smith, 1993; *J. Dairy Sci.* 76:3254; representatives from animal industry organizations.

agriculturalists and their critics interact to influence decision makers. Special interest or pressure groups are smaller versions of political parties with correspondingly narrower agendas and concerns. The ability of a pressure group to advance its interests depends on a bias for an exclusive interest. These groups tend to be skewed toward upper-middle-class membership and are unlikely to serve a balanced set of political or social interests. Correspondingly, the communications received by decision makers from these groups tend to carry a strong upper-class bias.

Conflicts over particular issues arise from the efforts of a special interest group to bring an agenda into the public arena while an opposing group(s) works to neutralize or avoid the confrontation. Federal and state agencies contribute to political conflict either by direct participation or by providing the forum in which conflict can occur. Regulatory agency participation is heightened by the increasing independence from legislative or executive control and the likelihood that each agency operates with a relatively narrow focus of societal need.

Policy is created by the ability of interest groups to define an issue, call attention to the group's position, and then pressure government to take a desired action (Fig. 23.1). Therefore, preventing the movement of an issue into the public arena is equally as powerful as creating awareness.

Animal, consumer, and environmental groups tend to be activist in nature, to be well funded, and to have a strong commitment to a narrow mission or cause; these groups are growing both in numbers of organizations and in membership. However, as the number of issue-oriented pressure groups continues to increase, there are signs of coalition formation designed to consolidate both political and economic power.

FIGURE 23.1   Animal rights groups protest in Washington, D.C. in an attempt to garner support. Photo by William E. Carnahan.

As the set of issues or concerns addressed by an organization broadens, the income-generating ability generally increases. Crucial to the fund-raising aspects of special interest group management is the need to maintain high media visibility, even to the extent of staging events or utilizing scant or misleading information.

Given this scenario, the animal agriculture industries must carefully monitor issues to determine appropriate responses while promoting caring **stewardship,** wholesome food production, and careful husbandry practices. Furthermore, agricultural organizations need to establish appropriate working coalitions with moderate special interest groups. Finally, organizations must improve their ability to motivate membership, to establish proactive positions with the media, government, and public, and to increase funding.

ASSESSING RISK

Assessing risk becomes important when evaluating most of the issues outlined in this chapter. Technology, shifts in societal attitudes, consumer perceptions, and the consequences of excessive regulations require the development of better systems of determining short- and long-term risk associated with applications of technology or the distribution of resources. Improvements in the precision of analytical measurement place additional pressure on the need for risk analysis.

The major issues facing the livestock industry tend to fall within one of four categories—animal welfare, diet-health, environmental impact, and food safety. Table 23.2 outlines the consumer issues facing the pork industry. Other industries are confronted with similar issues.

As residential and urban uses of land increase into traditional farming areas, the potential for conflict tends to increase. A survey of rural residents in Pennsylvania determined that approximately one-third of respondents had a complaint about a neighboring farm. Of those registering a complaint, 57% were concerned about odor, while an additional 18% complained about fly problems. No other topic elicited more than a 5% response rate from complainants.

The perceptions of farm neighbors in regards to various issues were quantified by livestock species and are presented in Table 23.3. These results point to the need for a proactive approach by the livestock industry to deal effectively with a breadth of issues and concerns.

Despite the concerns expressed by consumers, actual risk is oftentimes vastly different than perceived risk. For example, a resident of the United States has a 1 in 200 chance of being a victim of violent crime, a 1 in 375,000 chance of acquiring a foodborne disease from eating fruits or vegetables, a 1 in 700,000 chance of being attacked by a shark, and a 1 in 900,000 chance of acquiring a foodborne disease from eating beef.

TABLE 23.2   Consumer Concerns Relative to the Pork Industry

Issue	Ranking
Water quality	1.38
Pollution	1.48
Air quality	1.55
Environment	1.64
Food prices	1.93
Animal welfare	2.37
Confinement/production methods	2.44
Family farm	2.59
Structure of agriculture	2.94

1 = Very concerned, 5 = not concerned.

Source: Adapted from Kliebenstein and Hurley, 2001.

TABLE 23.3   Mean Perception of Rural Residents to Livestock Enterprises (scale of 1 to 8, with 1 = strong agreement)

Perception	Dairy	Swine	Poultry	Beef	Veal
Use too many pesticides	5.1	5.7	5.6	5.6	6.4
Dispose of animal waste properly	4.4	4.8	4.8	4.7	5.9
Harm surface or ground water	4.7	4.8	5.0	4.9	6.0
Provide pleasant scenery	2.9	4.9	5.1	3.3	5.4
Handle and care for animals humanely	3.5	4.3	4.7	3.9	5.7
Animals are generally healthy	3.3	4.3	4.7	3.8	5.9
Give too many antibiotics	6.0	6.1	6.0	5.9	6.6
Give too many growth promotants	6.0	6.1	5.9	5.9	6.6
Make good neighbors	2.8	4.4	4.3	3.3	5.4
Contribute significantly to local economy	2.7	3.8	3.3	3.3	5.4

Source: Adapted from Jones et al., 2000.

ANIMAL WELFARE

One of the common mistakes in evaluating the philosophical position of animal welfare/rights-oriented groups is to categorize them as all having the same agenda. It is much more logical to consider these groups along a spectrum that ranges from animal exploitation to animal liberation (Table 23.4).

Animal exploitation would be characterized as events or uses that are conducted without concern for the animals involved. Examples might include cockfighting or dogfighting. These activities typically occur in violation of existing laws. Animal use groups typically function under the premise that animals are for human use but that people have a responsibility to provide good animal care and to utilize management practices that minimize animal suffering. Examples of such groups might include livestock producers, breed associations, zoos, and equine sport enthusiasts. Animal control groups include regulatory agencies or organizations at all levels of government that deal with domestic or wild animal control. These groups typically focus on enforcement of existing laws.

Animal welfare groups, such as the traditional Humane Societies, go a step further to extend care for animals that are suffering or homeless due to neglect. Historically, these groups have focused on pets or companion animals. Animal rights groups advocate that animals have rights that include not being consumed, used in sport or research, or in some cases even used as pets or companions. These groups almost universally promote vegetarianism, are opposed to the taking of any life, and may even believe that spaying/neutering are violations of an animal's right to breed. Animal liberationists are extreme groups that

TABLE 23.4 Animal Welfare/Animal Rights Groups: A Summary of the Views

Category/Group	Viewpoint
Animal exploitation	These groups argue that animals exist for human use or even abuse; animals are human property. These groups advocate or conduct activities that are illegal in most states (e.g., dogfighting, cockfighting, live pigeon shooting). Most of their activities involve pain or death of the animals primarily for entertainment.
Animal use	These groups believe that animals exist primarily for human use (e.g., livestock production, hunting, fishing, trapping, rodeos, zoos). These organizations have guidelines for the responsible care of the animals. They believe that harvesting of animals for food should be as painless as possible.
Animal control	These organizations enforce the laws, ordinances, and regulations affecting animals. Animals may be supplied for research; surplus animals are destroyed. Many advocate spaying or neutering animals.
Animal welfare	These are national groups, humane societies, and welfare agencies that support humane treatment of animals. They work within existing laws to accomplish goals. They publicize and document animal abuses to get laws changed. They do not provide animals for research. They require spaying and neutering, and are willing to euthanize surplus pets rather than let them suffer.
Animal rights	These national and local groups believe animals have intrinsic rights that should be guaranteed like human rights. These rights include not being killed, eaten, used for sport or research, or abused in any way. Some hold that pets have the right to breed. Most require spaying or neutering.
Animal liberation	These groups believe that animals should not be forced to work or produce for human benefit. They call for animal liberation.

Source: Adapted from K. B. Morgan, *An Overview of Animal-Related Organizations with Some Guidelines for Recognizing Patterns* (Kansas City, MO: Community Animal Control, 1989).

TABLE 23.5   Colorado Cattlemen's Association Animal Welfare Code of Ethics

Item	Policy
Statement of position	The multi-billion dollar livestock industry in Colorado is dependent upon the welfare of the animals under its stewardship. It is the policy of the Colorado Cattlemen's Association to promote among its members good stewardship toward animals under their care. It is further the policy of the CCA to cooperate with the Colorado Department of Agriculture, the Colorado Federation of Animal Welfare Agencies and other organizations, agencies and individuals that share legitimate concerns about the humane treatment of animals.
General considerations in livestock raising	A. Livestock should be raised in conditions that meet their basic physical and behavioral needs. B. Handling facilities. Properly designed, well-kept facilities allow humane, efficient cattle movement. Facilities should be constantly evaluated to see if they can be modified to allow better and more humane animal handling. C. People with a good knowledge of working cattle and cattle behavior allow the best use of these facilities. Staff should be monitored to make sure they understand the best way to work cattle. Training should be available for those who need additional instruction in handling livestock. This applies especially to those who have not previously handled livestock. D. Inducements. Inducements of any sort (hot shot, whips, etc.) should be used as little as possible and should be used only to the extent that it is necessary to facilitate animal movement. They should never be used in a punitive or angry manner. E. Livestock should have access to professional veterinary care as required both to prevent and treat injuries and disease. Use of pharmaceuticals should be used based on an evaluation of the animal's need, not simply out of "habit."
Transport of animals	Density of the loading of livestock should be based upon careful consideration of the class of livestock and the planned duration of the trip. Under no circumstances should the animals be crowded to the point of causing undue stress during the transport. Length of time in the vehicle should be based upon the class and condition of the livestock. In no case should the animals be in the vehicle long enough to cause them inordinate amounts of stress.
Livestock auctions	A. Terminally sick or injured animals should be destroyed on the ranch and not subjected to the additional stress of being shipped to auction (i.e., nonambulatory cattle, severe cases of bovine ocular neoplasia). B. It is essential that auction management continually monitor their facilities and staff to make certain that conditions that may foster animal abuse do not exist.
Statement of duty	It is a livestock producer's duty to oppose inhumane treatment of livestock at any stage of the animal's life. Persons who willfully mistreat animals will not be tolerated in our business. We will provide any assistance necessary to proper officials during the investigation of prosecution of individuals who abuse livestock under their care.

extend the animal rights agenda in violent ways and believe that their cause is so just that any means used to advance that cause are therefore acceptable.

The debate over animal rights has generated discussion both external to and within U.S. animal agriculture sectors. This discussion has yielded a variety of viewpoints. For example, the Colorado Cattlemen's Association has adopted an animal welfare **code of ethics** that articulates the values of its membership (Table 23.5). This ethic might well represent the traditional perspective of animal husbandry. Tom Regan (1985) argues that "The rights view will not be satisfied with anything less than the total dissolution of the animal industry as we know it today." Philosopher Bernard Rollin (1993) articulates a reasoned perspective that "The ethic which has emerged in mainstream society does not say we should not use animals or animal product. It does say that the animals we use should live happy lives where they can meet the fundamental set of needs dictated by their natures and where they do not suffer at our hands."

A common assumption is that the animal issue can be conveniently categorized into two notions—animal welfare and animal rights. However, as Rollin (1995) points out, this assumption is flawed in that it (1) attempts to place the discussion into an us-versus-them context, and (2) fails to account for the truth that animals have rights that dictate the need for husbandry practices that assure their welfare. However, there are clear differences between the perspectives of livestock producers and the Animal Liberation Front or People for the Ethical Treatment of Animals.

Welfare concerns due to modern agricultural practices can be classified into three basic categories according to Rollin (1995): (1) Production diseases such as liver abscesses in feedlot cattle as a result of diets high in concentrates and low in roughages, (2) Scale effects from large animal units that yield less individual animal attention than traditionally sized livestock enterprises, and (3) Physical and psychological deprivation due to prolonged confinement. Specific concerns relative to each species are listed in Table 23.6, with some examples shown in Figure 23.2.

Animal protection groups have also expressed concern relative to the use of animals in research. Rollin (1995) has proposed the following model for assuring the well-being of animals in research settings:

1. Animals should not suffer pain or distress unless the pain or distress was the experimental objective, or unless control of pain/distress would invalidate the results.

2. Animals should not be used repeatedly for invasive experiments.

3. Drugs that cause paralysis while leaving the animal conscious should not be used without anesthesia.

4. Husbandry and housing should fit the nature of the animal.

5. Local committees of nonscientists and scientists should provide oversight of these principles via protocol review and facility inspection.

TABLE 23.6   Animal Welfare Concerns Relative to Animal Agriculture

Beef	Swine	Dairy	Veal	Poultry	Horses
Branding	Confinement	Calf management	White veal production	Housing/ Cages	Training techniques
Castration	Farrowing crates	Calf housing	Behavioral deprivation	Behavioral problems	Housing and stalling
Dehorning	Flooring systems	Stall systems	Flooring systems	Forced molting	Racing as two-year-olds
Cancer eye	Tail docking	Castration	Diet	Debeaking	Transport
Downers	Teeth clipping	Dehorning	Group housing	Toe trimming	Slaughter
Slaughter	Castration	Branding		Exercise	Wild horses
Gomer bulls	Early weaning	Tail docking		Nesting	Sports injuries
Feedlots	Halothene gene	Mastitis/lameness		Dust bathing	Insurance fraud
Handling/ transport	Handling/ transport	Downers		Boredom	Performance drugs

Source: Adapted from Rollin, 1995.

A

FIGURE 23.2 (A) Producers strive to assure the welfare of the animals under their care. Consumer concerns about management practices have created market niches for (B) eggs from free-ranging poultry enterprises.

B

The Food Marketing Institute (FMI) and National Council of Chain Restaurants (NCCR) have developed animal care guidelines to assure that their suppliers followed best practices in regards to transportation and harvest protocols, housing, and rearing. Examples of the recommendations include minimizing beak trimming in poultry, phasing out feed-withdrawal molting, increasing space allocations per bird, and eliminating tail docking in dairy cattle.

The transport and harvest guidelines adopted by FMI-NCCR include the following:

1. Animals should be transported and unloaded in a manner that prevents injury and distress.
2. Nonambulatory animals should never be loaded onto a transport vehicle destined for market or harvest facilities.
3. Nonambulatory animals as a result of transport should be unloaded appropriately.
4. Animals must be handled humanely and in accordance with applicable federal, state, and local laws.
5. Animals must be completely insensible prior to harvest (religious harvest will be addressed separately).

The animal welfare and rights issue is not likely to go away. The livestock industry will be best served by assuring that producers maintain a strong record of sound husbandry practices, continuing to educate consumers and policy makers, and continuing to research opportunities to improve facilities, management techniques, and animal handling.

BIOTECHNOLOGY

The term **biotechnology** elicits mixed reactions from both consumers and producers. Biotechnology has been defined as the application of physical, chemical, and engineering principles to biological systems. Therefore, biotechnology ranges from the use of vaccines to the manipulation of the genome. The application of biotechnology to the field of agriculture offers significant opportunities for the improvement of human health, food production, animal health, and modification of the impact of agricultural production on the environment. Biotechnical applications will help reduce dependence on agrichemicals by producing pest/disease-resistant plants and animals. Biotechnology will improve the efficacy of vaccines or even produce milk that boosts human immunity against bacterial infection. The use of biotechnology offers excellent benefits to human medicine via mass production of proteins such as blood factor VIII, tissue plasminogen activator, and antithrombin III, which serve to control bleeding common to hemophilia, to break up blood clots during cardiac arrest, and as a blood thinner, respectively. Table 23.7 provides a summary of the opportunities for biotechnology to add value to animal agriculture.

Major breakthroughs anticipated in the future include the ability to switch genes on and off, development of feeds that are capable of enhancing nutritional value and efficiency, the use of plants as a delivery mechanism for vaccines, and

TABLE 23.7   Benefits from the Application of Biotechnology to Animal Agriculture

Human Health
Decrease morbidity and mortality from disease
Biopharmaceuticals
Xenotransplantation

Food Production
Feeding livestock
- Improve efficiency of feed utilization
- Enhance nutritional value of feed grains and forages

Metabolic modifiers
- Enhance production efficiency
- Improve lean:fat ratio
- Increase milk yields
- Decrease production of animal waste

Transgenic animals[a]
- Enhance productivity of livestock, poultry, and aquaculture

Microbial genomic application to enhance food safety
- Development of diagnostic tools for rapid identification of pathogens
- Creation of vaccines against foodborne pathogens

Animal Health
Vaccine development
Selection of animals with enhanced disease resistance capability

Modification of Environmental Impact
Utilization of genetically modified crops on manure composition and quantity
- Decrease phosphorus excretion
- Decrease nitrogen excretion

Source: Adapted from Council for Agricultural Science and Technology, 2003.
[a]Transgenics will be slow to be incorporated into production systems.

TABLE 23.8   First Thought or Image When Thinking about Genetic Engineering

Genetic Engineering Category	Percent
Science/Technology	11.8
Test-Tube Baby	9.7
Plant/Animal/People	8.2
Negative/Frightened	7.5
Monster/Mutant	7.1
DNA/Chromosomes	5.8
Medicine	4.1
God/Creation	4.0
Progress	3.9
Crossbreeding	3.5
Neutral	2.6
Nazi/Hitler	2.3
Artificial/Tampering	1.8
Other	1.6
Don't know	26.1

Source: W. K. Hallman and J. Metcalfe, Rutgers University.

the widespread development of **xenotransplantation**—the use of transgenic animals as a source of organs, tissues, and cells for cross-species transplantation.

In the case of **genetic engineering,** the perception of people ranges from "progress" to "mutant/monster" (Table 23.8). Notice that almost 20% of those surveyed had negative perceptions about genetic engineering. Even management practices such as crossbreeding, which have been in effect for years, created some degree of disfavor among consumers (Table 23.9).

TABLE 23.9 First Thought or Image When Thinking about Genetic Engineering—General

Genetic Engineering Category	Percent
Do not approve of producing plants by cross-fertilization	20
Do not approve of producing hybrid animals by crossbreeding	62
Approve of bioengineering to produce plants	61
Approve of bioengineering to produce animals	28
Fruits/vegetables produced should be labeled	84
Would improve quality of life for people	69
Disagree that "scientists know what they are doing, so only moderate regulations on genetic engineering are probably necessary"	63

Source: W. K. Hallman and J. Metcalfe, Rutgers University.

Clearly consumers have contradictory notions about genetic engineering, as these techniques are viewed more favorably when applied to plants than animals, and are believed to have the potential to improve life but require strict regulations and labeling restrictions. Furthermore, consumers are very unsure as to whom to trust as a reliable source of information about genetic engineering. Several studies have found that scientists are not always seen as reliable sources of information about biotechnology. However, the 2000 Philip Morris survey of consumer opinions on agriculture reported that 57% of consumers supported biotechnology if it improved the taste of food, 69% if it increased food production, 65% if nutritional values were enhanced, and 73% if pesticide use was reduced as a result.

Bovine Somatotropin

Bovine somatotropin (BST) provides an excellent case study in terms of a biotechnical application being approved for agricultural use. In 1994, Monsanto's Posilac (trademark) became the first synthetic or recombinant BST product approved by the Food and Drug Administration (FDA) for commercial marketing. The use of BST is projected to increase milk production in dairy cows by 5–15 lb per day, accompanied by an increase in feed efficiency of 2–10%. BST is to be administered every 14 days over the last two-thirds of the lactation cycle. However, BST is not a substitute for good management.

BST is a naturally occurring complex protein that can be produced using recombinant techniques. All milk, whether from treated or nontreated cows, contains BST, and milk composition is not different between treated and untreated cows. Because BST is a protein, it is broken into amino acids by the digestive process. Furthermore, BST is not functional in humans as it will not bind to human cells. Extensive research (20,000 treated cows in hundreds of projects) has verified the safety of milk and meat from dairy cows supplemented with BST. Cows treated with BST exhibit reproductive and health performance similar to that of high-producing cows.

It has been suggested that BST is the most thoroughly studied substance in the history of food-related research (Table 23.10). BST is a technology that allows dairy producers the opportunity to provide a safe and wholesome product while utilizing fewer feed resources and fewer cows in the process.

It is important to recognize that biotechnology does not always offer simple, cost-effective solutions. Furthermore, the volume of genetic information carried in one cell limits the speed of breakthroughs arising from genetic engineering— it is estimated that there are nearly 3 billion pieces of genetic information in

TABLE 23.10   History of Bovine Somatotropin

Year	Event
1950s	Scientists conclude that BST is inactive in humans.
1982	Genetic splicing techniques used to mass-produce BST.
1985	FDA determines that milk from BST-treated cows is safe.
1987	FDA-ordered studies conclude that milk from BST-treated cows contains no more BST than other milk.
1989	FDA officials call BST "one of the safest products that we've ever administered."
1990	Monsanto receives approval for BST sales in Mexico, Brazil, Bulgaria, Nambia, and Zimbabwe.
1990	An NIH panel concludes BST is safe for human consumption.
1993	Government approves use of BST with a 90-day moratorium on commercial utilization.
1994	BST utilized on a commercial basis in the United States.

bovine cells. Additionally, evidence is accumulating that genetic markers may affect different traits in different breeds. Reproductive success may also be less than desirable. The oocyte maturation rate resulting from nuclear transfer is 70 and 80%, respectively, for cattle and hogs. Yet only 25 and 15% of embryos from these species develop to the blastocyst stage. Numerous problems with clones have become apparent including survival rate, accelerated aging, low reproductive efficiency, and increased birth weights. The cost of bovine cloning is a minimum of $15,000, which serves as a significant barrier to adoption.

Biotechnology brings with it a new set of moral and ethical dilemmas that will have to be resolved by society. Patenting of genetically engineered animals, regulation of the release of biotechnically altered species into the environment, and the use of biotechnology in human medicine are only a few of the issues that will face consumers and producers alike.

ENVIRONMENTAL ISSUES

Societal approaches to natural resource use have been transformed over time. In the early history of European settlement of North America, the prevailing attitude assumed that resources were limitless and environmental management focused almost exclusively on using resources from a short-term perspective without much regard for the future. Westward expansion was driven by the notion of "a land of milk and honey." Horace Greeley's exhortation to "Go west, young man" typified U.S. history prior to the late 1800s.

Gifford Pinchot, father of the U.S. Forest Service, advocated wise use with concerned planning for the future. Pinchot and other early federal lands managers believed that technically proficient professionals should manage public resources and that the goal of management should be to create the "greatest good for the greatest number." The resource agencies of the federal government were established based upon this philosophy.

Henry David Thoreau and other late-19th-century writers advanced the belief that wilderness has unique value and that certain resource or geographic areas ought to be protected from human influence. The National Wildlife Federation and the Nature Conservancy are organizations representative of these ideals.

Environmentalism initially emerged as a philosophy opposed to private use or control of public resources and typically viewed regulatory agencies as captured by commercial interests. Environmentalism has biosphere preservation as its first priority, with human needs secondary. Environmental philosophy runs a broad range of actual belief systems from those who believe in recycling as a way to conserve finite resources to activists who strive to develop very restrictive regulations to the use of natural resources.

Some have adopted a philosophy of retreatism, which is based on a belief that "humans are a cancer on the planet." Sacrificing standard of living and human needs are deemed necessary to protect the environment. Some followers of retreatism would go so far as to embrace terrorism as a legitimate means to advancing their cause.

Decisions regarding resource utilization and allocation will continue to result in conflict. Animal agriculture will likely face the greatest challenges in the areas of waste management, water utilization and quality, federal grazing land management, deforestation and land degradation, and wildlife habitat.

Waste Management

Waste management issues are of particular concern to intensive agricultural systems such as farrow-to-finish confinement facilities, dairies, and feedlots. Specific issues often involve **manure management** and odor control. Industry has shown an increasing willingness to overcome these challenges. The efforts of U.S. pork producers to improve environmental practices and management are outlined in Table 23.11. The stewardship beliefs of the National Cattlemen's Beef Association are outlined in Table 23.12.

Manure disposal is often accomplished by applying waste water and solids to land resources as a means of fertilization. Animal waste is rich in organic materials, nitrogen, phosphorus, and potassium. The use of animal waste as fertilizer is a classic example of creating both economic and environmental benefits by linking the animal and crop/forage system.

However, concerns about odor from livestock enterprises have led to the passage of restrictive legislation in some states. For example, Colorado voters in 1998 passed a ballot initiative that requires waste lagoons on swine farms to be covered as a method of odor control. The cost of covering these lagoons is prohibitive, and producers will not likely be able to stay in business and meet the letter of the law.

TABLE 23.11 Environmental Control Practices on U.S. Swine Farms

Waste Storage System	Small (<2,000 hd) %	Medium (2,000–9,999 hd) %	Large (>10,000 hd) %	All Farms
Above ground slurry storage	4	9	1	5
Deep pit ground slurry storage	55	67	49	57
Waste solids separated from liquids	16	8	20	15
Other systems	25	3	1	21
Anaerobic lagoon	17	45	65	23
Formal manure management plan	19	67	91	28
Manure applied to land	95	95	95	95
Diet manipulation for odor control[a]	47	62	84	50
Manure management to reduce odor[b]	25	43	54	29
Air quality management to reduce odor[c]	29	24	34	28

[a] Low crude protein, low phytase corn, pelleting, vegetable oil to control dust, etc.
[b] Chemical additives, composting, separate solids from liquids, etc.
[c] Biofiltering exhaust, windbreaks to reduce dust, etc.
Source: Adapted from NAHMS, 2000.

TABLE 23.12   Stewardship Beliefs of the National Cattlemen's Beef Association[a]

WHEREAS, productive natural resources are vital for the well-being, not only of the individual farmer, rancher, or feeder, but also for the local, state and national economy and society as a whole. Healthy natural resources provide a healthy watershed and a renewable source of feed for domestic animals and wildlife. Farming and ranching sustains open spaces and aesthetic features which contribute to recreational opportunities.

THEREFORE BE IT RESOLVED, that the National Cattlemen's Association promotes the prudent use of natural resources and offers the following Resource Stewardship recommendations. NCA further recognizes the value and benefit of periodic input and revision to keep the commitment to resource stewardship alive.

BE IT FURTHER RESOLVED, that NCA shall not be compelled to defend anyone in the beef cattle industry who has clearly acted to abuse grazing, water or air resources. To achieve these goals, the following environmental stewardship code is recognized by the industry:

1   Recognize the environment for its varying and distinct properties.
2   Manage for the whole resource, including climate, soil, topography, plant and animal communities.
3   Realize that natural resources are ever-changing and management must adapt.
4   Recognize and appreciate the interdependence of ecosystems.
5   Recognize that management practices should be site- and situation-specific, and must be locally designed and applied.
6   Recognize that successful management is an ongoing, long-term process and commit to stewardship, economic success and business continuity.
7   Strive to develop a management framework that involves family, employees, and business associates so that the entire team is committed to common goals.
8   Monitor and document for effective practices.
9   Never knowingly cause or permit abuses that result in permanent damage on public or private land.
10 Develop ways to communicate and share the vast practical experience of other resource stewards.
11 Become involved in organizations that provide an effective way to educate and support individuals.
12 Solicit input from a variety of sources on a regular basis as a means to improve the art and science resource management.
13 Help develop public and private research projects to enhance the current body of knowledge.
14 Recognize that individual improvement is the basis for any change.
15 Communicate with diverse interests to resolve resource management issues.

[a]Adopted by the membership of the National Cattlemen's Association, January 1995.

TABLE 23.13   Methodologies to Effectively Manage Manure

Method	Characteristics
Ambient digester	Methane gas from anaerobic digestion is captured beneath a plastic tarp for eventual use as a fuel.
Aerobic/anaerobic digester with microbial treatment	Solids are removed and effluent moves through both aerobic and anaerobic steps prior to recycled use to flush manure or wash floors, etc. Solid residue often used as fertilizer.
De-watering, drying, and desalinization	Water is removed, solids are dried, and mineral components are removed for use in related value-added products.
Thermophilic anaerobic digester	Solids are anaerobically digested at high temperatures to generate methane gas that can then be used to fuel boilers to produce the hot water used in the process.

Odor from manure can be a particular problem for livestock producers who manage animals in intensive systems (e.g., large feedlots, dairies, poultry and swine operations). Odor and noise regulations are currently handled by state and local regulations. As urban sprawl continues to move into traditional agricultural areas, odor and air-quality issues must be carefully monitored by animal agriculturalists. A variety of new technologies are being developed to better manage manure and animal waste (Table 23.13).

Federal regulations as defined in the Air Pollution Control Act affect animal agriculture primarily in regards to methane, reactive organic components, ammonia, and particulate matter. Managing these four pollutants is of interest as they can contribute to acid rain, ozone depletion, and excessive particulates in the atmosphere.

Water Utilization and Quality

Water-quality issues are often at the forefront of environmental debates. Specific concerns include nonpoint source pollution, direct ground or surface water contamination, and the amount of water utilized by various agricultural practices. Water is a critical natural resource for agriculturalists because of its vital importance to cultivation of crops and forages, and to maintenance of animal and human life, and its role in maintaining ecosystems. Particularly in the semiarid West and Southwest, conflicts over water rights and appropriation are frequent between agricultural interests, municipalities, water recreationists, and other users.

Several major water-quality issues that relate to animal production and agriculture in general are as follows:

1. The use of commercial fertilizers and manure increases the availability of nutrients needed by plants. However, excessive amounts of nitrogen and phosphorus from commercial fertilizers and manure, or pathogenic microorganisms from manure, may cause water-quality problems.

Based on survey data from the Centers for Disease Control, it has been estimated that waterborne infections caused by bacteria, viruses, or protozoan pathogens affect 994,000 people and are responsible for 900 deaths every year in the United States. Sources of waterborne microorganisms include home septic systems, cropland irrigated with sewage effluent, sanitary landfills, and land application of sludge, sewage effluent, and manure.

Commercial fertilizers and manure can increase nitrate/nitrite levels in water, resulting in health risks to both humans and farm animals. Proper fertilization practices, nitrification inhibitors, water-nitrogen management, and legume cropping rotations can be used to improve nitrogen utilization and reduce runoff and groundwater infiltration.

Transport of phosphorus into water is an issue when erosion moves soil materials into surface water. Using appropriate cropping systems and well-managed grazing programs with animals can prevent erosion.

2. Proper planning of animal facilities and drainage systems is needed to prevent animal wastes from contaminating water sources.

3. Pesticides include a number of chemicals that, if utilized inappropriately, may lead to water pollution. Following pesticide storage, mixing, and application protocols in conjunction with integrated pest management programs that minimize synthetic pesticides can be effective preventative measures to water contamination. As is the case with all environmental issues, risks to human health, environment, and economic systems must be evaluated and integrated into a balanced policy.

Federal Lands

The Louisiana Purchase of 1803 initiated nearly 200 years of **federal land** acquisitions, redistributions, and management policies. Various Homestead Acts, cash sales, and land grants to support schools and agricultural colleges

TABLE 23.14   Federal Lands in the West as a Percentage of State Land Mass

State	Percent
Nevada	86.0
Utah	64.1
Idaho	63.1
Oregon	52.2
Wyoming	47.6
California	44.6
Arizona	42.7
Colorado	35.9
New Mexico	33.1
Montana	29.4
Washington	28.8

Source: Public Lands Statistics, 1976, U.S. Bureau of Land Management.

disposed of some federal lands before the turn of the century. The first National Parks were created in 1872; the Federal Reserve Act of 1891 and the Weeks Act of 1911 allowed expansion of the lands managed as National Forests. Table 23.14 outlines the percentage of the 11 contiguous western states made up of federal lands.

The establishment of the open range livestock industry in the western states and territories followed the conclusion of the Civil War. Stimulated by the expansion of the railroad system complete with refrigerated cars, increasing demand for beef both domestically and abroad, and the ready availability of cattle in Texas to be driven northward, the open range livestock industry boomed in the 1870s and into the 1880s. The legends of the cattle barons were created by foreign investment, primarily from Scotland and England, which allowed for the development of large ranches such as the Swan Cattle Company which ran over 120,000 cattle at its zenith. Large tracts of land were controlled in the West by homesteading on or near water sources. A common thread in the history of the western states is that the control of water allowed for the domination of other resources as well.

Unfortunately, the cattle boom associated with the open range livestock industry was speculative and thus many of the management decisions of the time were shortsighted and resulted in damage to the rangelands. A devastating drought in 1886 and the blizzards of the winter of 1887 combined to cause cattle death losses as high as 60% across the Great Plains region. Just as quickly as it had boomed, the open range livestock industry went bust. It gave way to the forces of inclement weather, failed speculative ventures, and the continuous process of westward expansion and settlement. But significant damage to the western ecosystems had already been done. Restoration of the range resource yielded the first practitioners of range ecosystem management.

The 1905 grazing regulations were the first in a long line of policies that focused on improving and protecting the western rangelands. The Taylor Grazing Act helped establish the framework for the development of the Bureau of Land Management. Significant legislation that affects livestock grazing on federal lands is summarized in Table 23.15. Each of these legislative actions has yielded a degree of conflict and confrontation between groups who place differing values on the federal lands. The federal lands are host to a multitude of uses including livestock and wildlife grazing, timber production and harvest, mining,

TABLE 23.15   History of Legislative Actions Affecting Public Land Grazing

Homestead Acts of 1862, 1909, 1916
Weeks Act of 1911
Taylor Grazing Act of 1934
Multiple Use-Sustained Yield Act of 1960
Wilderness Act of 1964
Classification and Multiple Use Act of 1964
National Environmental Policy Act of 1969
Forest and Rangeland Renewable Resources Planning Act of 1974
National Forest Management Act of 1976
Federal Land Management and Policy Act of 1976
Public Rangeland Improvement Act of 1978
Rangeland Reform Proposal (Department of Interior) 1993–94

FIGURE 23.3   Multiple uses of federally owned/managed lands continue to be important. However, conflicting interests between user groups must be successfully managed to assure the viability of the resource.

a variety of recreation activities ranging from backpacking to skiing, and preservation of watersheds.

Multiple uses of federal lands will likely continue, as will disagreements relative to the management of the resources (Fig. 23.3). However, consensus-building models of decision making coupled with judicious management of these lands will continue to assure their productivity and sustainability.

Endangered Species

The **Endangered Species Act** of 1973 was passed to protect threatened and endangered species and their habitats from detrimental human activity. This legislation was reviewed in 1996. Once a plant or animal is placed on the endangered list, the Fish and Wildlife Service or National Marine Fisheries Service is delegated responsibility for drafting a recovery plan. More than 7,000 species have been listed, with 325-drafted recovery plans covering 411 species. Government species-recovery efforts have only allowed five delistings in the past two decades.

Meanwhile, it is estimated that better than three-quarters of the U.S. wildlife population depends on private land for at least part of the year. Conflict results when the government mandates the use of privately held resources because of the discovery of an endangered plant or animal. Some groups contend that domestic livestock and other agricultural enterprises compete for scarce resources and thus cause habitat loss for endangered species.

Examples of specific species-recovery plans that have raised issue with livestock producers include those for the desert tortoise, the sockeye salmon, and the timber wolf. Reintroduction of the wolf into Yellowstone National Park has stirred debate between those who believe the resurgence of predator species better balances ecosystems and those whose livestock may fall victim to predatory species.

These issues are difficult to resolve because they often become highly emotional and frequently end up in divisive positions instead of consensus solutions. As resources are demanded by divergent uses and as the human population multiplies, the need for balancing economic interests versus endangered species needs will be critical. The process of reaching consensus solutions is a difficult task. However, it is a responsibility that can no longer be ignored.

Global Warming

Global warming (sometimes called the *greenhouse effect*) becomes an issue with farm animals that are ruminants, because they emit methane (CH_4) into the atmosphere. When energy from the sun is radiated back from the earth into the atmosphere, certain gases—for example, methane, carbon dioxide, nitrous oxide, and chlorofluorocarbons—absorb it. Increased amounts of these gases absorb more infrared radiation, thus increasing the temperature of the atmosphere and causing global climate changes.

There is complete agreement as to whether global warming is occurring and, if so, at what speed. Yet there are some scientific studies that show the need for concern and more careful assessment. In 1994 a 1.8-mile Greenland Ice Core Project showed that there is an alarming volatility in the world climate—even in decades without changes brought about by humans, such as the greenhouse effect.

Methane is second in importance to carbon dioxide as a greenhouse gas. It is produced from wetlands, bogs, and rice paddies as a result of anaerobic fermentation. Ruminant animals produce methane by digesting plant material, which releases some of the carbon in the gas. This carbon is originally a result of photosynthesis; thus, it is not new gas but recycled carbon.

Accusations have been made that cattle and other ruminants are the major sources of greenhouse gases that are destroying the protective ozone layer. However, realistic calculations show that methane accounts for only 18% of the world's greenhouse gases, and only 7% of the world's methane is produced by all ruminants, domestic and wild (e.g., deer). U.S. beef cattle produce 0.5% of the world's methane, which is equivalent to 0.1% of the total greenhouse gases. Carbon dioxide emitted from automobiles contributes much more greenhouse gas than methane from ruminants. In fact, driving a few miles to purchase a hamburger produces more greenhouse gas than the methane required to produce the hamburger.

Additional charges against cattle production have been focused on the destruction of rain forests in Central America to increase forage ability for cattle used to import the beef to the United States. Consumption of hamburger in the United States has been blamed for the loss of rain forests in Central America. In reality, the United States imports less than 1% of its beef from Central America. The reason for the loss of the rain forests for cattle production results primarily from social, economic, and political pressures in the countries where rain forests are located.

Conversion of Agricultural Land

The conversion of agricultural land to nonagricultural uses is becoming an increasingly significant challenge to preserving wildlife habitat and assuring the security of food supplies for future generations (Fig. 23.4). Since 1978, Colorado has lost an average of 90,000 acres of agricultural land per year. This loss equates to an area 140 miles long and 1 mile wide.

Urban dwellers desire small acreage homesites, cities and communities grow, and growing recreational demand breaks up existing blocks of land. As large areas of land that once were managed as an ecosystem are divided into smaller parcels and then crisscrossed with infrastructure improvements such as roads and power lines, both wildlife populations and agricultural production capacity are negatively affected.

As the popularity of 5- to 35-acre homesite tracts grows, more pressure will be placed on traditional agricultural enterprises as they interface with urbanization. As large tracts of land are removed from production, not only agriculture is affected. Wildlife lose migrating routes, are exposed to increased predation from cats and dogs that accompany suburbanization, and lose vital habitat.

FIGURE 23.4 The loss of agricultural lands to urbanization is one of the most critical issues facing public policy makers. Continued loss of agricultural lands to other uses threatens to weaken the food security of the United States.

CONSUMER ISSUES

Diet and Health

Recent consumer surveys indicate that Americans are sensitive to health and nutrition relationships in the foods they purchase. Some people changed their diets during the 1980s and 1990s, primarily for health or nutritional reasons, although consumer surveys continually rank "taste" above "nutrition" as most important in food selection.

Most nutritionists agree that a healthy diet should contain all the required nutrients and enough calories to balance energy expenditure. Unfortunately, nutrient deficiencies or excesses exist in much of the world. Therefore, food choices, or lack of choices, determines the nutritional status of most people. Obesity caused by caloric excesses has become a leading nutritional problem in the United States with approximately one-third of the population classified as obese.

The relationship between diet and human health is a controversial and complex topic. The consumption of beef has been linked to two of the most dreaded human diseases—**coronary heart disease (CHD)** and cancer. Consumer perceptions have been influenced to accept these alleged relationships, and consumers have reduced their consumption of some animal products accordingly.

Before examining the root causes of human disease, it is worthwhile to understand the concept of risk analysis. American citizens are enjoying an ever-increasing life span, better health, and a higher standard of living than their ancestors could have imagined. Despite this evidence of well-being, widespread public worry about **diet-health relationships** persists. These worries are perpetrated when reports fail to account for the following:

■ the wholesale extrapolation of results obtained from lab animal models to humans,

■ the fact that natural compounds may contribute significantly more risk than man-made compounds, and

■ the effect of dose rate on disease incidence.

In the media-dominated environment where sensationalistic headlines are all too common, consumers are advised to be wary of **"junk science."** Warning signs that results of a study are being incorrectly represented are as follows:

1. The results recommend changes that offer quick-fix promises.
2. Foods are described as "good" versus "bad."
3. Simplistic conclusions are offered from a complex study.
4. The study was not peer reviewed.
5. Recommendations ignore differences between individuals or among groups.
6. Results are interpreted to offer significant negative consequences from a specific food item or diet selection.

Sound, well-based research work often appears to be lost in an emotionally charged issue. Many organizations and individuals have occasionally based judgments and decisions on emotion rather than on the best accumulated research facts. Some individuals feel that some self-proclaimed "diet and health experts" and a communications system that builds part of its readership on sensationalism direct consumer perceptions. Scientific principles should be identified so that decisions are based on true relationships.

The major known risk factors associated with coronary heart disease are genetics (a family history of CHD), high blood cholesterol, smoking, hypertension, physical inactivity, and obesity. Obesity caused by excess caloric intake is a major nutritional problem in the United States. Of the leading 10 causes of death in the United States, obesity is considered a risk factor in five (coronary heart disease, stroke, hypertension, type II diabetes, and some forms of cancer). More than 50% of U.S. adults exceed their recommended weights. Interestingly enough, survey results suggest that consumers are less concerned about caloric and fat intake than they were in 1990. In a 10-year span, the percentage of consumers who reported that they were always conscious of caloric intake fell from 40 to 25%. Consumers reporting that they were always cautious of caloric intake fell from 51 to 33% over the same time period. It is generally accepted that consumption of animal fat by humans causes an increase in the level of cholesterol in the blood, while consumption of vegetable oils (polyunsaturated fats) causes a decrease in blood cholesterol concentration. These relationships led to the theory that there is a relationship between consumption of animal fats and the incidence of **atherosclerosis** ("plugging" of the arteries with fatty tissue), which in turn results in an increased likelihood of death from coronary heart disease.

Evidence supporting the proposed blood cholesterol—heart disease relationship is still in question. Studies in which dietary fat intake has been modified, either in kind or amount, did not demonstrate significantly reduced mortality rates in test populations. Changing diets from animal fats to vegetable fats has not improved the heart disease record. There are data that suggest that poor health conditions can also result from people eating diets high in polyunsaturated fats.

Most consumers do not know the difference between saturated and polyunsaturated fats. Through many margarine and vegetable oil commercials, however, they have been informed that saturated is "bad" and **unsaturated** is "good." In a review of the diet and heart disease relationship, some medical doctors argue that the dietary-heart hypothesis became popular because a combination of the urgent pressure of special interest groups or health agencies, oil-food companies, and ambitious scientists had transformed that hypothesis into absolute fact.

Senator George McGovern's committee on Nutrition and Human Needs in the late 1960s and early 1970s shifted its emphasis from developing policy aided at eliminating malnutrition to dealing with the issue of consuming too many calories. Largely influenced by a self-proclaimed diet expert, McGovern's committee released a document entitled "Dietary Goals for the United States" based on two days of testimony. Written by a journalist with absolutely no background in science, nutrition, or health, the report would become the basis for a national policy focused on dietary fat. Twenty-five years later there is still not compelling and clear evidence as to the effect of dietary cholesterol and fat intake on human longevity.

The relationship between cholesterol intake and death from coronary heart disease is minimal. All **saturated fatty acids** are not equal in terms of their effect on serum cholesterol. For example, it is important to recognize that less than half of the fatty acids found in beef fat are saturated. The 18-carbon length fatty acid (stearic acid) has either a neutral effect on serum cholesterol or may actually lower serum cholesterol concentrations when substituted for other saturated fatty acids. Stearic acid comprises almost one-third of the saturated fatty acids in beef.

For example, a Select grade steak will have about 50% of its total fat in monounsaturated form of which nearly 90% is oleic acid (the beneficial fat found in olive oil). The remaining one-half of the total fat is saturated but one-third of that is stearic acid, which is potentially beneficial, but at the very worst neutral in its effect. In total, more than one-half, and as much as three-quarters, of the fat in the steak will lower cholesterol levels (Taubes, 2001).

Cholesterol is a naturally occurring substance in the human body. Every cell manufactures cholesterol on a daily basis. The average human turns over (uses and replenishes) 2,000 mg of cholesterol daily. Average dietary consumption of cholesterol is approximately 600 mg daily. Therefore, the body makes 1,400 mg each day to meet its needs.

The body cannot use cholesterol unless it is joined with a water-soluble protein, creating complexes known as lipoproteins. There are several different types of lipoproteins. Research workers have identified two of these lipoproteins: **HDL (high-density lipoprotein)** and **LDL (low-density lipoprotein).** High blood levels of the LDLs have been generally associated with increased cardiovascular problems, while some research data show that higher blood levels of HDLs may reduce heart attacks by 20%. Additional research with laboratory animals has shown that those fed beef had HDL levels 33% higher than animals fed soybean diets.

There is evidence of genetic differences in the proportion of HDLs and LDLs in individuals. People who are overweight, nonexercisers, and cigarette smokers have higher proportions of LDLs than those who are lean, exercisers, and nonsmokers.

Although the exact roles of dietary cholesterol and blood levels in the development of coronary heart disease are not known, it would appear logical from the existing information to use prudence in implementing drastic changes in dietary habits. Certainly, those individuals with high health risks primarily due to genetic background should take the greatest precautions.

Figure 23.5 shows that on an average per-capita basis, red meat and poultry supply less than 16 grams of fat per day. This level of fat (15.6 g) is less than 24% of the 67 grams of fat per day recommended by the American Heart Association for a 2,000-calorie-per-day diet. The 15.6 grams of fat represent 30% of the calories from fat where each gram of fat contains 9 calories.

Some individuals argue that consumers eat too much red meat and that excessive red meat consumption causes cancer. Evidence to support this statement is questionable, and excessive consumption must be defined. Figure 23.6 shows the per-capita red meat consumption at approximately 3.4 oz/day while a more recent report gives 2.6 oz/day. The 3.4 oz provides approximately 25 g of protein, which is less than half the average recommended daily allowance (RDA) for protein. The American Heart Association recommends 3.5 oz of cooked meat per person on a daily basis. Based on this recommendation, the average U.S. per-capita consumption of meat is not excessive.

Based on a review of several types of epidemiological studies, Austin et al. (1997) concluded that red meat consumption has not been clearly shown to be

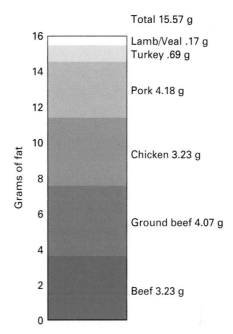

FIGURE 23.5 Average daily per-capita fat consumption from cooked meat. *Source:* USDA.

a risk factor for cancer. If an association is shown between red meat and cancer, one of three explanations is possible: (1) the association is due to chance or some bias in the study method; (2) the association is due to confounding evidence (i.e., meat and cancer are associated only because they both are related to some common underlying condition such as a low intake of cancer-protective fruits/vegetables); (3) the association is real. Whether or not it can be concluded that red meat is a risk factor for cancer depends on several criteria: consistency of the association, strength of the association (relative risk), specificity of the association, and congruence with existing knowledge (e.g., is there an explanation or biological mechanism).

The criteria of "consistency" require that the association be repeated under different circumstances, such as among various population groups and among individuals within a population. When a positive association between red meat and a specific cancer is demonstrated, it generally is weak. Moreover, the association between consumption of red meat of different types (beef, pork, lamb, and processed meats) and specific cancers is inconsistent. Although it has been suggested that meat components such as fat, protein, or iron, or chemicals formed during the cooking of meat might be carcinogenic, these hypotheses remain unproven. On the contrary, meat contains some components such as conjugated linoleic acid, a fatty acid, which may protect against cancer.

Inferring a relationship from epidemiological studies of diet and chronic disease is particularly difficult due to several characteristics of both diet and chronic disease. First, there are several problems with accurately quantifying dietary intake. This was a major problem in most of the epidemiological studies reviewed by Austin et al. (1997). Second, diseases such as cancer are caused by a variety of genetic and environmental factors. Diet (e.g., red meat intake) is only one of many lifestyle factors that may influence risk of developing a disease. Also, chronic diseases such as cancer tend to have a long latency period

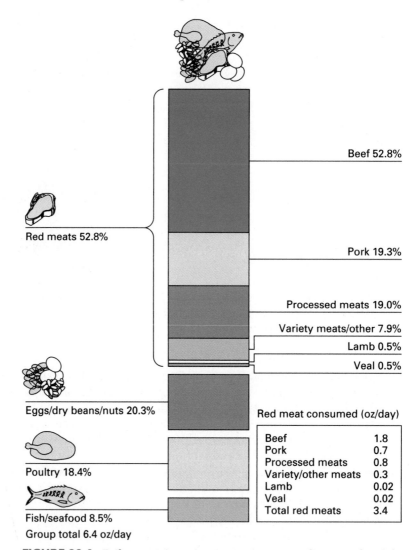

FIGURE 23.6 Daily per-capita meat consumption as part of meat, poultry, fish, dry beans, eggs, and nuts group (Food Guide Pyramid). *Source:* National Live Stock and Meat Board.

during which time changes in many factors may occur. For these reasons, it is difficult to determine if an association, such as one between red meat and cancer is real.

Epidemiological investigations can identify risk factors, not a cause-and-effect relationship. Only when epidemiological findings are supported by information from other types of scientific studies, such as experimental animal studies and human clinical trials, can a decision regarding a causal relationship be made on firmer ground.

The U.S. government has released several reports that outline dietary guidelines and goals for American citizens. Some argue that it is the government's responsibility to provide people with information about diet; others support scientifically based guidelines, but believe Americans should have freedom of choice.

The **Dietary Guidelines** published by the USDA and Department of Health and Human Services (HHS) include broad recommendations and are a reasonable attempt to initiate sound nutritional practices among individuals. The revised 1990 *Dietary Guidelines,* which draw heavily on two diet and health reports by the National Academy of Sciences and the U.S. Surgeon General, recommended that not more than 30% of daily calories come from fat and that less than 10% come from saturated fat. Some of the earlier versions of the *Guidelines* implied that red meat was a health risk by recommending that Americans eat less meat. The revised *Guidelines*, however, convey a more positive recommendation for lean red meat consumption. In 1992, the USDA began use of the Food Guide Pyramid (Fig. 23.7) in conjunction with the nutritional education programs. The pyramid replaced the wheel graphic used to display the four basic food groups that has been used in nutritional education programs since the 1950s.

In the period from 1970 to 1999, red meat consumption declined by 11% while per-capita fat and oil intake rose 31%. Soft drink consumption over the same time period jumped 109%. Moderating fat consumption is a challenge for many consumers because taste is associated with the fat content of foods. Many consumers rank taste as the most important criteria in their food selection. Because of this preference for fat in foods, the "actual consumption" pyramid is out of balance compared to the "recommended" Food Guide Pyramid (Fig. 23.7). The meat, poultry, fish, dry beans, eggs, and nuts group is in balance.

In many respects, the dietary guidelines, consumption goals, and nutritional information have not changed all that much over the past 20 years. Adherence to the suggested *Dietary Guidelines* can be easier by following these suggestions of the Dietary Guidelines Alliance:

1. Be realistic—make small, incremental changes in eating and exercise habits.

2. Be adventurous—expand your tastes to enjoy a variety of foods.

3. Be flexible—balance food consumption and exercise regime over several days instead of focusing on one meal or one day.

4. Be sensible—enjoy all foods—just don't overdo it. Choose sensible portion sizes.

5. Be active—exercise is a key to maintaining appropriate weight.

Vegetarian and vegan diets have been advocated as healthier than diets containing animal products. However, diets devoid of meat are relatively low in protein, saturated fat, long-chain n-3-fatty acids, retinol, vitamin B_{12}, and zinc while vegan diets are also low in calcium. A study by Key and coworkers in 2006 reported that there were no significant differences between meat-eaters and non-meat-eaters for incidence of hypertension, colorectal cancer, breast cancer, prostate cancer, and total mortality rate.

In the final analyses, balance and moderation are keys to appropriate and healthy dietary choices.

Food Safety

Although the food industry does not have a perfect record in regard to food safety, U.S. consumers enjoy a plentiful food supply that can arguably be called the safest in the world. Again, relative risk must be carefully evaluated when

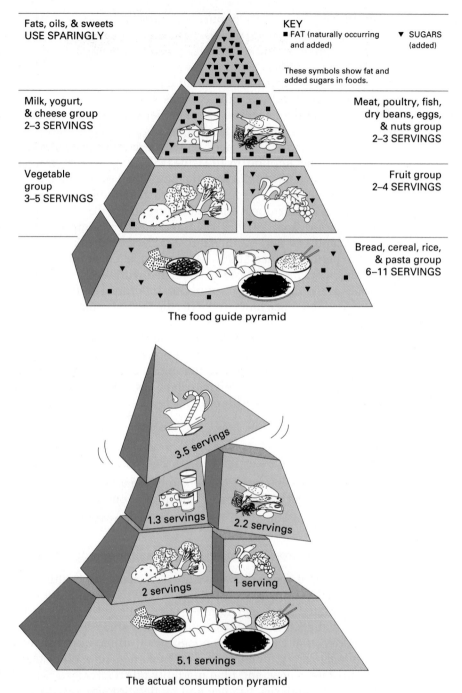

FIGURE 23.7 A comparison between the Food Guide Pyramid and actual consumption. *Source:* USDA and National Live Stock and Meat Board.

considering food safety issues. Assuring **food safety** is a joint responsibility shared by producers, processors, and consumers.

Food safety is a challenge for consumers, producers, processors, retailers, and foodservice outlets. Historically the food inspection system has been expected to assure food wholesomeness. The premise of **Total Quality Management** is that quality cannot be inspected into a product. Instead, systems must be developed to produce quality and safety at every key step in the process.

One of the most critical control points is management of food temperature. The danger zone—40 to 140°F—offers conditions under which bacterial pathogens can multiply. Therefore, foodservice establishments and consumers should strive to maintain raw and cooked foods outside the danger zone.

Additional steps that consumers should take include cooking and reheating food thoroughly, storing food at correct temperatures, avoiding cross-contamination between foods during storage and preparation, maintaining a high degree of personal sanitation, and keeping food preparation surfaces meticulously clean.

Food scientists rank food hazards as: (1) microbial contamination; (2) naturally occurring toxins; (3) environmental contaminants (e.g., heavy metals); (4) pesticide residues; and (5) food additives.

Examples of food safety issues for the livestock industry include infection by **E. coli 0157:H7** and other foodborne pathogens (Table 23.16). Concerns about *E. coli* 0157:H7 bacterial infection outbreaks peaked in 1993, when 1,000 cases were reported in the United States. Over three-quarters of the 1993 incidents were attributed to improperly prepared and cooked ground beef, most of which was traced to a single food-service supplier.

Data reported by the Centers for Disease Control in 2006 documented a decline in bacterial foodborne illnesses since 1996. Illness due to *Salmonella*, *Campylobacter*, *Shigella*, and *E. coli* 0157:H7 has declined by 9, 30, 43, and 29%, respectively. While the goal must be to eliminate foodborne illness outbreaks, these data show that large industries can make progress in food safety.

In fact, microbial contamination has resulted in a variety of new processing technologies designed to improve food safety. Some of the innovations include steam vacuuming, hot water washing, weak acid rinses, and steam pasteurization. Most processors are working to implement a multiple set of hurdles or blockades to microbial growth. Such systems have tremendous potential to enhance food safety.

Another food safety technology is **radiation pasteurization** or irradiation. Unlike the previously mentioned technologies, the use of radiation has been surrounded by a degree of controversy. The use of low doses of gamma rays, X-rays, and electron beams can safely control pathogenic microbes. The use of this technology doesn't make foods radioactive and essentially has little effect on food because its cells are no longer active. Small changes to vitamins, such as B_1 and C, do occur but are not dangerous to human health. As consumers learn more about the technology, their acceptance levels have increased. Irradiation of wheat and flour was approved in 1963 and since that time pork, poultry, red meat, and eggs have also been approved. Society's desire for an assured level of food safety will continue to create demand for new and improved technologies.

TABLE 23.16 An Overview of Foodborne, Illness-causing Organisms

Name of Organism	Description	Infective Dose[a]	Symptoms	Annual Occurrence in U.S.	Associated Foods
Salmonella	Rod-shaped, motile bacterium; widespread in animals, especially poultry and swine; gram-negative.	As few as 15–20 cells	Nausea, vomiting, abdominal cramps, fever.	2–4 million cases	Major: raw poultry, meats, eggs, dairy foods. Minor: coconuts, sauces, salad dressings, cake mixes, peanut butter, chocolate.
Clostridium botulism	Anaerobic, spore-forming rod; produces a pathogenic neurotoxin; 7 types of botulism are recognized; gram-positive.	A few ng of toxin	Weakness, vertigo; difficulty in breathing, swallowing; muscle weakness, constipation.	10–30 outbreaks	Most frequently associated with poorly processed, home-canned foods.
Staphylococcus aureus	Spherical bacteria; some strains produce entero toxins; gram-positive.	Less than 1 mg of toxin in contaminated food will yield symptoms	Nausea, vomiting, retching, abdominal cramping.	Relatively unknown due to mis-diagnosis	Foods that require a high degree of handling during preparation and that are kept at slightly elevated temperatures are typically involved.
Campylobacter jejuni	Slender, curvical, motile rod; gram-negative, relatively fragile.	400–500 bacteria	Diarrhea (watery or sticky), may contain blood and white cells; fever, abdominal pain, headache, muscle pain.	Leading cause of bacterial diarrhea	Major: raw chicken. Minor: unpasteurized milk, nonchlorinated water, seafood, hamburger, cheese, pork, eggs.
Listeria monocytogenes	Bacterium, non-spore forming, very hardy; gram-positive.	<1,000 total organisms	Flu-like symptoms may be followed by septicemia, meningitis, encephalitis, or cervical infections in pregnant women.	1,600, with 415 fatalities	Major: soft cheese and ground meat. Minor: poultry, dairy products, hot dogs, seafood, vegetables.
Escherichia coli 0157:H7	*E. coli* is a normal inhabitant of the gut of all animals. There are 4 classes of virulant *E. coli*. Majority of *E. coli* serve to suppress harmful bacteria and to synthesize vitamins. 0157:H7 produces a very potent toxin(s).	10 organisms	Severe cramping, diarrhea (may become bloody); typically self-limiting with 8-day duration; young children and the elderly may develop hemolytic uremic syndrome and renal failure.	Low and sporadic[b]	Major: undercooked ground beef, human-to-human contact, environmental exposure to organisms. Minor: poultry, apple cider, raw milk, vegetables, hot dogs, mayonnaise, salad bar items.

[a] Dose varies depending on immune state of the individual.
[b] This foodborne illness has received a high degree of coverage due to the impact on children.
Source: USDA.

Microorganisms

Microorganisms, such as molds, viruses, bacteria, yeasts, and parasites, are universally present. These microorganisms play potentially beneficial, harmful, or neutral roles in regard to food. Molds and yeasts are important in baking and cheese making, while some bacteria act as inhibitors to other harmful bacteria. Harmful microorganisms either cause spoilage or disease. Disease-causing microorganisms are referred to as **pathogens.**

Food spoilage microbes are usually visually apparent, while foodborne pathogens may be difficult to detect. The Centers for Disease Control reports that 77% of foodborne illness outbreaks originate due to improper handling and cooking in foodservice establishments, 20% due to improper food handling and cooking in the home, and 3% due to food manufacturing defects or failure.

Sanitation and temperature control are critical elements in controlling potentially harmful bacteria. Processors, retailers, and foodservice operators can minimize contamination via proper equipment sanitation, disciplined personnel hygiene practices, proper handling and storage, cooking to appropriate temperatures, and systematically assessing food safety efforts.

Consumers should never purchase outdated goods or those whose containers have been broken or dented. Furthermore, refrigerator temperatures should be maintained at or below 40°F and freezer temperatures at or below 0°F. Consumers are advised to follow label directions for preparation and serving, to avoid cross-contamination of foods, and to wash, rinse, and sanitize equipment, countertops, and cutting surfaces. Foods should never be thawed at room temperature, food should be cooked to appropriate temperatures (typically 140°F or greater, 155°F for ground meats, and 165°F for leftovers), and during storage or preservation foods should be kept outside the range of 40–190°F.

One consequence of a food safety failure is often the initiation of a national product recall. These product recalls are expensive from both the loss of product and the loss of consumer confidence. The five largest meat product recalls involved from 35 million to 19 million pounds of product resulting in millions of dollars in lost revenue. Prevention and early detection systems are being developed to better provide wholesome products to consumers.

Residues

Much of the U.S. public feels that chemical **residues** from pesticides, hormones, and food additives are the greatest food safety threats. There is a perception that the presence of these residues in any amount is extremely dangerous. In reality, large enough doses of many substances (including water) can be harmful and even fatal. What is critical to assess is the level of the residue in the food, the amount consumed, and the risks associated with that level of consumption.

Consumers may perceive that giving farm animals anabolic steroids/hormones will produce meat with residues that are harmful to health. Examples would be using a cattle ear implant containing growth promoting hormones such as estrogen to increase growth, and repartitioning feed energy to produce more muscle and less fat. As a result of this process, the average amount of estrogen in a 3-oz serving of beef is increased from 1.2 to 1.9 ng (a **nanogram** is one-billionth of a gram). Similar-size servings of potatoes, ice cream, and soybean oil contain more nanograms of estrogen (200, 500, and 1 million times more, respectively), compared to beef from implanted cattle. Considering that the average nonpregnant human female produces 480,000 nanograms of estrogen each day by normal physiological processes, the increased estrogen consumed in beef is of no physiological or medical consequence; that is, there are no problems associated with ingestion of meat from an animal.

In 1994, a jury awarded $42 million to 11 women whose mothers took diethylstilbestrol (DES), a synthetic estrogen, between 1947 and 1971 to prevent miscarriages during early pregnancy. This was the first time a trial was held claiming the hormone was related to reproductive problems and not to cancer. These recent awards raise additional perceptions that hormones are extremely dangerous health risks and should be avoided. However, in these situations, women were given massive therapeutic doses of DES, which is grossly different than the extremely small amounts of hormones found in meat and other foods.

There is a perception that so-called natural or **organic** animal products contain far fewer residues of antibiotics, hormones, drugs, and pesticides. Scientific tests of organic versus conventional beef have shown no evidence to confirm this perception. Monitoring by USDA shows that there are almost no ($< .0001$) violative residues of antibiotics in beef from fed steers and heifers. For nonfed cows and bulls the incidence rate of violative residues was higher ($< .003$) but still extremely low. Nonetheless, a segment of consumers desire the opportunity to purchase natural or organic foods.

Subtherapeutic **antibiotic** use in the livestock industry is predicated on the principle of facilitating the growth of resistant bacteria while restraining the growth of competing bacteria. Criticisms of the practice arise from concerns about the possibility of the resistant bacteria flourishing in farm environments and eventually infecting humans.

About 45% of U.S. antibiotic usage is by the animal industry (including pets) and nearly 90% of those go toward the treatment or prevention of specific diseases. Thus, only about 6% of all antibiotics are used at the subtherapeutic level to promote growth and efficiency.

A far graver concern is the dramatic overprescribing and misuse of antibiotics in human medicine. In fact, on a weight basis, humans use nearly 10 times the tonnage of antibiotics as compared to livestock. The leading predictor of a patient's likelihood of developing a drug-resistant infection is whether the patient was ingesting antibiotics in the previous 3 months. Add in the negative consequences of people not completing the full regimen of prescribed antibiotic therapy and the source of the problem is targeted toward human medicine.

Furthermore, human levels of antibiotic resistance are rising while resistance in livestock has remained historically stable. Proponents of banning subtherapeutic antibiotics fail to study the lessons in Europe where these products were banned in the early to mid-1990s. For instance, in Denmark the use of growth-promoting antibiotics was banned in 1997. As a result, illnesses requiring antibiotic therapy rose dramatically in swine herds with the end result being a dramatic rise in antibiotic use by the industry. Furthermore, pig productivity dropped significantly, mortality increased, and feed efficiency declined. Researchers at Iowa State predict that such a ban in the United States would increase the cost of production by $4.00 per pig.

The most logical approach to this issue is to assure responsible use of antibiotics, avoidance of the use of antibiotics of critical value to human medicine in livestock, continual antibiotic-resistance monitoring, and conducting research that evaluates the specific and direct cause-and-effect relationships.

Organic and Natural Products

Consumer demand for food that has been grown under organic conditions has grown at a slow but steady rate. As such, USDA developed the National Organic Program guidelines to provide uniform standards for the voluntary labeling of products grown under organic management.

Some of the guidelines include the following: Livestock must be maintained under continuous organic management from the last third of gestation or hatching with three exceptions—(1) edible poultry products must be from birds that have been managed organically beginning no later than the second day of life; (2) milk products must be from females managed continuously under organic conditions no later than one year prior to lactation; and (3) purchased breeding stock from nonorganic sources must be introduced into the herd prior to the last third of gestation. Further provisions are prescribed in regards to feedstuffs, health products, record keeping, and other management protocols.

Producers should carefully evaluate the cost and returns associated with aligning their enterprise with an organic supply chain. A careful evaluation of the requirements and restrictions along with a detailed market analysis is recommended.

Catastrophic Disease Outbreaks

Recent historical events such as the September 11 terrorist attacks, the foot and mouth disease outbreak in Europe, the occurrence of bovine spongiform encephalopathy (BSE) in cattle, and the concerns about avian influenza serve as evidence that an imperative priority of the livestock industry is to build a set of effective firewalls that either prevent or control an intentional or accidental outbreak. For example, to protect the U.S. cattle herd from BSE, a detailed feed rule has been implemented that prohibits the feeding of supplements to ruminants produced from the brains and spinal cords from cattle 30 months of age and older, tallow derived from prohibited materials if the tallow contains more than 0.15% impurities, and mechanically separated beef derived from prohibited materials.

The implementation of these firewalls is effective in maintaining consumer confidence when low-incidence disease is identified. For example, despite the identification of 3 head of U.S. cattle (1 imported, 2 native) being diagnosed with BSE starting with the first case in late 2003, consumer surveys found that the percent of U.S. consumers who were confident that U.S. beef was safe from BSE actually increased (Table 23.17). However, the loss of international markets to U.S. beef coupled with the costs of additional regulation and testing has resulted in an $8 billion loss to the American economy. The process of reopening

TABLE 23.17 Percent of Consumers Confident That U.S. Beef Was Safe from BSE

Date	% of Respondents
April 1996	80
April 2001	87
April 2003	85
April 2004	89
April 2005	92

Source: Adapted from NCBA, 2005.

international markets will take years to accomplish. BSE is a good example of the perfect storm created when politics, science, and international trade converge. It is staggering to consider the economic costs of three positive cases in the U.S. cowherd especially in light of the fact that no products from these three animals entered the food chain.

Nonetheless, increasing attention is being placed on developing a functional traceability system that would allow contamination to be traced to its source, early detection systems, and the creation of more effective intervention strategies. Furthermore, there is increasing pressure on governments to strengthen their regulatory efforts to prevent the introduction of catastrophic disease into national herds and flocks.

Development of biosecurity systems and protocols is of paramount importance to assuring food security.

Quality Assurance

Animal production began a new era in the 1990s as production practices were evaluated in light of their impact on the process of assuring that consumers received wholesome products of highly perceived value. Quality assurance programs have been instituted in the food and fiber industries as a means to coordinate genetic inputs, animal health care practices, and management protocols.

Quality assurance programs in the food animal industries have focused on (1) superior record keeping, (2) safe use of pharmaceutical products, with a special emphasis on product administration in accordance with label restrictions, (3) minimizing intramuscular injections, (4) assuring quality of animal feeds, and (5) avoiding residues.

Most quality assurance efforts are designed in accordance with **Hazard Analysis Critical Control Point (HACCP)** protocols. HACCP programs utilize a seven-step process to ensure that product quality is attained throughout the production process (Table 23.18).

Quality assurance efforts require an effective process of communicating critical information throughout the production and processing chain. Improved communication can go a long way toward adding value and strengthening positive consumer perceptions about animal products.

For example, contamination of wool with polypropylene products such as hay baling twine can significantly reduce the value of the clip. Contamination typically occurs when sheep bed down on areas where baling twine has been littered as a result of feeding hay. Wool producers are solving this problem by

TABLE 23.18 Seven Steps for Utilizing Hazard Analysis Critical Control Points

Number	Steps
1	Conduct a hazard analysis.
2	Identify critical control points in the process.
3	Establish preventative measure limits for each control point.
4	Establish monitoring process for critical control points.
5	Take corrective action.
6	Institute effective record-keeping system.
7	Monitor system on continuous basis and assure it works.

more carefully disposing of twine, avoiding the use of polypropylene feed tags, and communicating a price incentive for wool free from contamination.

Quality assurance efforts must be applied at every level of production, processing, and preparation to assure that food is safe and wholesome. For example, the following critical control points are important to the ability of foodservice establishments to assure food safety:

Food receiving

- Provide visual, packaging, and temperature inspection

Storage

- Hold refrigerated foods at 41°F or lower
- Store frozen foods at 0°F or lower
- Maintain dry storage temperatures between 50 and 70°F
- Implement first in–first out inventory control

Preparation

- Avoid cross-contamination
- Meats prepared separately from produce
- Poultry cooked to 165°F for 15 seconds
- Ground beef cooked to 155°F for 15 seconds
- Pork cooked to 145°F for 15 seconds
- Beef roasts heated to 145°F for 3 minutes
- Steaks heated to 145°F for 15 seconds
- Minimize human traffic in food prep areas

Cooking surfaces

- Scheduled cleaning throughout the business day
- Deep-cleaning at the close of business each day

Employee break rooms and wash areas

- Scheduled cleaning and sanitation
- Thorough and frequent handwashing
- Separate restrooms for employees and customers when possible
- Thorough cleaning at least daily

Dining areas

- Tables wiped and sanitized between customers
- Use separate cleaning tools for dining area, kitchen area, and restrooms to avoid cross-contamination

ISSUES AND OPPORTUNITIES

With challenging issues come opportunities. Jobs are created, recognition is given, and monetary rewards are available to successful problem solvers and decision makers. These successful individuals face the problems, meet the challenges, and bring reasonable resolution to the issues. The challenge is to separate the myths, perceptions, and emotions from the scientific truths and realities, then to effectively communicate these truths to the influencers and eventually to all concerned. U.S. consumers can be provided highly palatable animal products that are the cheapest and safest in the world. With the creativity of people, the environment can be sustained for generations for the economic, physical, and emotional well-being of the U.S. and world population—even with significant population increases.

CHAPTER SUMMARY

■ Issues facing the livestock industry can be categorized as animal welfare, biotechnology, environment, diet and health, food safety, and marketing.

■ Management of these issues in terms of public perception is important to the stability of the livestock industry.

■ Application of the principles of sound stewardship and husbandry practices is critical to the future of livestock enterprises.

KEY WORDS

public policy
stewardship
Assessing risk
animal welfare
code of ethics
biotechnology
xenotransplantation
genetic engineering
bovine somatotropin
  (BST)
environmentalism
manure management
federal lands
Endangered Species Act
global warming
coronary heart disease
  (CHD)

diet-health
  relationships
junk science
atherosclerosis
unsaturated fatty acids
saturated fatty acids
cholesterol
high-density
  lipoprotein (HDL)
low-density lipoprotein
  (LDL)
Dietary Guidelines
food safety
Total Quality
  Management
E. coli 0157:H7
Salmonella

Clostridium botulism
Staphylococcus aureus
Campylobacter jejuni
Listeria monocytogenes
radiation pasteurization
pathogens
residues
nanogram
organic
antibiotic
Hazard Analysis
  Critical Control Point
  (HACCP)

REVIEW QUESTIONS

1. *True or False:* Special interest groups represent the broad spectrum of societal opinion.

2. Animal agriculturalists would most likely fit into which three "animal groups"?

3. Rollin categorizes animal welfare concerns into three areas. What are they?

4. What is biotechnology?

5. What is BST?

6. *True or False:* Consumers universally understand and support application of genetic principles to agriculture.

7. To what important diseases of humans have the consumption of red meats, dairy products, and eggs been putatively linked?

8. *True or False:* Consumption of red meats, dairy products, and eggs is a major risk factor for coronary heart disease.

9. What are the major known risk factors for coronary heart disease?

10. What is the most significant food safety concern according to food scientists?

11. What are the keys to preventing illness from foodborne microorganisms?

12. *True or False:* Although perceived as a potential food safety threat, chemical residues in meat and dairy products are not a major problem in the United States.

13. What are some environmental issues facing livestock producers in the United States?

14. What are three specific water-quality issues that involve animal agriculture?

15. What are some marketing issues facing livestock producers?

16. What are the major issues associated with concentration and integration of livestock operations?

17. *True or False:* Value-based marketing has provided an effective incentive for livestock producers to improve carcass quality of livestock through genetic selection and nutrition.

18. What are some issues concerning use of public lands for livestock production?

SELECTED REFERENCES

Animal Industry Foundation. 1988. *Animal Agriculture: Myths and Facts.* Arlington, VA: Animal Industry Foundation.

Assessment of Marketing Strategies to Enhance Returns to Lamb Producers. 1991. College Station, TX: Texas Agricultural Market Research Center.

Austin, H. and L. D. McBean. 1997. Red meat and cancer. A review of current epidemiological findings. Chicago, IL: National Cattlemen's Beef Association.

Bauman, D. E. 1992. Bovine somatotropin: Review of an emerging animal technology. *J. Dairy Sci.* 75:3432–3451.

Bjerklie, S. 1990. Poultry's decade of issues. *Meat and Poultry* 36:24.

Cheeke, P. R. 1993. *Impacts of Livestock Production on Society. Diet/Health and the Environment.* Danville, IL: Interstate Publishers, Inc.

Council for Agricultural Science and Technology. 1998. *Food-borne Pathogens: Review of Recommendations.* Ames, IA.

Council for Agricultural Science and Technology. 2003. *Biotechnology in Animal Agriculture: An Overview.* Ames, IA.

Davidson, M. H., D. Hunningshake, K. C. Maki, P. O. Kwiterovich, and S. Kafonek. 1999. Comparison of the effects of lean red meat vs. lean white meat on serum lipid levels among free-living persons with hypercholesterolemia. *Arch. Intern. Med.* 159:1331.

Food Marketing Institute Animal Welfare Program. 2002. Washington, DC.

Fox, M. W. 1986. *Agricide.* New York: Schocken Books.

Gleick, P. H. (ed.). 1993. *Water in Crisis.* New York: Oxford University Press.

Hafs, H. D. (ed.). 1993. Genetically modified livestock: Progress, prospects and issues. *J. Anim. Sci.* 71(Suppl. 3).

Hallman, W. K., and J. Metcalfe. 1993. *Public Perceptions of Agricultural Biotechnology: A Survey of New Jersey Residents.* Ecosystem Policy Research Center, Rutgers University.

Jones, K., T. W. Kelsey, P. A. Nordstrom, L. L. Wilson, A. N. Maretski, and C. W. Pitt. 2000. Neighbors perceptions of animal agriculture. *Prof. An. Scientist* 16:105–110.

Key, T. J., P. N. Appleby, and M. S. Rosell. 2006. Health Effects of Vegetarian and Vegan Diets. Proc. of the Nutrition Society, 65:35–41. Oxford University.

Kliebenstein, J., and S. Hurley. 2001. Manure management: Consumers' perception levels. *PORK*, Sept., 32–34.

Meat industry: A sampling of the issues and controversies affecting meat and poultry industries all over the planet. 1989. *Meat and Poultry* 35:12–14, 41, 43.

National Animal Health Monitoring System. 1996. *Reference of 1996 Dairy Management Practices.* USDA: Veterinary Services.

National Animal Health Monitoring System. 2002. *Reference of 2000 Swine Health and Environmental Management.* USDA: Veterinary Services.

National Cattlemen's Foundation. 1990. *Special Interest Group Profiles.* Washington, DC: Hill and Knowlton.

National Research Council, Committee on Diet and Health. 1989. *Diet and Health. Implications for Reducing Chronic Disease Risk.* Washington, DC: National Academy Press.

Pearson, A. M., and T. R. Dutson. 1990. *Meat and Health.* Amsterdam: Elsevier Science Publishers.

Regan, T. 1985. *The Case for Animal Rights.* Berkeley: University of California Press.

Rollin, B. E. 1990. Animal welfare, animal rights, and agriculture. *J. Anim. Sci.* 68:3456.

Rollin, B. E. 1993. Animal Production and the New Social Ethic for Animals. Proc. Food Animal Well-Being: USDA and Purdue University.

Rollin, B. E. 1995. *Farm Animal Welfare: Social, Bioethical and Research Issues.* Ames, IA: Iowa State University Press.

Smidt, D. 1983. *Indicators Relevant to Farm Animal Welfare.* Boston: Martinus Nijhoff Publishers.

Smith, G. C. 1993. Anti-meat propaganda: Combatting myths with scientific facts. Presented to the International Meat Industry Convention. Chicago, IL.

Smith, G. C. 1993. Papers from AN 560 (Issues) class. Fort Collins, CO: Colorado State University.

Susskind, L., and P. Field. 1996. *Dealing with an Angry Public—The Mutual-Gains Approach to Dissolving Disputes.* New York: The Free Press.

Swinker, A. M., and J. C. Heird. 1994. Horse industry: Trends, opportunities, and issues. *Encyclopedia of Agricultural Science.* San Diego: Academic Press, Inc.

Taubes, G. 2001. The soft science of dietary fat. *Science* 291:2536–2545.

Taylor, R. E., and T. G. Field. 2002. *Beef Production and Management Decisions.* Upper Saddle River, NJ: Prentice Hall.

Thomas, J. A., and L. A. Myers. 1993. *Biotechnology and Safety Assessment.* New York: Raven Press.

CHAPTER 24

Making Effective Management Decisions

LEARNING OBJECTIVES

- Explain the profit formula
- Discuss the role of management in the livestock industry
- Discuss the human resource in the livestock industry
- Define management systems and explain the importance of the concept

Effective management of livestock operations implies that available resources are used to maximize net profit while the same resources are conserved or improved. Available resources include fixed resources (land, labor, capital, and management) and renewable biological resources (animals and plants). Effective management requires a manager who knows how to make timely decisions based on a careful assessment of management alternatives. Modern technology is providing useful tools to make more rapid and accurate management decisions.

Previous chapters have shown how biological principles determine the efficiency of animal production. However, it is important to identify other resources that, when combined with the efficiency of animal production, determine the profitability of an operation.

MANAGING FOR LOWER COSTS AND HIGHER RETURNS

Most livestock producers manage their operations with plans to make a profit. Simply stated, the **profitability formula** is:

$$\begin{array}{l} \text{profit} \\ \quad \text{or} \quad = (\text{production} \times \text{price} - \text{cost}) \\ \text{<loss>} \end{array}$$

The formula can be expanded to make management decisions more focused:

$$
\begin{matrix}
\text{profit} \\
\text{or} \\
\text{<loss>}
\end{matrix}
=
\left(
\text{production}
\begin{matrix}
\nearrow \text{offspring} \longrightarrow \text{pounds per} \\
\leftarrow \text{cull females} \longrightarrow \text{producing} \times \text{price} \\
\searrow \text{cull males} \nearrow \quad \text{female}
\end{matrix}
\right)
- \text{cost}
$$

Note: Costs include feed, labor, veterinary, repairs, fuel, interest, and other costs.

Obviously, profit occurs when output value exceeds input costs, and loss occurs when input costs exceed output value. Management decisions should focus on an optimum combination of output value and input costs to maximize profits while maintaining or improving resources.

The primary components of a long-term profitability formula include the following:

Costs—the primary production costs associated with livestock and poultry enterprises are feed, labor, and interest. However, many other costs, such as veterinary expense, fuel bills, and costs of repairs, should be carefully monitored.

Production *(output)*—output is usually measured in pounds and/or numbers sold. The outputs for the various species are shown in the enterprise budgets of the feeding and management chapters.

Price—the amount received per pound, per head, or per dozen (for eggs). Price is influenced primarily by supply and demand.

Maintaining or improving the resources—includes water, habitat, and land with the forage and crops produced from it. Maximizing short-term profits can easily deplete the land resource by overgrazing, erosion, and the like.

Each component of the profitability formula should initially be evaluated in terms of how it affects output value.

Price has a tremendous effect on output value, though individual producers have little influence over price. Producers accept the price the market offers at the time they sell their products (output). Therefore, the primary management focus for most livestock and poultry producers is reducing costs while increasing production and maintaining or improving resources.

The enterprise budget analyses in the species management contain the component parts of input costs. Managers can analyze these costs and make management decisions to reduce them. Throughout all of the chapters in the book, the biological principles affecting production are identified. Producers apply these principles in making cost-effective increases in production.

THE MANAGER

The manager is the individual responsible for planning and decision making. The management process in simple form is to plan, act, and evaluate. This process is described in more detail in Figure 24.1.

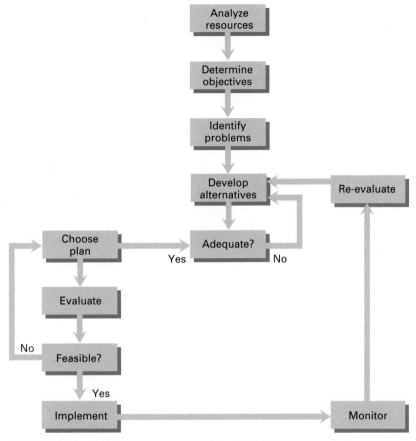

FIGURE 24.1 Major component parts of the planning process.

It is imperative that livestock and poultry producers manage their operations as businesses. Current economic pressures associated with keen competition are forcing more producers to manage their operations this way.

The manager may be an owner-operator with minimum additional labor, or the leader of a more complex organizational structure involving several other individuals or businesses (Fig. 24.2). An **effective manager,** whether involved in a one-person operation or a complex organizational structure, needs to (1) be profit-oriented; (2) identify objectives of the business and establish both short-term and long-range goals to achieve those objectives; (3) keep abreast of the current knowledge related to the operation; (4) know how to use time effectively; (5) attend to the physical, emotional, and financial needs of those employed in the operation; (6) incorporate incentive programs to motivate employees to perform at their full capacity each day; (7) be honest in business dealings; (8) effectively communicate responsibilities to all employees and make employees feel they are part of the operation; (9) know what needs to be done and at what time; (10) be a self-starter; (11) set priorities and allocate resources accordingly; (12) remove or alleviate high risks; and (13) set a good example for others to follow.

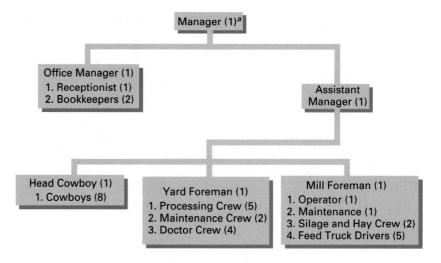

^aNumber in parentheses indicates number of employees in each position

FIGURE 24.2 Organizational structure of a large commercial feeding operation.

FINANCIAL MANAGEMENT

Costs, returns, and profitability of a livestock operation can only be assessed critically with a meaningful set of records. Table 24.1 identifies the financial records needed by most livestock and poultry operations. Each record is described and its purpose is given. Many producers keep only sufficient records to satisfy the IRS (Fig. 24.3). While this is important, additional records are necessary to make critical management decisions.

Profits can be determined by evaluating all cash costs of the operation, all inventory increases or decreases, and the value of opportunity costs. Opportunity costs represent the returns that would be forfeited if debt-free resources (such as owned land, livestock, and equipment) were used in their next best level of employment; for example, the value of pastureland if it were leased, returns if all capital represented by equipment and livestock were invested in a certificate of deposit (or similar investment), and returns if all family labor and management were utilized in other employment. By calculating this, a price can be established for living on a farm or ranch, and a goal can be established for increasing profits.

Credit and money management become crucial during periods of inflation, high interest rates, and relatively low livestock prices. Prudent use of credit can enable a livestock operation to grow more rapidly than it could through the use of reinvested earnings and savings, so long as borrowed funds return more over time than they cost. Thus, farmers and ranchers have to look to credit as a financial tool and learn to use it effectively.

INCOME TAX CONSIDERATIONS

Frequently, managers feel that they are ineffective if income taxes are overpaid. Paying little or no income tax should not be a major goal of the operation.

TABLE 24.1   Financial Records for Livestock and Poultry Operations

Financial Record	Description and Purpose
Cash transactions	Recording of all cash receipts and expenditures is the simplest yet most time-consuming of all financial tasks. This provides most of the information needed for filing income tax returns and in making loan applications.
Balance sheet	Provides a financial picture of the operation at one point in time—usually on the last day of the year. It reflects the net worth of the operation. Net worth = assets (what is owned) − liabilities (what is owed).
Income statement	A moving financial picture that describes most of the changes in net worth from one balance sheet to the next. Net income is calculated by subtracting the expenditures (cash, decrease in inventory, and depreciation) from income (cash receipts and increases in inventory).
Cash-flow statement	Shows cash generated and cash needed on a periodic basis (usually monthly) throughout the year. It assesses times when money must be borrowed and times when money is available for additional purchases or investment, or to retire existing debts. A cash-flow budget can be used to plan for the next calendar year. An active cash-flow statement tracks what is actually happening and evaluates the accuracy of the cash-flow budget.
Enterprise budget	Identifies costs and returns associated with a specific product or enterprise. It can aid in making financial decisions by identifying specific problem areas where management changes can be made. Enterprise budgets are also useful where operations include more than one enterprise or where additional enterprises are being considered. Components include production and marketing assumptions, operating receipts, direct costs, net receipts, and break-even analysis. Estimates of market weight and price can be used in assessing risk in production decisions.
Partial budget	Involves only those income and expense items that would change when implementing a proposed management decision.
Income tax forms	Form 1040 Schedule F is the primary income tax form for sole proprietors and individual partners in a partnership. Producers who have completed Schedule F have basically completed an income statement. There are numerous other forms and 1040 schedules (e.g., Asset Sales, Asset Purchases, Self-Employment Tax, Farm Rental Income and Expenses, Tax Withholding, Depreciation) that are also completed, depending on individual operations and circumstances.

Well-managed livestock operations pay income taxes if maximizing profitability is a goal. Therefore, it is not poor management to pay income tax but to pay more than is owed.

Tax laws are complex and constantly changing. Most cattle producers should consult a qualified tax adviser for tax reporting and longer-term tax management.

ESTATE AND GIFT-TAX PLANNING

Many livestock operations have a large amount of debt-free capital invested in land, livestock, buildings, and equipment. Adequate knowledge and proper planning are necessary for farmers and ranchers to pass on viable economic units to their heirs.

In recent years, significant changes in estate-tax policy have benefited family operations. For example, under a 1997 law, the unified credit exemption was phased up to $1 million in the year 2006. Effective in 1998, family farms and ranches were able to take advantage of a family business exemption of an additional $300,000. Producers can also make $10,000 annual cash gifts to children and grandchildren without paying federal gift tax. It is recommended that the services of an attorney and a tax adviser be employed for tax and estate planning.

SCHEDULE F
(Form 1040)

Department of the Treasury
Internal Revenue Service (O)

Profit or Loss From Farming

▶ Attach to Form 1040, Form 1041, Form 1065, or Form 1065-B.

▶ See Instructions for Schedule F (Form 1040).

OMB No. 1545-0074

1998

Attachment
Sequence No. **14**

Name of proprietor

Social security number (SSN)

A Principal product. Describe in one or two words your principal crop or activity for the current tax year.

B Enter NEW code from Part IV
▶

D Employer ID number (EIN), if any

C Accounting method: **(1)** ☐ Cash **(2)** ☐ Accrual

E Did you "materially participate" in the operation of this business during 1998? If "No," see page F-2 for limit on passive losses. ☐ Yes ☐ No

Part I Farm Income—Cash Method. Complete Parts I and II (Accrual method taxpayers complete Parts II and III, and line 11 of Part I.)
Do not include sales of livestock held for draft, breeding, sport, or dairy purposes; report these sales on Form 4797.

1	Sales of livestock and other items you bought for resale	1	
2	Cost or other basis of livestock and other items reported on line 1	2	
3	Subtract line 2 from line 1		3
4	Sales of livestock, produce, grains, and other products you raised		4
5a	Total cooperative distributions (Form(s) 1099-PATR)	5a	**5b** Taxable amount	5b
6a	Agricultural program payments (see page F-3)	6a	**6b** Taxable amount	6b
7	Commodity Credit Corporation (CCC) loans (see page F-3):		
a	CCC loans reported under election		7a
b	CCC loans forfeited	7b	**7c** Taxable amount	7c
8	Crop insurance proceeds and certain disaster payments (see page F-3):		
a	Amount received in 1998	8a	**8b** Taxable amount	8b
c	If election to defer to 1999 is attached, check here ▶ ☐	**8d** Amount deferred from 1997	8d
9	Custom hire (machine work) income		9
10	Other income, including Federal and state gasoline or fuel tax credit or refund (see page F-3)		10
11	**Gross income.** Add amounts in the right column for lines 3 through 10. If accrual method taxpayer, enter the amount from page 2, line 51 ▶		11

Part II Farm Expenses—Cash and Accrual Method. Do not include personal or living expenses such as taxes, insurance, repairs, etc., on your home.

12	Car and truck expenses (see page F-4—also attach **Form 4562**)	12		25	Pension and profit-sharing plans	25
13	Chemicals	13		26	Rent or lease (see page F-6):	
14	Conservation expenses (see page F-4)	14		a	Vehicles, machinery, and equipment	26a
15	Custom hire (machine work)	15		b	Other (land, animals, etc.)	26b
16	Depreciation and section 179 expense deduction not claimed elsewhere (see page F-5)	16		27	Repairs and maintenance	27
				28	Seeds and plants purchased	28
				29	Storage and warehousing	29
17	Employee benefit programs other than on line 25	17		30	Supplies purchased	30
18	Feed purchased	18		31	Taxes	31
19	Fertilizers and lime	19		32	Utilities	32
20	Freight and trucking	20		33	Veterinary, breeding, and medicine	33
21	Gasoline, fuel, and oil	21		34	Other expenses (specify):	
22	Insurance (other than health)	22		a	....	34a
23	Interest:			b	....	34b
a	Mortgage (paid to banks, etc.)	23a		c	....	34c
b	Other	23b		d	....	34d
24	Labor hired (less employment credits)	24		e	....	34e
				f	....	34f

35	**Total expenses.** Add lines 12 through 34f ▶		35
36	**Net farm profit or (loss).** Subtract line 35 from line 11. If a profit, enter on **Form 1040, line 18**, and ALSO on **Schedule SE, line 1.** If a loss, you MUST go on to line 37 (estates, trusts, and partnerships, see page F-6).		36
37	If you have a loss, you MUST check the box that describes your investment in this activity (see page F-6). • If you checked 37a, enter the loss on **Form 1040, line 18**, and ALSO on **Schedule SE, line 1.** • If you checked 37b, you MUST attach **Form 6198.**	37a ☐ All investment is at risk. 37b ☐ Some investment is not at risk.	

For Paperwork Reduction Act Notice, see Form 1040 instructions. Cat. No. 11346H Schedule F (Form 1040) 1998

FIGURE 24.3 Internal Revenue Service form (Schedule F) showing income and expense items for which documented records must be kept. *Source:* Department of the Treasury.

Livestock producers should review tax plans (estate, gift, and income) and consult with professionals who are knowledgeable about current tax laws.

PEOPLE

Human resource management and leadership are the most frequently overlooked topics associated with livestock production. Livestock and poultry managers often work with a variety of people including family, employees, business partners, clients, suppliers, and the general public. Failure to adequately prepare to effectively lead and interact with other people can lead to dismal financial performance in an enterprise even if all other elements of the business are well managed.

Effective communication skills (listening, writing, and speaking) are critical to the ability of a manager to gain the best efforts of employees and associates. The need for effective communication in an enterprise cannot be overstated. Listening to understand employees and other people is an important skill that is often overlooked by managers but is well understood and practiced by leaders.

Successful managers know how to accomplish the following:

1. Determine the optimum labor needs of an enterprise.
2. Identify, hire, and retain effective employees.
3. Motivate and reward employees with both financial and nonfinancial incentives.
4. Keep staff and management focused on the mission and goals of the organization.
5. Build and enhance effective teams and partnerships.

If the enterprise is family-owned or if family members are employed, the manager must also understand the following issues specific to the family business (Fig. 24.4).

1. Apply sound business principles rather than assuming things will work out simply because people are within the family. Involve all family members in financial decisions.
2. Evaluate other successful family operations. Determine why they are successful and how they resolve difficulties.
3. Include all family members in the written plan of responsibilities (e.g., who will make the decisions; how each family member will be paid; how vacation and other time away from the operation will be handled).
4. Hold weekly family councils for additional planning, evaluating, and problem solving. Create an environment that encourages open communication.
5. Recognize that family relationships have a higher priority than programs and profitability but that all can be compatible.
6. Have patience and tolerance with age differences in the family. Provide roles for family members and involve everyone (e.g., spouses, in-laws) in development of plans and goals.
7. Recognize that management changes can occur too fast or too slowly in how they affect family relationships and profitability of the operation.

FIGURE 24.4   A family-owned business often involves a significant number of people. Effective communication is critical to the success of the business.

MANAGEMENT SYSTEMS

Management systems analysis provides a method of systematically organizing information needed to make valid management decisions. It permits variables to be more critically assessed and analyzed in terms of their contribution to the desired end point—usually net profit. Individuals who have been educated to think broadly in the framework of management systems can often make valid management decisions without the use of data-processing equipment. A pencil, a hand calculator, and a well-trained mind are the primary components required for making competent management decisions. Without question, however, the use of the computer in synthesizing voluminous amounts of information enhances the management system.

Resources are different for each operation; no fixed recipe exists for successful livestock production. Each enterprise is unique as the result of such variables as different levels of forage production, varying marketing alternatives, varying energy costs, different types of animals, environmental differences, feed nutrients at various costs, and varying levels of competence in labor and management. All of these variables and their interactions pose challenges to the producer seeking to combine them into sound management decisions for a specific operation.

It is a common practice to increase or maximize production of animals by using known biological relationships. Animal production typically has been maximized without careful consideration of cost-benefit ratios and of how increased productivity relates to land, feed, and management resources. However, recognition is now being given to the need for optimization rather than maximization of animal productivity. Thus, there is an increased interest in management systems that attempt to optimize production with net profit being the primary (and possibly the only) goal involved. As a result, some valid biological relationships will not be useful or applicable because an economic analysis may prevent their inclusion in sound management decisions. For example, it is a well-known biological fact that calves born earlier in the calving season will have heavier weaning weights. Because of this relationship, some ranchers move the calving season to a period earlier in the year without a careful economic assessment. An economic evaluation for one ranch demonstrated that changing the calving season to match forage availability (in this case about 30 days to grazeable forage) increased profits significantly. Changing calving season alone was estimated to increase the ranch's carrying capacity by 30–40% in terms of animal units. In addition, the annual cow cost was reduced by 18% per cow because of reduction in winter feed requirements.

In dairy production, there is often a difference between maximum milk production and optimum milk production if producers want to maximize profit. The extra feed and management needed to maximize milk production may cost more than the value of the added milk.

Some farmers and ranchers are limited in utilizing the computer in management systems because of the lack of software. However, more management system software continues to be made available to producers. Learning to use information technologies may also be a limiting factor. The extension service or a local college typically offers training services.

CHAPTER SUMMARY

■ The profitability formula:

$$\text{profit or} \langle\text{loss}\rangle = (\text{production} \times \text{price}) - \text{cost}$$

identifies the components for animal producers to lower their costs and increase their revenue and profits.

■ Assessment of resources (human, financial, forage/feed, animals, etc.) and a decision-making process that integrates the resources is essential for the well-being of all resources.

■ Integrated resource management (management systems) is critical if future profits are to be sustained and resources are to be maintained or improved.

KEY WORDS

profit formula	effective manager	enterprise budget
costs	cash transaction	partial budget
production	balance sheet	income tax forms
price	income statement	management systems
resource improvement	cash-flow statement	

REVIEW QUESTIONS

1. What are the four components of the long-term profitability formula?
2. What is critical to assessing the costs, returns, and profitability of livestock operations?
3. What are five employee objectives successful managers know how to accomplish?

SELECTED REFERENCES

Bourdon, R. 1992. *The Systems Concept of Beef Production.* Beef Improvement Federation Fact Sheet.

Fetsch, R. J., and E. Paris. 1998. *Beyond the Basics and Estate Transfer Planning Workshop.* Colorado State University Extension Service.

Field, T. G. 2007. *Beef Production and Management Decisions.* Upper Saddle River, NJ: Prentice Hall.

Harl, N. E. 1997. *Estate Planning.* Iowa State University Extension Service. PM-993.

Libbin, J. D., and L. B. Catlett. 1987. *Farm and Ranch Financial Records.* New York: Macmillan.

Luft, L. D. Sources of credit and cost of credit. In *Great Plains Beef Cattle Handbook*, GPE-4351 and 4352.

Maddux, J. 1981. The man in management. *Proceedings: The Range Beef Cow, a Symposium on Production, VII.* Rapid City, SD, December.

NCA-IRM-SPA™. 2002. *Workbook for the Cow-Calf Enterprise.* College Station, TX: Texas A&M University.

CHAPTER 25

Beef Cattle Breeds and Breeding

LEARNING OBJECTIVES

- Explain the importance of breed type in the cattle industry
- Discuss the reasons for broadening the genetic base in the cattle business
- Outline the various approaches to improving cattle performance in the major economic traits
- Discuss the process of bull and replacement female selection
- Explain the role of sire summaries in genetic improvement
- Describe the various approaches to crossbreeding

A **breed** of cattle is defined as a race or variety, the members of which are related by descent and similar in certain distinguishable characteristics. More than 250 breeds of cattle are recognized throughout the world, and several hundred other varieties and types have not been identified with a breed name. The Texas Longhorn breed characterized the early history of the western cattle industry in the United States. The Longhorn was hardy, range-adapted, and disease-resistant but fell out of favor as the industry progressed with a focus on feedlot performance and superior carcass characteristics.

Some of the oldest recognized breeds in the United States were officially recognized as breeds during the middle to late 1800s. Most of these breeds originated from crossing and combining existing strains of cattle. When a breeder or group of breeders decided to establish a breed, distinguishing that breed from other breeds was of paramount importance; thus, major emphasis was placed on readily distinguishable visual characteristics, such as color, color pattern, polled or horned condition, and rather extreme differences in form and shape.

New cattle breeds, such as Brangus, Santa Gertrudis, Barzona, and Beefmaster, have come into existence in the United States during the past 50–75 years.

Attempting to combine the desirable characteristics of several existing breeds has helped to develop these breeds. Currently there are new breeds of cattle being developed. They are sometimes called **composite breeds** or **synthetic breeds.** In most cases, however, the same visual characteristics as previously mentioned are used to give the new breeds visual identity.

It was not long after some of the first breeds were developed that the word **purebred** was attached to them. Herd books and registry associations were established to assure the "purity" of each breed and to promote and improve each breed. *Purebred* refers to purity of ancestry, established by pedigree, which shows that only animals recorded in that particular breed have been mated to produce the animal in question. Purebreds, therefore, are cattle within the various breeds that have pedigrees recorded in their respective breed registry associations.

When viewing a herd of purebred Angus or Herefords, uniformity is noted, particularly uniformity of color or color pattern. Because of this uniformity of one or two characteristics, the word *purebred* has come to imply genetic uniformity (homozygosity) of all characteristics. Cattle within the same breed are not highly homozygous because high levels of homozygosity occur only after many generations of close inbreeding. This close inbreeding has not occurred in cattle breeds. If breeds were uniform genetically, they could not be improved or changed even if changes were desired.

WHAT IS A BREED?

The genetic basis of cattle breeds and their comparison is not well understood by most livestock producers. Often the statement is made, "There is more variation within a breed than there is between breeds." The validity of this claim needs to be examined carefully. Considerable variation does exist within a breed for most of the economically important traits. This variation is depicted in Figure 25.1, which shows the number of calves of a particular breed that fall

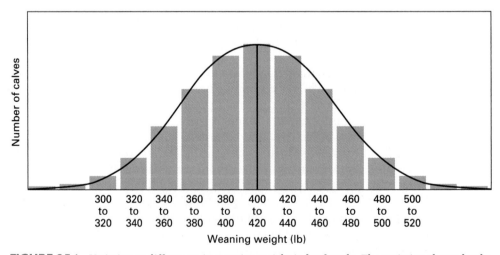

FIGURE 25.1   Variation or differences in weaning weight in beef cattle. The variation shown by the bell-shaped curve could be representative of a breed or a large herd. The vertical line in center is the average or the mean, which is 410 lb. Note that the number of calves is greater around the average and is less at the extremely light and extremely heavyweights. *Source:* Colorado State University.

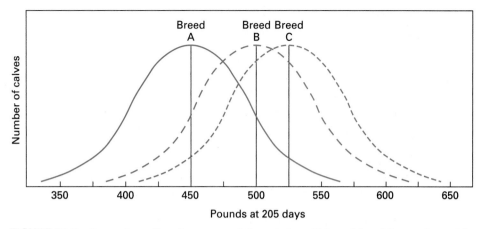

FIGURE 25.2   Comparison of breed averages and the variation within each breed for weaning weight in beef cattle. The vertical lines are the breed averages. Note that some individual animals in breeds B and C can be lower in weaning weight than the average of breed A. *Source:* Colorado State University.

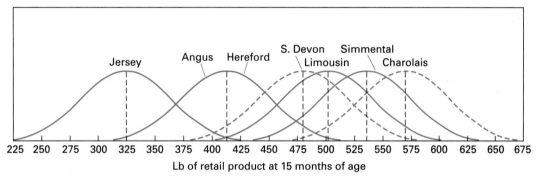

FIGURE 25.3   Breed differences in retail product when breeds are slaughtered at the same age. *Source:* U.S. Meat Animal Research Center.

within certain weight-range categories at 205 days of age. Connecting the high points of each bar forms a bell-shaped curve. The solid line that separates the bell curve into equal halves represents the breed average. Most of the calves are near the breed average; however, at the outer edge of the bell curve are high- and low-weaning weight calves. Note that there are fewer calves at these extremes.

Figure 25.2 allows us to compare three hypothetical breeds of cattle in terms of weaning weight. The breed averages are different; however, the variation within each breed is comparable among all three. The statement, "There is more variation within a breed than between breed averages," is more correct than the statement, "There is more variation within a breed than there is between breeds."

Figures 25.1 and 25.2 are hypothetical examples; however, they are based on realistic samples of data obtained from the various breeds of cattle. Figure 25.3 shows how breeds can differ greatly when only one trait is considered. Note how distinctly different the Charolais-sired calves are from the Jersey or Angus or Hereford calves. The statement, "There is more variation within the Charolais breed than between the Simmental and Jersey breeds" is not valid in this case. Rather, "There is more variation within the Charolais breed than between breed averages" is a more correct statement.

The information in Figure 25.3 supports Charolais as being the most superior breed based on breed averages. However, this is only for one trait, and other breeds show superiority to the Charolais breed when other economically important traits are considered.

MAJOR U.S. BEEF BREEDS

Shorthorn, Hereford, and Angus were the major beef breeds in the United States during the early 1900s. During the 1960s and 1970s, the number of cattle breeds remained relatively stable at between 15 and 20. Today, more than 60 breeds of cattle are available to U.S. beef producers. However, fewer than 20 have a widespread impact on the national herd. Why the large importation of the different breeds from several different countries? Following are several possible reasons:

1. Feeding more cattle larger amounts of grain, a practice started in the 1940s, resulted in many overfat cattle—cattle that had been previously selected to fatten on forage diets. Therefore, a need was established for grain-fed cattle that could produce a higher percentage of lean to fat at the desired slaughter weight.

2. Economic pressure to produce more weight in a shorter period of time demonstrated a need for cattle with more milk and more growth.

3. An opportunity was available for some promoters to capitalize on merchandising a certain breed as being the ultimate in all production traits. This was easily accomplished because there was little comparative information on breeds.

Color plates M through N show the leading breeds of beef cattle. Additional breed pictures can be accessed via www.ansi.okstate.edu/breeds/. Table 25.1 shows distinguishing characteristics and gives brief background information for more of these breeds.

The relative importance of the various breeds' contributions to the total beef industry is best estimated by the registration numbers of the breeds (Table 25.2). Although registration numbers are for purebred animals, they reflect the commercial cow-calf producers' demand for the different breeds. Registration numbers show that Angus, Limousin, Simmental, and Hereford are the most important breeds in the beef cattle industry in the United States. The numbers of cattle belonging to various breeds in this country have changed over the past years. No doubt some breeds will become more numerous in future years and others will decrease significantly in numbers. These changes will be influenced by economic conditions and how well the breeds meet the needs of the commercial beef industry.

IMPROVING BEEF CATTLE THROUGH BREEDING METHODS

Genetic improvement in beef cattle can be achieved by selection and by using a particular mating system. Significant genetic improvement by selection results when the selected animals are superior to the herd average and when the heritabilities of the traits are relatively high. It is important that the traits included in a breeding program be of economic importance and that they be measured objectively.

TABLE 25.1 Background and Distinguishing Characteristics of Major Beef Breeds in the United States

Breed	Distinguishing Characteristics	Brief Background
Angus	Black color; polled	Originated in Aberdeenshire and Angushire of Scotland. Imported into the United States in 1873.
Beefmaster	Various colors; horned	Developed in the United States from Brahman, Hereford, and Shorthorn breeds. Selected for ability to reproduce, produce milk, and grow under range conditions.
Brahman	Various colors, with gray predominant; they are one of the Zebu breeds, which have the hump over the top of the shoulder; most Zebu breeds also have large, drooping ears and excess skin in the throat and dewlap	Major importations to the United States from India and Brazil. Largest introductions in early 1900s. These cattle are heat tolerant and well adapted to the harsh conditions of the Gulf Coast region.
Brangus	Black and polled predominate, although there are Red Brangus which is a separate breed	U.S. breed developed around 1912—3/8 Brahman, 5/8 Angus.
Charolais	White color with heavy muscling; horned or polled	One of the oldest breeds in France. Brought into the United States soon after World War I, but its most rapid expansion occurred in the 1960s.
Chianina	White color with black eyes and nose; extremely tall cattle	An old breed originating in Italy. Acknowledged to be the largest breed, with mature bulls weighing more than 3,000 lb.
Gelbvieh	Golden colored; horned or polled	Originated in Austria and Germany. Developed as a dual-purpose breed used for draft, milk, and meat.
Hereford	Red body with white face; horned	Introduced in the United States in 1817 by Henry Clay. Followed the Longhorn in becoming the traditionally known range cattle.
Limousin	Golden color with marked expression of muscling; polled or horned	Introduced into the United States in 1969, primarily from France.
Longhorn	Multicolored with characteristically long horns	Came to West Indies with Columbus. Brought to the United States through Mexico by the Spanish explorers. Longhorns were the noted trail-drive cattle from Texas into the Plains states.
Polled Hereford	Red body with white face; polled	Bred in 1901 in Iowa by Warren Gammon, who accumulated several naturally polled cattle from horned Hereford herds.
Red Angus	Red color; polled	Founded as a performance breed in 1954 by sorting the genetic recessives from Black Angus herds.
Salers	Uniform mahogany red with medium to long hair; horned	Raised in the mountainous area of France, where they were selected for milk, meat, and draft.
Santa Gertrudis	Red color; horned	First U.S. breed of cattle that was developed on the King Ranch in Texas. Combination of 5/8 Shorthorn and 3/8 Brahman.
Shorthorn	Red, white, or roan in color; horned and polled	Introduced into the United States in 1783 under the name *Durham*. Most prominent in the United States around 1920.
Simmental	Yellow to red-and-white color pattern; polled and horned	A prominent breed in parts of Switzerland and France. First bull arrived in Canada in 1967. Originally selected as a dual-purpose for milk and meat.

TRAITS AND THEIR MEASUREMENT

The most economically important traits of beef cattle are classified as follows: (1) reproductive performance, (2) weaning weight, (3) postweaning growth, (4) feed efficiency, (5) carcass merit, (6) longevity (functional traits), (7) conformation, and (8) freedom from genetic defects.

TABLE 25.2   Annual Registration Numbers for Major U.S. Beef Breeds

	Annual Registrations (in thousands)						Year U.S. Association Formed
Breed[a]	2005	1995	1990	1985	1980	1975	
Angus	298.7	224.8	159.0	175.5	257.6	306.5	1883
Hereford[b]	68.8	120.0	170.5	180.1	353.2	419.2	1881
Simmental	43.9	71.0	79.3	73.0	66.1	75.5	1969
Red Angus	43.2	30.0	15.4	12.0	12.5	10.1	1954
Limousin	39.6	79.3	78.4	44.5	28.0	32.4	1968
Charolais	25.8	55.6	44.8	23.2	23.0	45.2	1957
Gelbvieh	27.5	33.8	22.8	16.1	NA	NA	1971
Brangus	22.9	31.0	32.1	30.3	24.5	13.4	1949
Beefmaster	19.0	47.1	38.4	32.1	30.0	12.0	1961
Shorthorn	18.0	20.0	18.0	16.7	19.4	29.2	1872
Brahman	8.0	15.4	13.0	29.9	36.4	27.5	1924

[a] Only breeds with more than 10,000 annual registrations are listed.
[b] Includes Polled Hereford.
Source: Beef breed associations.

Reproductive Performance

Reproductive performance has the highest economic importance of all the traits. Most cow-calf producers have a goal for percent calf crop weaned (number of calves weaned compared to the number of cows in the breeding herd) of 85% or higher. Beef producers also desire each cow to calve every 365 days or less and to have a calving season for the entire herd of less than 90 days. All of these are good objective measures of reproductive performance.

The heritability of fertility in beef cattle is quite low (less than 20%), so little genetic progress can be made through direct selection for that specific trait. Heritabilities for birth weight and **scrotal circumference** are high; therefore, selection for optimum levels of these traits will improve the percent calf crop weaned. The most effective way to improve certain reproductive traits (such as rebreeding after calving) is to improve the environment, for example through adequate nutrition and good herd health practices. Selecting bulls based on a reproductive soundness examination, which includes semen testing, will also improve reproductive performance in the herd.

Reproductive performance can be improved through breeding methods by crossbreeding to obtain heterosis for percent calf crop weaned; by using bulls with relatively light birth weights (heritability of birth weight is 40%), which decreases calving difficulty; and by selecting bulls that have a relatively large scrotal circumference. Scrotal circumference has a high heritability (40%); bulls with a larger scrotal size (over 32 cm for yearling bulls) produce a larger volume of semen and have half-sister heifers that reach puberty at earlier ages than heifers related to bulls with a smaller scrotal size.

Weaning Weight

Weaning weight, as measured objectively by weighing calves on scales, reflects the milking and mothering ability of the cow and the preweaning growth rate of the calf. Weaning weight is commonly expressed as the adjusted 205-day weight, where the weaning weight is adjusted for the age of the calf and age of the dam. This adjustment puts all weaning-weight records on a comparable basis since older calves weigh more than younger calves and mature cows (5–9 years of age) milk more heavily than younger cows (2–4 years of age) and older cows (10 years and older).

Weaning weights of calves are usually compared as a ratio expressed by dividing the calf's adjusted weight by the average weight of the other calves in the contemporary group. For example, a calf with a weaning weight of 440 lb, where the contemporary group average is 400 lb, has a ratio of 110. This calf's weaning-weight ratio is 10% above the herd average. Ratios can be used primarily for selecting cattle within the same herd in which they have had similar environmental opportunity. Comparing ratios between herds is misleading from a genetic standpoint because most differences between herds are caused by differences in the environment. Weaning weight will respond to selection because it has a heritability of 30%. Producers should avoid placing too much emphasis on increasing weaning weight due to the possibility of sacrificing economic performance via expenditure of excessive input costs to increase weaning weight.

Producers may opt to utilize a creep feeding system that allows calves access to a concentrate feed source but restricts access by the dams. While short-term creep feeding may be an effective means to prepare calves for feeding following weaning, it is typically expensive. Furthermore, creep feeding may cloud the ability to select accurately for increases in weaning weight that can be sustained on a forage-based diet.

Postweaning Growth

Postweaning growth measures the growth from weaning to a finished weight. Postweaning growth might take place on a pasture or in a feedlot. Usually, animals with relatively high postweaning gains make efficient gains at a relatively low cost to the producer.

Postweaning gain in cattle is usually measured in pounds gained per day after a calf has been on a feed test for 100–160 days. Weaning weight and postweaning gain are usually combined into a single trait; namely, adjusted 365-day weight (or yearling weight). It is computed as follows:

$$\text{Adjusted 365-day weight} = (160 \times \text{average daily gain}) + \text{adjusted 205-day weight}$$

Average daily gain and adjusted 365-day weight both have high heritabilities (40%); therefore, genetic improvement can be quite rapid when selection is practiced on postweaning growth or yearling weight.

Feed Efficiency

The pounds of feed required per pound of live weight gain measure **feed efficiency**. Only feeding each animal individually and keeping records on the amount of feed consumed can obtain specific records for feed efficiency. With the possible exception of some bull-testing programs, determining feed efficiency on an individual animal basis is seldom economically feasible. However, new feeding system technologies may allow more widespread collection of efficiency data.

Interpretation of feed-efficiency records can be somewhat difficult, depending on the endpoint to which the animals are fed. Feeding endpoint can be a certain number of days on feed (e.g., 140 days) to a specified slaughter weight (e.g., 1,200 lb) or to a carcass compositional endpoint (e.g., low choice quality grade or 0.4 in. of fat). Most differences in feed efficiency shown by individual animals are related to the pounds of body weight maintained through feeding periods and the daily rate of gain or feed intake of each animal. Cattle fed from a similar initial feedlot weight (e.g., 600 lb) to a similar slaughter weight (e.g., 1,100 lb) will demonstrate a high relationship between rate of gain and efficiency of gain. In this situation, cattle that gain faster will require fewer pounds of feed per pound of gain. Thus, a breeder can select for rate of gain and thereby make genetic improvement in feed efficiency. However, when cattle are fed to the same compositional endpoint (approximately the same carcass fat), there are small differences in the amount of feed required per pound of gain. This is true for different sizes and shapes of cattle, and even for various breeds that vary greatly in skeletal size and weight.

The heritability of feed efficiency at similar beginning and final weights is high (45%), so selection for more efficient cattle can be effective. It seems logical to use the genetic correlation between gain and efficiency where possible.

Carcass Merit

Carcass merit is presently measured by quality grades and yield grades, which are discussed in detail in Chapter 8. Many cattle-breeding programs have goals to produce cattle that will grade choice and have yield grade 2 carcasses. Objective measurements using ultrasound backfat thickness on the live animal and hip height measurement can assist in predicting the yield grade at certain slaughter weights. Visual appraisal, which is more subjective, can be used to predict amount of fat or predisposition to fat and skeletal size. These visual estimates can be relatively accurate in identifying actual yield grades if cattle differ by as much as one yield grade.

Because quality grade cannot be evaluated accurately in the live animal, it is necessary to evaluate the carcass. However, researchers are attempting to identify gene markers associated with marbling in the hope that evaluations of live animals may be possible in the future. Steer and heifer progeny of different bulls are slaughtered to identify the genetic superiority or inferiority of bulls for both quality grade and yield grade. The heritabilities of most beef carcass traits are high (over 30%), so selection can result in marked genetic improvement for these traits.

Longevity

Longevity measures the length of productive life. It is an important trait, particularly for cows, when replacement heifer costs are high or when a producer reaches an optimum level of herd performance and desires to stabilize it. Bulls are usually kept in a herd for 3–5 years, or inbreeding might occur. Some highly productive cows remain in the herd at age 15 years or older, whereas other highly productive cows have been culled before reaching 4 years of age. These cows may have been culled because of such problems as skeletal unsoundness, poor udders, eye problems (i.e., cancer eye), and lost or worn teeth. Little selection opportunity for longevity exists because few cows remain highly productive past the age of 10 years. Most cows that leave the herd early have poor reproduction performance.

Some producers need to improve their average herd performance as rapidly as possible rather than improve longevity. In this situation, a relatively rapid turnover of cows is needed. Some selection for longevity occurs because producers have the opportunity to keep more replacement heifers born to cows that are highly productive in the herd for a longer period of time than heifers born to cows that stay in the herd for a short time. Some beef producers attempt to identify bulls that have highly productive, relatively old dams. Stayability EPD estimates are being calculated for some breeds to assist breeders in selecting for longevity. Certain conformation traits, such as skeletal soundness and udder soundness, may be evaluated to extend the longevity of production.

Conformation

Conformation is the form, shape, and visual appearance of an animal. How much emphasis to put on conformation in a beef cattle selection program has been, and continues to be, controversial. Some producers believe that putting a productive animal into an attractive package contributes to additional economic returns. It is more logical, however, to place more selection emphasis on traits that will produce additional numbers and pounds of lean growth for a given number of cows. Placing some emphasis on conformation traits such as skeletal, udder, eye, and teeth soundness is justified. Conformation differences such as fat accumulation or predisposition to fat can be used effectively to make meaningful genetic improvement in carcass composition.

Most conformation traits are medium to high (30–60%) in heritability, so selection for these traits will result in genetic improvement.

Genetic Defects

Genetic defects, other than those previously identified under longevity and conformation, need to be considered in breeding productive beef cattle. Cattle have numerous known hereditary defects; most, however, occur infrequently and are of minor concern. Some defects increase in their frequency, and selection needs to be directed against them. A single pair of genes that are usually recessive determines most of these defects. When one of these hereditary defects occurs, it is a logical practice to cull both the cow and the bull.

Some commonly occurring genetic defects in cattle today are double muscling, syndactyly (mule foot), arthrogryposis (palate-pastern syndrome), osteopetrosis (marble bone disease), hydrocephalus, and dwarfism. An enlargement of the muscles evidences **double muscling** with large grooves between the muscle groups of the hind leg. Double-muscled cattle usually grow slowly and their fat deposition in and on the carcass is much less than that of the normal beef animal. **Syndactyly** is a condition in which one or more of the hooves are solid in structure rather than cloven. Mortality rate is high in calves with syndactyly. **Arthrogryposis** is a defect in which the pastern tendons are contracted, and the upper part of the mouth has not properly fused together. **Osteopetrosis** is characterized by the marrow cavity of the long bones being filled with bone tissue. All calves having osteopetrosis have short lower jaws, protruding tongue, and impacted molar teeth. A bulging forehead where fluid has accumulated in the brain area is typical of the defect of **hydrocephalus.** Calves with arthrogryposis, hydrocephalus, or osteopetrosis usually die shortly after birth.

The most common type of **dwarfism** is snorter dwarfism, in which the skeleton is quite small and the forehead has a slight bulge. Some snorter dwarfs exhibit a heavy, labored breathing sound. This defect was most common in the 1950s, and it has decreased significantly since that time.

BULL SELECTION

Bull selection must receive the greatest emphasis if optimum genetic improvement of a herd is to be achieved. Sire selection accounts for 80–90% of the genetic improvement in a herd over a period of several years. This does not diminish the importance of good beef females because genetically superior bulls have superior dams. However, most of the genetic superiority or inferiority of the cows will depend on the bulls previously used in the herd. Also, the accuracy of records is much higher for bulls than for cows and heifers because bulls produce more offspring and are used in more herds (especially with AI). While artificial insemination (Chapter 11) offers significant opportunities for increasing the rate of genetic change, the technology is mostly utilized by seedstock producers or commercial producers who are developing their own replacement heifers and desire to assure that young cows are bred to low-birth-weight sires. Less than 10% of all beef cows are artificially inseminated. This low rate of adoption is likely due to the inconvenience, high labor requirements, and need for animal handling facilities.

All commercial and purebred producers should identify seedstock suppliers who have honest, comparative records on their cattle. Most performance-minded seedstock producers record the birth weights, weaning weights, and yearling weights of their bulls. Some breeders also obtain feedlot and carcass data on their own bulls, or they use bulls from sires for whom these test data are known.

Purebred breeders should provide accurate performance data on their bulls, and commercial producers should request the information. Excellent performance records can be obtained and made available on the farm or ranch. The trait ratios are useful and comparative if the bulls have been fed and managed in similar environments.

Breeding Values

Phenotype is the appearance or performance of an animal determined by the *genotype* (genetic makeup) and the environment in which the animal was raised. Genotype is determined by two factors: (1) breeding value (what genes are present) and (2) nonadditive value (how genes are combined). Environmental factors influence the phenotype through known and unknown effects. Age of dam, resulting in different levels of milk production, is a known environmental effect adjusted for in an adjusted 205-day weight (phenotype). Unknown effects are things like injury or health problems for which it is difficult to adjust the phenotypic record.

An example depicting breeding values is shown in Figure 25.4. In the "brick wall concept," each brick represents a gene. Some genes have positive, additive effects, whereas other genes have negative effects. The different-sized bricks represent the magnitude of the gene's effect. In this example (Fig 25.4), the 10 pairs

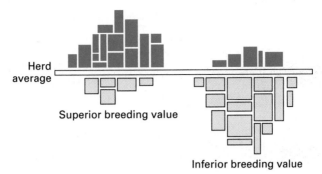

FIGURE 25.4 Breeding value as depicted by the "brick wall concept." *Source:* Gibb, Boggess, and Wagner, 1992.

of genes (20 bricks) are used to show a superior breeding value and a negative breeding value. The superior breeding value (higher brick wall) is where the sum of the additive gene effects (bricks) is above herd average.

Breed associations have developed computer programs that utilize performance records on the individual calf and its relatives. This information is used to calculate breeding values. Breeding values are most frequently reported on birth weight, maternal weaning weight, weaning weight, and yearling weight by using expected progeny differences (EPDs).

Sire Summaries

The development of sire summaries has made sire selection more effective. Most breed associations publish a sire summary that is updated annually or biannually.

Expected progeny difference (EPD) and **accuracy (ACC)** are the important terms used in understanding sire summaries. An EPD combines into one figure a measurement of genetic potential based on the individual's performance and the performance of related animals such as the sire, dam, and other relatives. EPD is expressed as a plus or minus value, reflecting the genetic transmitting ability of a sire. The most common EPDs reported for bulls, heifers, and cows are birth weight EPD, milk EPD, weaning-growth EPD, maternal EPD (includes milk and preweaning growth), and yearling-weight EPD. Some additional traits are included as well (Table 25.3).

Accuracy (ACC) is a measure of expected change in the EPD as additional progeny data become available. EPDs with ACC values of 0.90 and higher would be expected to change very little, whereas EPDs with ACC values below 0.70 might change dramatically with additional progeny data.

Table 25.4 shows sire summary data for several beef breed associations.

If bull B and bull C were used in the same herd (each on an equal group of cows), the expected average performance of their calves would be:

Bull B's calves would be 16 lb heavier at birth.

Bull B's calves would weigh 59 lb more at weaning.

Bull B's calves would weigh 78 lb more as yearlings.

Bull B's daughters would wean calves weighing 7 lb less.

Bull A and bull D have an optimum combination of calving ease (birth weight) and growth traits (weaning weight and yearling weight). Bull E is

TABLE 25.3   Sire Summary Data (EPDs or Ratios) for Selected Breeds

Traits	Angus[e]	Brahman[e]	Charolais[e]	Gelbvieh[e]	Hereford/Polled Hereford[e]	Limousin[e]	Red Angus[e]	Simmental[e]
Calving ease[a]								
first calf[b]	o	o	o	+	+	+	+	+
maternal[c]	o	o	o	+	+	+	+	+
Scrotal circumference	+	o	o	+	+	+	+	o
Birth weight	+	+	+	+	+	+	+	+
Gestation length	o	o	o	+	o	+	o	o
Weaning weight	+	+	+	+	+	+	+	+
Yearling weight	+	+	+	+	+	+	+	+
Milk[d]	+	+	+	+	+	+	+	+
Maternal (milk + 1/2 weaning weight)[d]	+	o	+	+	+	+	+	+
Disposition (docility)	o	o	o	o	o	+	o	o
Stayability	o	o	o	+	o	+	+	o
Heifer pregnancy	o	o	o	o	o	o	+	o
Mature cow energy requirements	o	o	o	o	o	o	+	o
Yearling hip height	+	o	o	o	o	o	o	o
Mature cow weight	+	o	o	o	o	o	o	+
Mature daughter height	+	o	o	o	o	o	o	+
Carcass weight	+	o	+	+	o	+	o	+
Marbling	+[f]	o	+	+	+[g]	+[f]	+[f]	+
Ribeye area	+[f]	o	+	+	+[g]	+[f]	+[f]	+
Fat thickness	+[f]	o	+	+	+[g]	+[f]	+[f]	+
Percent retail product	+[f]	o	o	o	o	o	o	+
Ultrasound rump fat	+	o	o	o	o	o	o	o
Grid value	+	o	o	+	o	o	o	o
Feedlot value	+	o	o	+	o	o	o	o
Beef value	+	o	o	o	o	o	o	o
Terminal Profit Index	o	o	+	o	o	o	o	o
Warner-Bratzler Shear Force	o	o	o	o	o	o	o	+

[a] Computed from birth weights.
[b] Expressed as a ratio, not an EPD. Measures calving ease of bulls' own calves.
[c] Expressed as a ratio, not an EPD. Measures calving ease of a bull's daughters.
[d] Computed from weaning weights.
[e] + = trait in sire summary; o = trait not in sire summary.
[f] Computed from both progeny and ultrasound data.
[g] Computed from ultrasound data.
Source: Breed sire summaries.

TABLE 25.4   Selected Data from an Angus Sire Evaluation Report

	Birth Weight		Weaning Weight Direct		Weaning Weight Maternal Milk[a]			Total Maternal[c]	Yearling Weight	
Sire	EPD	ACC	EPD	ACC	EPD	ACC	DTS[b]	EPD	EPD	ACC
A	−1	0.95	+20	0.95	+1	0.93	331	+11	+45	0.94
B	+9	0.95	+47	0.95	−17	0.89	88	6	+74	0.90
C	−7	0.83	−12	0.85	−10	0.79	29	−16	−4	0.85
D	+1	0.95	+22	0.95	+3	0.92	230	+14	+54	0.94
E	1	0.73	+15	0.70	+12	0.42	0	+20	+45	0.66

[a] Milk EPD is measured by comparing calf weaning weights from daughters of bulls.
[b] DTS is daughters.
[c] Calculated by taking $1/2$ of weaning weight (direct) + milk EPD.

a promising young sire with an excellent combination of EPDs. However, the ACC is relatively low and could significantly change with more progeny and as daughters start producing.

It should also be noted that while EPDs are very effective selection tools for within-breed comparisons, comparing expected progeny differences across breeds is more problematic. Conversion tables are available to assist breeders who desire to make across-breed EPD comparisons.

Bull selection becomes more complex in that desired genetic improvement is typically viewed as creating a combination of several traits. "Stacking pedigrees" for only one trait may result in some problems in other traits. A good example is selecting for yearling weight alone, which results in increased birth weight because the two traits are genetically correlated. Birth weight is associated with calving difficulty; so increased birth weight might be a problem. Large increases in birth EPD can occur and calving difficulty can be serious.

The most difficult challenge in bull selection is selecting bulls that will improve maternal traits. Frequently, too much emphasis is placed on yearling growth, frame size, and large mature size. These traits can be antagonistic to maternal traits such as birth weight, early puberty, and maintaining a cow size consistent with an economical feed supply. Figure 25.5 shows an example of stacking a pedigree for maternal traits for yearling bull selection where the bull is used naturally on heifers. If an older bull can be used through AI, then emphasis would be on accuracies higher than those available on young bulls.

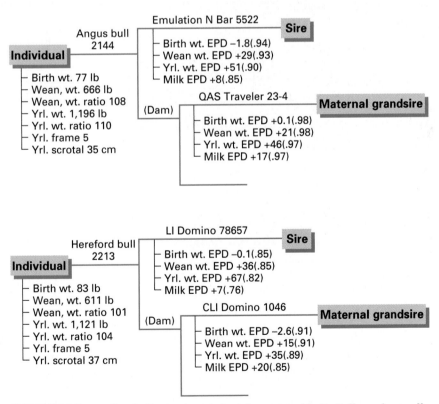

FIGURE 25.5 Yearling bull selection emphasizing maternal traits. Bulls used naturally on replacement heifers. *Source:* Colorado State University.

TABLE 25.5   Bull-Selection Criteria for Commercial Beef Producers

Goals and Maternal Traits	Bull-Selection Criteria for Maternal Traits[a]	Selection Criteria for Terminal Cross Bulls[b]
Breeding Program Goals 1. Select for an **optimum** combination of **maternal traits** to **maximize profitability** (avoiding genetic antagonisms and environmental conflicts that come with maximum production or single-trait selection) 2. Provide genetic input so **cows** can be **matched with their environment**—cows that wean more lifetime pounds of calf without overtaxing the forage, labor, or financial resources 3. Stack pedigrees for maternal traits **Maternal Traits** **Mature weight:** Medium-sized cows (1,000–1,250 lb under average feed supply) **Milk production:** Moderate (wean 500–550-lb calves under average feed) **Body condition** ("fleshing ability"): "5"condition score at calving without high-cost feeding **Early puberty/high conception:** Calve by 24 months of age **Calving ease:** Moderate birth-weights (65–75-lb heifers; 75–90-lb cows). Calf shape (head, shoulders, hips), which relates to unassisted births **Early growth and composition:** Rapid gains—relatively heavy weaning and yearling weights within medium (4–6) frame size. Yield grade 2 (steers slaughtered at 1,150 lb) **Functional traits (longevity):** Udders (shape, teats, pigment); eye pigment; disposition; structural soundness	**Yearling frame size:** 4.0–6.0 (smaller frame size will adapt better to harsher environments—e.g., less feed, more severe weather, less care) **Mature weight:** Under 2,000 lb in average condition (preference for future). Now we likely will have to consider bulls under 2,500 lb **Milk EPD:** 0 to +20 lb (prefer +10) **Maternal EPD:** +20 to +40 (prefer +25) **Body condition:** Backfat of 0.20 to 0.35 in. at yearling weights of 1,100–1,250 lb. Monitor reproduction of daughters **Scrotal circumference:** 34–40 cm at 365 days of age. Passed breeding soundness examination **Birth weight EPD:** Preferably under +1.0 lb. Evaluate calving ease of daughters **Weaning weight EPD:** Approximately +20 lb **Yearling weight EPD:** Approximately +40 lb **Accuracies:** All EPDs of 0.90 and higher (older, progeny-tested sires). Young bulls with accuracies below 0.90—evaluate trait ratios and pedigree EPDs to predict mature bulls with above-average EPDs and accuracies. Select sons of bulls that meet EPDs listed above **Functional traits:** Visual evaluation of bull and his daughters **Visual:** Sufficiently attractive to sire calves that would not be economically discriminated against in the marketplace Preference for "adequate middle" as medium-frame-size cattle need middle for feed capacity	**Yearling frame size:** 6.0–8.0 (should be evaluated with frame size of cows so slaughter progeny will average 5.0 and 6.0) **Mature weight:** No upper limit as long as birth weight and frame size are kept in desired range **Milk EPD:** Not considered **Maternal EPD:** Not considered **Body condition:** Evaluate with body condition of cows so that progeny will be between 0.25 and 0.45 in at 1,100–1,300-lb slaughter weights **Scrotal circumference:** 32–38 cm at 365 days of age. Passed breeding soundness exam **Birth weight EPD:** Preferably no higher than +5.0; want calves birth weights in 85–95-lb range with few, if any, over 100 lb **Weaning weight EPD:** +40 lb or higher **Yearling weight EPD:** +60 lb or higher **Accuracies:** All EPDs of 0.90 and higher (older, progeny-tested sires). Young bulls with accuracies below 0.90—evaluate trait ratios and pedigree EPDs to predict mature bulls with above average EPDs and accuracies. Select sons of bulls that meet EPDs listed above **Visual:** Functional traits: Disposition and structural soundness Sufficiently attractive to sire calves that would not be economically discriminated against in the marketplace

[a] EPDs are for British breeds.
[b] EPDs are for Continental breeds.

Table 25.5 identifies bull-selection criteria that would apply to a large number of commercial producers. First, the breeding program goals and the maternal traits are identified. Second, the bull-selection criteria are identified, giving emphasis to the maternal traits. These traits would receive emphasis if a commercial producer were using only one breed or a rotational crossbreeding program or in producing the cows used in a terminal crossbreeding program. Finally, Table 25.5 identifies the traits emphasized in selecting terminal cross

TABLE 25.6 Replacement Heifer Selection Guidelines at the Different Productive Stages

Stage of Heifer's Productive Life	Emphasis on Productive Trait	
	Primary	Secondary
Weaning (7–10 months of age)	Cull only the heifers whose actual weight is too light to prevent them from showing estrus by 15 months of age. Also consider the economics of the weight gains needed to have puberty expressed. Cull heifers that are too large in frame and birth weight.	Weaning weight ratio Weaning EPD Milk EPD Predisposition to fatness Moderate skeletal frame Skeletal soundness Docility
Yearling (12–15 months of age)	Cull heifers that have not reached the desired target breeding weight (e.g., minimum of 650–750 lb for small- to medium-sized breeds or cross, minimum of 750–850 lb for large-sized breeds and crosses). Cull heifers that are extreme in size and weight.	Milk and weaning weight EPDs Yearling weight ratio Yearling EPD Predisposition to fatness Moderate skeletal frame Skeletal soundness Docility
After breeding (19–21 months of age)	Cull heifers that are not pregnant and those that will calve in the latter one-third of the calving season.	Milk and weaning weight EPDs Yearling weight ratio Yearling EPD Predisposition to fatness Moderate skeletal frame Skeletal soundness Docility
After weaning First calf (31–34 months of age)	Cull to the number of first-calf heifers actually needed in the cow herd based on the weaning weight performance of the first calf. Preferably all the calves from these heifers have been sired by the same bull. Select for early pregnancy.	

bulls. In selecting terminal cross sires, little emphasis is placed on maternal traits as no replacement heifers from this cross are kept in the herd.

SELECTING REPLACEMENT HEIFERS

Heifers, as replacement breeding females, can be selected for several traits at different stages of their reproductive lives. The objective is to identify heifers that will conceive early in the breeding season, calve easily, give a flow of milk consistent with the feed supply, wean heavy calves, and make a desirable genetic contribution to the calves' postweaning growth and carcass merits.

Beef producers have found it challenging to determine which young heifers will make the most productive cows. Table 25.6 shows the selection process that producers use to select the most productive replacement heifers. This selection process assumes that more heifers will be selected at each stage of production than the actual numbers of cows to be replaced in the herd. The number of replacement heifers that producers keep is based primarily on their cost of production and their market value at various stages of production. More heifers than the number needed should be kept through pregnancy-check time. Heifers are selected on the basis that they become pregnant early in life primarily for economic reasons rather than expected genetic improvement from selection.

TABLE 25.7   Performance Data on High- and Low-Producing Cows in the Same Herd

Cow Number	Number of Calves	Weaning Weight (lb)	Weaning Weight Ratio	Yearling Weight Ratio
1	9	572	105	102
2	9[a]	464	89	94

[a] One calf died before weaning (not computed in averages).

COW SELECTION

Cows should be culled from the herd based on the productivity of their calves and additional evidence that they can be productive the following year, such as early pregnancy and soundness of udders, eyes, skeleton, and teeth. Productivity of cows is measured by pregnancy test, weaning and yearling weights (ratios) of their calves, and the EPDs of the cows. Table 25.7 shows the weaning and yearling values of the high- and low-producing cows in a herd. Cow 1 is a moderately high-producing cow, whereas cow 2 is the low-producing cow. Cow 2 should be culled and replaced with a heifer of higher breeding potential.

CROSSBREEDING PROGRAMS FOR COMMERCIAL PRODUCERS

Most commercial beef producers use **crossbreeding** programs to take advantage of **heterosis** in addition to the genetic improvement from selection. A crossbreeding system should be determined according to which breeds are available and then adapted to the commercial producers' feed supply, market demands, and other environmental conditions. A good example of adaptability is the Brahman breed, which is more heat and insect resistant than most other breeds. Because of this higher resistance, the level of productivity (in the southern and Gulf regions of the United States) is much higher for Brahmans and Brahman crosses.

Most commercial producers travel 150 or fewer miles to purchase bulls used in natural mating. Therefore, a producer should assess the breeders with excellent breeding programs in a 150-mile radius, as well as available breeds. This assessment, in most cases, should be determined before planning which breeds to use and in which combination to use them.

Breeds should be chosen for a crossbreeding system based on how well the breeds complement each other. Table 25.8 gives some comparative rankings of the major beef breeds for productive characteristics. Although the information in Table 25.8 is useful, it should not be considered the final answer for decisions on breeds to use. First, a producer needs to recognize that this information reflects breed averages; therefore, there are individual animals and herds of the same breed that are much higher or lower than the ranking given (Figs. 25.1–25.3). Producers need to use some of the previously described methods to identify superior animals within each breed. Second, it should be recognized that these average breed rankings can change with time, depending on the improvement programs used by the leading breeders within the same breed.

TABLE 25.8   Breed Crosses Grouped in Biological Types on Basis of Four Major Criteria

Breed Group[b]	Traits Used to Identify Biological Types[a]			
	Growth Rate and Mature Size	Lean-to-Fat Ratio	Age at Puberty	Milk Production
Jersey	X	X	X	XXXXX
Longhorn	X	XXX	XXX	XX
Hereford-Angus	XXX	XX	XXX	XX
Red Poll	XX	XX	XX	XXX
Devon	XX	XX	XXX	XX
Shorthorn	XXX	XX	XXX	XXX
Galloway	XX	XXX	XXX	XX
South Devon	XXX	XXX	XX	XXX
Tarentaise	XXX	XXX	XX	XXX
Pinzgauer	XXX	XXX	XX	XXX
Brangus	XXX	XX	XXXX	XX
Santa Gertrudis	XXX	XX	XXXX	XX
Beefmaster	XXX	XX	XXXX	XX
Sahiwal	XX	XXX	XXXXX	XXX
Brahman	XXXX	XXX	XXXXX	XXX
Nellore	XXXX	XXX	XXXXX	XXX
Brown Swiss	XXXX	XXXX	XX	XXXX
Gelbvieh	XXXX	XXXX	XX	XXXX
Holstein	XXXX	XXXX	XX	XXXXXX
Simmental	XXXXX	XXXX	XXX	XXXX
Maine Anjou	XXXXX	XXXX	XXX	XXX
Salers	XXXXX	XXXX	XXX	XXX
Piedmontese	XXX	XXXXXX	XX	XX
Limousin	XXX	XXXXX	XXXX	X
Charolais	XXXXX	XXXXX	XXXX	X
Chianina	XXXXX	XXXXX	XXXX	X

[a] Breed crosses grouped into several biological types based on relative differences (X = lowest, XXXXXX = highest).
[b] Breed group based on these breeds of sires mated to Hereford and Angus cows.
Source: Adapted from *MARC Research Progress Reports.*

Obviously, those traits that have high heritabilities would be expected to change most rapidly, assuming the same selection pressure for each trait. A careful analysis of the information in Table 25.8 shows that no one breed is superior for all important productive characteristics. This gives an advantage to commercial producers using a crossbreeding program if they select breeds whose superior traits complement each other. An excellent example of breed complementarity is shown by the Angus and Charolais breeds, which complement each other in both quality grade and yield grade.

Most of the heterosis achieved in cattle as a result of crossbreeding is expressed by weaning time. The cumulative effect of heterosis on pounds of calf weaned per cow exposed is shown in Figure 25.6, in which maximum heterosis is obtained when crossbred calves are born from crossbred cows. The traits that express an approximate heterosis of a 20% increase in pounds of calf weaned per cow exposed to breeding are early puberty of crossbred heifers, high conception rates in crossbred females, high survival rate of calves, increased milk production of crossbred cows, and a higher preweaning growth rate of crossbred calves.

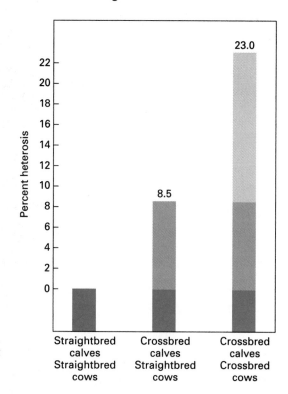

FIGURE 25.6 Heterosis resulting from crossbreeding for pounds of calf weaned per cow exposed to breeding. *Source:* USDA.

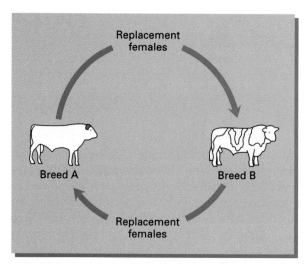

FIGURE 25.7 Two-breed rotation cross. Females sired by breed A are mated to breed B bulls, and heifers sired by breed B are mated to breed A bulls. This will increase the pounds of calf weaned per cow bred by approximately 15%. *Source:* Colorado State University.

Consistently high levels of heterosis (10–20%) can be maintained generation after generation if crossbreeding systems such as those shown in Figures 25.7–25.9 are used. The crossbreeding system shown in Figure 25.9 combines a two-breed rotation with a terminal cross. In this system, the two-breed rotation is used

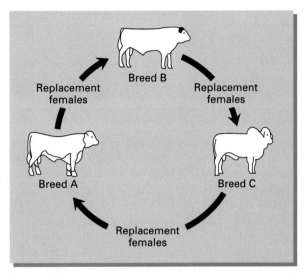

FIGURE 25.8 Three-breed rotation cross. Females sired by a specific breed are bred to the breed of the next bull in rotation. This will increase the pounds of calf weaned per cow by approximately 20%. *Source:* Colorado State University.

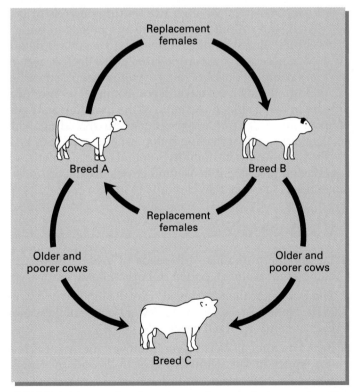

FIGURE 25.9 Two-breed rotation and terminal site crossbreeding system. Sires are used in the two-breed rotation primarily to produce replacement heifers. Terminal cross sires are mated to the less productive females. This system will increase the pounds of calf weaned per cow bred by more than 20%. *Source:* Colorado State University.

TABLE 25.9   Heritability and Heterosis for the Major Beef Cattle Traits

Traits	Heritability	Heterosis
Reproduction	Low[a]	High
Growth	Medium	Medium
Carcass	High	Low

[a] Exceptions are age at puberty, scrotal circumference, and birth weight (dystocia). They are highly heritable.

primarily to produce replacement females for the entire cowherd. In most cowherds, approximately 50% of the cows are bred to sires to produce replacement females, with the remaining 50% being bred to terminal cross sires. All terminal cross calves are sold. This crossbreeding system maintains heterosis levels as high as those in the three-breed rotation system.

In the rotational crossing, breeds with maternal-trait superiority (high conception, calving ease, and milking ability) would be selected. The terminal cross sire could come from a larger breed where growth rate and carcass cutability are emphasized. A primary advantage of the rotational-terminal cross system is that smaller- or medium-sized breeds can be used in rotational crossing and a larger breed could be used in terminal crossing. Terminal crossbreeding systems will maximize heterosis at approximately 25–30%. Some of the systems that will fit a one-breeding pasture program include (1) using a composite (synthetic) breed, (2) rotating two or three breeds of bulls (each breed every 3–5 years), and (3) putting multiple-sire breeds of bulls with crossbred cows. Each of these crossbreeding systems will maintain heterosis at approximately 10–15%. These systems may be more cost-effective and profitable than some of the more complex crossbreeding systems discussed earlier.

Table 25.9 shows the advantage a commercial producer has over a purebred breeder in being able to use more of the breeding methods for genetic improvement. Commercial producers can use crossbreeding, whereas purebred breeders cannot use crossbreeding if they wish to maintain breed purity.

Traits with a low **heritability** respond little to genetic selection, but they show a marked improvement in a sound crossbreeding program. The commerical producer needs to select sires carefully to improve the traits with a high heritability.

CHAPTER SUMMARY

- Phenotypic differences in breeds of beef cattle are primarily expressed as color (color pattern), polled or horned head, mature size, and muscling.

- There are more than 250 breeds of cattle in the world, with more than 60 breeds in the United States. However, fewer than 20 of these breeds make a significant impact on the national herd.

- Based on annual registration numbers, the Angus breed is the most numerous breed in the United States, followed by Hereford/Polled Hereford, Limousin, Beefmaster, and Charolais.

- Genetic improvement occurs by selecting for superiority (optimum combination) for the following traits: (1) reproductive performance, (2) weaning weight, (3) postweaning growth, (4) feed efficiency, (5) carcass merit (weight, yield grade, and quality grade), and (6) longevity.

- Effective bull selection, primarily through EPDs, accounts for 80–90% of the genetic improvement in a herd over a period of several years.

- It is economically important to select the biological type (mature weight, milk production, age at puberty, and carcass composition) that matches cows to a low-cost forage environment and their progeny to the marketplace.

- Optimum levels of heterosis are obtained by using crossbreeding or composite breeds.

KEY WORDS

breed
composite breeds
synthetic breeds
purebred
reproductive
 performance
scrotal circumference
weaning weight
postweaning growth

feed efficiency
carcass merit
longevity
conformation
genetic defects
double muscling
syndactyly
arthrogryposis
osteopetrosis

hydrocephalus
dwarfism
expected progeny
 difference (EPD)
accuracy (ACC)
crossbreeding
heterosis
heritability

REVIEW QUESTIONS

1. What is a breed of cattle?

2. What does *purebred* refer to?

3. *True or False:* A herd of purebred cattle, such as Angus or Herefords, which are particularly uniform in one or two characteristics, such as color or color pattern, will be highly homozygous.

4. What were the three major breeds of beef cattle in the United States during the early 1900s?

5. What are three reasons for the large increase in beef cattle breeds in the United States during the second half of the 20th century?

6. What are the four most important breeds of beef cattle in the United States today according to breed registration numbers?

7. When does the most genetic improvement in beef cattle occur?

8. What are the eight economically important traits of beef cattle?

9. Which trait of beef cattle has the greatest economic importance?

10. To what segment of the beef industry is reproductive performance especially important?

11. *True or False:* Reproductive performance is low in heritability so selection is relatively ineffective for making genetic progress in this trait.

12. What are the best ways to improve the reproductive performance of beef cattle?

13. What does weaning weight reflect?

14. To what segment of the beef industry is weaning weight especially important?

15. What does postweaning growth measure?

16. To which segments of the beef industry is postweaning growth especially important?

17. Producers can make genetic improvement in feed efficiency by selecting for what related trait?

18. How is carcass merit presently measured?

19. *True or False:* Producers can achieve genetic improvement by selecting for carcass merit because the heritability of the traits that affect carcass merit is high.

20. What does longevity measure?

21. To which segment of the beef industry is longevity especially important?

22. What is conformation?

23. *True or False:* Bull selection must receive the greatest emphasis for optimum genetic improvement of the herd to be achieved.

24. Breeding values of beef cattle are most frequently reported as what?

25. *True or False:* Most heterosis in crossbred beef cattle is expressed in traits measured after the age of weaning.

26. What type of crossbreeding system will maximize heterosis in a herd of beef cattle?

27. What types of traits respond most to crossbreeding?

SELECTED REFERENCES

Boggs, D. 1992. *Understanding and Using Sire Summaries.* Beef Improvement Federation, BIF-FS3.

Bull and Heifer Replacement Workshops (proceedings). 1990. Fort Collins, CO: Colorado State University.

Field, T. G. 2007. *Beef Production and Management Decisions.* Upper Saddle River, NJ: Prentice Hall.

Gibb, J., M. V. Boggess, and W. Wagner. 1992. *Understanding Performance Pedigrees.* Beef Improvement Federation, BIF-FS2.

Gregory, K. E., and L. V. Cundiff. 1980. Crossbreeding in beef cattle: Evaluation of systems. *J. Anim. Sci.* 51:1224.

Hickman, C. G. (ed.). 1991. *Cattle Genetic Resources.* Amsterdam: Elsevier Science Publishers.

Jarrige, R., and C. Berauger (eds.). 1992. *Beef Cattle Production.* Amsterdam: Elsevier Science Publishers.

Kress, D. D. 1989. Practical breeding programs for managing heterosis. *The Range Beef Cow Symposium XI.*

Legates, J. E. 1990. *Genetics of Livestock Improvement.* Englewood Cliffs, NJ: Prentice Hall.

Middleton, B. K., and J. B. Gibb. 1991. An overview of beef cattle improvement programs in the United States. *J. Anim. Sci.* 69:3861.

Silcox, R., and R. McGraw. 1992. *Commercial Beef Sire Selection.* Beef Improvement Federation, BIF-FS9.

CHAPTER 26

Feeding and Managing Beef Cattle

LEARNING OBJECTIVES

- Discuss the cyclic nature of profitability in the beef industry
- Describe the primary management approaches to optimizing percent calf crop and weaning weight
- List the primary factors affecting cost of production
- Explain the use of break-even price discovery
- Compare and contrast commercial and farmer-feeders
- Discuss profitability in the feedyard business
- Discuss the role of environmental management in cattle feeding

An overview of the beef industry and beef production is presented in Chapter 2. This chapter will identify the primary factors affecting beef cattle productivity and profitability.

COW-CALF MANAGEMENT

The effective and profitable management of a cow-calf enterprise is founded on the ability of a manager to allocate time and resources to critical activities. Each enterprise in the cattle industry has its own unique strengths and limitations thus making "one size fits all" recommendations impossible. However, a generalized management calendar for a cow-calf business is provided in Table 26.1.

Cow-calf producers are interested in managing their operations to generate profits while building a business that is capable of providing opportunities to others. One approach to assessing the profitability of a commercial cow-calf operation is by analyzing the following criteria: (1) **calf crop percentage** weaned (e.g., number of calves produced per 100 cows in the breeding herd), (2) average

TABLE 26.1　Management Calendar for a Selected Cow-Calf Operation with Spring and Fall Calving Herds[a]

Major Activities	Jan	Feb	Mar	Apr	May	June	July	Aug	Sept	Oct	Nov	Dec
Planning, assessing goals, record analysis	xxx x	xxx x	xxxx	xxxx	xxxx	xxxx	xxx x	xxx x	xxx x	xxx x	xxx x	xxx x
Calving		ss	ssss	ss				ff	ffff	f		
Breeding preparation (semen, equipment, bulls)			s						f			
Vaccinate cows (Leptospirosis & Vibriosis)	s						f					
Vaccinate heifers (*Clostridium c*, IBR, Lepto, BVD)				s					f			
Synchronize heifers				s					f			
Synchronize cows				s					f			
Breed heifers					sss	ssss					ffff	ff
Breed cows					ss	ssss	ss				fff	ff
Brand, castrate, dehorn calves				s					f			
Vaccinate calves (Blackleg, Malignant Edema)				s					f			
Pregnancy test		f							s			
Fertilize pastures				x								
Control pinkeye						xxxx	xxx x	xxx x				
Fly control tags						x						
Wean calves			f							s		
Process weaning weights			f							s		
Vaccinate at weaning (IBR, PI$_3$, Leptospirosis)			f							s		
Retag cows		f	f						s	s		
Grub control (cows & bulls)										x		
Grub control (calves)										s		
Vaccinate heifers for brucellosis				f							s	
Cull cows (sell or feed)					f					s		
Vaccinate cows (*Clostridium c*)	s						f					
Select replacement heifers (weaning, yearling, and pregnancy test)		f	s	f					sf	s		
Increase precalving level of feed	s											
Keep facilities and equipment repaired	x	x	x	x	x	x	x	x	x	x	x	x
Income tax preparation and filing	x											
Income tax evaluation and adjustment										x		
Cash-flow update and evaluation	x	x	x	x	x	x	x	x	x	x	x	x
Check salt and mineral supply	x	x	x	x	x	x	x	x	x	x	x	x
Evaluation target weights for replacement heifers		s				f		f				s
Evaluate management plan	x	x	x	x	x	x	x	x	x	x	x	x

TABLE 26.1 *(continued)*

Major Activities	Jan	Feb	Mar	Apr	May	June	July	Aug	Sept	Oct	Nov	Dec
Spend time planning, both short- and long-term	x	x	x	x	x	x	x	x	x	x	x	x
Check water supply	xxx	xxx	xxxx	xxxx	xxxx	xxxx	xxx	xxx	xxx	xxx	xxx	xxx
	x	x					x	x	x	x	x	x
Attend educational conference	x	x									x	x
Plan next year's budget and cash-flow statement										x		

ªs = spring calving herd, f = fall calving herd, x = applies to both herds. Placement of letter within the month shows the week designation. Type of activities and timing of activities may be different for other operations.

TABLE 26.2 Break-Even Price (per 100 lb) for Commercial Cow-Calf Operations with Varying Calf Crop Percentages, Annual Cow Costs, and Weaning Weights

Calf Crop Percent Weaned	Annual Cow Cost	Break-even Prices for Calves of Different Weaning Weights		
		400 lb	500 lb	600 lb
95	$350	$92.10	$73.68	$61.40
95	300	78.95	63.16	52.63
95	250	65.79	52.63	43.85
85	350	102.94	82.35	68.63
85	300	88.24	70.59	58.82
85	250	73.53	58.25	49.02
75	350	116.67	93.33	77.78
75	300	100.00	80.00	66.67
75	250	83.33	66.67	55.56

weight of calves at weaning (7–9 months of age), and (3) annual cow cost (dollars required to keep a cow each year).

An example of the economic assessment of a commercial cow-calf producer who has an 85% calf crop weaned, 500-lb weaning weights, and a $300 annual cost would be as follows:

Calf crop % (0.85) × weaning weight (500 lb) = 425 lb of calf weaned per cow in the breeding herd

Annual cow cost ($300) ÷ lb of calf weaned (425 lb) = $70.59 per hundredweight

A break-even price of $70.59 per hundredweight means that the producer would have to receive more than $0.70 per pound, or $70.59 per 100 lb of calf sold, to cover the yearly cost of maintaining each cow in the breeding herd. Table 26.2 illustrates the break-even price for several levels of calf crop percentage, weaning weight, and annual cow cost. This information reflects different management levels in which the break-even price ranges from more than $1.00 per pound to less than $0.50 per pound. Profitability of the commercial cow-calf operation is determined by comparing the market price of the calves at the time they are sold with the break-even price.

Cow-calf operations are managed best by operators who know the factors that affect calf crop percentage (Fig. 26.1), weaning weight, and annual cow cost. The primary management objective should be to generate profits via improvement in pounds of calf weaned per cow and reduction or control of the annual cow costs.

Costs and Returns

Figure 26.2 shows the dollars returned over cash costs for an average U.S. cow-calf operation. Note the wide deviation in net returns over the 20-year period. Also notice the general inverse relationship between cattle inventory numbers and returns over costs. During 1986 to 1993, higher prices for calves resulted in higher levels of profitability; however, returns for 1995 to 1998 are negative. As supplies came back into line with demand, cow-calf profits increased beginning in 1999.

FIGURE 26.1   Percent calf crop, as measured by a live calf born and raised per cow, is the most economically important trait for the cow-calf producer. The bull, cow-calf, and producer each make a meaningful contribution to the level of productivity for this trait.

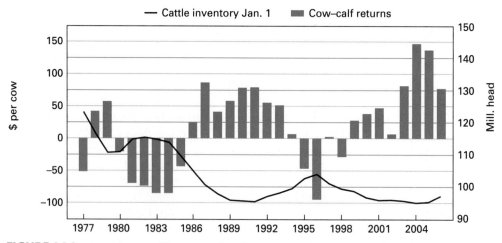

FIGURE 26.2   Annual cow–calf returns and cattle inventory. *Source: Livestock Marketing Information Center.*

Costs and returns are computed from an **enterprise budget** (sometimes called an **enterprise analysis**). The enterprise budget is one of the most useful financial forms upon which to make management decisions.

The component of receipts and expenses for an average cow-calf enterprise are shown in Table 26.3. Notice that as weaning percent increases, the amount of returns needed to cover costs (break-even price) is reduced.

TABLE 26.3   Cost-Return Projection—Beef Cow-Calf Enterprise (per cow)

	Weaning Percentage		
	82%	88%	94%
RETURNS PER COW:			
1. Steers: 560 lbs	$ 190.52	$ 204.46	$ 218.40
2. Heifers: 540 lbs	168.62	180.96	193.29
3. Cull Cows: 1,050 lbs × 16%	74.78	74.78	74.78
4. Other			
A. Gross Returns per Cow	$ 433.92	$ 460.20	$ 486.47
5. Summer Pasture (6 months)	$ 119.23	$ 119.23	$ 119.23
6. Crop Residue	7.50	7.50	7.50
7. Hay—Forage	102.38	102.38	102.38
8. Grain			
9. Protein and Mineral	25.50	25.50	25.50
10. Labor	50.00	50.00	50.00
11. Veterinary, Drugs, and Supplies	16.00	16.00	16.00
12. Marketing Costs	10.00	10.00	10.00
13. Utilities, Fuel, and Oil	22.20	22.20	22.20
14. Facilities and Equipment Repairs	29.00	29.00	29.00
15. Breeding Charge			
a. Capital Replacement (16% of Heifer Calves)	65.80	65.80	65.80
b. Annual Bull Cost or A.I. Charge	12.00	12.00	12.00
c. Interest on Breeding Stock	61.12	61.12	61.12
d. Insurance on Breeding Stock	7.64	7.64	7.64
16. Professional Fees (legal, accounting, etc.)	1.00	1.00	1.00
17. Miscellaneous	6.00	6.00	6.00
18. Depreciation on Facilities and Equipment	$ 12.39	$ 12.39	$ 12.39
19. Interest on Facilities and Equipment[1]	13.95	13.95	13.95
20. Insurance and Taxes on Facilities and Equipment	4.52	4.52	4.52
B. Subtotal	$ 566.24	$ 566.24	$ 566.24
21. Interest on 1/2 Operating Costs @ 8%	15.55	15.55	15.55
C. Total Costs per Cow	$ 581.79	$ 581.79	$ 581.79
D. Return Over Total Costs (A–C)	$ −147.88	$ −121.60	$ −95.32
22. Cwt. Weaned	4.51	4.84	5.17
E. Average Gross Return Needed/Cwt.			
23. To Cover Total Costs	$ 112.42	$ 104.76	$ 98.07
24. To Cover Feed Costs	$ 56.45	$ 52.62	$ 49.25
F. Asset Turnover (D ÷ Investment)[2]	42.44%	45.01%	47.58%
G. Net Return on Investment ((D + 15c + 9 + 21) ÷ Investment)[2]	−5.60%	−3.03%	−0.46%

[1]Original cost of facilities and equipment plus salvage value divided by 2, times at an interest rate of 8%.
[2]Investment equals total value of breeding stock and facilities-equipment.

MANAGEMENT FOR OPTIMUM CALF CROP PERCENTAGES

The primary management factors affecting calf crop percentages are as follows:

1. Heifers need to be fed adequate levels of a balanced ration to reach puberty at 15 months of age if they are to calve at the desired age of 2 years. Medium-frame-sized heifers of English breeds and crosses (e.g., Angus and Hereford) should weigh 650–750 lb at 15 months of age. Cattle of larger-frame-sized exotic breeds or crosses should weigh 100–150 lb more to ensure a high percentage of heifers cycling at breeding.

2. Heifers should be bred to calve early in the calving season. Heifers calving early are more likely to be pregnant as 2- and 3-year-olds, whereas heifers calving late will likely not conceive during the next breeding season. Some producers save more heifers as potential replacements at weaning and as yearlings so that selection for early pregnancy can be made at pregnancy test time.

3. Heifers typically have a longer postpartum interval than cows. This interval becomes shorter if first-calf heifers (heifers with their first calves) are separated from mature cows 60 days prior to calving and after calving. This separation allows the heifers to obtain their share of the feed essential for a rapid return to estrus.

4. Feeding programs are designed to have cows and heifers in a moderate body condition (visually estimated by the fat over the back and ribs) at calving time. Table 26.4 illustrates a **body condition scoring (BCS)** system. Cows that are thin at calving usually have longer postpartum intervals (Fig 26.3). Cows that are too fat reflect higher feed inputs and thus higher costs than is necessary for profitable production.

5. Cows, particularly first-calf heifers, should be observed every few hours at calving time. Some females will have difficulty calving **(dystocia)** and will need assistance in delivery of the calf. Calving difficulties should be kept to a minimum to prevent potential death of calves and cows. Calving difficulty is also undesirable because cows given assistance will usually have longer postpartum intervals.

6. Calving difficulty should be minimized but usually cannot be eliminated. A balance should be maintained between the number of calves born alive and the weight of the calves at weaning. Calves that are heavier at birth usually have heavier weaning weights. When calves are too heavy at birth, however, the death loss of the calves increases. Conversely, calves that are too light at birth often have increased death rates as well. Birth weight is the primary cause of calving difficulty; therefore, management decisions should be made to keep birth weights moderate. Bulls of the larger breeds or larger frame sizes should not be bred to heifers, and large, extremely high growth bulls, even in the breeds known for calving ease, should not be bred to heifers. Birth weight within a herd is influenced by genetics. Genetic differences are more important than certain environmental differences such as amount of feed during gestation. Bulls to be used artificially should have extensive progeny test records for birth weight and calving ease in addition to an individual birth-weight record.

TABLE 26.4 System of Body Condition Scoring (BCS) for Beef Cattle

Group	BCS	Description
Thin condition	1	*Emaciated*—Cow is extremely emaciated, with no palpable fat detectable over spinous processes, transverse processes, hip bones, or ribs. Tail-head and ribs project quite prominently.
	2	*Poor*—Cow still appears somewhat emaciated but tail-head and ribs are less prominent. Individual spinous processes are still rather sharp to the touch, but there is some tissue cover over dorsal portion of ribs.
	3	*Thin*—Ribs are still individually identifiable but not quite as sharp to the touch. There is obvious palpable fat along spine and over tail-head, with some tissue cover over dorsal portion of ribs.
Borderline condition	4	*Borderline*—Individual ribs are no longer visually obvious. The spinous processes can be identified individually on palpation but feel rounded rather than sharp. There is some fat cover over ribs, transverse processes, and hip bones.
Optimum moderate condition	5	*Moderate*—Cow has generally good overall appearance. On palpation, fat cover over ribs feels spongy and areas on either side of tail-head now have palpable fat cover.
	6	*High moderate*—Firm pressure now needs to be applied to feel spinous processes. A high degree of fat is palpable over ribs and around tail-head.
Fat condition	7	*Good*—Cow appears fleshy and obviously carries considerable fat. There is very spongy fat cover over ribs and around tail-head. In fact, "rounds" or "pones" are beginning to be obvious. There is some fat around vulva and in crotch.
	8	*Fat*—Cow is very fleshy and overconditioned. Spinous processes almost impossible to palpate. Cow has large fat deposits over ribs and around tail-head, and below vulva. "Rounds" or "pones" are obvious.
	9	*Extremely fat*—Cow is obviously extremely wasty and patchy and looks blocky. Tail-head and hips are buried in fatty tissue and "rounds" or "pones" of fat are protruding. Bone structure is no longer visible and barely palpable. Animal's mobility might even be impaired by large fatty deposits.

Source: Richards et al., 1986. *J. Anim. Sci.* 62:300.

FIGURE 26.3 The cow on the left is a body condition score (BCS)4, whereas the cow on the right is a BCS 6. Cows with BCSs below 5 have a longer interval (days) between calving and pregnancy compared to cows having BCSs of 5 and higher. Courtesy of Oklahoma State University.

7. The bull's role in affecting pregnancy rate has a marked influence on calf crop percentage. Before breeding, bulls should be evaluated for breeding soundness by addressing physical conformation and skeletal soundness, palpating the genital organs, measuring scrotal circumference, and testing the semen for motility and morphology. **Libido** (sex drive) and mating capacity are additional important factors affecting the bull's influence on pregnancy rate. These traits are not easily measured in individual bulls

before breeding or even after breeding in multiple-sire herds. The typical cow-to-bull ratio utilized by most cattle producers is 30:1. However, some bulls can settle more than 50 cows in a 60-day breeding season. In some large pastures with rough terrain, the cow-to-bull ratio might have to be lower to assure a high calf crop percentage.

8. Crossbreeding affects calf crop percentage in several ways. Crossbred heifers usually cycle earlier and have higher conception rates than their straight bred counterparts. Crossbred calves are more vigorous and have a higher survival rate. An effective crossbreeding program can increase the calf crop by 8–12%.

9. The primary nutritional factor influencing calf crop percentage is adequate intake of energy, which can be expressed in pounds of total digestible nutrients (TDN). The quantity of TDN intake is important in helping to initiate puberty, maintaining proper body condition at calving, and keeping the postpartum interval relatively short. Other nutrients of major importance are protein, calcium, and phosphorus. Additional vitamins and minerals are important only in areas where the soil or feed is deficient.

10. Calf losses during gestation are usually low (2–3%) unless certain diseases are present in the herd. Serious reproductive diseases such as brucellosis, leptospirosis, vibriosis, and infectious bovine rhinotracheitis (IBR) can cause abortions, which may markedly reduce the calf crop percentage. These diseases can be managed by blood-testing animals entering the herd or by vaccinating. Herd health programs vary for different enterprises, depending on the incidence of the diseases in the region. Details of these programs should be worked out with the local veterinarian.

11. Calf losses after 1–2 days following birth are usually small (2–3%) in most cow-calf operations. Severe weather problems, such as spring blizzards, can cause high calf losses where protection from the weather is limited. In certain areas and in certain years, health problems can also cause high death losses. Infectious calf scours and secondary pneumonia can occasionally reduce the potential calf crop by 10–30%.

MANAGEMENT FOR OPTIMUM WEANING WEIGHTS

The primary management factors affecting calf weaning weights are as follows:

1. Calves born early in the calving season are heavier at weaning primarily because they are older. Calves are typically born over a several-week period but are weaned together on one specified day. Every time the cow cycles during the breeding season and fails to become pregnant, the weaning weight of her calf is reduced by 30–40 lb. Most commercial producers limit the breeding season to 90 days or less so the calves are heavier at weaning and can be managed in uniform groups (Fig 26.4).

2. The amount of forage available to the cow and the calf has a marked influence on weaning weights. The cow needs feed to produce milk for the calf. The calf, after about 3 months of age, will consume forage directly in addition to the milk it receives.

FIGURE 26.4 This calf is approximately 7 months old and soon will be weaned from its dam. The calf is the result of mating a terminal sire bull to an F1 cross female to capture a high degree of heterosis. Courtesy of the American-International Charolais Association.

FIGURE 26.5 The squeeze chute and head catch is an essential piece of equipment for humanely restraining cattle. Insertion of ear tags, administration of medicine and growth promotants, and other management tasks are more easily accomplished as a result of the squeeze chute. Note that this animal is being restrained to allow the use of an ultrasound machine to measure backfat and rib-eye area.

3. **Growth stimulants**, commonly given to nursing calves, will increase the weaning weight by 5–15% (Fig. 26.5). Ralgro (Zeranol), Synovex C, and Compudose, common growth stimulants are implanted as pellets under the skin of the ear. The pellets dissolve over a period of several weeks and

supply the growth-stimulating substance, which is absorbed into the bloodstream. This implant, however, should not be used on breeding replacement bulls and heifers, because it sometimes interferes with the proper development and functioning of the reproductive organs. Recent studies have shown that use of certain implants and repeated use of implants may lower percent of cattle grading USDA choice and may decrease palatability of the final product. Producers should communicate with buyers of their cattle to determine the appropriate implant protocol.

4. Providing supplemental feed to the calves where it is inaccessible to the cows will increase the weaning weight of the calves. This practice of **creep feeding** should be used with caution because it is not always profitable. It helps calves make the transition of the weaning process and is a feasible practice under drought or marginal feed-supply conditions. Creep feeding of breeding heifer calves can impair the development of their mammary system and subsequently reduce milk production. This impairment is caused by fat accumulating in the udder and crowding the secretory tissue.

5. Diseases that affect the milk supply of the cow or growth of the calf will cause a reduction in the weaning weight of the calf (see Chapter 21).

6. Genetic selection for milk production and calf growth rate will typically increase calf weaning weight. Selection based on EPDs (weaning weight and milk) is the most effective way to change weaning weights in a herd genetically. Effective bull selection will account for 80–90% of the genetic improvement in weaning weight, although weaning-weight information can also be used in culling cows and in selecting replacement heifers. It has been well demonstrated that effective selection can result in an increase in weaning weight of 4–6 lb per year on a per-calf basis.

7. Crossbreeding for the average cow-calf producer can result in a 10–30% (average 20%) increase in pounds of calf weaned per cow exposed in the breeding herd. Most of the increase occurs due to improved reproductive performance; however, one-fourth to one-third of the 20% increase is due to the effect of heterosis on growth rate of the calf and increased milk production of the crossbred cow. The lifetime performance of crossbred females compared to straightbreds is considerably higher as a result of weaning a greater number of calves that tend to also be heavier.

MANAGEMENT FOR LOW ANNUAL COW COSTS

In times of inflationary economic conditions, it may be difficult for a producer to lower annual cow costs. Some producers manage to lower costs, while others keep costs at a level similar to past years. Adequate income and expense records must be maintained so that cost areas can be carefully analyzed (Fig. 26.6). An enterprise budget (Table 26.3) is needed to assess cow costs and make management decisions to lower cow costs.

Table 26.5 shows cow productivity, annual cow costs, and calf break-even prices for average, low-cost, and high-cost producers in the NCA-IRM-SPA database.

FIGURE 26.6 Cattle producers must keep accurate financial, forage, and herd productivity records to assess annual cow costs and the profitability of an enterprise.

TABLE 26.5   Cow-Calf Producer Profiles by Cost Group

	Cost Group		
	Average	Low One-Third	High One-Third
Percent calf crop	85%	83%	84%
Weaning weight	514 lb	527 lb	498 lb
Calf weight per cow exposed	438 lb	441 lb	418 lb
Cost per cow (financial) [a]	$377	$268	$490
Calf break-even price [a]	$0.86	$0.61	$1.17

[a]Before noncalf revenue adjustment.
Source: NCA-IRM-SPA database (1995 NCA Cattlemen's College).

TABLE 26.6   Comparing Major Differences in Net Income

Item	High Cost	Low Cost	Difference
Financial cost per cow	$490	$268	$222
Pounds weaned per cow	441	418	18[a]

[a]Weaning weight valued at $0.80 per pound.
Source: Adapted from NCBA-IRM-SPA database.

There is a difference of approximately $200 per cow in net income when comparing low-cost producers ($61/cwt breakeven) with high-cost producers ($117/cwt breakeven). Table 26.6 illustrates that most of this $200 difference is due to financial cost per cow. While pounds weaned per cow are important, its contribution to net income is far overshadowed by cost per cow. This difference shows producers where management decisions should first be focused.

Low-cost producers in the NCA-IRM-SPA database were requested to identify the primary factors that determined their low break-even prices (Table 26.7).

Even though these "top five" are ranked as most important, the other, or miscellaneous, costs should not be overlooked. These smaller expense items

TABLE 26.7   Top Five Ways Low-Cost Producers Reduce Costs

1. Reduced supplemental feed costs (40%[a])
2. Rotational grazing (better pasture management) (30%[a])
3. Right genetics (27%[a])
4. Reduced labor costs (25%[a])
5. Strong herd health program (19%[a])

[a]Percent of respondents.
Source: NCBA-IRM-SPA database (1995 NCBA Cattlemen's College).

FIGURE 26.7   Cornstalk fields are available to cattle after the gain has been harvested. There are millions of acres of stalk fields and other crop-aftermath feeds that can be grazed by cattle and other livestock. Courtesy of *BEEF*.

(e.g., fuel, machinery, repair, supplies, utilities, and taxes) account for less than 20% of an average budget. Yet, they make up over 40% of the difference in cow cost between high-cost and low-cost producers. These smaller expense items should be evaluated on a regular basis.

The greatest consideration should be given to **feed costs,** as they constitute the largest part of annual cow costs, usually 50–70%. The period from the weaning of a calf to the last one-third of gestation in the next pregnancy is the time when cows can be maintained on comparatively small amounts of relatively cheap, low-quality feeds. Cow-calf operations having available crop aftermath feeds (e.g., cornstalks, grain stubble, and straw) usually have the greatest opportunity to keep feed costs lower than other operations (Fig. 26.7). The most economical feed resource must be matched to the right biological type of cows—e.g., usually those that display early puberty, calving ease, moderate milk, moderate mature weight, and fleshing ability (BCS) for early rebreeding. In many operations, the most economical feed resource comes from maximizing grazed forage with cows receiving minimal amounts of harvested and purchased feed.

Labor costs usually compose 15–20% of the annual cow costs. Labor costs are usually lower on a per-cow basis as herd size increases and in areas where moderate weather conditions prevail. Typically 15–20 hours of labor per cow unit per year are required. Operations that use labor inefficiently require twice as much labor per cow.

Interest charges on operating capital account for another 10–15% of the annual cow cost. Producers can reduce interest charges by carefully analyzing the costs of different credit sources.

Cows and heifers should be palpated at approximately 45 days after the end of the breeding season to determine if pregnancy has occurred. The producer should consider all marketing alternatives to maximize profits when selling open cows. Failing to check cows for pregnancy contributes to higher annual cow costs, lower calf crop percentages, and higher break-even costs of the calves produced.

STOCKER-YEARLING PRODUCTION

Several alternate stocker-yearling production programs are identified in Chapter 2, Figure 2.5. Good producers understand the factors that affect the productivity and profitability of these various programs.

The primary factors affecting the costs and returns of stocker-yearling operations are marketing (both purchasing and selling the cattle), the gaining ability of the cattle, the amount and quality of available forage and roughage, and the health of the cattle. Table 26.8 shows that management decisions affecting purchase price and sale price have the greatest impact on profitability.

Stocker-yearling producers need to be aware of current market prices for the cattle they purchase or sell. They also need to understand the loss of weight of the cattle from the time of purchase to the time the cattle are delivered to their farm or ranch. This loss in weight, called **shrink,** can sometimes reflect the difference in the profit or loss of the stocker-yearling operation. It is common for calves and yearlings to shrink 3–12% from on-farm weight to delivered weight. For example, yearlings purchased at 700 lb that shrink 4% will have a delivered weight of 672 lb. It typically takes 2–3 weeks to recover the weight loss.

The gaining ability of most stocker-yearling cattle is estimated visually. Cattle that are lightweight for their age, thin but healthy, with a relatively large skeletal frame size, usually have a high gain potential. Cattle that are light for their age are typically most profitable for the stocker-yearling operator, whereas heavier cattle are most profitable for the cow-calf producer.

TABLE 26.8 Effect of a 10% Change on the Break-Even Price ($78/cwt) of a Stocker-Yearling Budget

Factor	Change (%)	Decrease in Break-even Price ($/cwt)	Increase in Profit ($/head)
Sale price	+10%	$ 0.00	$ 56.01
Purchase price	−10	6.95	48.63
Average gain	+10	2.44	18.01
Pasture cost	−10	0.59	4.10
Interest cost	−10	0.33	2.29
Death loss	−10	0.20	1.42
All factors	10	10.02	130.59

Source: Cattle-Fax, Englewood, CO.

Stocker-yearling cattle that are purchased and sold several times often encounter stress situations such as fatigue, hunger, thirst, and exposure to multiple disease organisms. The most common diseases include shipping fever complex and other respiratory diseases. These stress conditions make it necessary for stocker-yearling producers to have effective health programs for newly purchased cattle. Producers who have poor herd health programs typically experience higher costs of gain and higher death losses.

The primary objective of the stocker-yearling operation is to obtain the most pounds of cattle gain within economic reason while ensuring that high-quality forage yields can be obtained consistently each year. Forage management to obtain efficient production and consumption of nutritious feed is another essential ingredient of a successful stocker-yearling operation. Time of grazing and intensity of grazing (number of animals per acre) are important considerations if maximum forage production and utilization are to be maintained.

TYPES OF CATTLE FEEDING OPERATIONS

There are two basic types of cattle feeding operations: the commercial feeder and the farmer-feeder. The two types are generally distinguished by type of ownership and size of feedlot. **Commercial feedlots** are usually defined as having more than 1,000-head capacity and **farmer-feeder** feedlots as having less than 1,000-head one-time capacity.

The farmer-feeder operation is usually owned and operated by an individual or family. The commercial feedlot may be owned by an individual, a partnership, or a corporation—the last type is most common, especially as feedlot size increases. Commercial feedlots may own the cattle, feed the cattle owned by someone else (called *custom cattle feeding* or *custom feedlots*), or engage in some combination of the two. Custom-fed cattle are owned by other cattle feeders, investors, cattle producers, or packers and are fed on a contractual basis.

Even though the number of cattle fed in feedlots having more than 1,000-head capacity is increasing while feedlots less than 1,000 head are decreasing, there are some advantages to a farmer-feeder operation. Consider the following advantages and disadvantages:

1. The farmer-feeder can utilize cattle as a market for homegrown feeds.
2. The farmer-feeder can effectively utilize high roughage feeds in a backgrounding or warmup operation.
3. The farmer-feeder can distribute available labor over several different enterprises with cattle feeding being only one of those enterprises.
4. The farmer-feeder may have advantages in flexibility. When feeding cattle is unprofitable, the farmer-feeder can totally close out the cattle feeding and divert time and dollars into other phases of the farming operation. The large commercial feedlot has much higher overhead costs, which are fixed regardless of whether the pens are filled with cattle or are empty.

TABLE 26.9   **The Major Component Parts of a Feedlot Business Analysis**

Major Component, with Primary Factors Influencing Them				
Investment in Facilities	Cost of Feeder Cattle	Feed Cost per Pound of Gain	Nonfeed Cost per Pound of Gain	Total Dollars Received
Land	Grade	Ration	Death loss	Market alternative
Pens	Weight	Rate of gain	Labor	Transportation
Equipment	Shrink	Feed efficiency	Taxes	Shrink
Feed mill	Transportation	Length of time on feed	Insurance	Dressing percentage
Office	Gain potential		Utilities	Quality grade
			Veterinary expenses	Yield grade
			Repairs	Manure value

5. Commercial feedlots usually obtain and analyze more records and information and make more effective management decisions. Commercial feedlots typically utilize more professional expertise (consultants) in managing nutrition, health, and marketing.

6. Custom cattle feeding greatly reduce the operating capital requirement for the commercial feedlot and shift some of the risk to the customer. However, one of the greatest risks for commercial feedlots is keeping the lots full of cattle. Customers will stop putting cattle into custom lots when financial losses occur over a period of time, creating the problem identified in item 4. Farmer-feeders usually feed just one group of cattle per year, which requires leaving the facilities vacant for several months. Commercial feedlots attempt to keep their lots full year-round, so they have a cattle turnover rate of 2–2.5 times per year.

FEEDLOT CATTLE MANAGEMENT

The primary factors needed to analyze and properly manage a feedlot operation are investment in facilities, cost of feeder cattle, feed cost per pound of gain, nonfeed costs per pound of gain, and marketing. A more detailed analysis of these factors is shown in Table 26.9.

Facilities Investment

The investment in facilities varies with type and location of feedlots. In the United States, larger commercial feedlots are quite similar regardless of where they are located. The general layout is an open lot of dirt pens, with pen capacities varying from 100–500 head. The pens are sometimes mounded in the center to provide a dry resting area for cattle. The fences are constructed of pole, cable, or pipe. A feed mill to process grains and other feeds is usually a part of the feedlot. Special trucks distribute feed to fenceline feed bunks. Bunker trench silos hold corn silage and other roughages. Grains might be stored in these silos; however, they are more often stored in steel bins above the ground. The investment cost per head of capacity for this type of feedlot is approximately $150.

Feedlots for farmer-feeder operations vary from unpaved, wood-fenced pens to paved lots with windbreaks, sheds, or total confinement buildings. The latter might have manure collection pits located under the cattle, which stand on slotted floors. The feed might be stored in airtight structures. In farmer-feeder operations, however, feeds are typically stored in upright silos and grain bins, particularly where rainfall is high. Feeds are typically processed on the farm and distributed with tractor-powered equipment to feed bunks located either inside or outside the pens. Investment costs for these feedlots vary from $200–$500 per head.

Cost of Feeder Cattle

Before buying feeder cattle, the feedlot operator first estimates anticipated feed costs and the price the fed slaughter cattle will bring. These figures are then used to project the cost of feeder cattle or what the operator can afford to pay for them. Feeder cattle are priced according to weight, sex, **fill** (content of the digestive tract), skeletal size, thickness, and body condition. Most commercial feeders prefer to buy cattle with **compensatory gain.** These cattle are thin and relatively old for their weight. They have usually been grown out on a relatively low quality of feed. When placed on feedlot rations, they gain rapidly and compensate for the previous lack of energy in their diet.

The feeder cattle buyer typically projects a high gain potential in cattle that have a large skeletal frame and little finish or body condition. However, not all cattle of this type will gain rapidly.

Heifers are usually priced a few cents a pound under steers of similar weight. The primary reason is that heifers gain more slowly, the cost per pound of gain is higher, and some feeder heifers are pregnant.

Feeder cattle of the same weight, sex, frame size, and body condition can vary several cents a pound in cost. This value difference is usually due to differences in fill. The differences in fill can amount to 10–40 lb in live weight of feeder cattle. Feed and water consumed before weighing, distance and time of shipping, temperature, and the manner in which cattle are loaded and transported are some major factors affecting the amount of shrink. Shrink results primarily from loss of fill, but weight losses can occur in other parts of the body as well.

Feed Costs

Feed costs per pound of gain form the major costs of putting additional weight on feeder cattle. Typically, feed costs are 60–75% of the total costs of gain.

Feed costs per pound of gain are influenced by several factors, and the knowledge of these factors is important for proper management decision-making. The choice of feed ingredients and how they are processed and fed are key decisions that affect feed costs.

Cattle that gain more rapidly and efficiently on the same feeding program have lower feed costs. Some of these differences are genetic and can sometimes be identified with the specific producer of the cattle. Most feeder cattle receive feed additives (e.g., Rumensin) and ear implants (e.g., Ralgro or Synovex) that improve gain and efficiency and eventually the cost of gain.

Feed cost per pound of gain gets progressively higher as days on feed increase. Therefore, cattle feeders should avoid feeding cattle beyond their optimum combination of slaughter weight, quality grade, and yield grade.

Nonfeed Costs

Nonfeed cost per pound of gain is sometimes referred to as **yardage** cost. Yardage cost includes costs of gain other than feed. These costs can be expressed as either cost per pound of gain or cost per head per day. Obviously, cattle that gain faster move in and out of feedlots sooner and accumulate fewer total dollar yardage costs.

Death loss and veterinary expenses caused by feeder cattle health problems can increase the nonfeed costs significantly. Most cattle feeders prefer to feed yearlings rather than calves because the death loss and health problems in yearlings are significantly lower than in calves.

Gross Receipts

The total dollar amount received for slaughter cattle emphasizes the need for the cattle feeder to be aware of marketing alternatives and the kinds of carcasses the cattle will produce. Nearly 80% of the fed cattle are sold directly to a packer. This marketing alternative requires the cattle feeder to be aware of current market prices for the weight and grade of cattle being sold.

Many slaughter cattle at large commercial feedlots are sold on a standard shrink (pencil shrink) of 4%, with the cattle being weighed at the feedlot without being fed the morning of weigh day. Feeders who ship their cattle some distance before a sale weight is taken should manage their cattle to minimize the shrink loss.

Approximately 58% of fed cattle are currently sold on a carcass basis, an increase from the approximately 20% in 1971. Nebraska, Iowa, and Colorado account for about two-thirds of all carcass grade and weight marketings of steers and heifers. Packers in Texas, Iowa, Nebraska, Minnesota, and Wisconsin account for approximately three-fifths of all cows and bulls purchased on a carcass basis. Some slaughter steers and heifers are sold on a live weight basis, with the buyer estimating the carcass weight, quality grade, and yield grade. Slaughter cattle of yield grades 4 and 5 usually have large price discounts. The price spread between the select and choice quality grades can vary considerably over time. Some marketing alternatives will not show a price differential between select and choice if the cattle have been well fed for a minimum number of days (100–120 days for yearling feeder cattle).

Cattle feeders who manage their cattle consistently for profitable returns know how to purchase high-performing cattle at reasonable prices. These cattle feeders formulate rations that will optimize cattle performance with feed costs. Also, their cattle have minimum health problems with a low death loss. The feeder feeds cattle for the minimum number of days required to assure carcass acceptability and palatability, and develops a marketing plan that will yield the maximum financial returns.

Some cattle are sold on a grade and yield basis (carcass weight and carcass grade), where total value is determined after the cattle have been slaughtered. This method of marketing is most useful to producers who know the carcass characteristics of their cattle.

TABLE 26.10   Budget Showing Costs and Return for a Feedlot

Budget Assumptions			
In-weight	750 lb	Conversion rate (AF)	7.9 lb
Out-weight	1,175 lb	In-price ($/cwt)	$84.14
Net gain	450 lb	Out-price ($/cwt)	$68.00
Days on feed	133 days	Operating interest rate	10.0%
Average daily gain	3.3 lb	Death loss	1.0%

Operating Budget		
	Total Cost ($/head)	Percent of Total
Feeder cost	$638.55	74.3%
Feed	170.00	19.8
Interest	26.93	3.1
Death loss	6.39	0.7
Vet-med	6.00	0.7
Overhead	12.00	1.4
Total cost/head sold	$859.87	100.0%
Total income/head	$804.24	
Net profit (loss)/head	($ 55.62)	
Break-even ($/cwt)	$ 71.66	
Total cost of gain ($/cwt)	$ 42.54	

Source: Cattle-Fax.

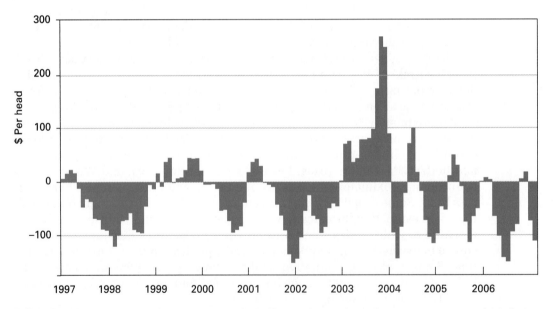

FIGURE 26.8 Average monthly returns to cattle feedlots in the Southern Plains. *Source: Livestock Marketing Information Center.*

COSTS AND RETURNS

The costs and returns for average fed cattle production are shown in Table 26.10. The enterprise budget is calculated on a dollar-per-head basis. The two major cost categories were feeder cattle purchase (74.3%) and feed (19.8%). Because the market was below the break-even price, a loss of almost $56 per head was realized. Figure 26.8 shows average monthly returns for cattle feeders in the southern Great Plains. Similar returns would be expected throughout the United States. Observe the contrasts in highly profitable versus highly unprofitable years.

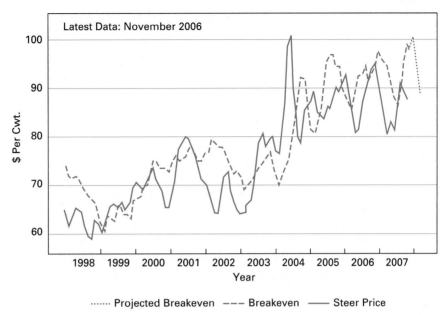

FIGURE 26.9   Choice steer prices compared to breakeven prices for the Southern Plains, 1998–2007. *Source: Livestock Marketing Information Center.*

PRODUCTION AND PRICES

Beef production tonnage and cattle prices are related. When production is low, prices are usually high. Producers typically save more replacement heifers and cull fewer cows during times of high prices. As a result the increased number of breeding females eventually increases beef production beyond consumer demand and prices decline. When prices decrease, low-cost producers with break-even prices below industry average have the profitability advantage (Figure 26.9).

ENVIRONMENTAL MANAGEMENT

The four primary environmental management issues for feedyards are dust, odor, flies, and water quality. Dust management is best accomplished via regular pen maintenance with timely manure removal in late spring. The regular use of box scrapers reduces manure accumulation without cratering the pen hardpan. The use of overhead sprinklers or water truck sprayers can be utilized to maintain the recommended moisture rate of 25–35% in 1 in. of manure requires approximately 14 gal of water per head of pen capacity. Achieving the same moisture increase in 2 in. of manure would require roughly 28 gal per head of pen capacity.

Control of odor requires regular pen maintenance, the use of correctly constructed and maintained runoff holding ponds; and proper nutritional management. The use of a more precisely balanced ration to avoid overfeeding of phosphorus is a key strategy.

Research is being conducted to use plant-oil extracts or fat extracts as a treatment for pen surfaces to control both odor and dust. Plant-oil extracts

TABLE 26.11   Nutrients in Manure from Four Sources (lb/hd/yr)

Source	N	P_2O_5	K_2O
Solid manure from open lots	65	41	65
Solid manure from deep-bedding building	132	66	132
Liquid runoff from open lots	5	2	11
Liquid manure from deep pit	89	55	79

Source: Adapted from Iowa State University (2001).

may be useful in inhibition of manure fermentation as a means to preserve nutrient value and suppress odor. These extracts may also have antimicrobial properties that help to inhibit the growth of organisms that contribute to foodborne illness.

Another strategy is to compost manure. The benefits of **composting** are to reduce volume and weight, kill weeds, concentrate nutrients, reduce odor, and to reduce fly populations. The disadvantages are labor and equipment costs, storage space, and nutrient loss. The nutrients in manure vary by type of lot construction (Table 26.11).

Fly control can be achieved via chemical, biological, or combination strategies. Chemical controls used alone are typically costly, short term in effect, and increase risk of human or environmental chemical exposure. The use of biological controls via fly-parasites has been successfully adopted in the industry.

Fly-parasites, also known as parasitic wasps, lay their eggs inside fly pupae. The fly pupae then become a food source for the fly-parasites. The use of early releases of fly-parasites before fly season followed by scheduled weekly releases is recommended. The cost of biological control ranges from $.20–$1.00 per head of cattle.

Water quality issues are likely to have a significant impact on feedlots and other intensive animal management facilities. Waste management is a primary concern as the decomposition of manure can negatively affect water quality via pathogens, nitrate, ammonia, phosphorus, salts, and organic salts. Incorrect handling, storage, or land application of manure can result in contamination of groundwater or surface water.

In an active feedyard, a layer of soil and manure becomes sufficiently compacted to create a seal that serves as a barrier to seepage. If the integrity of this layer is ensured, then water infiltration can be kept to less than 0.05 in. per day. It is critical that this layer be left undisturbed during pen cleaning.

Applying manure or wastewater directly to agricultural lands requires thorough knowledge about the following factors associated with a particular site— soil type, slope, irrigation practices, precipitation levels, crop nutrient requirements, nutrient levels of manure, and proximity to waterways or sells. The goal of manure management is essentially to collect, store, and apply wastes to lands at appropriate agronomic rates with the objectives of optimizing crop growth rates, economic returns, and protecting water quality.

Runoff problems can be minimized via employment of **best management practices (BMP)** such as building up-gradient ditches, dams, grass filter strips, filter fences, and other appropriate controlled drainage systems. Lagoons or ponds may be required to contain wastewater and runoff. BMP for manure handling are provided in Table 26.12.

TABLE 26.12 Summary of Best Management Practices for Manure Handling, Storage, and Application

Analyze for nutrient content
Account for available N from the total system
Apply to land areas large enough to accommodate manure volume
Calculate long-term manure loading rates
Maintain records of manure and soil analysis; application volume, timing, and methodology; additional fertilizer applications; and plant yields
Base manure application rates upon site-specific nutrient plans
Incorporate manure soon after application to avoid runoff
Determine application protocol based on soil composition and risk of aquifer contamination
Apply manure uniformly utilizing correct calibration
Utilize buffer zones to prevent water contamination
Use grass strips to catch and filter nutrients and sediments from runoff
Use rotational application schemes when planting high N use crops or forages
Locate manure stockpiles away from wells
Divert run-off via ditches, terraces, etc.
Maintain integrity of manure-soil seal when cleaning feedlot pens

Source: Adapted from Colorado State University (1994).

CHAPTER SUMMARY

■ Well-managed commercial cow-calf producers keep cost-effective records to achieve low break-even prices (BEs) by implementing the following formula:

$$BE = \frac{(\% \text{ calf crop} \times \text{weaning weight})}{\text{annual cow cost}}$$

■ Visual scores of body condition (BCS) in cows are used to minimize costs and obtain optimum levels of percent calf crop and weaning weights.

■ The primary factors affecting costs and returns of stocker-yearling and feedlot operations are (1) marketing (purchase price and sale price), (2) gaining ability of the cattle, (3) amount and quality of forage/feed, and (4) the health of the cattle.

KEY WORDS

calf crop percentage
enterprise budget
  (enterprise analysis)
body condition score
  (BCS)
dystocia
libido

growth stimulants
creep feeding
feed costs
labor costs
shrink
commercial feedlots
farmer-feeders

fill
compensatory gain
yardage
composting
best management
  practices (BMP)

REVIEW QUESTIONS

1. What are three criteria for assessing profitability in commercial cow-calf operations?

2. At what age should beef heifers be first bred and first calved?

3. *True or False:* Heifers should be bred at the same time as cows.

4. *True or False:* Feeding programs should be designed to have cows and heifers in a moderate body condition.

5. What is the typical cow-to-bull ratio for pasture breeding?

6. How can crossbreeding improve calf crop weaned?

7. *True or False:* Calves born early in the calving season are heavier at weaning primarily because they are older.

8. How does the amount of forage available influence weaning weights of calves?

9. *True or False:* Growth stimulants given to nursing calves will increase weaning weights by 5–15%.

10. *True or False:* Creep feeding can increase weaning weights and is always economical.

11. How can genetic selection be used to increase weaning weight?

12. What constitutes the largest part of annual cow costs?

13. What are the primary factors affecting profitability in stocker-yearling operations?

14. What is the primary objective of stocker-yearling operations?

15. What are the primary factors affecting profitability of feedlot operations?

16. What are nonfeed costs per pound of gain referred to as?

17. How are most slaughter steers and heifers sold?

SELECTED REFERENCES

Albin, R. C., and G. B. Thompson. 1996. *Cattle Feeding: A Guide to Management.* Amarillo, TX: Trafton Printing, Inc.

Duckett, S. K., D. G. Wagner, F. N. Owens, H. G. Dolezal, and D. R. Gill. 1996. Effects of estrogenic and androgenic implants on performance, carcass traits and meat tenderness in feedlot steers: A review. *The Professional Animal Scientist* 12:205–214.

Field, T. G. 2007. *Beef Production and Management Decisions.* Upper Saddle River, NJ: Prentice Hall.

Harner, J. P., and J. P. Murphy. 1998. *Planning Cattle Feedlots.* Kansas State University Ext. Publ. MF-2316. Manhattan, KS.

Iowa State University. 2001. *Beef Feedlot Systems Manual.* ISU Ext. Publ. 1867. Ames, IA.

Jarrige, R., and C. Beranger (eds.). 1981. *Beef Cattle Production.* Amsterdam: Elsevier Science Publishers.

National Cattlemen's Beef Association. 2001. *Cattle and Beef Handbook.* Englewood, CO.

National Research Council. 1996. *Nutrient Requirements of Beef Cattle.* Washington, DC: National Academy Press.

Richards, M. W., J. C. Spitzer, and M. B. Warner. 1986. Effect of varying levels of nutrition and body condition at calving on subsequent reproductive performance. *J. Anim. Sci.* 62:300.

Ritchie, H. D. 1992. *Calving Difficulty in Beef Cattle.* Beef Improvement Federation, BIF-FS6a and FS6b.

USDA. 2000. *Baseline References of Feedlot Management Practices.* National Animal Health Monitoring System. APHIS, VS.

CHAPTER 27

Dairy Cattle Breeds and Breeding

LEARNING OBJECTIVES

- Compare and contrast the performance of the primary dairy breeds
- Explain the need for dairy type description and linear classification
- Discuss the selection of dairy cattle for improved productivity
- Compare and contrast the tools available to dairy breeders

Production of milk per cow has been increased markedly in the past 50 years by improvements in breeding, feeding, sanitation, and management. The application of genetic selection, coupled with the extensive use of artificial insemination, has contributed to a successful dairy herd improvement program.

CHARACTERISTICS OF BREEDS

Six major breeds of dairy cattle—**Holstein, Ayrshire, Brown Swiss, Guernsey, Jersey,** and **Red and White**—are used for milk production in the United States. These breeds are shown in color plates O and P (following Chapter 25) where production characteristics and other information about the breeds are also given.

It is important to recognize the variation for milk yield and percent milk components that exist between breeds and within a breed. Table 27.1 shows the variation in milk production, percent fat, and percent protein for the major breeds. While the breeds are distinctly different based on breed averages, it is possible, for example, to identify Holstein cows that have higher fat percentages than Jersey cows.

TABLE 27.1   Average DHI Herd Performance by Breed (all test categories)

Breed	Herds (n)	Milk (lb)	Fat (%)	Protein (%)
Ayrshire	122	15,625	3.85	3.16
Brown Swiss	270	17,925	4.04	3.37
Guernsey	180	15,101	4.51	3.37
Holstein	21,659	22,347	3.64	3.05
Jersey	1,148	16,099	4.60	3.57
Milking Shorthorn	43	14,299	3.64	3.11
Red and White	29	19,594	3.65	3.02

Source: Adapted from Animal Improvement Programs Laboratory, 2006.

TABLE 27.2   Annual Registration Numbers for Major U.S. Breeds of Dairy Cattle

Breed	Annual Registrations (in thousands)						Year Association Formed
	2005	1995	1990	1985	1980	1970	
Holstein	293.5	329.9	395.9	394.5	535.9	281.6	1870
Jersey	73.0	63.4	53.6	65.4	61.0	37.1	1818
Brown Swiss	10.7	10.8	11.7	11.9	12.9	16.4	1880
Guernsey	5.0	7.4	13.9	25.1	20.9	43.8	1877
Ayrshire	6.0	6.4	7.8	11.1	10.9	15.0	1875
Red and White	3.7	4.4	6.7	5.3	4.8	—	1964
Milking Shorthorn	2.9	3.2	2.3	3.4	4.9	5.4	1912

Source: Various dairy cattle breed associations.

Some Holstein cows produce extremely large amounts of milk—more than 100 lb/day at peak lactation. Thus, in Holsteins and other cows with high milk production, great stress is placed on udder ligaments, which can break down and no longer support the udder. If an udder breaks down, it is more suscepti- ble to injury and disease, often necessitating culling the cow.

Ayrshires and Brown Swiss produce milk over a greater number of years than Holsteins, but their production levels are lower. Efficiency of milk pro- duction per 100 lb body weight is similar for the major breeds of dairy cows.

Guernsey and Jersey cows produce milk having high percentages of milk fat and solids-not-fat, but the total amount of milk produced is relatively low. The efficiency of energy production of different breeds varies less among the breeds than do either total quantity of milk produced or percentage of fat in the milk.

Registration Numbers

Breed popularity can be estimated from registration numbers, as shown in Table 27.2. Most dairy cows are grade cows, which mean they do not have pedi- grees (with individual registration numbers) recorded by a breed association. However, producers with registered cattle are a source of breeding cattle for the total dairy industry. Breeds with the largest registration numbers reflect the demand from the commercial dairy industry.

DAIRY TYPE

Some descriptive terms describing **ideal dairy type** are *stature, angularity, level rump, long and lean neck, milk veins,* and *strong feet and legs.* In the past, dairy producers have placed great emphasis on dairy type, but research studies indicate that some components of dairy type may have little or no value in improving milk

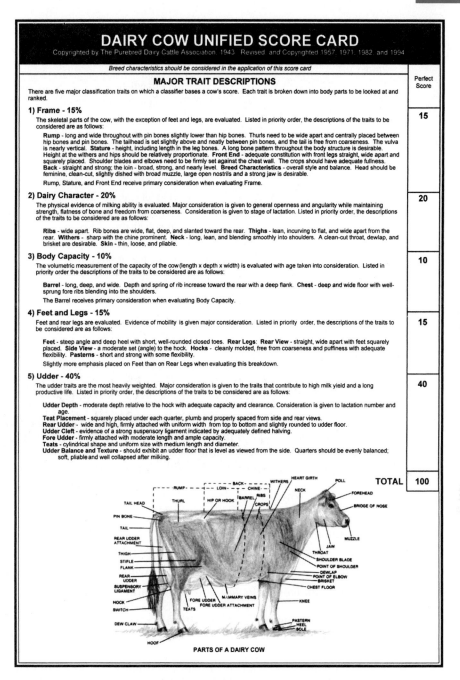

DAIRY COW UNIFIED SCORE CARD

Copyrighted by The Purebred Dairy Cattle Association. 1943 Revised. and Copyrighted 1957, 1971, 1982. and 1994

Breed characteristics should be considered in the application of this score card

MAJOR TRAIT DESCRIPTIONS

	Perfect Score

There are five major classification traits on which a classifier bases a cow's score. Each trait is broken down into body parts to be looked at and ranked.

1) Frame - 15% — **15**

The skeletal parts of the cow, with the exception of feet and legs, are evaluated. Listed in priority order, the descriptions of the traits to be considered are as follows:

Rump - long and wide throughout with pin bones slightly lower than hip bones. Thurls need to be wide apart and centrally placed between hip bones and pin bones. The tailhead is set slightly above and neatly between pin bones, and the tail is free from coarseness. The vulva is nearly vertical. **Stature** - height, including length in the leg bones. A long bone pattern throughout the body structure is desirable. Height at the withers and hips should be relatively proportionate. **Front End** - adequate constitution with front legs straight, wide apart and squarely placed. Shoulder blades and elbows need to be firmly set against the chest wall. The crops should have adequate fullness. **Back** - straight and strong; the loin - broad, strong, and nearly level. **Breed Characteristics** - overall style and balance. Head should be feminine, clean-cut, slightly dished with broad muzzle, large open nostrils and a strong jaw is desirable.

Rump, Stature, and Front End receive primary consideration when evaluating Frame.

2) Dairy Character - 20% — **20**

The physical evidence of milking ability is evaluated. Major consideration is given to general openness and angularity while maintaining strength, flatness of bone and freedom from coarseness. Consideration is given to stage of lactation. Listed in priority order, the descriptions of the traits to be considered are as follows:

Ribs - wide apart. Rib bones are wide, flat, deep, and slanted toward the rear. **Thighs** - lean, incurving to flat, and wide apart from the rear. **Withers** - sharp with the chine prominent. **Neck** - long, lean, and blending smoothly into shoulders. A clean-cut throat, dewlap, and brisket are desirable. **Skin** - thin, loose, and pliable.

3) Body Capacity - 10% — **10**

The volumetric measurement of the capacity of the cow (length x depth x width) is evaluated with age taken into consideration. Listed in priority order the descriptions of the traits to be considered are as follows:

Barrel - long, deep, and wide. Depth and spring of rib increase toward the rear with a deep flank. **Chest** - deep and wide floor with well-sprung fore ribs blending into the shoulders.

The Barrel receives primary consideration when evaluating Body Capacity.

4) Feet and Legs - 15% — **15**

Feet and rear legs are evaluated. Evidence of mobility is given major consideration. Listed in priority order, the descriptions of the traits to be considered are as follows:

Feet - steep angle and deep heel with short, well-rounded closed toes. **Rear Legs: Rear View** - straight, wide apart with feet squarely placed. **Side View** - a moderate set (angle) to the hock. **Hocks** - cleanly molded, free from coarseness and puffiness with adequate flexibility. **Pasterns** - short and strong with some flexibility.

Slightly more emphasis placed on Feet than on Rear Legs when evaluating this breakdown.

5) Udder - 40% — **40**

The udder traits are the most heavily weighted. Major consideration is given to the traits that contribute to high milk yield and a long productive life. Listed in priority order, the descriptions of the traits to be considered are as follows:

Udder Depth - moderate depth relative to the hock with adequate capacity and clearance. Consideration is given to lactation number and age.
Teat Placement - squarely placed under each quarter, plumb and properly spaced from side and rear views.
Rear Udder - wide and high, firmly attached with uniform width from top to bottom and slightly rounded to udder floor.
Udder Cleft - evidence of a strong suspensory ligament indicated by adequately defined halving.
Fore Udder - firmly attached with moderate length and ample capacity.
Teats - cylindrical shape and uniform size with medium length and diameter.
Udder Balance and Texture - should exhibit an udder floor that is level as viewed from the side. Quarters should be evenly balanced; soft, pliable and well collapsed after milking.

TOTAL — **100**

PARTS OF A DAIRY COW

FIGURE 27.1 Parts of the dairy cow and a description of the preferred dairy type. © Purebred Dairy Cattle Association.

production and might even be deleterious if overstressed in selection. However, a properly attached udder and strong feet and legs are good indicators that a cow will remain a high producer for a long time.

Figure 27.1 shows the parts of a dairy cow and also gives a description of the component parts (general appearance, dairy character, body capacity, and udder) of preferred dairy type. The points on the scorecard show greatest emphasis for general appearance (especially feet and legs) and udder, with each receiving 35 points

TABLE 27.3 Linear Classification Scoring System to Assess Type for Selected Traits

Trait Class	Trait	Description	Score
Form	Stature	Extremely tall	45
		Intermediate	25
		Extremely short	5
Form	Body depth	Extremely deep	45
		Intermediate	25
		Extremely shallow	5
Rump	Rump angle/width	Extremely sloped/extremely wide	45
		Slight slope/intermediate	25
		High pins/extremely narrow	5
Legs and feet	Rear legs (side view)	Extremely sickled	45
		Intermediate	25
		Extremely posty	5
Udder	Fore udder attachment	Extremely strong	45
		Intermediate	25
		Extremely loose	5
Teats	Front teat placement	Extremely close	45
		Centrally placed	25
		Extremely wide	5

Source: Adapted from Linear Classification System, Holstein Association, USA, 1999.

out of a possible 100. Table 27.3 illustrates the scores assigned to descriptions or assessments of a sample of traits evaluated in a linear classification system.

Some dairy breed associations have a classification program to evaluate the type traits. Breeders use the **linear classification system** as an evaluation tool to enhance selection for high-producing cows with the durability to stay productive. The Holstein Association offers five participation programs:

Classic—All cows in the herd are evaluated; eligible for herd recognition awards.

Standard—Most, but not all, cows are evaluated; eligible for herd awards.

Limited—Focused on evaluation of first-calf heifers.

Introductory—Offered to breeders who want to get started in the program.

Breeders' Choice—Selected cows are evaluated.

A classifier approved by the association scores each cow or bull for the several traits and gives a final score if these scores are to be official records recognized by the association.

In the Holstein Association program a final score is calculated by rating the four major categories of classification traits: general appearance, dairy character, body capacity, and mammary system. The emphasis for each category for cows and bulls is shown in Table 27.4.

The final score represents the degree of physical perfection of any given animal. It is expressed in the following numbers and words:

Excellent (EX): 90–100   Good (G): 75–79

Very good (VG): 85–89   Fair (F): 65–74

Good plus (G+): 80–84   Poor (P): 50–64

TABLE 27.4   Classification Traits Used for Final Scores in Holstein Cattle

Trait	Emphasis (%)	
	Cows	Bulls
Frame	15	30
Dairy character	20	25
Body capacity	10	20
Feet and legs	15	25
Udder	40	—

Source: Holstein Association, Linear Classification Program.

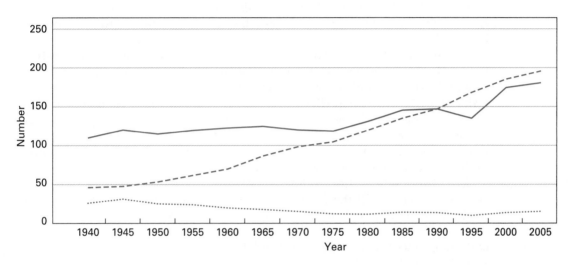

.... Number of cows (mil)   – – – Average milk per cow (100 lbs)   ——— Total milk (bil lbs)

FIGURE 27.2   Changes in milk production in the United States, 1940–2005. *Source:* Adapted from USDA data.

The final score is used in computing predicted transmitting ability for type (PTAT). PTAT identifies genetic differences in sires that can be considered in selection programs to improve dairy cattle type.

Type has value from a sales standpoint. Although some components of type score are negatively related to milk production, type is important as a measure of the likelihood that a cow will sustain a high level of production over several years.

IMPROVING MILK PRODUCTION

Great strides have been made over the past 50 years in improving milk production through improved management and breeding. For example, the total amount of milk produced in the United States continues to increase; yet the number of dairy cows in 2001 was less than half the number in 1950 (Fig. 27.2). The average production per cow in 2005 was more than three times greater than the average production in 1940. In the past 10 years alone, annual milk production in the United States has increased by approximately 320 lb per cow per year.

SELECTION OF DAIRY COWS

The average productive life of a dairy cow is short (approximately 3–4 years). Many cows are culled primarily because of reproductive failure, low milk yield, udder breakdown, foot and leg weaknesses, and mastitis.

Heifers whose ancestral records indicate they will be high producers should be used to replace cows that are **culled** for low production. Such heifers should also be used to replace cows that leave the herd because of infertility, mastitis, or death, although the improvement gained thereby is generally modest.

A basis for evaluating a dairy cow is the quantity of milk (pounds) and quality (total solids) that she produces. For a dairy operator to know which cows are good producers and which cows should be culled, a record of milk production is essential.

Nearly one-half of U.S. dairy cows are enrolled in the **National Cooperative Dairy Herd Improvement Program (NCDHIP),** a national, industry-wide production-testing and record-keeping program. It is often referred to as the **Dairy Herd Improvement (DHI)** program. DHI is delivered via a cooperative association of 30 affiliates, 6 regional data processing centers, and the USDA Animal Improvement Programs Laboratory, which handles the genetic evaluation. DHI is designed to collect and process raw data into a systematic format useful in making dairy management decisions.

DHI also facilitates the development of a national database that is utilized in genetic evaluation. Of the nearly 4.5 million enrolled cows, about 55% contribute data to the genetic evaluation effort.

DHI offers numerous recording programs. Four of the most common data collection procedures are described as follows:

1. **Supervised test**—the DHIA technician weighs and samples milk for all cows from each milking during a 24-hour period.

2. **Partially supervised test**—milk weights and samples are taken alternately at A.M. and P.M. milkings by the DHIA technician and one other person.

3. **Owner conducted test**—someone other than the DHIA technician records test-day production data.

4. **Supervised electronic test**—supervised test whereby data collection is conducted electronically with the technician certifying procedures and accuracy.

Some restricted, unsupervised data collections are allowed. DHI data collection and processing is performed under strict quality standards to ensure the integrity of the information. Records obtained allow producers to compare cows within the herd and to compare herd performance to that of other herds in the region.

TriStar is the program offered by the Holstein Association for providing production records, cow and herd genetic performance reports, and recognition programs, such as the Gold Medal Dam and Dam of Merit recognition. TriStar offers four service options to meet the needs of its members.

Figure 27.3 shows an individual cow's record used in the DHIR program. Especially note the milk and milk fat recorded under the heading "EXT. 305 DAYS." Records are standardized to a lactation length of 305 days, 2 milkings per day (2x), and to a mature age of cow (ME = mature equivalent). Adjustments are made to records so they can be more accurately compared on a standardized basis.

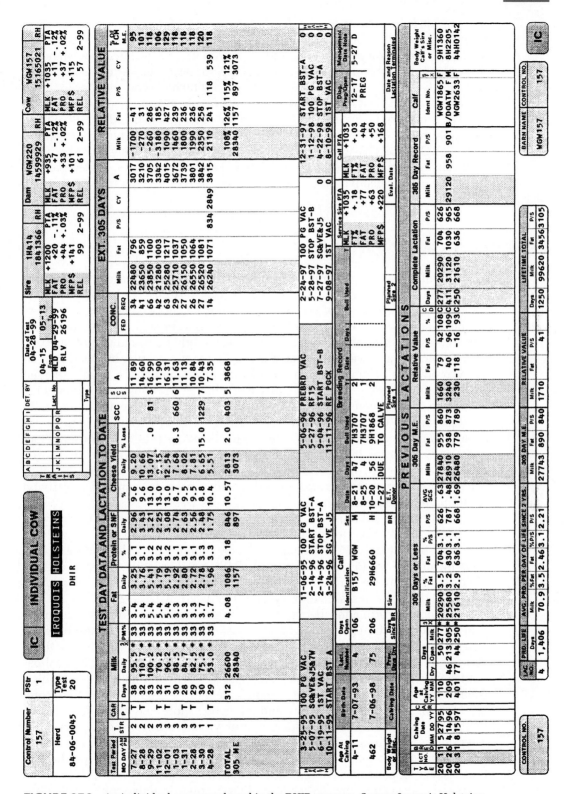

FIGURE 27.3 An individual cow record used in the DHIR program. *Source:* Iroquois Holsteins.

A registry association for purebred animals exists for each dairy cattle breed in the United States. In addition, some breed associations honor cows with outstanding production performance and bulls with daughters that are outstanding in production. The Holstein selective registry, for example, has the Gold Medal Dam and Gold Medal Sire awards. To win a Gold Medal a cow must produce an average of 24,881 lb of milk and 500 lb of milk fat per year during her productive life; she must also have a minimum type score of 83 points based on 100 points. The Gold Medal cow should also have three daughters that meet the requirements. A Gold Medal–winning sire must have 10 or more daughters that meet high standards in milk and fat production and also in type score. Other breed associations have selective registries as a means of encouraging improvement in production.

BREEDING DAIRY CATTLE

The time of breeding is an important phase in dairy cattle management. Because they are milked each day, dairy cows are more closely observed than beef cows, allowing visual detection of estrus. When in estrus, dairy cows may show restlessness, enlarged vulvas, and a temporary decline in milk production. Also, when cows are in standing heat they will permit other cows to mount them.

Technicians are available to artificially inseminate cattle, but well-trained dairy producers or employees are excellent inseminators. The use of semen from genetically proven sires is highly desirable, even though this semen may cost more than semen from an average bull. Considering the additional milk production that can be expected from heifers sired by a good bull, the extra investment can return high dividends. Semen costing $25–$150 per unit from genetically superior bulls may be a better investment than semen from less desirable bulls at $10–$20 per unit. Bulls should not run with the milking cows, because bulls are often dangerous. Bulls that provide semen for artificial insemination should be handled with caution as well.

The heritability of traits is indicative of the progress that can be made by selection. Susceptibilities to cystic ovaries, ketosis, mastitis, and milk fever are all low in heritability (5–10%). Percentages of fat, protein, and solids-not-fat are all high in heritability (50%). Yearly milk (ME), protein, solids-not-fat, and fat yields are medium in heritability (25–30%). Heritability estimates for traits assessed by the linear classification systems are listed in Table 27.5.

The genetic correlation between two traits is indicative of the amount of genetic change in trait A that might be expected from a certain amount of selection pressure applied to trait B. The most important genetic correlations are those that might be associated with milk yields during the first lactation. Fat, solids-not-fat, protein yield, lifetime milk yields, and length of productive life have high genetic correlations with first lactation milk yield (0.70–0.90). Overall type score, levelness of rump, udder texture, and strength of fore and rear udder are all negatively correlated with first lactation milk yield (−0.20 to −0.40). Dairy character and udder depth are positively correlated (0.35–0.40) with first lactation yield. Inbreeding tends to increase mortality rate and to reduce all production traits except fat percentage of milk and mature body weight.

There are several **inherited abnormalities** known in dairy cattle. This does not mean that dairy cattle have a higher number of inherited abnormalities than

TABLE 27.5   Heritabilities of Holstein Association Type Traits

Trait	h^2
Stature	0.42
Strength	0.31
Body depth	0.37
Dairy form	0.29
Rump angle	0.33
Thurl width	0.26
Rear leg score	0.21
Foot angle	0.15
Fore attachment	0.29
Rear udder height	0.28
Rear udder width	0.23
Udder cleft	0.24
Udder depth	0.28
Teat placement	0.26
Teat length	0.26
Final Score	0.29

Source: Adapted from Holstein Association Sire Summary.

other farm animals, but that more is known about dairy cattle than about most other farm animals because dairy cattle are observed more closely. No breed of dairy cattle is free from all inherited abnormalities. Some of these abnormalities include achondroplasia (short bones), weavers, limber limbs, rectal-vaginal constriction, dumps, flexed pasterns (feet turned back), fused teats (teats on same side of udder are fused), hairlessness (almost no hair on calf), and syndactylism (only one toe on a foot). Many of these inherited abnormalities are lethal and most are recessive in their mode of inheritance.

Usually the occurrence of inherited abnormalities is infrequent. However, occasionally an outstanding sire might carry an abnormality or a specific dairy herd may have several genetically abnormal calves. As most inherited abnormalities result from recessive genes, both the sire and dam of genetically abnormal calves carry the undesirable gene. Breeding stock should not be kept from either parent.

One reason such rapid progress has been made in improving milk production is that genetically superior sires are identified and used widely in artificial insemination programs. Genetically superior bulls today might become the sires of 100,000 calves or more each in their productive lives by use of artificial insemination. Generally, the better sires will sire more calves than ordinary sires because dairy producers know the value of semen from outstanding bulls. Thus, selection is enhanced markedly for greater milk production by use of artificial insemination (Fig. 27.4).

Traits of importance in selection of dairy cattle include milk production, milk composition, longevity, and structural correctness. These traits are usually emphasized according to their relative heritability and economic importance. Fertility is extremely important but low in heritability; so marked improvement in this trait is more likely to be accomplished environmentally through good nutrition and management rather than through selection.

Most dairy cattle in the United States are straightbred because crossing breeds has failed to lead to a significant improvement in milk production. No combination of breeds, for example, equals the straightbred Holstein in total milk production. Because many genes control milk production and the effect of each of the genes is unknown, it is impossible to manipulate genes that control milk

FIGURE 27.4   Cows being artificially inseminated with semen from a genetically superior bull. An effectively managed AI program can make a significant impact on efforts to increase milk production in dairy herds. Courtesy of Select Sires.

production by the same method that would be used in the case of simple inheritance. Also, milk production is a sex-limited trait expressed only in the female. Furthermore, milk production is highly influenced by the environment. In order for milk production to be improved by genetic selection, the environment must be standardized among all animals present to ensure, insofar as possible, that differences between animals are due to inheritance rather than environment.

In addition to a high level of milk production, characteristics of longevity, regularity of breeding, ease of milking, and quiet disposition are important in dairy cows.

Bulls are evaluated for their ability to transmit the characteristic of high-level milk production both by considering the production level of their ancestors (pedigree) and by considering the production level of their daughters (progeny testing). An index is used as a predictive evaluation of the bull's ability to transmit the characteristic of high-level milk production.

Sire Genetic Evaluations, computed by the USDA, are based on comparing daughters of a given sire with their contemporary herd mates. Each sire is assigned a **predicted transmitting ability** based on the superiority or inferiority of his daughters to their herd mates. Many sires have daughters in 50–100 different herds, so the predicted transmitting ability (PTA) for milk, fat, and type is generally highly reliable and provides a sound basis for the selection of semen. The USDA publishes PTAs among sires semiannually.

SIRE SELECTION

The dairy industry (breed associations, National Association of Animal Breeders, and the USDA) uses the **best linear unbiased prediction (BLUP)** method for estimating predicted transmitting ability (PTA) among sires. Dr. C. R. Henderson developed this method at Cornell University. The BLUP procedure accounts for genetic competition among bulls within a herd, genetic progress of the breed over generations, pedigree information available on young bulls, and differing numbers of herd mate's sires, and partially accounts for the differential culling of daughters among sires. In the BLUP method, direct comparisons are made among bulls that have daughters in the same herd. Using bulls that have daughters in two or more herds makes indirect comparisons. An example presented by the Holstein Association (1980) depicts the use of three different sires in two herds:

Herd 1	*Herd 2*
Daughters of sire A	Daughters of sire B
Daughters of sire C	Daughters of sire C

Direct comparisons can be made between sires B and C in herd 2. Using the common sire C as a basis for comparison can make indirect comparisons between sires A and B. PTAs are calculated for milk, protein, fat, type, and dollars returned.

The next step in the evolution of selection is the utilization of indices that combine both biological performance data on several traits as well as economic weightings for each trait as a means to calculate estimates that describe the genetic potential of an animal for a particularly important set of economically valuable traits. **Net merit index (NM$)** is an excellent example of a bioeconomic index. Net merit index is calculated by the following formula:

$$\text{NM\$} = 0.7\,(\text{MFP\$}) + \$11.30\,(\text{PTA PL}) - \$28.22\,(\text{PTA SCS} - \text{breed ave. SCS})$$

Where:

MFP\$ (milk-fat-protein) = \$0.031 (PTA milk lbs) + \$0.80 (PTA fat lbs) + \$2.00 (PTA protein lbs)

PTA PL = Predicted transmitting ability productive live (number of months in milk until the cow is 84 months old). PTA PL is a measure of ability to avoid culling and is highly correlated with production and with type classification score, especially udders.

PTA SCS = Predicted transmitting ability somatic cell score. This estimate is *not* a replacement for excellent management and preventative care. It is useful as a tool to balance selection for increased yield without incurring excessive mastitis.

The **Type Production Index (TPI)** combines differences for milk production traits and differences for type traits into a single value. TPI is calculated via the following formula:

$$\text{TPI} = \left[3\left(\frac{\text{PTAP}}{19}\right) + \left(\frac{\text{PTAF}}{22.5}\right) + \left(\frac{\text{PTAT}}{.7}\right) + \left(\frac{\text{UDC}}{.8}\right) \right] 50 + 576$$

Where:

PTAP = Predicted Transmitting Ability Protein
PTAF = Predicted Transmitting Ability Fat
PTAT = Predicted Transmitting Ability Type
UDC = Udder Composite

FIGURE 27.5 Paradise-R Cleitus Mathie is an outstanding Holstein bull. More than 900,000 units of semen have been collected from this sire. He has 19,399 daughters in 5,469 herds. His PTA is +1,768 (milk) with a net merit of +$163.

Figure 27.5 identifies a Holstein bull that is genetically superior in combination of type and milk traits.

CHAPTER SUMMARY

■ Based on registration numbers, the Holstein is the major breed of dairy cattle, followed by Jersey, Brown Swiss, Guernsey, Ayrshire, Red and White, and Milking Shorthorn.

■ Selection for milk production has more than doubled the pounds of milk produced per cow from 1960 to 2001.

■ The dairy industry has implemented effective production testing programs and record-keeping systems for a longer period of time compared to most other livestock industries.

KEY WORDS

Holstein
Ayrshire
Brown Swiss
Guernsey
Jersey
Red and White
ideal dairy types
linear classification
  system

culled
National Cooperative
  Dairy Herd
  Improvement Program
  (NCDHIP)
Dairy Herd
  Improvement (DHI)
TriStar
inherited abnormalities

Sire Genetic Evaluation
predicted transmitting
  ability
best linear unbiased
  prediction (BLUP)
net merit index (NM$)
Type Production Index
  (TPI)

REVIEW QUESTIONS

1. What are the six major breeds of dairy cattle in the United States?
2. Which breed produces the greatest amount of milk per day?
3. Which breed of dairy cattle is most popular?
4. *True or False:* Efficiency of milk production per 100 pounds of body weight is about the same for the major breeds of dairy cattle.

5. *True or False:* Most dairy cattle are registered.

6. What are the major reasons many dairy cattle are culled?

7. What is the Dairy Herd Improvement Association?

8. What is the most reliable sign that a cow is in estrus?

9. *True or False:* Expensive semen from genetically superior bulls is more desirable than less expensive semen from average bulls.

10. What traits are most important to select for in order to improve milk production in dairy cattle?

11. Why are most dairy cattle in the United States purebred rather than crossbred?

12. How are bulls evaluated for their ability to transmit the characteristic of high milk production?

SELECTED REFERENCES

Bath, D. L., F. N. Dickinson, H. A. Tucker, and R. D. Appleman. 1985. *Dairy Cattle: Principles, Practices, Problems, Profits.* Philadelphia, PA: Lea & Febiger.

Holstein Association USA. 2006. *Linear Classification Program.* Brattleboro, VT.

Holstein Association USA. 2006. *Holstein Type-Production Sire Summaries.* Brattlesboro, VT.

Legates, J. E. 1990. *Breeding and Improvement of Farm Animals.* 4th ed. New York: McGraw Hill.

Schmidt, G. H., et al. 1988. *Principles of Dairy Science.* Englewood Cliffs, NJ: Prentice Hall.

Trimberger, G. W., W. E. Etgen, and D. M. Galton. 1987. *Dairy Cattle Judging Techniques.* Englewood Cliffs, NJ: Prentice Hall.

Wiggins, G. R. 1991. National genetic improvement programs for dairy cattle in the United States. *J. Anim. Sci.* 69:3853.

CHAPTER 28

Feeding and Managing Dairy Cattle

LEARNING OBJECTIVES

- Explain the nutrient and milk yield relationships during lactation and gestation
- List the elements of a successful dairy nutrition program
- Describe how to adjust a dairy ration for heat stress
- Compare the nutritional program for dry cows and replacement heifers
- Describe the appropriate management of dairy bulls
- Describe the various types of dairy housing
- Explain the importance of a nutrient management plan
- Describe the milking process
- List the primary reasons dairy females are culled
- Discuss mastitis prevention and preventative health practices
- Compare and contrast profitable versus unprofitable dairies
- Discuss the value of performance records benchmarks

The U.S. dairy industry has changed greatly from the days of the family milk cow. In 1850, one-third of all cattle were dairy cows. By 1940, nearly 40% of U.S. cattle were dairy animals. Currently, fewer than 10% of the cattle herds are utilized as dairy cows. The contemporary dairy industry is a highly specialized industry that produces, processes, and distributes milk. A large investment in cows, machinery, barn, and milking parlor is necessary. Dairy operators who produce their own feed need additional money for land on which to grow the feed. They also need machinery to produce, harvest, and process the crops.

Although the size of dairy operations varies from 30 or less to more than 5,000 milking cows, the average dairy has approximately 100 milking cows, 30 dry cows, and 100 replacement heifers. The average dairy producer farms 200–300 acres of land, raises much of the forage, and markets the milk through coopera-tives of which he or she is a member. These producers sell about 3 tons of milk daily, or about 2.2 million pounds annually, worth about $230,000. Their average total capital investment may exceed a half million dollars. The average producer has a partnership with a family member or another person to make management of time and resources easier.

The dairy operator must provide feed and other management inputs to keep animals healthy and at a high level of efficient milk production. Feed and other inputs must be provided at a relatively low cost compared with the price of milk if the dairy operation is to be profitable.

NUTRITION OF LACTATING COWS

The average milk production per cow in the United States for a lactation period of 305 days is approximately 16,870 lb. Some herd averages exceed 25,000 lb, and some top-producing cows yield more than 40,000 lb of milk per year. Thus, some lactating cows may produce more than 150 lb of milk, more than 5 lb of milk fat, and more than 4.5 lb of protein per day. A great need for energy and total amount of feed is created by lactation. For example, a cow weighing 1,400 lb that produces 40 lb of milk daily needs one and a quarter times as much energy for lactation as for maintenance. If she is producing 80 lb of milk daily, she needs 2.5 times more energy for milk production than for maintenance.

Providing adequate nutrition to lactating dairy cows is challenging and com-plex. Some reasons for this complexity are shown in Figure 28.1. The dairy cow's nutritional needs vary widely during the production cycle. Dairy pro-ducers must formulate rations to match each period of the production cycle so as to optimize milk yield and reproduction, prevent metabolic disorders, and increase longevity. Water and energy are the two most limiting nutrients in a dairy ration.

It is difficult during the first 2–4 months after calving to provide adequate nutrition because milk yield is high and intake is limited. As the nutrient intake is less than the nutrient demand for milk, the cow uses her body fat and protein reserves to make up the difference. Thus the cow is in a negative energy balance and usually loses body weight during the months of heavy milk production (Fig. 28.1). This same period of time can also be challenging to reproduction, as conception rates are usually lower when cows are losing weight.

Body condition scores (1 = thin, 3 = average, 5 = fat) can be used to monitor nutrition, reproduction, and health programs for dairy herds. Condition scores in the low 3s and high 2s are acceptable during the first few weeks after calv-ing, but body condition scores in the low 4s are necessary as the cows move into the dry period. Scores lower or higher than the aforementioned benchmarks suggest potential management problems. To provide adequate nutrition at ef-fective costs, cows can be grouped by condition score and stage of lactation.

At 2–3 months into the lactation period, daily milk production peaks and then starts to decline; feed intake is adequate or higher than milk production demands. This contributes to body-weight gain (Fig. 28.1). The energy content of the ration should be monitored after approximately 5 months of lactation to prevent the cow from becoming too fat.

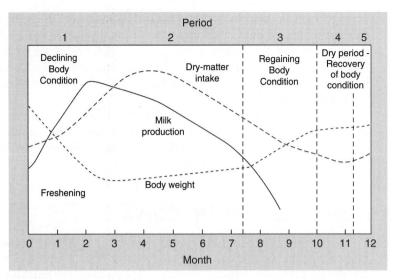

FIGURE 28.1   Nutrient and milk yield relationships during lactation and gestation. Courtesy of Hoffmann-LaRoche, Inc. Adapted from *Nutritional Needs of Dairy Cows* (Growthlines). Fall 1989.

Different types and amounts of feeds can be used to provide the needed energy, protein, vitamins, and minerals to lactating cows. The availability, palatability, and relative costs of different feeds are the primary factors influencing ration composition. Basic nutrition principles, presented in Chapters 15, 16, and 17, provide the foundation for developing dairy cattle feeding programs.

The elements of a successful dairy nutrition program can be described as follows:

1. Assure a constant, high-quality water supply.
2. Utilize high-quality feedstuffs.
3. Maintain long fiber size in forages (20% of fibers >1.5 in. to ensure normal rumen function).
4. Maintain an optimal level of grain feeding (<60% of DM as concentrate).
5. Monitor cud-chewing to assess rumen function and sufficient fiber length (60–75% of cows chewing cud during observation at rest).
6. Assure sufficient bunk space (75% of cows need to be able to eat at once).
7. Maintain consistent, high levels of DM intake (3.8–4.0 lb of DM per 100 lb of body weight).
8. Ensure that feed is available 24 hours per day and especially when cows return from the milking parlor.
9. Do not offer free-choice ration components as this makes it impossible to control the total mixed ration.
10. Manage floors, ventilation, stalls, animal handling practices, and temperature control to assure cow comfort.

11. Monitor concentrate and forage quality to minimize variation in feedstuffs.

12. Manage to minimize occurrence of displaced abomasums and ketosis.

Monitoring cow activity is one means to head off nutritional and reproductive problems. Ideally, cows should spend their time in the following activities to assure optimal performance—lying down (50%), eating (21%), milking (13%), social (6%), lockups (6%), and drinking (4%). A charting spreadsheet is available from ABS Global (www.absglobal.com) to calculate cow activity.

Most dairy cattle rations are based on roughages (hay and silages). Roughages are usually the cheapest source of nutrients; often they are produced by dairy farms, but increasing herd size, greater managerial demands, higher land-tax costs, and cheaper feeds are changing this traditional role. Silages and concentrates are typically mixed together before being fed to the cows; this is known as a **total mixed ration,** and is the preferred way of feeding dairy cows. Chopped hay and silage can be delivered to mangers on each side of an open alleyway.

Lactating dairy cows cannot obtain an adequate supply of nutrients from an all-roughage ration. Concentrates are supplied in amounts consistent with the level of milk production, body weight of the cow, amount of nutrients in the roughage, and nutrient content of the concentrates. Concentrates are usually provided to cows while they are being milked. However, high-producing cows requiring large amounts of concentrates do not have sufficient time in the milking parlor to consume all the necessary grain; therefore, additional concentrates are usually fed at another location. Another innovation that enhances management of cows in negative energy balance during the first third of lactation is the use of by-pass proteins and protected fats. These approaches protect nutrients from the action of the microbes in the rumen. For example, protein produced by the activity of the microbes in the rumen has a less desirable amino acid balance than many proteins provided in the diet, so bypassing the rumen offers the opportunity for absorption of a more desirable amino acid mix in the small intestine.

There are economic advantages in feeding concentrates in relation to quantity of milk produced by each cow. However, this method tends to feed a cow that is declining in production too generously and to feed a cow that is increasing in production inadequately. It is a poor economic practice to allow cows in their last 2 months of lactation to have all the concentrates that they can consume. Heavy feeding at this stage of lactation does not result in increased milk production.

Young (2-year-old) cows that are genetically capable of high production should be fed large amounts of concentrates to provide the nutrition they need to grow as well as to produce milk. Without adequate nutrition, their subsequent breeding and lactation may be hindered. The dairy cow transitioning from dry to lactating, and the first-calf heifer moving into her first lactation, need to be on a rising plane of energy and protein in the 30–60 days prior to parturition, with the most rapid increase occurring in the 14 days prior to calving.

Forages vary immensely in nutrient concentrations. Legumes (alfalfa) have high calcium, protein, and potassium relative to animal nutrient requirements, whereas other forages such as grasses and corn silage are considerably lower. Concentrates (grains) usually are low in calcium and high in phosphorus. The

TABLE 28.1   Examples of Rations Needed to Meet the Nutrient Requirements of a 1,350-lb Dairy Cow with 3.8% Fat Test

Period[a]	Daily Milk Production (lb)	Daily Dry-Matter Intake (lb)	Ration (lb/day; as fed)
1	90	49	Alfalfa hay (28 lb)[b]; corn-oats grain mix (21 lb); soybean oil meal (5 lb); dical (0.5 lb); salt, vitamin, trace mineral mix (0.3 lb)
3	50	38	Alfalfa hay (27 lb)[b]; corn-oats grain mix (16 lb); soybean oil meal (4.5 lb); dical (0.3 lb); salt, vitamin, trace mineral mix (0.25 lb)
4	dry	24	Alfalfa hay (12 lb)[b]; corn silage (43 lb); monosodium phosphate (0.1 lb); salt, vitamin, trace mineral mix (0.1 lb)

[a]Refer to Fig. 28.1.
[b]Contains 20% crude protein.
Source: Adapted from Hoffmann-LaRoche, Inc. *Nutritional needs of dairy cows.* Growthlines. Fall 1989.

amounts of protein supplement (soybean meal) and mineral supplement that must be added to a concentrate formulation obviously depend on the type and amount of forage being fed. Trace mineralized salt and vitamins A, D, and E are usually added to concentrates.

Table 28.1 shows examples of rations formulated to meet the nutrient requirements of dairy cows in various periods of production. It is recommended that higher producing cows receive rations specifically formatted to their needs. Higher producing females may require greater amounts of fiber, energy, and bypass protein. Feed typically accounts for about one-half of the cash costs associated with a dairy. As such, producers should allocate a comparable level of attention to the feeding aspect of the dairy.

ADJUSTING FOR HEAT STRESS

Heat stress may lead to milk production losses of 8 to 10 lb per day. Heat stress may occur at temperatures in excess of 77°F or at humidity indices of greater than 72°F. Symptoms of heat stress include a body temperature of more than 102.5°F (101.5°F is normal), excessive panting (more than 80 breaths per minute), feed intake losses which may be as high as 15%, decreased milk yields, and declining pregnancy rates of 20% or more. Cows under heat stress may also experience higher incidences of acidosis.

Therefore, managers need to implement environmental adaptation strategies. Management protocols to minimize the effects of heat stress include provision of shade, assuring adequate air movement, providing cooling from fans or sprinklers, reformulating diets to match intake, provision of ample sources of clean water, and minimizing cow movement to avoid increasing core body temperature.

NUTRITION OF DRY COWS

How dry cows are fed and managed may influence their milk production level and health in the next lactation. Suggested nutrient densities for dry cow rations are provided in Table 28.2.

A common practice in drying off lactating cows is abruptly to stop milking the cow; with high producers, however, this may be traumatic and dry-off may need to be more gradual (intermittent milking). The buildup of pressure in the

TABLE 28.2   Nutrient Density for Dry Cow Rations

Nutrient	Early Dry Period	Pre-Fresh Period
Dry matter intake (lb)	26–28	23–26
Crude protein (%)	12.5–13.0	14.0–15.0
NE^2 (meal/lb)	0.55–0.62	0.68–0.70
Calcium (%)	>0.60	0.45–1.50
Phosphorus (%)	0.30–0.35	0.35–0.40
Vitamin A (IU/day)	100,000	130,000
Vitamin D (IU/day)	20,000	30,000
Vitamin E (IU/day)	800	1,000–3,000

Source: Adapted from Perkins, 1998. Colorado Dairy Nutrition Conference.

mammary gland causes the secretory tissues to stop producing milk. At the last milking, the cow should be infused with a treatment for preventing mastitis.

Dairy producers plan for a 50- to 60-day dry period. Short dry periods usually reduce future milk yield because the cow has not adequately improved body condition and the mammary tissue has not properly regenerated. Long dry periods can lower milk yield because the cow may become overly fat, and profitability may be less because feed costs are increased.

Dry cows should be separated from lactating cows so they can be fed and managed consistent with their needs. Dry cows need fewer concentrates than lactating cows. If dry cows overeat, they will likely become fat. This excessive body fat may lower future milk yield and cause health problems, such as fatty liver, ketosis, mastitis, retained placenta, metritis, milk fever, or even death.

NUTRITION OF REPLACEMENT HEIFERS

Young heifers (5 months of age and older) can meet most of their nutritional needs from good-quality legume or grass-legume pasture in addition to 2–3 lb of grain. If the pasture quality is poor, they will need good-quality legume hay and 3–5 lb of grain. In winter, good-quality legume hay or silage and 2–3 lb of grain are needed. If the forage is grass (hay or silage), the heifers will need 3–5 lb of grain to keep them growing well.

Heifers should be large enough to breed at about 15 months of age and calve as 2-year-olds. Weight recommendations for heifers from different breeds are shown in Table 28.3. Weight is important because it affects when the heifer reaches puberty and can be bred. Also, heifers that are heavier at first calving produce more milk. For example, the optimum weight at calving for Holstein heifers is 1,200–1,250 lb. Lighter heifers produce less milk and heavier heifers are too costly for the increased production.

TABLE 28.3   Recommended Weights at Breeding for Replacement Heifers

Breed	Weight (lb)
Ayrshire	720–750
Brown Swiss	800–875
Guernsey	720–750
Holstein	800–875
Jersey	630–650

Source: Adapted from Hoffmann-LaRoche, Inc. *Nutritional needs of dairy cows.* Growthlines. Fall 1989.

MANAGEMENT OF BULLS

It is generally recommended that dairy bulls not be kept on the farm and that all breeding be done by AI. By not keeping bulls on the farm, producers (1) decrease the risk of a dangerous bull and (2) increase genetic progress by using superior AI bulls. A 1996 NAHMS survey of dairy producers found that almost half of dairy farms had no on-farm mature bulls, and another one-third had only a single sire on the farm.

Some commercial producers do keep bulls on their farms to breed cows naturally, primarily because bulls detect estrus more accurately and keep conception rates high. Clean-up bulls are typically 2 years of age or younger due to the aggressive nature of mature males. Nose rings are sometimes placed in the noses of bulls to provide a means of restraint and control. Typically these devices are nonrusting and self-piercing with application usually occurring when the bull is between 9 and 12 months of age.

Often breeding bulls are housed with lactating cows and are thus fed the same ration. Diets appropriate for the high producing cow will exceed the nutritional requirements of young sires. The high calcium content of a diet appropriate for lactating cows may lead to calcification of muscle in the mature bull if they are fed under this regime for an extended period of time.

CALVING OPERATIONS

Dairy cows that are close to calving should be separated from other cows, and each cow should be placed in a maternity stall that has been thoroughly cleaned and bedded with clean bedding. The cow in a maternity stall can be fed and watered there. A cow that delivers her calf without difficulty should not be disturbed; however, assistance may be necessary if the cow has not calved by 4–6 hours from the start of labor. Extreme difficulties in delivering a calf may require the services of a veterinarian, who should be called as early as possible.

As soon as a calf arrives, it should be wiped dry. Any membranes covering its mouth or nostrils should be removed, and its **navel** should be dipped in a tincture of iodine solution to deter infection. Producers should be sure newborn calves have an adequate amount of colostrum because colostrum contains antibodies to help the calf resist any invading microorganisms that might cause illness. The cow should be milked to stimulate her milk production.

Many commercial dairy operators dispose of bull calves shortly after the calves are born. Some bull calves are fed for veal, while others are castrated and fed for beef. Heifer calves are grown, bred, and milked for at least one lactation to evaluate their milk-producing ability.

Dairy calves rarely nurse their dams. Usually they are removed from their dams and fed milk or milk replacer for 4 to 8 weeks. Calves that are separated from one another usually have fewer health problems and thus a higher survival rate (Fig. 28.2).

Milk that is not salable or milk substitutes are fed to the young calves. Grain and leafy hay are provided to young calves to encourage them to start eating dry

FIGURE 28.2 An example of a housing system designed to maintain calf separation to minimize disease transfer. This system allows ease of monitoring calf health and feeding as well as providing protection for the calf.

feeds and to stimulate rumen development. As soon as calves are able to consume dry feeds (45–60 days of age), milk or milk replacer is removed from the diet because of its high cost.

MILKING AND HOUSING FACILITIES FOR DAIRY COWS

Dairy farming is labor-intensive, so labor-saving machines are common parts of the facilities. Dairy cows usually are managed in groups of 20–50 head; they may be housed in free-stall (cows have access to stalls but are not tied) or loose housing systems. These facilities reduce cleaning, feeding, and handling time as well as labor. In tie-stall barns, cows are tied in a stanchion and remain there much of the year; feeding and milking are done individually in the stanchion. In this case, a pipeline milking system is used. Cows are milked in place and milk is carried via a stainless-steel pipe to the bulk tank, where it is cooled and stored. In loose housing systems, such as the loaf shed or free stall (Fig. 28.3), cows are brought to a specialized milking facility, called a **parlor,** for milking. The percentages of dairy farms using various **housing systems** are listed in Table 28.4.

Many housing systems require daily or twice-daily removal of manure and waste feed; this may be done mechanically (with a tractor blade) or by flushing with water. Loose housing systems without free stalls usually have cows on a manure pack, where the pack is typically removed once or twice a year. Manure,

FIGURE 28.3   A water flush free-stall barn with individual lock-in stanchions. These facilities provide for inside feeding of a totally mixed ration. Courtesy of Colorado State University.

TABLE 28.4   Housing Type by Production Stage

Type	Percent of Enterprise			
	Unweaned Dairy Heifers	Weaned Dairy Heifers	Lactating Cows	Maternity
Freestall	2.5	9.7	24.4	5.6
Individual animal area	29.7	6.6	2.3	38.3
Multiple animal area	40.0	73.9	17.9	26.3
Tie stall or stanchion	10.5	11.5	61.4	26.3
Drylot	9.1	38.1	47.2	28.9
Pasture	7.4	51.4	59.6	41.9
Calf hutch	32.5	—	—	—

Source: Adapted from NAHMS, 1996.

urine, and other wastes usually are applied directly to land and/or stored in a lagoon, pit, or storage tank for breakdown before being sprayed or pumped (irrigated) onto land.

WASTE MANAGEMENT

The increased attention of consumers and environmental protection groups toward confined animal enterprises has resulted in a multitude of local, state, and federal regulations. As a result, many dairy operators have taken a proactive stance to better manage their environmental impact. The creation of nutrient management plans coupled with the use of rations that are more precisely formulated to minimize generation of problematic levels of nitrogen and phosphorus in animal waste can effectively improve environmental impact.

A South Carolina study found that almost three-quarters of dairies evaluated were feeding excess phosphorus in the diet. This excess nutrient was not correlated with improved milk production. As such, the use of precision ration balancing had the potential to reduce the amount of P fed by about 20% on average.

Developing a **nutrient management plan** for a dairy is a critical step in assuring conformance to environmental regulations and dealing with complaints from neighbors. Nutrient management plans should be accomplished via a thorough development process and include site information (names and contact information, physical enterprise description, and emergency plans); production data (number of animals, estimated amount of waste and waste water produced); permit information as specified by law; manure application data; test results and dead animal disposal protocols.

Records that should be recorded and retained in regards to nutrient management planning and monitoring include informational records such as soil tests, manure tests, and manure application agreements. Also, activity records such as manure storage, field application sites and quantities, and equipment calibration records.

When manure is applied to land, then the following pieces of information should be logged—date of application, amount applied, number of acres, how it was applied, who performed the application, the wind speed and direction, air temperature, soil conditions, and sky conditions. Finally, the cropping history should be recorded including crop yields, nutrient credits taken, use of commercial fertilizers, level of precipitation, and tillage practices.

MILKING OPERATIONS

An example of a modern milking parlor is a concrete platform raised about 30 in. above the floor (pit) of the parlor (Fig. 28.4). It is designed to speed the milking operation, reduce labor, and make milking easier for the operator. Cows enter the parlor in groups or individually, depending on the facility. The udder and teats are washed clean, dried, and massaged for about 20–30 seconds. Milking begins about 1 minute later and continues for about 6–8 minutes. After milking is completed, the milking unit is removed manually or by computer, and the teats are dipped in a weak iodine solution. Fig. 28.5 illustrates a rotary milking parlor design that allows cows to flow through the parlor in a circular pattern with the milking crew located centrally.

FIGURE 28.4 A trigon milking parlor equipped with computer units and automatic milker takeoffs. Two sides of the milking parlor will accommodate six cows each, while the third side holds four cows. Courtesy of Colorado State University.

FIGURE 28.5   A rotary milking parlor that facilitates continuous flow of cows to and from the milking area and optimizes convenience for milking personnel.

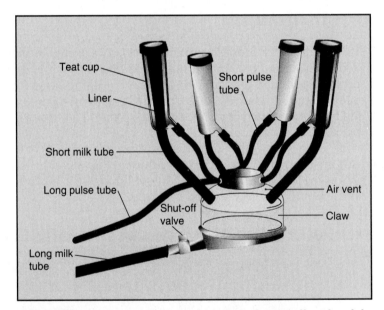

FIGURE 28.6   Modern milking machines have dramatically reduced the labor intensity of dairy farming. However, close supervision of the milking process is required to ensure that the udder and teats are clean prior to milking, that the suction cups are applied correctly, and that the milking machine is removed from the cow as soon as milking is complete. Courtesy Managing Milk Quality, copyright 1998. ITA LaPocaitiere, www.hoards.coms.

Modern milking machines (Fig. 28.6) are designed to milk cows gently, quickly, and comfortably. Some have a takeoff device to remove the milking cup once the rate of flow drops below a designated level. Removal of the milking unit as soon as milking is complete is important because prolonged exposure to the machine can cause teat and udder damage and lead to disease (mastitis) problems.

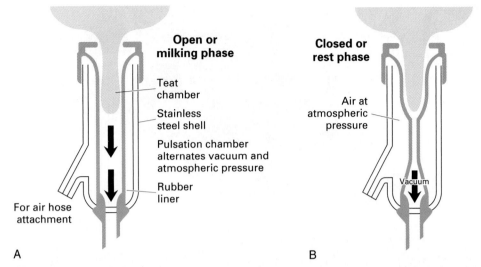

FIGURE 28.7   Longitudinal section of the teat cup of a milking machine. (A) During the open or milking phase, the vacuum level is the same on both sides of the liner, permitting it to be in an open position. (B) During the closed or rest phase, atmospheric air is admitted into the pulsation chamber by the pulsator. The pressure differential between the inside of the liner and the outside causes the liner to collapse. Courtesy of De Laval Agricultural Division. Alfa-Laval, Inc., Poughkeepsie, NY.

The milking machine operates on a two-vacuum system: one vacuum is located inside the rubber liner and the other outside the rubber liner of the teat cup (Fig. 28.7). There is a constant vacuum on the teat to remove the milk and keep the unit on the teat. The intermittent vacuum in the pulsation chamber causes the rubber liner to collapse around the teat. This assists blood and lymph to flow out of the distal teat into the upper part of the teat and udder.

The milking parlor and its equipment must be kept sanitary. The pipes are cleaned and sanitized between milkings. The total milking parlor is illustrated in Figure 28.8.

Regular milking times are established with equal intervals between milkings. Most cows are milked twice per day; however, high-producing cows produce more milk if they are milked three times as day. The latter is labor-intensive and economically feasible only in herds with high levels of productivity.

CONTROLLING DISEASES

The same diseases afflict dairy and beef cattle, but some diseases are more serious in dairy cattle than in beef cattle. Certain disease organisms can be transmitted through milk and can be a major problem for people who consume unpasteurized milk. Disease organisms are carefully monitored. The goal levels for bacterial cultures in on-farm milk tanks are listed in Table 28.5.

Several diseases, such as tuberculosis and brucellosis, can affect the health of dairy cattle and of humans consuming their milk. Requirements for the production of grade A milk include that the herd must be (1) checked regularly with the ring test for tuberculosis and (2) bled and tested for brucellosis.

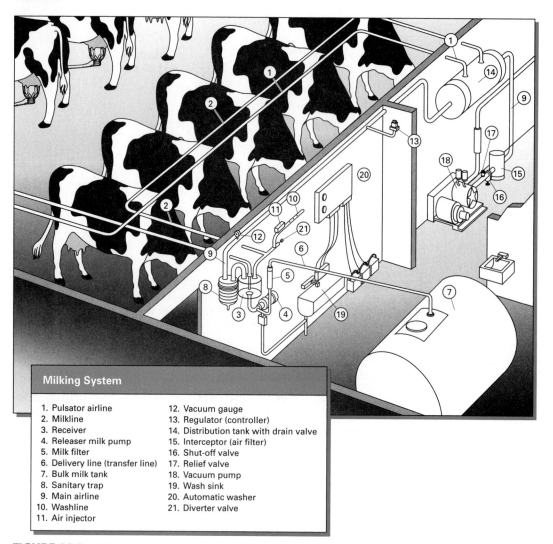

Milking System

1. Pulsator airline
2. Milkline
3. Receiver
4. Releaser milk pump
5. Milk filter
6. Delivery line (transfer line)
7. Bulk milk tank
8. Sanitary trap
9. Main airline
10. Washline
11. Air injector
12. Vacuum gauge
13. Regulator (controller)
14. Distribution tank with drain valve
15. Interceptor (air filter)
16. Shut-off valve
17. Relief valve
18. Vacuum pump
19. Wash sink
20. Automatic washer
21. Diverter valve

FIGURE 28.8   Schematic of the integrated milking system. Courtesy of Managing Milk Quality, copyright 1998, ITA LaPocaitiere, www.hoards.com.

TABLE 28.5   Goal Levels for Herd Tank Cultures (colony-forming units)

Type of Bacteria	Goal Levels	Moderate Levels	High Levels	Very High Levels
Strep. ag.	0	50–200	200–400	>400
Staph. Aureus	0–50	50–150	150–250	>250
Coliforms	<100	100–400	400–700	>700
Environ. Staph.	<300	300–500	500–750	>750
Strep. non-ag.	<700	700–1,200	1,200–2,000	>2,000

Source: Adapted from Colorado State University.

Further, milking and housing facilities must be clean and meet certain specifications. Grade A milk must have a low bacteria count and somatic cell count (low mastitis). Grade A milk sells at a higher price than lower grades of milk; therefore, most dairy producers strive to produce grade A milk.

Bang's disease (brucellosis) is caused by the *Brucella abortus* organism. It can markedly reduce fertility in cows and bulls and humans can contract it as a disease known as *undulant fever.* Heifers should be vaccinated in calfhood to provide immunity from disease. Since most dairy cows are bred by artificial insemination, there is little danger of a clean herd becoming contaminated from breeding. Persons drinking unpasteurized milk are taking a serious risk unless they know that the milk comes from a clean herd.

Perhaps the most troublesome disease in dairy cattle is **mastitis**—inflammation and infection of the mammary gland. The disease destroys tissue, impedes milk production, and lowers milk quality. Mastitis costs the U.S. dairy industry more than $1.5 billion each year or approximately $200 per cow annually. Reasons for culling dairy cows are listed in Table 28.6. Mastitis and reproductive failure are the leading reasons for culling.

In its early stage of development, subclinical mastitis is undetectable by the human eye. This type of mastitis can be detected only with laboratory equipment (Somatic Cell Counter at the DHI testing laboratory) or with a cow-side test called the California Mastitis Test (CMT). In the latter test, a reagent is added with a paddle to small wells, and then milk is squirted from each quarter into a specific well. If a reaction occurs, then the relative degree of subclinical mastitis infection can be determined. Small white clots appear in the milk of cows with advanced mastitis. At this stage it is called **clinical mastitis** and is easily observed. Milk can also be collected using a strip cup, which has a fine screen on a dark background such that white flakes, strings, or blood (from infected areas) can be seen. As mastitis progresses, the milk shows more clots and is watery, and acute infection is seen. At this stage, the signs of mastitis in the cow are obvious: the udder is swollen, red, and hot and gives pain to the cow when touched. In the final stages of mastitis, the gel formation becomes dark and is present in a watery fluid. At this stage, the udder has been so badly damaged that it no longer functions properly.

Susceptibility to mastitis may be genetically related to a certain degree, but environmental factors such as bruises, improper milking, and unsanitary conditions are more prevalent causes. High-producing animals may be more prone to stress and mastitis than low-producing animals. Also, cows with **pendulous** udders or oddly shaped teats are more likely to develop mastitis. Housing facilities and milking procedures are other important factors.

The best approach to controlling mastitis is using good management techniques to prevent its outbreak. These include using routine mastitis tests (DHI or CMT), treating infected animals, preventing infection from spreading, dipping teats after milking, using dry-cow therapy, and practicing good husbandry

TABLE 28.6   Dairy Cow Culling Practices

Reason	% Dairy Cows Culled
Udder/mastitis problems	26.9
Reproductive problems	26.5
Poor production	19.3
Lameness/injury	16.3
Disease	6.0
Other	4.1
Poor temperament	0.9

Source: Adapted from NAHMS, 2002.

for cleanliness. The development of a milking parlor quality assurance program coupled with an effective employer training effort can yield positive results. Such a program should include written protocols coupled with routine audits designed to monitor conformance to the prescribed standards. Suggested protocols that should be evaluated include the following:

1. Udder prep should result in a uniformly clean surface including the teat ends.
2. Milking units should be attached to udder within one minute of the start of udder preparation.
3. Monitor teat cup attachment to minimize air leaks during milking.
4. Postmilking teat dipping should result in thorough surface coverage.

Mastitis prevention protocols and use on dairy farms are described in Table 28.7. Chronically infected cows may have to be sold due to poor performance. For example, one study found that cows with mastitis required 10 more days to first breeding, required .5 more services per conception, and were open 25 days longer.

Most treatments involve infusing an antibiotic into the udder through the teat canal. Depending on the antibiotic used, milk from treated cows should not be sold for human use for 3–5 days after the final treatment is given.

Vaccination and preventative health management practices for replacement heifers and cows are described in Tables 28.8 and 28.9.

Other diseases and health practices are discussed in chapter 21.

TABLE 28.7   Mastitis Preventative Management on U.S. Dairy Farms

Treatment	% Enterprises Using Treatment
Dry cow mammary infusions at final milk-out:	
Treated all quarters on all cows	76.5
Treated all quarters on selected cows	14.6
Did not treat in all four quarters	8.9
Lactating cows at milking time:	
Premilking wash pen	7.0
Premilking test dip/spray	58.3
Postmilking teat dip/spray	88.9

Source: Adapted from NAHMS, 1996.

TABLE 28.8   Dairy Heifer and Cow Vaccination Protocols

Vaccine	% Enterprises Vaccinating Heifers	% Enterprises Vaccinating Cows
Bovine viral diarrhea	71.5	74.2
Infectious bovine rhinotracheitis	67.0	69.3
Parainfluenza (Type 3)	60.0	62.2
Bovine respiratory syncytial virus	58.2	61.1
Hemophilus somnus	31.4	32.4
Leptospirosis	65.1	70.1
Clostridia	32.8	25.0
Brucellosis	51.0	—

Source: Adapted from NAHMS, 2004.

TABLE 28.9 Preventative Practices in Replacement Heifers

Practice	% of Enterprises—Heifers	% of Enterprises—Cows
Deworm	67.3	53.4
Coccidiostats in feed	46.5	—
Vitamin A-D-E (injection)	16.3	15.5
Vitamin A-D-E (in feed)	76.9	81.4
Selenium (injection)	12.7	—
Selenium (in feed)	70.8	—
Ionophores	42.2	—
Probiotics	13.1	16.7

Source: Adapted from NAHMS, 1996.

COSTS AND RETURNS

Assessment of profit opportunities requires that sufficient records be available to evaluate the farm's overall financial position, to monitor cost controls, to assure high labor efficiency, to measure production level, and to determine appropriate herd size.

Table 28.10 shows enterprise averages for 100 dairy farms. Comparisons for 26 profitable farms and 74 unprofitable farms are made for financial efficiency, total costs per cow, and cost per hundred pounds of milk. To be considered profitable, the dairy farm had to earn at least $30,000 of labor and management income per operator. Profitable farms averaged $47,111 per operator. Their return

TABLE 28.10 Comparison of Profitable and Unprofitable Dairy Farms[a]

Item	26 Profitable Farms	74 Unprofitable Farms
Financial Efficiency (percent of gross income)		
Operating expense	62.1%	67.4%
Interest expense	4.3	6.3
Depreciation expense	6.3	9.6
Net farm income	27.3	16.7
Total	100.0	100.0
Total Costs per Cow		
Hired labor	$257	$226
Feed	620	663
Machinery	294	300
Livestock	391	420
Crops	211	246
Real estate	202	198
Other	170	190
Total	$2,145	$2,243
Cost per Hundredweight of Milk		
Purchased feed and crop expenses	$4.17	$4.75
Breeding	0.19	0.21
Veterinarian and medicine	0.41	0.51
Milk marketing	0.42	0.43
Other expenses	0.69	0.76
Total	$5.88	$6.61

[a]Profitability not related to herd size or level of milk production—e.g., profitable farms' range in herd size was 34–270 cows, while milk production per cow ranged from 13,040–27,278 lb.
Source: Adapted from University of Wisconsin.

TABLE 28.11 Performance Records Summary

Management Area	Trait	Goal	Rationale
Lactation distribution	Days in milk (DIM)	180 days and even calving pattern	Average milk production declines with prolonged DIM
	Days dry	>80% of cows plus or minus 20% of the option for days dry	Dry periods less than 30 days or greater than 60 days result in economic loss
	Herd distribution and calving interval	365 days	Average calving intervals beyond 365 days yield a high percentage of cows in late lactation and below break-even production
Herd distribution	Percentage of DIM	Describes the average percentage of the month that cows spent in milk	Estimates the proportion of the month that cows contribute to cash flow
	Annual % DIM	85–87%	Each 1% increase through 91% yields a 100-lb increase in average annual lactation
Daily milk (DM) production	DM production	Depends on herd	Directly related to cash flow—trait is easily influenced by management and climate
	Average lb peaked 1st 90 days	Depends on herd	For each 1-lb increase in peak yield, the 305-day yield increases by about 220 lb
DIM at 1st breeding	Voluntary waiting period following calving prior to breeding	45–50 days	Allows for time to complete uterine involution and maintenance of a 13-month or less calving interval
	Average age at 1st calving	22.5–25.5 mos.	>25 months yields lower relative net income per cow, profit per day of life, and lifetime profit per cow

Management Area	Trait	Interval (days)	Herd Distribution (%)	Possible Problems
Heat detection status	Intervals between breeding	≤18	10	Cystic ovary, poor heat detection
		18–24	65	Normal
		25–35	10	Abnormal cycle, error in cow ID or heat detection
		36–48	10	Missed heat
		>48	5	2 missed heats or embryonic death

Management Area	Trait	Goal	Rationale
	Heats detected	>80%	Approximately $40 loss for each missed heat
Breeding performance	Services per conception	<1.5	Each 0.1 service over 1.5 services per conception costs $1.50 per cow
	First service conception	>75%	Low values indicate extended calving interval in the future
	Service per cow	>2.0	If services per cow are greater than service per conception by ≥0.4, immediately look for cause of problem

Management Area	Trait	SCS	% Herd	High SCS
Somatic cell summary	Somatic cell score	Low	>90	Indicate possible subclinical mastitis, problems with quality control or milking equipment
		Med.	<7	
		High	<3	

TABLE 28.11 *(continued)*

Management Area	Trait	Goal	Rationale
Culling and herd turnover	Culling guidelines or do not breed (DNB)	Depends on herd	Decisions depend on a comparison of expected income from cow and expected income from a replacement heifer
	Selective culling	15–35%	Assure genetic rate of change by removing poor performers
	Nonselective culling	*Annual culling rates* Reproductive culls: 5–10% Udder health culls: <5–10% Disease/injury culls: <5%	Removal of reproductive failures, diseased or injured cows, and poor udders

Source: Adapted from Western Regional Extension Program 117.

compared to that of unprofitable farms was 13.8% versus −0.8% for return on equity and 11.2% versus 2.1% for return on assets. Interestingly, as dairies increase in size, economies of size and scale are not always realized. One Wisconsin study found that cash costs per cwt. equivalent for enterprises with fewer than 50 cows, 76 to 100 cows, 151–250 cows, and more than 250 head were $7.73, $7.81, $7.64, and $7.96, respectively.

Fortunately, the dairy industry has superior herd production and financial record-keeping systems available. Herd managers should measure their herds' productivity against regional or national **benchmarks.** For example, the competitive benchmarks for milking parlor turns per hour are 4.5, parlor labor cost per cwt. of milk at $0.50, cows milked per worker per hour at 100 with 3,000 cwts. of milk per hour, and a hourly labor cost per milking stall of $0.50. A summary of typical dairy production records and the recommended goals is provided in Table 28.11.

CHAPTER SUMMARY

- Dairy farming is labor-intensive, so facilities are important in saving labor, feeding cattle, and producing wholesome, clean milk.

- Nutrition and health programs are extremely important in maintaining high milk production in cows.

- The most troublesome disease in many dairy operations is mastitis—the inflammation and infection of the mammary gland.

- Forage and concentrates must be high in nutrient content and must be provided in large amounts to express the high genetic potential for milk production.

KEY WORDS

body condition	housing systems	mastitis
total mixed ration	nutrient management	clinical mastitis
heat stress	plan	pendulous
navel	(Bang's disease)	benchmarks
parlor	brucellosis	

REVIEW QUESTIONS

1. What is the standard length of lactation period in dairy cattle in the United States?

2. *True or False:* It is often difficult to meet the nutrient requirements of high-producing dairy cattle during the first 2 to 4 months postpartum because nutrient demand is high but feed intake is limited.

3. A ration in which silages and roughages are mixed together and fed to dairy cattle is called a _____ .

4. Why are concentrates fed to lactating dairy cattle?

5. What nutrient requirement do young (2-year-old) cattle have in addition to those for maintenance and lactation?

6. Why are dry periods of 50–60 days necessary?

7. When should heifers be first bred?

8. What are two advantages of purchasing bull semen for use in artificial insemination?

9. Where is milking of dairy cattle most commonly performed?

10. Why is it important to remove the milking machine as soon as milking is complete?

11. How are dairy cows typically housed?

12. What is the most troublesome disease of dairy cattle?

13. What does the best approach to controlling mastitis include?

SELECTED REFERENCES

Bath, D. L., F. N. Dickinson, H. A. Tucker, and R. D. Appleman. 1985. *Dairy Cattle: Principles, Practices, Problems, Profits.* Philadelphia: Lea & Febiger.

Bertrand, J. A., J. C. Freck, and J. C. McConnell. 1999. Phosphorus intake and excretion on South Carolina dairy farms. *Prof. Ani. Sci.* 15:264–267.

Dairy Info Base. 2006. USDA:ADDS. www.adds.org.

Fiez, E., D. Grusenmeyer, E. Thomason, and B. Wiesen. 1992. Dairy Record Templates. *Western Regional Extension Publ. 117.*

Hoffmann-LaRoche, Inc. 1989–1990. *Nutritional Needs of Dairy Cows.* Nutley, NJ: Hoffmann-LaRoche.

Meyer, D. 2000. Dairying and the environment. *J. Dairy Sci.* 83:1419–1427.

National Animal Health Monitoring System. 1996. Reference of Dairy Management Practice. USDA:VS.

National Animal Health Monitoring System. 2004. Animal Disease Exclusion Practices on U.S. Dairy Operations. USDA:VS.

National Research Council. 1988. *Nutrient Requirements of Dairy Cattle.* Washington, DC: National Academy Press.

Pennsylvania State University Extension Service. 1995. Proc. Managing Dairy Farms into the 21st Century.

Perkins, B. L. 1998. Managing the 30,000 Pound Herd. Proc. Colorado Dairy Nutrition Conference, Fort Collins, CO.

CHAPTER 29

Swine Breeds and Breeding

LEARNING OBJECTIVES

- List and describe the characteristics of the major swine breeds
- Describe the traits that impact profitability in the swine industry
- Discuss the effective use of performance records in the swine industry
- Explain selection of replacement gilts and boars
- Overview the use of crossbreeding in swine production

Breeds of swine are genetic resources available to swine producers. Most market pigs in the United States today are crossbreds, but these crossbreds derive their origin and continued perpetuation from seedstock herds that maintain pure breeds of swine. Knowledge of the breeds, their productive characteristics, and their crossing abilities is an important component of profitable swine production.

CHARACTERISTICS OF SWINE BREEDS

In the early stages of the formation of swine breeds, breeders distinguished one breed from another through assessment of visual characteristics. These visual distinguishing characteristics for swine were, and continue to be, hair color and erect or drooping ears. Figure 29.1 shows the major breeds of swine and their primary distinguishing characteristics.

Relative importance of swine breeds is shown by registration numbers recorded by the different breed associations (Table 29.1). Although the registration numbers are for registered purebreds, the numbers do reflect the demand

Berkshire
(black with
white on
legs, snout,
and tail;
erect ears)

Chester
White
(white;
drooped
ears)

Duroc
(red;
drooped
ears)

Hampshire
(black with
white belt;
erect ears)

Landrace
(white;
large,
drooped
ears)

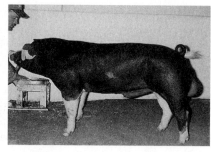

Poland
China
(black with
white on
legs, snout,
and tail;
drooped
ears)

Spotted
(black and
white spots;
drooped
ears)

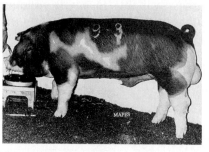

Yorkshire
(white;
erect ears)

FIGURE 29.1   The major breeds of swine in the United States. Courtesy of Mapes Livestock Photos (Berkshire, Chester White, Landrace, Poland China, Spotted, Duroc, and Hampshire).

for breeding stock by commercial producers who produce market swine primarily by crossbreeding two or more breeds. Note how the popularity of breeds has changed with time. This change has resulted, in part, from use of more objective measurements of productivity and the breeds' ability to respond to the demand and selection for higher productivity and profitability.

Competitors of purebred breeders are large corporate seedstock suppliers who offer boars and gilts of specialized bloodlines. These lines of breeding, usually

TABLE 29.1 Number of Litters Recorded for Major U.S. Breeds of Swine

Breed	2005	1995	1990	1985	Year Association Formed
Berkshire	5,458	1,934	2,071	2,158	1875
Chester White	1,824	3,077	5,554	5,158	1930
Duroc	84,111[a]	13,709	22,179	21,852	1883
Hampshire	45,806[a]	16,244	18,925	14,117	1893
Landrace	31,650[a]	4,405	4,365	3,501	1950
Poland	1,047	1,600	1,848	2,318	1876
Spotted	2,080	3,251	6,443	9,310	1914
Yorkshire	111,137[a]	19,007	23,861	24,335	1935

[a]Reported in number of head instead of number of litters.
Source: Adapted from National Swine Registry and other associations.

called **hybrids,** originated from crossing two or more breeds, then applying some specialized selection programs. Farmers Hybrid Company was the first swine-breeding corporation; it formed in the 1940s. Since 1960 DeKalb, Land O'Lakes, Pig Improvement Company (PIC), Newsham Hybrids, Swine Genetics International, Ltd., and others have joined it as genetic providers in the industry.

TRAITS AND THEIR MEASUREMENTS

Table 29.2 gives an overview of the traits preferred in market barrows and gilts. These are specific targets for measurable traits.

Sow Productivity

Litter size, number weaned per litter, 21-day litter weight, and number of litters per sow per year measure **sow productivity,** which is of extremely high economic importance. A combination of number of pigs (born alive) and litter weight at 21 days best reflects sow productivity (Fig. 29.2). Litter weight at 21 days is used to measure milk production because the young pigs have consumed little supplemental feed before that time. Sow productivity traits are low in heritability, but economically important. They can be best improved through managing the environment (e.g., feed, health), rather than through selective breeding. Eliminating sows low in productivity while keeping highly productive sows is still a logical practice for economic reasons, but not when making major genetic change. It is important for commercial producers with effective crossbreeding programs to use breeds that rank high in sow and maternal productivity traits. Producers should carefully evaluate the cost of developing replacement females and understand the effect of pigs sold per litter on the number of farrowings required before a sow returns positive value. The impact of pigs marketed per litter and cost of sow development are illustrated in Table 29.3. For example, a female that is culled after only 1 farrowing returned no profit, while those females developed at the lowest cost and with the longest sustained production per litter returned the most profit to the enterprise.

Growth

Growth rate is economically important to most swine enterprises and has a heritability of sufficient magnitude (35%) to be included in a selection

TABLE 29.2   Symbol: A Standard of Performance

Symbol	Is the graphic representation of a standard of performance and carcass composition that provides a basis for the continuing improvement of the modern hog.
Symbol	Symbolizes the commitment of the entire pork industry to making lean pork the true meat of choice of the twenty-first century.
Symbol I	**A Market Barrow—adopted in 1983** Was the target or goal for all pork producers. These figures probably represented the top 10% of all hogs marketed in 1983. • 250-lb market barrow • 180-lb, 32-in.-long carcass yields 105 lb of lean pork • Feed conversion efficiency of 2.5 from birth to a market age of 150 days • Demonstrates a lean gain of ¾ lb per day of age • Last rib fat depth measurement at slaughter is 0.7 in. • Loin muscle area is 5.8 sq in. • Average of three backfat measurements is 1.0 in.
Symbol II	**A Market Gilt—adopted in 1996** • 195-lb carcass from a live weight of 260 lb • Intramuscular fat level greater than or equal to 2.9% • High health production system • Produced by an environmentally assured producer on PQA Level III • Result of a terminal crossbreeding program • From a maternal line capable of weaning 25 pigs per year • Marketed at 164 gilts days of age • Live weight feed efficiency of 2.40 • Fat-free lean gain efficiency of 5.90 • Fat-free lean gain of 0.78 lb per day • Standard Reference backfat of 0.6 • Fat-Free Lean Index is 52.2
Symbol III	**A Market Goal—adopted in 2005** • 205-lb carcass from a live weight of 270 lb • Intramuscular fat level greater than or equal to 3.0% • High health production system, free of parasites and Halothane gene • Produced by an environmentally assured producer on PQA program with animal care assessment • Result of a terminal crossbreeding program • From a maternal line capable of weaning 25 pigs per year after multiple parities • Marketed at 156 (164 gilts) days of age • Live weight feed efficiency of 2.4 (2.40) • Fat-free lean gain efficiency of 5.9 (5.80) • Fat-free lean gain of 0.95 lb per day • Fat-Free Lean Index of 53.0 (54.7) • Loin muscle area of 6.5 sq. in. (7.1) • Belly thickness of 1.0 in. • 10th rib fat of 0.7 in. (0.6) • Muscle color score of 4.0

Numbers in parentheses represent goals for gilts.
Source: Adapted from National Pork Producers Council.

program. Seedstock producers need a scale to measure weight in relation to age (Fig. 29.3). Growth rate can be expressed in several ways and typically is adjusted to some constant basis, such as days required for swine to reach 250 lb.

Feed Efficiency

Feed efficiency measures the pounds of feed required per pound of gain and has an associated heritability of 0.30. This trait is economically important because feed costs account for 60–70% of the total production costs for commercial

FIGURE 29.2 The number of pigs raised per sow and the weight of the litter at 21 days of age effectively measure sow productivity. Courtesy of the National Swine Registry.

TABLE 29.3 Effects of Pigs Sold per Litter and Replacement Gilt Development Cost on Number of Farrowings Required to Return Positive Value to the Enterprise

Pigs Sold per Litter	Farrowing and Positive Return[a]					
	1	2	3	4	5	6
7.5	—	—	—	—	—	$ 17
8.0	—	—	$ 10	$ 60	$110	159
8.5	—	$ 7	82	156	229	300
9.0	—	56	155	252	348	442
Replacement Female Price ($/hd)						
150	—	$ 63	$138	$212	$284	$356
175	—	35	110	184	256	328
200	—	7	82	156	229	300
225	—	—	54	128	200	273
250	—	—	27	100	173	245

[a]Measured as Net Present Value. Assumes a $46/cwt market price.
Source: Adapted from Stadler and Lacy, 1998.

producers. Obtaining feed-efficiency records requires keeping individual or group feed records. However, feed efficiency can be improved without keeping feed records, such as by testing and selecting individual pigs for gain and back-fat. This improvement occurs because fast-gaining lean pigs tend to be more efficient in their feed utilization. Feed-efficiency records are obtained in some central boar-testing stations; however, costs and other factors need to be considered before similar records are taken on the farm.

FIGURE 29.3 Using scales to measure the growth rate of individual boars is an essential part of a sound breeding system. Courtesy of Iowa State University.

Carcass Traits

Carcass traits are used to estimate the pounds or percentage of acceptable-quality lean pork (10% fat) in the carcass. Percent lean measured on a carcass basis has a heritability of 0.48. The traits used to predict carcass composition are, in order of importance, (1) fat depth over the loin at the 10th rib, (2) loin muscle area, and (3) carcass muscling score (see Chapter 8). Fortunately, measures of fatness can be predicted reliably by backfat measurements taken on live pigs by ultrasonics (Fig. 29.4). Experienced persons can visually predict significant differences in lean-to-fat composition in live pigs with a fairly high degree of accuracy.

Structural Soundness

Structural soundness—the capacity of breeding and slaughter animals to withstand the rigors of confinement rearing and breeding—is vital to the swine industry today. Breeders feel that most unsoundness results from intensive swine production; actually, though, intensive rearing only makes unsoundness more noticeable. Some seedstock producers raise their breeding stock in modern facilities (similar to the manner in which most commercial producers raise their offspring) and then cull the unsound animals.

Recent studies report unsoundness to be medium in heritability; therefore, improvement can be made through selection. Soundness can be improved through visual selection if breeders decide to cull restricted, too straight-legged, unsound boars lacking the proper flex at the hock, set of the shoulder, even size of toes, or proper curvature and cushion to the forearm and pastern.

Inherited defects and abnormalities generally occur with a low degree of frequency across the swine population. It is important to know that they exist

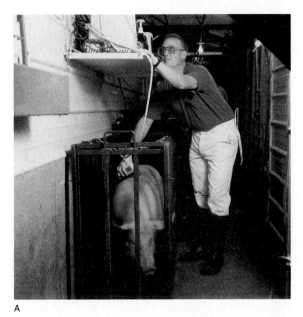

A

B

FIGURE 29.4 (A) Ultrasound measurements being taken on a pig. (B) Screen showing ultrasound image of a loineye and fat deposition (above the loineye muscle) from the pig shown in (A). Courtesy of Monsanto Choice Genetics.

and that certain herds may experience a relatively high incidence. Certain structural unsoundness characteristics might be categorized as inherited defects. **Cryptorchidism** (retention of one or both testicles in the abdomen), umbilical and scrotal **hernias** (rupture), and **inverted nipples** (nipples or teats that do not protrude but are inverted into the mammary gland), and **pale, soft,**

and exudative (PSE) carcasses are considered the most commonly occurring genetic defects and abnormalities. The PSE condition and the **porcine stress syndrome (PSS)** have occurred in recent years due to selection for extreme muscling.

The PSS gene in its homozygous mutant form *(nn)* results in either death from physical stress or PSE meat. However, the presence of the normal allele *(N)* nearly always prevents stress death or the rigidity response to testing using the anesthetic halothane.

Research studies have shown the carrier animals to be 1 to 4% higher than normal pigs *(NN)* in lean percentage and to be equal, if not superior, in growth rate. All of these studies, however, have suggested that the muscle quality of these pigs *(Nn)* is generally undesirable and that the frequency of PSE is in the 30–50% range.

There is evidence that the PSS gene does not account for all of the pork industry's PSE problems. However, because the gene offers only a minimal increase in lean percentage and significantly increases the PSE frequency, the gene should be eliminated from the U.S. pig population.

If pork is to be elevated to the "meat of choice" as noted in industry literature, then carcass leanness and muscling cannot be the entire focal point. In addition to eliminating the PSS gene, selection methods to increase pork quality appear advisable. Consumer preference studies on pork suggest that an increase in the lipid level of meat would make it more competitive with chicken and possibly other meats.

EFFECTIVE USE OF PERFORMANCE RECORDS

Commercial producers recognize that the performance levels in their herds are influenced by two factors: genetics and environment (e.g., feed, health, and housing).

One of the most important decisions in the swine enterprise is the selection of an individual, company, or cooperative to supply genetics. Determination of a seedstock supplier should be based on an assessment of individual herd performance and need, freedom from disease and genetic defects in the seedstock herd, availability of reliable data, and availability of associated herd management services. A sound genetic improvement program should include four features: (1) accurate, complete performance records including animal identification, consistent measurement of all boars and gilts (not on-again, off-again or limited, partial performance testing), and ranking of animals within defined contemporary groups; (2) assessment of the genetic merit of economically important traits (growth rate, feed efficiency, carcass merit, and reproductive performance) based on the individual's performance relative to its contemporary group and incorporating the performance of relatives; (3) indexes weighting traits relative to their economic importance in commercial pork production (the indexes should correctly rank the individuals relative to their intended use in crossbreeding and marketing systems); and (4) selection of the highest-ranking boars and gilts based on selection indexes. Seedstock producers should utilize selection indexes as their primary selection criteria.

The rate of genetic improvement in a commercial pork producer's herd will parallel the rate of genetic progress made by the seedstock supplier the commercial producer chooses (Fig. 29.5). Therefore, the commercial producer

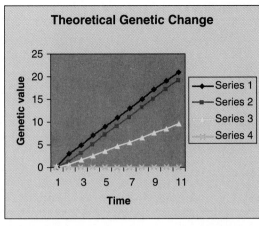

Series 1   Genetic trend of seedstock herds using available selection tools

Series 2   Genetic trend of commercial herds buying superior breeding stock from superior seedstock herds

Series 3   Genetic trend of commercial herds buying average breeding stock from superior seedstock herds

Series 4   Genetic trend of commercial herds using unimproved breeding stock and no selection pressure

FIGURE 29.5   Comparison of genetic trends under various selection strategies. *Source:* Adapted from *Pork Industry Handbook* (PIH-9).

should select seedstock producers whose improvement programs have the four features previously mentioned.

There are several excellent genetic evaluation programs available to seedstock breeders (e.g., SWINE-EBV, Nebraska SPF, and STAGES). **STAGES** (the Swine Testing and Genetic Evaluation System), which is a series of computer programs implemented by breed associations, will be highlighted here. STAGES compute genetic evaluations for several traits within a single herd. Reproductive traits include 21-day litter weight and number born alive. Postweaning traits with EPDs reported are backfat, loineye area, pounds of fat-free lean adjusted to 250 lb, days to 250 lb, and feed per pound of gain. A series of bioeconomic indexes are also reported and include the **sow productivity index (SPI),** the **terminal sire index (TSI),** and the **maternal line index (MLI).** These traits are evaluated in a multibreed national sire evaluation system for the Duroc, Hampshire, Yorkshire, and Landrace breeds.

The genetic evaluation that results from STAGES is expressed as an expected progeny difference (EPD).

Several breeds each have an across-herd sire summary that is published on an annual basis. Sire summaries provide unbiased estimates of the genetic merit of boars that have a legitimate performance record or have progeny with performance records. These estimates are in the form of EPDs.

The EPD is a prediction of the progeny performance of an animal compared to the progeny of an average animal in the breed based on all information currently available. This allows for the direct comparison of all boars evaluated in the sire summary based on the predicted differences between the performance of their progeny. These EPDs are expressed as plus or minus values, with the average EPD for the breed being approximately zero. Table 29.4 compares two boars assuming the EPDs have equal accuracies.

Progeny of boar A are expected to reach 250 lb 3.5 days sooner than progeny of boar B and to have 0.10 in. less backfat. This assumes the boars are bred to sows of similar genetic merit.

An accuracy value is listed with each EPD. Accuracies range from 0–1.0 and give a confidence estimate of the EPD. Accuracies that approach 1.0 indicate the EPD is a good estimate of the boar's true genetic potential for that trait.

TABLE 29.4   Comparison of EPDs for Two Boars

	Expected Progeny Difference	
Boar ID	Days to 230 lb	Backfat
A	−3.5	−0.05
B	0	+0.05

TABLE 29.5   Comparison of EPD Index for Two Boars

	Expected Progeny Difference		
Boar ID	SPI	MLI	TSI
C	92	106	136
D	118	105	94

The three bioeconomic indexes allow producers to focus selection pressure on specific phases of the production cycle. Sow productivity index (SPI) ranks individuals for reproductive traits using a composite of traits including number born alive, number weaned, and litter weight. Each unit of SPI represents $1 per litter produced by each daughter of a particular parent.

The maternal line index (MLI) is used to evaluate replacement gilts for cross-breeding systems. Both maternal and terminal traits are included, but the maternal traits are weighted twice as much as the terminal.

The terminal sire index (TSI) is useful in selection of animals for terminal crossing programs and includes only postweaning traits (backfat, days to 250 lb, pounds of lean, and feed per pound of gain). Each TSI unit is the equivalent of $1 for every 10 pigs marketed.

Table 29.5 compares the indexes of two sires. Each litter by sire D would be expected to be worth $26 more than those of C due to heavier weaning weights and greater numbers of live pigs. Sires C and D are virtually equal in terms of maternal line index. However, every 10 pigs out of C would be expected to be worth $42 more than those sired by D due to superior growth and leanness.

The results of the National Pork Producers Council Terminal Line National Genetic Evaluation Program (NGEP) were published in 1995. This was the largest, most comprehensive genetic evaluation program ever done in the swine industry. A large representative sample of boars (795) from nine different sire lines (breeds and hybrids) were each mated to a similar genetic base of sows to produce 1,990 litters. These progeny test pigs were from matings to give 100% heterosis.

More than 3,200 test pigs were evaluated for growth and carcass traits. Evaluations were also made for loin meat quality traits and loin eating-quality traits, with more than 30 traits measured for these three major categories. Index 1 and Index 2 were calculated, with each index combining several traits into an economic value expressed as dollars per pig. Index 1 establishes the value of a sire line to a producer selling hogs on a backfat-based carcass merit buying system. Index 2 gives the value of a sire line if retail values can be realized in addition to the values achieved in Index 1. Thus, realizing these value differences per pig is dependent on the marketing alternative utilized. Table 29.6 gives a summary of several selected traits and Indexes 1 and 2.

The differences in sire lines in the NGEP are useful in planning effective selection programs. However, individual boar selection within a breed or line using EPDs and other performance data should receive the greater emphasis.

TABLE 29.6   Selected Trait Summary

				Trait					
Sire Line	Days to Reach 250 lb	Feed Conversion (lb feed/ lb gain)	Backfat (in.)	Loin Area (sq. in.)	Loin Drip Loss (%)[a]	Loin pH[b]	Tender- ness (lb)[c]	Index 1 ($/pig)[d]	Index 2 ($/pig)[e]
Berkshire	175	3.07	1.25	5.74	2.43	5.91	12.63	−$4.14	−$4.05
Danbred HD	176	2.88	0.98	6.75	3.34	5.75	12.78	2.12	1.52
Duroc	170	2.91	1.13	6.14	2.75	5.85	12.43	0.64	10.51
Hampshire	175	2.92	1.01	6.58	3.56	5.70	12.89	1.34	1.55
NGT Large White	173	2.94	1.17	5.62	2.92	5.84	13.40	−0.97	−8.87
Nebraska SPF Duroc	168	2.89	1.11	6.35	2.81	5.88	12.72	1.50	8.25
Newsham Hybrid	172	2.83	0.98	6.45	2.99	5.82	13.46	3.53	0.89
Spotted	174	3.14	1.24	5.83	2.88	5.83	13.02	−4.77	−8.20
Yorkshire	175	2.93	1.05	6.17	2.85	5.84	13.49	−0.74	−1.70

[a]Important because 1% drip loss equals a loss of 1 lb for every 100 lb meat.
[b]Higher pH associated with low drip loss, darker color, more firmness, increased tenderness—all positive attributes.
[c]Pounds of pressure to push probe into cooked loin (Instron machine).
[d]Index 1 combines days to 250 lb, feed conversion, and backfat at 10th rib.
[e]Index 2 combines intramuscular fat in loin, pH, tenderness, drip loss, and loin muscle area that were estimated using results of consumer preference study, then added to traits in Index 1.
Source: Adapted from National Genetic Swine Evaluation Program.

National consumer preference studies throughout the United States were made on samples of pork from the NGEP to evaluate the pork and also compare it to chicken breast. The major findings were as follows:

1. Price (58%) was the most important attribute when comparing chicken to pork. Next most important was lipid level (24%).

2. Higher lipid levels and greater tenderness were preferred when taste-testing pork.

3. Increases in lipid level have more than twice the economic value to these consumers as increased tenderness or more preferred pH levels.

4. The most preferred visual characteristics must be balanced against the most preferred taste characteristics to identify the pork chop with the greatest long-term consumer appeal.

SELECTING REPLACEMENT FEMALES

Sow productivity is the foundation of commercial pork production. The sow herd also contributes half of the genetic composition to the growing-finishing pigs. These two factors show the importance of replacement gilt selection so that highly productive gilts can be retained in the swine herd. Fast-growing, sound, moderately lean replacement gilts with good body capacity should be selected from litters where 8–14 excellent pigs were weaned (Fig. 29.6). Among sows that have farrowed and will rebreed, those that have physical problems, bad dispositions, extremely small litters (two pigs below the herd average), and poor mothering records should be culled.

FIGURE 29.6 The Hampshire gilt has many of the visual characteristics and performance records identified in Table 29.2. Courtesy of Mapes Livestock Photos.

Balance between sow culling and gilt selection needs to be established to maintain productive individuals while assuring a profitable herd size. Replacement gilts should be available in sufficient numbers to replace culled sows. Gilts replacing sows represent a major opportunity for genetic change in the sow herd. As sows generally produce larger litters of heavier pigs than do gilts, replacing large numbers of sows with gilts may reduce production levels. This production differential and the low relationship among the performances of successive litters argue for low rates of culling based on sow performance to maintain high levels of production.

The higher productive rate of sows over gilts must be balanced against the genetic change made possible by bringing gilts into production. A total gilt replacement level of 20–25% is suggested for each farrowing. Pork producers may find economic advantages in timing the culling of sows to take advantage of high prices for sows.

The gilt selection and sow-culling scheme suggested assumes that there are no major genetic antagonisms between maternal traits (litter size and milking ability) and rate of gain and low backfat thickness. Some evidence indicates that the so-called meaty gilt does not make a good sow. There is, however, no documented evidence that selecting fast-growing, low-backfat gilts that are not too extreme in muscling will adversely affect sow performance. Guidelines for gilt selection by age and weight are given in Table 29.7.

BOAR SELECTION

Boar selection is extremely important in making genetic change in a swine herd. Over several generations, selected boars will contribute 80–90% of the genetic composition of the herd; however, this contribution does not diminish the importance of female selection. Boars should have dams that are highly productive sows. Productivity of replacement gilts is highly dependent on the level of

TABLE 29.7   Gilt Selection Calendar

When	What
Birth	Identify gilts born in large litters. Hernias, cryptorchids, and other abnormalities should disqualify all gilts in a litter from which replacement gilts are to be selected.
	Record birth dates, litter size, identification.
	Equalize litter size by moving boar pigs from large litters to sows with small litters. Pigs should nurse before moving.
	Keep notes on sow behavior at time of farrowing and check on disposition, length of farrowing, any drugs (such as oxytocin) administered, condition of udder, and extended fever.
3–5 weeks	Take 21-day weight of litter. Wean litters. Feed balanced, well-fortified diets for excellent growth and development.
	Screen gilts identified at birth by examining underlines and reject those with fewer than 12 well-spaced teats. If possible, at this time select and identify about two to three times the number of gilts needed for replacement.
180–200 lb	Weigh gilts and obtain ultrasound measurement of backfat thickness. Evaluate for soundness.
	Select for replacements the fastest-growing, leanest gilts that are sound and from large litters.
	Save 25–30% more than needed for breeding.
	Remove selected gilts from market hogs. Place on restricted feed.
	Allow gilts to have exposure to a boar along the fence that separates them.
	Observe gilts for sexual maturity. If records are kept, give advantage to those gilts that have had several heat cycles prior to final selection.
Breeding time	Make final cull when the breeding season begins and keep sufficient extra gilts to offset the percentage nonconception in the herd.
	Make sure all sows and gilts are earmarked, ear-tagged, or otherwise identified.

Source: Adapted from *Pork Industry Handbook,* PIH-27, Guidelines for Choosing Replacement Females, 1992.

TABLE 29.8   Suggested Selection Standards for Replacement Boars

Trait	Standard
Litter size	10 or more pigs farrowed; 8 or more pigs weaned
Underline	12 or more fully developed, well-spaced teats
Feet and legs	Wide stance both front and rear; free in movement; good cushion to both front and rear feet; equal-sized toes
Age at 230 lb	155 days or less
Feed/gain, boar basis (60–230 lb)	240 lb of feed/100 lb (cwt) of gain or less
Daily gain (60–230 lb)	2.0 lb/day or more
Backfat ultrasound (adjusted to 230 lb)	0.8 in. or less

Source: Adapted from *Pork Industry Handbook* (PIH-9), 1988 Boar Selection for Commercial Production.

sow productivity passed on by the boars. Selection standards for replacement boars are shown in Table 29.8. An example of a genetically superior boar is shown in Figure 29.7.

CROSSBREEDING FOR COMMERCIAL SWINE PRODUCERS

The primary function of seedstock breeders is to provide breeding stock for commercial producers. Breeding programs for seedstock producers are designed to genetically improve the economically important traits. Within a specific breed, rate of improvement will be dependent on heritability of the traits, how much selection is practiced, and how quickly generation turnover occurs. The most rapid genetic changes that have occurred over the past few decades are in growth rate and leanness.

FIGURE 29.7 "Ulf 166," a Yorkshire boar who demonstrated exceptional genetic superiority as evidenced by the following data:

■ During a 3-year period (1991–1994), Ulf was either first-, second- or third-ranking boar in the Yorkshire breed for number born alive (NBA), 21-day litter weight (LTWT), backfat (BF), and maternal line index (ML1).

■ "Ulf's" domination of the Yorkshire National Sire Summary was most evident in the January 1994 edition:
—"Ulf" ranked 1st for ML1, while his sons ranked 2nd through 7th, 9th, and 10th.
—"Ulf" ranked 3rd for NBA, while his sons ranked 1st and 4th.
—"Ulf" ranked 3rd for LTWT, while his sons ranked 1st, 2nd, 4th, 6th, and 8th.
—"Ulf" ranked 2nd for SPI, while his sons ranked 1st, 3rd, and 5th through 8th.

Courtesy of Swine Genetics International Ltd.

Commercial producers and seedstock producers select for the same economically important traits. Commercial producers have an added advantage over seedstock producers in making genetic change, because crossing two or more swine breeds will increase productivity. This genetic phenomenon of heterosis or hybrid vigor is shown in Table 29.9. There is a 40% increase in total litter market weight of crossbreds, which is a cumulative effect of larger litters, high pig survival, and increased growth rate of individual pigs. This marked increase in productivity tells why more than 90% of the market hogs in commercial production are crossbreds. Crossbreeding allows genetic improvement in some traits with low heritabilities, such as sow productivity. There is little hybrid vigor for traits with high heritabilities, such as carcass traits.

An effective crossbreeding program takes advantage of using hybrid vigor and selecting genetically superior breeding animals from breeds that best complement one another. A producer needs to know the relative strengths and weaknesses of the available breeds shown in Table 29.10. These breed comparisons can change over time as a result of genetic changes made in the individual breeds. Thus, breed comparisons must be made on a regular basis. In addition, the variation within a breed must be considered; individual pig selection could vary from the performance evaluations shown in Table 29.10.

TABLE 29.9 Heterosis Advantage for Production Traits

Item	First-Cross Purebred Sow	Multiple-Cross Crossbred Sow	Crossbred Boar
Percentage of Advantage over Purebred			
Reproduction			
Conception rate	0.0%	8.0%	10.0%
Pigs born alive	0.5	8.0	0.0
Litter size at 21 days	9.0	23.0	0.0
Litter size, weaned	10.0	24.0	0.0
Production			
21-day litter weight	10.0	27.0	0.0
Days to 220 lb	7.5	7.0	0.0
Feed/gain	2.0	1.0	0.0
Carcass composition			
Length	0.3	0.5	0.0
Backfat thickness	−2.0	−2.0	0.0
Loin muscle area	1.0	2.0	0.0
Marbling score	0.3	1.0	0.0

Source: Adapted from *Pork Industry Handbook* (PIH-39), 1987.

TABLE 29.10 Comparative Performance Evaluations for Swine Breeds

Trait	Berkshire	Chester White	Duroc	Hampshire	Landrace	Poland China	Spotted	Yorkshire
Conception rate (%)	+	+	A	A	−−	A	A	−
Litter size (no. pigs raised to weaning)	−	++	A	−	++	−−	−	++
21-day litter weight (lb)	−	−	A	A	++	−	−−	+
Rate of gain (lb/day)	−	−	++	A	A	−	−	A
Feed efficiency (lb feed/lb gain)	A	A	+	+	A	−	A	A
Backfat (in.)	A	A	A	+	A	A	A	A
Loin eye area (sq. in.)	A	A	A	+	A	+	A	A

Evaluation system:
++ = substantially superior to average
+ = superior to average
A = average or near average
− = below average
−− = substantially below average

Source: Adapted from *Pork Industry Handbook* (PIH-39), 1987; and *National Hog Farmer* (June and October 1990).

The crossbred female is an integral and important part of an effective cross-breeding program. Even though crossbreeding provides an opportunity to reap the benefits of many genetic sources, an unplanned crossing program will not yield success or profit for the pork producer. A well-planned crossbreeding system capitalizes on heterosis, takes advantage of breed strengths, and fits the producer's management program.

The two basic types of crossbreeding systems are the **rotational cross** system and the **terminal cross** system. The rotational cross system combines two or more breeds, with a different breed of boar being mated to the replacement crossbred females produced by the previous generation (Fig. 29.8). In a simple terminal cross system, a two-breed single or rotational cross female is mated to a boar of the third breed. A more complex terminal crossing system, shown in Figure 29.9, uses four breeds. In this example, crossbred boars are mated to crossbred females.

Figure 29.10 shows a **rota-terminal crossbreeding** system, which combines the three-breed rotational system and the terminal system of crossbreeding. Thus, all

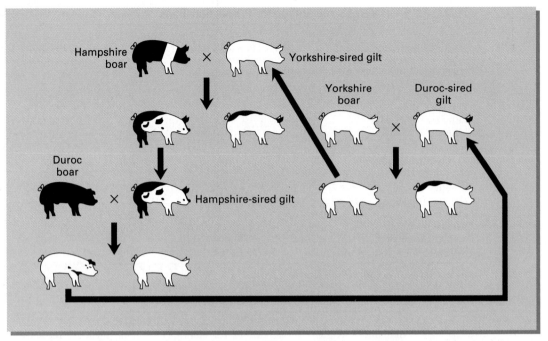

FIGURE 29.8 Three-breed rotational crossbreeding system. Adapted from *Pork Industry Handbook* (PIH-39).

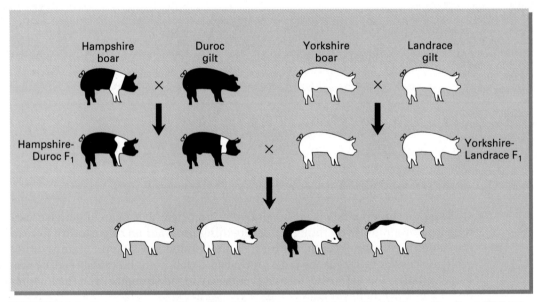

FIGURE 29.9 Four-breed terminal crossbreeding system. Adapted from *Pork Industry Handbook* (PIH-39).

the pigs produced by terminal cross boars are sold because the gilts would lack the superiority needed in maternal traits (reproduction and milk). The female stock can be purchased through a system primarily emphasizing reproductive performance. This latter system maximizes heterosis; however, purchasing replacement females increases the risk of introducing disease into the herd.

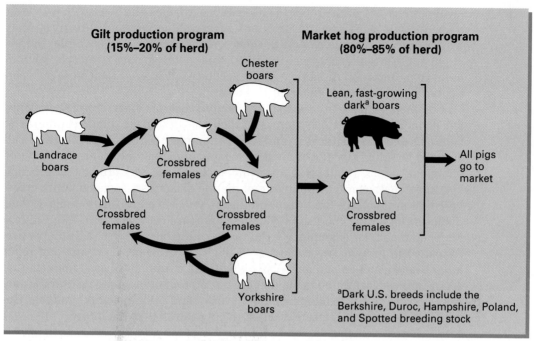

FIGURE 29.10   A rota-terminal crossbreeding system that combines both rotational crossing and terminal crossing. Adapted from *Pork Industry Handbook* (PIH-106), Genetic Principles and Their Applications.

A two- or three-breed rotational crossbreeding system may better fit a producer's management program even though heterosis is less than with a terminal cross system. A two-breed rotation results in 67% of the heterosis of a terminal cross, whereas a three-breed rotation gives 86% of the heterosis of a terminal crossing system.

The three-breed rotational cross is a popular crossbreeding system, although many of the larger commercial operations use terminal crossbreeding. The three-breed rotational cross combines the strong traits of a third breed not available in the other two breeds. Sires from three breeds are systematically rotated each generation and replacement crossbred females are selected each generation. These females are mated to the sire breed to which they are least related. Because reproductive performance is to be stressed in the initial two-breed cross, growth, feed efficiency, and superior carcass composition may be lower than desired. These traits can be emphasized in the individual boars selected from the third breed, although the reproduction and maternal traits should also receive attention.

Although limited comparative research information is available on the use of crossbred and hybrid boars, these boars are apparently more aggressive breeders, have fewer problems in leg soundness, and improve conception rate (Table 29.9) in comparison to purebred boars. Crossbred or hybrid boars can combine superior traits that may not be available in one purebred breed.

Hybrid boars sold by some commercial companies should not be confused with crossbred boars sold by private breeders. Hybrid boars are developed from specific line crosses. These lines have been selected and developed for specific traits. When specific crosses are made, the hybrid boar is used on specific cross females to obtain the recommended breed combination and to realize a high level of heterosis in the offspring. Genetic companies and seedstock providers often include very specific programs of crossbreeding and directives as to which lines to include as part of their service package. For example, DeKalb might suggest that boars from the DK97 line, noted for heavy muscle, high libido, soundness and durability, might best be used as a terminal cross on a particular maternal line or maternal-line cross.

Commercial pork producers have many selection tools, crossing systems, and genetic breeding-stock sources for their use. Producers should capitalize on heterosis and breed strengths, and require complete performance records on all selected breeding stock. Although there is no one best system, breed, or source of breeding stock, producers must evaluate their total pork production program and integrate the most profitable combination of factors associated with a crossbreeding program. In the future, producers will have even more tools at their disposal, including the use of genetic-marker-assisted selection. The Chinese Meishan breed, for example, is extremely prolific but also has undesirable performance in growth and carcass data. Pig Improvement Company and other large swine genetic companies have identified a number of gene markers that assist in the selection of swine that are more productive, more resistant to disease, and capable of producing more flavorful and desirable pork products. Genetic innovations will create even more opportunities in the future.

CHAPTER SUMMARY

- The most important visual characteristics in identifying breeds of swine are hair color, color pattern, and erect or drooping ears.
- Based on registration number, Yorkshire is the most numerous breed of swine, followed by Duroc and Hampshire.
- Traits of high economic importance that are included in most selection programs are (1) sow productivity (number in letter and litter weight at 21 days), (2) growth rate, (3) feed efficiency, (4) carcass traits (pounds of acceptable-quality pork), and (5) skeletal soundness.
- The use of selection indexes such as SPI, MLI, and TSI are excellent approaches to making progress in the swine herd.
- Crossbreeding in swine is widely used whereby the outstanding traits of two or more breeds are combined and heterosis increases the pounds of market pigs sold per litter.

KEY WORDS

hybrids
sow productivity
growth rate
Symbol I
Symbol II
Symbol III
feed efficiency
cryptorchidism
hernias

inverted nipples
pale, soft, and
  exudative (PSE)
porcine stress syndrome
  (PSS)
STAGES
sow productivity index
  (SPI)
terminal sire index (TSI)

maternal line index
  (MLI)
rotational cross
terminal cross
rota-terminal
  crossbreeding

REVIEW QUESTIONS _____

1. What are the three most popular breeds of swine in the United States today?
2. What are measures of sow productivity?
3. What two measures combined are the best indication of sow productivity?
4. How can sow productivity best be improved?
5. *True or False:* Pigs sold per litter has minimal effect on profit.
6. Why is feed efficiency such an economically important trait for most commercial swine producers?
7. What are three quality measures of carcass traits?
8. What is meant by PSE?
9. Why are structural soundness traits of importance to commercial swine producers?
10. What is a major disadvantage of replacing large numbers of sows with gilts?
11. What is a major advantage of replacing large numbers of sows with gilts?
12. What are the four features that should be included in a sound program of genetic improvement?
13. What is the primary purpose of seedstock producers?
14. What are breeding programs of seedstock producers designed to do?
15. Why are more than 90% of all hogs in commercial production crossbred?
16. What are the two basic types of crossbreeding systems for swine?
17. A _____ crossbreeding system combines aspects of the three-breed rotational cross and terminal cross systems.
18. Which three breeds of swine are strongest in reproductive traits?
19. Which breed of swine is strongest in growth performance traits?
20. Which breed is superior in carcass traits?

SELECTED REFERENCES _____

Boar selection guidelines for commercial pork producers. 1988. *Pork Industry Handbook,* PIH-9. Swine Extension Specialists: various state universities.

Crossbreeding systems for commercial pork production. 1988. *Pork Industry Handbook.* Swine Extension Specialists: various state universities.

Genetic evaluation programs. 1990. National Swine Improvement Federation Fact Sheet No. 12. West Lafayette, IN: Purdue University Cooperative Extension Service.

Guidelines for choosing replacement females. 1992. *Pork Industry Handbook,* PIH-27. Swine Extension Specialists: various state universities.

Legates, J. E. 1990. *Breeding and Improvement of Farm Animals.* 4th ed. New York: McGraw-Hill.

Plain, R., J. Lawrence, and G. Grimes. 2006. The structure of the U.S. pork industry. Fact Sheet 15-01-01. Des Moines, IA: Pork Information Gateway.

Pond, W. G., J. H. Maner, and D. L. Harris. 1991. *Pork Production Systems*. Westport, CT: AVI Publishing Co.

Porcine Stress Syndrome. 1992. *Pork Industry Handbook*, PIH-26. Swine Extension Specialists: various state universities.

Stadler, K., and C. Lacy. 1998. How profitable is sow longevity? *Pork '98*, February.

Stewart, T. S., et al. 1991. Genetic improvement programs in livestock: Swine testing and genetic evaluation system (STAGES). *J. Anim. Sci.* 69:3882.

CHAPTER 30

Feeding and Managing Swine

LEARNING OBJECTIVES

- Describe the four phases of pork production and the five types of enterprises
- Discuss breeding female management strategies
- Outline nutritional management across the swine industry
- List the reasons for preweaned pig death losses
- Describe animal identification in the swine industry
- Compare diets as pigs grow and develop
- List the problems associated with gossypol and mycotoxins
- Explain the use of contracting in the swine industry
- Discuss the major environmental issues in the swine industry
- Outline the cyclic nature of swine profitability

Swine production in the United States is concentrated heavily in the nation's midsection, known as the Corn Belt. This area produces most of the nation's corn, which is the principal feed used for swine.

TYPES OF SWINE OPERATIONS

Swine production occurs in four basic steps: (1) breeding and gestation; (2) farrowing to weaning; (3) nursery (care of pigs from weaning until 30–70 pounds); and (4) finishing by feeding hogs to market weights of 223–300 pounds. There are five primary types of swine operations: **farrow-to-finish,** in which a breeding herd is maintained where pigs are produced and fed to slaughter weight on the same farm (steps 1 through 4); **farrow-to-feeder pig production,** in which the producer is involved in steps 1, 2, and 3; **feeder pig finishing,** in which feeder pigs are purchased and then fed to slaughter weight (step 4 only); **farrow-to-weaner,** in which pigs are managed to weaning age and then sold or moved off site; and pure-bred or **seedstock** operations, which are similar to farrow-to-finish except that their salable product is primarily breeding boars and gilts.

FARROW-TO-FINISH OPERATIONS

The selecting and breeding practices used to produce highly productive swine are presented in Chapters 29. The primary objective in the production phase where a resident breeding herd is maintained is to feed and manage sows, gilts, and boars economically to ensure a large litter of healthy, vigorous pigs from all breeding females.

Boar Management

Boars should be purchased at least 60 days before the breeding season so they can adapt to their new environment. A boar should be purchased from a herd that has an excellent herd health program (vaccinated for leptospirosis, erysipelas, porcine parvovirus, and other appropriate pathogens) and then isolated from the new owner's swine for at least 30 days. During this time, the boar can be treated for internal and external parasites if the past owner has not recently done so. The boar should be revaccinated for erysipelas and leptospirosis. Boars not purchased from a validated brucellosis-free herd should have passed a negative brucellosis test within 30 days of the purchase date. An additional 30-day period of fence-line exposure to the sow herd is recommended to develop immunities before the boar is used.

Boars should be managed to assure excellent quality and quantity of semen without overfeeding. Boars that are fed excessive nutrients, especially energy, may become too large for useful service and often exhibit poor libido and produce poor quality semen. Young boars should receive 6–8 lb of a balanced ration up to 260 lb. Boars that are servicing females should receive a balanced diet of 3–6 lb. The amount of feed should be adjusted to keep the boar in a body condition that is neither fat nor thin. Assuring that breeding males are not heat stressed is critical, as those boars in heat stress will experience substantial declines in fertility 1 to 2 months later. Spermatozoa produced under heat stress will be of poor quality at the conclusion of spermatogenesis and maturation 30 to 50 days later.

Young, untried boars should be test-mated to several gilts before the breeding season, as records show that approximately 1 of 12 young boars is infertile or sterile. The breeding aggressiveness, mounting procedure, and ability to get the gilts pregnant should be evaluated.

Successful breeding of a group of females requires an assessment of the number of boars required and the breeding ability of each boar. Generally, young boars can pen-breed 8–10 gilts during a 4-week period. A mature boar can breed 10–12 females. **Handmating** (in which producers individually mate the boar to each female) takes more labor but extends the breeding capacity of the boar. Young boars can be handmated once daily and mature boars can be used twice daily or approximately 10–12 times a week. Handmating can be used to mate heavier, mature boars with the use of a breeding crate. The breeding crate takes most of the boar's weight off the female. Breeding by handmating should correspond with time of ovulation (approximately 40 hours from the beginning of estrus) or litter size will be reduced.

Boars should be considered dangerous at all times and handled with care. Facilities can be designed that allow handlers a convenient escape route from the mating pen. It is unsafe for a boar to have tusks, as he may inflict injury on the handler or on other pigs. All tusks should be removed prior to the breeding season and every six months thereafter.

Management of Breeding Females

Proper management of females in the breeding herd is necessary to achieve optimum reproductive efficiency. **Gilts** may start cycling as early as 5 months of age. It is recommended that breeding be avoided during their first heat cycle as the number of ova released is usually less than in subsequent heat cycles. Mixing pens of confinement-reared gilts and regrouping them with direct boar contact initiates early puberty.

Breeding gilts should weigh 250–260 pounds at 8 months of age when they are exposed to mating for the first time. Prebreeding gilts should be fed individually or in groups of no more than 10 head. Guidelines for developing breeding gilts are listed in Table 30.1.

Hot weather is detrimental to a high reproductive rate in the breeding females. High temperatures (above 85°F) will delay or prevent the occurrence of estrus, reduce ovulation rate, and increase early embryonic deaths. The animals will suffer heat stress when they are sick and have elevated body temperatures as well as when the environmental temperature is high.

Besides reducing litter size, diseases such as leptospirosis, pseudorabies, and those associated with stillbirths, mummified fetuses, and embryonic deaths may also increase the problems in getting the females pregnant. Attention to a good herd health program will assure a higher reproductive performance over the years.

The swine industry is rapidly moving to the utilization of artificial insemination with nearly two-thirds of all market hogs resulting from **AI.** As expected, as the sizes of the farms increase, use of AI becomes more common. In 2003, more than 90% of hogs marketed by producers, with over 10,000 annual marketings, were the result of AI. The success of an AI program is highly

---

TABLE 30.1   Guidelines for Gilt Development Programs

Expose gilts to first-litter or older cull sows once they reach 50–60 lb to allow adaptation to herd-level pathogens.

Segregate pigs from first-litter gilts in the nursery and finishing barns.

Feed a vitamin and mineral fortified ration.

Following first breeding, feed intake should be restricted to approximately 4 lb/day for 3–5 days.

Feed intake should then be increased to 5 lb/day to day 90 of pregnancy.

From day 90 until 2–3 days prior to parturition, feed intake should be boosted to 6.5 lb/day.

In the 48-to-72-hour timeframe in advance of farrowing, intake should be limited to 5 lb/day.

During lactation, the feeding program should target maximum consumption with particular attention given to lycine, calcium, and phosphorus.

Source: Multiple sources.

TABLE 30.2   Impact of AI Technician[a] on Productivity

Technician	Farrowing Rate (%)	Pigs Born Alive per Litter	Total Pigs Born per Litter	Pigs Produced
1	90.6	10.3	11.0	2,348
2	85.9	10.5	11.2	2,310
3	81.6	10.3	11.0	2,153
4	89.1	10.2	10.8	2,346
5	89.8	10.4	11.1	2,413
6	67.8	8.5	9.3	1,377

[a] Each technician bred 256 sows.
Source: Flowers, 2002.

dependent on the timing of insemination and the ability of the technician. Mating should occur 6 to 8 hours prior to ovulation to allow for best conception rates. Appropriate training and monitoring is very important to achieving a high reproductive rate (Table 30.2).

It is important that sows and gilts obtain adequate nutrients during gestation to assure optimum reproduction. Feeding in excess is not only wasteful and costly but is also likely to increase embryonic mortality. A limited feeding system using balanced, fortified diets is recommended to ensure that each sow gets her daily requirements of nutrients without consuming excess energy. As a rule of thumb, 4–5 lb of a balanced ration provides adequate protein and energy; however, during cold weather an additional pound of feed may prove beneficial, especially for bred gilts. With limited feeding, it is important that each sow get her portion of feed and no more. The use of individual feeding stalls for each sow is best because it prevents the "boss sows" from taking feed from slower eating or timid sows. During gestation, gilts should be fed so they will gain about 100–120 lb (Fig. 30.1), and sows should gain approximately 75–100 lb. This allows gilts to farrow at 350–400 lb and sows to weigh 425–550 lb at farrowing.

Table 30.3 shows several sow rations that utilize different grains. Note how the rations are balanced for protein, specific amino acids, minerals, and vitamins. Breeding females should be closely monitored in regards to body condition score to assure optimal performance. Body condition scores of 1–5, based on visual appraisal, are recommended (1 = extremely thin, 5 = extremely fat). Females should have a body condition score of 3–4 at time of birth.

Pregnancy Detection

Determining the gestational status of the breeding female is important to good management decisions. Many producers depend on observations of sows returning to estrus as a primary test for failure to attain pregnancy. However, this practice is labor-intensive, often requires the ability to provide boar exposure, and not absolutely conclusive such as is the case with sows experiencing cystic ovaries.

Pregnancy tests that measure the progesterone concentration in a serum or urine sample provide another alternative. While accurate, the need to collect samples limits the commercial applicability of this protocol. The use of ultrasound evaluation provides a useful pregnancy test in many situations and this

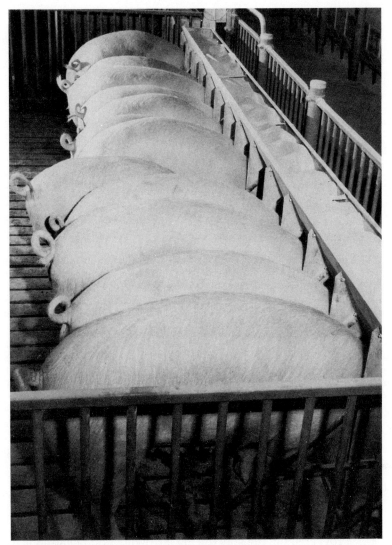

FIGURE 30.1 Bred gilts should be fed a diet that allows them to reach to specified target weight at farrowing (usually at approximately 350 to 400 lb). Courtesy of University of Illinois.

technology has become the preferred method of evaluation in the industry, particularly for larger herds.

Management of the Sow during Farrowing and Lactation

Farrowing is the process of the sow or gilt giving birth to her pigs. On the average, there are approximately 10 pigs farrowed per litter, with approximately 9 pigs weaned per litter. Well-managed swine herds typically have three to four more pigs per litter than these average figures. Extremely large litters are usually undesirable because the pigs are small and weak at birth and death losses can be high.

TABLE 30.3   Selected Sow Rations (with corn, grain, sorghum, or wheat as the grain source)

Ingredient	Gestation	Pre-farrowing	Lactation 10 pigs .33 lb ADG #1	Lactation 10 pigs .33 lb ADG #2	Lactation 10 pigs .44 lb ADG #1	Lactation 10 pigs .44 lb ADG #2
Corn, yellow	1,619	1,257	1,412	—	1,323	618
Grain, sorghum	—	—	—	1189	—	—
Wheat	—	—	—	—	—	618
Soybean meal, 44%	—	—	—	535	590	—
Soybean meal (dehulled)	245	465	470	—	—	588
Dehydrated alfalfa meal, 17%	60	100	—	100	—	—
Fat (choice grease)	—	100	—	100	—	100
Calcium carbonate	16	18	22	18	28	29
Dicalcium phosphate	37	37	38	35	36	32
Salt	10	10	10	10	10	10
Vitamin-trace mineral mix	8	8	8	8	8	8
Other	5	5	5	5	5	5
Total	2,000	2,000	2,000	2,000	2,000	2,000
Protein, %	13.1	17.2	17.2	19.0	18.5	19.9
Lysine, %	0.60	0.90	0.90	0.98	1.00	1.07
Tryptophan, %	0.14	0.20	0.19	0.24	0.22	0.29
Threonine, %	0.48	0.65	0.65	0.71	0.70	0.75
Methionine + cystine, %	0.48	0.58	0.59	0.60	0.63	0.67
Calcium, %	0.80	0.90	0.91	0.89	1.02	1.00
Available phosphorus, %	0.38	0.40	0.40	0.41	0.40	0.41
Metabolizable energy, kcal/lb	1,473	1,557	1,490	1,529	1,460	1,575

Source: Adapted from *Pork Information Gateway*, Swine Diets, 2004.

Proper care of the sow during farrowing and lactation is necessary to ensure large litters of pigs at birth and at weaning. Sows should be dewormed about 2 weeks before being moved to the farrowing facility. The sow should be treated for external parasites at least twice, a few days prior to farrowing time. The farrowing facility should be cleaned completely of organic matter, disinfected, and left unused for 5–7 days before sows are placed in the unit.

Before the sow is put in the farrowing pen or stall, her belly and teats should be washed with a mild soap and warm water. This helps eliminate bacteria that can cause diarrhea in nursing pigs.

Sows should be in the farrowing facility at the right time. Breeding records and projected farrowing dates should be used to determine the proper time. If farrowing is to occur in a crate or pen, the sow should be placed there no later than the 110th day of gestation. This gives some assurance that the sow will be in the facility at farrowing time, as the normal gestation is 111–115 days in length. Farrowing time can be estimated by observing an enlarged abdominal area, swollen vulva, increased respiration rate, and filled teats. The presence of milk in the teats indicates that farrowing is likely to occur within 24 hours.

Before farrowing, while the sows are in the farrowing facilities, they can be fed a ration similar to the one fed during gestation. A laxative ration can be helpful at this time to prevent constipation. The ration can be made laxative by the addition of 10 lb/ton of Epsom salts or potassium chloride, part of the protein as linseed meal, or the addition of 10% wheat bran, alfalfa meal, or beet pulp.

Sows need not be fed for 12–24 hours after farrowing, but water should be continuously available. Two or three pounds of a laxative feed may be fed at the first postfarrow feeding; the amount fed should be gradually increased until the optimum feed level is reached by 10 days after farrowing. Sows that are thin at farrowing may benefit from generous feeding early after farrowing. Sows nursing fewer than 8 pigs may be fed a basic maintenance amount using a rule of thumb of 3 lb for the sow and 1 lb for each pig nursed on a daily basis. Sows and first-lactation gilts raising more than 8 pigs should be fed free choice. It is unnecessary to reduce feed intake before weaning. Regardless of level of feed intake, milk secretion in the udder will cease when pressure reaches a certain threshold level.

Induced Farrowing

The practice of stipulating the time of farrowing has been utilized by many swine operations for the past decade or more. The advantages of this approach include reduced pig mortality, heavier weaning weights, more efficient use of labor resources, and more cross-fostering. Induced farrowing provides the opportunity for timely observation of the farrowing sow and can lead to reduced perinatal mortality of 0.5 pigs per litter or 1 additional pig per annum.

The standard recommendation is to induce parturition on day 112 of gestation. Typically, induction should not occur prior to 2 days before due date and as such knowing the date of breeding and the average gestation length of the sow population is required for success. Furthermore, the producer must be able to accurately compute the benefits in light of the relatively high cost of this practice. If drugs are given too early, the producer risks increased rates of abortion and stillbirth.

The preferred drugs include prostaglandin and cloprostenol (synthetic prostaglandin). Oxytocin may be administered to those sows not farrowing within 25 hours of the original injection.

Baby Pig Management from Birth to Weaning

Most sows and gilts are farrowed in farrowing stalls or pens to protect baby pigs from the female lying on them (Fig. 30.2). Producers who give attention at farrowing will decrease the number of pigs that die during birth or within the first few hours after birth. Approximately one-third of pre-weaning pig deaths occur within 72 hours of birth. The primary causes of preweaning **mortality** in pigs are outlined in Table 30.4. Pigs can be freed from membranes, weak pigs can be revived, and other care can be given that will reduce baby pig death loss. Manual assistance in the birth process should not be given unless obviously needed. Duration of labor ranges from 30 minutes to more than 5 hours, with pigs being born either head or feet first. The average interval between births of individual pigs is approximately 15 minutes, but this can vary from almost simultaneously to several hours in individual cases. An injection of oxytocin can be given (sows only) to speed the rate of delivery after 5 to 6 pigs have been

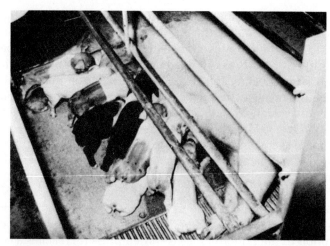

FIGURE 30.2 Sow and litter of pigs in a farrowing crate. The farrowing crate is a pen (approximately 5 × 7 ft) that restricts the sow to a fixed area so that she does not lay on her babies. A panel several inches off the floor separates the sow from the baby pigs while allowing them access for nursing. Courtesy of the University of Illinois.

TABLE 30.4 Reasons for Preweaning Pig Mortality

Cause	%
Crushing	51
Starvation	19
Various-known	8
Scours	11
Respiratory problems	2
Unknown	9

Source: NAHMS, 2000.

born. This step should be preceded by palpation of the tract to assure that there is not an obstruction of the birth canal.

Manual assistance, using a well-lubricated gloved hand, can be used to aid in difficult deliveries. The hand and arm should be inserted into the reproductive tract as far as is needed to locate the pig; the pig should be grasped and gently but firmly pulled to assist delivery. Difficult births are often associated with the occurrence of lactation failure where **mastitis** (inflammation of the udder), **metritis** (inflammation of the uterus), and **agalactia** (inadequate milk supply) may occur. Antibacterial solutions such as nitrofurazone are infused into the reproductive tract after farrowing. Sometimes intramuscular injections of antibiotics are used to decrease or prevent some of these infections.

It is important that the newborn pig receive colostrum within 4 to 6 hours postfarrowing to give immediate and temporary protection against common bacterial infections. Within 6 hours following birth, a sow's immunoglobulin concentration in colostrum declines by 50%. Pigs from extremely large litters can be transferred to sows having smaller litters to equalize the size of all litters. Care should be given to ensure that baby pigs receive adequate colostrum before the transfer takes place.

TABLE 30.5　Target Air Temperatures for Swine

Weight/Age	Optimum (°F)	Acceptable Range (°F)
Newborn	95	90–100
21 days	80	75–85
10–30 lbs	80	75–85
30–50 lbs	75	70–80
50–75 lbs	65	60–70
75–180 lbs	60	55–70
180 lbs–market	60	55–70

Source: Adapted from *PORK*, July 2002.

Air temperature within the creep area for baby pigs equipped with zone heat (heat lamps, gas brooders, or floor heat) should have a temperature of 90–100°F. Other target air temperatures for swine of known weights are listed in Table 30.5. Otherwise the pigs will chill and die or become susceptible to serious health problems. Moisture should be controlled or removed from the farrowing facility without creating a draft on the pigs.

Management of environmental temperature from birth to 2 weeks of age is critical. At 2 weeks of age, the baby pig has developed the ability to regulate its own body temperature.

Soon after birth, the navel cord should be cut to 3–4 in. from the body and then treated with iodine. This disinfects an area that could easily allow bacteria to enter the body and cause joints to swell and abscesses to occur.

The producer should also clip the eight sharp needle teeth of the baby pig. This prevents the sow's udder from being injured and keeps facial lacerations from occurring when the baby pigs fight one another. Approximately one-half of each tooth can be removed by using side-cutting pliers or toenail clippers.

Good records are important to the producer interested in obtaining optimum production efficiency. The basis of good production records is pig identification. Ear notching pigs at 1–3 days of age using the system shown in Figure 30.3 provides positive identification for the rest of the pig's life.

Baby pig management from 3 days to 3 weeks of age includes controlling anemia and scours, castration, and tail docking. Sow's milk is deficient in iron, so iron dextran shots are given intramuscularly at 3–4 days of age (Fig. 30.4).

Cross-fostering is the practice of transferring piglets from one sow to another within a farrowing group to reduce within litter weight variation among pigs and to assure that each sow is nursing the number of pigs best suited to her ability. This practice requires that pigs be left on their natural mother for a period of 4 to 6 hours to assure adequate colostrum intake, that the transfer occurs within 24 to 30 hours of birth, and that the foster dam farrowed within 24 hours of the donor sow. The predominate reason for this timing is that pigs establish a preferred teat to nurse within 48 hours of birth and disrupting this preference often leads to increased fighting among piglets.

Baby pig scours are an ongoing problem for the swine producer. Colostrum and a warm, dry, draft-free, and sanitary environment are the best preventative measures against scours. Orally administered drugs determined to be effective against specific bacterial strains have been found to be the best control measures for scours. Serious diarrhea problems resulting from diseases such as

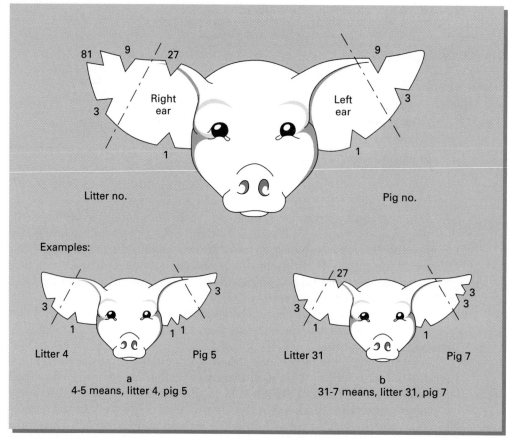

FIGURE 30.3   An ear-notching system for swine. Individual pig identification is necessary for meaningful production and management records. *Source:* Colorado State University.

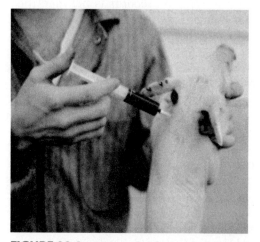

FIGURE 30.4   An iron shot being administered to a baby pig. The preferred site of injection is the neck muscle. Injection should not be administered in the ham. Courtesy of Michigan State University.

transmissible gastroenteritis (TGE) and swine dysentery should be treated under the direction of a local veterinarian.

Castration is usually done before the pigs are 2 weeks old. Pigs castrated at this age are easier to handle, heal faster, and are not subjected to as much stress as those castrated later. The use of a clean, sharp instrument, low incisions to promote good drainage, and antiseptic procedures make castration a simple operation.

Tail docking has become a common management practice to prevent tail biting during the confinement raising of pigs. Tails are removed 0.5 to 0.75 inches from the body, and the stump should be disinfected along with the instrument after each pig is docked.

Pigs should be offered feed at 1–2 weeks of age in the creep area. These starter feeds can be placed on the floor or in a shallow pan. By 3–4 weeks after farrowing, the sow's milk production has likely peaked; however, the baby pigs should be eating the supplemental feed and growing rapidly (Fig. 30.5).

Rations for baby pigs (weighing 10–25 lb, 14–20 lb, or 25–40 lb) are shown in Table 30.6. Note the components of the rations, especially when milk products (dried whey or dried skim milk) are included in the rations of 10- to 25-lb pigs. Milk by-products are beneficial because of their desirable protein and lactose content that boosts pig performance. Table 30.7 outlines the recommended levels of feed intake for lactating gilts and sows. Females should be kept cool to assure sufficient feed intake to maintain sufficient milk production to allow pigs rapid and healthy early growth.

Pigs can be weaned from their dams at a time that is consistent with the available facilities and the management of the producer. Pigs weaned at young ages should receive a higher level of management. General weaning guidelines include weaning only pigs over 10 lb, weaning pigs over a 2- to 3-day period, weaning the heavier pigs in the litter first, grouping pigs according to size in pens of 30 pigs

FIGURE 30.5 Weaned pigs in a modern nursery unit. The pigs were weaned from the sow and placed into these pens when they weighed 10 to 20 Ib. The pigs will be moved into pens for growing and finishing when they weigh approximately 40 to 45 Ib. Courtesy of the University of Illinois.

TABLE 30.6   Starter Diets for Early Weaned Pigs

| | Ration Number and lb/Ration | | | | | |
| | Pigs 10–14 lb | | Pigs 14–20 lb | | Pigs 20–45 lb | |
Ingredient	1	2	1	2	1	2
Corn, yellow	882	682	945	766	1,156	1,229
Oat groats	—	200	—	200	—	—
Soybean oil	80	80	40	40	—	—
Fat (choice grease)	—	—	—	—	—	40
Soybean meal, 44%	265	300	480	460	590	670
Fish meal	80	80	100	100	—	—
Lactose	50	100	—	—	—	—
Dried skim milk	100	—	—	—	—	—
Dried whey	400	400	400	400	200	—
Dried blood meal	50	50	—	—	—	—
Dried plasma protein	50	60	—	—	—	—
Lysine	1	2	2	2	2	2
Methionine	2	2	1	1	—	—
Calcium carbonate	14	16	10	10	18	19
Dicalcium phosphate	11	13	7	6	19	25
Copper sulfate	2	2	2	2	2	2
Salt	5	5	5	5	7	7
Vitamin-trace mineral	8	8	8	8	8	8
Total	2,000	2,000	2,000	2,000	2,000	2,000
Protein, %	20.33	20.36	20.07	20.28	19.03	19.87
Lysine, %	1.40	1.40	1.30	1.30	1.15	1.19
Tryptophan, %	0.27	0.27	0.24	0.25	0.23	0.24
Threonine, %	0.90	0.89	0.83	0.83	0.75	0.76
Methionine + cystine, %	0.81	0.79	0.74	0.73	0.65	0.66
Lysine:calorie ratio	4.02	4.05	3.89	3.87	3.57	3.56
Calcium, %	0.86	0.86	0.75	0.75	0.71	0.73
Available phosphorus, %	0.45	0.43	0.41	0.40	0.32	0.32
Metabolizable energy, kcal/lb	1,580	1,573	1,517	1,521	1,467	1,513

Source: Adapted from *Pork Information Gateway*, Swine Diets, 2004.

TABLE 30.7   Feed Intake in Lactating Sows

| | | Total Feed Intake/hd/day (lb) | | |
	Day of Lactation	1st = Litter Gilts	Sows	Feeding Frequency
	Farrowing	4	6	1 ×
Day	2	6	8	1 ×
	3	8	10	1 ×
	4	10	12	1 ×
	5	12	14	2 ×
	6	14	16	2 ×
	7	14	16	2 ×
	8	16	18	2 ×
	9	16	20	2 ×
	10 +	16	20	2 ×

Source: Adapted from D. Lewis, 2002.

or less, providing 1 feeder hole for 2 pigs and 1 waterer for each 8–10 pigs, limiting feed for 48 hours, and using medicated water if scours develop.

Feeding and Management from Weaning to Market

A dependable and economical source of feed is the backbone of a profitable swine operation. Since approximately 60–70% of the total cost of pork production is feed, the swine producer should be keenly aware of all aspects of swine nutrition.

The pig is an efficient converter of feed to meat. Modern meat-type hogs can be produced with a feed efficiency of less than 3.0 lb of feed per pound of gain from 40 lb to market. To obtain maximum feed use, it is necessary to feed well-balanced rations designed for specific purposes.

Swine rations for growing-finishing market pigs as well as rations for breeding stock are formulated around cereal grains, the largest component of swine rations. The most common cereal grains—those that provide the basic energy source—are corn, milo, barley, wheat, and their by-products. Because of its abundance and readily available energy, corn is used as the base cereal for comparing the nutritive value of other cereal grains. Milo (grain sorghum) is similar in nutritional content to corn and can completely replace corn in swine rations; however, milo needs more processing (grinding, rolling, steaming, etc.). Its energy value is about 95% the value of corn. Barley contains more protein and fiber than corn, but its relative feeding value is 85–95% of corn and it is less palatable than corn. Wheat is equal to corn in feeding value and is highly palatable. However, for best feeding results, wheat should be mixed half-and-half with some other grain. Which grain or combination of grains to use can best be determined by availability and relative cost.

Grinding the grains improves all grains for feeding, especially those high in fiber such as barley and oats. Caution needs to be applied as feed that is too finely ground may lead to increased gastric ulcers. Pelleting the ration may increase gains and efficiency of gains from 5–10%. However, the advantages of pelleting are usually offset by its higher cost.

Cereal grains usually contain lesser quantities of proteins, minerals, and vitamins than swine require; therefore, rations must be supplemented with other feeds to increase these nutrients to recommended levels. Soybean meal has been proven to be the best single source of protein for pigs when palatability, uniformity, and economics are considered. Other feeds rich in protein (meat and bone meal or tankage) can compose as much as 25% of the supplemental protein if economics so dictate. Cottonseed meal is not recommended as a protein source unless the gossypol has been removed. **Gossypol** in relatively high levels is toxic to pigs.

Proper harvest, storage, and handling of corn is critical to minimize the occurrence of molds that may produce undesirable **mycotoxins** such as **aflatoxin, zearalenone,** and **vomitoxin.** These mycotoxins can result in poor growth performance, abortion, and other symptoms.

If economics dictate, a producer can supplement specific amino acids in place of some of the protein supplement. Typically, lysine may be supplemented separately because lysine is the first limiting amino acid in most swine diets.

Calcium and phosphorus are the minerals most likely to be deficient in swine diets, so care should be exercised to ensure adequate levels in the diet.

The standard ingredients for supplying calcium and phosphorus in the swine diet are limestone and either dicalcium phosphate or defluorinated rock phosphate. If an imbalance of calcium and zinc exists, a skin disease called **parakeratosis** is likely to occur.

Swine producers may formulate their swine diets in several ways: (1) by using grain and a complete swine supplement; (2) by using grain and soybean meal plus a complete base mix carrying all necessary vitamins, minerals, and antibiotics; or (3) by using grain, soybean meal, vitamin premix, trace-mineral premix, salt, calcium and phosphorus sources, and an antibiotic premix. Which ration is used depends on the producer's expertise, the cost, and the availability of ingredients. The trace-mineral premix typically includes vitamins A, D, E, K, B_{12}, and other B vitamins (niacin, pantothenic acid, riboflavin, and choline).

Table 30.8 shows example diets for growing pigs. Diets for finishing pigs are noted in Table 30.9. These diets are formulated using the common grains (corn, grain sorghum, barley, or wheat).

Most swine producers use feed additives because the additives have the demonstrated ability to increase growth rate, improve feed efficiency, and reduce death and sickness from infectious organisms. Feed additives available to swine producers fall into three classifications: (1) **antibiotics** (compounds coming from bacteria and molds that kill other microorganisms), (2) **chemotherapeutics** (compounds that are similar to antibiotics but are produced chemically rather than microbiologically), and (3) **anthelmintics** (dewormers). Antibiotics and chemotherapeutics should not be substituted for poor management (e.g., unsanitary conditions and poor health care).

TABLE 30.8   Selected Growing Rations for Average and High-Lean Barrows (45–75 lb) and High-Lean Gilts (45–110 lb)

Ingredient	Ration Number and lb/Ration			
	1	2	3	4
Corn, yellow	1,407	1,494	—	—
Grain sorghum	—	—	1,470	1,275
Soybean meal, 44%	542	451	565	572
Fat (choice grease)	—	—	—	100
Calcium carbonate	18	18	19	18
Dicalcium phosphate	20	21	20	20
Salt	7	7	7	7
Trace mineral-vitamin	6	6	6	6
Lysine	—	3	3	2
Total (lb)	2,000	2,000	2,000	2,000
Protein, %	17.71	16.22	17.30	18.48
Lysine, %	0.95	0.95	0.95	1.02
Tryptophan, %	0.20	0.18	0.21	0.23
Threonine, %	0.67	0.61	0.63	0.69
Methionine + cystine, %	0.60	0.56	0.56	0.59
Calcium, %	0.65	0.65	0.65	0.65
Available phosphorus, %	0.27	0.27	0.28	0.28
Metabolizable energy, kcal/lb	1,482	1,484	1,456	1,558

Source: Adapted from *Pork Information Gateway*, Swine Diets, 2004.

Table 30.9   Selected Finishing Rations for Gilts and Barrows (175–265)

Ingredient	Gilt–1	Gilt–2	Barrow–1	Barrow–2
	Ration Number and lb/Ration			
Corn, yellow	1,660	—	1,719	—
Grain sorghum	—	1,721	—	1,779
Soybean meal, 44%	300	235	245	181
Soybean meal, 48%	—	—	—	235
Lysine	—	3	—	3
Calcium carbonate	16	16	16	16
Dicalcium phosphate	13	14	9	10
Salt	7	7	7	7
Trace mineral-vitamin mix	4	4	4	4
Total, lb	2,000	2,000	2,000	2,000
Protein, %	13.46	13.20	12.50	12.29
Lysine, %	0.64	0.64	0.57	0.57
Tryptophan, %	0.14	0.16	0.12	0.14
Threonine, %	0.50	0.47	0.46	0.43
Methionine + cystine, %	0.50	0.45	0.47	0.42
Calcium, %	0.50	0.50	0.45	0.45
Available phosphorus, %	0.18	0.20	0.14	0.16
Metabolizable energy, kcal/lb	1,504	1,473	1,510	1,479

Source: Adapted from *Pork Information Gateway*, Swine Diets, 2004.

Feed additive selection and the level to be used will vary with the existing farm environment, management conditions, and stage of the production cycle. Disease and parasite levels will vary from farm to farm. Also, the first few weeks of a pig's life are far more critical in terms of health protection. Immunity acquired from colostrum diminishes by 3 weeks of age, and the pig does not begin producing sufficient amounts of antibodies until 5–6 weeks of age. Also during this time, the pig is subjected to stress conditions of castration, vaccinations, and weaning. Therefore, at this age the additive level of the feed is much higher than when the pig is more than 75 lb and growing rapidly toward slaughter weight. Producers should be cautious because some additives have withdrawal times, which mean the additive should no longer be included in the ration. This is necessary to prevent additives from occurring as residues in meat.

Market pigs that are fed properly prepared diets will likely have rapid and efficient liveweight gains. However, rapid and low-cost gains are also associated with excellent health programs and well-constructed facilities (Fig. 30.6). Providing adequate and consistent water supplies is also critical to the successful management of a swine enterprise. Watering system specifications are listed in Table 30.10.

The swine industry has adapted two innovative feeding approaches to more efficiently feed growing hogs—phase feeding and split-sex feeding protocols. **Phase feeding** simply means the use of specific rations for the specific needs of pigs differing in growth rate. Phase feeding typically involves the use of a sequence of four rations balanced for lysine—the most limiting amino acid for swine. **Split-sex feeding** recognizes the different nutritional requirements of barrows and gilts. The benefits of split-sex management include improved growth performance, lower feed costs, and improved market timing. Barrows typically

FIGURE 30.6 A group of pigs in a growing and finishing production unit. The building is enclosed and the floor is slatted. The manure from the pigs drops through the slats into a 36-in.-deep pit. The waste is removed from the pit via flushing with water. Management of waste and air quality is a priority for swine enterprises. Courtesy of University of Illinois.

TABLE 30.10 Guidelines for Providing Water to Swine

Stage	Head Per Nipple Waterer	Waterer Span (ft)	Minimum Flow Rate Per Minute (cups)
Nursery	10	12	1–1.5
Grower	10–15	18	2–3
Finisher	15	24–36	3–4
Gestation	15	36	3–4
Lactation	1	NA	3–4

Source: Adapted from T. See, 2002.

require more feed to attain similar growth rates to their female contemporaries but need less lysine.

Technological breakthroughs achieved through active research and development programs have yielded feed additives that enhance the ability of hogs to more efficiently produce lean. For example, ractopamine, marketed as Paylean™, is a nutrient that yields a repartitioning response in which feed is shifted from production of fat to production of lean. Another supplement approved for use in swine diets to enhance lean production is chromium picolinate at 200 parts per billion. It is suspected that chromium enhances carbohydrate metabolism by stimulating insulin function.

MANAGEMENT OF PURCHASED FEEDER PIGS

Feeder pigs marketed from one producer to another are subjected to more stress than pigs produced in a farrow-to-finish operation. These stresses are fatigue, hunger, thirst, temperature changes, diet changes, different surroundings, and social problems. Almost every group has a shipping-fever reaction. Care of newly arrived pigs must be directed to relieving stresses and treating shipping fever correctly and promptly.

Management priorities for newly purchased feeder pigs should include (1) a dry, draft-free, well-bedded barn or shed with no more than 50 pigs per pen; (2) a specially formulated starter ration having 12–14% protein, more fiber, and a higher level of vitamins and antibiotics than a typical starter ration; (3) an adequate intake of medicated water containing electrolytes or water-soluble antibiotics and electrolytes; and (4) prompt and correct treatment of any sick pigs.

MARKET MANAGEMENT DECISIONS

Slaughter weights of market hogs and which market is selected can have a marked effect on income and profitability of the swine enterprise. For many years the recommended weight to sell barrows and gilts was, in most instances, 200–220 lb. The primary reasons for selling these animals at a maximum weight of 220 lb were increased cost of gain and increased fat accumulations at heavier weights. Market discounts were common on pigs weighing over 220 lb because of the increased amount of fat. Today market hogs average 265 pounds with some contracts providing incentives for market hogs weighing up to 300 pounds. These heavier pigs are leaner, and they are not subject to the same market discounts as fatter pigs marketed several years ago. Decisions to market pigs at heavier weights should be based on feed costs and the gaining ability of the pigs. Most industry forecasts suggest that the trend for increased market weights will continue into the future.

Prices for slaughter pigs can vary among markets. Marketing costs, such as selling charges, transportation, and shrink (loss of live weight) can also vary. If more than one market is available, producers should occasionally patronize different markets as a check against their usual marketing program. A single market is seldom the best market.

The increase in the number of hogs sold on contract with final price determined on a carcass merit basis has been dramatic in the past decade. Approximately 80% of all hogs are sold on a contract basis. The most common approach is a **formula contract** where price is based on the cash market plus a predetermined premium. Other contract types include a **forward-cash contract,** a **risk-share contract** where the base price is determined from a formula that accounts for fluctuations in feed costs, and a risk-share contract where cash prices are paid within a predetermined range and adjustments are made if the cash market falls outside the range (producer and packer split gains and losses).

Several branded product initiatives have also been introduced into the market place in an attempt to capture more value as the result of creating product differentiation. These programs typically establish a specific set of market criteria such as lean color, lean pH, percent lean, and specific genetics such as breed or line. The American Berkshire Association has created a targeted supply chain approach that focuses on adding value to Berkshire and Berkshire cross pigs for both the domestic and export market.

COSTS AND RETURNS

Enterprise records for farrow-to-finish hog operations are shown in Table 30.11. Compare the costs and returns for the high-profit versus the low-profit farms. Excellent management is reflected in lower costs and higher profits (compare items 2 and 8). Recognize that 1998 was one of the poorest profit years on record for swine producers.

TABLE 30.11 Enterprise Records for Farrow-to-Finish Swine Operations

	Top 1/3	Bottom 1/3	Average
Sorted by Margin Over All Costs Including Inventory			
1 Return to capital, unpaid labor and management, $	$ 85,684	$ 27,873	$ 71,524
2 Net profit and return to management this period, $	$ 53,919	$ (28,182)	$ 25,642
3 Return per hour for all hours of labor and management, $/hr	$ 29.25	$ 4.80	$ 17.31
4 Percent return on capital, %	37.79%	1.33%	20.14%
7 Average price per cwt of market hogs, $	$ 34.72	$ 33.62	$ 33.96
8 Average price per cwt of cull breeding stock sold, $	$ 23.34	$ 22.76	$ 23.09
9 Average price per cwt of all market animals sold excluding newly weaned pigs, $	$ 33.89	$ 33.14	$ 33.30
10 Feed cost per cwt of pork produced, $	$ 19.22	$ 20.57	$ 19.54
11 Other operating cost (except hired labor) per cwt of pork produced, $	$ 4.59	$ 6.67	$ 5.54
12 Utilities, fuel, electric, and telephone per cwt, $	$ 1.37	$ 1.55	$ 1.42
13 Veterinary services and medicine per cwt, $	$ 1.19	$ 1.70	$ 1.36
14 Depreciation, taxes and insurance costs per cwt of pork produced, $	$ 2.58	$ 3.20	$ 2.78
15 Capital charge on fixed capital per cwt of pork produced, $	$ 1.46	$ 2.01	$ 1.57
16 Capital charge on operating capital per cwt of pork produced, $	$ 0.89	$ 1.09	$ 0.94
17 Value of labor (all) per cwt of pork produced, $	$ 4.15	$ 4.68	$ 4.48
18 Total cost per cwt of pork produced, $	$ 32.89	$ 38.22	$ 34.84
19 Margin over all costs per cwt of pork sold, not including inventory, $	$ 2.87	$ (4.21)	$ (0.64)
20 Margin over all costs per cwt of pork produced, including inventory, $	$ 8.60	$ (2.23)	$ 3.41
21 Margin over all costs per head sold, including inventory, $	$ 20.03	$ (5.44)	$ 8.26
22 Fixed costs per period per female maintained, $	$ 160	$ 205	$ 177
23 Fixed costs per period per crate maintained, $	$ 747	$ 1,102	$ 897
24 Fixed costs per pig weaned, $	$ 9.39	$ 12.23	$ 10.49
25 Net profit per period per female maintained, $	$ 339	$ (79)	$ 141
26 Net profit per period per crate maintained, $	$ 1,464	$ (483)	$ 600
27 Net profit per pig marketed, $	$ 20.80	$ (5.72)	$ 8.56
32 Total number of market hogs sold this period	2,359	4,521	3,610
33 Average weight of market hogs sold, lb	253	252	254
34 Pig death loss, birth to weaning (% of number farrowed live)	13.29%	11.57%	12.60%
35 Pig death loss, weaning to feeder (% of number weaned)	3.02%	3.10%	2.86%
36 Pig death loss, feeder to market (% of number of feeders started)	6.30%	6.11%	6.20%
37 Breeding stock death loss, (% of number maintained)	5.68%	5.35%	5.79%
38 Average breeding female inventory, number of head	180	307	251
39 Number of litters weaned per female per year	1.96	1.94	1.91
40 Number of pigs weaned per litter	8.52	8.94	8.73
41 Number of pigs weaned per female per year	16.8	17.5	16.8
42 Number of litters weaned per crate per year	9.0	10.7	9.6
43 Number of pigs weaned per crate per year	76.8	97.5	84.8
44 Total pounds of feed per cwt of pork produced, lb	349	354	348
45 Average cost of diets per cwt, $	$ 5.53	$ 5.82	$ 5.62
46 Hours of labor per cwt of pork produced, hours	0.47	0.52	0.49
47 Hours of labor per female maintained per year, hours	18.28	20.50	19.54
48 Hours of labor per litter weaned, hours	9.34	11.01	10.43

Source: Iowa State University, *Swine Business Record.*

Table 30.12 outlines a sample of the performance **benchmarks** for a farrow-to-finish enterprise. Careful monitoring of these benchmarks is critical to assess progress and to identify problem areas before excessive costs are incurred.

Costs and returns over several years are noted for farrow-to-finish operations (Fig. 30.7). Excellent managers achieve larger positive returns and lower losses

TABLE 30.12 Performance Benchmarks for a Farrow-to-Finish Enterprise

Trait	Goal
Litters farrowed/sow/year	2.0 (pen mating)
	2.2 (handmating—confinement)
Farrowing rate	>80%
Pigs weaned/sow/year	>20
Mortality rate	
Breeding herd	<4% (confinement)
	<6% (outside)
Preweaning	<10%
Nursery	<2%
Growing/finishing	<3%
Days to market	<190[a]
Feed efficiency	
Growing/finishing	<3.2
Nursery	<2.0
Whole herd	<3.3

[a] 45-lb nursery pig @ 56 days, 1.6-lb gain to 260 lb = 134 days.
Source: Adapted from Baas, 1999.

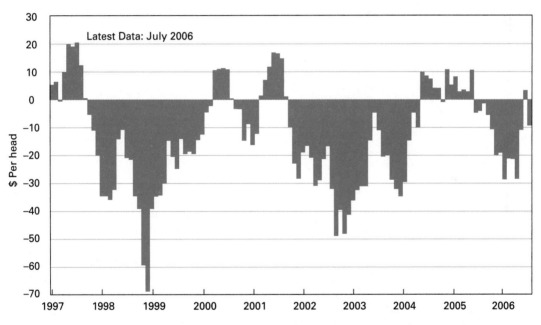

FIGURE 30.7 Average monthly returns for farrow-to-finish enterprises. *Source:* Livestock Marketing Information Center.

than shown in this figure. These top managers have low-cost production (especially low feed costs), while buying and selling pigs below and above average market prices. Even with excellent management, shifts in the market price for hogs (Fig. 30.8) can make it difficult to generate profits.

ENVIRONMENTAL MANAGEMENT

Management of waste and air quality factors are an important consideration for swine producers. The goals of an effective manure management plan should be to avoid or minimize (1) direct discharge, runoff, or seepage into a stream,

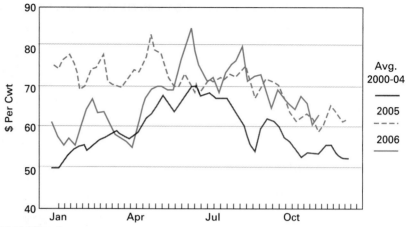

FIGURE 30.8 Market barrow and gilt prices. *Source:* Livestock Marketing Information Center.

TABLE 30.13 Feed-Related Odor Control Practices in the Swine Industry

Protocol	%
Finely ground grain	27.3
Vegetable oil or fat to control dust	24.0
Low crude protein and/or synthetic amino acids	19.8
Pelleting	15.3
Phytase	11.0
Other feed additives	10.1
Add 10% fiber	8.5
Other	8.5

Source: NAHMS, 2000.

(2) release of ammonia into the atmosphere, (3) detectable odor beyond the farm borders, (4) release of pathogens into the environment, and (5) contamination of soil and groundwater.

The most frequently utilized strategies to **control odor** include dietary manipulation (50.2%), manure management (28.9%), air quality interventions (28.2%), and the addition of chemical-biological agents to lagoons or waste storage sites (16.0%) according to the USDA. The specific feed-related odor control practices are described in Table 30.13.

Swine managers can also affect water and odor impacts by minimizing feed and water spills, improving utilization with feed processing technologies, adjusting diets for specific animal requirements, and adjusting diets for climatic variation. The specific effects of management practices on the nitrogen and phosphorus levels of waste are described in Table 30.14.

Consumers are increasingly vocal in expressing their concerns about animal handling in confinement and transportation situations. The Food Marketing Institute and National Council of Chain Restaurants have adapted sow-housing guidelines that include the following:

1. The pregnant sow should be able to lie on her side with her teats extending into the adjacent stall.

Table 30.14   Nutritional Management to Reduce Excess Nitrogen and Phosphorus (combined effects are not additive)

Method	Potential Decrease in: (%)	
	Nitrogen	Phosphorus
Precision ration formulation	10–15	10–15
Limit protein and amino acid supplements	20–40	—
Include highly digestible feeds	5	20–30
Include phytase in a diet with reduced phosphorus supplementation	2–5	20–30
Incorporate phase-feeding program	5–10	5–10

Source: Multiple sources.

TABLE 30.15   Swine Handling and Transport Guidelines

Handlers should move slowly among pigs in a pen for 1 minute per day or 5 minutes per week during the finishing period.

Feed should be withheld for 4–6 hours prior to loading or 12–18 hours prior to harvest. Ready access to water should be provided.

Eliminate shadows and assure even distribution of light.

Panels, paddles, or flags should be used to move hogs. Avoid any contact that might injure or stress the animal.

Move groups of 5–6 pigs on 3-ft.-wide alleys and only 3 pigs in a 2-ft.-wide alley.

Load pigs in the evening or early morning during periods when mid- or late-day heat and humidity are high. Use partitions to divide the load.

Avoid sharp turns in alleys; loading chutes should have less than 20° angles.

Minimize fighting among pigs by avoiding co-mingling of large groups of pigs in alleyways or holding areas.

The number of pigs per-running-foot of truck floor space (92 in. truck) should be 2.2 for 200-lb hogs, 1.8 for 250-lb hogs, and 1.6 for 300-lb hogs.

Avoid leaving pigs standing on a stationary, fully loaded truck.

Drivers should operate smoothly to avoid injuring pigs.

Schedule arrival at the packing plant so that hogs can be promptly off-loaded.

Source: National Institute for Animal Agriculture.

2. The sow's head should not have to rest on a raised feeder box.

3. The sow's hindquarters should not be in contact with the back portion of the stall.

4. The sow should be able to stand up unimpeded.

Additional transport and swine handling protocols are listed in Table 30.15.

CHAPTER SUMMARY

- Swine operations are classified into (1) farrow-to-finish, (2) farrow-to-feeder pig production, (3) feeder pig finishing, (4) farrow-to-weaner, and (5) purebred or seedstock.

- Management of the sow during farrowing and lactation and of the baby pigs from birth to weaning are the most critical periods in a total swine management program.

- Sow rations and baby pig rations must have the proper combinations of amino acids, vitamins, and minerals (in addition to protein and energy) to assure adequate and cost-effective gains.

- Herd health programs must be well planned to prevent reproductive diseases, assure adequate lactation, and prevent scours and other diseases in baby pigs.

KEY WORDS

farrow-to-finish	mastitis	antibiotics
farrow-to-feeder pig	metritis	chemotherapeutics
production	agalactia	anthelmintics
feeder pig finishing	cross-fostering	phase feeding
farrow-to-weaner	transmissible	split-sex feeding
purebred seedstock	gastroenteritis (TGE)	formula contract
boar	dysentery	forward-cash contract
handmating	gossypol	risk-share contract
gilt	mycotoxins	(2 types)
artificial insemination	aflatoxin	benchmarks
(AI)	zearalenone	odor control
farrowing	vomitoxin	
pig mortality	parakeratosis	

REVIEW QUESTIONS

1. What are the five primary types of swine operations?

2. Which is the major type of swine operation?

3. Why should boars be purchased at least 60 days before the breeding season?

4. Why is it best to feed pregnant sows and gilts in individual feeding stalls?

5. Approximately what percent of market hogs are conceived by artificial insemination?

6. Why are gilts and sows farrowed in farrowing stalls?

7. Why is it important for newborn pigs to receive colostrum?

8. At what temperature should newborn pigs be maintained during the first 2 weeks?

9. Why is it important to maintain baby pigs at 85–95°F during the first 2 weeks after birth?

10. What is the basis of good production records?

11. What is the most common form of permanent identification used on pigs?

12. Why is clipping of needle teeth performed on baby pigs?

13. Why is docking of tails performed on baby pigs?

14. *True or False:* Cereal grains such as corn, milo, barley, and wheat form the basis of pig rations because they are important protein sources.

15. *True or False:* Swine rations can be supplemented with lysine because it is usually the first limiting amino acid in most swine diets.

16. What are the three classes of feed additives for swine rations?

17. What is the reason that heavy pigs exceeding market weight are often discounted by the packer?

18. What marketing alternative is used to market nearly 60% of market hogs?

SELECTED REFERENCES

Aherne, F. 2006. Feeding the gestating sow. Fact sheet 07-01-06. Des Moines, IA: Pork Information Gateway.

Aherne, F. 2006. Feeding the lactating sow. Fact sheet 07-01-05. Des Moines, IA: Pork Information Gateway.

Baas, T. J. 1999. Key production factors to monitor the pork production business. Iowa State University (personal correspondence).

Flowers, B. 2002. Value of a good A.I. technician. *PORK,* Nov., 42.

Flowers, W. L., and H. D. Alhusen. 1991. Reproductive performance and estimates of labor requirements associated with combinations of artificial insemination and natural service in swine. *J. Anim. Sci.* 70:615.

Iowa Pork Industry Center. 1999. *Iowa Swine Business Record Summary.* Ames, IA: Iowa State University.

Lewis, D. 2002. Is your sow management costing you pigs. *PORK,* August, 20–22.

Miller, E. R., D. E. Ullrey, and A. J. Lewis. 1992. *Swine Nutrition.* Boca Raton, FL: CRC Press, Inc.

National Research Council. 1988. *Nutrient Requirements of Swine.* Washington, DC: National Academy Press.

Pond, W. G., J. H. Maner, and D. L. Harris. 1991. *Pork Production Systems.* Westport, CT: AVI Publishing Co.

Pork Industry Handbook (various years). Contains Fact Sheets (e.g., Management and Nutrition of the Newly Weaned Pig; Swine Diets; Swine Growing-Finishing Units; Pork Production Systems with Business Analyses). Cooperative Extension Services of several universities.

See, T. 2002. Water guidelines for swine. *PORK,* June, 3.

Sheep Breeds and Breeding

LEARNING OBJECTIVES

- Compare and contrast wool-type, meat-type, and dual-purpose sheep breeds
- Characterize the major sheep breeds
- Discuss the primary traits of economic importance in the sheep industry
- Contrast the management of the sheep breeding season with other livestock
- Compare and contrast the purebred and commercial sheep production segments
- Outline the use of crossbreeding in the sheep industry
- List the major inherited abnormalities in sheep

Sheep apparently evolved in the dry mountainous areas of southwest and central Asia. Present-day domesticated sheep were derived from wild animals that existed in Asia some 8,000–10,000 years ago. It is believed that domestication occurred during the early civilizations of southwest Asia. The Romans introduced a white, woolen sheep into Western Europe, and these sheep contributed heavily to the British breeds. Some 100–200 years ago, Europeans took the Merino and British breeds of sheep to South America, Australia, South Africa, and New Zealand. Additional information on the world and U.S. sheep industry can be found in Chapter 2.

MAJOR U.S. SHEEP BREEDS

Sheep have been bred for three major purposes: production of fine wool for making high-quality clothing; production of long wool for making heavy clothing, upholstering, and rugs; and production of mutton and lamb. In recent years, dual-purpose sheep breeds (which produce both wool and meat) have been developed by crossing fine-wool breeds with long-wool breeds and then selecting for both improved meat and wool production. Some breeds serve specific purposes; an example is the Karakul breed, which supplies pelts for clothing items such as caps and Persian coats.

Sheep breeds can be divided into three major categories:

1. **Ewe breeds** are usually white-faced and have fine or medium wool, long wool, or crosses of these types. They are noted for reproductive efficiency, wool production, size, milking ability, and longevity. The Rambouillet, Debouillet, Merino, Corriedale, Targhee, Finnsheep, and Border Leicester are breeds in the ewe breed category.

2. **Ram breeds** are meat-type breeds raised primarily to produce rams for crossing with ewes of the ewe breed category. They are noted for growth rate and carcass characteristics. The Suffolk, Hampshire, Shropshire, Oxford, Southdown, Montadale, and Cheviot are classed as ram breeds. Several recently imported breeds such as the Texel offer a wide variety of genetics to choose from. The Texel expresses a high degree of muscularity and lean yield.

3. **Dual-purpose breeds** are used either as ewe breeds or as ram breeds. The Dorset and Columbia are good examples of breeds that are often used as ewe breeds to be crossed with a good ram breed. The Dorset, Lincoln, and Romney are often used as ram breeds to cross on other breeds of ewes to improve milking ability and fertility of the ewe flock.

Characteristics

Breeds of sheep and their characteristics are listed in Table 31.1. Most breeds are distinguished by wool characteristics, color of face, size, and horned or polled condition. Inheritance of the horned or polled condition is sex-influenced and generally controlled by a single pair of genes. In the heterozygote, the genes are usually expressed as recessive in the ewe and dominant in the ram (Table 31.2).

The sheep breeds commonly used in the United States are shown in Figures 31.1 and 31.2. The Merino and Rambouillet (fine-wool) breeds and the dual-purpose breeds that possess fine-wool characteristics have a herding instinct; so one sheepherder with dogs can handle a band of 1,000 ewes and their lambs on the summer range. Winter bands of 2,500–3,000 ewes are common. Ram breeds, by contrast, tend to scatter over the grazing area. They are highly adaptable to grazing fenced pastures where feed is abundant. Rams of the ram breeds (meat-type) are often used for breeding ewes of fine-wool breeding on the range for the production of market lambs. Crossbreeding in sheep has been used for more than a half century.

The U.S. Sheep Experiment Station in Dubois, Idaho, developed the Columbia and Targhee breeds. Both breeds have proven useful on western ranges because they have the herding instinct and, when bred to a meat breed of ram, they raise better market lambs than do fine-wool ewes. The U.S. Sheep Experiment Station has made great strides in improving the Rambouillet, Columbia, and Targhee breeds.

TABLE 31.1 Characteristics of Sheep Breeds within Type Classification

Breed	Mature Size	Growth Rate[a]	Degree of Muscularity[a]	Wool Type	Fiber Diameter (micron)	Staple Length (in.)	Grease Fleece[b] Weight (lb)	Color	Origin (Imported to U.S.)
Ewe breeds									
Corriedale	R[e] 175–275 E[f] 130–180	M	M	Crossbred	31.5–24.5	3.5–6	10–17	White	New Zealand (1914)
Finnsheep	R 150–200 E 120–140	L	L	Medium	31–23.5	3–6	4–8	White	Finland (late 1960s)
Merino[c]	R 150–225 E 110–150	LM	L	Fine to very fine	22–19	2.5–4	8–14	White	France/Spain
Rambouillet	R 250–300 E 150–200	MH	M	Fine	24.5–18.5	2–4	8–18	White	France
Targhee	R 200–300 E 125–200	MH	M	Medium	26.5–21.5	3–4.5	10–14	White	U.S. (1926)
Ram breeds									
Cheviot	R 160–200 E 120–160	M	H	Medium	33–27	3–5	5–10	White with black nose	Scotland (1838)
Hampshire	R 225–325 E 175–225	H	H	Medium	33–25	2–3.5	6–10	White with wool cap and black face	England (1880)
Oxford	R 200–300 E 150–200	MH	H	Medium	34.5–30	3–5	8–12	White with brown face and legs, wool cap	England (1846)
Shropshire	R 225–250 E 150–180	MH	M	Medium	32.5–24.5	2.5–4	6–10	White with black face, wool cap	England (1855)
Southdown	R 190–230 E 130–180	M	H	Medium	29–23.5	1.5–2.5	5–8	White with grey face	England (1803)
Suffolk	R 250–350 E 180–250	H+	H	Medium	33–25.5	2–3.5	5–8	White with black face	England (1888)
Dual-purpose breeds									
Columbia	R 225–300 E 150–225	H	M	Medium	31–24	3.5–5	10–16	White	U.S. (1912)
Dorset[d]	R 210–250 E 140–180	MH	MH	Medium	33–27	2.5–4	5–9	White	England (1855)
Lincoln	R 250–350 E 200–250	M	M	Long	41–33.5	8–15	12–20	White	England (1825)
Romney	R 225–275 E 150–200	M	M	Long	38–31	4–6	8–12	White	England (1904)

[a] L = low, M = medium, H = high.
[b] Ewe basis.
[c] Rams, horned; ewes polled.
[d] Either horned or polled.
[e] Ram.
[f] Ewe.

TABLE 31.2　Horn Inheritance in Sheep

Genotype	Phenotype of Ewes	Phenotype of Rams
HH	Horned	Horned
Hh	Polled	Horned
hh	Polled	Polled

The most important trait in meat breeds of sheep is weight at 90 days of age, with some emphasis also given to conformation and finish of the lambs. Weight at 90 days of age can be calculated for lambs that are weaned at younger or older ages than 90 days as follows:

$$\textbf{90-day weight} = \frac{\text{weaning weight} - \text{birth weight}}{\text{age at weaning}} \times 90 + \text{birth weight}$$

If a lamb weighs 10 lb at birth and 100 lb at 100 days, the 90-day weight is:

$$\frac{100 \text{ lb} - 10 \text{ lb}}{100 \text{ days}} \times 90 \text{ days} + 10 \text{ lb} = 91 \text{ lb}$$

If birth weight was not recorded, an assumed birth weight of 8 lb can be used.

Approximately 85–90% of the total income from sheep of the meat breeds is derived from the sale of lambs, with only 10–15% coming from the sale of wool. By contrast, the sale of wool accounts for 30–35% of the income derived from fine-wool and long-wool breeds. Even with wool breeds, the income from lambs produced constitutes the greater portion of total income.

The U.S. Sheep Experiment Station in Dubois, Idaho, has developed a new synthetic breed of sheep, called the **Polypay,** which has superiority in lamb production and in carcass quality. Four breeds (Dorset, Targhee, Rambouillet, and Finnsheep) provided the basic genetic material for the Polypay. Targhees were crossed with Dorsets, and Rambouillets were crossed with Finnsheep, after which the two crossbred groups were crossed. The offspring produced by crossing the two crossbred groups were intermated and a rigid selection program was practiced. This population was closed to outside breeding. The Rambouillet and Targhee breeds contributed hardiness, herding instinct, size, a long breeding season, and wool of high quality. The Dorset contributed good carcass, high milking quality, and a long breeding season. The Finnsheep contributed early puberty, early postpartum fertility, and high lambing rate.

The Polypay breed has shown outstanding performance in conventional once-a-year lambing under range conditions and superior performance in twice-a-year lambing when compared with other breeds or breed crosses.

There are several U.S. breeds of sheep that are not listed in Table 31.1. Some of these breeds possess certain characteristics that may be useful for improving the most popular U.S. breeds either through systematic crossbreeding or in the establishment of new breeds.

Based on annual registration numbers, the Suffolk, Dorset, Rambouillet, and Hampshire are the most numerous breeds of sheep (Table 31.3). The Suffolk and Hampshire are used extensively in crossbreeding programs to produce market lambs both in farm flocks and on the range. The most popular ewe breeds on the range are the Rambouillet, Columbia, and Corriedale. In a farm flock lamb-production system, the Dorset and Shropshire are important ewe breeds.

Cheviot

Cheviot

Hampshire

Hampshire

Southdown

Southdown

Texel

FIGURE 31.1 Breeds of sheep commonly used as straightbreds or for crossbreeding to produce market lambs—Cheviot (*Source:* American Sheep Industry Association); Dorset; Hampshire; Shropshire (*Source:* American Sheep Industry Association); Southdown (*Source:* American Southdown Breeders Association); Suffolk; Texel.

Columbia

Corriedale

Polypay

Rambouillet

Targhee

FIGURE 31.2 Breeds of sheep commonly used as straightbreds or for crossbreeding to improve wool production or maternal traits—Columbia (*Source:* American Sheep Industry Association); Corriedale (*Source:* American Sheep Industry Association); Polypay (*Source:* American Sheep Industry Association); Rambouillet (*Source:* American Sheep Industry Association); Targhee (*Source:* U.S. Targhee Sheep Association).

BREEDING SHEEP

Reproduction

Sheep differ from many farm animals in having a breeding season that occurs mainly in the fall of the year. The length of the breeding season varies with breeds. Ewes of breeds with a long breeding season show heat cycles from mid- to late

TABLE 31.3   Major U.S. Sheep Breeds Based on Annual Registration Numbers (in thousands), 1975–2005

Breed	2005	1998	1995	1990	1985	1980	1975	Year U.S. Association Formed
Suffolk[a]		23.5	29.8	41.8	49.1	60.3	31.8	1935
Dorset	11.6	12.2	12.6	19.5	12.8	15.2	8.7	1898
Hampshire	55.6	9.4	11.2	16.4	16.9	60.4	17.2	1889
Rambouillet		5.0	12.4	17.1	12.2	11.9	10.0	1889
Polypay		2.5	2.9	1.4	3.2	0	0	1979
Columbia		4.3	5.0	7.8	8.0	10.0	6.0	1942
Southdown		5.5	5.8	5.9	4.8	4.4	3.6	1882
Corriedale		2.7	3.1	5.0	4.4	6.5	6.1	1816
Montadale		3.2	2.9	3.8	3.1	3.1	1.9	1945
Shropshire		2.5	2.6	2.9	3.4	4.4	3.8	1884
Cheviot		2.1	2.4	2.8	2.5	2.5	2.3	1891

Note: Only breeds with more than 2,100 annual registrations are listed.
[a] Registrations in the National Suffolk Sheep Association and American Suffolk Sheep Association combined.
Source: Various breed associations.

summer until midwinter. The breeds in this category are the Rambouillet, Merino, and Dorset. Breeds with an intermediate breeding season (Suffolk, Hampshire, Columbia, and Corriedale) start cycling in late August or early September and continue to cycle until early winter. Ewes of breeds having long or intermediate breeding seasons are more likely to fit into a program of three lamb crops in 2 years.

Ewes of breeds with short breeding seasons (Southdown, Cheviot, and Shropshire) do not start cycling until early fall and discontinue cycling at the end of the fall period.

Puberty is reached at 5–12 months of age and is influenced by breed, nutrition, and date of birth. The average length of **estrous cycles** is slightly over 16 days, and the length of estrus (when the ewe is receptive to the ram) is about 30 hours. The length of gestation (time from breeding until the lamb is born) averages 147 days but varies; medium-wooled and meat breeds have shorter gestation periods, whereas fine-wooled breeds have longer gestation periods.

Several factors affect **fertility** in sheep. Lambing rates vary both within and between breeds, with some ewes consistently producing only one lamb per year and others producing three or four lambs each year. To improve lambing rate, producers keep replacements from ewes that consistently produce two to four lambs each year.

Examining the semen should confirm fertility of rams used in a breeding program. Semen can be collected by use of an artificial vagina and examined with a microscope. Ram fertility is evaluated by (1) checking for abnormal sperm (tailless, bent tails, no heads, and so on), (2) observing the percentage of live sperm (determined by a staining technique using fresh semen), and (3) checking the sperm motility and concentration in freshly collected semen. Two semen collections 3–4 days apart should be examined. When rams have not ejaculated for some time, there may be dead sperm in the **ejaculate.**

Other Factors Affecting Reproduction

Crossbreeding is one of the most advantageous tools available to enhance reproductive performance. Crossbred ewe lambs, when adequately fed, are usually bred to lamb at 1 year of age. Generally, crossbred lambs are more likely to conceive as lambs than are purebred lambs.

Age impacts reproductive rate as mature (3- to 7-year-old) ewes are more fertile and raise a higher percentage of lambs born than do younger or older ewes.

Light, temperature, and relative humidity also affect reproduction in sheep. Sheep respond to decreased day lengths both by showing a greater proportion of ewes in estrus and by higher conception rates. *Temperature* has a marked effect on both ewes and rams. High temperatures cause heat sterility in rams because the testicles must be at a temperature below normal body temperature for viable sperm production. Embryo survival in the ewe is also influenced by temperature. When the ambient temperature exceeds 90°F, embryo survival decreases during the first 8 days after breeding.

Environmental factors such as disease, parasites, and insufficient feed resources impact the health and well-being of sheep and when not correctly managed may reduce the number of lambs produced. Producers can increase their income and profit from a sheep operation by controlling diseases and parasites and by providing the sheep with an adequate supply of good feed. Sheep in moderate body condition are usually more productive than fat sheep. Ewes that are in moderate condition and that are gaining weight before and during the breeding season will have more and stronger lambs than ewes that are either overfat or **emaciated** at breeding.

Estrus synchronization and AI are tools that can be used to influence reproduction. In some intensively managed sheep operations, hormones can be used to synchronize estrus. Hormones can also be used to bring ewes into estrus at times other than during their normal breeding season.

Estrus can be synchronized if progesterone is given for a 12- to 14-day period in the feed, as an implant, via an intravaginal device, or by daily injections followed by injections of pregnant mare serum (PMS) the day progesterone is discontinued. Conception rates at this estrus are low; a second injection of pregnant mare serum 15–17 days later will bring the ewes into estrus and fertile matings will occur. There must be sufficient ram power to breed a large number of ewes naturally if they are synchronized, or artificial insemination can be used.

This system of synchronizing estrus can also be used to obtain pregnancies out of the normal breeding season and to obtain a normal lamb crop from breeding ewe lambs. Also, producers can use estrus synchronization for accelerated lambing if it is desired to raise two lamb crops per year.

If the ewes are to be artificially inseminated, high-quality fresh semen should be diluted with egg-yolk-citrate diluter shortly before insemination. Ram semen has been frozen, but conception rates have been low.

Estrogen content in feeds may result in a flock of sheep expressing very low fertility because of high estrogen content of the legume pasture or hay that the sheep are consuming. Producers are advised to have the legume hay or legumes in the pasture checked for estrogen if they are experiencing fertility problems in their sheep.

The Breeding Season

Sheep may be handmated or pasture mated. If they are **handmated,** some **teaser rams** are needed to locate ewes that are in heat. Either a **vasectomized** ram (each vas deferens has been severed so sperm are prevented from moving from the testicles) can be used, or an apron can be put on the ram such that a strong cloth prevents copulation.

Tagging ewes, or removing wool from the dock and vulva region before breeding, will result in a larger percentage of lambs as well as a lamb flock of more uniform age.

Ewes should be checked twice daily for **heat.** Ewes normally stay in heat for 30 hours and ovulate near the end of heat. It is desirable to breed ewes the morning after they are found in heat in the afternoon. Ewes that are found in heat in the morning should be bred in the late afternoon.

If ewes are to be handmated, the ram should breed them once. Recently bred ewes should be separated from other ewes for 1 or 2 days so that the teaser ram does not spend his energies mounting the same ewe repeatedly.

If ewes are to be **pasture mated,** they are sorted into groups according to the rams to which they are to be mated. It is best to have an empty pasture between breeding pastures so as not to entice rams to be with ewes in another breeding group.

Using numbered branding irons to apply scourable paint is a common means to identify sheep. The brisket of the breeding ram can be painted with scourable paint so that he marks the ewe when he mounts her. A light-colored paint should be used initially, and then a dark color can be used about 14 days later to detect ewes that return in heat. An orange paint can be used initially, followed, successively, with green, red, and black. The paint color should be changed each 16–17 days. Because the ewes are numbered, they can be observed daily and the breeding date of each can be recorded. Also, a ram that is not settling his ewes can be detected and replaced. Occasionally, a ram may be low in fertility even though his semen was given a satisfactory evaluation before the breeding season.

A breeding season of 40 days results in a lamb crop of uniform age and identifies ewes for culling that have an inherent tendency for late lambing.

The Purebred Breeder

The goal of purebred breeders (Fig. 31.3) is to make genetic changes in economically important traits. These breeders use selection as their method for genetic improvement, as crossbreeding is limited primarily to commercial producers. Purebred breeders with large operations may find it desirable to close their flocks and select ewe and ram replacements from within the closed flocks.

Normally, purebred breeders have selection programs to produce superior rams for commercial producers. At the same time, commercial producers prefer rams that contribute outstanding performance in a commercial crossbreeding program.

In general, great progress in sheep improvement can be made by selection within a breed for traits that are highly heritable and economically important. Crossbreeding can give significant genetic improvement in traits of low

FIGURE 31.3 Selection for economically important traits is key to genetic improvement. Improvement in wool quality and production can be achieved because wool traits are typically highly heritable and can be measured objectively and accurately.

heritability such as fertility. Traits that are only moderately heritable can be improved by selecting genetically superior breeding animals and also by using crossbreeding.

Highly heritable traits (40% or higher) include mature body size, yearling type score, face cover, **skin folds,** clean fleece yield, yearling staple length, gestation length, loin eye area, fat weight, and retail cut weight. Traits having low heritability (below 20%) include weaning type score, weaning condition score, multiple births, number of lambs weaned, fat thickness over loin, carcass weight per day of age, carcass grades, and dressing percentage. Moderately heritable traits (20–39%) include birth weight, 90-day weight, rate of gain, neck folds, grease fleece weight, fleece grade, lambing date, milk production, carcass length, and bone weight.

The National Sheep Improvement Program was initiated in 1986. Genetic estimates are calculated for number of lambs born, litter weight at 60 days, individual weights at 30, 60, 90, 120, 180, or 360 days (3 weights analyzed per flock), weight of the wool clip, wool fiber diameter, and staple length.

Principles applicable to breeding and improving sheep are presented in Chapters 12 (genetics), 13 (selection), and 14 (systems of mating).

Commercial Sheep Production

Producers should consider the following issues when designing a breeding system for flock:

1. What are the available pasture and management resources?
2. What is the desired income percentage from the sale of lambs versus wool? (Wool and growth traits tend to be antagonistic.)
3. What are the potential benefits from heterosis?

TABLE 31.4 Effects of Heterosis in Crossbred Lambs

Trait	Level of Heterosis (%)
Birth weight	3
Weaning weight	5
Postweaning gain	7
Survival of weaning	10
Pounds of lamb weaned per exposed ewe	18

Source: Adapted from Schoenian, 2007.

4. How can breed differences best be utilized to meet the breeding and marketing goals of the flock?

5. What is the anticipated season of marketing?

6. How large is the flock? (Smaller flocks have fewer options than larger flocks.)

Producers must carefully balance the desire to take maximum advantage of heterosis (Table 31.4) with the need to optimize the breeds utilized to assure that the management system requirements of the flock are met. Despite its advantages, crossbreeding may not be desirable for small flocks or in flocks where the intensity of management is not sufficient to assure correct implementation of the crossbreeding plan.

Terminal Crossing

For example, in some extensive management conditions of the semiarid West, the use of the Rambouillet is quite common. The Rambouillet's fine-wool production, durability, and mature size are so desirable that the use of a ram from a complementary ewe breed is difficult to identify. However, the use of a black-faced breed such as the Hampshire or Suffolk is an effective terminal cross for these conditions. Under the more intensive management conditions of many farm flocks, the use of crossbreeding systems to produce ewe replacements is more common.

Most market lambs are crossbreds (produced either from crossing breeds or from mating crossbred ewes with purebred rams), but the key to the success of **crossbreeding** is the improvement in production traits made by the purebred breeders in their selection programs. Crossbreeding can be used to combine meat conformation of the sire breeds with lambing ability and wool characteristics of the ewe breeds, and also to obtain the advantage of the fast growth rate of lambs that result from heterosis. The additive as well as the heterotic effects is evident in crossbred lambs. Maximum **heterosis** is obtained in three- or four-breed crosses, either rotational or terminal.

A clear understanding of the market target and time of marketing is also useful. For example, if the spring or Easter lamb market is the target, the use of a sire breed with moderate mature size and excellent muscularity (such as the Southdown) bred to a highly fertile maternal breed (such as the Dorset) may be a desirable mating system. This system is effective because the resulting lambs finish at relatively low weights (70–80 lb) and this meets the requirements of the spring market. While the spring market has historically been a target, its importance as a niche market is declining.

Three-Breed Terminal Crossbreeding

The use of terminal crossing systems is particularly valuable in those production systems where fine-wool production is important. For example, in Australia, the Merino is the predominant wool-producing breed. While the Merino produces an outstanding quality of wool clip, Merino wethers are lightly muscled and poor in growth rate. A common crossbreeding system to produce better-quality market lambs has evolved (Fig. 31.4). In such a system, poorer-performing Merino ewes or older ewes in the flock are allocated to the process of producing F_1 females for mating to a terminal sire.

Utilizing breed differences in crossing systems has led producers to experiment with a variety of breed types. For example, the Finnsheep is highly prolific, often producing three to six lambs per lambing. Crossbred ewes of 25–50% Finnsheep genetics produce significantly more lambs than straightbred or crossbred ewes of the meat breeds. The use of Finnsheep in crossing programs was initiated by a desire to attain more rapid reproductive efficiency. Unfortunately, high-percentage Finnsheep market lambs lack the growth rate and muscularity required to be profitable. These producers utilizing Finnsheep have attempted to gain the efficiency improvements without the loss of performance by utilizing $\frac{1}{4}$ blood Finnsheep rams.

Some research shows that two lamb crops can be raised per year or, if feed and other conditions do not support such intensive production, three lamb crops every 2 years can be produced. To achieve intense production of this type, producers need ewes that will breed throughout the year. At present, the Polypay seem to fit into such a program fairly well. Also, the Rambouillet and Dorset have a tendency to breed out of season.

To raise three lamb crops in 2 years, ewes should be bred in late summer, probably for a 30-day period (August), for lambs to arrive in January (lamb crop 1). The ewes and lambs need to be well fed so the lambs are sufficiently finished to go to market at 90 days of age (in April). The ewes are then bred in April (30-day breeding period) to lamb in August (lamb crop 2), and the ewes and lambs need good feed so the lambs are well finished and sufficiently large to be marketed at 90 days

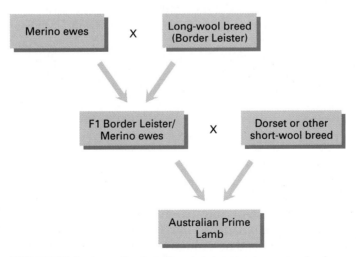

FIGURE 31.4 Australian breeding system to produce prime lambs.

of age (November). The ewes are bred in November to lamb in April (third lamb crop) and the lambs are marketed at 90 days of age in July.

INHERITED ABNORMALITIES

It is important to guard against certain genetic abnormalities when managing sheep. Although exceptions occur, sheep showing obvious genetic defects are usually culled. Some inherited abnormalities include the following.

Cryptorchidism is inherited as a simple recessive trait; therefore, a ram with only one testis descended into the scrotum should never be used for breeding. In addition, producers should cull rams that sire lambs having cryptorchidism, as well as the ewes that produced the lambs.

Dwarfism is inherited as recessive and is lethal; therefore, ewes and rams producing dwarf offspring should be culled.

The mode of inheritance of **entropion** (turned-in eyelids) has not been determined, but it is known to be under genetic control. A record should be made of any lamb having entropion so that the lamb can be marketed.

Sheep have lower front teeth but lack upper front teeth. They graze by closing the lower teeth against the dental pad of the upper jaw. If the lower jaw is either too short **(overshot** or **parrot mouth)** or too long **(undershot),** the teeth cannot close against the dental pad and grazing is difficult. The mode of inheritance of these jaw abnormalities is unknown, but the conditions are under genetic control. Sheep having abnormal jaws should be culled.

Rectal prolapse is common in black-faced sheep. It is a serious defect, and both inheritance and the environment are influential in its occurrence. Lambs on heavy feeding or lush pastures are more likely to show rectal prolapse. If surgery is used to correct the condition, the animal should not be used as a breeding animal.

Skin folds, open-faced, closed-faced, and **wool blindness** are inherited and selection against these traits has been effective.

Spider syndrome is a severe skeletal deformity occurring in Suffolk sheep. The most striking feature is an outward bending of the forelimbs from the knees. Angular deformities of the hind limbs are generally present. The genetic abnormality is due to a recessive gene and can be effectively selected against.

Many wool defects are known, including black fibers, black-tipped fibers, hairiness, fuzziness, and "high belly wool" in which wool that is typical of wool on the belly is present on the sides of sheep. Selection against all of these fleece defects should be practiced as a means of reducing the frequency of their occurrence.

Genes with Positive Effects on Important Traits

A single allele (CLPG) is responsible for the expression of a condition known as callipyge. Sheep expressing this condition are significantly more muscular, leaner, and have higher dressing percentages than those without the trait. Unfortunately, the resulting meat from these animals is tougher and thus less desirable.

A prolific strain of Merino from Australia is known as the Booroola Merino and its exceptional reproductive performance has been tied to an autosomal allele designated as Fec^B. Ewes with one or two copies of the allele have higher production of offspring per birth and fine wool. However, the resulting offspring have not performed well.

Both of these conditions are examples where the ability to identify and isolate a gene with significant effects may allow producers the ability to make dramatic changes in particular traits. It is important to recognize that resulting antagonisms may limit the widespread utilization of such an approach.

CHAPTER SUMMARY

- Breeds of sheep are classified primarily by wool production and ewe traits (e.g., Rambouillet), meat production (e.g., Suffolk), or a combination of the two (dual-purpose—e.g., Columbia).
- Based on registration numbers, the most numerous sheep breed is the Suffolk, followed by the Dorset, Hampshire, Southdown, and Rambouillet.
- Under most conditions, multiple births (i.e., twinning) are highly desirable, with other economically important traits being growth rate, wool production, and carcass merit (combination of quality grades and yield grades).
- The breeding program for many large range sheep flocks involves using wool breeds or dual-purpose breeds of ewes to mate to ram breeds excelling in growth rate and carcass characteristics.

KEY WORDS

ewe breeds	teaser ram	dwarfism
ram breeds	vasectomized	entropion
dual-purpose breeds	tagging	parrot mouth (overshot)
90-day weight	heat	undershot
Polypay	pasture mated	rectal prolapse
puberty	skin folds	wool blindness
estrous cycles	crossbreeding	spider syndrome
fertility	heterosis	
ejaculate	three-breed terminal	
emaciated	cross	
handmated	cryptorchidism	

REVIEW QUESTIONS

1. What are the three categories of sheep breeds?
2. What are the production characteristics of ewe breeds?
3. What characteristics are ram breeds noted for?
4. What is the most important trait of meat breeds?
5. What is the average length of the estrous cycle in ewes?
6. What is the average length of gestation in sheep?
7. At what age does puberty occur in sheep?
8. What factors influence the age of puberty in sheep?

9. How is ram fertility initially evaluated before the breeding season?

10. What four characteristics of ram semen are examined during fertility evaluation?

11. What factors affect reproduction in sheep?

12. What is tagging?

13. What is the primary goal of purebred sheep breeders?

14. What is the key to success in a crossbreeding program?

15. What is the relationship between growth and wool traits?

16. Describe the typical Australian crossbreeding system?

17. Which three ewe breeds of sheep are notable for breeding throughout the year and fit well into a program of three lamb crops in 2 years?

18. What characteristic of ewe breeds makes them highly suitable for range production?

SELECTED REFERENCES

Botkin, M. P., R. A. Field, and C. L. Johnson. 1988. *Sheep and Wool: Science, Production, and Management.* Englewood Cliffs, NJ: Prentice Hall.

Dickerson, G. E. 1978. *Crossbreeding Evaluation of Finnsheep and Some U.S. Breeds for Market Lamb Production.* North Central Regional Publication #246. ARS, USDA, and University of Nebraska.

Lasley, J. F. 1987. *Genetics of Livestock Improvement.* Englewood Cliffs, NJ: Prentice Hall.

Maijala, K. 1991. *Genetic Resources of Pig, Sheep, and Goat.* Amsterdam: Elsevier Science Publishers.

Neimann-Sorenson, A., and D. E. Tube (eds. in chief). 1982. World Animal Science, Series C: Production System Approach. I. E. Coop (ed.). *Sheep and Goat Production.* New York: Elsevier.

Ross, C. V. 1989. *Sheep Production and Management.* Englewood Cliffs, NJ: Prentice Hall.

Schoenian, S. 2007. Breeding systems. www.sheep101.info. *The Sheepman's Production Handbook.* 1988. Denver, CO: Sheep Industry Development Program.

Wilson, D. E., and D. G. Morrical. 1991. The national sheep improvement program: A review. *J. Anim. Sci.* 69:3872.

CHAPTER 32

Feeding and Managing Sheep

LEARNING OBJECTIVES

- Describe the distribution of sheep on various types of sheep enterprises
- List the primary facilities and equipment required for sheep production
- Discuss the types of farm flock enterprises
- Describe the nutritional management of the ewe flock
- Discuss management during lambing, shearing, and weaning
- Contrast range flock management with farm flock management
- Discuss the factors affecting profitability of the sheep enterprise

Sheep feeding and management are essential for the success of an operation. These areas must be integrated with knowledge of how breeding and environmental factors affect sheep productivity and profitability. Several areas of feeding and management that affect efficient production are discussed in this chapter. Sheep enterprises can generally be classified into one of three categories—farm flocks, range flocks, or lamb feeders. Refer to Table 32.1 for a comparison of the size of farm versus range flocks.

PRODUCTION REQUIREMENTS FOR FARM FLOCKS

Pastures

Good pastures are essential to the typical farm flock operator (Fig. 32.1). Grass-legume mixtures, such as rye grass and clover or orchard grass and alfalfa, are ideal for sheep. Sheep can also graze on temporary pastures of such plants as Sudan grass or rape, which are often used to provide forage in the dry part of

TABLE 32.1 Percent of Sheep on Enterprises by Flock Type and Region

	Total U.S.	Herded Range Flock	Fenced Range Flock	Farm Flock	Intensive Confinement	Multiple
West Coast[a]	16.7	36.9	35.9	23.6	1.1	2.5
Mountain[b]	33.6	38.0	32.6	14.9	10.0	4.5
West North Central[c]	17.7	0.6	18.8	65.1	13.8	1.7
West South Central[d]	20.6	0.4	77.9	18.9	1.0	1.8
East North Central[e]	5.4	0.1	7.6	87.7	3.5	1.1
East South Central[f]	0.9	0.0	5.3	90.9	0.5	3.3
Northeast[g]	2.8	0.3	5.3	90.4	2.4	1.6
Southeast[h]	2.3	0.9	7.5	87.2	2.5	1.9

[a] OR, WA, CA.
[b] MT, ID, WY, NV, UT, CO, AZ, NM.
[c] SD, ND, NE, KS, MN, IA, MO.
[d] OK, TX.
[e] MI, OH, IN, WI, IL.
[f] AR, LA, AL, MS, TN, KY.
[g] PA, NY, and New England.
[h] FL, GA, SC, NC, VA, WV, MD.
Source: NAHMS, USDA, 1996.

FIGURE 32.1 Effective management of forages and pastures is critical to the success of the sheep enterprise. Courtesy of Colorado State University.

summer when permanent pastures may show no new growth. Crops of grain and grass seed may also provide pasture for sheep during autumn and early spring. Sheep do not trample wet soil as severely as do cattle; pasturing sheep on grass-seed and small-grain crops in winter and early spring when the soil is wet does not cause serious damage to the plants or soil.

Fencing

A woven-wire or electric fence is necessary to contain sheep in a pasture. Forage can be best utilized by rotating (moving) sheep from one pasture to another, a practice that also assists in the control of internal parasites. Some sheep operators use temporary fencing (such as electric fencing) to divide pastures and for predator control. It is necessary to use two electrified wires, one located low enough to prevent sheep from going underneath the fence, and the other located high enough to prevent them from jumping over it.

Corrals and Chutes

It is occasionally necessary to put sheep in a small enclosure to sort them into different groups or to treat any sick animals. Proper equipment is helpful in this task because sheep are often difficult to drive, particularly when ewes are being separated from their lambs. A **cutting chute** can be constructed to direct sheep into various small lots. A well-designed chute is sufficiently narrow to keep sheep from turning around and can be blocked so that sheep can be packed together closely for such purposes as treating diseases and reading ear tags (Fig. 32.2). The chute can be constructed of lumber that is nailed to wooden posts set into the ground and properly spaced so that the correct width (14–16 in.) is provided when the boards are nailed on the inside. Pens used to enclose small groups of sheep can be constructed of woven-wire fencing. A loading chute is

FIGURE 32.2 Moving sheep into a catch pen to be worked through a chute to be vaccinated. Courtesy of American Sheep Industry Association.

FIGURE 32.3 Unloading sheep off a truck onto a high country pasture.

used to place sheep onto a truck for hauling. A portable loading chute (Fig. 32.3) is ideal because it can be moved to different locations where loading and unloading sheep are necessary.

Shelters

Sheep do not normally suffer from cold because they have a heavy wool covering; therefore, open sheds are excellent for housing and feeding wintering ewe lambs, pregnant ewes, and rams. Although the ewes can be lambed in these sheds, the newborns need an enclosed and heated room when the weather is cold. However, sheep are vulnerable to changes in the weather following shearing. Allowances should be made to minimize stress following shearing.

Lambing Equipment

Small pens, about 4×4 ft, can be constructed for holding ewes and their lambs until they are strong enough to be put with other ewes and lambs. Four-foot panels can be constructed from 1×4-in. lumber, and the two panels can be hinged together. The lambing pens (called **lambing jugs**) can be made along a wall by wiring these hinged panels. Heat lamps are extremely valuable for keeping newborn lambs warm. A heat lamp above each lambing jug can be located at a height that provides a temperature of 90°F at the lamb's level (Fig. 32.4).

It is advisable to identify each lamb and to record which lambs belong to which ewes. Often, a ewe and her lambs are branded with a scourable paint to identify them. Numbered ear tags can be applied to the lamb at birth. If newborn lambs are to be weighed, a dairy scale and a large bucket are needed. The lamb is placed in the bucket, which hangs from the scale, for weighing. It is advisable to immerse the umbilical cords of newborn lambs in a tincture of iodine.

FIGURE 32.4 Lambing jugs are designed to ensure that newborn lambs can be kept warm and bonded to their dams.

Feeding Equipment

Usually sheep are given hay during the winter feeding period. Feeding mangers should be provided because hay is wasted if it is fed on the ground. Some sheep may require limited amounts of concentrates. Concentrates can be fed in the same bunk as hay if the hay bunk is properly constructed, or separate feed troughs (Fig. 32.5) may be preferable.

Additional concentrates can be provided for lambs by placing the concentrates in a **creep**, which is constructed with openings large enough to allow the lambs to enter but small enough to keep out the ewes. In addition to concentrates, it is advisable to provide good-quality legume hay and a mineral mix containing calcium, phosphorus, and salt. Creep diets are balanced for needed nutrients, but need not be complex in nature. Wheat or barley could replace up to half of the corn or sorghum (milo) in these diets. Lambs normally eat 1.5 lb of creep diet daily from 10–120 days of age (0.10 lb at 3 weeks and 3.0 lb at 120 days).

Water is essential for sheep at all times. It can be provided by tubs or automatic waterers. Tubs should be cleaned once each week. Large buckets are usually used for watering ewes in lambing jugs.

A separate pen equipped with a milk feeder may be needed for orphan lambs. Milk or a milk replacer can be provided free-choice if it is kept cold. Heat lamps kept some distance from the milk feeder can be provided.

Feed Storage

Areas should be provided for storage of hay and concentrates so that feeds can be purchased in quantity or so homegrown feeds can be stored in a dry place.

FIGURE 32.5   Sheep eating corn and pellets from a feed bunk. Courtesy of American Sheep Industry Association.

An open shed is satisfactory for hay storage. Some operators prefer a feed bin for concentrates; it is designed so that the feed can be put in at the top and removed from the bottom.

TYPES OF FARM FLOCK PRODUCERS

Some producers raise purebred sheep and sell rams for breeding. Commercial producers whose pastures are productive can produce slaughter lambs on pasture, whereas those whose pastures are less desirable produce feeder lambs. Lambs that are neither sufficiently fat nor sufficiently large for harvest at weaning time usually go to feedlot operators where they are fed to market weight and condition. However, most feedlot lambs are obtained from producers of range sheep.

Purebred Breeder

Purebred sheep are usually given more feed than commercial sheep because the **purebred breeder** is interested in growing sheep that express their growth potential. Records are essential to indicate which ram is bred to each of the ewes and which ewe is the mother of each lamb. All ram lambs are usually kept together to compare and identify those most desirable for sale. The purebred breeder usually retains the ram lambs until they are a year of age, at which time they are offered for sale. Buyers usually seek to obtain rams from purebred breeders in early summer.

Purebred breeders have major responsibilities to the sheep industry. Because purebred breeders determine the genetic productivity of commercial sheep, they should rigidly select animals that are kept for breeding and should offer for sale only those rams that will contribute improved productivity for the commercial producer.

Commercial Slaughter Lamb Producers

The lamb producer whose feed and pasture conditions are favorable aims for lambs to be born in late winter and early spring and to be weaned at 60–70 days of age. Weaned lambs from this system typically weigh 60 lb. These lambs are often penned and fed to gain 0.7 lb per day, with a market weight target of 120 lb. Access to a creep-fed diet in the preweaning phase is advisable.

Lambs can be finished on pasture or in a combined pasture-supplement feeding scenario. However, gain will be less than it would be for lot-fed lambs. Example diets for feeding early-weaned lambs are provided in Table 32.2.

Commercial producers can gain some assurance of raising heavy, well-finished market lambs at weaning by using good ram selection and crossbreeding programs. Rams that are heavy at 90 days of age are more likely to sire lambs that are heavy at this age than are rams that are light in weight at 90 days of age. The breeds used in a three-breed rotation should include one that is noted for milk-producing ability, one that is noted for rapid growth, and one that is noted for ruggedness and adaptability. The Dorset could be considered for milk production, either the Hampshire or Suffolk for rapid growth, and the Cheviot for ruggedness. All of these breeds make desirable carcasses.

Commercial producers castrate all ram lambs. Young ewes that are selected to replace old ewes should be large and growthy and should be daughters of ewes that produce relatively many lambs. Ewes that have started to decline in production and poorly productive ewes should be culled.

Commercial Feeder Lamb Producers

Some commercial sheep are produced under pasture conditions that are insufficient for growing the quantity or quality of feed needed for heavy slaughter lambs at weaning. Lambs produced under such conditions may be either fed out in the summer or carried through the summer on pasture with their dams and finished in the fall. Sudan grass or rape can be seeded so that good pasture is available in the summer, and lambs can be finished on pasture by giving them some concentrates. If good summer pasture cannot be made available, it is advisable to wait until autumn, at which time the lambs are put on full feed in a feedlot.

TABLE 32.2   Example Diets for Finishing Early-Weaned Lambs in Drylot

Ingredient	Ration Number and % of Diet	
	1	2
Ground corn (shelled)	37.3	—
Cracked corn	37.4	—
Ground corn (ear)	—	76.9
Cottonseed hulls	10.0	—
Soybean meal, 48%	12.8	15.6
Molasses	—	5.0
Ground limestone	1.0	1.0
Trace mineral + selenium	1.0	1.0
Ammonium chloride	0.5	0.5
Vitamin premix[a]	+	+

[a]According to manufacturer's instructions.
Source: Adapted from *Sheep Production Handbook.*

TABLE 32.3 Example Rations for Growing and Finishing Lambs

| | Ration Number and % of Diet | | | | | |
| | Up to 70 lb | | 70–90 lb | | 90 lb to Market | |
Ingredient	1	2	1	2	1	2
Corn (ground/cracked)	52.0	—	62.0	30.0	72.0	60.5
Corn (ground ear)	—	60.5	—	30.5	—	—
Corn (ground cobs)	20.0	—	10.0	—	—	—
Soybean meal, 48%	10.5	5.5	10.5	5.5	10.5	5.5
Dehydrated alfalfa	10.0	—	10.0	—	10.0	—
Alfalfa hay (ground)	—	27.5	—	27.5	—	27.5
Molasses	5.0	5.0	5.0	5.0	5.0	5.0
Dicalcium phosphate	1.0	—	1.0	—	1.0	—
Trace mineral salt + selenium	1.0	1.0	1.0	1.0	1.0	1.0
Ammonium chloride	0.5	0.5	0.5	0.5	0.5	0.5

Source: Adapted from *Sheep Production Handbook.*

Commercial Feedlot Operator

Lambs that come to the feedlot for finishing are treated for internal **parasites** and for certain diseases, particularly **overeating disease.** They are provided water and hay initially, after which concentrate feeding is allowed and increased until lambs receive all the concentrates they will consume. Death losses may be high in unthrifty lambs.

The feedlot operator hopes to profit by efficiently increasing the lambs' weight. Lambs on feed should gain about 0.5–0.8 lb per head per day. Feeder buyers prefer feeder lambs weighing 80–90 lb over larger lambs because of the need to put 30–40 lb of additional weight on the lambs to finish them. Many feeders target a final weight of 130 lb. Table 32.3 shows some example diets for growing-finishing lambs.

Feeding lambs is riskier than producing them because of death losses, price fluctuations of feed and sheep, and the necessity of making large investments.

FEEDING EWES

Figure 32.6 shows the expected weight changes in a 160-lb ewe raising twin lambs. A ewe of similar weight, raising a single lamb, would have about two-thirds of the weight changes shown in Figure 32.6. Economical feeding programs are implemented to correspond with these expected weight changes.

Mature, pregnant ewes usually need nothing more than lower-quality roughages (e.g., hay, wheat straw, or corn stover) during the first half of pregnancy, after which some concentrates and good-quality legume hay should be fed. Any grains such as corn, barley, oats, milo, and wheat are satisfactory feeds. Sheep usually chew these grains sufficiently so that grinding or rolling is not essential. Rolled or cracked grains may be digested slightly more efficiently and may be more palatable, but finely ground grains are undesirable for sheep unless the grains are pelleted.

Sheep need energy, protein, salt, iodine, phosphorus, and vitamins A, D, and E. In some areas selenium is deficient and must be supplied either in the feed or by injection. Mature, ruminating sheep have little need for quality protein or B vitamins,

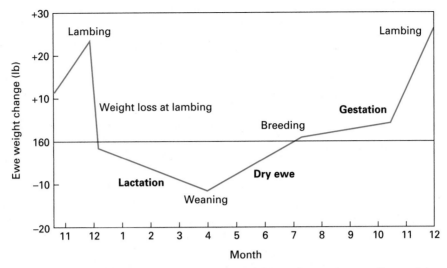

FIGURE 32.6 Weight changes normally expected during the year for a 160-lb ewe giving birth to and raising twin lambs. *Source: Sheep Production Handbook.*

but immature sheep have variable requirements. Rumen microorganisms can synthesize protein from nonprotein nitrogenous substances in the ration.

Sheep will perform poorly or die quickly when the water supply is inadequate, so the importance of supplying water cannot be overstated. Water that is not clean will not be well accepted by sheep. Water intake is influenced by amount of food eaten, protein intake, environmental temperature, mineral intake, water temperature, pregnancy, water content of feed eaten (including rain and dew on pastures), and the odor and taste of the water.

Energy is perhaps the most common limiting nutrient for ewes. Underfeeding is the primary cause for a deficiency of energy because most feed materials are high in energy. Dry range grasses and mature forages such as grain straws may be high in gross energy but so low in digestibility that sheep cannot obtain all their energy needs from them.

The amount of protein in the ration for sheep is of greater importance than the quality of protein because sheep can make the essential amino acids by the action of microorganisms in the rumen. The oil meals (soybean, linseed, peanut, and cottonseed) are all high in protein. Soybean meal is the most palatable and its use in a ration encourages a high feed intake. Nonprotein sources of nitrogen such as urea and **biuret** can be used to supply a portion, but not all, of the nitrogen needs of sheep. Not more than one-third of the nitrogen should be supplied by urea or biuret, and these materials are not recommended for young lambs with developing rumens or for range sheep on low-energy diets. If protein intake is limited by mixing with salt, adequate water must be provided. The mineral needs for sheep are calcium, phosphorus, sulfur, potassium, sodium, chlorine, magnesium, iron, zinc, copper, manganese, cobalt, iodine, molybdenum, and selenium. All of these minerals are found in varying amounts in different tissues of the body. For example, 99% of the calcium, 80–95% of the phosphorus, and 70% of the magnesium occur in the skeleton and more than 80% of the iodine is found in the thyroid gland.

Breed, age, sex, and growth rate of young animals; reproduction and lactation of the ewe; level and chemical form injected or fed; climate; and balance and adequacy of the ration influence the requirement for minerals.

Salt is important in sheep nutrition and it can become badly needed when sheep are grazing on lush pastures. Sheep may consume too much salt when they are forced to drink brackish water.

In some areas, iodine and selenium are deficient. The use of iodized salt in an area where iodine is deficient may supply the iodine needs. Selenium can be added to the concentrate mixture that is being fed or it can be given by injection.

The most critical periods of nutritional needs for the ewe are at breeding and just before, during, and shortly after lambing. A lesser but still important period for the ewe is during lactation. After the lamb is weaned, the ewe normally can perform satisfactorily on pasture or range without any additional feed. If accelerated lamb production is practiced, the ewe should be well fed after the lamb is weaned so the ewe can be bred again.

Example diets for ewes are shown in Table 32.4. Note how the amount of feed changes from maintenance to gestation to lactation. In both the gestation and lactation periods, the maintenance requirement must be met first.

Ewes are normally run on pasture or range where they obtain all their nutritional needs from grasses, browse, and forbs, except during the winter or dry periods when these plants are not growing. During these periods when there is no plant growth, the sheep must be given supplemental feed. Monitoring **body condition score** is an efficient means to determine if ewes are receiving the appropriate level of feed (Table 32.5). Table 32.6 outlines target body condition scores for various phases of the production process.

Occasionally lambs are raised on milk replacer or on cow's milk if the ewe dies during lambing or if her udder becomes nonfunctional. During these two periods, frozen colostrum should be available for the lambs. If milk or milk replacer is used, it can be bottle-fed twice daily or it can be self-fed if it is kept cold. It is important to keep milk cold when it is self-fed to prevent over

TABLE 32.4   Example Diets (as fed) for 155-lb Ewes at Different Stages of Production

Stage of Production	Ration No.	Alfalfa Hay (midbloom/lb)	Corn Silage (mature/lb)	Corn Grain (lb)	Soybean Meal (44% CP/lb)	Salt/Trace Mineral Mix (lb)[a]
Maintenance	1	3.0	—	—	—	0.05
	2	—	6.0	—	0.20	0.05
Gestation	1	3.5	—	—	—	0.05
(first 15 weeks)	2	—	6.0	—	0.25	0.05
Gestation (last 4 weeks)						
130–150%						
lambing	1	3.5	—	0.75	—	0.05
	2	—	6.0	0.75	0.40	0.05
180–225%						
lambing	1	3.5	—	1.25	—	0.05
	2	—	7.0	1.00	0.50	0.05
Lactation						
(first 6–8 weeks)	1	4.0	—	2.00	—	0.05
Suckling single	2	—	9.0	1.00	0.85	0.05

[a] Contains 50% trace mineral salt (for sheep) and 50% dicalcium phosphate.
Source: Sheep Production Handbook, 1988.

TABLE 32.5   Body Condition Scores for Ewes

Score	Spinous Process[a]	Transverse Process[a]	General Description
1	Very prominent	Clearly protrude	Very thin, no fat cover
2	Prominent, but smooth	Rounded	Thin, but skeletal features do not protrude
3	Rounded, but smooth	Rounded, but smooth	Evidence of fat along rib, shoulder, back; hips protrude
4	Barely evident	Cannot be palpated	Moderate fat deposition throughout
5	Not detectable	Not detectable	Excessively fat

[a]Should be palpated down the spine, along the loin and over the rib.

TABLE 32.6   Desired Body Condition Scores at Various Production Phases

Production Phase	Desired Body Condition Score
Dry ewe (116–176 days)	1.5–2.0
Breeding (35–52 days)	2.5–3.0
Early gestation (first 15–17 weeks)	2.0–2.5
Late gestation[a] (last 4–6 weeks)	2.5–3.0
Early lactation[a] (first 6–8 weeks)	3.0–3.5
Late lactation, weaning (last 4–6 weeks)	2.0–2.5

[a] Add 0.5 for ewes pregnant with or nursing twins.
Source: Adapted from *Sheep Production Handbook.*

consumption. There should be heat lamps not far from where the lambs consume the cold milk so they can go to the heated area to become warm and to sleep.

Rams should be fed to keep them healthy but not fat. A small amount of grain along with good-quality hay satisfies their nutritional needs in winter. Young bred ewes should be fed some grain along with all the legume hay they will consume because they are growing and also need some nutritional reserve for the subsequent lactation. Ewe lambs should be grown out but not fattened in the winter. Limited grain feeding along with legume hay satisfies their nutritional needs.

At lambing time, the grain allowance of ewes needs to be increased to assist them in producing a heavy flow of milk. Also, the lambs need to be fed a high-energy ration in the creep. Lambs obtain sufficient protein in the milk given by their mothers, but they need more energy. Rolled grains provided free choice in the creep are palatable and provide the energy needed.

Lambs on full feed are allowed some hay of good quality and all the concentrates they will consume. Some feedlot operators pellet the hay and concentrates, whereas others feed loose hay and grains. After the grasses and legumes start to grow in the spring, all sheep generally obtain their nutritional needs from the pasture.

CARE AND MANAGEMENT OF FARM FLOCKS

The handling of sheep is extremely important. A sheep should never be restrained by its wool because the skin is pulled away from the flesh, causing a bruise. When a group of sheep is crowded into a small enclosure, the sheep will face away from the person who enters the enclosure. When the sheep's rear flank is grasped with one hand, the sheep starts walking backward; this allows the handler to reach out with the other hand and grasp the skin under the sheep's chin. When a sheep is being held, grasping the skin under the chin with one hand and grasping the top of the head with the other hand enables the holder to pull the sheep forward so its brisket is against the holder's knee. If a sheep is to be moved forward, the skin under the chin can be grasped with one hand and the **dock** (the place where the tail was removed) can be grasped with the other hand. Putting pressure on the dock makes the sheep move forward, while holding its chin with the other hand prevents it from escaping.

Lambing Operations

Before the time the ewes are due to lamb, wool should be clipped from the dock, udder, and vulva regions. This process is called **crutching** or *tagging*. If weather conditions permit, the ewes may be completely shorn. Also, all **dung tags** (small pieces of dung that stick to the wool) should be clipped from the rear and flank of pregnant ewes. Young lambs will try to locate the teat of the ewe and may try to nurse a dung tag if it is present.

Ewes should be checked periodically to locate those that have lambed. The ewe and newborn lamb should be placed in a lambing jug. A ewe that is about ready to lamb can be watched carefully but with no interference if delivery is proceeding normally. In a normal presentation, the head and front feet of the lamb emerge first. If the rear legs emerge first **(breech presentation),** assistance may be needed if delivery is slow because the lamb can suffocate if deprived of oxygen for too long. If the front feet are presented but not the head, the lamb should be pushed back enough to bring the head forward for presentation.

As soon as the lamb is born, membranes or mucus that may interfere with its breathing should be removed. When the weather is cold, it may be necessary to dry the lamb by rubbing it with a dry cloth. If a lamb becomes chilled, it can be immersed in warm water from the neck down to restore body temperature, after which it should be wiped dry. The lamb should be encouraged to nurse as soon as possible. A lamb that has nursed and is dry should survive without difficulty if a heat lamp is provided. An ear tag, a tattoo, or both can identify the lamb.

Some ewes may not want to claim their lambs. It may be necessary to tie a ewe with a rope halter so she cannot butt or trample her lamb.

Castrating and Docking

Because birth is a period of stress for lambs, it is best to wait 3–4 days before castrating and **docking** them. To use the elastrator method of castration and docking, a tight rubber band is placed around the scrotum above the testicles (for castration) and around the tail about an inch from the buttocks (for docking). Some death losses can occur when tetanus-causing bacteria invade the tissue where the elastrator was applied. Another castration and docking practice is to remove the testicles and tail surgically. The emasculator is also useful for docking; the skin of the tail is pulled toward the lamb, the emasculator is applied about an inch from the lamb's buttocks, and the tail is cut loose next to the emasculator. A fly repellent should be applied around any wound to lessen the possibility of **fly strike** (fly eggs are deposited during warm weather).

Occasionally, ewes develop a vaginal or uterine **prolapse** (protrusion of the reproductive tract to the outside through the vulva). This condition is extremely serious and leads to death if corrective trained personnel do not take appropriate measures. The tissue should be pushed back in place, even if it is necessary to hoist the ewe up by her hind legs as a means of reducing pressure that the ewe is applying to push the tract out. After the tract is in place, the ewe can be harnessed so that external pressure is applied on both sides of the vulva. In some cases, it may be necessary to suture the tract to make certain that it stays in place. Once a ewe has prolapsed her reproductive tract, the tract shows a weakness that is likely to recur; therefore, a record should be kept so that the ewe can be culled after she weans her lamb.

Shearing

Sheep are usually shorn in the spring, before the hot weather months. Sheep should be kept off feed and water for 6–12 hours prior to shearing to reduce gut fill and to help reduce skin cuts. Only dry sheep should be shorn. Professional personnel typically do shearing, though shearing classes are available to teach the operator. The usual method of shearing involves clipping the fleece from the animal with power-driven shears, leaving sufficient wool covering to protect the sheep's skin. In the shearing operation, the sheep is set on its dock and cradled between the shearer's legs, which are used to maneuver the sheep into the positions needed to ease the shearing.

A fleece that has been properly clipped will remain in one large piece. It is spread with the clipped side out, rolled with the edges inside, and tied with paper twine. It is best to remove dung tags and coarse material that is clipped from the legs and put these items in a separate container. The tied fleece is put into a huge sack (Fig. 32.7). When buyers examine the wool, they can obtain core samples from the sack. The core sample is taken by inserting a hollow tube that

FIGURE 32.7 Raw wool being hydraulically packed into bales. Wool of a specific grade range is packed together to help ensure the uniformity of each bale.

is sharp at the end into the sack of wool. The sample obtained is examined to evaluate the wool in the sack rather than having to remove the fleeces to examine them. If undesirable material is obtained in the core sample, the price offered will be much lower than if only good wool is found.

If the wool is kept for some time before it is sold, it should be stored in a dry place and on a wooden or concrete floor to avoid damage from moisture.

Sheep that are nicked during shearing should be treated to avoid infection. Immediately following shearing, sheep should be placed into clean, dry pens and given protection from cold, wet, windy, or hot conditions.

FACILITIES FOR PRODUCTION OF RANGE SHEEP

Sheep differ from cattle in that they more readily graze weedy plants and brush as well as grasses and legumes. Because of their different grazing patterns, cattle and sheep can be effectively grazed together in some range areas. Total pounds of live weight produced can be higher than when the species are grazed separately on the same range.

Range sheep are produced in large flocks primarily in arid and semiarid regions. Sheep of fine-wool breeding tend to stay together as they graze, which makes herding possible in large range areas. Range sheep are moved about either in trucks or by trailing so they can consume available forage at various elevations. Requirements for the production of range sheep are usually different from those for farm flock operations. One type of range sheep operation is described here, but it must be noted that variations exist.

Range sheep are usually bred to lamb later than sheep in farm flocks; therefore, they can be lambed on the range. Few provisions are needed for lambing when sheep are lambed on the range, but under some conditions a tent or a lambing shed may be used to give range sheep protection from severe weather at lambing time. Temporary corrals can be constructed using snow fences and steel posts when it is necessary to contain the sheep at the lambing camp or at lambing sheds.

Sheep on range are usually wintered at relatively low elevations in areas where little precipitation occurs. Wintering sheep are provided **feed bunks** if hay is to be fed, and windbreaks to give protection from cold winds. Some producers of range sheep provide pelleted feed to supplement the winter forage. Pellets are usually placed on the ground but are sometimes dispersed in grain troughs.

A *sheep camp* or *sheepwagon* is the mobile housing used by the sheepherder. The camp is moved by truck or by horses because the sheep need to be moved over large grazing areas. A sheepherder usually has a horse and dogs to assist in herding the sheep. The sheep are brought together to a night bedding area each evening.

MANAGING RANGE SHEEP

Range sheep are grazed in three general areas: (1) the **winter headquarters,** which is a relatively low and dry area and sometimes provides forage for winter grazing; (2) the **spring-fall range,** which is a somewhat higher elevation area and receives more precipitation; and (3) the **summer grazing** area, which is at high mountainous elevations and receives considerable precipitation, resulting in lush feeds.

The Winter Headquarters

The forage on the winter range, where there is usually less than 10 in. of precipitation annually, is composed of sagebrush and grasses. The grasses are cured on the ground from the growth of the previous summer; consequently, winter forage is of lower quality than green forage because the plants in the winter forage are mature and because they have lost nutrients. The soil in these areas is often alkaline and the water is sometimes alkaline.

Because forage in the wintering area is of poor quality, supplemental feeding that provides needed protein, carotene, and minerals (such as copper, cobalt, iodine, and selenium) is usually necessary. A pelleted mixture made by mixing sun-cured alfalfa leaf meal, grain, solvent-extracted soybean or cottonseed meal, beet pulp, molasses, bone meal or dicalcium phosphate, and trace-mineralized salt is fed at the rate of 0.25–2.0 lb per head per day, depending on the condition of the sheep. The feed may be mixed with salt to regulate intake so that feed can be placed before the sheep at all times. If the intake of feed is to be regulated through the use of salt, trace-mineralized salt should be avoided. Adequate water must also be provided at all times, because heavy salt intake is quite harmful if sheep do not have water for long periods of time.

The ewes are brought to the winter headquarters about the first of November. Rams are turned in with the ewes for breeding in November if lambing is to take place in sheds or if a spring range that is not likely to experience severe weather conditions is available for lambing. Otherwise, the rams are put with the ewes in December for breeding.

January, February, and March are critical months because the sheep are then in the process of exhausting their body stores and because severe snowstorms can occur. If sheep become snowbound, they should each be given 2 lb of alfalfa hay plus 1 lb of pellets per day containing at least 12% protein. Adequate feeding of ewes while they are being bred and afterward results in at least a 30% increase in lambs produced and about a 10% increase in wool produced. In addition, death losses are markedly reduced. Sheep that are stressed by inadequate nutrition, either as a result of insufficient feed or a ration that is improperly balanced, are highly susceptible to pneumonia and resulting heavy death losses.

The Spring-Fall Range

Pregnant ewes are shorn at the winter headquarters (usually in April). They are then moved to the spring-fall range, where they are lambed. If they are lambed on the range, a protected area is necessary. An area having scrub oak or big sagebrush on the south slopes of foothills and ample feed and water is ideal for range lambing. Portable tents can be used if the weather is severe.

Ewes that have lambed are kept in the same area for about 3 days until the lambs become strong enough to travel. The ewes that have lambs are usually fed a pelleted ration that is high in protein and fortified with trace-mineralized salt and either bone meal or dicalcium phosphate. Feeding at this time can help prevent sheep from eating poisonous plants.

Ewes with lambs are kept separate from those yet to lamb until all ewes have borne their lambs. In addition, ewes that are almost ready to lamb are separated from those that will not lamb for some time yet. Thus, after lambing gets under way, three separate groups of ewes are usually present until lambing is completed.

If the ewes are bred to lamb earlier than is usual for range lambing and a crested wheat-grass pasture is available, ewes may be lambed in open sheds. If good pasture is unavailable, the ewes may be confined in yards around the lambing sheds, starting a month before lambing. In this event, the ewes must be fed alfalfa or other legume hay and 0.50–0.75 lb of grain per head per day. The ewes should have access to a mixture of trace-mineralized salt and bone meal or dicalcium phosphate.

Although shed lambing is more expensive than range lambing, higher prices for lambs marketed earlier have made shed lambing advantageous. Fewer lambs are lost in shed lambing, and lambing can take place earlier in the year. The heavy market lambs that result produce enough income to offset the costs of shed lambing.

Summer Grazing

The Summer Range. Sheep are moved to the summer range shortly after lambing is completed if weather conditions have been such that snow has melted and lush plant growth is occurring. The sheep are put into bands of about 1,000–1,200 ewes and their lambs. In some large operations, the general practice is to put ewes with single lambs in one band and ewes with twins in another. The ewes with twin lambs are given the best range area so the lambs will have added growth from the better forage supply. Sheep are herded on the summer range to assist them in finding the best available forage.

In mid-September to October, prior to the winter storms, the lambs are weaned. Colorado producers on the range expect to exceed a 130% lamb crop. Lambs that carry sufficient finish are sent to slaughter and other lambs are sold to lamb feeders. It is the general practice among producers of Rambouillet, Columbia, and Targhee sheep to breed some of the most productive and best wooled ewes to rams of the same breed to raise replacement ewe lambs. Most of these ewe lambs are kept and grown out, and only the less desirable ones are culled. The remainder of the ewes are bred to meat-type rams, such as the Suffolk or Hampshire, and all their lambs are marketed for slaughter as feeders or as stockers (animals used in the flock for breeding).

The Fall Range. As soon as the lambs are weaned, the ewes are moved to the spring-fall range. Later they go to the winter headquarters for wintering.

The number of ewes that can be bred per ram during the breeding period of about 2 months is 15 for ram lambs, 30 for yearling rams, and 35 for mature, but not aged, rams. These numbers are general and depend greatly on the type of conditions existing on the range.

CONTROLLING DISEASES AND PARASITES _____

Sheep raised by most producers are confronted with a few serious diseases, several serious internal parasites, and some external parasites. Some common diseases of sheep include the following types.

Enterotoxemia (overeating disease) is often serious when sheep are in a high nutritional state (e.g., lambs in the feedlot), though it can also affect sheep that are on lush pastures. The disease can be prevented by administering type D toxoid. Usually, three treatments are given: two about 4 weeks apart and a booster treatment 6 months later. Vaccinating pregnant ewes can prevent losses from this disease among young lambs.

E. coli **complex**—*Clostridium E* is another disease common in sheep. At least three organisms are involved: (1) *E. coli,* (2) *Clostridium perfringens*, and (3) a virus called *rotavirus*. This disease affects lambs from birth to a few days of age. Vaccinating all pregnant ewes twice in later pregnancy with type C and D *Clostridium perfringens* toxoid can prevent the disease. The lambing pen should be thoroughly cleaned. Broad-spectrum **antibiotics** given to afflicted lambs help. The ewe flock should be vaccinated with *Clostridium perfringens* type C and D toxoid. Enterotoxemia may be caused by *Clostridium perfringens* type D or C. The *Clostridium perfringens* type CD toxoid given to the ewe will give the lamb protection from both types of enterotoxemias.

Lamb dysentery, caused by *C. perfringens* type B, occurs very early in life and often during wet weather. Vaccinating ewes can prevent dysentery with type BCD vaccine.

Footrot is one of the most serious and common diseases affecting the sheep industry. The disease can be treated with systemic medication. It can be cured by severe trimming so that all affected parts are exposed, treating the diseased area with a solution of one part formalin solution to nine parts of water, and then turning the sheep onto a clean pasture so that reinfection does not occur. Formaldehyde must be used with caution because the fumes are damaging to the respiratory system of both the sheep and the person applying the formaldehyde.

Once all sheep in the flock are free of footrot, making sure that it is not reintroduced into the flock can best prevent it. Rams introduced for breeding should be isolated for 30–60 days for observation. If footrot develops, rams should continue in isolation until free of the disease. A vaccine is available for footrot.

Sore mouth usually affects lambs rather than adult sheep. It is caused by a virus and can be contracted by humans. It can be controlled by vaccination, and sheep should be vaccinated as a routine practice.

Sheep are subject to a nutritional disease known as **white muscle disease.** To prevent it, pregnant ewes should be given an injection of selenium during the last one-third of pregnancy and the lambs should be given an injection of selenium at birth. Selenium can be added to the feed of pregnant ewes to prevent white muscle disease in their lambs. If trace-mineralized salt is provided, it should contain sufficient selenium to supply the needs of the sheep.

Shipping fever is a highly infectious and contagious disease complex that usually affects lambs after the stress of transportation. Antibiotics and

sulfonamides are usually effective as treatment. Care in transporting lambs helps prevent this disease.

Caseous lymphadenitis is a disease that occurs with greater frequency as sheep increase in age from lambs to old animals. The pathogenic organism grows in lymph glands and causes a large development of caseous material (a thick, cheeselike accumulation) to form. It is a serious disease and one that is difficult to control. The abscesses can be opened and flushed with a solution of equal parts of 0.2% nitrofurazone solution and 3% hydrogen peroxide. One should not open an abscess and let the thick pus go onto the floor or soil where well sheep will be traveling because this may cause the disease to spread. All infected animals should be culled and should be isolated from noninfected sheep immediately on appearance of being infected.

Milk fever is due to hypocalcemia. Afflicted animals can be treated with an injection of calcium salts. A 20% solution of calcium borogluconate given **intravenously** at the rate of 100 ml per sheep should have an afflicted animal up and in good condition within 2 hours.

Urinary calculi (kidney stones) occur when the salts in the body that are normally excreted in the urine are precipitated and form stones that may lodge in the kidneys, ureters, bladder, or urethra. Providing a constant supply of clean water helps immensely in preventing formation of calculi. The ideal ratio of calcium to phosphorus is 1.6:1.0. Ammonium chloride can be included in the ration to help prevent urinary calculi.

Pregnancy disease (ketosis) is a metabolic disease that affects ewes in late pregnancy, particularly if they are carrying twins or triplets. The problem is that ewes carrying twins or triplets must break down body fat to provide their energy needs in later pregnancy. It is possible that fat breakdown may not be sufficient for the glucose needs of these ewes, resulting in hypoglycemia. Feeding some high-energy grain such as corn, barley, or milo may prevent pregnancy disease, but one or more large lambs reducing rumen space in the mother aggravate the condition.

Grass tetany (or **grass staggers**) is due to a deficiency of magnesium at a particular time, most frequently in spring when lactating ewes are put onto lush pasture where there is insufficient magnesium available. An injection of 50–100 ml of 20% calcium borogluconate or an injection of magnesium sulfate should give rapid recovery.

Annual death loss in adult sheep is approximately 5–6% of inventory while lamb death loss averages 9–10% of the lamb crop on an annual basis. About two-thirds of these losses result from nonpredator causes and losses due to predation account for the remaining one-third in adult sheep. The percent of death loss in lambs due to predators increases to approximately 43% of total losses.

The leading three causes of death loss (nonpredator) in adult sheep are old age, lambing problems, and digestive disorders. Nonpredator death losses in lambs most often result from respiratory disease, digestive problems, weather-related issues, and dystocia. These four categories account for three-quarters of nonpredator lamb mortalities.

Sheep have internal and external parasites, though internal parasites are the more serious of the two. The most important internal parasites are coccidiosis, stomach worms, nodular worms, liver flukes, lungworms, roundworms, and tapeworms.

Common external parasites of sheep include blowfly maggots, **keds** (sheep ticks), lice, mites, screwworms, and sheep bots. Making two applications of an effective insecticide that is not harmful to sheep can control external parasites. The two applications should be spaced so that eggs hatched after the first application will not result in egg-laying adults prior to the second application.

DETERMINING THE AGE OF SHEEP BY THEIR TEETH

Teeth can determine the age of a sheep. Lambs have four pairs of narrow lower incisors called *milk teeth* or *baby teeth*. At approximately a year of age the middle pair of milk teeth is replaced by a pair of larger, permanent teeth. At 2 years, a second pair of milk teeth is replaced. This process continues until, at 4 years of age, the sheep has all permanent incisors. The teeth start to spread apart and some are lost at about 6–7 years of age. When all the permanent incisors are lost, the sheep has difficulty grazing and should be marketed.

COSTS AND RETURNS

The objective of farm flock operators is the efficient production of slaughter lambs ready for market when prices are usually highest (Fig. 32.8).

The items used to evaluate costs and returns for ewes are shown in Table 32.7. Because price is only part of the profit formula, understanding cost of production is important.

Note that only in the highest-production flocks did returns over variable cost reach profitable levels. Profit is only realized when break-even prices are carefully managed and the market is sufficiently strong.

FIGURE 32.8 Timing of the flock's production schedule with cyclic patterns of climate and markets is critical to success in the sheep enterprise. Courtesy of American Sheep Industry Association.

TABLE 32.7 Production Costs and Returns per Range Ewe

	2002		
	Production Levels		
	1	**2**	**3**
Production			
Lamb crop	120%	140%	160%
Culling rate	20%	20%	20%
Retention rate	21%	21%	21%
Variable costs per ewe			
1. Pasture (1.12 aum × $17/aum)	$ 19.40	$ 19.40	$ 19.40
2. Sorghum silage (730 lb × $17.36)	6.50	6.50	6.50
3. Alfalfa hay (457 lb × $79/ton)	18.85	18.85	18.85
4. Grain sorghum (629 lb × $4.42/cwt)	22.02	27.80	26.92
5. Protein (42 lb × $202.06/ton)	3.24	4.24	3.96
6. Vitamins-minerals (18.6 lb × $0.23/lb)	4.19	4.19	4.19
7. Feed processing	2.74	3.45	3.33
8. Labor (4 hours × $10.00/hour)	40.00	40.00	40.00
9. Veterinary, drugs, and supplies	3.00	3.00	3.00
10. Breeding costs			
11. Marketing costs	3.00	3.00	3.00
12. Shearing	2.50	3.00	3.50
13. Utilities, fuel, oil	5.00	5.20	5.40
14. Building and equipment repairs	4.37	4.37	4.37
15. Miscellaneous	0.50	0.50	0.50
16. Interest on ½ variable costs @ 8%	5.41	5.82	5.88
A. Total variable costs	**$140.72**	**$151.42**	**$153.00**
Fixed costs per ewe			
17. Depreciation on buildings and equipment	$ 11.54	$ 11.54	$ 11.54
18. Depreciation on rams	0.93	0.93	0.93
19. Interest on buildings and equipment	7.43	7.43	7.43
20. Insurance-taxes on building and equipment	3.25	3.25	3.25
21. Interest on breeding flock	7.74	7.74	7.74
22. Insurance on breeding flock	0.97	0.97	0.97
B. Total fixed costs	**$ 31.87**	**$ 31.87**	**$ 31.87**
C. Total costs per ewe (A + B)	**$172.90**	**$183.28**	**$184.87**
Returns per ewe			
23. Market lambs (120 lb × $72.00 cwt × cwt produced)	$84.68	$101.79	$118.90
24. Cull ewes (0.2 × 140 lb × $31 cwt)	8.68	8.68	8.68
25. Wool (8.5 lb × $1.00/lb)	8.50	8.50	8.50
26. Ewe replacement (retention rate × ewe value)			
D. Gross returns per ewe	**$101.86**	**$118.97**	**$136.08**
E. Return over variable cost (D − A)	**$ −38.86**	**$ −32.45**	**$ −16.93**
F. Return over total costs (D − C)	**$ −70.73**	**$ −64.32**	**$ −48.79**
G. Average selling price needed per hundredweight			
27. To cover variable costs (lines A − 24 − 25 + 26) ÷ 29	$105.04	$ 94.95	$ 82.25
28. To cover total costs (lines C − 24 − 25 + 26) ÷ 29	$132.14	$117.49	$101.55
H. Total feed costs (lines 1 through 6)	**$ 76.94**	**$ 86.53**	**$ 87.35**
29. Hundredweight produced (marketed)	1.17	1.41	1.65
30. Feed cost per hundredweight lamb marketed (H ÷ 29)	$ 0.65	$ 0.61	$ 0.53
I. Asset turnover (D ÷ investment)	**36%**	**42%**	**48%**
J. Net return on investment [(F + 16 + 19 + 21) ÷ investment]	**−17.7%**	**−15.3%**	**9.8%**

Source: KSU Extension, 2002.

CHAPTER SUMMARY

- Sheep need good facilities for lambing and working them for sorting, branding, and treatment.

- Farm flocks and range flocks utilize different resources and management strategies to achieve their enterprise goals.

- Sheep are usually shorn in the spring, prior to the hot weather months. Late spring storms sometimes challenge the sheep flock, especially under range conditions where there is limited protection.

- Sheep are challenged by health problems when compared to most other livestock. Therefore, preventative herd health and treatment programs must be implemented in a timely manner.

KEY WORDS

cutting chute	docking	white muscle disease
lambing jugs	fly strike	shipping fever
creep	prolapse	sulfonamides
purebred breeders	feed bunk	caseous lymphadenitis
commercial producer	winter headquarters	milk fever
parasites	spring-fall range	intravenously
overeating disease	summer grazing	urinary calculi
biuret	enterotoxemia	ketosis (pregnancy
body condition score	*E. coli* complex	disease)
dock	antibiotics	grass tetany
crutching	lamb dysentery	(grass staggers)
dung tags	footrot	keds
breech presentation	sore mouth	

REVIEW QUESTIONS

1. *True or False:* Although sheep normally do not suffer from cold weather because they have a heavy wool covering, newborn lambs should be sheltered in an enclosed and heated room when the weather is cold.

2. In what region of the United States are most range flocks located?

3. Under what conditions will farm flocks produce a larger proportion of feedlot lambs?

4. At what age and weight are commercial slaughter lambs typically finished?

5. What disease of growing lambs is of particular concern to commercial feedlot operators?

6. *True or False:* Mature, pregnant ewes usually need high-quality roughage and concentrate feeds during the first half of gestation.

7. What factors influence water intake of sheep?

8. _____ is the most common limiting factor in the diet of ewes.

9. When do the most critical periods of nutritional needs occur for ewes?

10. In what three general areas are range sheep grazed?

11. When does the breeding season of most sheep occur?

12. *True or False:* Because forage on winter range is of poor quality, supplemental feeding of sheep is usually needed to meet nutritional requirements.

13. What is the number of ewes that can be bred per ram during a 60-day breeding period?

SELECTED REFERENCES

Battaglia, R. A., and V. B. Mayrose. 1981. *Handbook of Livestock Management Techniques.* New York: Macmillan.

Botkin, M. P., R. A. Field, and C. L. Johnson. 1988. *Sheep and Wool: Science, Production, and Management.* Englewood Cliffs, NJ: Prentice Hall.

Ensminger, M. E., and R. O. Parker. 1986. *Sheep and Goat Science.* Danville, IL: Interstate Printers and Publishers.

National Animal Health Monitoring System. 1996. *U.S. Regional Sheep and Health Management Practices.* Washington, DC: USDA: APHIS.

National Animal Health Monitoring System. 2006. *Sheep and Lamb Nonpredator Death Loss in the United States—2004.* Washington, DC: USDA: APHIS.

National Research Council. 1985. *Nutrient Requirements of Sheep.* Washington, DC: National Academy Press.

Sheep Housing and Equipment Handbook. 2005. Ames, IA: Midwest Plan Service.

Sheep Industry Council. 2004. *The Sheep Production Handbook.* Denver, CO: Sheep Industry Development Program.

Stillman, R., T. Crawford, and L. Aldrich. July 1990. *The U.S. Sheep Industry.* USDA/ERS, Staff Report no. AGES 9048.

Warmann, G. W. 1998. *Farm Ewe Flocks—Once a Year Lambing.* Kansas State University Extension Service. Publ. MF-421.

CHAPTER 33

Horse Breeds and Breeding

LEARNING OBJECTIVES

- List and describe the primary light and draft horse breeds
- Contrast selection in the horse industry with the other livestock enterprises
- Describe the ideal conformation of the horse
- List and describe the unsoundnesses and blemishes affecting the horse
- Describe the various gaits
- Discuss how to determine a horse's age by evaluating its teeth

HORSES AND HUMANS

The horse was domesticated approximately 5,000 years ago, and thus began the relationship between people and horses. The role of the horse through history ranges from a source of food to the focus of artisans and sculptors. The horse has played an important part in human history by changing military strategies and the division of labor in agriculture and transportation, serving as a focal point in sport, and providing satisfying recreational activities for humans both young and old.

BREEDS OF HORSES

Horses have been used for so many different purposes that many breeds have been developed to fill specific needs. The major breeds of horses and their primary uses are listed in Table 33.1. No attempt has been made in Table 33.1 to indicate all the ways each breed is used.

TABLE 33.1 Characteristics and Uses of Selected Breeds of Horses

Breed	Color	Origin	Height (in hands)[a]	Weight (lb)	Uses
Riding and Harness Horses					
American Quarter Horse	All colors except paint	United States	14.2–15.2	1,000–1,250	Short racing, showing, stock work
American Saddlebred	Chestnut, bay, brown, black, gray	Kentucky	15.0–16.0	1,000–1,150	Showing, pleasure riding, 3 and 5 gaited
Arabian	Bay, chestnut, white, gray, black	Arabia	14.2–15.2	850–1,000	Pleasure riding, showing
Morgan	Bay, chestnut, brown, black	New England	14.2–15.1	950–1,150	Pleasure riding, driving, showing
Standardbred	Bay most common, but all solid colors	United States	14.2–16.2	850–1,200	Harness racing
Tennessee Walking Horse	Black, bay, chestnut	Tennessee	15.0–16.0	1,000–1,200	Pleasure riding, showing
Thoroughbred	Bay, brown, chestnut, or any other solid color	England	15.2–17.0	1,000–1,300	Long racing
Ponies					
Hackney	Bay, chestnut, black, brown	England	11.2–14.2	450–850	Light harness, showing
Pony of America	Appaloosa	Iowa	11.5–13.5	700–800	Riding by children, showing
Shetland	Bay, chestnut, brown, black, gray	Shetland Islands	9.2–10.0	300–400	Riding by children, showing
Welsh	Any color except piebald and skewbald	Wales	11.0–13.0	350–850	Riding by children, showing
Draft[b]					
Belgian	Chestnut, roan usually	Belgium	15.2–17.0	1,900–2,400	Heavy pulling
Clydesdale	Bay, brown, black, roan	Great Britain	15.2–17.0	1,700–2,000	Heavy pulling
Percheron	Black, gray usually	France	15.2–17.0	1,600–2,200	Heavy pulling
Shire	Bay, brown, usually black, gray	England	16.2–17.0	1,800–2,200	Heavy pulling
Suffolk	Chestnut	England	15.2–16.2	1,500–1,900	Heavy pulling
Color Registries					
Appaloosa	Leopard, blanket, roan	Pacific Northwest	14–16	900–1,250	Pleasure riding, showing, stock work
Buckskin	Buckskin, dun, grulla	United States	14–16	900–1,250	Pleasure riding, showing, stock work
Paint	Tobiano, overo	United States	14–16	900–1,250	Pleasure riding, showing, stock work
Palomino	Palomino		14–16	900–1,250	Pleasure riding, showing, stock work
Pinto	Pinto	United States (via horses brought by Spanish conquistadors)	14–16	900–1,250	Pleasure riding, showing, stock work
White and cremes	White and cream		14–16	900–1,250	Pleasure riding, showing, stock work

[a] Height is measured in inches but reported in *hands*. One hand equals 4 in.
[b] Draft horses are heavy horses used for pulling; other horses are called *light horses* and are used primarily as pleasure horses.
More detailed information can be found at www.ansi.okstate.edu/breeds/.

The **"light" breeds of horses** provide pleasure for their owners through activities such as racing, riding, and exhibition in shows. Color Plates Q and R show examples of these breeds. Quarter Horses are excellent as cutting horses (horses that separate individual cattle out of a herd) and for running short races. Ponies, such as Shetland and Welsh, are selected for their friendliness and safety with children. The American Saddlebred and the Tennessee Walking Horse have been selected for their comfortable gaits and responsive attitudes. The American Saddlebred has regional popularity and is considered "the peacock of the show ring." The Palomino, Appaloosa, and Paint horses are color breeds developed for showing and for working livestock.

Beauty in color and markings is important in horses used for show and breeding purposes, and colored breeds have been developed accordingly. The Appaloosa, for example, has color markings of three patterns: leopard, blanket, and roan. It appears that these color patterns are under different genetic controls. In Appaloosas, Paints, and Palominos, coloration can be affected or eliminated by certain other genes, such as the gene for roaning and the gene for graying. The gene for gray can eliminate both colors and markings, as shown by gray horses that turn white with age.

Another group of horses are the **draft horses**—large and powerful animals that are used for heavy work. The Percheron, Shire, Clydesdale, Belgian, and Suffolk are examples of draft horses (Table 33.1).

Popularity of Breeds

Registration numbers are a measure of breed popularity. Table 33.2 shows the registration numbers for draft, light, and pony breeds. Based on these numbers, the Quarter Horse is the most popular horse breed in the United States.

TABLE 33.2 Major U.S. Draft, Light Horse, and Pony Breeds Based on Annual Registration Numbers (in thousands), 1980–2002

Breed	2002	1998	1995	1990	1985	1980	Year Association Formed
Light Horse							
Quarter Horse	158.0	125.3	110.0	110.6	157.4	137.1	1940
Paint	60.0	55.4	34.0	22.4	12.7	9.7	1962
Thoroughbred	35.6	36.5	35.2	43.6	50.4	39.4	1894
Tennessee Walking	16.0	14.0	6.7	8.0	7.6	6.8	1935
Standardbred	11.0	10.8	11.9	16.6	18.4	15.2	1938
Arabian	9.7	11.3	13.0	17.7	30.0	19.7	1908
Appaloosa	9.3	9.1	11.2	10.7	16.2	25.4	1938
Pinto	4.6	4.0	3.8	2.9	3.9	—	1956
Anglo and Half-Arab	4.2	4.0	4.2	4.3	10.1	14.3	1950
Morgan	3.6	2.5	3.0	3.6	4.5	4.5	1907
American Saddlebred	3.1	3.0	3.2	3.6	4.4	3.9	1891
Draft							
Belgian	4.0	3.9	3.9	3.7	4.2	.7	1887
Percheron	2.4	2.2	2.0	1.3	1.5	.3	1876
Clydesdale	0.6	0.5	0.5	0.4	0.2	.2	1879
Pony							
Miniature	—	8.3	11.0	11.2	—	—	1971
Welsh	—	0.8	0.7	0.7	0.5	0.9	1906

Sources: Adapted from the American Horse Council, *Horse Industry Directory*, 2006; and various breed associations.

BREEDING PROGRAM

Reproduction

Mares of the light breeds reach sexual maturity at 12–18 months of age, whereas draft mares are 18–24 months of age when sexual maturity is reached. Mares come into estrus every 21 days during the breeding season if they do not become pregnant. Heat lasts for 5–7 days with ovulation occurring toward the end of heat. Because of the relatively long duration of heat and because ovulation occurs toward the end of heat, horse owners often delay breeding a mare for 2 days after she has first been observed in heat. Some breeders have the mare bred every other day while she is in heat.

Although about 10% of all ovulations in mares are multiple ovulations, twinning occurs in only about 0.5% of the pregnancies that carry to term. The uterus of the mare apparently cannot support twin fetuses; consequently, most twin conceptions result in the loss of both embryos. The length of gestation is about 340 days (approximately 11 months), with usually only one foal being born. Mares usually come into heat 5–12 days following foaling, and fertile matings occur at this heat if the mare has recovered from the previous delivery.

Improving reproductive efficiency has been one of the major areas of focus in equine research over the past 25 years. Improved technologies and management have yielded both improved reproduction performance and also the ability to allow horses with inherent fertility problems to reproduce.

Selection

An effective breeding program is a completely objective one. There is no place in a breeding program for emotional attachment toward an animal that results in a loss of selection pressure. Every attempt should be made to see that the environment is the same for all animals in a breeding program. A particularly appealing foal that is given special care and training may develop into a desirable animal. However, many foals less appealing in early life might also develop into desirable animals if given similar attention. Environmental effects may mislead breeders resulting in the propagation of less than desirable animals whose weaknesses are masked by nutrition or other favorable management conditions. Thus, selecting a horse for breeding that has had special care and training may in fact be selecting only for special care and training. Certainly these environmentally produced differences in horses are not inherited. In fact, it is often wise to select animals that were developed under the type of environment in which they are expected to perform. If stock horses are being developed for herding cattle in rough, rugged country, selection under such conditions is more desirable than where conditions are less rigorous. Horses that possess inherited weaknesses tend to become unsound in a rugged environment and, as a result, are not used for breeding. Such animals might never show those inherited weaknesses in a less rugged environment.

An ideal environment for most horse-breeding programs has quality forage distributed over an area that requires horses to exercise as they graze. Where access to land is limited, horses may have to be kept in a small area and forced to exercise a great deal. Forced exercise tends to keep the animals from becoming too fat and gives strength to the feet and legs. Horses need regular, not sporadic, exercise. Regular exercise, even if quite strenuous, is healthy for animals that are genetically sound and may reveal the weaknesses of those that are not.

Strenuous exercise can, however, be harmful to animals that have not previously exercised for a considerable period of time.

An environment should be provided that identifies horses having genetic superiority for their intended performance. For example, horses that are being bred for endurance in traveling should be made to travel long distances on a regular basis to determine if they can remain sound. Horses bred for jumping should be trained to jump as soon as they are physically mature so that those lacking the ability to jump or those that become unsound from jumping can be removed from the breeding program. Draft horses should be trained to pull heavy loads early in life (3 or 4 years of age) to determine their ability to remain sound and their willingness to pull. Performance should be measured before animals are used in a breeding program.

Any horse that is unsound should not be used for breeding regardless of the purposes for which the horse is being bred. Such abnormalities as toeing-in or toeing-out, sickle hocks, cow hocks, and contracted heels will likely lead to unsoundness and difficulties or lack of safety in traveling. Interference and forging actions are extremely objectionable because they can cause the horse to stumble or fall. Eye, mouth, and respiration defects should be selected against in all horses.

CONFORMATION OF THE HORSE

Purchasing the Horse

Buying a horse often involves a significant investment and a situation where both buyer and seller may be amateurs. However, a disciplined approach to purchasing a horse will yield favorable results (Table 33.3). Protection of the interest of both parties is an important consideration. One way to accomplish this task is through the use of the prepurchase exam. While there is variation in the way that these exams are conducted, the exam should be performed by an experienced equine veterinarian with both buyer and seller present.

The prepurchase exam usually focuses on an assessment of the horse's general health, physical condition, and structural soundness. Furthermore, it may be important to assess the horse's athleticism, temperament, and appropriateness for the horsemanship skills of the buyer. These topics are covered in more detail later in this chapter.

Body Parts

To understand the conformation of the horse, one should be familiar with the body parts (Fig. 33.1). (A more detailed drawing of the skeletal structure of the horse is shown in Chapter 18.)

TABLE 33.3 The Process of Making a Horse Purchase

1. Establish your wants and needs in advance.
2. Shop around but only with reputable sellers—take your time.
3. Have a set of predetermined questions, take notes, use videotape if possible.
4. Always get a prepurchase exam.
5. Use a thorough sales contract to protect buyer and seller.
6. Keep all records and notes relative to the sale.
7. Do not buy on impulse, and stay on budget.

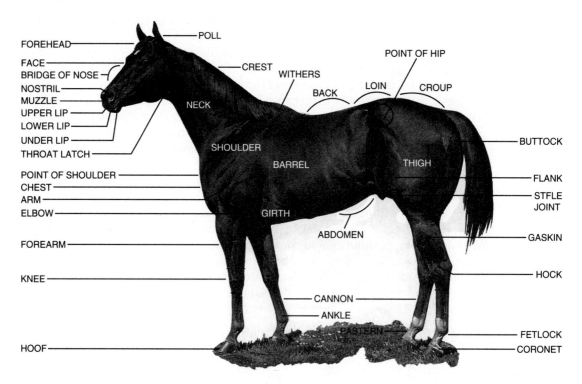

FIGURE 33.1   Parts of the horse. *Source:* Reprinted with the permission of the American Quarter Horse Association © American Quarter Horse Association.

Feet and Legs.   Major emphasis is placed on feet and legs in describing conformation in the horse—for identifying both correctness and conditions of unsoundness. The old saying, "No feet, no horse" is still considered valid by most horse producers. As foot and leg structure is important, a review of skeletal structure of the front leg and hind leg is given in Figures 33.2 and 33.3.

Figure 33.4 shows the front legs from a front view. From this view, a vertical line from the point of the shoulder should fall in the centers of the **knee, pastern, cannon,** and **foot.** Each leg is divided into two equal halves. Deviations from this ideal position are shown with the common terminology association with them.

The front legs are shown from a side view in Figure 33.5. A vertical line from the shoulder should fall through the center of the elbow and the center of the foot. The angle of the pastern in the ideal position is 45–50°.

Figure 33.6 shows the correct hind leg position from a rear view; the less desirable position of the feet and legs are also shown. In the ideal position, a vertical line from the point of the buttocks should fall through the centers of the **hock,** cannon, pastern, and foot.

The ideal position of the hind legs from a side view is shown in Figure 33.7. Deviations from the ideal position are shown and described. The ideal position is shown where a vertical line from the point of the buttocks touches the rear edge of the cannon from the hock to the fetlock and meets the ground behind the heel.

The Hoof.   Care of the horse's feet is essential to keep the horse sound and serviceable. Regular cleaning, trimming, and shoeing (depending on frequency of

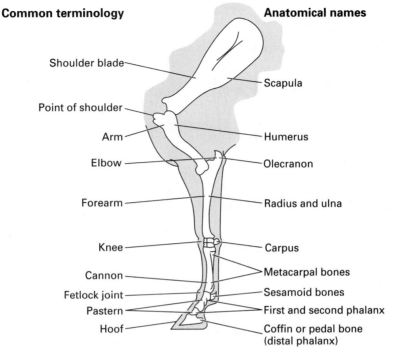

FIGURE 33.2 Skeletal front leg with common terminology and anatomical names. *Source:* Colorado State University.

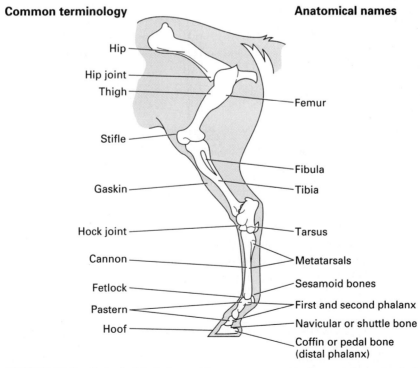

FIGURE 33.3 Skeletal hind leg with common terminology and anatomical names. *Source:* Colorado State University.

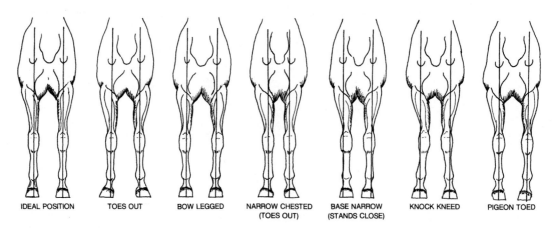

IDEAL POSITION   TOES OUT   BOW LEGGED   NARROW CHESTED (TOES OUT)   BASE NARROW (STANDS CLOSE)   KNOCK KNEED   PIGEON TOED

FIGURE 33.4 Correct and faulty conformation of front feet and legs from a front view. *Source:* Adapted from William R. Culbertson.

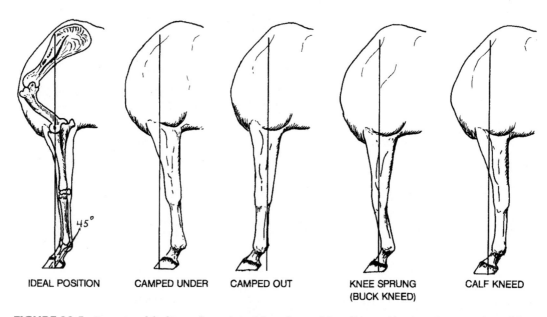

IDEAL POSITION   CAMPED UNDER   CAMPED OUT   KNEE SPRUNG (BUCK KNEED)   CALF KNEED

FIGURE 33.5 Correct and faulty conformation of front feet and legs from a side view. *Source:* Adapted from William R. Culbertson.

use) are needed. The hoof will, in mature horses, grow $\frac{1}{4}$–$\frac{1}{2}$ in. per month, so trimming is typically needed every 6–8 weeks.

The external parts of the hoof are shown in Figure 33.8.

UNSOUNDNESS AND BLEMISHES OF HORSES

Two terms, *unsoundness* and *blemish*, are used in denoting abnormal conditions in horses. **Unsoundness** is any defect that interferes with the usefulness of the horse. It may be caused by an injury or improper feeding, be inherited, or

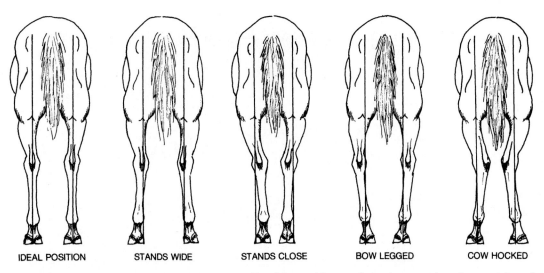

IDEAL POSITION   STANDS WIDE   STANDS CLOSE   BOW LEGGED   COW HOCKED

FIGURE 33.6   Correct and faulty conformation of hind feet and legs as shown in a rear view. *Source:* Adapted from William R. Culbertson.

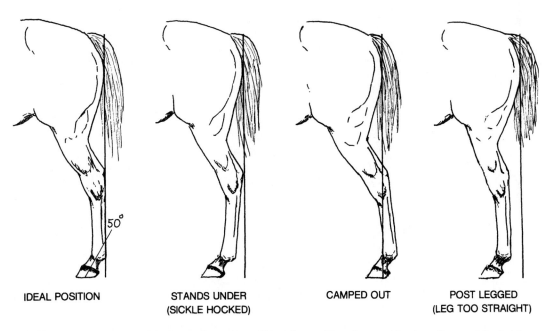

IDEAL POSITION   STANDS UNDER   CAMPED OUT   POST LEGGED
                       (SICKLE HOCKED)   (LEG TOO STRAIGHT)

FIGURE 33.7   Correct and faulty conformation of hind feet and legs from a side view. *Source:* Adapted from William R. Culbertson.

develop as a result of inherited abnormalities in conformation. A **blemish** is a defect that detracts from the appearance of the horse but does not interfere with its usefulness. A wire cut or saddle sore, for instance, may cause a blemish without interfering with the usefulness of the horse.

Horses may have anatomical abnormalities that interfere with their usefulness. Many of these abnormalities are either inherited directly or develop

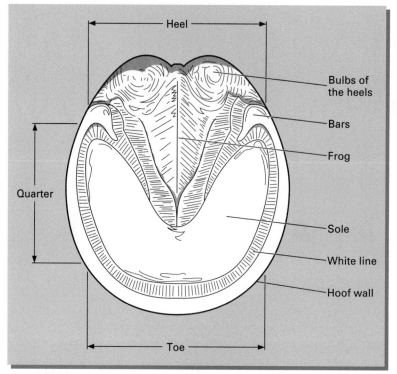

FIGURE 33.8 The conformation of the hoof (viewed from the bottom side) with several parts identified. *Source:* Colorado State University.

because of an inherited condition. Abnormalities of the eyes, respiratory system, circulatory system, and conformation of the feet and legs are all important.

Some of the major unsoundness and blemishes include the following (several of which are identified by location on the body in Fig. 33.9).

Bog spavin is a soft swelling on the inner, anterior aspect of the hock. Although unsightly, the condition usually does not cause lameness. **Bone spavin** is a bony enlargement on the inner aspect of the hock. Both bog and bone spavin arise when stresses are applied to horses that have improperly constructed hocks. Lameness usually accompanies this condition, but most animals return to service after rest and, in some cases, surgery.

Capped hock is a thickening of the skin at the point of the hock. **Curb** is a hard swelling above the cannon bone (a hand's width below the hock). The plantar ligament becomes inflamed and swollen, usually owing to poor conformation or a direct blow.

Cataract is inherited as a dominant trait. It can be eliminated if the afflicted horse does not produce several foals before the cataract develops.

Contracted heels are commonly an inherited conformation defect. Contracted heels may result when the frog or cushion of the foot is damaged and shrinks, allowing the heels to come together. The bottom surface of the foot becomes smaller in circumference than at the coronet band.

The term *cow hocked* (Fig. 33.6) indicates that the points of the hocks turn inward. Such hocks are greatly stressed when the horse is pulling, running, or jumping.

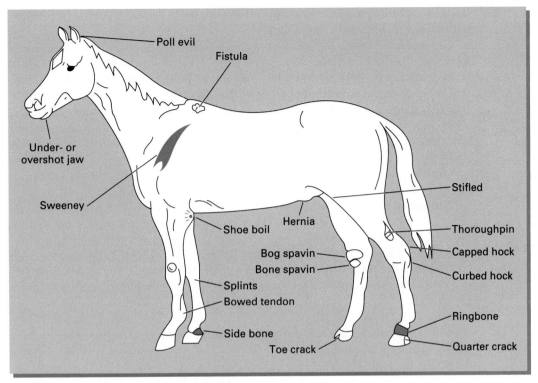

FIGURE 33.9   Locations of several potential unsoundnesses and blemishes. *Source:* Colorado State University.

Heaves is a respiratory disease in which the horse experiences difficulty in exhaling air. The horse can exhale a certain volume of air normally, after which an effort is exerted to complete exhalation. A horse with a mild case of heaves can continue with light work, but horses with moderate to severe heaves have a limited ability to work.

Hyperkalemic periodic paralysis (HYPP) is a muscle disease that is genetically controlled. HYPP affects some lines of Quarter Horses, Paints, and Appaloosas. Affected horses experience tremors as a result of abnormal muscle fiber activity. A mutation in the gene responsible for sodium and potassium regulation is apparently the cause. Ideally, those horses that carry the trait should not be allowed to breed.

Laminitis, also called **founder,** is an inflammation of the laminae of the foot causing severe pain and lameness. The front feet are affected more frequently than the hind feet. Overeating of grain, consumption of large amounts of water by overheated horses, and overworking of horses on hard surfaces are frequent causes of laminitis.

Moon blindness is periodic ophthalmia in which the horse is blind for a short time, regains its sight, and then again becomes blind for a time. Periods of blindness may initially be spaced as much as 6 months apart. The periods of blindness become progressively closer together until the horse is continuously blind. This condition received the name *moon blindness* because the trait is first noticed when the periods of blindness occur about a month apart, which originally led to the thought that the periods of blindness were associated with changes in the moon. More recently this condition has been associated with infections, parasites, and riboflavin deficiency.

Navicular disease is a disease complex manifested in pain in the heels of the front foot. Navicular pain may result from a variety of hoof faults including contracted heels, underslung heels, or lack of hoof symmetry. While the navicular bone was commonly considered the affected tissue, new studies suggest that this disease may be linked to damage of the tendons, ligaments, or coffin joint. Diagnosis is complicated by the multiple factors that may be involved. Early treatment, including medication, surgery, or rest is critical to assure recovery.

Quittor is a deep sore that drains at the coronet. The infection is caused by puncture wounds, corns, and the like, and results in severe lameness.

Ring bone is a condition in which the cartilage around the pastern bone ossifies. Ring bone shows as a hard bony enlargement encircling the areas of the pastern joint and coronet.

Ruptured blood vessels are a defect of circulation in which the blood vessels in the lungs are fragile and may rupture when the horse is put under the stress of exercising. Some racehorses have died due to hemorrhage from these fragile blood vessels.

Shoe boil, or **capped elbow,** is a soft swelling on the elbow. Common causes are injury to the elbow while the horse is lying down or getting up or injury from a long heel on a front shoe.

Sickle hocked (Fig. 33.7) is the term used to describe the hock when it has too much set or bend. As a result, the hind feet are set too far forward. The strain of pulling, jumping, or running is much more severe on a horse with sickle hocks than on a horse whose hocks are of normal conformation.

Side bones is an abnormality that occurs when the lateral cartilages in the foot ossify. During the ossification process, lameness can occur, but some horses regain soundness with proper rest and shoeing.

A **stifled** horse is one in which the patella (the kneecap in humans) has been displaced. Older horses seldom become sound once they are stifled, while younger horses usually recover. There is a surgical operation for this condition.

String halt is an involuntary flexion of the hock during movement. It is considered a nerve disorder. Surgery can improve the condition.

Sweeney refers to atrophied muscles at any location, although many people use it to refer only to shoulder muscles. In the shoulder sweeney, the nerve crossing the shoulder blade has been injured.

A **thoroughpin** is a soft enlargement of the tendon sheath of the large tendon (tendon of Achilles) of the hock and the fleshy portion of the hind leg. Stress on the flexor tendon allows synovial fluid to collect in the depression of the hock. Lameness rarely occurs.

Toeing-in, or **pigeon-toed** (Fig. 33.4), refers to the turning in of the toes of the front feet. **Toeing-out** (Fig. 33.4) refers to the turning out of the toes of the front feet. These conditions influence the way in which the horse will move its feet when traveling. Toeing-out or moving the front feet inward is considered the more serious defect because it can lead to further interference and faults.

Tying-up, or **exertional rhabdomyolysis,** is a condition categorized as either early-form or late-form. Affecting the unconditioned or overexerted horse, tying up is characterized by a shuffling gait, heavy sweating, and overly contracted rump and thigh muscles. Severe forms may result in muscular trauma. Afflicted horses usually recover without long-term consequences. Prevention involves feeding according to a horse's workload and assuring that horses are prepared for strenuous work.

Windgalls, sometimes referred to as **wind puffs** or *puffs*, occur when the joint capsules on tendon sheaths around the pastern or fetlock joints are enlarged. The disease is common, but not serious, in hardworking horses.

GAITS OF HORSES

The major gaits of horses, along with their modifications, are as follows:

1. **Walk** is a four-beat gait in which each of the four-feet strikes the ground independently.

2. **Trot** is a diagonal, two-beat gait in which the right front and left rear feet hit the ground in unison, and the left front and right rear feet hit the ground in unison. The horse travels straight without swaying sideways when trotting.

3. **Pace** is a lateral two-beat gait in which the right front and rear feet hit the ground in unison and the left front and rear feet hit the ground in unison. There is a swaying from right to left when the horse paces.

4. **Gallop** is the fastest gait with four beats.

5. **Canter** is a three-beat gait. Depending on the lead, two diagonal legs hit the ground at the same time, with the other hind leg and lead leg hitting at different times.

6. **Rack** is a snappy four-beat gait in which the joints of the legs are highly flexed. The forelegs are lifted upward to produce a flashy effect. This is an artificial gait, whereas the walk, trot, pace, gallop, and canter are natural gaits. The rack is popular in the show ring for speed and animation.

7. **Running walk** is the fast ground-covering walk unique to the Tennessee Walking Horse. It is faster than the normal walk. The horse moves with a gliding motion as the hind leg oversteps the forefoot print by 12–18 inches or more.

EASE OF RIDING AND WAY OF GOING

When a horse's foot strikes the ground, a large shock is created that would be objectionable to the rider if no shock absorption occurred. There are several shock-absorbing mechanisms existing in horses' feet and legs. The horse has lateral cartilages on all four feet that expand outward when the foot strikes the ground. This absorbs some of the shock. The pastern on each leg absorbs some of the shock when the foot strikes the ground by bending somewhat. A pastern that is too straight will not absorb much of the shock and one that is too long and weak will let the leg go to the ground. These kinds of pasterns will soon result in unsound horses. Thus, it is very important that a pastern have the proper slope so it can absorb the optimal amount of shock without the leg going to the ground or causing too much concussion on the joints and, ultimately, the rider.

The front legs each have two joints that allow movement and absorb shock: the joint between the ulna and the humerus, and the joint between the humerus and the scapula. Also, the hind legs each have two joints that bend and thus absorb some shock. These joints are located between the metatarsus and tibia and between the tibia and femur.

If the horse's feet and legs have proper conformation, a pleasant ride can be enjoyed. If there are abnormalities due to inheritance, injury, improper nutrition, or disease, the horse will give the rider a less pleasurable ride.

Abnormalities in Way of Going

A horse that toes out with its front feet tends to dish or swing its feet inward (wing in) when its legs are in action. Swinging the feet inward can cause the striding foot to strike the supporting leg so **interference** to forward movement results. A horse that toes-in (pigeon-toed) tends to swing its front feet outward, giving a **paddling** action.

Some horses overreach with the hind leg and catch the heel of the front foot with the toe of the hind foot. This action, called **overreaching,** can cause the horse to stumble or fall. **Forging** occurs when the hind foot hits the shoe on the front foot.

Figure 33.10 shows the way of going well as the horse moves straight and true, each foot moving in a straight line. The other illustrations show the path of flight of each foot when the structure of the foot and leg deviates from the desired norm. Figure 33.11 shows how the length and slope of the hoof affects way of going.

Since few horses move perfectly true, it is important to know which movements may be unsafe. A horse that wings in (interferes) is potentially more unsafe than a horse that wings out (paddles), as the former horse may trip itself.

DETERMINING THE AGE OF A HORSE BY ITS TEETH

The age of a horse can be estimated by its teeth (Figs. 33.12, 33.13, and 33.14). A foal at 6–10 months of age has 24 baby or milk teeth (12 incisors and 12 molars). The incisors include three pairs of upper and three pairs of lower incisors.

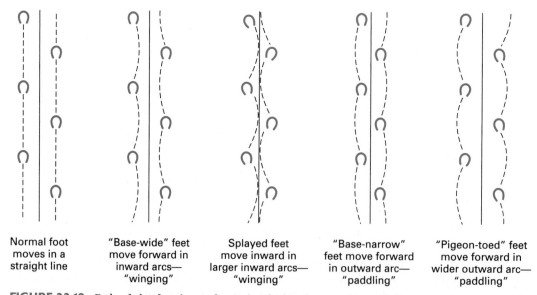

| Normal foot moves in a straight line | "Base-wide" feet move forward in inward arcs— "winging" | Splayed feet move inward in larger inward arcs— "winging" | "Base-narrow" feet move forward in outward arc— "paddling" | "Pigeon-toed" feet move forward in wider outward arc— "paddling" |

FIGURE 33.10 Path of the feet (way of going), which relates to foot and leg structure, as seen from above. *Source:* Colorado State University.

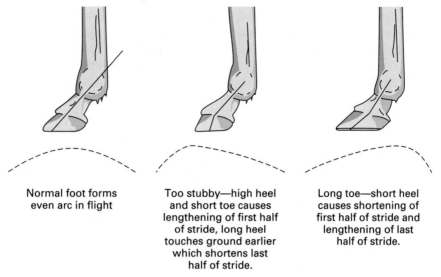

Normal foot forms
even arc in flight

Too stubby—high heel
and short toe causes
lengthening of first half
of stride, long heel
touches ground earlier
which shortens last
half of stride.

Long toe—short heel
causes shortening of
first half of stride and
lengthening of last
half of stride.

FIGURE 33.11 Illustration of how length and slope of the hoof affects way of going. *Source:* Colorado State University.

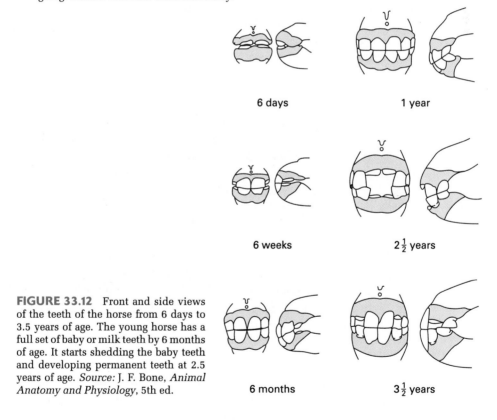

6 days

1 year

6 weeks

2½ years

FIGURE 33.12 Front and side views of the teeth of the horse from 6 days to 3.5 years of age. The young horse has a full set of baby or milk teeth by 6 months of age. It starts shedding the baby teeth and developing permanent teeth at 2.5 years of age. *Source:* J. F. Bone, *Animal Anatomy and Physiology*, 5th ed.

6 months

3½ years

Chewing causes the incisors to become worn. The wearing starts with the middle pair and continues laterally. At 1 year of age, the center incisors show wear; at 1.5 years, the intermediates show wear; and at 2 years, the outer, or lateral, incisors show wear. At 2.5 years, shedding of the baby teeth starts.

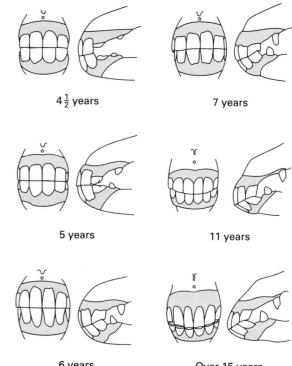

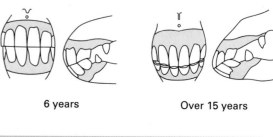

FIGURE 33.13 Front and side views of the teeth of the horse from 4.5 to 15 years of age. The horse's mouth changes in shape as it becomes older, such that the front teeth protrude somewhat forward. *Source:* J. F. Bone, *Animal Anatomy and Physiology*, 5th ed.

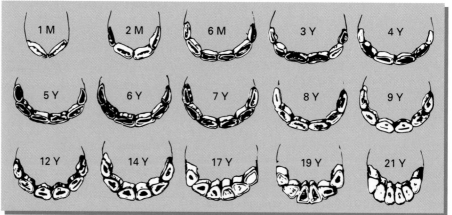

FIGURE 33.14   Surface of the lower incisors of the horse from 1 month (M) to 21 years (Y) of age. *Source:* J. F. Bone, *Animal Anatomy and Physiology*, 5th ed.

The center incisors are shed first. Thus, at 2.5 years, the center incisors become permanent teeth; at 4 years, the intermediates are shed; at 5 years, the outer, or lateral, incisors are shed and replaced by permanent teeth.

A horse at 5 years of age is said to have a **full mouth,** because all the teeth are permanent. At 6 years, the center incisors show wear; at 7 years, the intermediates show wear; and at 8 years, the outer, or lateral, incisors show wear. Wearing is shown by a change from a deep groove to a rounded dental cup on the grinding surface of a tooth.

CHAPTER SUMMARY

- Horses are typically classified as light, draft, or pony breeds.
- Based on registration numbers, the Quarter Horse is the most popular horse breed.
- Mares have a 21-day estrous cycle, an estrus of 5–7 days, and a gestation length of 340 days.
- Unsoundness in horses leads to diminished usefulness and should be strictly selected against.
- Gaits of horses include the walk, trot, pace, gallop, canter, and rack. The Tennessee Walker also exhibits the running walk.

KEY WORDS

light horse breeds
draft horses
ponies
color registries
knee
pastern
cannon
foot
hock
unsoundness
blemish
bog spavin
bone spavin
capped hock
curb
cataract
contracted heels
heaves

hyperkalemic periodic
 paralysis (HYPP)
laminitis (founder)
moon blindness
navicular disease
quittor
ring bone
ruptured blood vessels
shoe boil (capped
 elbow)
sickle hocked
side bones
stifled
string halt
sweeney
thoroughpin
toeing-in
pigeon-toed

toeing-out
tying-up (exertional
 rhabdomyolysis)
windgalls (wind puffs)
walk
trot
pace
gallop
canter
rack
running walk
interference
paddling
overreaching
forging
full mouth

REVIEW QUESTIONS

1. What is the most popular breed of horse in the United States today?
2. What is the length of the estrous cycle of mares?
3. What is the duration of gestation in horses?
4. *True or False:* Stallions for a breeding program should always be selected based on their ability to correct and improve upon a weakness in a herd, without consideration to new weaknesses which they may possess and introduce into the herd.
5. What is unsoundness?
6. What is a blemish?
7. What are the seven major gaits of horses?
8. How are a trot and pace similar? How do they differ?

9. *True or False:* The age of a horse can readily be estimated by examining its teeth.

10. When is a horse said to have a full mouth?

SELECTED REFERENCES

Bone, J. F. 1988. *Animal Anatomy and Physiology*. 5th ed. Corvallis, OR: Oregon State University Book Stores.

Bowling, A. 1996. *Horse Genetics*. Oxon, UK: CAB International.

Butler, D. 1985. *Principles of Horseshoeing II*. Grand Prairie, TX: Equine Research Inc.

Ensminger, M. E. 1990. *Horses and Horsemanship*. Danville, IL: Interstate Publishers, Inc.

Evans, J. W. 1989. *Horses: A Guide to Selection, Care and Enjoyment*. 4th ed. San Francisco, CA: W. H. Freeman.

Evans, J. W. (ed.). 1992. *Horse Breeding and Management*. Amsterdam: Elsevier Science Publishers B.V.

Evans, J. W., A. Borton, H. F. Hintz, and L. D. Van Vleck. 1990. *The Horse*. 2d ed. San Francisco: W. H. Freeman.

Harris, S. 1993. *Horse Gaits, Balance and Movement*. Grand Prairie, TX: Equine Research Inc. *Horse Industry Directory*. 2006. Washington, DC: American Horse Council.

Jones, W. E. 1982. *Genetics and Horse Breeding*. Philadelphia, PA: Lea & Febiger.

Lasley, J. F. 1987. *Genetics of Livestock Improvement*. Englewood Cliffs, NJ: Prentice Hall.

Rich, G. A. 1981. *Horse Judging Guide*. Colorado State University Extension Service Publication MOOOOOG.

Stashak, T. S. (ed.). 1987. *Adam's Lameness in Horses*. 4th ed. Grand Prairie, TX: Equine Research Inc.

CHAPTER **34**

Feeding and Managing Horses

LEARNING OBJECTIVES

- Discuss the use of body condition scoring in horse management
- Discuss the role of exercise on nutritional requirements of the horse
- Outline the nutritional and feeding practices used in the horse industry
- Describe reproductive management of the horse
- Describe the facilities and equipment required in horse management
- Discuss health management of the equine

Humans and horses share a unique and mutually beneficial bond. Although many people enjoy horses, most own only a few; however, whether a person keeps only one horse for personal recreation or manages a large breeding farm, basic horse knowledge is vital. Information on managing horses, such as feeding, facilities, disease prevention, and parasite control, is important. Additional material on management is discussed in Chapter 24.

FEEDS AND FEEDING

The equine digestive tract is well suited to a forage-based diet comprised of grazed grasses/legumes, as well as mechanically harvested forages. There are significant regional differences in both availability and type of **forage** (Table 34.1). Grains also provide an important ingredient in equine rations. **Concentrate** mixtures containing grains, protein supplements, and vitamin and mineral additives are prepared and sold by commercial feed companies, especially to owners with only one or two horses. Wet molasses can be added to these concentrate mixtures to make sweet feed. The forage and concentrate mixtures may be mixed and pelleted to make a complete, higher-priced convenience feed for horse owners.

TABLE 34.1   Forages for Various Regional Equine Pastures

Northeast, Midwest, Upper South, and Pacific Northwest:
alfalfa
bird's foot trefoil
brome grass
Kentucky bluegrass
orchard grass
red clover
timothy
white clover

Southeast:
alfalfa
Bahia grass
Bermuda grass
Dallis grass

Southwest:
alfalfa
Bermuda grass

Great Plains and Intermountain West:
alfalfa
wheat grass

Some forage varieties, such as endophyte infected fescue, can be harmful to horses—particularly broodmares. Fescue toxicity is problematic in the southeastern region of the United States and may result in increased abortion rates, reduced rebreeding rates, diminished milk production, and significantly higher rates of stillbirth. Management approaches to solving this dilemma are multifaceted and often variable in effectiveness. Certainly, pastures with fescue should be tested for the presence of endophytes, pregnant mares should be kept off infected pastures and hay in the last 3 months of pregnancy, and reproductive rates should be carefully monitored.

Although horses spend less time chewing than ruminants, the normal, healthy horse with a full set of teeth can grind grains such as oats, barley, and corn so that cracking or rolling these feeds is unnecessary. Wheat and milo, however, should be cracked to improve digestibility. The horse's stomach is relatively small, composing only 10% of the total digestive capacity. Only a small amount of digestion takes place in the stomach, and food moves rapidly to the small intestine. From 60–70% of the protein and soluble carbohydrates are digested in the small intestine, and about 80% of the fiber is digested in the cecum and colon. The large intestine has about 60% of the total digestive capacity, with the colon being the largest component. Bacteria that live in the cecum aid digestion there. Minerals, proteins as amino acids, lipids, and readily available carbohydrates such as glucose are absorbed in the small intestine.

Owners of pleasure horses may liberally feed horses in an attempt to attain a pleasing physical appearance. Perhaps more horses are overfed than are underfed. Proper condition of the horse can be monitored via a numerical scoring system. The range of **body condition scores** is provided in Table 34.2. Also, many people want to be kind to their animals, keeping them housed in a box stall when weather conditions are undesirable. This may not be best for the

TABLE 34.2 Description of Equine Body Condition Scores

Condition Score	Description
1	Poor—extremely emaciated; ribs and bone structure easily discernable
2	Very thin—emaciated; ribs prominent; bone structure somewhat noticeable
3	Thin—slight fat cover over ribs
4	Moderately thin—faint outline of ribs, neck, shoulder, and withers not obviously thin
5	Moderate—ribs not visually apparent, but easily palpated; back level over loin; withers rounded
6	Moderately fleshy—fat over ribs and tailhead feels spongy; fat deposits generally apparent
7	Fleshy—noticeable filling of space between ribs with fat; crease down back over loin
8	Fat—deposition of fat along inner buttocks; thickening of neck; fat withers, tailhead, and behind shoulders
9	Obese—bulging fat, flank filled in flush; patchy fat appearing over ribs

Source: Adapted from Henneke et al. (1983).

physiological state of the horse. Certainly, if any deficiency exists in the feed provided, such a deficiency is much more likely to affect horses that are not running on good pasture, where they can forage for themselves.

Young, growing foals should be fed correctly to allow for proper growth, but overfeeding and obesity are discouraged. Quality of protein and amounts of protein, minerals, and energy are important for young growing horses. Soybean meal or dried milk products in the concentrate mixture provide the amino acids and minerals that might otherwise be deficient or marginal in the weanling ration.

Good-quality pasture or hay supplemented with grain can provide the nutrition needed by young horses. An appropriate salt-mineral source should be provided at all times, and clean water is essential.

During the first 8 months of gestation, pregnant mares perform well on good pastures or on good-quality hay supplemented with a small amount of grain and an appropriate source of salt-minerals. As the fetus grows during the last trimester of pregnancy, the mare requires more concentrate and less fibrous, bulky hay. Oats, corn, or barley make excellent feed grains for pregnant mares. Pregnant mares should be in good body condition but not obese. The pregnant mare should not be allowed to drop below a body score of 5 as a means to assure high fertility rates. Animals used for riding or working, whether they are pregnant or not, need more energy than those not working (Table 34.3). Horses being exercised heavily should be fed appropriate amounts of concentrates or high-energy forages to replace the energy used in the work. Body condition of the working horse should be carefully monitored to assure optimal health and performance. A thin horse is depicted in Figure 34.1. Horses may have insufficient body condition for a variety of reasons including insufficient feed intake, parasite infestation, or as a result of an illness.

Generally, lactating mares require more grain feeding than do geldings and pregnant or nonpregnant mares. Lactation is the most stressful nutritional period for a mare. If lactating mares are exercised, they must be fed additional grain and hay to meet the nutrient demand of the physical activity.

Stallions need to be fed as working horses during the breeding season and given maintenance ration during the nonbreeding season. Feeding good-quality hay with limited amounts of grain is usually sufficient for the stallion.

TABLE 34.3   Megacalories Burned as a Result of Various Forms of Exercise

Exercise—1 hour	Megacalories Burned per 1,000 lb of Body Weight
Walking	0.2
Slow trot	2.3
Fast trot–slow canter	5.7
Canter–full gallop	10.5
Strenuous (racing, reining, etc.)	17.7

Source: NRC, *Nutrient Requirements of Horses.*

FIGURE 34.1   Body condition of horses should be carefully monitored to ensure that appropriate feed rations are provided. This horse is thin as the result of not receiving sufficient nutrients.

Feed companies provide properly balanced rations for horse farms of all sizes. An owner who has only one or two horses may find it highly advantageous to use a prepared feed, since it is difficult and laborious to prepare balanced rations for only a few animals. Using commercially prepared feeds or custom-blended rations can prevent nutritional errors, save time, and, for larger farms, be more cost-effective.

Table 34.4 shows some examples of horse rations and describes when and how these rations should be fed. Feed delivery to horses varies from region to region for both forages and concentrates (Table 34.5 and Table 34.6). According to the NAHMS study (1998), horse owners fed their horses once per day, twice per day, three times or more daily, or less than daily at rates of 25%, 48%, 19%, and 8%, respectively.

Selection of feeds for inclusion in a ration also requires that managers make a careful assessment of the **quality characteristics** of a particular ingredient. For example, when selecting hay, it is important to evaluate the plant type composition, degree of leaf retention, stage of plant maturity at time of harvest, degree of foreign matter contamination, as well as aroma, color, and texture. Ideally,

TABLE 34.4 Sample Rations for Horses of Different Ages and in Various States of Production

Creep feed for nursing foals. The grain should be fed at a rate of 0.5–0.75 lb of grain/100 lb body weight.

	Percentage in Grain Mix
Corn, rolled or flaked	34.0%
Oats, rolled or flaked	34.0
Soybean meal	22.0
Molasses	6.0
Dicalcium phosphate	2.0
Limestone	1.5
Trace-mineral salt	0.5

Grain mix for weanlings. The grain should be fed at a rate of 0.75–1.5 lb of grain/100 lb body weight. Select the grain according to the type of roughage fed. Allow free-choice consumption of either roughage type.

	Percentage in Grain Mix	
	Alfalfa Hay	Grass Hay
Corn, rolled or flaked	40.0%	34.0%
Oats, rolled or flaked	40.0	34.0
Soybean meal	12.0	23.0
Molasses	5.0	5.0
Dicalcium phosphate	2.5	3.0
Limestone	0	0.5
Trace-mineral salt	0.5	0.5

Grain mix for yearlings and mares. The grain should be fed at a rate of 0.5–1.0 lb of grain/l00 lb of body weight. Select the grain according to the type of roughage fed. Allow free-choice consumption of either roughage.

Mares during late gestation and lactation can be fed the same grain mixes as yearlings. The grains should be fed at a rate of 0–0.5 lb/100 lb of body weight. Roughage consumption can vary from 1.5–2.5 lb/100 lb of body weight.

	Percentage in Grain Mix	
	Alfalfa Hay	Grass Hay
Corn, rolled or flaked	46.5%	38.0%
Oats, rolled or flaked	46.5	38.0
Soybean meal	0	15.5
Molasses	5.0	5.0
Dicalcium phosphate	1.5	2.0
Limestone	0	1.0
Trace-mineral salt	0.5	0.5

Grain mix for horses at maintenance and at work, dry mares during the first 8 months of gestation, or stallions. The grain should be fed as needed (to maintain body condition). Roughage consumption can vary from 1.5–2.5 lb/100 lb of body weight.

	Percentage in Grain Mix	
	Alfalfa Hay	Grass Hay
Corn	46.5%	46.5%
Oats	46.5	46.5
Molasses	5.0	5.0
Dicalcium phosphate	0	1.5
Monosodium phosphate	1.5	0
Trace-mineral salt	0.5	0.5

Source: Ginger Rich, *Horse Judging Guide,* Colorado State University Extension Service.

TABLE 34.5 Percent of Operations That Fed Various Forages in 1997 by Region

Type	Region				
	Southern	Northeast	Western	Central	All
Small bales (< 200 lbs)	82.9	92.7	92.8	83.2	86.6
Grass hay	69.5	54.9	43.8	36.0	54.2
Alfalfa hay	18.9	11.7	51.3	24.7	26.9
Grass/alfalfa mixed hay	15.3	53.8	40.4	55.9	35.1
Corn stalks, oat straw, other	5.1	4.6	12.7	8.0	7.5
Large bales (> 200 lbs)	37.4	21.1	17.5	36.2	30.3
Grass hay	34.6	14.8	6.2	19.9	22.1
Alfalfa hay	2.9	1.2	5.6	4.5	3.7
Grass/alfalfa mixed hay	3.6	7.6	6.0	16.1	7.5
Corn stalks, oat straw, other	1.9	0.0	4.1	1.8	2.2
Nonbaled dried forage (hay cubes, etc.)	5.9	5.4	8.8	7.6	6.9
Nondried forage	1.4	0.0	1.2	0.6	1.0

Forage delivery systems were troughs/racks (41.4%), loose on the ground (41.1%), other individual feeders (8.5%), rubber tires (2.1%), hay nets (2.4%), and other (4.5%).
Source: NAHMS, 1998.

TABLE 34.6 Percent of Operations Feeding Various Grains/Concentrates by Region

Source	Region				
	Southern	Northeast	Western	Central	All
Unpelleted sweet feed	60.7	63.2	51.1	53.4	57.2
Unpelleted grain	41.9	29.2	47.5	48.1	42.9
Geriatric feed	5.5	9.6	10.8	6.9	7.6
Complete feed pellets/cubes	21.1	17.0	20.5	13.3	18.7
Grain mix with pellets	24.3	27.7	15.6	20.9	21.9
Other	8.1	7.4	11.1	6.2	8.3
None	2.2	6.2	11.9	5.0	5.6

Source: NAHMS, 1998.

feeds should be tested at a qualified laboratory to quantify the nutrient content as well as to determine the presence of contaminants.

One of the most potentially devastating digestive disorders of the horse is the condition known as **colic.** Colic is a term utilized to describe a broad range of abdominal pain. Abrupt changes in diet, feeding schedule, exercise regime, or housing type may lead to colic. The pain can be attributed to gas distension, decreased gut motility, parasitic infestation, ulcers, bowel displacement or twisted gut, or ingestion of sand or other foreign objects. While most cases of colic are relatively mild, affected horses typically exhibit pawing, pacing, or rolling in response to the discomfort. Treatment varies, but immediate veterinary care is advised.

The digestive system of the horse is adapted to a consistent diet of grass or hay. Table 34.7 outlines the steps managers can take to avoid colic.

MANAGING HORSES

Proper management of horses is essential at several critical times: during the breeding season, foaling, weaning of foals, castration, and strenuous work. Enterprises that own horses are highly variable in type with about half serving as recreational experiences, one-quarter are farms and ranches, 16% are breeding establishments, and 10% are focused on competitive showing according to the USDA.

TABLE 34.7 Management Steps to Prevent Colic

Feed appropriate to the horse's need.

- High fiber, low carbohydrate feedstuffs of 8 to 10% protein are preferred.
- Feed no more than $^1/_2$ of the ration in concentrate form.
- Concentrate feedings should be spaced out in several small meals.

Do not make sudden or frequent dietary changes.
Keep in mind that horses are nibblers often grazing for as much as 20 hours per day. The feeding regime should mimic this natural behavior.
Keep horses active. Those equine that spend most of their time in stalled conditions are most at risk to colic.
Avoid feeding on sandy ground to avoid sand impactions of the gut.
Maintain regular internal parasite prevention practices by assuring good sanitation in pens, paddocks, or stalls, and via regular administration of dewormers.

TABLE 34.8 Percent of Operations, Horses, and Type of Enterprise in the United States

	Small (< 9 hd)	Medium (10–19 hd)	Large (> 19 hd)
Percent of enterprises	66	26	8
Percent of inventory	36	34	30
Boarding/training	3	10	17
Breeding	9	22	34
Farm/ranch	40	42	32
Pleasure	46	22	10
Other	2	3	6

Source: USDA, 2006.

Equine enterprises tend to have relatively small numbers of horses on inventory with the smallest category (less than 9 head) dominated by recreational or farm/ranch enterprises (86%) but controlling only one-third of all horses. The larger enterprises account for less than 10% of all horse operations but control almost one-third of the inventory (Table 34.8).

Breeding Season

Fertility rates can be enhanced by dealing effectively with chronic uterine infection, identification of uterine damage due to **foaling,** waiting to breed mares until after the transition period following an estrus, and utilizing a good teasing program.

Mares should be teased daily with a stallion to determine the stage of their estrous cycle (Fig. 34.2). When the mare is in standing heat (estrus), she is ready to be bred. When the stallion approaches the front of the mare, the mare reacts violently against the stallion if not in heat, but squats and urinates with a winking of the vulva if in heat. Cleanliness is paramount for both natural breeding and artificial insemination (AI) programs. Mares bred naturally should have the vulva washed and dried and the tail wrapped before being served by the stallion. If the mare can be bred twice during heat without overusing the stallion, breeding 2 and 4 days after the mare is first noticed in heat is desirable. A mature stallion can serve twice daily over a short time and once per day over a period of 1 or 2 months. A young stallion should be used lightly at about three or fewer services per week. In an AI program, the stallion's semen can be collected every other day to cover as many mares as possible, depending on the stallion's sperm numbers and motility. AI programs allow better management of the stallion.

FIGURE 34.2   Chute teasing. A stallion is led along a chute holding several mares. The behavior of each mare is observed to detect expression of estrus. Courtesy of Colorado State University.

FIGURE 34.3   A normal delivery presentation of a foal is front feet and head first.

Foaling Time

Mares normally give birth to foals in early spring, at which time the weather can be unpleasant. A clean box stall that is bedded with fresh straw should be made available. If the weather is pleasant, mares can foal on clean pastures. When a mare starts to foal, she should be observed carefully but not disturbed unless assistance is necessary. If the head and front feet of the foal are being presented, it should be delivered without difficulty (Fig. 34.3). If necessary, however, a qualified person can assist by pulling the foal as the mare labors. Do not pull when the mare is not laboring and do not use a tackle to pull the foal. If the front feet are presented but not the head (Fig. 34.4), it may be necessary to push the foal back enough to get the head started along with the front feet. **Breech presentations** can endanger the foal if delivery is delayed; therefore, assistance should be given to help the mare make a rapid delivery if breech presentation

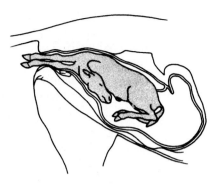

FIGURE 34.4 Malpresentation of foal for delivery; head and neck back. *Source: R. A. Battaglia and V. B. Mayrose, Handbook of Livetock Management Techniques.*

occurs. If it appears that difficulties are likely to occur, a veterinarian should be called as soon as possible.

As soon as the foal is delivered, its mouth and nostrils should be cleared of membranes and mucus so it can breathe. If the weather is cold, the foal should be wiped dry and assisted in nursing. As soon as the foal nurses, its metabolic rate increases helping it stay warm. The **umbilical cord** should be dipped in a tincture of iodine solution to prevent harmful microorganisms from invading the body.

In general, it is highly desirable to exercise pregnant mares up to the time of foaling. Mares that are exercised properly while pregnant have better muscle tone and are likely to experience less difficulty when foaling than those that get little or no exercise while pregnant. During the horsepower era, many mares used for plowing and similar work foaled in the field without difficulty. The foal was usually left with the mare for a few days, after which the foal was left in a box stall while the mare was worked.

Weaning the Foal

When weaning time arrives, it is best to remove the mare and allow the foal to remain in the surroundings to which it is accustomed. Since the foal will make every attempt to escape to find its mother, it should be left in a box stall or a secure and safe fenced lot. Fencing other than barbed wire should be used. Prior to weaning, the foal should become accustomed to eating and drinking on its own. High-quality hay or pasture and a balanced concentrate feed should be provided for the foal.

Castration

Colts that are not kept for breeding can be castrated any time after the testicles have descended; however, the stress of castration should not be imposed at weaning time. Some people prefer to delay castration until the colt has reached a year of age, whereas others prefer an earlier time. Genetic potential and nutrition, not time of castration, determine mature height and weight. Castration is performed because geldings can be handled more safely than stallions by owners and are less dangerous around other people.

The colt can be put onto a turntable or manually restrained with ropes for **castration.** Painkillers and muscle relaxants can be used. A qualified person opens the scrotum on each side and the membrane around each testicle is split to expose the testicle. The testicle is pulled from the body cavity far enough to expose the cord for clamping. The cord is crushed by the clamp to prevent excessive bleeding and then severed. It may be desirable to give the colt an injection of an antibiotic to prevent infection. If the scrotum was cleaned with a mild disinfectant before castration, the wound need not be washed with an antiseptic. Harsh disinfectants should not be applied to the wound. After the colt is castrated, he should be placed in a small, clean pasture for close observation. If fly infestation is a problem, a fly repellant should be applied about the scrotal area.

Identification

Identification of horses varies considerably as some horses are not formally documented (29%) while others have multiple forms of identification. About one-half of equines are identified by registration paper, one-quarter via a Coggins Test document, 10% by brand, and 4% with a lip tattoo according to the USDA.

Tattooing them on the inside of the upper lip can permanently identify horses. This does not disfigure the animals and can be easily read by raising the upper lip. Other methods of identification include freeze branding, hot-iron branding, and electronic implants.

Care of Hardworking Animals

Hardworking horses that are sweating and have elevated temperature, pulse, and respiration rates should be washed (cooled down) before receiving water and feed. The animals should be given only small amounts of water and walked until they have stopped sweating and their pulse and respiration are within 10% of resting levels. At that time, water, hay, and grain can be given. This handling process is important; otherwise, horses may become colicky or founder.

Hardworking horses (Fig. 34.5) do require extra energy, usually in the form of grain. However, care must be taken to avoid overfeeding grain to horses. Excess grain, and sometimes lush green pasture, can cause a horse to founder. Founder can result in death or severe lameness.

HOUSING AND EQUIPMENT

Barns should be located on a higher elevation than the surrounding area to assure good drainage. They should be accessible to utilities and vehicles and, preferably, have a southeast exposure. Design should also take into account safety, air quality, labor intensity, and durability.

Although many styles of barns exist, those constructed with an aisle between two rows of stalls provide easy access and efficient use of space. Feed can be placed in the stalls on either side as the feed cart goes down the aisle. If the stalls can be opened from the outside and cleaned with mechanical equipment, labor also is saved. The hay manger in stalls should be constructed at chest height to the horse. Hayracks placed above the horse's head force the horse to inhale hay dust when reaching for the hay, and health problems can result.

To alleviate mixing hay with grain, a separate grain feeder should be placed several feet from the hayrack. The waterer or water bucket should be located along the outside wall for drainage purposes. The water source should also be placed some distance from the hay and grain so feed will not drop in the water.

A   B

FIGURE 34.5 Many horses are utilized in (A) athletic events or as (B) working animals on farms or ranches. The working horse requires additional feed to ensure that the horse is able to perform at peak levels. Reprinted with permission of the American Quarter House Association © American Quarter Horse Association.

Electric float-controlled waterers save labor but are prone to malfunction. Although water buckets require more labor, they allow water intake to be easily monitored.

Box stalls are used for foaling and for the mare and the foal when the weather is severe. The stall should be constructed to allow complete cleaning and proper drainage. Minimum dimensions for a foaling stall are 14×14 ft, whereas a regular box stall should be at least 10×10 ft.

Feed should be stored in an area connected with or adjacent to the barn. Truck access to the feed storage area is vital. The floor should be solid (e.g., concrete) and the rodent-proof bins should be easily cleaned. Overhead (loft) storage of hay requires a great deal of labor and expensive construction. Large amounts of hay stored in the barn where the horses are housed increases dust and danger of fire.

Fences should be constructed based on assessment of cost, durability, safety, and strength. Desirable fence materials include PVC, wood, rubber/nylon, and diamond-mesh wire. If wood is used, it is important to protect material from horses that crib or chew the fence.

A dry, dust-free **tack room** is needed to house riding equipment, including saddles, blankets, bridles, and halters. In addition, it may be advisable to have a washroom where horses can be washed.

A **stock** is essential to care for horses that have been injured and need attention. If the horse is in a stock, it is unlikely that a person working around the horse can be kicked. The stock can be used for AI, teasing, palpation, or treating health problems. The stock is also useful for injections or when blood samples are being taken. Often, horses strike with a front foot when a needle is inserted for vaccinations or blood samples; a properly constructed stock will protect people as well as the horse.

CONTROLLING DISEASES AND PARASITES

Sanitation is of vital importance in controlling diseases and parasites of horses. Horses should have clean stalls and be groomed regularly. A horse that is to be introduced into the herd should be isolated for a month to prevent exposing other horses to diseases and parasites. Horses may have illnesses caused by bacteria or viruses, internal or external parasites, poisonous plants, nitrates, and/or moldy grains.

Regardless of the size of the equine enterprise, it is critical to establish a trusted working relationship with an equine veterinarian. Veterinarians can help prevent diseases as well as treat animals that are diseased. Most veterinarians prefer to assist in preventing health problems rather than treating the animals after they are ill. A variety of diseases are vaccinated against by horse owners (Table 34.9). A veterinarian should be contacted for assistance in establishing a sound preventative health plan.

Horse manure is an excellent medium for microorganisms that cause **tetanus.** Horses should be vaccinated to prevent tetanus, which can develop if an injury allows tetanus-causing microorganisms to invade through the skin. Usually two shots are given to establish **immunity,** after which a booster shot is given each year. Because the same microorganisms that cause tetanus in horses also affect humans, those who work with horses should also have tetanus shots.

Strangles, also known as *distemper,* is a highly contagious bacterial disease that affects the upper respiratory tract and associated lymph glands. High fever, nasal discharge, swollen lymph glands, and a rattling sound associated with breathing are signs of strangles. The disease is spread by contamination of feed and water. Afflicted horses must be isolated and provided clean water and feed. A strangles bacterin is available. The injections should be given 3–4 weeks apart followed by an annual booster.

Broodmares are subject to many infectious agents that invade the uterus and cause **abortion;** examples include *Salmonella* and *Streptococcus* bacteria and the viruses of **rhinopneumonitis** and **arteritis.** Should abortion occur, professional assistance should be obtained to determine the specific cause and to develop a preventative plan for the future.

Sleeping sickness, or **equine encephalomyelitis,** is caused by viral infections that affect the brain of the horse. Different types, such as the eastern, western, and Venezuelan, are known. Vectors such as mosquitoes transmit encephalomyelitis.

TABLE 34.9 **Percent of Equine Operations Administering Vaccines**

Vaccine Type	% of Operations
Flu	54
Strangles	27
Rhinopneumonitis	47
Rabies	33
West Nile virus	64
Equine encephalitis	
Western/Eastern/	56
Venezuelan	18
Tetanus	61
Equine viral arteritis	12
Potomoc horse fever	11

Source: Adapted from USDA, 2006.

Horses rubbing noses together or sharing water and feed containers can also spread it. Vaccination against the disease consists of two intradermal injections spaced 1 week to 10 days apart. These injections should be given in April on an annual basis. Horses in the southern United States should be vaccinated twice a year.

Influenza is a common respiratory disease of horses. The virus that causes influenza is airborne, so frequent exposure may occur where horses congregate. The acute disease causes high fever and a severe cough when the horse is exercised; rest and good nursing care for 3 weeks usually gives the horse an opportunity to recover. Severe aftereffects are rare when complete rest is provided. Horse owners who plan for shows should vaccinate for influenza each spring. Two injections are required the first year, with one annual booster thereafter.

Unfortunately, new diseases may be introduced into a region with potentially devastating consequences. A case in point is **West Nile virus,** a disease that is transmitted via mosquito or bird vectors to both equines and humans. The introduction of the virus into the United States followed outbreaks in Africa, Asia, and Europe. Affected animals exhibit depression, loss of hindquarter coordination, tremors, and paralysis. Symptoms may progress rapidly and spread widely as a result of mosquito proliferation and bird migration. A vaccine is available but its efficacy is not well proven.

Horses can become infected with internal and external parasites. Control of internal parasites consists of rotating horses from one pasture to another, spreading manure from stables on land that horses do not graze, and treating infested animals.

Pinworms develop in the colon and rectum from eggs that are swallowed as the horse consumes contaminated feed or water. These parasites irritate the anus, which causes the horse to rub the base of its tail against objects even to the point of wearing off hair and causing skin abrasions. Pinworms are controlled by oral administration of proper vermifuges.

Bots are the larval stage of the botfly. The female botfly lays eggs on the hairs of the throat, front legs, and belly of the horse. The irritation of the botfly causes the horse to lick itself. The eggs are then attached to the tongue and lips of the horse, where they hatch into larvae that burrow into the tissues. The larvae later migrate down the throat and attach to the lining of the stomach, where they remain for about 6 months and cause serious damage. A paste wormer is available to control bots.

Adult **strongyles** (bloodworms) firmly attach to the walls of the large intestine. The adult female lays eggs that pass out with the feces. After the eggs hatch, the larvae climb blades of grass where they are swallowed by grazing horses. The larvae migrate to various organs and arteries where severe damage results. Blood clots, which form where arteries are damaged, can break loose and plug an artery. Strongyles are treated with Ivermectin and other wormers.

Adult **ascaris worms** are located in the small intestine. The adult female produces large numbers of eggs that pass out with the feces. The eggs become infective if they are swallowed when the horse eats them while grazing. The eggs hatch in the stomach and small intestine, and the larvae migrate into the bloodstream and are carried to the liver and lungs. The small larvae are coughed up from the lungs and swallowed. When they reach the small intestine, they mature and produce eggs. These large worms may cause an intestinal blockage in young horses. The same chemicals used for the control of strongyles are effective in the control of ascarids.

Horses should be dewormed at least twice a year and up to four times a year. The type of wormer should be rotated every second time.

An important component of a sound health management plan is the development of an effective record-keeping system. Table 34.10 outlines the important records that should be maintained on individual horses.

Horses are afflicted by a variety of diseases and disorders. The rate of morbidity and mortality varies by the age of the affected horse and by the particular disease in question. Table 34.11 outlines rate of disease incidence while Table 34.12 documents mortality from the most common causes of death.

TABLE 34.10 Important Records to Be Kept on Individual Horses[a]

Vaccinations
Parasite control
Medical treatments
Disease and/or injury
Dental exams/floating schedules
Shoeing and hoof trimming
Feeding schedule and ration quantity and composition
Baseline vital statistics (temperature, pulse, respiration rate)
Breeding records
Competition and training schedule

[a]Dates, products used, professionals involved should be recorded.

TABLE 34.11 The Incidence of Various Diseases and Disorders in the Horse

Disorder	Incidence in Foals (< 6 mo.) %	Incidence in Horses (> 6 mo.) %
Colic	4	2
Other digestive disorder	6	—
Respiratory disease	4	2
Eye conditions	1	1
Skin problems	1	1
Reproductive disease	—	1
Injury	9	5
Lameness or hoof problems	3	3
Neurological disorder	6	—

Source: USDA, 2006.

TABLE 34.12 Mortality Rate Originating from Various Diseases and Disorders in the Horse

Cause of Death	Incidence in Foals (< 6 mo.) %	Incidence in Horses (> 6 mo.) %
Old age	—	30
Injury	24	16
Colic	3	15
Lameness	8	8
Neurological condition	—	3
Noncolic digestive disorder	8	3
Strangles	2	1
Other respiratory disease	5	2
Dystocia	—	2

Source: USDA, 2006.

CHAPTER SUMMARY

- The major digestive structures of the equine include the stomach, small intestine, cecum, and colon.

- Nutrients should be provided to the horse in accordance with its requirements as determined by age, work schedule, lactation status, and gestation status.

- Key management areas for horses include reproduction, weaning, housing, and health.

KEY WORDS

forage
concentrates
body condition scores
quality characteristics
colic
foaling
breech presentation
umbilical cord
castration
identification

barns
box stalls
tack room
stock
tetanus
immunity
strangles
abortion
rhinopneumonitis
arteritis

sleeping sickness
 (equine
  encephalomyelitis)
influenza
West Nile virus
bots
strongyles
ascaris worms

REVIEW QUESTIONS

1. Where is fiber digested in the gastrointestinal tract of a horse?

2. *True or False:* Most horses are overfed.

3. Horses used for riding or working need more _____ in their diets than those not working.

4. At what critical times is proper management of horses essential?

5. How is the stage of the estrous cycle determined in mares?

6. When do mares normally give birth to foals?

7. *True or False:* Mares should not be worked or exercised during gestation to prevent injury of the fetus.

8. Why is castration of males performed on horses that are not to be used in a breeding program?

9. Describe how horses are typically identified?

10. Describe the percent of equine operations and the percent of inventory controlled by each of the three primary categories (small, medium, and large).

11. Compare and contrast the use of horses on small-, medium-, and large-sized enterprises.

12. Hardworking horses that are sweating and have elevated temperature, pulse, and respiration rates should be _____ before receiving water and feed.

13. What disorder may result from overfeeding grain or lush green pasture to horses?

14. _____ is of vital importance in controlling diseases and parasites of horses.

15. *True or False:* Because horse manure is an excellent medium for the microorganism that causes tetanus, and the tetanus-causing microorganism is the same for horses and humans, horses and those who work with horses should be immunized against tetanus.

16. How can internal and external parasites of horses be controlled?

17. What are four common internal parasites of horses that should be controlled?

18. Describe the common vaccinations given to horses.

SELECTED REFERENCES _____

Ambrosiano and Harcourt. 1989. *Complete Plans for Building Horse Barns Big and Small.* Grand Prairie, TX: Equine Research, Inc.

Battaglia, R. A., and V. B. Mayrose. 1981. *Handbook of Livestock Management Techniques.* New York: Macmillan.

Cunha, T. J. 1991. *Horse Feeding and Nutrition.* 2d ed. San Diego: Academic Press, Inc.

Ensminger, M. E. 1990. *Horses and Horsemanship.* Danville, IL: Interstate Publishers, Inc.

Ensminger, M. E., J. E. Oldfield, and W. W. Heinemann. 1990. *Feeds and Nutrition.* Clovis, CA: Ensminger Publishing Co.

Evans, J. W. 1989. *Horses: A Guide to Selection, Care, and Enjoyment.* San Francisco: W. H. Freeman.

Evans, J. W., A. Borton, H. F. Hintz, and L. D. Van Vleck. 1990. *The Horse.* 2d ed. San Francisco: W. H. Freeman.

Henneke, O. R. 1983. An objection method for judging a horse's body condition. *Equine Veterinary Journal,* 371–372.

Hickman, J. 1987. *Horse Management.* San Diego: Academic Press, Inc.

Loving, N. S. 1993. *Veterinary Manual for the Performance Horse.* Grand Prairie, TX: Equine Research, Inc.

National Research Council. 1989. *Nutrient Requirements of Horses.* Washington, DC: National Academy Press.

Shideler, R. K., and J. L. Voss. 1984. *Management of the Pregnant Mare and Newborn Foal.* Colorado State University Expt. Sta. Spec. Series 35.

USDA. 2006. *Equine 2005—Baseline Reference of Equine Health and Management.* Washington, DC: National Animal Health Monitoring Systems: APHIS: VS.

Voss, J. L., and B. W. Pickett. *Reproductive Management of the Broodmare.* Colorado State University.

CHAPTER 35

Poultry Breeding, Feeding, and Management

LEARNING OBJECTIVES

- Describe the traditional breeds of poultry and their distinguishing characteristics
- Describe the productivity of turkeys and broilers
- Compare the use of genetic selection and crossing in the poultry industry to other livestock enterprises
- Describe incubation management
- List the primary facility and environmental management concerns in poultry enterprises
- Discuss management of laying hens
- Discuss the factors affecting profitability of poultry production
- Explain vertical integration in the poultry industry

The poultry industry in the United States has been transformed into a dynamic, fully coordinated system of production, processing, and marketing. The distribution of cash receipts from the production of poultry is 62%, 21%, 14%, and 3% for broilers, eggs, turkeys, and others (geese, ducks, etc.).

BREEDS AND BREEDING

The term *poultry* applies to chickens, turkeys, geese, ducks, pigeons, peafowls, and guineas. The head and neck characteristics that distinguish several poultry types are shown in Figure 35.1. Turkeys have some unusual identifying

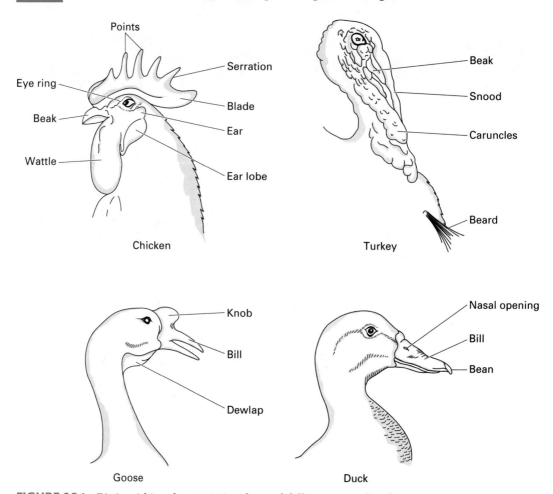

FIGURE 35.1 Distinguishing characteristics of several different types of poultry.

characteristics including a beard (a black lock of hair on the upper chest of the male turkey). They also have a **caruncle**—a red-pinkish fleshlike covering on the throat and neck, with the **snood** hanging over the beak.

Characteristics of Breeds

Chickens. Chickens are classified according to class, breed, and variety. A **class** is a group of birds that has been developed in the same broad geographical area. The four major classes of chickens are American, Asiatic, English, and Mediterranean. A **breed** is a subdivision of a class composed of birds of similar size and shape. Some important breeds, strains, lines, and synthetics are shown in Figure 35.2. The parts of a chicken are shown in Figure 35.3. A **variety** is a subdivision of a breed composed of birds of the same feather color and type of comb.

Factors such as egg number and size, eggshell quality, efficiency of production, fertility, and hatchability are most important to commercial egg producers. Broiler producers consider such characteristics as white plumage color and picking quality, egg production, fertility, hatchability, growth rate, carcass quality,

FIGURE 35.2 Traditional chicken breeds included layers such as the (A) Rhode Island Red and (B) Plymouth Rock, as well as broilers such as the (C) New Hampshire and (D) Cornish. Modern breeding methods have created specialized lines such as the (E) Ross Male, (F) Ross Female, and (G) Ross Broiler. These modern methods have yielded a more consistent product that better meets the needs of consumers. A–D courtesy of Watt publishing Company. E–G courtesy of Ross Breeders, Inc.

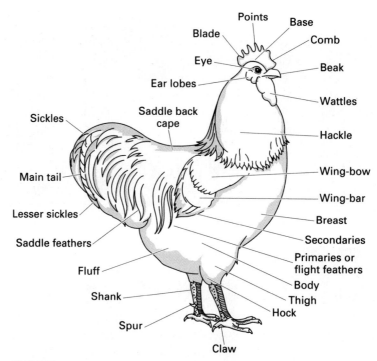

FIGURE 35.3 The external parts of a chicken.

TABLE 35.1   Certain Breeds of Chickens and Their Main Characteristics

Breed	Purpose	Type of Comb	Color of Egg
American Breeds			
White Plymouth Rock	Eggs and meat	Single	Brown
Wyandotte[a]	Eggs	Rose	Brown
Rhode Island Red	Eggs	Single and rose	Brown
New Hampshire	Eggs and meat	Single	Brown
Asiatic Breeds			
Brahma[a]	Meat	Pea	Brown
Cochin[a]	Meat	Single	Brown
English Breeds			
Australorp	Eggs	Single	Brown
Cornish	Meat	Pea	Brown
Orpington[a]	Meat	Single	Brown
Mediterranean Breed			
Leghorn	Eggs	Single and rose	White

[a] These breeds are of minor importance to the U.S. poultry industry.

feed efficiency, livability, and egg production. Also, breeds, strains, and lines that cross well with each other are important to broiler breeders.

The breeds of chickens listed in Table 35.1 were developed many years ago. These specific breeds are not easily identified in the commercial poultry industry because the breeds have been crossed to produce different varieties and strains (see Fig. 35.2). Some breeds are exhibited at shows or propagated for specialty marketing and are more a novelty than part of today's poultry industry.

FIGURE 35.4 Male and female white turkeys that are nearing maturity. Courtesy of Hubbard ISA.

Turkeys.   Turkeys are categorized into eight varieties, but most turkeys grown for meat are white. The eight varieties are (1) Bronze, (2) Narragansett, (3) White Holland, (4) Black, (5) Slate, (6) Bourbon Red, (7) Small Beltsville White, and (8) Royal Palm. Two types of turkeys are commercially important in the United States today: the Small White and the large Broad White. Since the 1950s, the emphasis has been on producing larger, white turkeys of the type shown in Figure 35.4. A corresponding decrease in the number of Small Whites has occurred.

Ducks.   Ducks are classified into many domestic breeds but only a few are of economic importance. The most popular breed in the United States is the White Pekin. It is most valuable for its meat, and produces excellent carcasses at 7–8 weeks of age. Another breed, the Khaki Campbell, is used commercially in some countries for egg production.

Geese.   Breeds of geese such as the Embden, Toulouse, White Chinese, and Pilgrim are all satisfactory for meat production. The characteristics preferred by most commercial goose producers include a medium-sized carcass, good livability (low mortality), rapid growth, and a heavy coat of white or nearly white feathers. The Embden and White Chinese breeds meet these requirements. One variety of the Toulouse is gray; another variety is buff. The Pilgrim gander is white; the female is grayish. The White Chinese breed is pure white.

Breeding Poultry

Turkeys.   There is a specific breeding cycle in the production of baby poults for distribution to growers who raise them to market weight. The pedigree flocks (generation 1) or parent stocks are pure lines. The pure lines are crossed with one another to produce generation 2. Generation 2 lines are crossed with one another to produce generation 3. Eggs from these line crosses are selected into female or male lines and sent to hatcheries. Males from the male lines are selected for such meat traits as thicker thighs, meatier drumsticks, plumper breasts, faster rate of growth, and higher feed efficiency. The females of the female line are selected for greater fertility, hatchability, egg size, and meat conformation. The male-line males are crossed with female-line females; the eggs produced are hatched and the poults are sold for the production of market turkeys.

FIGURE 35.5 A flock of young pullet turkey hens in the breeding program. Courtesy of Hubbard ISA.

Laying hens are approximately 30 weeks of age when they reach sexual maturity (Fig. 35.5). They are put under controlled lighting to stimulate the start of laying. The average hen usually lays for about 25 weeks and produces 88–93 eggs. The hen is considered "spent" at the end of the 25-week laying cycle, and most hens are marketed at that time for meat. Hens can be molted and stimulated to lay for another 25-week cycle. They require 90 days of molting and produce only 75–80 eggs during this later laying cycle; therefore, producers should compare the costs of growing new layers with that of recycling the old ones before deciding if it is economical to recycle old layers. The spent hens usually go into soup products.

Turkey eggs are not produced for human consumption because it costs much more to produce food from turkey eggs than from chicken eggs. It would cost about 50 cents each to produce turkey eggs for this purpose.

All turkey hens are artificially inseminated to obtain fertile eggs. As turkeys have been selected for such broad breasts and large mature size, it is difficult for toms to mate successfully with hens. Semen is collected from the toms and used for inseminating the hens. Usually 1 tom to 10 hens is sufficient for providing the semen needed.

The present-day turkey is much different from the turkeys of many years ago. Today's turkey grows more rapidly, converts feed into meat more efficiently, has more edible meat per unit of weight, and requires less time to reach market. Most turkeys grown today are white-feathered, so there are no dark pin feathers to discolor the carcass. Breeders have made substantial genetic progress in the turkey industry (Fig. 35.6). Over the past 30 years, the market weight of turkeys has nearly doubled while substantially decreasing the time to market and the feed required to achieve a pound of weight gain (Table 35.2).

Chickens.   Sophisticated selection and breeding methods have been developed in the United States for increased productivity of chickens for both eggs and meat. For example, since the late 1950s, breeders have increased 56-day body weight by over 4 lb. Improvements in feed efficiency, total yield, and breast yield have also been remarkable. The discussion here is directed primarily at improvement

FIGURE 35.6 Efficient feed conversion is an economic advantage for the turkey industry. *Source:* Courtesy of Grant Heilman Photograph, Inc.

TABLE 35.2 Average Live Weights and Feed Efficiencies of Large White Toms and Hens

Year	Live Weight at 18 Weeks		Feed/Gain 0–18 Weeks		Days to 35 lb Market Weight	
	Toms	Hens	Toms	Hens	Toms	Hens
1970	16.9	9.0	3.10	3.10	235	151
1975	21.3	11.7	2.80	2.40	194	144
1980	22.4	12.2	2.70	2.30	185	135
1985	23.0	12.3	2.74	2.45	175	130
1990	27.1	14.2	2.80	2.44	156	107
1995	29.3	15.1	2.69	2.35	149	103
2000	31.9	15.5	2.62	2.36	143	100
2002	32.2	15.5	2.61	2.30	136	99

Source: Multiple sources.

of chickens; however, methods described for chickens can also be applied to improve meat production in turkeys, geese, and ducks. The genome of the chicken is carried on 39 chromosomes with more than 1 million base pairs. The sequencing of the chicken genome in 2004 paves the way for breeders to more effectively select for traits that have been difficult or costly to measure such as disease resistance. Furthermore, breeders will be able to better match populations of birds to specific environmental and management conditions.

Early poultry breeding and selection were concentrated on qualitative traits, which, from a genetic standpoint, are more predictable than quantitative traits. **Qualitative traits,** such as color, comb type, abnormalities, and sex-linked characteristics, while important, are less important than **quantitative traits,** such as egg production, egg characteristics, growth, fertility, and hatchability.

Quantitative traits are more difficult to select for than are qualitative traits because the mode of inheritance is more complex and the role of the environment is greater. Quantitative traits differ greatly as to the amount of progress that can be attained through selection. For example, increased body weight is much easier to attain than increased egg production. Furthermore, a relationship usually exists between body size and egg size. Generally, if body size increases, a corresponding increase in egg size will occur. Most quantitative traits of chickens fall in the low-to-medium range of heritability, whereas qualitative traits fall in the high range.

Progress in selecting for egg production has been aided by the **trap nest**; a nest equipped with a door that allows a hen to enter but prevents her from leaving. The trap nest enables accurate determination of egg production of individual hens for any given period and helps to identify and eliminate undesirable egg traits and broodiness (the hen wanting to sit on eggs to hatch them). Furthermore, it allows the breeder to begin pedigree work within a flock.

The U.S. broiler industry is an example of the phenomenal changes that can result from the application of genetic selection pressure and improved management (Table 35.3). The dramatic improvements in live weight, feed conversion, survivability, and time to market have been largely responsible for the remarkable growth of chicken's market share. From 1945 to 1995 the cost (adjusted for inflation) of producing live broilers declined from $2 per pound to approximately $0.25 per pound. This dramatic improvement can be attributed to the broiler multiplication process as described in Table 35.4. The global value of the grandparent, parent, and broiler inventories is estimated at $275 million, $1 billion, and $60 billion, respectively. The foundation and great grandparent stock are so valuable as to elude estimation.

TABLE 35.3   Average Performance of Broilers

Year	Average Live Wt.	Feed Conversion	% Mortality	Age in Days
1925	2.2	4.7	18	112
1935	2.6	4.4	14	98
1945	3.1	4.0	10	84
1955	3.3	3.0	7	70
1965	3.5	2.4	6	63
1975	3.7	2.1	5	56
1985	4.2	2.0	5	49
1995	4.7	1.95	5	47
2000	5.0	1.95	5	46
2005	5.2	1.9	4.5	46

Source: USDA.

TABLE 35.4   Broiler Multiplication Process

Generations	Population Size	Generation Goal
Pedigree (F)	90,000 hens	Purelines subjected to heavy culling pressure, foundation stock that is never sold.
Great grand-parents (GGP)	250,000 hens	Some selection pressure but mass production is goal. These stocks are also not sold to protect competitive advantage.
Grandparents (GP)	10,000,000 hens	Sold as baby chicks in global market.
Parents (P)	400,000,000 hens	Day-old parents are sold by a variety of companies in a broad market.
Broilers (B)	40,000,000,000/yr	Produced for consumption.

Source: Adapted from multiple sources.

TABLE 35.5   Big 4 Broiler Genetics Companies

Company	% Market Share
Aviagen[a]	44
Cobb-Ventress (owned by Tyson)	33
Hubbard/ISA	10
Hybro	5
Others	8

[a] Formed by merger of Ross, Arbor Acres, and Lohmenn Indian River.
Source: Adapted from multiple sources.

FIGURE 35.7 A broiler breeding pen containing 1 male and 12 females. This is a typical breeding unit. Courtesy of Hubbard ISA.

The competitive nature of the genetics business has led to a high degree of market concentration. The top two genetic providers control more than 75% of the market while the top four combine to control just over 90% (Table 35.5).

Genetic providers in the future will have to refocus their emphasis on those traits that directly affect consumer demands and perception.

Consumers will continue to demand products that are affordable but also deliver superior palatability while meeting societal expectations in regards to food safety, environmental impact, and animal welfare. For example, selection emphasis will be placed on improved bird health, reduced metabolic dysfunction, improved skeletal structure, and limiting pecking behavior to meet welfare concerns.

The two most important types of selection applied primarily to chickens and, to a lesser extent, to other poultry today are mass selection and family selection. In **mass selection,** the older method, the best-performing males are mated to the best-performing females (Fig. 35.7). The program is effective in improving traits of high heritability. **Family selection** is a system whereby all offspring from a particular mating are designated as a family, and selection and culling within a population of birds are based on the performance level of the entire family. This type of selection is most adaptable for traits of low heritability. This is not to say that progress in traits of high heritability cannot be made by using family selection—quite the opposite is true. However, mass selection is effective for traits of high heritability and certainly easier.

Progress through selection is rather slow for the reproductive traits because they are low in heritability. Most breeding programs are geared toward the improvement of more than one trait at a time. For example, to improve egg-laying lines, it might be necessary to attempt simultaneously to improve egg numbers, shell thickness, interior quality of the egg, and feed efficiency.

Outcrossing. Defined as mating unrelated breeds or strains, **outcrossing** is probably more adaptable to the modern broiler industry than to the egg-production industry, though it can be used to improve egg production. Numerous experiments have established fairly accurately which breeds or strains cross well with each other. Knowledge today is sophisticated to the extent that breeders know which strain or line to use as the male line and which to use as the female line.

Crossing Inbred Lines. Development of **inbred lines** for crossing is used largely for increasing egg production. *Inbreeding* is defined as the mating of related individuals or as the mating of individuals more closely related than the average of the flock from which they originated. The mating of brother with sister is the most common system of inbreeding for poultry; however, mating parents with offspring gives the same results. Both of these types of inbreeding increase the degree of inbreeding at the same rate per generation.

Hybrid chickens are produced by developing inbred lines and then crossing them to produce chickens that exhibit hybrid vigor. The technique employed in production of hybrid chickens is essentially the same as that used in development of hybrid plants. The breeder works with many egg-laying strains or breeds while producing hybrids. The strains or breeds are inbred (mostly brother or sister) for a number of generations, preferably five or more. In the inbreedng phase, many undesirable factors are culled out—selection is most rigid at this stage.

In all phases of hybrid production, the strains are completely tested for egg production and for important egg-quality traits. After the birds are inbred for the necessary number of generations, the second step—crossing the remaining inbred lines in all possible combinations—is initiated. These crossbreds are tested for egg production and for egg-quality traits. Crossbreds that show improvement in these traits are bred in the third phase, in which the best test crosses are crossed into three-way and four-way crosses in all possible combinations. Most breeders prefer the four-way cross. The offspring that result from three-way and four-way crosses are known as **hybrid chickens.**

Strain Crossing. **Strain crossing** is more easily done than inbreeding and crossing inbred lines because it entails only the crossing of two strains that possess similar egg-production traits. It is definitely a type of crossbreeding if the two strains are not related to each other.

Selection is oftentimes based on the computation of a net merit index utilizing the heritability of each trait considered, the relative economic importance of each trait, and genetic correlations among the traits. Birds having the most desirable index are used for breeding.

In production of broiler or layer hybrid chicks, computational software is used to determine which lines or strains are most likely to yield superior chickens. In broiler hybrids, hatchability, growth rate, economy of feed use, and carcass desirability are important. In layer hybrids, hatchability, egg production, egg size, egg and shell quality, and livability are important.

Practically all line-cross layers are now free from **broodiness** because attempts have been made to eliminate genes for broodiness, and records are available on which lines produce broody chicks when crossed. Lines that produce broody chicks in a line cross are no longer used for producing commercial layers.

FEEDING AND MANAGEMENT

The success of modern poultry operations depends on many factors. Hatchery operators must care for breeding stock properly so that eggs of good quality are available to the hatchery, and they must incubate eggs under environmental conditions that ensure the hatching of healthy and vigorous birds (Fig. 35.8). Feeding, health, and financial programs are essential to sound poultry management. The broiler and egg production cycles are provided in Figures 35.9 and 35.10, respectively.

Incubation Management

Each type of poultry has an **incubation** period of definite length (Table 35.6), and incubation management practices are geared to the needs of the eggs throughout that time. The discussion here of incubation management applies specifically to chickens. Although the principles of incubation management generally apply to other types of poultry as well, specific information should be obtained from other sources. Egg production varies by species with egg-type hens producing 240 per year, broiler-type hens yielding 170 eggs annually, and turkey hens producing 105 eggs per year.

FIGURE 35.8 Removal of a batch of broiler chickens from the incubator. *Source:* Tyson Foods, Inc.

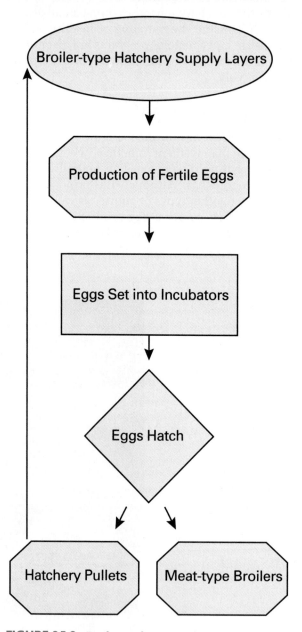

FIGURE 35.9 Broiler production cycle.

All modern commercial hatcheries have some form of forced-air incubation system. Today's commercial incubator has a forced-air (fan) system that creates a highly uniform environment inside the incubator. Forced-air incubators are available in many sizes. All sizes are equipped with sophisticated systems that control temperature and humidity, turn eggs, and bring about an adequate

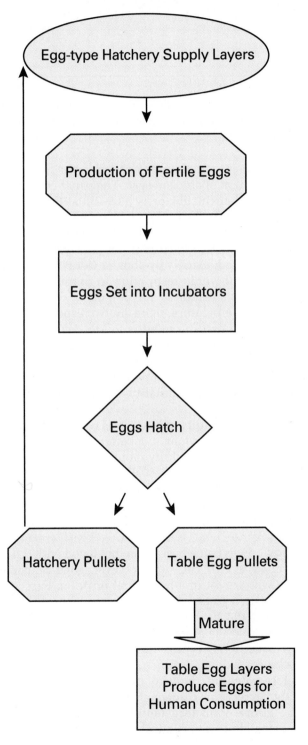

FIGURE 35.10 Egg production cycle.

TABLE 35.6 Incubation Times for Various Birds

Type	Incubation Period (days)
Chicken	21
Turkey	28
Duck	28
Goose	28–32
Pheasant	24
Bobwhite quail	23
Guineas	27

exchange of air between the inside and outside of the incubator. The egg-holding capacity of forced-air incubators ranges from several hundred eggs to 100,000 or more.

Temperature. Proper temperature is probably the most critical requirement for successful incubation of chicken eggs. The usual beginning incubation temperature in the forced-air incubator is 99.5–100.0°F. The temperature should usually be lowered slightly (by 0.25–0.5°F) at the end of the fourth day and remain constant until the eggs are to be transferred to the hatching compartments (approximately 3 days before hatching). Three days before hatching, the temperature is again lowered, usually by about 1.0–1.5°F. Just before hatching, the chicks switch from embryonic respiration to normal respiration and give off considerable heat, which results in a high incubator temperature.

With reference to temperature, there are two especially critical periods during incubation—(1) the first through the fourth day and (2) the last portion of incubation. Higher-than-optimum temperatures usually speed the embryonic process and result in embryonic mortality or deformed chicks at hatching. Lower-than-optimum incubation temperatures usually slow the embryonic process and also cause embryonic mortality or deformed chicks.

Humidity. A relative humidity of 60–65% is needed for optimum **hatchability.** All modern commercial forced-air incubators are equipped with sensitive humidity controls. The relative humidity should usually be raised slightly in the last few days of incubation, as it has been shown that successful hatching rates are greater when the relative humidity is close to 70% in the last few days of incubation. Providing optimum relative humidity is essential in reducing evaporation from eggs during incubation.

Position of Eggs. Eggs hatch best when incubated with the large end (the area of the space called the *air cell*) up; however, good hatchability can also be achieved when eggs are set in a horizontal position. Eggs should never be set with the small end up because a high percentage of the developing embryos will die before reaching the hatching stage. The head of the developing embryo should be situated near the air cell. In the last few days of incubation, the eggs are transferred to a different type of tray—one in which the eggs are placed in a horizontal position.

Shortly before hatching, the beak of the chick penetrates the air cell. The chick is now able to receive an adequate supply of air for its normal respiratory processes. The horny point (egg tooth) of the beak eventually weakens the eggshell until a small hole is opened, at which point the egg is said to be *pipped.*

The chick is typically out of the shell within a few hours. The hatching process varies considerably among different species.

Turning of Eggs.   Modern commercial incubators are equipped with time-controlled devices that permit eggs to be turned periodically. Eggs that are not turned enough during incubation have little or no chance of hatching because the embryo often becomes stuck to the shell membrane.

Commercial incubators are equipped with setting trays or compartments that allow eggs to be set in a vertical position. They also have mechanisms that rotate these trays so that the chicken eggs rotate 90° each time they are turned. The number of times the eggs are turned daily is most important. If the eggs are rotated once or twice daily, the rate of hatchability will be much lower than if the eggs are rotated five or more times daily.

Oxygen Content.   The air surrounding incubating eggs should be 21% oxygen by volume. At high altitudes, however, the available oxygen in the air may be too low to sustain the physiological needs of developing chick embryos, many of which may therefore die. Good hatchability can often be attained at high altitudes if supplemental oxygen is provided. This is necessary for commerical turkey hatching, but hatchability can be improved by selecting breeding birds that hatch well in environments in which the supply of oxygen is limited.

Carbon Dioxide Content.   It is vital that the incubator be properly ventilated to prevent excessive accumulation of carbon dioxide (CO_2). The CO_2 content of the air in the incubator should never be allowed to exceed 0.5% by volume. Hatchability is lowered drastically if the CO_2 content of air in the incubator reaches 2.0%. Levels of more than 2.0% almost certainly reduce hatchability to near zero.

Sanitation.   The incubator must be kept as free of disease-causing microorganisms as possible. The setting and hatching compartments must be thoroughly washed or steam-cleaned and fumigated between settings. In many cases, they should be fumigated more than once for each setting. An excellent schedule is to fumigate eggs immediately after they have been set and again as soon as they have been transferred to the hatching compartment.

Candling of Eggs.   In maintaining a healthy and germ-free environment, eggs must be **candled**—examined by shining a light through each egg to see if a chick embryo is developing—at least once during incubation so that infertile and dead-germ eggs (eggs containing dead embryos) can be identified. Many operators candle chicken eggs on the fourth or fifth day and turkey and waterfowl eggs on the seventh to tenth day. Some operators candle eggs a second time when transferring eggs to the hatching compartment.

Managing Young Poultry

The main objective in managing young poultry is to provide a clean and comfortable environment with sufficient feed and water (Fig 35.11).

A

B

FIGURE 35.11   (A) A commercial poultry house containing young broilers. (B) Environmental conditions are well controlled for these many hundred birds by virtue of an automated system. Photo A courtesy of Big Dutchman, Inc.

Housing for Broilers

Broilers raised in the South are typically housed in long (> 500 ft long × 50 ft wide) housing arrangements. Designs vary from totally enclosed walls to siding provided via curtains that can be moved in accordance with weather conditions. Flooring is usually dirt, but concrete or asphalt may be used. Most are designed to use large fans for increased ventilation. Broiler houses located in northern regions tend to be fully insulated, windowless structures with concrete floors.

Brooders, water dispensers, and feeding systems are usually suspended from the ceiling so that they can be lifted out of the way for bird penning or house cleaning. Feed and water is provided via a fully or semiautomated system on a continuous basis. Vaccinations and medications are typically provided via water or feed.

A typical broiler farm will consist of 159 acres with a small portion allotted to the broiler enterprise. The average farm has 3.2 houses with a total bird capacity of 64,000 birds housed at a density of 0.5 to 1 sq ft per bird. The broiler flock turns over 6.5 times per annum on average for a total annual production of more than 410,000 birds per farm.

House Preparation.   The brooder house should be thoroughly cleaned, disinfected, and dried several days before it receives young birds. All necessary brooding equipment should tested ahead of time and be in the proper place. Such advance preparation is probably as essential to success as any management practice applied to young birds. It is imperative that the disinfectant have no effect on the meat or eggs produced by birds. Food and Drug Administration regulations that govern the use of disinfectants should be strictly observed.

Litter.   **Litter** should be placed in the house when equipment is checked. Some commonly used substances are planer shavings, sawdust, wood chips, peat moss, ground corncobs, peanut hulls, rice hulls, and sugarcane fiber. However, the use of recycled paper litter is replacing some of the more traditional types as it provides improved control of ammonia, fecal caking, litter compaction, and enhances bird welfare. The entire floor should be covered by at least 2 in. of litter, which must be perfectly dry when the birds enter.

The primary purpose of litter is to absorb moisture. Litter that gets extremely damp should be replaced. Moisture levels in litter should not exceed 35%.

Floor Space.   The availability of too much or too little **floor space** can adversely affect growth and efficiency of production. If floor space is insufficient, young birds have difficulty finding adequate feed and water. This can lead to feather picking and actual cannibalism. Too much space can cause birds to become bored, which can lead to problems similar to those caused by overcrowding. However, birds should be allowed freedom of movement in the growing area.

Broilers below 4.5 lb of live weight should have a maximum stocking density of 6.5 lb per square foot. Birds weighing between 4.5 to 5.5 lb and those exceeding 5.5 lb should have a stocking density that doesn't exceed 7.5 and 8.5 lb per square foot, respectively.

Broilers, turkeys, and waterfowl should have an area of 110–145 in.² for at least their first 8 weeks. In certain situations, such birds could be allowed as much as 145–220 in.² before they are 8 weeks old.

Feeder Space. Chick trough feeders should be placed within the hover guard area or partially under the hover (0.8 in. per chick or 80 in. per 100 chicks to 6 weeks of age). Producers should be sure the 10-watt attraction lights under the hover are working. One linear inch of **feeder space** per bird is sufficient for most species from the age of 1 day to 3 weeks; 2 in. for 3–6 weeks; and 3 in. beyond 6 weeks. Quail require less space, whereas turkeys, geese, and ducks require slightly more. Most feeders are designed so that birds can eat on either side of them.

The operator can determine how much feeder space is needed through observation. If relatively few birds are eating at one time, the feeding area is probably too large. If too many are eating at once, the area is probably too small. Young birds are usually fed automatically after the brooder stove enclosure is removed.

Water Requirements. Three fountain-type waterers should be placed within the hover guard area (one fountain-type waterer per 100 chicks). Generally, two 1-gal water fountains are adequate for 100 birds that are 1 day old. More waterers can be added as necessary. Most commercial producers switch from water fountains to an automatic watering system when birds are put on automatic feeding. If trough-type waterers are used, allow at least 1 in. of water space per bird until they reach 10–12 weeks of age. More space might be needed thereafter. Trough-type waterers designed so that birds can drink from both sides are available. A space of at least 0.5 in. per bird is needed if pan-type waterers are used.

Lighting Requirements. Manipulation of lighting is an important management tool to help control the growth rate of broilers. With the intense selection for improved muscularity of broilers, birds early in the growth curve partition nutrients to muscle growth at the expense of other tissues especially skeletal and internal organ systems. Lighting systems coupled with low-density feed and moderate feed restrictions are useful in overcoming this problem.

Two lighting systems are typically used in the industry—16 hours light with 8 hours of dark or increasing light protocols. The lighting system must be matched with housing design, feeding program, market weight goals, and grower contract requirements. Most programs are initiated on day 3 following receipt of chicks. Except for the first week and the final 14 days of growout, at least 4 hours of darkness should be provided in every 24-hour period during which illumination at the bird level does not exceed 50% of the light level in the remaining hours.

Other Management Factors. Temperatures in the hatchery holding room should be at least 72°F. Young birds should have access to the proper feed as soon as they are placed in the brooder house. It is important that birds such as turkey poults start eating feed early. Some young poults die for lack of feed; although feed is readily available, they have not learned to eat it.

Birds should be vaccinated and given health care according to a prescribed schedule; a veterinarian should be consulted.

Managing 10- to 20-Week-Old Poultry

Management of poultry from approximately 10–20 weeks of age is quite different from managing younger birds. Improper practices in this most critical period could adversely affect subsequent production.

Confinement rearing is used by commercial producers of replacement birds (the birds that will be kept for egg production). In the three basic systems of confinement rearing, birds are raised on solid floors, slatted floors, or wire floors or cages.

Birds raised on floors in confinement should have from 1–2 ft^2 of floor space per individual. The amount of space required depends on the environmental conditions and on the condition of the house. Caged birds should be allowed 0.5 ft^2 to not more than 0.75 ft^2. Air quality is also of concern and ammonia levels should not be allowed to exceed 25 parts per million. Dust is also of concern to both the health of the flock and the people responsible for their care. The use of electrostatic space charge systems are one new approach to improving air quality in poultry houses.

Replacement chickens raised in confinement are generally fed a completely balanced growing, or developing, ration that is 15–18% protein. Most birds are kept on a full-feeding program, but in certain conditions some producers restrict the amount of feed provided. Restricted feeding can take different forms: total feed intake, protein intake, or energy intake may be limited. The main purposes of a restricted feeding program are to slow growth rate (thus delaying the onset of sexual maturity and reducing the number of small eggs produced) and to lower feeding cost. Feed intake must not be restricted whenever a restricted lighting system (which is also used to delay sexual maturity) is in effect.

Automatic feeding and watering devices are used in most confinement operations. Watering- and feeding-space requirements are practically the same as those for birds that are 6–10 weeks old. As long as birds are not crowding the feeders and waterers, there is no particular need to increase the feeding and watering space.

Water consumption is of critical importance in assuring that feed consumption and bird growth rates are optimized. Most commercial growers monitor water consumption on a house-by-house basis if not on a pen-by-pen level. Water consumption becomes of particular concern as ambient temperature rises. For reasons of water conservation and labor efficiency, the use of nipple-type watering systems is most prevalent.

Lighting conditions are of particular concern in regards to poultry. Light plays three important roles in poultry management—stimulation of physiological cycles that are light dependent, initiation of hormone release, and impact on bird vision.

Feed intake can be enhanced with a variety of specific protocols:

1. Lighting practices of 23 hr light: 1 hr dark or an intermittent lighting of 1 hr light: 3 hr dark with an intensity of 4–5 ft-candles will reduce bird activity and raise consumption.

2. Maintain sufficient floor space per bird to enhance intake.

3. Assure that feed grains are free of mold and other contamination.

4. Monitor the house for unproductive birds and remove them to improve overall flock efficiency.

The lighting regime used in confinement rearing is very important. Birds raised in window-type housing receive the normal light of long-day periods unless the house is equipped with some type of light-check. During short-day periods, supplemental lighting can be used to meet the requirements of growing chickens. Replacement chickens should receive approximately 14 hours of light daily up to 12 weeks of age. To delay sexual maturity, the amount should then be reduced to about 8–9 hours daily until the birds have reached 20–22 weeks of age. The light is then either abruptly increased to 16 hours per day, or it is increased by 2–3 hours in weekly increments of 15–20 minutes added until 16 hours of light per day are reached.

Regardless of which system replacement pullets are reared under, they should be placed in the laying house when they are approximately 20 weeks of age so they can adjust to the house and its equipment before beginning to lay.

Management of Laying Hens

Prior to World War II, large numbers of small flocks of hens were the foundation of egg production. Average production per year was 100 eggs or fewer, hens suffered mortality rates approaching 40%, and the flock was subjected to disease, climatic variability, predation, and a host of other stressors. Management of the laying hen has changed significantly with the increased level of specialization in agriculture. For example, historically hens laid a clutch of eggs and then set with them until hatching. The hen and her chicks ate the same diet and by virtue of shared living space the offspring developed immunity quickly. Under modern production practices, the hen lays eggs into a clean nest, the eggs are gathered, disinfected before being set into a sterile hatchery environment, and then moved into relatively clean poultry houses as chicks.

Consolidation of hatcheries has been significant with fewer than 310 enterprises in business today as compared to approximately 5,000 in the late 1950s. However, the capacity of the U.S. hatchery complex has grown during the same time period.

The requirements of laying hens for floor space varies from 1.5–2 ft² per bird for egg-production strains and from 2.5–3.5 ft² for dual-purpose and broiler strains. Turkey breeding hens require 4–6 ft², but less area is needed if hens are housed in cages. If turkey hens have access to an outside yard, 4 ft² of floor space is best. Game birds such as quail and pheasants require less floor space than egg-production hens. In general, 3 ft² or more of floor space is adequate for ducks, whereas geese need approximately 5 ft².

Breeder hens are housed on litter or on slatted floors, whereas birds used for commercial egg production are typically kept in cages (Fig. 35.12). Free-ranging laying hens are housed less intensively (Fig 35.13).

Floor-type houses for breeder hens usually have 60% of the floor space covered with slats. Slatted floors are usually several feet above the base of the building and manure is often allowed to accumulate for a long time before being removed. Many slatted-floor houses are equipped with mechanical floor scrapers that remove the manure periodically. Frequent removal of manure lessens the chance that ammonia will accumulate.

Gathering of eggs in floor-type houses can be done automatically if some type of roll-away nesting equipment is used (Fig. 35.14). In houses equipped with individual nests, the eggs are gathered manually. Some operators prefer colony nests (open nests that accommodate several birds at a time).

FIGURE 35.12 Laying hens in a modern facility that utilizes stacked cages. Courtesy of Big Dutchman, Inc.

Some type of litter or nesting material must be placed on the bottom of individual and colony nests if eggs are to be gathered manually. Birds should not be allowed to roost in the nests in darkness because dirty nests result.

Cage operations are used by most commercial egg-producing farms. Young chickens may be brooded in colony cages, whereas laying hens may be housed in 4–6 tiers of decked laying cages. A range of 67–86 square inches of space per hen is recommended. The lower end of the range is appropriate for housing smaller Leghorn strains in shallow design cages while Brown hens in deep cages require more space. Protection from predation and environmental extremes are the primary objectives of poultry housing, and management should focus on minimizing the spread of pathogens and parasites.

All cages are equipped with feed troughs that usually extend the entire length of the cage allowing all of the hens to eat at one time. Some troughs are filled with feed manually, others automatically. It is sometimes advantageous to dub (remove the combs and wattles from) caged layers so they can obtain feed from automatic feeders by reaching their heads through the openings. Additional considerations include assuring that manure cannot drop from one cage onto others, cage floor slopes should not exceed 8 degrees, and hens should be able to stand comfortably upright. To assure favorable conditions for visual inspection of birds, light intensity at 0.5 to 1-ft candle should be provided at feeding levels of the facility.

Several types of watering devices (trough, nipple-type, or individual cup-type waterers) are available for use in individual and colony cages. The watering

FIGURE 35.13 Laying hens in a free-ranging facility.

system should be equipped with a metering device that makes it possible to medicate the birds quickly when necessary by mixing the exact dosage of medication required with the water. Recommendations for watering space are provided in Table 35.7.

The arrangement of cages within a house varies greatly. Some producers have a step-up arrangement, such as a double row of cages at a high position, with a single row at a low position on either side. Droppings from birds caged in the double row fall free of the birds in the single row. An aisle approximately 3 ft wide is usually located between each group of cages. Rather than aisles, some systems have a movable ramp that can travel above the birds from one end of the house to the other. In other systems, several double cages are stacked on top of each other.

In cage operations, manure is typically collected into pits below the cages and removed by mechanical pit scrapers or belts. In stacked housing arrangements, the manure is moved via a conveyor system from each stack to a central collection point (Fig. 35.15). Manure may also drop into pits that are slightly sloped from end to end and partially filled with water. The manure can be flushed out with the water into lagoons.

An alternative to cage housing is the use of free-ranging barns that allow birds more freedom of movement and the opportunity to use "brood houses" for laying. Feed and water are delivered via automatic systems. Some consumers prefer to purchase products from farms and processors employing these systems.

A

FIGURE 35.14 Egg collectors designed for (A) a stacked housing layer barn and (B) a free-ranging facility. Photo A courtesy of Big Dutchman, Inc.

B

TABLE 35.7 Water Space Requirements for Layer Hens

Age of Bird (wks)	Linear Trough Space/Bird (in.)	Maximum Number of Birds per Cup or Nipple[a]
0–6	0.6	20
6–18	0.8	15
> 18	1.0	12

[a] Perimeter space required for round-type watering systems can be calculated by multiplying linear trough space by 0.8.
Source: Adapted from United Egg Producers' Animal Husbandry Guidelines.

FIGURE 35.15 A manure collection system from a stacked housing system. Manure is continuously removed to help ensure proper ventilation, humidity, and air quality for the birds. Courtesy of Big Dutchman, Inc.

Housing Poultry

Factors such as temperature, moisture, ventilation, and insulation are given careful consideration in planning and managing poultry houses. An important consideration is to assure that standby generators and alarm systems are in place to allow normal house operation during emergency situations.

Temperature. Most poultry houses are built to prevent sudden changes in house temperature. A bird having an average body temperature of 106.5°F usually loses heat to its environment except in extremely hot weather. Chickens perform well in temperatures between 35 and 85°F, but 55–75°F is optimal.

House temperature can be influenced by such factors as the prevailing ambient temperature, solar radiation, wind velocity, and heat production of the birds. A 4.5-lb laying hen can produce nearly 44 **British thermal units (Btu)** of heat per hour (1 Btu is the quantity of heat required to raise the temperature of 1 lb of water 1°F at or near 39°F). The amount of Btu produced varies with activity, egg-production rate, and the amount of feed consumed. Heat production by birds must be considered in designing poultry houses (about 40 of the 44 Btu produced per hour by a laying hen are available for heating). Although heat produced by birds can be of great benefit in severe cold, the house must be designed so that excess heat produced by birds in hot weather can be dissipated.

Moisture. Excess house moisture, especially in cold weather, can create an environment that is extremely uncomfortable, that can lead to a drop in production by laying birds, and that if allowed to continue for long, can cause illness.

Much of the moisture in a poultry house comes from water spilled from waterers, water in inflowing air, and water vapor from the birds themselves and their droppings. Every effort should be made, of course, to minimize spillage from watering equipment. The amount of moisture in a poultry house can be reduced by increasing the air temperature or by increasing the rate at which air is removed. Because air holds more moisture at high temperatures than at low temperatures, raising the air temperature causes moisture from the litter to enter the air and thus be dissipated. Air temperature in the house can be increased by retaining the heat produced by the birds themselves and by supplemental heat.

If incoming air is considerably colder than the air in the house, it must be warmed or it will fail to aid in moisture removal. In cold weather, most exhaust ventilation fans are run more slowly than normal, so most moisture removal is accomplished by increasing the air temperature in the house.

Ventilation. A properly designed **ventilation** system provides adequate fresh air, aids in removing excess moisture, and is essential in maintaining a proper temperature within the house. Type and amount of **insulation** and heat produced by the birds themselves must be considered in planning for ventilation.

Ventilation is accomplished by positive pressure, through which air is forced into the house to create air turbulence, or by negative pressure, through which air is removed from the house by exhaust fans. The positive pressure system is accomplished by having fans in the attic so that air is forced through holes. The negative pressure system is accomplished by locating exhaust fans near the ceiling. Some houses are ventilated so that the air is kept fairly warm and dry by the use of open walls. Also, when outside air is cold and dry, air can enter at low portions of the house and leave through vents near the ceiling as it warms. The heat created by the chickens warms the colder air, which then takes moisture from the house.

Two integral and necessary parts of the most common ventilation systems are the exhaust fan and the air-intake arrangement. The number of exhaust fans varies with size of the house, number of birds, and capacity of the fans. Most fans are rated on the basis of their cubic feet per minute (cfm) capacity, that is, how many cubic feet of air they move per minute. Fans in laying houses should operate at 4–4.5 cfm per bird when the temperature is moderate. In summer, cfm per bird could be as high as 10.

Fans also have a static pressure (water pressure for low house temperatures and a high speed for warm house temperatures). Others operate at only one speed; the amount of air removed is controlled by shutters that open, so that more air can be removed when the temperature of the house exceeds the thermostatically fixed temperature. These shutters then close at lower temperatures.

An air-intake area must be provided for the ventilation system. It is usually a slotted area in the ceiling or near the top of the walls. There should be at least 100 in.[2] of inlet space for each 400 cfm of fan capacity.

Feeds and Feeding

Rations fed to poultry today are complex mixtures that should include, in a balanced proportion, all ingredients that have been found to be necessary for body maintenance, maximum production of eggs and meat, and optimum reproduction (fertility and hatchability).

Maximum growth potential cannot be obtained in broilers and turkeys unless maximum feed intake is coupled with correct diet formulation. Dietary intake is affected by a variety of factors including dietary energy, protein, amino acids, vitamins and minerals, antinutritional factors, and water consumption.

Antinutritional factors such as alkaloids, phytates, and other natural feed components can inhibit nutrient availability, reduce intake, and suppress growth. Meat-type poultry will consume twice as much water as compared to feed on a weight basis. Thus, adequate access to water is critical to feed consumption.

Management factors may also have significant influence on the level of feed intake. The primary management protocols of note are assuring access to water and feed, environmental stress levels, and the degree of disease challenge. Managers must assure adequate feeder and waterer space so that even more submissive birds will have free access to nutrients. Feed and water must be available *ad libitum*. Feed availability that is disrupted for more than four hours can lead to increased susceptibility to enteric disease.

Increased levels of stress will limit feed intake. Careful attention to dealing with heat stress, poor air quality, and poor litter quality will enhance feed intake and consequent growth rate. Disease management is also critical as any factor, including vaccination, that elicits an immune response will also depress feed consumption.

Energy requirements are supplied mainly by cereal grains, grain by-products, and fats. Some important grains are yellow corn, wheat, sorghum grains (milo), barley, and oats. Most rations contain high amounts of grain (60% or higher, depending on the type of ration). A ration containing a combination of grains is generally better than a ration having only one type. Animal fats and vegetable oils are excellent sources of energy. They are usually incorporated into broiler rations or any high-energy rations.

Protein is so highly essential that most commercial poultry feeds are sold on the basis of their protein content. The types of amino acids present determine the nutritional value of protein (see Chapter 15). Excellent protein can be derived from both plants and animals. Most rations contain both plant and animal protein so that each source can supply amino acids that the other source lacks. The most common sources of plant protein are soybean meal, cottonseed meal, peanut meal, alfalfa meal, and corn gluten meal. Cereal grains contain insufficient protein to meet the needs of birds. The best sources of animal protein are fish meal, milk by-products, meat by-products, tankage, blood meal, and feather meal. The protein requirements of birds vary according to species, age, and purpose for which the birds are being raised. The protein requirements of certain birds are shown in Table 35.8. Where variable values are listed, the highest level is to be fed to the youngest individuals. Recommended nutrients for layers are provided in Table 35.9.

TABLE 35.8   Protein Requirements of Poultry

Type	Age (weeks)	Percentage of Protein Required in Diet
Chickens		
Broilers	0–6	20.0–23.0%
Replacement pullets	0–14	15.0–18.0
Replacement pullets	14–20	12.0
Laying and breeding hens		14.5
Turkeys		
Starting	0–8	26.0–28.0
Growing	8–24	14.0–22.0
Breeders		14.0
Pheasant and quail		16.0–30.0
Starting and growing		
Ducks		
Starting and growing		16.0–22.0

Source: National Research Council, *Nutrient Requirements of Poultry*, 8th ed. (Washington, DC: National Academy of Sciences, 1984).

TABLE 35.9   Nutrient Recommendations for Egg-Type Layers

Nutrient (%)	Nutrient Needs (mg per day)	Daily Intake (g per hen)		
		80	100	120
CP	15,000	18.8	15.0	12.5
Arginine	700	0.88	0.70	0.58
Histidine	170	0.21	0.17	0.14
Isoleucine	650	0.81	0.65	0.54
Leucine	820	1.03	0.82	0.68
Lysine	690	0.86	0.69	0.58
Methionine + Cystine	580	0.73	0.58	0.48
Phenylalanine + Tyrosine	830	1.04	0.83	0.69
Threonine	470	0.59	0.47	0.39
Tryptophan	160	0.20	0.16	0.13
Valine	700	0.88	0.70	0.58
Calcium	3,250	4.06	3.25	2.71
Nonphytate Phosphorus	250	0.31	0.25	0.21
Sodium	150	0.19	2.15	0.13
Chlorine	130	0.16	0.13	0.11

Source: NRC, 1994.

Whatever ration is adequate for turkeys is generally suitable for game birds. Some game bird producers feed a turkey ration at all times. Others feed a complete game bird ration. The actual nutrient requirements of game birds are not as well known as are the requirement of chickens and turkeys.

A considerable number of minerals are essential, especially calcium, phosphorus, magnesium, manganese, iron, copper, zinc, and iodine. Calcium and phosphorus, along with vitamin D, are essential for proper bone formation. A deficiency of either of these elements can lead to a bone condition known as **rickets.** Calcium is also essential for proper eggshell formation.

The amount of calcium required varies somewhat with age, rate of egg production, and temperature. Chickens and turkeys up to 8 weeks of age require 0.8–1.2% of calcium in their diets. Calcium can be reduced from 0.6–0.8% in 8- to 16-week-old birds. Laying hens require at least 3.4% calcium. The amounts of phosphorus required for chickens of different ages are as follows: 0–6 weeks, 0.4%; 6–14 weeks, 0.35%; and mature, 0.32%. Turkeys require from

TABLE 35.10 Recommended Supplemental Vitamin and Mineral Levels for Various Poultry Classes

	Starter Chicks (0–8 wks)	Growing Chickens (8–18 wks)	Egg-Type Layers	Starter Turkeys (0–8 wks)	Growing-Finishing Turkeys (8 wks to mkt)
Vitamin[1]					
A (miu)[2]	7.0	7.0	6.0	9.0	7.0
D_3 (miu)[2]	2.0	2.0	2.0	3.0	2.5
E (tiu)[3]	6.0	6.0	5.0	11.0	8.0
B_{12} (mg)	10.0	10.0	6.0	6.0	6.0
Riboflavin (g)	6.0	5.0	4.0	6.0	4.0
Niacin (g)	30.0	30.0	15.0	65.0	45.0
Pantothenic Acid (g)	10.0	10.0	6.0	14.0	10.0
Choline (g)	450.0	450.0	250.0	600.0	550.0
Folic Acid (g)	0.6	0.6	0.2	1.0	0.7
Thiamine (g)	1.0	1.0	1.0	1.0	1.0
Pyridoxine (g)	3.0	3.0	1.0	3.0	2.0
Biotin (mg)	50.0	50.0	30.0	100.0	50.0
Minerals[4]					
Manganese (mg)	25.0	25.0	50.0	50.0	50.0
Zinc (mg)	25.0	25.0	50.0	50.0	50.0
Iron (mg)	50.0	50.0	50.0	50.0	50.0
Copper (mg)	5.0	5.0	5.0	5.0	5.0
Iodine (mg)	0.2	0.2	0.2	0.2	0.2
Selenium (mg)	0.05	0.05	0.05	0.1	0.1

[1] Supplement per ton.
[2] Million International units.
[3] Thousand International units.
[4] Supplemental per pound.
Source: Adapted from Waldroup, 2002.

0.3–0.6% phosphorus; game birds, approximately 0.5%. Requirements for other minerals vary greatly depending on the stage of growth and production.

The vitamins that are most important to poultry are A, D, K, and E (fat soluble), and thiamin, riboflavin, pantothenic acid, niacin, vitamin B_6, choline, biotin, folacin, and vitamin B_{12} (water soluble). Recommended supplemental vitamin and mineral supplementation for various classes of poultry are listed in Table 35.10.

Issues Management

Affluent cultures have the luxury of choosing amongst a large array of food choices. Due to the wealth of resources enjoyed by consumers in developed nations, food security is not a daily concern and as such the market has created opportunities for differentiation on a number of levels such as animal handling, types of production systems, and organic/natural labels, for example. Simultaneously, activists and governmental regulators have pressured the poultry industry on topics such as bird housing, litter management, and poultry handling.

Environmental Impact

The 7 billion broilers in the United States produce about 1 ton of manure per each 1,000 birds. In the context of the average broiler farm previously discussed

COSTS AND RETURNS

The U.S. poultry industry has undergone significant structural and market change over the past 30 years. Over 80% of the farms producing poultry are located in the Northeast, Southeast, mid-South, Gulf Coast, and Corn Belt states. More than 75% of the total broiler production occurs in the top 10 states (GA, AR, AL, NC, MS, TX, MD, DE, VA, CA). In the turkey industry, 80% of the numbers are produced in the top 10 states (NC, MN, AR, VA, MO, CA, IN, PA, SC, IA) with the top six accounting for more than two-thirds of the total. Almost two-thirds of the eggs are produced in the top 10 states (OH, CA, IA, PA, IN, GA, TX, AR, MN, NE).

Prior to the integration and consolidation of the poultry industry, broiler production was a secondary farm enterprise uniformly spread across the country (Fig. 35.16). However, by the early 1990s, the geographic shift had occurred (Fig. 35.17). The primary reasons for this change were the development of improved transportation and packaging systems that allowed consolidation to occur in the southern states plus California. Furthermore, the climatic and socioeconomic conditions of the South were favorable to the growing of broilers under contract for a rapidly integrating industry. The location of a majority of the broiler processing facilities in the South and Southeast hastened the

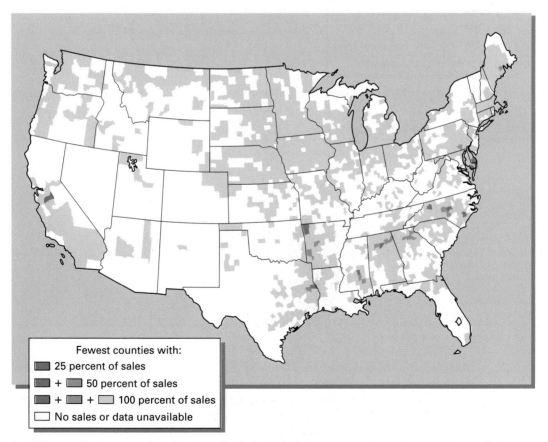

Fewest counties with:
- 25 percent of sales
- + 50 percent of sales
- + + 100 percent of sales
- No sales or data unavailable

FIGURE 35.16   Concentration of broiler sales in the United States, 1969. *Source:* Economic Research Service.

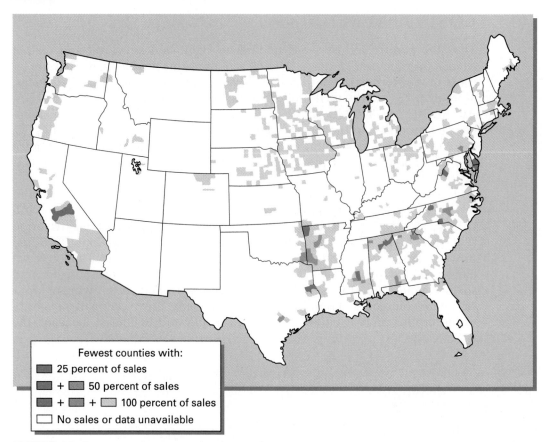

FIGURE 35.17 Concentration of broiler sales in the United States, 1992. *Source:* Economic Research Service.

TABLE 35.12 **Top 10 Broiler Companies**

Firm	Harvest Plants	N. Harvested (mil hd)	Average Live Wt.	Million lb (ready-to-cook)	Headquarters
Tyson Foods, Inc.	39	42.6	4.9	150	AR
Pilgrims Pride Corp.	27	29.2	5.0	113	TX
Gold Kist, Inc.	11	14.3	5.6	62	GA
Perdue Farms, Inc.	10	10.6	5.6	40	MD
Wayne Farms, LLC	NA	5.4	6.6	30	GA
Sanderson Farms, Inc.	7	5.3	6.8	30	MS
Mountaire Farms, Inc.	3	4.3	7.2	25	GA
Foster Farms	6	5.7	5.3	18	CA
Peco Foods, Inc.	4	3.1	7.3	17	AL
House of Raeford Farms	4	3.0	6.6	16	NC

Source: Adapted from multiple sources.

consolidation of the industry (Table 35.12). The top 3 firms have 47% market share in broilers (up from 35% in 1990) while the top 10 control almost 75%.

The poultry integrators also understood the value of a coordinated production system aligned with a processing sector focused on improving packaging, value-added product innovations, brand-name marketing, and convenience attributes of poultry products.

A result of these structural changes was the development of contract grower arrangements that now account for more than 95% of the value of poultry and egg production. Of the remainder, a vast majority of the production occurs on farms directly owned by the integrator-processor. The widespread use of **contract growing** offers farmers the opportunity to reduce market and production risk, gain market access, stabilize income, and gain access to technical and management innovation. The drawbacks for farmers include loss of upside market potential, potential loss of incentive to improve as managers and decision-makers, and loss of bargaining power. The benefits to processors include the ability to standardize input supplies, enhance quality control, and allow for more rapid response to shifts in consumer demand.

The allocation of responsibility to growers or processors is listed in Table 35.13. In general, growers provide land, labor, and housing in exchange for high-quality genetics, management expertise, veterinary care, and marketing power.

Contracts are structured to include a base payment, or fixed price per pound produced, an incentive-discount adjustment designed to reward better performance, and a disaster payment clause that allows for compensation if losses occur due to fire, flood, etc. Broiler production under contract is most prevalent in the Southeast and Gulf Coast states, where the majority of broiler processing and further-processed plants are located. This integrated production and marketing system has allowed the broiler industry to steadily increase both bird numbers and total production (Fig. 35.18).

Representative characteristics of poultry farms are as follows:

Broiler Production
- $^2/_3$ of farm dedicated to other enterprises
- 160-acre farm
- 3 houses per farm
- 65,000 birds per farm

Turkey Production
- $^3/_4$ of farm dedicated to other enterprises
- 225-acre farm
- 3 houses per farm
- 27,000 birds per farm

The cost of constructing housing on a farm is a significant investment with construction of a standard broiler house (50 × 500 ft) approaching $200,000. Thus, a typical farm with three houses has an investment of $600,000 just in the cost of building materials and construction. The life span of a new broiler house

TABLE 35.13 Distribution of Responsibility Under Contract Poultry Grower Arrangements

Grower	Processor
Day-to-day care of poultry	Chicks[b]
Land	Feed[b] and ration formulation
Housing	Veterinary supplies/service
Utilities	Management services
Labor	Transport to and from farm
Facility sanitation[a]	Determines capacity and design of housing
Manure disposal[a]	Production strategy
	Flock rotation schedule
	Genetics

[a] May vary by contract.
[b] The largest costs in broiler production (in order) are feed and birds.
Source: USDA, 1999.

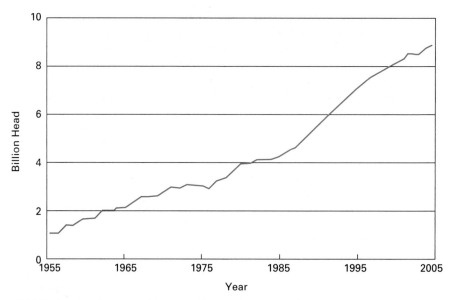

FIGURE 35.18 Changes in broiler production (1955–2005). *Source:* USDA.

TABLE 35.14 Poultry Meat and Egg Production Costs and Returns

Item	Production Cost (cents per doz/lb)		Net Return (cents per doz/lb)
	Feed	Total	
Market eggs	28.8	47.0	8.9
Broilers	16.0	26.4	7.6
Turkeys	21.9	35.6	5.4

Source: USDA.

TABLE 35.15 Industry Performance Targets

Trait	Target
Feed conversion—broilers	<2800 calories/lb of liveweight
Feed conversion—breeders	<7.5 lb per settable dozen
Average daily gain	>0.09 lb/day
Labor efficiency	>185 broilers/hour of labor[a]
Hatchability	>83%
Chicks/hen/week	>3.2

[a] Receiving through the chiller.

is approximately 35–45 years and gross income of $2.10 to $2.25 per square foot would be expected prior to debt retirement. Most loans are 15 years in duration and cash flows would improve significantly once the debt is settled. Annual operational costs of a house include litter management ($900–$1,000), electricity ($3,500), fuel ($5,000), repairs plus equipment replacement ($1,800), and insurance plus taxes ($2,800).

Table 35.14 shows the production costs for eggs, broilers, and turkeys. Note that feed costs comprise 60–70% of the total costs. Management practices that monitor and lower feed costs usually have a positive effect on net returns.

Table 35.15 outlines a few of the **performance targets** for profitable poultry production. It is important to recognize that the high degree of integration in the poultry industry allows for an extremely coordinated level of communication within the production and marketing chain. The use of objective goals and measurements has allowed dramatic growth in the poultry industry.

While the industry made significant strides in profitability via improved cost controls, particularly in regards to feed, an overemphasis on production costs can lead to reduced profitability. For example, if the changes made to a ration to lower costs result in lower liveability or reduced gains, the net effect may be lost net income as a result of poorer productivity albeit at a lower cost. The poultry industry in the future will continue to monitor costs but will ultimately succeed because of the ability of managers to optimize costs and production tonnage.

As the animal protein market has differentiated into a variety of specialty or niche markets, some poultry producers have moved into production systems that do not utilize chemical feed additives such as antibiotics or hormones as a means to create a unique brand attribute for their products. Poultry and egg products originating from these systems often sell at a premium price, typically double that of conventionally produced poultry and eggs. However, there are significant costs associated with additive-free or natural production systems. For example, the costs associated with ration formulation, additional clean-out requirements, added payments to the grower, and loss of feed efficiency add production costs per bird of 6, 1.5, 8, and 3%, respectively. While capturing market premiums are attractive to producers, the additional costs associated with these systems must be carefully evaluated in terms of net profit.

The poultry industry has also become increasingly dependent on the export of chicken meat as a primary strategy to increase profitability. While exports have shown steady growth, the ability of the United States to capture additional share of the global market has been limited by the growth of the Brazilian broiler industry spurred by significant cost of production advantages. Concerns over avian influenza add instability to the export market. Nonetheless, growth of the poultry industry will require access to global markets and the ability of the industry to remain competitive on both a price and value basis.

CHAPTER SUMMARY

- ◼ The most important poultry in the United States are chickens and turkeys.
- ◼ Breeds of chickens are relatively unimportant, as the commercial broiler and egg producers utilize hybrid chickens resulting from sophisticated breeding methods.
- ◼ The production of broilers and eggs is very intensively managed in an integrated system involving breeding, hatching, feeding, processing, and marketing name-brand products to the consumer.

KEY WORDS

caruncle	qualitative traits	outcrossing
snood	quantitative traits	inbred lines
class	trap nest	hybrid chickens
breed	mass selection	strain crossing
variety	family selection	broodiness

incubation	feeder space	rickets
hatchability	British thermal unit	debeaking
candling	(Btu)	Avian influenza
litter	ventilation	contract growers
floor space	insulation	performance targets

REVIEW QUESTIONS

1. How are breeds of chickens classified?
2. What are the four major classes of chickens?
3. What characteristics are most important to commercial egg producers?
4. What Mediterranean breed of chicken is of major importance to the U.S. poultry industry?
5. What three American breeds of chickens are important to the U.S. poultry industry?
6. What English breed of chicken is important to the U.S. broiler industry?
7. What is the incubation period for chicken eggs?
8. What is the incubation period for turkey eggs?
9. What factors must be regulated during incubation of poultry eggs for hatching?
10. What are the two critical periods for temperature regulation of hatching eggs?
11. What problems may result from incubation temperatures that are too high or too low?
12. Why is maintaining humidity near 60% important in hatching incubators?
13. How should eggs be positioned in an incubator for good hatchability?
14. Why will hatchability be greatly reduced if eggs are not turned often enough during incubation?
15. *True or False:* The oxygen content of incubators should be maintained at 21%.
16. *True or False:* It is vital to maintain the carbon dioxide content of incubators at or below 0.5% to maintain good hatchability of eggs.
17. What is the primary purpose of placing litter on the floors of brooder houses?
18. What is debeaking or beak trimming, and why is it performed?
19. Replacement laying hens should be managed to reach puberty and begin laying eggs at _____ of age.
20. What factors are carefully considered when planning and managing poultry houses?
21. What are the two integral parts of a ventilation system for poultry?
22. What mineral is especially important in the diets of laying hens?
23. What accounts for 60–70% of total costs of poultry production?

SELECTED REFERENCES

American Poultry History. 1996. Mount Morris, IL: Watt Publishing Co.

Carpenter, E. 2001. Nutrient management comes of age. *Poultry USA*, Dec., 60–74.

Lundeen, T. 2002. Feed intake most important factor in meat-type poultry growth. *Feedstuffs*, Oct. 7, 9.

National Research Council. 1994. *Nutrient Requirements of Poultry.* 9th ed. Washington, DC: National Academy Press.

North, M. O., and D. D. Bell. 1990. *Commercial Chicken Production Manual.* Westport, CT: AVI Publishing Co.

Perry, J., D. Banker, and R. Green. 1999. *Broiler Farms' Organization, Management, and Performance.* ERS: USDA. Bulletin No. 748.

Scanes, C. G., G. Brant, and M. E. Ensminger. 2004. *Poultry Science.* Upper Saddle River, NJ: Pearson Prentice Hall.

Waldroup, P. W. 2002. Dietary nutrient allowances for chickens and turkeys. Feedstuff, 74:28.

Watt Poultry Yearbook. 2006. Mount Morris, IL: Watt Publishing Co.

CHAPTER 36

Goat Breeding, Feeding, and Management

LEARNING OBJECTIVES

- Describe the production characteristics of goats
- Learn the external parts of the goat
- Discuss the timing of breeding and kidding
- Describe the nutritional management of goats
- List the primary diseases of concern and their management
- Compare and contrast dairy, meat, and fiber goats

Geological records reveal that goats existed five million years ago, and bones of goats closely associated with people have been found dating back 12,000 years. It appears that people at that time existed in small groups, moving to where food and water could be found for themselves and the goats. The goats were particularly important to people because they provided food (milk and meat), skins for making certain items of clothing, and mohair for many uses.

Goats thrive on **browse** (brushy plants) and **forbs** (broadleaf plants), but they also eat grass extensively, depending on forage species and stage of maturity. Choice of plants and plant parts is important to goats because they are more likely to select the more nutritious parts. Because of their browsing over a wide territory, goats are less likely to suffer from internal parasites.

Goats have a large impact on the economy and food supply for people of the tropical world and also several Mediterranean countries. In many countries, people consume more goat's milk than cow's milk, and goats are an important source of meat. In the United States, the consumption of both goat's milk and goat meat is increasing, and the production of mohair is an important industry in Texas.

There are five major kinds of domesticated goats: (1) the **meat goat** is used for brush clearance, for the production of meat, and, in some parts of the world, for skins and fine leather; (2) the **Angora goat** is used mainly for the production of mohair and, to a lesser extent, for brush clearance and the production of meat; (3) the **dairy goat** is used largely for the production of milk and, to a lesser extent, for meat; (4) the **Cashmere goat** is noted for the soft cashmere fibers used in producing high-quality clothing; and (5) the **pygmy goat** is used as a laboratory ruminant animal and pet in the United States but is an important disease-resistant meat and milk producer in West Africa (where it is known as the West African Dwarf) and other countries.

In some parts of the world, goat meat production is extremely important and, in fact, meat goats outnumber dairy goats worldwide. Goat numbers in the United States are modest compared to those for cattle and sheep because of the highly specialized and effective dairy industry built around the dairy cow, as well as the specialized beef cattle and sheep industries. Yet in some parts of the world, including India, Iran, Pakistan, and Sudan, large quantities of dairy products are produced from goats.

Milk produced by dairy goats differs from cow's milk in that all carotene has been converted into vitamin A in goat's milk. The type of curd formed from goat's milk is different from the curd from cow's milk because of differences in the major caseins; milk fat in goat's milk is in smaller globules than in cow's milk, does not rise or coalesce as readily, and contains much greater amounts of short-chain fatty acids. Goat's milk is more readily digested and assimilated by people and animals because of these differences. The dairy goat is a desirable animal for providing milk for the family because it is small and less expensive to feed than the cow. Goats can consume large quantities of browse, which is not very palatable to cattle.

U.S. GOAT INDUSTRY

The inventory of goats in the United States is almost 3.5 million head comprised of 2.9 million head of meat goats, 320,000 dairy goats, and 285,000 Angoras. Approximately one-half million meat goats are harvested annually under federal inspection programs. The geographic distribution of goats is provided in Table 36.1. The meat goat industry in particular is expanding, driven by ethnic demand for **cabrito** (goat meat) and the value of goats as brush and invasive plant control agents.

TABLE 36.1 Geographic Distribution of Meat, Dairy, and Angora Goats in the United States

State	Meat Goats (1,000)	State	Dairy Goats (1,000)	State	Angora Goats (1,000)
TX	942	CA	37	TX	230
TN	107	WI	26	AZ	28
OK	73	TX	23	NM	7
GA	66	OK	14	CA	5
KY	62	NY	12.8	MO	2.5
CA	61	PA	12.6	OH	2.2
NC	59	OR	9.2	PA	2.1
AL	47	MI	8.9	NC	1.6
SC	38	MO	8.6	MI	1.4
MO	37	IA	8.5	VA	1.2

Source: USDA.

TABLE 36.2   Comparison of Cabrito to Competing Proteins for Nutritional Value

3 Oz. Roasted	Calories	Fat (g)	Saturated Fat (g)	Protein (g)	Iron (g)
Goat	122	3	0.8	23	3.2
Beef	245	16	6.8	23	2.2
Pork	310	24	8.7	21	2.7
Lamb	235	16	7.3	22	1.4
Chicken	120	3.5	1.1	21	1.5

Source: USDA.

Demand for cabrito is strongly tied to the growth of ethnic populations and ethnic themed foodservice establishments. A 3-ounce serving of cabrito is comparable to other animal proteins for protein content, has more iron than beef, pork, lamb, or chicken; and has less fat than its competitors (Table 36.2). Because of its low fat content, cabrito must be cooked slowly under moist conditions and overcooking should be avoided to preserve palatability.

The largest ethnic markets are Hispanic, Muslim, and Caribbean consumers. Goat demand increases in accordance with holidays and festivals such as Easter, Independence Day, Ramadan, and Eid Al Fitr. The spring market tends to favor kid goats weighing 20–50 lb while fall and winter markets prefer goats weighing 60 lb and heavier.

PHYSIOLOGICAL CHARACTERISTICS

Goats have many characteristics that are similar to those of sheep. Some of the production traits of goats are summarized in Table 36.3.

MEAT GOATS

Meat goats are found in limited numbers in the southwestern and western United States, though they are raised primarily in Africa and the Middle East, where their main uses are for meat and skins. In West Africa there has been a

TABLE 36.3   Production Characteristics of Goats

Trait	Time/lb
Gestation length	144–155 days (average 150 days)
Length of estrous cycle	15–18 days (average 16 days)
Length of estrus	1–3 days (average 2 days)
Age at puberty	120 days to over 1 year[a]
Normal breeding season	Late summer, early fall, or late winter[b] (breeds near the equator are less seasonal)
Size of kids at birth	1.5–11.0 lb[c] (average 5.5 lb)
Adult size	
Does	40–190 lb (average 130 lb)
Bucks	50–300 lb[d] (average 160 lb)

[a] Age of puberty depends on the breed and nutrition. Pygmy goats tend to reach puberty at an early age; some kid at 9 months of age.
[b] Normal breeding season depends on the breed. Pygmy does are capable of breeding over a long breeding season but are more fertile when bred in the fall or early winter. Angora goats usually have a highly restrictive breeding season during the fall.
[c] Pygmy and Angora goats are smaller at birth than are dairy goat kids.
[d] Pygmy goats are the smallest in size, Angora goats are intermediate, and dairy goats are the largest. Bucks of all breeds are larger than corresponding does.

FIGURE 36.1 Boer goats have been imported from South Africa to improve meat production. *Source:* The American Boer Goat Association.

preference for a small, well-muscled, agile goat—the West African Dwarf. These goats can obtain most of their feed from brushy plants and low trees. As many of the areas away from cities and towns are lacking in refrigeration, small goats that are slaughtered can be consumed by the family before the meat spoils.

The predominant meat-type goat is the Spanish goat. Spanish goats are able to breed out of season and make excellent range animals due to their excellent hardiness and survival rates under low input management systems. Kid crops range from 75% in extensive, low-input management systems to over 150% for intensive, spring/fall kidding, and high-input management scenarios. Spanish goats are highly variable in phenotypic characteristics such as color and ear shape.

The Boer breed of goat (Fig. 36.1) has been imported from New Zealand. Developed in South Africa, the Boer is superior in growth rate and lean meat yield. Compared to other livestock species, meat goats have a lower carcass dressing percent, higher lean yield, and less carcass fat. Yearling male and female Boer goats average 100 and 85 lb, respectively. Adult male and female goats have average weights of 250 and 140 lb, respectively.

Boer goats with a white body, cherry red head and neck, as well as a blaze face are preferred. They are short haired, have long, droopy ears, and a convex shape to their face.

DAIRY GOATS

The external parts of a goat are shown in Figure 36.2. The dairy goat and dairy cow are about equal in efficiency of converting feed into milk, even though the dairy goat produces much more milk in relation to its size and weight than does the dairy cow. Feed needed for maintenance is higher per unit of body weight for goats than for dairy cows.

Gall (1981) compared the efficiency of Holstein dairy cows with that of Alpine Beetel dairy goats. The cows weighed 1,325 lb and produced 13,260 lb

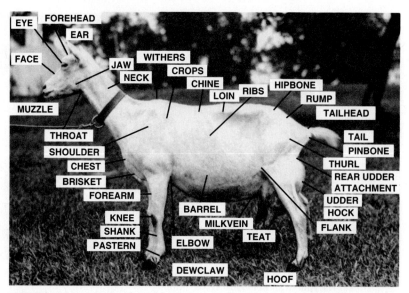

FIGURE 36.2   The external parts of the dairy goat. *Source:* R. F. Crawford and Ted Edwards. *Emerald Dairy Goat Association Newsletter.* Photograph by Ole Hoskinson.

of milk, whereas the dairy goats weighed 108 lb and produced 1,083 lb of milk per lactation. The dry-matter intakes per lactation and per quart of milk for the Holstein cows were 13,300 lb and 2.3 lb, and for the goats they were 1,173 lb and 2.4 lb per quart of milk, respectively.

The six major dairy goat breeds registered in the United States—Toggenburg, Saanen, Alpine, Nubian, LaMancha, and Oberhasli—are all capable of high milk productivity. These breeds are shown in Figure 36.3 and described in Table 36.4. Milk production characteristics are listed in Table 36.5.

The Toggenburg breed is medium in size, is vigorous, and is capable of high milk production. The Saanen is a large, all-white breed. The Alpine breed is a large, somewhat rangy goat. The Nubian is a tall goat breed that differs from the other five breeds in being a better meat producer and having long, wide, pendulous ears and a Roman nose. However, Nubian milk is distinctive for its higher milk fat content (7.4% versus 3.5% for Swiss breeds). The Nubian is also considered a dual-purpose breed. The LaMancha goat breed of California origin is smaller and differs from all other breeds in having almost no external ears ("gopher" ears) or extremely short ears ("elf" or "cookie" ears). The Oberhasli is a medium-sized dairy goat breed, also of Swiss origin, with usually solid red or black colors.

ANGORA GOATS

Angora goats (Fig. 36.4) are raised chiefly to produce mohair, which is made into fine clothing. It is estimated that world mohair production is approximately 15,000 tons per year. These goats are produced in Turkey, South Africa, and, in the United States, the Southwest (Texas has the most Angora goats of any state), the Ozarks of Missouri and Arkansas, and the Pacific Northwest.

Alpine

LaMancha

Nubian

Oberhasli

Saanen

Toggenburg

FIGURE 36.3 The six major goat breeds registered in the United States. Courtesy of American Dairy Goat Association. www.adga.org.

They do best on browse, so they are most commonly kept on rough, brushy areas. They are not as well adapted for intensive grazing pastures because they prefer broadleaf plants over some grasses, and because they are more likely to become parasitized when grazing near the soil level in overgrazed pastures.

Some Angora goats have a tendency to shed their mohair in the spring, but selection against this trait has been successful. Some Angora goats are not shorn for 2–3 years, which allows the mohair to grow to lengths of 1–2 ft. This special

TABLE 36.4 Characteristics of Dairy Goat Breeds

Breed	Origin	Body Structure	Color	Ear and Face Structure
Alpine	France	Doe—110–200 lb; Buck—175–220 lb	Multiple	Erect, straight or dished
LaMancha	U.S.	Doe—110–140 lb; Buck—140–180 lb	Multiple	Very short, straight
Nubian	Orient	Doe—110–140 lb; Buck—140–180 lb	Multiple	Long and droopy, roman nose
Oberhasli	Switzerland	Doe—110–130 lb; Buck—145–175 lb	Bay with black markings	Erect, straight
Saanen	Switzerland	Doe—110–200 lb, Buck—175–275 lb	White	Erect, straight or dished
Toggenburg	Switzerland	Doe—110–130 lb, Buck—140–180 lb	Light fawn to dark chocolate	Erect, straight or dished

TABLE 36.5 Milk Production Characteristics of Dairy Goat Breeds

Breed	Milk (lb)	Fat (lb)	Protein (lb)
Alpine	2,085	69	62
LaMancha	1,799	66	55
Nubian	1,760	66	52
Oberhasli	1,830	64	52
Saanen	1,936	71	57
Toggenburg	1,665	57	47

Source: USDA.

FIGURE 36.4 A group of Angora goats showing a full fleece of mohair. Angora goats are useful for meat production, brush clearing, and mohair production. *Source:* Mohair Council of America.

mohair brings a high price and is used for making wigs and doll hair. Shearing is usually done in spring or twice a year. The clip from does weighs 4–6 lb, whereas bucks and wethers may shear 5–8 lb. Highly improved goats often shear 10–15 lb of mohair per year. Three classes of mohair are produced: the **tight lock** with ringlets the full length of the fibers; the **flat lock,** which is wavy; and the **fluffy** or **open fleece.** The tight-lock fleeces are fine in texture but low in yield. The flat lock lacks the fineness of the tight lock but is satisfactory for making cloth, and its yield is high. The fluffy fleece is of low quality, often quite coarse, and is easily caught in brush and thereby lost.

Fleeces should be sorted at shearing time. Coarse fleeces and those with burrs should be kept separate from good fleeces. The clip from each fleece should be tied separately so that it can be properly graded.

Fertility can be low in Angora goats. Losses of kids on the open range or in scattered timber areas of the West Coast states are often high, especially because of predators such as coyotes, bobcats, and wild dogs. Generally, the kid crop that is raised to weaning is about 80% of the number conceived, as estimated on the basis of the number of does bred. Losses of 20% or more may occur when proper care is lacking at the time kids are born.

Angora goats usually have horns that are curved sideways. Swiss (dairy) goats have straight sickle horns. When they are in good condition on the range, mature bucks and wethers may weigh 150–200 lb. Mature does weigh 90–110 lb, but well-fed show animals usually weigh more. Kids are usually weaned at 5 months of age, which allows the does to gain in weight at breeding time. Breeding over a period of days is usually done in late September and October. Since the gestation is 147–152 days, this time of breeding results in the kids arriving in the spring after the weather has moderated. Kidding may take place in open sheds when the weather is severe, or outside when the weather is favorable. Young does are usually first bred to kid at the age of 2 years but, if they are properly developed, they may be bred to kid when they are yearlings.

Angora goats need shelter during rainy periods or wet snows, especially after shearing; an open shed is sufficient. The long mohair coats of Angoras keep them warm during cold weather, but they may nevertheless crowd together and smother in severe snowstorms if no shelter is provided.

Angora goats are subject to the same parasites and diseases as dairy goats, and the same treatments are effective for both. In most other respects, though, the care given Angora goats resembles the care given sheep rather than that given dairy goats.

CASHMERE GOATS

Cashmere goats originated in Central Asia; they are white in color and have erect ears and long, twisted horns. In recent years, there has been an increase in Cashmere goats and producers in the United States.

Cashmere goats produce the Cashmere fiber which is extremely fine and highly valued. The average yield per animal is 0.05 to 2 lb of unscoured fiber, with 60% yield as clean fiber. The estimated world cashmere production is 3,000–4,000 tons per year. The fine undercoat is combed out with scrupulous care, the process taking 8–10 days. The majority of the world's production of cashmere is processed into the most luxurious, expensive cloth and knitwear in Scotland.

PYGMY GOATS

Pygmy goats were developed in Africa. Those brought to the United States came from the Cameroon area to the Caribbean and were then imported to the United States. Early pygmy goats were brought to the United States as exotic zoo animals. Zoos sold pygmy animals not only to other zoos but also to research scientists. A colony of pygmy goats was established in Oregon and several other research herds were established, including herds at the University of California. Because pygmies are small in size, the cost of keeping them is minimal. Pygmy goats are also raised for meat and milk.

Breeders of pygmy goats enjoy showing their goats just as do dairy goat and Angora goat breeders. There are over 50 pygmy goat shows held annually in the United States, where several hundred goats are shown.

Pygmy goats vary in color from black to white with a dorsal stripe down the back and dark on the legs. On all pygmy goats except the black ones, the muzzle, forehead, eyes, and ears are accented in tones lighter than the dark portion of the body.

The pygmy goat is also becoming popular as an animal for 4-H projects. The goats are small and do not require a lot of space or feed. All goats are very friendly animals, which make them ideal for young people to handle; they are also easily trained.

CARE AND MANAGEMENT

Simple housing for goats is adequate in areas where the weather is mild because goats do best when they are outside on pasture or in an area where they can exercise freely. If the weather is wet at times, an open shed with a good roof is necessary. Lactating goats can be fed in an open shed in stormy weather and taken to the milking parlor for milking. If the weather is severely cold, an enclosed barn with ample space is needed. At least 20 square feet per goat is needed if goats are to be housed in a barn, but 16 square feet per goat may suffice in open sheds.

Fencing

It has been said that it is not a question of if goats will escape confinement but when. Thus good fencing is particularly important. Fences must be properly constructed with woven wire approximately 6 ft high and with 6-in. stays every 10–12 in. Particular attention should be given to those areas where uneven terrain makes it difficult to prevent gaps of sufficient size to allow goats to crawl underneath. Fencing for goats is different from fencing for cows; a woven-wire fence is best for containing goats because they can climb a rail fence. In addition, a special type of bracing at corners of the fence is necessary because goats can walk up a brace pole and jump over the fence.

With electric fences, two offset electric wires should be used at 8 in. high and 6–8 in. away from the fence. Fences should be charged to a minimum of 4,000 volts. However, an electric fence may not contain bucks of any breed; some may go through the fence despite the shock.

Another approach is to use an 8-strand barbed wire fence or a 5-strand electrified high tensile fence system. Four by four woven wire also makes an acceptable barrier. Woven-wire fences larger than these are not acceptable as they are big enough to allow goats to get their heads stuck.

Four-foot-tall fenced catch pens are useful for handling goats for routine procedures such as sorting, vaccinations, and foot trimming. Working chutes, alleys, and a head catch are useful if the number of animals to be worked is relatively large. A good website for facility information is www.clemson.edu/agronomy/goats/handbook.

Equipment

Several pieces of equipment are needed for normal care: dehorners, a tattoo set, hoof shears and hoof knife, a grooming brush, an emasculator, and a **balling gun** for administering boluses (large pills for dosing animals).

Dehorning

The polled condition has advantages because a polled goat is less likely to catch its head in a woven-wire fence than is a horned goat. Horned animals should be disbudded during the first week following birth by use of an electric dehorning iron. Goat breeders have been plagued by the genetic linkage between hermaphrodites and polled expression. However, polled goats with good fertility can be found.

Hoof Trimming

Goats' feet should be kept properly trimmed to prevent deformities and footrot. In rough country, the terrain will likely keep hooves at a correct length. Under smooth conditions or under pen management, hoof trimming should be scheduled every 6–8 weeks. A good pruning shear is ideal for leveling and shaping the hoof, but final trimming can be done with a hoof knife. Footrot should be treated with formaldehyde, copper sulfate, or iodine. Affected goats are best isolated from the others and placed on clean, dry ground after treatment.

Identification

All kids can be identified by a tattoo in the ear, except for LaMancha and pygmies, where identification is placed in the tail-web. Ear tags are also useful and are easy to read. The tattoo serves as permanent identification in case an ear tag is lost. All registered goats must be tattooed because the goat registry associations do not accept ear-tag identification.

Castration

Male kids not acceptable for breeding should be castrated at 4–6 weeks of age. This can be done by constricting the blood circulation to the testicles by use of a rubber band elastrator or by surgically removing the testicles. Some breeders nonsurgically crush the cords to the testicles with an **emasculator.** After the blood supply to the testicles is discontinued, they usually atrophy, but exceptions occur. After the emasculator is used, the blood circulation to the testicles may continue, and although the male is sterile, he continues to produce testosterone and has normal sex drive. A problem with the elastrator is that tetanus can occur when the tissue dies below the elastrator band. Surgical removal of the testicles creates a wound that attracts flies in the warm season; therefore a fly repellent should be applied around the wound. It is recommended that a veterinarian castrate older goats. Whatever method is used, caution must be taken to prevent infection, gangrene, and other complications.

Milking

It is important that the udders of dairy goats have strong fore and rear attachments and that the two teats be well spaced. Goats are milked by hand (Fig. 36.5) or by milking machines. The milking stanchion should be elevated to place the goat at a convenient level for the person doing the milking. Clean, sanitary conditions of the milking stanchion, the milking machine, the goat, and her udder and teats are very important prior to milking. After milking, the teats should be dipped in a weak iodine solution.

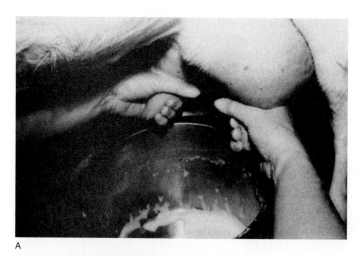

A

B

FIGURE 36.5 Goats are milked either (A) by hand or (B) by machine. *Source:* George Haenlein, University of Delaware.

Time of Breeding

The female goat comes into heat at intervals of 15–18 days until she becomes pregnant. Young does can first be bred when they weigh 85–95 lbs at about 6–10 months of age, which means that does in good general condition may be bred to have kids at 1 year of age. Most goats are seasonal breeders; the normal breeding season occurs in September, October, and November. If no effort is made to breed does at other times, most of the young will be born in February, March, or April. Normal goat lactations are 7–10 months in duration. A period of at least 2 months when no goats are lactating might occur, but with staggered breeding, the kidding and milking seasons can be extended. Housing goats in the dark for several hours each day in the spring and summer months (to simulate the onset of shorter days) causes some to come into estrus earlier than usual. Conversely, artificial additional light in the goat barn may delay estrus in the autumn.

One service is all that is needed to obtain pregnancy, but it is generally wise to delay breeding for a day after the goat first shows signs of heat. The doe stays in heat from 1–3 days, but the optimum period of standing heat may last only a few hours at the end of estrus. A female in heat is often noisy and restless, and her milk production may decrease sharply. She may disturb other goats in the herd, so it is advisable to keep her in a separate stall until she goes out of heat.

Male goats during the breeding season usually have a rank odor, especially when confined to a pen. Housing the males downwind apart from the does will reduce some of the nuisance of male odors. Cleanliness of the male's long beard and shaggy hair helps reduce "bucky" odors.

Careful records should be kept showing breeding dates and the bucks used. From breeding records, it is possible to calculate when does should deliver kids, as the gestation period is about 150 days. Detailed breeding records are essential for knowing ancestors, which in turn helps in the selection of superior males to produce especially desirable offspring.

Delivery of twins is common and delivery of triplets occurs on occasion, particularly among mature does. More males are conceived and born than are females; the ratio is approximately 115:100. Mature does average 2 kids, whereas younger does average 1.5 kids. Kids at birth weigh 5–8 lbs, but singles may be heavier than twins or triplets, and males are usually heavier than females.

Time of Kidding

A doe almost ready to deliver young should be placed in a clean pen that is well bedded with clean straw or shavings. The doe should have plenty of clean water, some laxative feed (such as wheat bran), and fresh, soft legume hay. She should be carefully observed but not disturbed unless assistance is necessary, as indicated by excessive straining for 3 hours or more. If the kid presents the front feet and head, delivery should be easy without assistance. If only the front feet, but not the head, are presented, the kid should be pushed back enough to bring the head forward in line with the front feet. Breech presentations are not uncommon, but these deliveries should be rapid to prevent the kid from suffocating. If assistance is needed, a qualified attendant with small, well-lubricated hands should pull when contractions occur. Gentleness is essential; harsh or ill-timed pulling can cause severe internal damage.

As soon as the kid arrives, its mouth and nostrils should be wiped clean of membranes and mucus. In cold weather, it is necessary to take the newborn to a heated room for drying. A chilled kid must be helped to regain its body temperature with a heat lamp or even by immersing its body up to the chin in water that is as hot as can be tolerated when the attendant's elbow is immersed in it for 2 minutes. The navel should be kept out of the water. Kids should be encouraged to nurse as soon as they are dry, as nursing helps the newborn kid to keep warm.

Difficulty in kidding may be caused by a low selenium supply, diseases, and internal parasites. It is advisable to use disposable rubber or plastic gloves when assisting with delivery and to wash and disinfect everything diligently. Some diseases (including brucellosis) are transmissible to people; however, brucellosis has not been recently reported in U.S. goats.

Feeding

Because goats are ruminants, they can digest roughage effectively. The ability of goats to consume lower-quality forage is illustrated in Table 36.6. However, the types and proportions of feed should be related to the functions of the goats. For example, dry does and bucks that are not actively breeding perform satisfactorily on ample browse, good pasture, and good-quality grass and legume hay. If grass is short and hay is of poor quality, and goats are milking well, feeding of supplemental concentrates is necessary. Overfeeding of supplemental concentrates can cause diarrhea or obesity, which interferes with reproduction and subsequent lactation.

A summary of feed requirements for meat goats is provided in Table 36.7 and for dairy goats in Table 36.8.

Heavily lactating does and young does that must grow while lactating should be given good-quality hay *ad libitum* and relatively high levels of concentrates. Good-quality pasture should supplement hay well, but even then lactating goats require some roughage in the form of hay or other source of long fiber to prevent scouring.

Pregnant does should be fed to gain weight in order to ensure adequate nutrition of the kids. A doe should be in good flesh, but not fat, when she kids, because she draws from her body reserves for milk production.

Grains such as corn, oats, barley, and milo may be fed whole because goats crack grains by chewing. If protein supplements are mixed with the grain or if the feed mix is being pelleted, the grain may be rolled, cracked, or coarsely ground. Some people prefer to mix the hay and grain and prepare the ration in

TABLE 36.6   Comparison of Goat Diets by Forage Type

Location	% Composition of Diet		
	Grass	Forbs	Browse
College Station	18	5	77
Kerrville	16	9	75
Sonora	46	10	44
Tilden	37	5	58

Source: Adapted from Texas Agricultural Experiment Station, 92.

TABLE 36.7   Protein and Intake Requirements for Meat Goats at Various Stages of Production

Production Phase	Protein %	Pounds per Day as Fed
Preweaning/creep	18	0.25–0.33
Weanlings	16	0.5–0.75
Finishing	14–16	1
Flushing (30 days prior and 30 days after breeding)	14–16	1–3
Gestation	14–16	0.5–1 (2nd and 3rd month)
		0.75–2 (last 1.5 months)
Lactation (single)	14–16	0.75–1.25
Lactation (twins)	14–16	2
Replacement does	16	0.5–1
Bucks (mature)	14	< 0.5

Free-choice forage and mineral is assumed.
Source: Adapted from ADM Alliance Nutrition.

TABLE 36.8   Protein and Intake Requirements for Dairy Goats at Various Stages of Production

Production Phase	Protein %	Pounds per Day as Fed
Preweaning/starter (2–4 mo)	18	Free choice
Growing (4 mo to 6 to 8 wks before kidding)	14–16	1–1.5
Dry does (6 to 8 wks before kidding)	14–16	1–2
Lactating does	14–16	1 per 3 lb of milk production
Bucks (mature)	14–16	1–2

Free-choice forage and mineral is assumed.
Source: Adapted from ADM Alliance Nutrition.

a total pellet form. There is added expense in pelleting, but much less feed is wasted.

In some areas, deficiencies in minerals such as phosphorus, selenium, and iodine may exist. The use of iodized or trace-mineralized salt along with dicalcium phosphate or steamed bonemeal usually provides enough minerals if legume hay or good pastures are available. Calcium is usually present in sufficient quantity in legume hay and phosphorus is adequately provided in grain feeds. To be safe, it is best to provide a mixture of trace-mineralized salt and bonemeal free-choice.

Young, growing goats and lactating goats need more protein than do bucks or dry does. A ration containing 12–15% protein is desirable for bucks and dry does, but 15–20% protein may be better for young goats and for does that are producing much milk.

Generally, kids must be allowed to nurse their dams to obtain the first milk (colostrum). After 3 days, kids may be removed from their mothers and given milk replacer by means of a lamb feeder or hand-held bottle until they are large enough to eat hay and concentrates. It is advisable to encourage young kids to eat solid feed at an early age (2–3 weeks) by having leafy legume hay and palatable fresh concentrates such as rolled grain available at all times. Solid feeds are less expensive than milk replacers, and when the kids can do well on solid feeds, milk replacers should not be fed. Small kids need concentrates until their rumens are sufficiently developed to digest enough roughage to meet all their nutritional needs. At 5–6 months of age, young goats can do well on good pasture or good-quality legume hay alone.

Goats of any kind are usually run with sheep outside of the United States; this practice may be advantageous in the United States as well. Goats normally eat more browse and forbs than sheep.

Rams should never be run with doe goats, and buck goats should never be run with ewes. These animals will mate when the females come in heat (estrus); if they conceive, the pregnancy will usually terminate at about 3 or 4 months. There are reports of rare cases of sheep-goat crosses delivered alive at term.

CONTROLLING DISEASES AND PARASITES

Major diseases affecting goats are described in this section. The normal temperature of a goat is 102–103° F. Observing goats' feeding habits is a good means to monitor health status since goats not eating or grinding their teeth should be evaluated closely. As goats are highly resistant to **bluetongue,** there is no reason for alarm even if sheep that have bluetongue are grazing in the same pasture.

Johne's disease also affects sheep, and it is a serious health problem in goats. Affected goats become unthrifty, emaciated, and unproductive.

A troublesome disease, **caseous lymphadenitis,** is characterized by nodules in the lymph area (throat). When one of the infected nodes is opened, a thick, caseous material is exposed. The disease is difficult to control. An isolation program is desirable for purchased animals to see if they develop the disease. Infected animals should be isolated; then infected nodes are opened and flushed with H_2O_2. Lancing abscesses should be performed under the direction of a veterinarian as this disease is transmissible to humans.

Caprine arthritis encephalitis (CAE) affects the central nervous systems of young kids; head tremors, loss of coordination, and partial paralysis can be observed. The arthritis form of the disease occurs in mature goats and affects the feet and other limb joints. Other symptoms are loss of condition and "hard udder" (a firm, swollen udder in freshly kidded goats). There is no effective treatment for the disease. Preventative methods include reducing contact with infected animals (e.g., at shows and sales), blood-testing suspects, and removing infected animals from the herd.

Caprine pleuropneumonia results in high mortality. The services of a veterinarian should be obtained to help control this disease.

Contagious ecthyma, also called **soremouth,** is contagious to sheep and humans. Animals that recover from it are immune for several years. While mildly problematic in adult goats, young goats may stop nursing due to mouth lesions and may develop lameness as the result of foot abscesses. A vaccine can be administered to kids on premises that are infected.

Enterotoxemia is caused by *Clostridium perfringens* types C and D. Animals of all ages are susceptible. *C. perfringens* antitoxin given intravenously or subcutaneously will obtain a dramatic response. Toxoid vaccination given to month-old kids followed by a second dose in 2 weeks and booster doses each year will control the disease. This disease, sometimes called *overeating* disease, can be prevented by avoiding feeding excess milk and grain.

Urinary calculi results from the formation of stones in the urinary tract of male goats, especially wethers due to an incorrect balance of calcium and phosphorus in the diet. Inability to urinate, belly kicking, excessive stretching, and general restlessness are recognizable symptoms. Diets must be carefully balanced and goats should have continuous access to fresh water at all times.

Tetanus is caused by *Clostridium tetani,* which infects wounds and often causes death. The causative organism is prevalent in soil contaminated with horse feces. Two doses of toxoid to goats over 1 month of age and then an annual booster injection will control the disease.

Mastitis is an inflammation of the udder predisposed by bruising, lack of proper sanitation, and improper milking. The milk becomes curdled and stringy. It is advisable to use a milk-culture sensitivity test and engage the services of a veterinarian for treating severe cases. An udder may be treated with a suitable antibiotic by sliding a special dull plastic needle up the teat canal into the udder cistern and depositing the antibiotic, or by systemic antibiotic treatment intramuscularly or intravenously. Hot packs may help reduce the edema and, in severe cases, frequent milking is necessary and most helpful. Dry goat treatments can effectively prevent mastitis.

Brucellosis is transmissible to humans. All goats should be tested for goat brucellosis, and reactor animals or suspects must be slaughtered. Goats do not get the same brucellosis that affects cows.

Footrot occurs when goats are kept on wet land. It is contagious, but the bacterial organism causing it does not live long in the soil. Any animal to be introduced into a herd should be isolated for several days to see if footrot is present. Severe hoof trimming of infected feet, followed by treatment with formaldehyde (as described in Chapter 32), cures this disease. Generally, dairy goats dislike wet land and footrot is not a common problem in U.S. dairy goats.

Ketosis, or pregnancy disease, rarely occurs in goats. Affected goats are in pain and cannot walk. Exercising goats that are pregnant, reducing the amount of feed in the latter part of the pregnancy to avoid fat conditions, and making certain that the goats have clean, fresh water at all times helps prevent ketosis.

Milk fever can occur, but rarely does, in goats that are lactating heavily. Some heavy-milking goats may deplete their calcium stores. They cannot stand and may die of progressive paralysis. An intravenous injection of calcium gluconate results in rapid recovery. The goat may be up and normal in less than an hour after a calcium gluconate injection.

Goats are subject to many severe external and internal parasites. The external parasites include **lice** and **mites.** The animals may be sprayed, dipped, or dusted with appropriate insecticides similar to those used on dairy cows. Dairy feed stores are usually good sources of materials and advice for treatment.

Internal parasites of goats include **stomach worms** and **coccidia.** The same treatment for stomach worms that is effective in sheep or horses can be used. **Coccidiosis,** caused by a protozoan organism in the intestinal tract, is found mostly in young goats. **Drenching** the animal with sulfa drugs or certain antibiotics helps control coccidiosis. Routine treatment with recommended anthelmintics controls internal parasites in goat herds. Professional help in administering these drugs is advisable.

BRUSH CONTROL

Angora goats, Spanish meat goats, or cull dairy goats are used in some areas to kill brush. One method is to stock the area heavily with goats and give them little feed as long as their health is not impaired. Keeping the goats hungry for green feed causes all brush to be destroyed in 2 years because the goats nip off the buds of new growth, thus starving the plant. It usually requires 2 years to starve hardy plants, but some of the less hardy ones will succumb the first year. Another method is to mix goats with cattle for 6–10 years. The goats tend to eat leaves of brushy plants while the cattle consume the grasses. Some goat owners offer their herds for hire to help other landowners control undesirable plants. Stocking rates for goats are 4 goats per cow equivalent on improved pastures and 10 goats per cow equivalent on brushy range.

CHAPTER SUMMARY

■ Goats have a large impact on the economy and food supply of people in several countries of the world. The consumption of both goat's milk and goat meat is increasing in the United States.

■ The major kinds of domesticated boats are (1) dairy goat, (2) meat goat, (3) Angora goat, (4) Cashmere goat, and (5) pygmy goat.

■ Besides the Angora goat, the most numerous goats are the leading breeds of dairy goats: Toggenburg, Saanen, Alpine, Nubian, LaMancha, and Oberhasli.

■ Goat management involves understanding the primary principles and production practices affecting reproduction, feeding, and health.

KEY WORDS

browse	*ad libitum*	mastitis
forbs	bluetongue	brucellosis
meat goat	Johne's disease	footrot
dairy goat	caseous lymphadenitis	ketosis
Angora goat	caprine arthritis	milk fever
Cashmere goat	encephalitis (CAE)	lice
pygmy goat	caprine	mites
cabrito	pleuropneumonia	stomach worms
tight lock	contagious ecthyma	coccidia
flat lock	(soremouth)	coccidiosis
open fleece (fluffy)	enterotoxemia	drenching
emasculator	urinary calculi	
balling gun	tetanus	

REVIEW QUESTIONS

1. What are the five kinds of goats and what are they used for?
2. Describe the reasons for the growth of the meat goat industry.
3. What is the length of the estrous cycle in goats?
4. What is the duration of gestation in goats?
5. *True or False:* Dairy goats are about equal to dairy cows in efficiency of converting feed into milk.
6. *True or False:* The milk of dairy goats is more easily digested by humans than is cow's milk.

7. What are the six major breeds of dairy goats in the United States?

8. Why is woven wire preferred to rail for fencing of goats?

9. How are goats permanently identified?

10. When is the normal breeding season for goats?

11. *True or False:* It is acceptable to run does with rams and bucks with ewes because goats and sheep will not interbreed.

12. Describe the major diseases affecting goats.

SELECTED REFERENCES

ADM Alliance Nutrition. 2005. *The Goat Guide.* Quincy, IL.

Coffey, L. 2006. *Meat Goats: Sustainable Production.* Fayetteville, AR: National Sustainable Agriculture Information Service.

Gall, C. 1981. *Doing Goats.* London: Academic Press.

Haenlein, G. F. W. 1993. Producing quality goat milk. *Int. J. of Anim. Sci.* 8:79.

Haenlein, G. F. W. 1996. Status and prospects of the dairy goat industry in the United States. *J. Anim. Sci.* 74:1173.

Mackenzie, D., and R. Goodwin. 1993. *Goat Husbandry.* 5th ed. Boston, MA: Faber and Faber.

National Research Council. 1981. *Nutrient Requirements of Goats.* Washington, DC: National Academy Press.

National Symposium on Goat Meat and Marketing. 1991. Langston, OK: Caprine Development Foundation.

Paschal, J. C., B. F. Craddock, C. W. Honselka, and D. Rollins. 1992. *Spanish Goat Management.* Texas Ag. Ext. Service Bul. B-5021.

CHAPTER 37

Careers and Career Preparation in the Animal Sciences

LEARNING OBJECTIVES

- Describe the career paths of Animal Science program graduates
- Explain the diverse nature of careers in animal-related fields
- Compare the salary opportunities in agribusiness, production, food processing, and veterinary medicine
- Explore the process of career preparation
- Write an initial career plan

The millions of domestic animals that provide food, fiber, and recreation for people create a multitude of diverse career opportunities. The results of a placement survey of approximately 10,000 agriculture graduates in the United States are shown in Table 37.1. Data on placement in careers in the animal sciences

TABLE 37.1   Post Graduation Careers of Bachelor Degree Recipients in Agriculture

Post Graduation Career/Activity	Percent of Graduates
Agribusiness (industry and business)	34%
Graduate and professional study	20
Farming and ranching	11
Education, extension, communications	10
Scientists, engineers, related specialists	9
Miscellaneous	16

Source: Adapted from the Food and Agricultural Education Information System Placement Report.

follow a similar pattern because workers in those fields comprise a high percentage of the graduates shown in these tables.

Beef cattle, dairy cattle, horses, poultry, sheep, and swine are the animals of primary importance in animal science curricula; reproduction, nutrition, breeding (genetics), meats, and management are the specialized topics typically covered in animal science courses. A few animal science programs include studies of pet and companion animals. Most college and university curricula in the animal sciences are designed to assist students in the broad career areas of production, science, agribusiness, and the food industry. Some animal science departments combine production and agribusiness into an industry concentration or a double-major program. The major careers in these areas are shown in Table 37.2.

TABLE 37.2   A Sample of Animal Sciences Careers in Production, Science, Agribusiness, Food Industry, and Equine-Related Fields

Animal Sciences				
Production	Science	Agribusiness	Food Industry	Equine
Feedlot positions	Graduate school for Master of Science (M.S.) and Doctor of Philosophy (Ph.D.) degrees	Sales and management positions with feed companies, packing companies, health and pharmacy companies, equipment companies, etc.	Food-processing plants	Management (stalls, farm, breeding barn, foals, etc.)
Livestock production operations (beef, dairy, swine, sheep, and poultry)	Research (university or industry) in nutrition, reproduction, breeding and genetics, products, and production management	Livestock publications	Food-ingredient plants Food-manufacturing plants	Facility design and construction Trainer—riding instructor
Ranch positions	University or college teaching	Advertising and promotion	Government—protection and regulatory agencies	Writer, artist, photographer
Breed associations	University extension and area extension	Finance (PCAs, banks, etc.)	Government—Department of Defense (food supply and food service)	Feed, pharmaceuticals, tack sales, marketing, and service companies
AI studs and breeding		Public relations		
Livestock buyers for feeders and packers	Management positions in industry	Meat grading (federal government)	Government—Department of Agriculture (research and information)	Farrier
County extension agents/specialists	Government work International opportunities	International marketing opportunities	University research, teaching, and extension	Recreation related (guide, outfitter, guest ranch, etc.)
Meat grading and distribution	Laboratory technicians	Graduate school for Master's in Business Administration (M.B.A.)	Positions with food companies	Insurance agent Veterinarian or vet-technician
Marketing (auctions, Cattle Fax, livestock sales management, alliances, etc.)	Veterinary school for Doctor of Veterinary Medicine (D.V.M.): private practice, consulting, university teaching and research	Foreign agriculture	Research and development with food companies	Massage therapist Association management, marketing, public relations
International opportunities		Technical sales and service	Quality control management	Horse buyer, appraiser
Livestock and meat market reporting (government)	Meat inspection	Companion animals		Horse-related real estate sales
Feeding manufacturing	Industry/commercial			
Companion animals: boarding, breeding, training	Companion animal research			
Game farm management				

FIGURE 37.1   Many opportunities exist for people to work in animal production. (A) Herdsman. (B) Livestock marketing specialist. (C) AI technician. (D) Horse trainer. (E) Riding instructor. Photo D courtesy of the American Quarter Horse Association.

PRODUCTION

Many individuals are intrigued with animal production because of the possibility of being their own boss and working directly with animals (Fig. 37.1). These ambitions should be held in light of the financial investment required in land, which is typically hundreds of thousands of dollars. Beef cattle, dairy, poultry, sheep, and swine production operations continue to become fewer in number, more specialized, and larger in terms of the amount of capital invested.

TABLE 37.3   Livestock Enterprise Salaries and Benefits, 2006

Enterprise/Position	Average[a] ($1,000)	Range ($1,000)	% Partial or Full Benefit in Addition to Salary	
			Housing	Insurance
Swine				
Manager	42.8	58–33	41	81
Assistant manager	32.1	43–26	22	83
Herdsman	30.2	35–27	10	81
Dairy				
Manager	44.0	48–35	21	68
Assistant manager	32.7	43–25	8	81
Herdsman	35.7	40–30	68	42
Beef				
Manager	44.3	68–30	52	82
Assistant	32.6	38–25	42	68

[a] Does not include performance bonus/incentive.
Source: Adapted from Agri-Careers, Inc.

Some who find a career in the production area have a family operation to which they can return. A few others have the needed capital to invest. Others find opportunities via lease arrangements, in managing for non-family-owned enterprises, or with investors.

Sometimes students who seek employment in animal production are disillusioned with the base salary, benefits, and working hours when production work is compared with agribusiness careers. Salary and benefits in the production sector are becoming more competitive (Table 37.3). However, some graduates are anxious and willing to obtain experience and prove they can work, and eventually locate permanent employment in production operations. Some producers of cattle, sheep, swine, and horse operations allow an employee to buy into an operation on a limited basis or own some of the animals after working a year or two.

Certain careers require education and experience in the production area even though the individual will not be producing animals directly. Career opportunities such as those for breed association and publication positions, field persons, extension agents, associated product and service sales, and livestock marketing require an understanding of production, as one works directly with livestock producers.

SCIENCE

An animal science student who concentrates heavily in science courses is usually preparing for further academic work with a goal of achieving one or more advanced degrees. A minimum grade average of 3.0 and relatively high scores on the Graduate Record Exam are usually required for admission to graduate school or a professional veterinary medicine program. Advanced degrees are usually obtained after entrance into graduate school or after being admitted to a professional veterinary medicine program (Fig. 37.2) where a Doctor of Veterinary Medicine (D.V.M.) degree is awarded. Table 37.4 shows where graduates of veterinary medical colleges are employed and their average starting salaries.

FIGURE 37.2 Careers in the veterinary profession require admission to a professional school of veterinary medicine.

TABLE 37.4 Professional Incomes and Employment for U.S. Veterinarians

Career Area	1995 Avg. Income[a]	2003 Avg. Income[a]	Avg. Starting Salary in 1995	Avg. Starting Salary in 2006
Large animal exclusive	$63,547	$101,435	$39,535	$61,028
Large animal predominant	56,500	80,689	34,290	53,397
Mixed animal	50,862	83,656	31,892	52,254
Small animal predominant	55,179	92,236	30,970	56,241
Small animal exclusive	57,991	103,108	31,882	57,117
Equine predominant	64,240	117,799	31,924	40,130
Average (private practice)	$57,557	101,140	NA	55,031
College or university	$63,749	$96,788	NA	NA
Federal government	59,626	90,163	NA	NA
State/local government	57,430	74,997	NA	NA
Uniformed services	57,619	86,587	NA	NA
Corporate	94,046	131,748	NA	NA
Average (public or corporate)	$67,043	103,750	NA	NA

[a] Annual salary after several years of employment.
Source: Adapted from *JAVMA* 229:1087 (Oct. 1, 2006); 226:208 (Jan. 15, 2005).

One of the major emerging opportunities for veterinarians is in the area of food animals. The shortage of well-trained food animal veterinarians is of concern to the livestock industry, and a number of programs have been initiated to attract people into the field of food animal medicine.

FIGURE 37.3 Advances in science and biotechnology offer new career opportunities to students with excellent preparation in both theory and application.

FIGURE 37.4 Opportunities to work as executives in agricultural organizations offer job satisfaction to those with an interest in working with people.

The Master of Science (M.S.) and Doctor of Philosophy (Ph.D.) degrees are the advanced degrees received by the science-oriented student (Fig. 37.3). Although the 2-year college certificate or Bachelor of Science (B.S.) degree allows breadth of education, advanced degrees are generally directed toward a specialization. These specialized animal science areas are typically in nutrition, reproduction, breeding (genetics and statistics), animal products, and management systems.

AGRIBUSINESS

Although the number of individuals employed in livestock and poultry production has decreased, the number of individuals and businesses serving producers has greatly increased. Positions in sales, management, finance, advertising, public relations, and publications are prevalent in feed, animal health, equipment, meat packing, marketing, and livestock organizations (Fig. 37.4).

Agribusiness careers that relate to the livestock industry require a student to have a foundation of knowledge of livestock along with an excellent comprehension of business, economics, computer science, and effective communication. A graduate should understand human interaction, know how to effectively communicate, and enjoy working with people. Extracurricular activities that give students experience in leadership and working with people are invaluable for meaningful career preparation.

Some animal science students who concentrate heavily in business and economics courses will pursue an advanced degree. This is typically a Master's in Business Administration (M.B.A.). A combination of a B.S. degree in Animal Science with an M.B.A. is an excellent preparation for management positions in livestock-oriented businesses.

FOOD INDUSTRY

Red meats, poultry, milk, natural fibers, and eggs—the primary end products of animal production—provide basic nutrition and eating enjoyment for millions of consumers. Career opportunities are numerous in providing one of our basic needs—food. Processing, packaging, and distribution of food are important components of the food-production chain. New food products are continually being produced (Fig. 37.5), and new methods of manufacturing and fabricating food are being developed. Electrical stimulation of carcasses for increased tenderization; vacuum packaging of primal cuts for a longer shelf life, and preparation of convenience products are examples of innovations in meat processing. A vital and continuing challenge over the next several decades will be to provide a food supply that is nutritious, safe, convenient, attractive, and economical—which is also satisfying to the consumer.

Middle and senior management positions in the food processing sector are financially rewarding and provide a high degree of professional challenge and fulfillment (Table 37.5). Attainment of at least a B.S. degree is an important prerequisite for these positions.

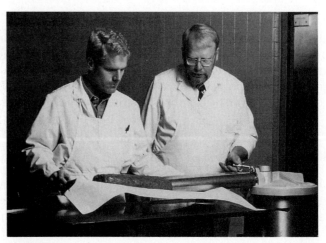

FIGURE 37.5 Improving value-added characteristics of animal products is essential to success.

TABLE 37.5   Meat and Poultry Processing Company Salaries—2005

Position	Median Total[a] Cash Compensation	Percent with a Four-Year Degree or Higher
Corporate Chief Executive Officer	$298,700	55
Corporate Chief Financial Officer	$153,700	100
Corporate Marketing Executive	160,600	60
Plant General Manager	123,300	72
Plant R&D Manager	77,000	100
Plant HACCP Manager	74,250	74
Plant Quality Assurance Manager	70,000	72
Plant Engineer	79,050	60

[a] Salary plus profit sharing, etc.

Source: Adapted from *Meat & Poultry,* 2006.

HORSES

While horses continue to play an important role in many agricultural settings, their greatest use is recreation and sport. The list of employment options related to horses is exceptionally diverse. Many of the opportunities are similar to those in other livestock industries (Table 37.2). However, students will find careers in the horse industry that range from artist to trail ride operator (Table 37.2).

People seeking career options in the horse industry will typically find it easier to get started in positions related to sales of products and services utilized by horse owners. It is more difficult to gain entry as a horse facility manager or trainer. Flexibility, tenacity, and a willingness to work up the ladder are essential elements of success. The value of internships and mentors cannot be overstated (Fig. 37.6).

COMPANION ANIMALS

The most obvious career related to companion animals is that of the veterinarian. However, jobs are available in boarding, breeding, training, and grooming of companion animals. Furthermore, the animal health care products and pet nutrition industries offer opportunities for those desiring to work with companion animals (Table 37.2). Very limited opportunities are available in working with animals at zoological gardens and wildlife parks (Fig. 37.7).

INTERNATIONAL OPPORTUNITIES

Much has been written during the past decade of the challenge to provide adequate nutrition to an ever-expanding world population. Many countries have tremendous natural resources for expanded food production but lack technical knowledge and adequate capital to develop these resources. Federal government programs, designed to help foreign countries help themselves, offer several career opportunities in organizations such as the Peace Corps, Vista, and USAID. Many of these opportunities of assisting people in agricultural production in developing countries are open to individuals educated and experienced in the animal sciences.

The Foreign Agricultural Service of the USDA employs people in animal economics, marketing, and administration as attachés and international secretaries.

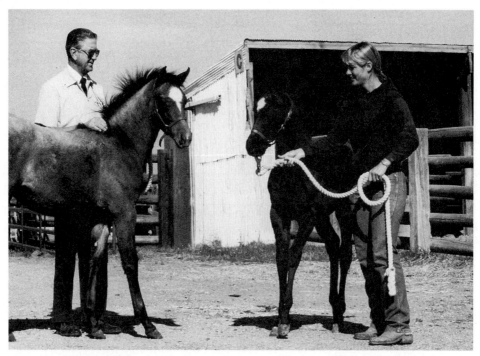

FIGURE 37.6 Internships and working with mentors are critical for students desiring to obtain meaningful career opportunities.

FIGURE 37.7 Limited opportunities exist to work with zoo animals.

Animal science students take several courses in economics, marketing, foreign languages, and business administration if they wish to qualify for these positions.

Certain private industries offer opportunities in animal production and related businesses in foreign countries. Multinational firms such as livestock feed companies, animal health and pharmaceutical companies, companies that export and import animals and animal products, and consulting companies are some examples.

MANAGEMENT POSITIONS

Management and administrative positions exist in the production, science, agribusiness, and food industry areas of animal science. Because management sometimes implies more responsibility and more pay than other positions that require more physical labor, young men and women typically desire management positions. Nevertheless, although management and administrative salaries are usually higher, the workload and pressures are usually much greater.

Many college graduates have the impression that once the degree is in hand, it automatically qualifies them for management positions. In actuality, management positions are usually earned based on a person's proven ability to work with people, solve problems, and make effective decisions. After being employed for a few years, individuals having previously learned the academic principles must also experience the component parts of an operation or business. For example, presidents or vice presidents of feed companies have typically started their initial careers in feed sales. Most successful managers have had a continual series of learning experiences since the end of the formal education. Managers need to understand all aspects of the business they are managing, particularly the products that are produced and sold. Students can begin the process of preparing for management opportunities by seeking appropriate internships, developing a professional network, reading and studying business and entrepreneurial issues, and fine-tuning communication and presentation skills.

The importance of involvement in student clubs, judging teams, and other extracurricular activities cannot be overstated when evaluating the career paths of people successful in animal agriculture (Fig. 37.8).

FUTURE CAREER OPPORTUNITIES

A 2005 USDA report (Table 37.6) suggests that during 2005–2010, average annual employment opportunities for agricultural graduates will be 52,000 with only 49,300 graduates (agriculture, natural resources, veterinary medicine) available. The shortage will largely be concentrated in marketing, financial, and sales positions. However, excellent opportunities will exist in value-added technologies and processes, as well as in the science and engineering.

In contrast, consumer services professionals and agricultural production managers will find a more competitive job market. This is not to say that there are no opportunities in these fields. However, students seeking employment in these areas will have to assure that they are exceptionally well prepared and able to make significant contributions to potential employers.

CAREER PREPARATION

An education should provide opportunities for people to earn a living, continue learning, and live a full, productive life even beyond the typical retirement years. A broad-based education is important because a person may be preparing for a career that does not presently exist or a career not recognized at the present time. People tend to change positions 6–8 times during their careers,

FIGURE 37.8 Participation in a judging team allows students the opportunity to travel, expand their professional networks, and become more familiar with the challenges and enjoyment that come from working in the livestock industry.

TABLE 37.6   Projected Average Annual Employment Opportunities for College Graduates in the Food and Agricultural Sciences, United States, 2005–2010

Job Category	Opportunities (N)	Percent of Total	Qualified Graduates Available (N)
Science and engineering occupations	12,916	25	7,423
Management and business occupations	24,125	46	12,198
Communication–education occupations	6,967	13	6,323
Agriculture production occupations	8,022	16	6,381

Source: Adapted from USDA, 2005.

so a person needs to be flexible and adaptable and willing to continue learning throughout life. Students should be aware of industry trends affecting employment opportunities (Table 37.7).

Occasionally someone chooses a career for which he or she has little real talent or for which she or he is not qualified or properly motivated. These individuals may spend years preparing for or achieving partial success when they could have been outstanding in another area. In the animal sciences, students preparing for acceptance into veterinary medicine, graduate school, or another professional school seem to experience this frustration frequently. Individuals are capable of success in several different careers if they receive the appropriate

TABLE 37.7 Four Major Trends That Significantly Affect the U.S. Employment Opportunities for Agricultural Graduates

Consumers and their preferences for food, services, and experiences.
Changing structure of the U.S. and global food system driven by markets, consolidation, and policy.
Innovations in science and technology.
Public policy relative to agriculture and the food system including food security issues.

Source: Adapted from USDA, 2005.

education and experience. A career choice should be consistent with a person's interest, desire, and motivation. A career goal should be established with flexibility for change if it proves to be unrealistic.

After tentatively choosing a career area, it should be pursued vigorously with sustained personal motivation. Students should write to and visit with people currently working in the student's chosen career area. Every personal visit or written letter should request another lead or source of further information. Students should continually evaluate the reality of a chosen career. They should think realistically about the facts rather than getting caught up in the perceived glamour of a career choice. Career interest tests can be helpful in identifying areas to pursue in more detail. However, these tests are not infallible; they can provide direction but they are not the final answer. Students should always temper the results with additional information from several sources before making final decisions.

Preparing for a career is a lifelong process that is founded on several fundamental concepts—there are no shortcuts, first impressions count, improvement is a continuous process, and development of an array of skills is typically more valuable than having a narrow specialization. Furthermore, there is absolutely no substitute for hard work, tenacity, and enthusiasm regardless of the career area. Students desiring insight into a particular career should practice a three-step process—observe, evaluate, and emulate. Opportunities to observe professionals should be pursued, the traits demonstrated by successful professionals should be critically evaluated, and then the true principles of success should be emulated.

The effectively prepared student maintains a balance of good grades, development of teamwork and leadership skills via participation in extracurricular activities, and attainment of professional experiences via internships, independent studies, and summer employment. Students seeking careers related to the livestock, poultry, and food industries should develop appropriate technical skills in animal sciences and then couple those experiences with study in related fields. This process allows individuals to create unique arrays of skills that enhance the number of options available.

Employers typically list hard work, communication skills, people skills (management, coaching, leadership, and teamwork), job experience, business background, willingness to relocate, technical knowledge appropriate to the position, curiosity, and enthusiasm as important traits. Too often entrants into a profession overfocus on the initial salary instead of seeking unique experiences that offer more upside potential once job performance is established. A combination of useful courses, practical experience, effective communication, and excellent work habits is the best assurance of finding and keeping a meaningful career in the animal sciences.

CHAPTER SUMMARY

- Livestock production and the associated organizations and agribusinesses provide numerous and varied career opportunities for students majoring in the animal sciences.

- Students concentrating on science courses are usually preparing for advanced degrees, such as doctor of veterinary medicine (D.V.M.) or the research-oriented degrees of master of science (M.S.) and doctor of philosophy (Ph.D.).

- Students concentrating on economics, agricultural business, and business courses are usually preparing for careers in livestock production and the agribusiness careers in the meat, animal health, and feed industries.

- In addition to an excellent course preparation, employers emphasize skills in effective communication, computers, and work ethic.

- Meaningful internships make a significant contribution to career preparation.

REVIEW QUESTIONS

1. *True or False:* Agribusiness employs more agricultural degree recipients than does farming and ranching.

2. *True or False:* Career options for animal science graduates are limited.

3. *True or False:* Managers in the swine industry tend to have higher average salaries than managers in the beef industry.

4. What are four major trends affecting employment of agricultural graduates?

SELECTED REFERENCES

Agri-Careers, Inc. 2006. Highways 148 and 92 West, Wassena, Iowa 50853, (800) 633-8387.

Annual M&P Salary Survey. 2006. *Meat & Poultry*, July.

Employment starting salaries and educational indebtedness of 2006 graduates of U.S. veterinary medical colleges. 2006. *JAVMA* 229:1087.

Food and Agricultural Education Information System. 2005. Placement for Agricultural and Natural Resources Graduates. Washington, DC: USDA.

Goecker, A. D., J. L. Gilmore, E. Smith, and P. G. Smith. 2005. *Employment Opportunities for College Graduates in the Food and Agricultural Sciences.* Washington, DC: USDA, Higher Education Program.

Income of U.S. veterinarians. 2005. *JAVMA* 226:208.

CHAPTER 38

Aquaculture

LEARNING OBJECTIVES

- Compare and contrast the role of aquaculture in the United States and the world
- Overview the primary aquatic species managed in U.S. aquaculture
- Describe the steps in catfish production
- Overview the process of trout production
- Discuss tilapia production
- Overview the crawfish production cycle
- Discuss health management in aquaculture

OVERVIEW

Aquaculture production involves both plants and animals. However, this chapter will focus entirely upon the raising of animals in water for the benefit of people. Given the vast quantities of water that cover the earth's surface, it is not surprising that global aquaculture produces nearly 35 million tons of raw product, valued at more than 46 billion dollars (Fig. 38.1). While capture of nonfarmed fish has stabilized since the mid-1980s, aquaculture has experienced steady growth over the same time period. On a worldwide scale both aquaculture and capture fishery production is most extensive in Asia with China as the significant producer. Per-capita consumption is highest in Japan with the United States comparable to that of China.

There are five major fish species raised in U.S. aquaculture—catfish, trout, salmon, tilapia, and hybrid striped bass. Crawfish, mollusks, and shrimp are also economically significant.

The United States faces the following competitive challenges relative to foreign producers: expense of compliance with regulation, high costs of disease management, limited access to investment capital, and opposition to new marine leases.

A

B

C

D

E

FIGURE 38.1 Aquaculture is becoming an increasingly important food supplier. The (A) harvests of (B) catfish, (C) crawfish and (D) tilapia are important commercial ventures in the United States. A growing demand for specialty dishes such as crawfish (E) has increased production of farmed fish.

CATFISH OVERVIEW

Catfish production is dominated by four states—Alabama, Arkansas, Louisiana, and Mississippi (Tables 38.1 and 38.2). Catfish was produced on 156,000 acres of water located on 1,160 U.S. farms in 2005 with a sales production of nearly $462 million. Processors paid average monthly prices ranging from $0.73 per lb to $0.84 per lb in 2006. Processors sold fresh and frozen catfish products for an average price of $2.54 per lb in the same year. Fresh and frozen catfish is merchandised by processors as fillets (53% of volume), whole (36%), or further processed as nuggets or steaks (11%).

Growth of the catfish industry can be attributed to increasing numbers of health-conscious consumers and successful marketing that expanded demand for catfish beyond the traditional southern regional market. The production of catfish is expected to remain concentrated in the South due to the abundance of land/water resources, favorable climatic conditions, and the availability of technical expertise and infrastructure.

Catfish Production

Seven species of catfish have been successfully cultured (Table 38.3). The external features of the channel catfish are illustrated in Figure 38.2. Cultured catfish **spawn** at weights greater than 2 pounds, eggs will hatch within a 5- to 10-day period with water temperature influencing the speed of hatch. Warm temperatures increase speed of the hatch, while cooler temperatures reduce the time to hatch. Temperatures outside the range of 65–85°F negatively affect the hatch.

Catfish prefer to spawn in environments that are relatively dark with protected surroundings. Therefore, the use of submerged containers in commercial

TABLE 38.1   Number of Farms and Gross Sales from Catfish—2005 and 1998

	Farms—2005	Sales ($1,000)	Farms—1998	Sales ($1,000)
Alabama	192	98,413	250	58,222
Arkansas	142	77,852	156	55,307
Louisiana	33	14,998	100	28,427
Mississippi	386	243,122	404	285,366
United States	1,160	461,885	1,370	450,710

Source: USDA.

TABLE 38.2   Catfish Numbers and Live Weight of Brood Fish, Food-size Fish, Stockers, and Fingerlings by State—2005

State	Broodstock		Food-size Fish		Stockers		Fingerlings	
	N (1,000)	Live Wt. (1,000 lb)	N (1,000)	Live Wt. (1,000 lb)	N (1,000)	Live Wt. (1,000 lb)	N (1,000)	Live Wt. (1,000 lb)
Alabama	298	2,189	78,258	142,186	13,580	3,603	35,941	NA
Arkansas	70	331	65,061	104,140	8,313	1,177	121,286	NA
Louisiana	NA	NA	14,841	21,130	NA	NA	NA	NA
Mississippi	68	317	222,508	313,441	8,241	1,100	507,268	NA

Source: USDA.

TABLE 38.3   Cultured Species of Catfish

Channel catfish
Blue catfish
White catfish
Black bullhead
Brown bullhead
Yellow bullhead
Flathead catfish

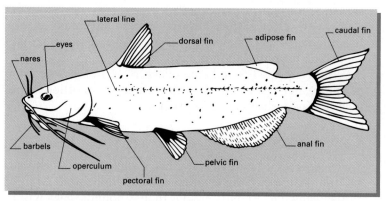

FIGURE 38.2   External parts of the channel catfish. From Southern Regional
Aquaculture Center, Publication #180.

ponds greatly enhances the spawning rate. Growers use cans, buckets, or boxes
that are large enough to house an adult brood pair. A reasonable production goal
in a correctly managed catfish enterprise is >1,000 **fingerlings** per pound of
brooder (female).

Catfish are commercially spawned in ponds by the use of one of the following methods:

1. Fry transfer—freshly hatched fry are transferred from spawning containers to nursery ponds. Equalization of temperature and quality of
   water between the two ponds is required for successful transfer.

2. Egg transfer—following spawning, eggs are removed for incubation in
   a hatchery. While the most effective, this system requires the greatest
   expertise and intensity of management.

3. Pen spawning—brood pairs are penned together but carefully monitored to
   assure that the male isn't hurting the female. Just following spawning, the
   female is removed; eggs can then either be pond-hatched or transferred to
   an incubator.

Regardless of methodology, providing an optimum environment for hatching is critical (Table 38.4).

As **sac-fry** hatch, they swim through the screen of the incubation baskets
to cluster together in the bottom of the trough. As sac-fry supply their own
nourishment via the attached yolk sac, they do not have to be fed for the first
24–48 hours. Once the sac-fry have absorbed the yolk sac, they are classified as
swim-up-fry and they begin to search for food. At this stage, they are moved to a
rearing trough. Optimal growing conditions are specified in Table 38.5. Following

TABLE 38.4   Optimal Conditions for Hatching

Water temperature	75°–82°F
Oxygen levels	≥6 ppm
Water pH	6.5–8.5
Water quality	Disease free, free of algae, decaying plant material
Water flow	2 gallon/minute for 100 g

Source: Adapted from Southern Regional Aquaculture Center, 2002.

TABLE 38.5   Optimal Conditions for Growing Catfish

Water temperature	85°
Oxygen concentrations	≥4 ppm
Stocking rate	
4–6 in. fingerlings	3,000–4,000 per acre
Ponds with limited water flow	<1,500 per acre
Harvest weight/age	1.25 lb/18 months

Source: Adapted from Southern Regional Aquaculture Center, 2002.

TABLE 38.6   Length-Weight Relationship for Fingerling Catfish

Length (in.)	Weight (lbs/1,000 fingerlings)
1.0	0.7
2.0	3.1
3.0	8.8
4.0	19.1
5.0	35.3
6.0	58.8
7.0	91.0
8.0	133.3
9.0	187.1

Source: Adapted from Parker, 1995.

the growing phase, catfish are rotationally harvested from ponds once they have achieved a desired weight and then the pond is restocked based on a 1:1 basis for those harvested. Thus, restocking is a nearly continuous process.

Stocking rates vary based on the production system and management expertise. Fingerlings (4–6 inches) can be stocked at 3,000–4,000 per acre. However, as stocking rate increases, the potential for health and nutritional problems is enhanced. Stocking rate assessment is enhanced by knowledge of the length-weight relationship in catfish (Table 38.6).

As is the case with other livestock enterprises, the nutritional management of the catfish enterprise is the most critical function. Least-cost balanced rations are required in catfish production. Palatability is also a critical issue to assure sufficient intake of supplemental feeds.

Catfish feed is available in meal or crumble forms and may be sinking, slow sinking, or floating. Water temperature is a major consideration in determining feed type. Floating feeds are typically more expensive but preferred for water temperatures in excess of 65°F as feeding behavior is more easily observed. Sinking and slow-sinking feeds are most frequently utilized when water temperatures fall below 65°F, as catfish stay deeper at these temperatures.

Because ponds are periodically topped (harvested), the variation in size of fish within the pond is substantial. However, daily feedings of multiple feed types and sizes is not practical. Therefore, most producers use $^5/_{32}$- to $^3/_{16}$-inch

TABLE 38.7   **Examples of Typical Feeds for Catfish**

Ingredient (percent of protein)	% of Ration			
	32%	32%	28%	28%
Soybean meal	36.5	34.5	26.3	21.2
Cottonseed meal	10.0	12.0	10.0	12.0
Merhaden meal	4.0	—	4.0	4.0
Meat/bone/blood meal	4.0	8.0	4.0	4.0
Corn grain	22.9	22.4	30.6	51.4
Wheat middlings	20.0	20.0	22.5	4.0
Dicalcium phosphate	1.0	1.0	1.0	1.0
Vitamin mix (specific)	include	include	include	include
Mineral mix (specific)	include	include	include	include
Catfish oil[a]	1.5	2.0	1.5	1.5

[a] Sprayed onto finished pellets to reduce dust.
Source: Adapted from Southern Regional Aquaculture Center, 2002.

TABLE 38.8   **Feed Consumption and Feed Conversion Ratio for Varying Catfish Sizes at Optimal Water Temperatures**

Fish Size (lbs/1,000 fish)	Feed Consumption (% of body weight)	Feed Conversion (ratio)
60	4.0–4.5	1.1–1.2
100	3.5–4.0	1.3–1.4
300	2.5–3.0	1.4–1.6
600	2.0–2.5	1.6–1.8
1,000	1.3–1.5	1.9–2.0
3,000	1.0–1.1	2.2–2.4

Source: Southern Regional Aquaculture Center, 2002.

feed pellets. Most feeding regimens are based on once-a-day feeding to satiety. Satiation is feeding to demand for a period of 10–15 minutes. Unfortunately, this system is highly subjective and may adversely affect water quality.

Examples of representative catfish diets are provided in Table 38.7. Most growers utilize a once-per-day feeding schedule with feed delivered to the ponds via mechanical blowers that scatter the feed widely across the pond surface to maximize feeding opportunity. Estimated feed consumption and feed conversion ratios at optimal water temperature are provided in Table 38.8. Note that feed consumption as a percent of body weight declines as fish size increases. Two-thirds of catfish farm operators utilize a 32% protein ration for food-size fish with the remainder choosing the 28% protein option.

Fish are typically fed early in the morning. Peak oxygen demand generally occurs approximately 6 hours postfeeding. Therefore, it is critical to assure sufficiently high dissolved oxygen levels at and beyond the time of feeding. Failure to do so may result in stress and/or death. Once water temperatures begin to drop below 70°F with the onset of winter, feeding schedules and rates should be adjusted downward.

Feeding practices are typically variable by season with a majority of farms feeding every other day in the spring and fall months and daily during the summer months. About one-third of producers have adopted limit feeding programs during the winter months.

TROUT OVERVIEW

Trout have been cultivated in the United States for nearly a century and a half. Originally, production was focused on supplementing wild populations. Rainbow trout are the most popular farm-raised trout with Brook and Brown trout also utilized in production systems. Idaho is the leading producer of farmed trout, although there is wide geographic distribution of enterprises (Table 38.9).

U.S. producers sold products valued at approximately $74 million in 2005. Food-size sales were valued at just over $63 million with an average value per pound of $1.05. The three largest market outlets were processors, fee and recreational fishing enterprises, and retailers/restaurants.

Stocker sales (fish 6–12 inches long) generated $5.2 million in revenue at an average price per pound of $2.82 in 2005. Primary market outlets include recreational fishing businesses, governmental agencies, and other producers (Table 38.10). Management of disease is critical to the financial health of an enterprise as it is the leading cause of trout loss (Table 38.11).

TABLE 38.9   Trout Operations and Sales by State—2005

State	Number of Operations	Total Sales (1,000 $)
CA	16	6,860
CO	8	1,474
ID	26	35,520
MI	21	1,011
MO	10	2,469
NY	24	640
NC	42	6,607
OR	16	806
PA	40	4,819
UT	7	537
VA	16	1,276
WA	14	9,127
WI	46	1,580
WV	16	345
Total	410	79,282

Source: USDA.

TABLE 38.10   Market Outlets for Food-size and Stocker Trout as Percent of Value—2001

Outlet	Food-size (%)	Stocker (%)
Live haulers	1	7
Fee/recreational fishing	19	59
Other producers	2	13
Government	1	11
Direct to consumer	3	3
Processors	68	1
Retail and restaurants	5	1
Other	1	5

Source: USDA.

TABLE 38.11   Trout Losses by Cause—2001

Cause	%
Disease[1]	74
Theft	<1
Chemicals	<1
Drought	2
Flood	<1
Predators[2]	13
Other	9

[1] Losses from both bacterial and parasitic causes.
[2] Mink, otters, birds, and others.
Source: USDA.

Trout Production

The majority of cultured trout are rainbow trout. On trout farms, eggs are typically produced on brood-fish production units and then transferred to separate units where the production of food fish occurs. Unlike catfish, trout egg incubation and hatching occurs at cooler temperatures (~55°F). Trout eggs are usually shipped once they reach the "eyed" stage or about midway in the incubation period. Shipping of trout eggs from one region of the country to another is relatively common. Because of the disease risk associated with transferring eggs from one production unit to another, an effective disinfectant plan should be in place. Eggs are disinfected typically by bathing them for 10 minutes in an iodine solution. The iodine should be concentrated at 100 ppm. It is very important that the concentration level be quantified. Furthermore, in acidic waters or where the alkalinity is less than 30 ppm sodium bicarbonate should be added at 0.05%. Once the eggs have been gently and thoroughly bathed in the disinfectant, they can be transferred to incubators.

Three types of **incubator systems** are most common. These include vertical trays, "California" trays, and upwell incubators (Fig. 38.3). "California" trays are flat-bottomed, which fit horizontally into rearing troughs. Vertical trays are arranged in stacks to make more efficient use of space when incubating large numbers of eggs. The stacking also allows re-aeration of water as it moves through the stack. The upwell incubator partially suspends eggs in the flow of water. In all these systems, eggs should not receive direct sunlight and dead eggs should be removed within 3-day intervals. Following hatching, sac-fry are moved to rearing troughs. At approximately 14 days posthatching, the fry swim up to the surface of the trough.

Once a majority of fry swim up, feeding is initiated. Automatic feeders are used to provide a frequent feeding schedule (Table 38.12). Overfeeding should be avoided and unused feed should be removed daily. Once the fry grow to 200 to 250 fish per pound, transfer to larger grow-out tanks or **raceways** takes place.

Fish grown in raceways require a large amount of high-quality water. A flow rate of 2–3 gallons per minute is required to prevent accumulation of solid wastes and to dilute liquid wastes. Ammonia concentration should be kept low to assure optimal fish health. Stocking rates for the raceway are approximately 4.5 lb per cubic foot. Maintaining fish in comparable sized groups is important to assure uniformity at time of marketing. Optimal temperatures for trout growth are 55–65°F. Harvest should occur when trout are 7 to 14 inches in length and weighing $\frac{1}{2}$ to 1 pound at 7 to 14 months of age.

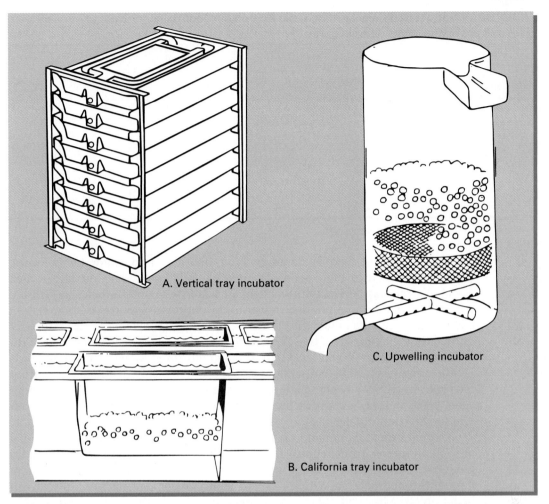

A. Vertical tray incubator

C. Upwelling incubator

B. California tray incubator

FIGURE 38.3   Trout incubator alternatives. From Southern Regional Aquaculture Center, Publication #220.

TABLE 38.12   **Recommended Feeding Rate for Small Rainbow Trout**

Fish Size (number/pound)	Feed (size)	Feedings/Day (N)	Amount of Feed (% of fish weight/day)
2,500–2,000	starter meal	8	6–9
2,000–800	#1 granule	8	5.2–7.5
800–250	#2 granule	6	5.1–7.2
250–100	#3 granule	4	4.2–6.1
100–30	#4 granule	3	3.2–4.9

Source: Southern Regional Aquaculture Center, 2002.

Nutritional management of trout is based primarily on high protein feeds. Fry and fingerling trout need feeds that are approximately 50% protein and 20% fat. As fish grow, feed composition should be approximately 38–45% protein and 10–18% fat. The transition between feeds is accomplished via moving from a "crumble"-type feed to a pelleted feed. Fish feeds are typically manufactured commercially and utilize fishmeal as the primary protein source.

TABLE 38.13 Suggested Daily Feeding Ratio (% body wt.) for Rainbow Trout at Varying Sizes and at Varying Water Temperatures (°F)

Number of Fish/lb	Temperatures				
	38–41	46–48	52–54	58–60	64–67
1,200	3.6	4.5	5.2	6.5	7.1
600	3.5	4.3	5.1	6.2	6.8
300	3.4	4.2	5.0	5.8	6.3
100	2.9	3.9	4.6	5.0	5.5
60	2.6	3.3	4.0	4.4	4.7
30	1.8	2.7	3.3	3.8	4.0
15	1.4	2.1	2.7	2.9	3.2
5	0.9	1.2	1.6	1.9	1.8
1	0.5	0.8	1.1	1.4	1.4

Source: Southern Regional Aquaculture Center, 2002.

The amount of feed is dependent on fish size and water temperature. Smaller fish require more feed per unit of body mass than do larger fish. Metabolic rates also vary with changes in water temperature because fish are cold-blooded. Trout maintained in warmer water require more feed than fish in cooler environments (Table 38.13). At water temperatures below 38°F, trout require only a maintenance diet. When water temperatures exceed 68°F, trout digest feed inefficiently and fecal matter may contribute to heavy nutrient loads in the water. As water temperatures rise, oxygen levels decline. Thus, trout managed in warmer waters may experience more respiratory challenge. Management of water temperature and water quality are crucial to assuring excellent growth performance.

Feeding protocols depend on the age and weight of the trout and the size of the enterprise. Trout require feed 7–10 days posthatching and are initially fed a finely ground feed offered up to 10 times per day. Automatic feeders are frequently incorporated with 2–3 hand feedings per day so that the fish can be frequently observed. As the fry increase in mass, feedings can be reduced to 5 times per day. It is important to assure that the feed is widely distributed along the water surface to ensure adequate feeding opportunity.

After fingerlings are moved to tanks or ponds, the use of demand feeders becomes desirable. A demand feeder (Fig. 38.4) consists of a series of bins containing pelleted feed that releases feed via activation of a pendulum that extends into the water. These feeders are located every 25 to 30 feet along the tanks. Trout over 5 inches in length are readily trained to use a demand feeder.

The benefits of demand feeders include enhanced growth rates and feed efficiency, elimination of sharp oxygen declines when feedings are scheduled for only several per day, and reduction in labor costs. Disadvantages include the tendency to use excess feed, overfeeding, and release of feed into a relatively small portion of the tank. Fish should not be fed prior to handling and transport. Prior to harvest, fish should be kept off feed for 3–4 days.

TILAPIA OVERVIEW

Tilapia were cultured by the Egyptians more than 3,000 years ago. Currently, tilapia are second only to carp as the world's most important cultured freshwater fish.

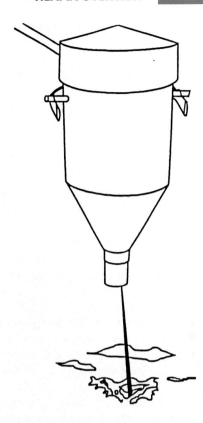

FIGURE 38.4 Demand feeder used in trout production. From Southern Regional Aquaculture Center, Publication #223.

Tilapia were introduced into the United States in the 1950s from Africa and the Middle East. Several species and hybrids are suitable for aquaculture in the United States. However, cold tolerance should be considered when choosing species. Nonetheless, tilapia are very adaptive to poor water quality and generally exhibit a high level of hardiness.

In 2006, 156 farms in the United States sold $31.3 million worth of tilapia. California and Idaho are the leading states.

Tilapia Production

Ideal water temperatures for tilapia range from 79–90°F. Tilapia are tolerant of salinity levels up to 30 ppt and low levels of dissolved oxygen so long as they have access to the surface. Tilapia must be wintered in heated water in more temperate climates.

Tilapia are **mouth-brooders** whereby the female deposits eggs into a prepared nest on the floor of the pond. The eggs are subsequently fertilized by the male at which time the female gathers the fertilized eggs in her mouth where they are held until they hatch. The number of eggs spawned is related to the size of the female with females at 2 pounds producing up to 1,500 eggs. Tilapia are highly fertile and can spawn year-round. Because of their high reproductive rate, excess numbers of fingerlings in a pond may become a constraint to production.

Tilapia are most frequently pond-cultured to allow the utilization of natural food sources. In pond-culture systems, it is critical to not overpopulate with fry and fingerlings or the percentage of marketable fish will be diminished.

TABLE 38.14   Comparison of Mixed-sex and Mono-sex Culturing of Tilapia

	Mixed-sex	Mono-sex
Stocking rate	1 gram fry (1 month old) 2,000–6,000/acre	4,000–20,000/acre
Harvest weight	8 oz/fish, 1,400 lb/acre	18 oz/fish, 2.2 tons/acre
Survival rate	70%	90%

Source: Adapted from Parker, 1995.

At harvest time, ponds are drained to allow a 100% harvest rate. This process also assists with removal of fry and fingerlings that would interfere with the next production cycle. There are two predominant pond production systems—mixed-sex culture and mono-sex culture (Table 38.14).

In mixed-sex culture, male and female fry are cultured together and harvested at about the time they reach sexual maturity. In mono-sex systems, fry are nurseried together and then separated by gender at the fingerling stage. This system allows males to maximize their growth rate.

Tilapia utilize a varied diet that includes natural food sources such as plankton, enval fish, and decomposing organic matter. Even with high levels of supplemental feeds, tilapia will achieve one-third to one-half of their growth from natural food sources. Tilapia have grown in prominence due to their hardiness, high reproductive rate, low cost of production, and the desirability of the texture and flavor of tilapia meat.

Alternatives to pond rearing of tilapia include **cage** or **tank production systems.** Tilapia can be raised in high densities in cages so long as water is freely circulating. The primary advantages of cage-rearing are that the breeding cycle is disrupted. Cages also allow convenience in harvesting, ease of treating disease, and relatively low capital investment. The disadvantages include greater risk of disease occurrence, complete dependence as supplemental feed, and lowered tolerances to poor water quality.

Tank culture of tilapia is superior to pond culture in that they disrupt the reproductive cycle, allow more rigorous control over environmental conditions, and require less labor input. The disadvantages include total reliance on supplemental feed, costs are increased due to water pumping and aeration, and the risk of mechanical failure requires backup electrical and mechanical systems to avert disaster.

CRAWFISH OVERVIEW

Crawfish are crustaceans and their production involves the culture of two primary species—the Red Swamp crawfish and the White River crawfish. The primary advantage of crawfish farming is the simplicity of the production method as no hatcheries are required to produce the young. Furthermore, because crawfish consume decaying plant material there is no requirement for the use of formulated supplementary feedstuffs. Disadvantages include the high water requirements of 70 to 100 gallons per minute per surface acre. Market disadvantages include the regional nature of domestic demand and the cost competitiveness of foreign producers, most notably China. In 2005, 648 farms in the United States produced sales of just over $21 million with Louisiana accounting for nearly 90% of the farms and 96% of the sales.

Crawfish Production

Crawfish are highly adaptable and their production is based on the mirroring of natural conditions that include a cycle of wet and dry conditions. The wet cycle is the time of growth and maturation, while during the dry cycle, crawfish burrow into pond banks.

Crawfish are managed in either **single-crop** (Table 38.15) or **rotational production systems.** Under single-crop systems, ponds are designed only for crawfish production. In rotational systems, crawfish production is rotated with other crop options or the site of production is rotated.

Single-crop ponds are typically one of three types—cultivated forage ponds, naturally vegetated ponds, or wooded ponds (Table 38.16). Cultivated forage ponds are the most intensively managed.

In rotational-pond systems, crawfish production is most typically rotated with rice and/or soybeans. In a rice-crawfish-rice system, two crops are harvested each year. In a rice-crawfish-soybean system, three crops are taken each two-year cycle. A comparison of production schedules is provided in Table 38.17.

Harvesting crawfish is labor intensive and nearly three-quarters of the production cost is incurred in harvest season. Harvest is a 4- to 5-month season lasting from November to June under optimal conditions. Crawfish are trapped in mesh wire enclosures set at a rate of 20 to 40 traps per acre. Traps are baited and emptied daily during harvest season. Bait purchases account for more than 40% of variable costs with labor accounting for an additional 20%.

TABLE 38.15 Crawfish Culture Cycle for Single-Crop Ponds

Season	Activity
April–May	Stock 50–60 pounds of adults per acre into new ponds
May–June	Drain pond over a 14–28-day cycle
June–August	Plant crawfish forage or manage natural vegetation
October	Re-flood ponds
November–May/June	Harvest crawfish
May/June	Drain pond to restart cycle (no restocking required)

Source: Adapted from Southern Regional Aquaculture Center, 2002.

TABLE 38.16 Comparison of Pond Types Used in Single-Crop Crawfish Production Systems

Factor	Cultivated Forage	Naturally Vegetated	Wooded
Ease of tillage	High	No tillage	Built on heavy clay soils in wooded swamps
Predominant crop	Rice	Natural vegetation	Leaf litter from cypress-tupelo swamps provide most of forage
Water circulation	Baffle levies/ recirculation systems	Minimal, water quality problematic	Intense water management needed
Yield/acre	Highest	High organic matter in soil may lower yields which are typically inconsistent	Lowest

Sources: Adapted from Southern Regional Aquaculture Center, 2002.

TABLE 38.17 Comparison of Management Calendar for Three Rotational Crawfish Systems

Production Phase	Production System (month of implementation)		
	Rice-Crawfish-Rice	Rice-Crawfish-Soybeans	Rice-Crawfish-Fallow
Plant rice	March–April	March–April	March–April
Ponds flooded, rice 6–8 in. high, stock adult crawfish	June	June	June
Drain field and harvest rice	August	August	August
Re-flood rice field	October	October	October
Harvest crawfish	December–April	December–May	December–June/July
Drain pond and replant rice— restocking may or may not be required	March–April	NA	NA
Drain pond and plant soybeans	NA	Late May–June	NA
Harvest soybeans	NA	October–November	NA
Re-flood field and harvest crawfish	NA	November–March	NA
Plant rice and restock crawfish	NA	March–April	NA
Drain pond and fallow	NA	NA	July–March
Plant rice	NA	NA	March–April

Source: Adapted from Southern Aquaculture Center, 2002.

DISEASE MANAGEMENT

Careful management of the aquatic environment is critical to assuring healthy populations of fish. Water quality, oxygen availability, correct pH, and avoidance of pollutants are critical factors in assuring a healthy environment. Furthermore, fishery managers should strive to maintain optimum stocking rates to avoid stress due to overcrowding. Even under excellent management, disease-causing organisms can impact a production system.

The most common causes of disease in aquatic cultures include bacteria, viruses, fungi, and parasites. Examples of specific diseases are provided in Table 38.18 through Table 38.20.

TABLE 38.18 Parasitic Diseases of Fish

Disease	Agent	Symptoms	Complicating Factors	Treatment
Tapeworm[a]	*Parasite*	Listlessness, weight loss, lowered fertility	Introduction via purchased brood-fish, avian fecal material	None available
Trichodiniasis[b]	*Tricodina protozoa*	Diminished consumption, listlessness	Poor water quality, malnutrition	Formalin, potassium permanganate, copper sulfate
Ichthyophthiriasis[b]	*Ciliated protozoa*	Small raised spots on body surface, fish congregate at pond intake or outlet	Poor water quality	Formalin, table salt, copper sulfate
Anchor parasites[a]	*Parasitic crustacean*	Reddish lesions on the external surface, parasite is attached via a small barb	Cross-contamination by wildlife	Avoid stocking with infected fish
Fish lice	*Parasitic branchiurans*	Infected fish rub against tank sides, red spots on skin	Stocking with infected fish	Avoid stocking with infected fish

[a] Catfish.
[b] Freshwater fish.

Sources: Adapted from Parker, 1995; Walker and Crummett, 1996; and Southern Regional Aquaculture Center, 1999.

TABLE 38.19   Bacterial Diseases of Fish

Disease	Agent	Symptoms	Complicating Factors	Treatment
Hemorrhagic Septicemia[a]	Aeromonas hydrophila Pseudomonas fluorescens	Irregular ulcerations, swollen belly, raised scales, red streaking of the fins	Typically occur in the spring, associated stress due to transport, handling, overcrowding	Terramycin in the diet
Columnaris[a]	Flexibacter columneris	Discoloration resembling fungal condition, fin atrophy and tail loss	Injury due to mishandling, overcrowding, shifts in water temperature	Potassium permanganate or Diquat
Enteric Redmouth[b]	Yersinia ruckeri	Inflammation and atrophy of jaws, palate, reddening of the intestine and body fat	Often linked with transport or handling	Obtain fish and eggs from disease-free sources, vaccine is available
Enteric Septicemia[c]	Edwardsiella ictaluri	Lesion between the eyes, raised lesions on body, hemorrhagic body cavity, off-feed, rapidly swimming in circles	Low oxygen levels, high ammonia and nitrate concentration	Good management, antibiotics in feed
Chilodonelliasis[a]	Chilodonella cyprini	Bright red, bleeding gills	Poor water quality, crowded conditions	Good management, Formalin, potassium permanganate

[a] All fish.
[b] Trout.
[c] Catfish.
Sources: Adapted from Parker, 1995; Walker and Crummett, 1996; and Southern Regional Aquaculture Center.

TABLE 38.20   Viral Diseases of Fish

Disease	Symptoms	Complicating Factors	Treatment
Channel Catfish Viral Disease	Swollen abdomen, clear yellow fluid in body cavity, whirling activity, skin hemorrhage	Low oxygen, high ammonia, excessive water temperature	Minimize handling and overcrowding, no therapeutic agents available
Infectious Hematopoietic Necrosis	Pale gills, dark coloration, hemorrhages at base of fins	High death loss in young fish, survivors become carriers	Avoid infected eggs, feed and fish; no known treatment
Infectious Pancreatic Necrosis	Popeyes, abdominal swelling, spiraling activity	Young fish most susceptible, spread through feces, eggs, seminal fluids	No known treatment

Sources: Adapted from Parker, 1995; Walker and Crummett, 1996; Southern Regional Aquaculture Center.

Treatment protocols for fish include dipping, tank flushing, bathing, feeding, and injections. Dipping involves catching fish in a net and dipping them into a prepared treatment solution. Tank flushing is the addition of a solution to the tank, which is then quickly spread through the tank. Bathing is the addition of the drug to the holding unit for a specified time period followed by rapid flushing with fresh water. With bathing systems the even distribution of the agent is of particular concern.

Medications may be distributed via feed or by using a balling gun for application to specific individuals. Injections may be used for larger fish. Injections may occur via intraperitoneal or intramuscular application.

Because treatment is typically expensive, it is important to minimize stress. Factors that increase stress for cultured fish include low levels of diffused oxygen, dramatic shifts in water temperature, poor nutritional practices, water quality problems, handling, overstocking, and pollutants. Fish show a variety of signs that they are experiencing stress. These indicators include listlessness, poor feed intake, erratic swimming behavior, loss of equilibrium, crowding inflow areas, excessive resting at the bottom of the tank or pond, and **flushing** (rubbing) against objects.

CHAPTER SUMMARY

- Aquaculture is an important global industry with annual production valued at more than $46 billion.
- Catfish, crawfish, and trout are the most significant enterprises in U.S. aquaculture.
- Assuring water quality, nutrition, and freedom from disease are the most critical management functions.

KEY WORDS

spawn	stocking rate	cage (tank) systems
fingerlings	incubator systems	single-crop system
brooder	raceway	rotational production
sac-fry	mouth-brooders	flushing

REVIEW QUESTIONS

1. What are the major species raised in U.S. aquaculture?
2. Why is catfish and crawfish production concentrated in the South?
3. Describe the ideal environment for catfish spawning.
4. Compare and contrast the three systems of commercial spawning.
5. What are the ideal conditions for catfish eggs to hatch?
6. What are the ideal conditions for catfish to grow?
7. What are the leading states for trout production?
8. What types of incubator systems are used in trout production?
9. What are optimal conditions for trout growth?
10. What is the relationship between feeding rates, trout size, and water temperature?
11. Define the term "mouth-brooder."
12. Compare and contrast mixed-sex versus mono-sex cultivating of tilapia.
13. Describe the pond type used in single-crop crawfish production systems.
14. Compare the management schedules of rotational crop crawfish production.

15. Describe the factors that lead to fish stress and the symptoms of stress.

16. Discuss the symptoms, complicating factors, and treatment of selected parasitic, bacterial, and viral diseases of fish.

SELECTED REFERENCES

Cain, K., and D. Ganling. 1993. *Trout Culture in the North Central Region*. North Central Reg. Aquaculture Center. Fact sheet #108.

de la Bretonne, L. W., and R. P. Romaire. 1990. *Crawfish Production*. South. Reg. Aquaculture Center. Publ. No. 242.

Hinshaw, J. M. 1990. *Trout Farming*. South. Reg. Aquaculture Center. Publ. No. 222.

Hinshaw, J. M. 1999. *Trout Production: Feeds and Feeding Method*. South. Reg. Aquaculture Center. Publ. No. 223.

Hinshaw, J. M. 1990. *Trout Production: Handling Eggs and Fry*. South. Reg. Aquaculture Center. Publ. No. 220.

Hinshaw, J. M., L. E. Rogers, and J. E. Easley. 1990. *Budgets for Trout Production*. South. Reg. Aquaculture Center. Publ. No. 221.

McClain, W. R., J. L. Avery, and R. P. Romaire. 1998. *Crawfish Production*. South. Reg. Aquaculture Center. Publ. No. 241.

McGinty, A. S., and J. E. Rakocy. 1999. *Cage Culture of Tilapia*. South. Reg. Aquaculture Center. Publ. No. 281.

Parker, R. 1995. *Aquaculture Science*. Albany, NY: Delmar Publishers.

Popma, T., and M. Masser. 1999. Tilapia: *Life History and Biology*. South. Reg. Aquaculture Center. Publ. No. 283.

Rakocy, J. E. 1989. *Tank Culture of Tilapia*. South. Reg. Aquaculture Center. Publ. No. 282.

Robinson, E. H., M. H. Li, and M. W. Brunson. 1998. *Feeding Catfish in Commercial Ponds*. South. Reg. Aquaculture Center. Publ. No. 181.

Southern Regional Aquaculture Center. 1998. *Analysis of Regional and National Markets for Aquacultural Food Products in the Southern Region*. Publ. No. 601.

USDA. 1995. *Overview of Aquaculture in the United States*. Center for Epidemiology and Animal Health, Fort Collins, CO.

USDA. 2003. *Reference of 2003 U.S. Catfish Health and Production Practices: Part I*. National Animal Health Monitoring System. Animal and Plant Health Inspection Service, Fort Collins, CO.

USDA. 2003. *Reference of 2003 U.S. Catfish Management Practices: Part II*. National Animal Health Monitoring System. Animal and Plant Health Inspection Service, Fort Collins, CO.

Walker, S. S., and D. Crummett. 1996. *Aquaculture*. Multistate Academic and Vocational Curriculum Consortium, Stillwater, OK.

Wellborn, T. L. 1998. *Channel Catfish*. South. Reg. Aquaculture Center. Publ. No. 180.

Glossary

abomasum The fourth stomach compartment of ruminant animals that corresponds to the true stomach of monogastric animals.

abortion Delivery of fetus between conception and a few days before normal parturition.

abscess Localized collection of pus in a cavity formed by disintegration of tissues.

absorption The passage of liquid and digested (soluble) food across the gut wall.

accessory organs The seminal vesicles, prostate, and Cowper's glands in the male. These glands add their secretions to the sperm to form semen.

accuracy (ACC) of selection Numerical value, ranging from 0–1.0, denoting the confidence that can be placed in the EPD (expected progeny difference); e.g., high ($\geq$0.70), medium (0.40–0.69), low ($\leq$0.40).

adipose Fat cells or fat tissue.

ad libitum Free choice; allowing animals to eat all they want.

afterbirth The membranes attached to the fetus that are expelled after parturition.

AI Abbreviation for artificial insemination.

air dry Refers to feeds in equilibrium with air; they would contain approximately 10% water or 90% dry matter.

albumen The white of an egg.

alimentary tract Passageway for food and waste products through the body.

alleles Genes occupying corresponding loci in homologous chromosomes that affect the same hereditary trait but in different ways.

allelomimetic behavior Doing the same thing. Animals tend to follow the actions of other animals.

alveolus (plural *alveoli*) A hollow cluster of cells. In the mammary gland, these cells secret milk.

amino acid Any of a class of 20 molecules that are combined to form proteins in living things.

amnion A fluid-filled membrane located next to the fetus.

ampulla The dilated or enlarged upper portion of the vas deferens in bulls, bucks, and rams, where sperm are stored for sudden release at ejaculation.

anabolic A constructive, or "building up," process.

anaerobic Able to survive or function where there is no oxygen.

anatomy Science of animal body structure and the relation of the body parts.

androgen A male sex hormone, such as testosterone.

anemia Deficiency of hemoglobin, often accompanied by a reduced number of red blood cells. Usually results from an iron deficiency.

anestrous Period of time when female is not in estrus; the nonbreeding season.

antemortem Before death.

anterior Situated in front of, or toward the front part of, a point of reference. Toward the head of an animal.

anterior pituitary (AP) The part of the pituitary gland, located at the base of the brain, that produces several hormones.

anthelmintic A drug or chemical agent used to kill or remove internal parasites.

antibiotic A product produced by living organisms, such as yeast, which destroys or inhibits the growth of other microorganisms, especially bacteria.

antibody A specific protein molecule that is produced in response to a foreign protein (antigen) that has been introduced into the body.

antigen A foreign substance that, when introduced into the blood or tissues, causes the formation of antibodies. Antigens may be toxins or native proteins.

anti-inflammatory An agent that acts to decrease inflammation and associated pain, heat, and swelling.

antiseptic A chemical agent used on living tissue to control the growth and development of microorganisms.

antitoxin An antibody that is capable of neutralizing poisons from animal and vegetable sources.

AP See *anterior pituitary*.

APCS A popular milk testing system where weight is recorded at two milkings.

arteriosclerosis A disease resulting in the thickening and hardening of the artery walls.

artery Vessel through which blood passes from the heart to all parts of the body.

artificial insemination The introduction of semen into the female reproductive tract (usually the cervix or uterus) by a technique other than natural service.

artificial vagina   A device used to collect semen from a male when he mounts in a normal manner to copulate. The male ejaculates into this device, which simulates the vagina of the female in pressure, temperature, and sensation to the penis.

ascaris   Any of the genus *(Ascaris)* of parasitic roundworms.

as fed   Refers to feeding feeds that contain their normal amount of moisture.

assimilation   The process of transforming food into living tissue.

atherosclerosis   A form of arteriosclerosis involving fatty deposits in the inner walls of the arteries.

atrophy   Shrinking or wasting away of a tissue or organ.

auction   A market facility where an auctioneer sells animals to the highest bidder.

autopsy   A postmortem examination in which the body is dissected to determine the cause of death.

avian   Refers to birds, including poultry.

balance sheet   A statement of assets owned and liabilities owed in dollar terms that shows the equity or net worth at a specific point in time (e.g., net worth statement).

band   (1) a relatively large group of range sheep; (2) method of identification (e.g., put a band around the leg of a chicken).

Bang's disease   See *brucellosis*.

barren   Not capable of producing offspring.

barrow   A male swine that was castrated before reaching puberty.

basal metabolism   The chemical changes that occur in an animal's body when the animal is in a thermoneutral environment, resting, and in a postabsorptive state. It is usually determined by measuring oxygen consumption and carbon dioxide production.

base pair   Two nitrogenous bases (adenine and thymine or guanine and cytosine) held together by weak bonds. Two strands of DNA are held together in the shape of a double helix by the bonds between base pairs.

beef   The meat from cattle (bovine species) other than calves (the meat from calves is called *veal*).

beri-beri   A disease caused by a deficiency of vitamin B_1.

biologicals   Medicinal products used primarily to prevent disease, including serums, vaccines, antigens, antitoxins, etc.

biotechnology   The use of microorganisms, plant cells, animal cells, or parts of cells (such as enzymes) to produce industrially important products or processes.

birth weight expected progeny difference (EPD)   The expected average increase or decrease in the birth weight of a beef bull's calves when compared to the other bulls in the sire summary.

blemish   Any defect or injury that mars the appearance of, but does not impair the usefulness of, an animal.

bloat   An abnormal condition in ruminants characterized by a distention of the rumen, usually seen on an animal's upper left side, owing to an accumulation of gases.

blood spots   Spots in the egg caused by a rupture of one or more blood vessels in the yolk follicle at the time of ovulation.

BLUP   Best linear unbiased prediction, method for estimating breeding values of breeding animals.

boar   (1) A male swine of breeding age. (2) Denotes a male pig, which is called a *boar pig*.

bog spavin   A soft enlargement of the anterior, inner aspect of the hock.

bolus   (1) Regurgitated food. (2) A large pill for dosing animals.

bone spavin   A bony (hard) enlargement of the inner aspect of the hock.

bots   Any of a number of related flies whose larvae are parasitic in horses and sheep.

bovine   A general family grouping of cattle.

boxed beef   Cuts of beef put in boxes for shipping from packing plant to retailers. These primal and subprimal cuts are intermediate cuts between the carcass and retail cuts.

boxed lamb   See *boxed beef*. Similar process except lamb instead of beef.

break joint   Denotes the point on a lamb carcass where the foot and pastern are removed at the cartilaginous junction of the front leg.

bred   Female has been mated to the male. Usually implies the female is pregnant.

breech   The buttocks. A breech presentation at birth is where the rear portion of the fetus is presented first.

breed   Animals of common origin with characteristics that distinguish them from other groups within the same species.

breeding value   A genetic measure for one trait of an animal, calculated by combining into one number several performance values that have been accumulated on the animal and the animal's relatives.

brisket disease   A noninfectious disease of cattle characterized by congestive right heart failure. It affects animals residing at high altitudes (usually above 7,000 ft).

British breeds   Breeds of beef cattle originating in England. Examples are Angus, Hereford, and Shorthorn.

British thermal unit (Btu)   The quantity of heat required to raise the temperature of 1 lb of water 1°F at or near 39.2°F.

brockle-faced   White-faced with other colors splotched on the face and head.

broiler   A young meat-type chicken of either sex (usually up to 6–8 weeks of age) weighing 3–5 lb. Also referred to as a *fryer* or *young chicken*.

broken-mouth   Some teeth are missing or broken.

brooder   Fish that have reached reproductive maturity.

broodiness The desire of a female bird to sit on eggs (incubate).

browse Woody or brushy plants. Livestock feed on tender shoots or twigs.

brucellosis A contagious bacterial disease that results in abortions; also called *Bang's disease.*

buck A male sheep or goat. This term usually denotes animals of breeding age.

bulbourethral (Cowper's) gland An accessory gland of the male that secretes a fluid which constitutes a portion of the semen.

bull A bovine male. The term usually denotes animals of breeding age.

buller-steer syndrome A behavior problem where a steer has a sexual attraction to other steers in the pen. The steer is ridden by the other steers, resulting in poor performance and injury.

bullock A young bull, typically less than 20 months of age.

buttermilk The fluid remaining after butter has been made from cream. By use of bacteria, cultured buttermilk is also produced from milk.

buttons May refer to cartilage or dorsal processes of the thoracic vertebrae. Also see *cotyledons.*

by-product A product of considerably less value than the major product. For example in U.S. meat animals, the hide, pelt, and offal are by-products, whereas meat is the major product.

C-section See *cesarean section.*

calf A young male or female bovine animal under a year of age.

calorie The amount of heat required to raise the temperature of 1 g of water from 15°C to 16°C.

calve Giving birth to a calf. Same as parturition.

calving interval The amount of time (days or months) between the birth of a calf and the birth of a subsequent calf, both from the same cow.

candling The shining of a bright light through an egg to see if it contains a live embryo.

canter A slow, easy gallop.

capon Castrated male chicken. Castration usually occurs between 3 and 4 weeks of age.

capped hocks Hocks that have hard growths that cover, or "cap," their points.

carbohydrates Any foods, including starches, sugars, celluloses, and gums, that are broken down to simple sugars through digestion.

carcass merit The value of a carcass for consumption.

carnivorous Subsisting or feeding on animal tissues.

carotene The orange pigment found in carrots, leafy plants, yellow corn, and other feeds, which can be broken down to form two molecules of vitamin A.

caruncle (1) The red and blue fleshy, unfeathered area of skin on the upper region of the turkey's neck. (2) The "buttons" on the ruminant uterus where the cotyledons on the fetal membranes attach.

casein   The major protein of milk.

cash-flow statement   A financial statement summarizing all cash receipts and disbursements over the period of time covered by the statement.

castrate   (1) To remove the testicles. (2) An animal that has had its testicles removed.

cattalo   A cross between domestic cattle and bison.

cecum (ceca)   Large, sock-shaped pouch between the horse's small and large intestines; important in cellulose digestion.

cervix   The portion of the female reproductive tract between the vagina and the uterus. It is usually sealed by thick mucus except when the female is in estrus or delivering young.

cesarean section   Delivery of fetus through an incision in abdominal and uterine walls.

chalaza   A spiral band of thick albumen that helps hold the yolk of an egg in place.

chemotherapeutics   Chemical agents used to prevent or treat diseases.

chevon   Meat from goats.

chick   A young chicken that has recently been hatched.

chorion   The outermost layer of fetal membranes.

chromosome   The self-replicating genetic structure of cells containing the cellular DNA that bears in its nucleotide sequence the linear array of genes.

chyme   The thick, liquid mixture of food that passes from the stomach to the small intestine.

chymotrypsin   A milk-digesting enzyme secreted by the pancreas into the small intestine.

class   A group of animals categorized primarily by sex and age.

clip   One season's yield of wool.

clitoris   The ventral part of the vulva of the female reproductive tract that is homologous to the penis in the male. It is highly sensory.

cloaca   Portion of the lower end of the avian digestive tract that provides a passageway for products of the urinary, digestive, and reproductive tracts.

closed face   A condition in which sheep cannot see because wool covers their eyes.

clostridium   Genus of anaerobic bacteria. Many produce potent toxins that cause such diverse diseases as tetanus, botulism, and gas gangrene. Some of these anerobic bacteria inhabit the soil and feces.

clutch   Eggs laid by a hen on consecutive days.

coccidia   A protozoan organism that causes an intestinal disease called *coccidiosis*.

coccidiosis   A morbid state caused by the presence of organisms called coccidia, which belong to a class of sporozoans.

cock   A male chicken; also called a *rooster*.

cockerel   Immature male chicken.

cod   Scrotal area of steer remaining after castration.

coefficient of determination   Percentage of variation in one trait that is accounted for by variation in another trait.

colic   A nonspecific pain of the digestive tract.

colon   The large intestine from the end of the ileum and beginning with the cecum to the anus.

colostrum   The first milk given by a female after delivery of her young. It is high in antibodies that protect young animals from invading microorganisms.

colt   A young male of the horse or donkey species.

comb   The fleshy outgrowth on the top of a chicken's head, usually red in color, with varying sizes and shapes.

commercial   (1) a carcass grade of cattle; (2) livestock that are not registered or pedigreed by a registry (e.g., breed) association.

compensatory gain   A faster-than-normal rate of gain after a period of restricted gain.

compensatory growth   See *compensatory gain*.

complete feed   (compound) A nutritionally adequate feed for animals other than humans by specific formula compounded to be fed as the sole ration and capable to maintaining life and/or promoting production without any additional substance, except water, being consumed.

composite breed   A breed that has been formed by crossing two or more breeds.

concentrate   A feed used with another to improve the nutritive balance of the total and intended to be further-diluted and mixed to produce a supplement or a complete feed.

conception   Fertilization of the ovum (egg).

conditioning   The treatment of animals by vaccination and other means before putting them in the feedlot.

conformation   The physical form of an animal; its shape and arrangement of parts.

contagious disease   Infectious disease; a disease that is transmitted from one animal to another.

contemporaries   A group of animals of the same sex and breed (or similar breeding) that have been raised under similar environmental conditions (same management group).

Continental breeds   Breeds of beef cattle originating in countries other than England. Sometimes called European or exotic breeds. Examples are Charolais, Limousin, and Simmental.

contracted heels   A condition in which the heels of a horse are pulled in so that expansion of the heel cannot occur when the foot strikes the ground.

core samples   Samples (wool, feed, or meat) taken by a coring device to determine the composition of the sample.

coronary band (coronet)   Boundary between the top of the hoof wall and the skin at the bottom of the pastern where hoof growth begins.

corpus luteum   A yellowish body in the mammalian ovary. The cells where follicular cells develop into the corpus luteum, which secretes progesterone. It becomes yellow in color from the yellow lipids that are in the cells.

correlation coefficient   A measure of the association of one trait with another.

cost of gain   The total cost divided by the total pound gained; usually expressed on a per-pound basis.

cotyledon   An area of the placenta that contacts the uterine lining to allow nutrients and wastes to pass from the mother to the developing young. Sometimes referred to as *button*.

cow   A sexually mature female bovine animal—usually one that has produced a calf.

cow-calf operation   A management unit that maintains a breeding herd and produces weaned calves.

cow hocked   A condition in which the hocks are close together but the feet stand apart.

creep   An enclosure in which young can enter to obtain feed but larger animals cannot enter. This process is called *creep feeding*.

crimp   The waves, or kinks, in a wool fiber.

crossbred   An animal produced by crossing two or more breeds.

crossbreeding   Mating animals from genetically diverse groups (i.e., breeds) within a species.

crutching   See *tagging*.

cryptorchidism   The retention of one or both testicles in the abdominal cavity in animals that typically have the testicles hanging in a scrotal sac.

cud   Bolus of feed a ruminant animal regurgitates for further chewing.

cull   To eliminate one or more animals from the beeding herd or flock.

curb   A hard swelling that occurs just below the point of the hock.

curd   Coagulated milk.

cutability   Fat, lean, and bone composition of meat animals. Used interchangeably with yield grade. (See also *yield grade*.)

cutting chute   A narrow chute where animals proceed through gates in single file such that animals can be directed into pens along the side of the chute.

cwt   An abbreviation for hundredweight (100 lb).

cycling   Infers that nonpregnant females have active estrous cycles.

dam   Female parent.

dark cutter   Color of the lean (muscle) in the carcass has a dark appearance, usually caused by stress (excitement, etc.) to the animal before slaughter.

debeaking   To remove the tip of the beak of chickens.

dehorn   To remove the horns from an animal.

deoxyribonucleic acid (DNA)   A complex molecule consisting of deoxyribose (a sugar), phosphoric acid, and four nitrogen bases (a gene is a piece of DNA).

depreciation   An accounting procedure by which the purchase price of an asset with a useful life of more than 1 year is prorated over time.

dewclaws   Hard horny structures above the hoof on the rear surface of the legs of cattle, swine, and sheep.

dewlap   Loose skin under the chin and neck of cattle.

DHIA   Dairy Herd Improvement Association, an association which dairy producers participate in keeping dairy records. Sanctioned by the National Cooperative Dairy Herd Improvement Program.

DHIR   Dairy Herd Improvement Registry, a dairy record-keeping plan sponsored by the breed associations.

diet   Feed ingredients or mixture of ingredients (including water) which are consumed by animals.

digestibility   The quality of being digestible. If a high percentage of a given food taken into the digestive tract is absorbed into the body, that food is said to have *high digestibility*.

digestion   The reduction in particle size of feed so that the feed becomes soluble and can pass across the gut wall into the vascular or lymph system.

diploid   Having the normal, paired chromosomes of somatic tissue as produced by the doubling of the primary chromosomes of the germ cells at fertilization.

disease   Any deviation from a normal state of health.

disinfect   To kill, or render ineffective, harmful microorganisms and parasites.

disinfectant   A chemical that destroys disease-producing microorganisms or parasites.

distal   Position that is distant from the point of attachment of an organ.

DM   See *dry matter.*

DNA   (deoxyribonucleic acid) The molecule that encodes genetic information. DNA is a double-stranded molecule held together by weak bonds between base pairs of nucleotides.

DNA fingerprint   Pattern of DNA fragments unique to an individual. Often found by using restriction enzymes to cut the DNA into fragments. These fragments can be sorted and documented, forming a unique "fingerprint." This technology, the same used as a forensic tool at crime scenes, is also used to DNA parentage test animals.

DNA sequence   The relative order of base pairs, whether in a fragment of DNA, a gene, a chromosome or an entire genome.

dock   (1) To cut off the tail. (2) The remaining portion of the tail of a sheep that has been docked. (3) To reduce or lower in value.

doe   A female goat or rabbit.

dominance   (1) A situation in which one gene of an allelic pair prevents the phenotypic expression of the other member of the allelic pair. (2) A type of social behavior in which an animal exerts influence over one or more other animals.

dominant gene   A gene that overpowers and prevents the expression of its recessive allele when the two alleles are present in a heterozygous individual.

dorsal   Of, on, or near the back of an animal.

double muscling   A genetic trait in cattle where muscles are greatly enlarged rather than duplicate muscles.

down   Soft, fluffy type of feather located under the contour feathers. Serves as insulating material.

drake   Mature male duck.

drench   To give fluid by mouth (e.g., medicated fluid is given to sheep for parasite control).

dressing percentage   The percentage of the live animal weight that becomes the carcass weight at slaughter. It is determined by dividing the carcass weight by the live weight, then multiplying by 100.

drop   Body parts removed at slaughter—primarily hide (pelt), head, shanks, and offal.

drop credit   Value of the drop.

dry (cow, ewe, sow, mare)   Refers to a nonlactating female.

dry matter   Feed after water (moisture) has been removed (100% dry).

dubbing   The removal of part or all of the soft tissues (comb and wattles) of chickens.

dung   The feces (manure) of farm animals.

dwarfism   The state of being abnormally undersized. Two kinds of dwarfs are recognized; one is proportionate and the other is disproportionate.

dysentery   Severe diarrhea.

dystocia   Difficult birth.

ectoderm   The outermost layer of the three layers of the primitive embryo.

edema   Abnormal collection of fluid in body tissues that causes soft swelling.

ejaculation   Discharge of semen from the male.

emaciation   Thinness; loss of flesh where bony structures (hips, ribs, and vertebrae) become prominent.

embryo   Very early stage of individual development within the uterus. The embryo grows and develops into a fetus. In poultry, the embryo develops within the eggshell.

embryo transfer   The transfer of fertilized eggs from a donor female to one or more recipient females.

endocrine gland   A ductless gland that secretes a hormone into the bloodstream.

endoderm   The innermost layer of the three layers of the primitive embryo.

enterotoxemia   A disease of the intestinal tract caused by bacterial secretion of toxins. Its symptoms are characteristic of food poisoning.

entropion Turned-in eyelids.

environment The sum total of all external conditions that affect the well-being and performance of an animal.

enzyme A complex protein produced by living cells that causes changes in other substances in the cells without being changed itself and without becoming a part of the product.

EPD See *expected progeny difference*.

epididymis The long, coiled tubule leading from the testis to the vas deferens.

epididymitis An inflammation of the epididymis.

epiphysis A piece of bone separated from a long bone in early life by cartilage, which later becomes part of the larger bone.

epistasis A situation in which a gene or gene pair masks (or controls) the expression of another nonallelic pair of genes.

equine Refers to horses.

equine encephalomyelitis An inflammation of the brain of horses.

eruction (or eructation) The elimination of gas by belching.

esophageal groove A groove in the reticulum between the esophagus and omasum. Directs milk in the nursing young ruminant directly from the esophagus to the omasum.

essential nutrient A nutrient that cannot be synthesized by the body and must be supplied in the diet.

estrogen Any hormone (including estradiol, estriol, and estrone) that causes the female to come physiologically into heat and to be receptive to the male. Estrogens are produced by the follicle of the ovary and by the placenta.

estrous An adjective meaning "heat," which modifies such words as "cycle." The estrous cycle is the heat cycle, or time from one heat to the next.

estrous synchronization Controlling the estrous cycle so that a high percentage of the females in the herd express estrus at approximately the same time.

estrus The period of mating activity in the female mammal. Same as heat.

ET Abbreviation for *embryo transfer*.

ethology Study of animal behavior in the animal's natural environment.

eukaryote Cell or organism with membrane-bound, structurally discrete nucleus and other well-developed subcellular compartments. Eukaryotes include all organisms except viruses, bacteria, and blue-green algae.

European breeds See *Continental breeds*.

eviscerate The removal of the internal organs during the slaughtering process.

ewe A sexually mature female sheep. A ewe lamb is a female sheep before attaining sexual maturity.

exocrine gland Gland that secretes fluid into a duct.

exotic breeds See *Continental breeds*.

expected progeny difference (EPD) One-half of the breeding value; the difference in performance to be expected from future progeny of a sire, compared with that expected from future progeny of an average bull in the same test.

family selection Selection based on performance of a family.

farrow To deliver, or give birth to, pigs.

fat Adipose tissue.

FDA See *Food and Drug Administration*.

feather picking The picking of feathers from one bird by another.

feces Bowel movements, excrement from the intestinal tract.

feed additive Ingredient (such as an antibiotic or hormone-like substance) added to a diet to perform a specific role (e.g., to improve gain or feed efficiency).

feed bunk A trough or container used to feed farm animals.

feed efficiency (1) The amount of feed required to produce a unit of weight gain or milk; for poultry, this term can also denote the amount of feed required to produce a given quantity of eggs. (2) The amount of gain made per unit of feed.

feeder Animals (e.g., cattle, lambs, pigs) that need further feeding prior to slaughter.

feeder grades Visual classifications (descriptive and/or numerical) of feeder animals. Most of these grades have been established by the USDA.

felting The process of pressing wool fibers together in conjunction with heat and moisture to produce a fabric.

femininity Well-developed secondary female sex characteristics, udder development and refinement in head and neck.

feral Domesticated animals that return to nature to survive and reproduce.

fertility The capacity to initiate, sustain, and support reproduction. With reference to poultry, the term typically refers to the percentage of eggs that, when incubated, show some degree of embryonic development.

fertilization The process in which a sperm unites with an egg to produce a zygote.

fetus Later stage of individual development within the uterus. Generally, the new individual is regarded as an embryo during the first half of pregnancy, and as a fetus during the last half.

fill The contents of the digestive tract.

filly A young female horse.

fineness A term used to describe the diameter of wool fibers.

fingerlings Usually 1–6 inches long.

finish The degree of fatness of an animal.

Finnsheep A highly prolific breed of sheep introduced into the United States in 1968.

fistula A running sore at the top of the withers of a horse, resulting from a bruise followed by invasion of microorganisms.

flank firmness   Firmness of the flank muscle in lamb carcass evaluation.

flank streaking   Streaks of fat in the flank muscle of lamb carcasses.

fleece   The wool shorn at one time from all parts of the sheep.

flehmen   A pattern of behavior expressed in some male animals (e.g., bull, ram, stallion) during sexual activity. The upper lip curls up and the animal inhales in the vicinity of the vulva or urine.

flock   A group of sheep or poultry.

flushing   Placing females (typically sheep and swine) on a gaining level of nutrition before breeding to stimulate greater ovulation rates; also, a behavior in fish whereby diseased fish rub against objects in tanks or ponds.

fly strike   An infestation with large numbers of blowfly maggots.

foal   A young male or female horse (noun) or the act of giving birth (verb).

follicle   A blisterlike, fluid-filled structure in the ovary that contains the egg.

follicle-stimulating hormone (FSH)   A hormone produced and released by the anterior pituitary that stimulates the development of the follicle in the ovary.

Food and Drug Administration (FDA)   A U.S. government agency responsible for protecting the public against impure and unsafe foods, drugs, veterinary products, and other products.

food-size   Commercially grown fish produced for food, usually ranging from .75 to 1.0 pounds and over 1 foot in length.

footrot   A disease of the foot in sheep and cattle. In sheep it causes rotting of tissue between the horny part of the foot and the soft tissue underneath.

forb   Weedy or broad-leaf plants, as contrasted to grasses, that serve as pasture for animals.

forging   The striking of the heel of the front foot with the toe of the hind foot by a horse in action.

founder   Nutritional ailment resulting from overeating. Lameness in front feet with excessive hoof growth usually occurs.

frame score   A numerical rating of frame size.

frame size   A measure of skeletal size. It can be visual or by measurement (usually taken at the hips).

freemartin   Female born twin to a bull (approximately 9 of 10 will not conceive).

freshen   To give birth to young and initiate milk production. This term is usually used in reference to dairy cattle.

fry   Stage from hatching until fish reach 1 inch in length.

fryer   See *broiler*.

FSH   See *follicle-stimulating hormone*.

full-mouth   Animal has all permanent teeth fully exposed.

full sibs   Animals having the same sire and dam.

gallop   A three-beat gait in which each of the two front feet and both of the hind feet strike the ground at different times.

gametes   Male and female reproductive cells. The sperm and the egg.

gametogenesis   The process by which sperm and eggs are produced.

gander   Mature male goose.

gelding   A male horse that has been castrated.

gene   The fundamental physical and functional unit of heredity.

gene expression   The process by which a gene's coded information is converted into the structures present and operating in the cell.

gene mapping   Determination of the relative positions of genes on a DNA molecule (chromosome or plasmid) and of the distance, in linkage units or physical units, between them.

general combining ability   The ability of individuals of one line or population to combine favorably or unfavorably with individuals of several other lines or populations.

generation interval   Average age of the parents when offspring are born.

generation turnover   Length of time from one generation of animals to the next generation.

genetic code   The sequence of nucleotides, coded in triplets (codons) along the mRNA, that determines the sequence of amino acids in protein synthesis.

genetic engineering   The technique of removing, modifying, or adding genes to a DNA molecule.

genome   The sum total of a living organism's genetic material. The genome is divided into chromosomes, which contain genes, and genes are made of DNA.

genomics   The study of genes and their function.

genotype   The genetic constitution, or makeup, of an individual. For any pair of alleles, three genotypes (e.g., *AA, Aa,* and *aa*) are possible.

gestation   The time from breeding or conception of a female until she gives birth to her young.

gilt   A young female swine prior to the time that she has produced her first litter.

goiter   Enlargement of the thyroid gland, usually caused by iodine-deficient diets.

gonad   The testis of the male; the ovary of the female.

gonadotrophin   Hormone that stimulates the gonads.

gossypol   A toxic product contained in cottonseed.

grade   (1) a designation of live or carcass merit (e.g., Choice grade); (2) livestock not registered with registry (e.g., breed) association.

grading up   The continued use of purebred sires of the same breed in a grade herd or flock.

grass tetany   A disease of cattle and sheep marked by staggering, convulsions, coma, and frequently death, caused by a mineral imbalance (magnesium) while grazing lush pasture.

grease wool   Wool as it comes from the sheep and prior to cleaning. It contains the natural oils from the sheep.

gross energy The amount of heat, measured in calories, produced when a substance is completely oxidized. It does not reveal the amount of energy that an animal could derive from eating the substance.

growth The increase in protein over its loss in the animal body. Growth occurs by increases in cell numbers, cell size, or both.

habituation The gradual adaptation to a stimulus or to the environment.

half sib Animals having one common parent.

hand mating Same as hand breeding—bringing a female to a male for service (breeding), after which she is removed from the area where the male is located.

hank A measurement of the fineness of wool. A hank is 560 yards of yarn. More hanks of yarn are produced from fine wools than coarse wools.

haploid One-half of the diploid number of chromosomes for a given species, as found in the germ cells.

hatchability A term that indicates the percentage of a given number of eggs set from which viable young hatch, sometimes calculated specifically from the number of fertile eggs set.

hay Harvested forage such as alfalfa hay.

heat See *estrus*.

heat increment The increase in heat production after consumption of feed when an animal is in a thermoneutral environment. It includes additional heat generated in fermentation, digestion, and nutrient metabolism.

heaves A respiratory defect in horses during which the animal has difficulty completing the exhalation of inhaled air.

heifer A young female bovine cow before the time that she has produced her first calf.

heiferette A heifer that has calved once, after which the heifer is fed for slaughter; the calf has usually died or been weaned at an early age.

hemoglobin The iron-containing pigment of the red blood cells. It carries oxygen from the lungs to the tissues.

hen An adult female domestic fowl, such as a chicken or turkey.

herbivorous Subsisting or feeding on plants.

herd A group of animals. Used with beef, dairy, or swine.

heritability The portion of the total variation or phenotypic differences among animals that is due to heredity.

hernia The protrusion of some of the intestine through an opening in the body wall (also commonly called *rupture*). Two types of hernias, umbilical and scrotal, occur in farm animals.

heterosis Performance of offspring that is greater than the average of the parents. Usually the amount of superiority of the crossbred over the average of the parental breeds. Also referred to as *hybrid vigor*.

heterozygous A term designating an individual that possesses unlike genes for a particular trait.

hides   Skins from animals such as cattle, horses, and pigs; beef hides weigh more than 30 lb each as contrasted to calf skins which weigh less.

hinny   The offspring that results from crossing a stallion with a female donkey (jenny).

hobble   To tie two of an animal's legs together. An animal is hobbled to prevent it from kicking or moving a long distance.

homeotherm   A warm-blooded animal. An animal that maintains its characteristic body temperature even though environmental temperature varies.

homogenized   Milk that has had the fat droplets broken into very small particles so that the milk fat stays in suspension in the milk fluids.

homologous   Corresponding in type of structure and derived from a common primitive origin.

homologous chromosomes   Chromosomes having the same size and shape that contain genes affecting the same characters. Homologous chromosomes occur in pairs in typical diploid cells.

homology   Similarity in DNA or protein sequences between individuals of the same species or among different species.

homozygous   A term designating an individual whose genes for a particular trait are alike.

hormone   A chemical substance secreted by a ductless gland. Usually carried by the bloodstream to other places in the body where it has its specific effect on another organ.

hybrid vigor   See *heterosis*.

hydrocephalus   A condition characterized by an abnormal increase in the amount of cerebral fluid, accompanied by dilation of the cerebral ventricles.

hyperkalemic periodic paralysis (HYPP)   An inherited muscle disorder characterized by muscle tremors, weakness and, in severe cases, collapse and death.

hypertension   High blood pressure.

hypothalamus   A portion of the brain found in the floor of the third ventricle. It regulates reproduction, hunger, and body temperature and has other functions.

hypoxia   A condition resulting from deficient oxygenation of the blood.

ileum   Distal portion of the small intestine.

immunity   The ability of an animal to resist or overcome an infection.

impaction   Obstructive lodging of food in the intestine.

implant   To graft or insert material to intact tissues.

implantation   The attachment of the fertilized egg to the uterine wall.

imprinting   Learning associated with maturational readiness.

inbreeding   The mating of individuals who are more closely related than the average individuals in a population. Inbreeding increases homozygosity in the population but it does not change gene frequency.

incisor   A front tooth.

incubation period   The time that elapses from the time an egg is placed into an incubator until the young is hatched.

independent culling level   Selection method in which minimum acceptable phenotypic levels are assigned to several traits.

index   (1) An overall merit rating of an animal. (2) A method of predicting the milk-producing ability that a bull will transmit to his daughters.

infection   Invasion of the body tissues by microbial agents or parasites other than insects.

infectious   Capable of invading and growing in living tissues. Use to describe various pathogenic microorganisms such as viruses, bacteria, protozoa, and fungi.

influenza   A virus disease characterized by inflammation of the respiratory tract, high fever, and muscular pain.

ingest   Anything taken into the stomach.

inheritance   The transmission of genes from parents to offspring.

insemination   Deposition of semen in the female reproductive tract.

instinct   Inborn behavior.

insulin   Hormone secreted by the pancreas to control blood sugar level and utilization of sugar in the body.

integration   The bringing together of all segments of a livestock or poultry production program under one centrally organized unit.

intelligence   The ability to learn to adjust successfully to situations.

interference   The striking of the supporting leg by the foot of the striding leg by a horse in action.

interstitial cells   The cells between the seminiferous tubules of the testicle that produce testosterone.

intravenous   Within the vein. An intravenous injection is an injection into a vein.

in vitro   Outside the living body; in a test tube or other artificial environment.

jack   A male donkey.

jackass   See *jack*.

jennet   A female donkey.

jenny   A female donkey.

Karakul   A breed of fat-tailed sheep having coarse, wiry furlike hair. Used to produce Persian lambskins.

ked   An external parasite that affects sheep. Although commonly called *sheep tick*, it is actually a wingless fly.

kemp   Coarse, opaque, hairlike fibers in wool.

ketosis   A condition (also called *acetonemia*) that is characterized by a high concentration of ketone bodies in the body tissues and fluids.

kid   Young goat.

kilocalorie (kcal, Kcal)   An amount of heat equal to 1,000 calories. (See also *calorie.*)

kosher meat   Meat from ruminant animals with split hooves where the animals have been slaughtered according to Jewish law.

lactalbumin   A nutritive protein of milk.

lactation   The secretion and production of milk.

lactose   Milk sugar. When digested, it is broken down into one molecule of glucose and one of galactose.

lamb   (1) A young male or female sheep, usually less than a year of age. (2) To deliver, or give birth to, a lamb.

lamb dysentery   See *dysentery.*

lambing   Act of giving birth. Same as parturition.

lambing jug   A small pen in which a ewe is put for lambing. It is also used for containing the ewe and her lamb until the lamb is strong enough to run with other ewes and lambs.

laminitis   Inflammation of the sensitive plates of soft tissue (laminae) within the horse's foot caused by physical or physiologic injury. Severe cases of laminitis may result in founder, an internal deformity of the foot. *Acute* laminitis sets in rapidly and usually responds to appropriate, intensive treatment, while *chronic* laminitis is a persistent, long-term condition that may be unresponsive to treatment.

lard   The fat from pigs that has been produced through a rendering process.

layer   A hen that is kept for egg production.

legume   Any plant of the family *leguminosae*, such as pea, bean, alfalfa, and clover.

leucocytes   White blood cells.

LH   See *luteinizing hormone.*

libido   Sex drive or the desire to mate on the part of the male.

lice   Small, flat, wingless insect with sucking mouth parts that is parasitic on the skin of animals.

linebreeding   A mild form of inbreeding that maintains a high genetic relationship to an outstanding ancestor.

line crossing   The crossing of inbred lines.

linkage   The proximity of two or more markers on a chromosome. The closer together the markers are, the lower the probability that they will be separated during DNA repair or replication processes and hence the greater the probability that they will be inherited together.

linkage map   A map of the relative positions of genetic loci on a chromosome, determined on the basis of how often the loci are inherited together. Distance is measured in centimorgans.

lipid   An organic substance that is soluble in alcohol or ether but insoluble in water; used interchangeably with the term *fat.*

litter The young produced by multiparous females such as swine. The young in a litter are called *littermates*.

liver flukes A parasitic flatworm found in the liver.

locus The place on a chromosome where a gene is located.

longevity Life span of an animal. Usually refers to a long life span.

luteinizing hormone (LH) A protein hormone, produced and released by the anterior pituitary, which stimulates the formation and retention of the corpus luteum. It also initiates ovulation.

lymph Transparent, nutritive yellow liquid that exudes from blood vessels into tissue spaces and is drained back into the veins through lymph vessels. Lymph plays an important role in fighting infection and maintaining the body's fluid balance.

macroclimate The large, general climate in which an animal exists.

macromineral A mineral that is needed in the diet in relatively large amounts.

maintenance A condition in which the body is maintained without an increase or decrease in body weight and with no production or work being done.

mammal Warm-blooded animals that suckle their young.

mammary gland Gland that secretes milk.

management The act, art, or manner of managing, handling, controlling, or directing a resource or integrating several resources.

marbling The distribution of fat in muscular tissue; intramuscular fat.

mare A sexually developed female horse.

marker An identifiable physical location on a chromosome whose inheritance can be monitored. Markers can be expressed regions of DNA (genes) or some segment of DNA with no known coding function but whose pattern of inheritance can be determined.

market class Animals grouped according to the use to which they will be put, such as slaughter, feeder, or stocker.

market grade Animals grouped within a market class according to their value.

masticate To chew food.

mastitis Inflammation of the mammary gland.

masturbation Ejaculation by a male by some process other than sexual intercourse.

mean (1) Statistical term for average. (2) Term used to describe animals having bad behavior.

meat The tissues of the animal body that are used for food.

meat spots Spots in the egg that are blood spots which have changed color or tissue sloughed off from the reproductive organs of the hen.

meiosis A special type of cell nuclear division that is undergone in the production of gametes (sperm in the male, ova in the female). As a result of meiosis, each gamete carries half the number of chromosomes of a typical body cell in that species.

melengestrol acetate   (MGA) A feed additive that suppresses estrus in heifers and is widely used in the feedlot industry.

mesoderm   The middle layer of the three layers of the primitive embryo.

messenger RNA   (mRNA) RNA that serves as a template for protein synthesis.

metabolism   (1) The sum total of chemical changes in the body, including the "building up" and "breaking down" processes. (2) The transformation by which energy is made available for body uses.

metabolizable energy   Gross energy in the feed minus the sum of energy in feces, gaseous products of digestion, and energy in urine. Energy that is available for metabolism by the body.

metritis   Inflammation (infection) of the uterus.

MGA   See *melengestrol acetate.*

microclimate   A small, special climate within a macroclimate created by the use of such devices as shelters, heat lamps, and bedding.

microcomputer   A small computer that has a smaller memory capacity than a larger or mainframe computer.

micromineral   A mineral that is needed in the diet in relatively small amounts. The quantity needed is so small that such a mineral is often called a *trace mineral.*

milk EPD   A genetic estimate of the milking ability of a beef bull's daughters when compared to the daughters of other bulls.

milk fat   The fat in milk; synonymous with butterfat.

milk fever   See *parturient paresis.*

milk letdown   The release of milk into the teat cisterns.

Milk Only records   Dairy record system similar to DHI except no milk fat samples are taken.

minimum culling level   A selection method in which an animal must meet minimum standards for each trait desired in order to qualify for being retained for breeding purposes.

mites   Very small arachnids that are often parasitic upon animals.

mitosis   A process in which a cell divides to produce two daughter cells, each of which contains the same chromosome complement as the mother cell from which they came.

modifying genes   Genes that modify the expression of other genes.

mohair   Fleece of the Angora goat.

monogastric   Having only one stomach or only one compartment in the stomach. Examples are swine and poultry.

monoparous   A term designating animals that usually produce only one offspring at each pregnancy. Horses and cattle are monoparous.

monotocous   Producing a single offspring at a birth.

moon blindness   Periodic blindness that occurs in horses.

morbidity   Measurement of illness; morbidity rate is the number of individuals in a group that become ill during a specified time.

mortality rate   Number of individuals that die from a disease during a specified time, usually 1 year.

mouth-brooder   Fish that hold eggs or newly hatched young in their mouths.

mouthed   The examination of an animal's teeth.

mule   The hybrid that is produced by mating a male donkey with a female horse. They are usually sterile.

mulefoot   Having one instead of the expected two toes, on one or more of the feet.

multiparous   Having had two or more pregnancies which resulted in viable fetuses.

mutation   A change in a gene.

mutton   The meat from a sheep that is over 1 year old.

muzzle   The nose of horse, cattle, or sheep.

myofibrils   The primary component part of muscle fibers.

navel   The area where the umbilical cord was formerly attached to the body of the offspring.

necropsy   Perform a postmortem examination.

net energy   Metabolizable energy minus heat increments. The energy available to the animal for maintenance and production.

nicking   The way in which certain lines, strains, or breeds perform when mated together. When outstanding offspring result, the parents are said to have *nicked* well.

nipple   See *teat*.

nodular worm   An internal parasitic worm that causes the formation of nodules in the intestines.

nonruminant   Simple-stomached or monogastric animal.

NPN (nonprotein nitrogen)   Nitrogen in feeds from substances such as urea and amino acids, but not from preformed proteins.

nucleotide   The subunit of DNA composed of a five carbon sugar, a nitrogenous base, and a phosphate group.

nutrient   (1) A substance that nourishes the metabolic processes of the body. (2) The end product of digestion.

nutrient density   Amount of essential nutrients relative to the number of calories in a given amount of food.

obesity   An excessive accumulation of body fat.

offal   All organs and tissues removed from inside the animal during the slaughtering process.

omasum   One of the stomach components of ruminant animals that has many folds.

omnivorous   Feeding on both animal and vegetable substances.

on full feed   A term that refers to animals that are receiving all the feed they will consume. See also *ad libitum*.

oogenesis   The process by which eggs, or ova, are produced.

open   Refers to nonpregnant females.

open-faced   Face of sheep that is free from wool, particularly around the eyes.

opportunity costs   Returns given up if debt-free resources (e.g., land, livestock, equipment) are used in their next-best level of employment.

optimum level of performance   The level at which a trait or traits maximizes net profit. Resources are managed to achieve a combined balance of traits that sustains high levels of profitability.

osteopetrosis   Abnormal thickening, hardening, and fragility of bones, making them weaker.

osteoporosis   An abnormal decrease in bone mass with an increased fragility of the bones.

outbreeding   The process of continuously mating females of the herd to unrelated males of the same breed.

outcrossing   The mating of an individual to another in the same breed that is not related to it. Outcrossing is a specific type of outbreeding system.

ova   Plural of ovum, meaning *eggs*.

ovary   The female reproductive gland in which the eggs are formed and progesterone and estrogenic hormones are produced.

overeating disease   A toxic condition caused by the presence of undigested carbohydrates in the intestine, which stimulates harmful bacteria to multiply. When the bacteria die, they release toxins. Called *enterotoxemia* in some animals.

overshot jaw   Upper jaw is longer than lower jaw. Also called *parrot mouth*.

oviduct   A duct leading from the ovary to the horn of the uterus.

ovine   Refers to sheep.

ovulation   The shedding, or release, of the egg from the follicle of the ovary.

ovum   The egg produced by a female.

Owner-sampler record   Dairy record system similar to DHI except milk weights and samples are taken by the dairy producer instead of a DHIA supervisor.

pace   A lateral two-beat gain in which the right rear and front feet hit the ground at one time and the left rear and front feet strike the ground at another time.

paddling   The outward swinging of the front feet of a horse that toes in.

pale, soft, exudative (PSE)   A genetically predisposed condition in swine in which the pork is very light colored, soft, and watery.

palpation   Feeling by hand.

parasite An organism that lives a part of its life cycle in or on, and at the expense of, another organism. Parasites of farm animals live at the expense of the farm animals.

parity Number of different times a female has had offspring.

parrot mouth Upper jaw is longer than lower jaw.

parturient paresis Partial paralysis that occurs at or near time of giving birth to young and beginning lactation. The mother mobilizes large amounts of calcium to produce milk to feed newborn, and blood calcium levels drop below the point necessary for impulse transmission along the nerve tracks. Commonly called *milk fever.*

parturition The process of giving birth.

pasteurization The process of heating milk to 161°F and holding it at that temperature for 15 seconds to destroy pathogenic microorganisms.

pasture rotation The rotation of animals from one pasture to another so that some pasture areas have no livestock on them in certain periods.

pathogen Biologic agent (i.e., bacteria, virus, protozoa, nematode) that may produce disease or illness.

paunch Another name for *rumen.*

pay weight The actual weight for which payment is made. In many cases it is the shrunk weight (actual weight minus pencil shrink).

pedigree The record of the ancestry of an animal.

pelt The natural, whole skin covering, including the wool, hair, or fur (e.g., a sheep pelt has the wool left on).

pencil shrink An arithmetic deduction (percent of liveweight) from an animal's weight to account for fill.

pendulous Hanging loosely.

penis The male organ of copulation. It serves both as a channel for passage of urine from the bladder as an extension of the urethra, and as a copulatory organ through which sperm are deposited into the female reproductive tract.

per capita Per person.

performance test The evaluation of an animal according to it performance.

pernicious anemia A chronic type of mycrocitic anemia caused by a deficiency of vitamin B_{12} or a failure of intestinal absorption of vitamin B_{12}.

pharmaceuticals Medicinal products (drugs) used primarily to treat disease.

phenotype The characteristics of an animal that can be seen and/or measured (e.g., the presence or absence of horns, the color, or the weight of an animal).

pheromones Chemical substances that attract the opposite sex.

photoperiod Time period when light is present.

physiology The science that pertains to the functions of organs, organ systems, or the entire animal.

picking The removal of feathers in dressing poultry.

pigeon-toed See *toeing-in.*

pin bones   In cattle, the posterior ends of the pelvic bones that appear as two raised areas on either side of the tail head.

pink tooth   Congenital porphyria, teeth are pink gray and the animals tend to sunburn easily.

pinworms   A small nematode worm with unsegmented body found as a parasite in the rectum and large intestine of animals.

pituitary   Small endocrine gland located at the base of the brain.

placenta   The vascular organ that unites the fetus to the uterus.

pneumonia   Inflammation or infection of alveoli of the lungs caused by either bacteria or viruses.

poikilotherm   A cold-blooded animal; one whose body temperature varies with that of the environment.

polled   Naturally or genetically hornless.

poll evil   An abscess behind the ears of a horse.

Polypay   A synthetic breed of sheep developed in the United States by combining the Dorset, Targhee, Rambouillet, and Finnsheep breeds.

polytocous   Giving birth to several offspring at one time.

porcine stress syndrome   A genetic defect in swine inherited as a simple recessive. It is associated with heavily muscled animals that may suddenly die when exposed to stressful conditions. Their muscle is usually pale, soft, and exudative (PSE).

pork   The meat from swine.

posterior   Toward the rear end of an animal.

postgastric fermentation   The fermentation of feed that occurs in the cecum, behind the area where digestion has occurred.

postnatal   See *postpartum*.

postpartum   After birth.

postpartum interval   The length of time from parturition until the dam is pregnant again.

poult   A young turkey of either sex, from hatching to approximately 10 weeks of age.

poultry   This term includes chickens, turkeys, geese, pigeons, peafowls, guineas, and game birds.

predicted difference   Dairy bull record based on superiority or inferiority of the bull's daughters compared to their herd mates.

predicted transmitting ability (PTA)   Estimate of genetic transmitting ability (i.e., one-half of the breeding value) of dairy bulls. Estimated amount by which daughters of a bull will differ from the breed average.

pregastric fermentation   Fermentation that occurs in the rumen of ruminant animals. It occurs before feed passes into the portion of the digestive tract in which digestion actually occurs.

pregnancy disease A metabolic disease in late pregnancy affecting primarily ewes carrying twins or triplets. A form of ketosis. Also called *pregnancy toxemia*.

pregnancy testing Evaluation of females for pregnancy through palpation or using an ultrasound machine.

premix A uniform mixture of one or more micro-ingredients with diluent and/or carrier. Premixers are used to facilitate uniform dispersion of the micro-ingredients in a large mix.

prenatal Prior to being born; before birth.

primiparous Bearing or having borne but one offspring.

probe A device used to measure backfat thickness in pigs and cattle.

production testing An evaluation of an animal based on its production record.

progeny testing An evaluation of an animal on the basis of performance of its offspring.

progesterone A hormone produced by the corpus luteum that stimulates progestational proliferation in the uterus of the female.

prokaryote Cell or organism lacking a membrane-bound, structurally discrete nucleus and other subcellular compartments. Bacteria are prokaryotes.

prolapsed Turned inside out.

prostaglandins Chemical mediators that control many physiological and biochemical functions in the body. One prostaglandin ($PGF_2\alpha$) can be used to synchronize estrus.

prostate A gland of the male reproductive tract that is located just back of the bladder. It secretes a fluid that becomes part of semen at ejaculation.

protein A large molecule composed of one or more chains of amino acids in a specific order; the order is determined by the base sequence of nucleotides in the gene coding for the protein.

protein supplement Any dietary component containing a high concentration (at least 25%) of protein.

proximal Nearest. The position that is closest to the point of attachment for a limb or bone.

PSE See *pale, soft, and exudative*.

PSS See *porcine stress syndrome*.

PTA See *predicted transmitting ability*.

puberty The age at which the reproductive organs become functionally operative.

pullet Young female chicken from day of hatch through onset of egg production; sometimes the term is used through the first laying year.

purebred An animal eligible for registry with a recognized breed association.

qualitative trait A trait expressed categorically because of a sharp distinction between phenotypes (e.g., black and red). Usually only one or a few pairs of genes are involved in the expression of a qualitative trait.

quality grades Animals grouped according to value as Prime, Choice, etc., based on conformation and fatness of the animals.

quantitative trait A trait expressed on a continuous/numerical scale because of a gradual variation from one phenotype to another (e.g., weaning weight). Usually many gene pairs and environmental influences are involved in the expression of such traits.

rack (1) A rapid four-beat gait of a horse. (2) A wholesale cut of lamb located between the shoulder and loin.

ram A male sheep that is sexually mature.

ration The amount of total feed fed to an animal over a 24-hour period.

reach See *selection differential.*

realized heritability The portion obtained of what is reached for in selection.

realizer A feeder animal (usually cattle) that has serious health problems or injury. Economics dictate the animal be sold rather than continue the duration of the feeding program.

reasoning The ability of an animal to respond correctly to a stimulus the first time the animal encounters a new situation.

recessive gene A gene that has its phenotype masked by its dominant allele when the two genes are present together in an individual.

reciprocal recurrent selection The selection of breeding animals in two populations based on the performance of their offspring after animals from two populations are crossed.

recombinant DNA (rDNA) Isolated DNA molecules that can be inserted into the DNA of another cell. rDNA is used in the genetic-engineering process.

rectal prolapse Protrusion of part of large intestine through the anus.

recurrent selection Selection for general combining ability by selecting males that sire outstanding offspring when mated to females from varying genetic backgrounds.

red meat Meat from cattle, sheep, swine, and goats, as contrasted to the white meat of poultry.

registered Recorded in the herdbook of a breed.

regurgitate To cast up digested food to the mouth as is done by ruminants.

reinforcement A reward for making the proper response to a stimulus or condition.

replicate To duplicate, or make another exactly like, the original.

reproduction The production of live, normal offspring.

retained placenta Placenta remains within the reproductive tract after parturition has occurred.

reticulum One of the stomach components of ruminant animals that is lined with small compartments, giving a honeycomb appearance.

rhinitis Inflammation of the mucous membranes lining the nasal passages.

rhinopneumonitis Equine herpes virus-1. It produces acute catarrh upon primary infection.

ribonucleic acid (RNA)   An essential component of living cells, composed of long chains of phosphate, ribose sugar, and several bases.

rickets   A disease of disturbed ossification of the bones caused by a lack of vitamin D or unbalanced calcium/phosphorus ratio.

ridgling   Another term for cryptorchid.

ringbone   An ossification of the lateral cartilage of the foot of a horse all around the foot.

riparian   An area next to water (stream, river, or lake) where more vegetation grows (compared to a greater distance from the water source) because of the added moisture from the water. Grazing animals usually inhabit this area more frequently than others, thus increasing the possibility of overgrazing.

RNA   (ribonucleic acid) A chemical found in the nucleus and cytoplasm of cells. RNA plays an important role in protein synthesis and other chemical activities of the cell.

roman nose   A nose having a prominent bridge (e.g., a roman-nosed horse).

roughage   A feed that is high in fiber, low in digestible nutritents, and low in energy. Such feeds as hay, straw, silage, and pasture are examples.

rumen   The large fermentation pouch of the ruminant animal in which bacteria and protozoa break down fibrous plant material that is swallowed by the animal, sometimes referred to as the *paunch*.

ruminant   A mammal whose stomach has four parts (rumen, reticulum, omasum, and abomasum). Cattle, sheep, goats, deer, and elk are ruminants.

rumination   The regurgitation of undigested food and chewing it a second time, after which it is again swallowed.

sac-fry   Fish with an external yolk sac.

salmonella   Gram-positive, rod-shaped bacteria that cause various diseases such as food poisoning in animals.

scale   (1) Size. (2) Equipment on which an animal is weighed.

scoured wool   Wool that has been cleaned of grease and other foreign material.

scours   Diarrhea; a profuse watery discharge from the intestines.

screwworms   Larvae of several American flies that infest wounds of animals.

scrotal circumference   A measurement (usually cm or in.) of the circumference of both testicles and the scrotal sac that surrounds them.

scrotum   A pouch that contains the testes. It is also a thermoregulatory organ that contracts when cold and relaxes when warm, thus tending to keep the testes at a lower temperature than that of the body.

scurs   Small growths of hornlike tissue attached to the skin of polled or dehorned animals.

scurvy   A deficiency disease in humans that causes spongy gums and loose teeth. It is caused by a lack of vitamin C (ascorbic acid).

seedstock   Breeding animals; sometimes used interchangeably with *purebred*.

selection   Differentially reproducing what one wants in a herd or flock.

selection differential   The difference between the average for a trait in selected animals and the average of the group from which they come; also called *reach*.

selection index   Selection method in which several traits are evaluated and expressed as one total score.

semen   The fluid containing the sperm that is ejaculated by the male. Secretions from the seminal vesicles, the prostate gland, the bulbourethral glands, and the urethral glands provide most of the fluid.

seminal vesicles   Accessory sex glands of the male that provide a portion of the fluid of semen.

seminiferous tubules   Minute tubules in the testicles in which sperm are produced. They comprise about 90% of the mass of the testes.

service   To breed or mate.

settle   To become pregnant.

sex-limited   Existing in only one sex, such as milk production in dairy cattle.

shearing   The process of removing the fleece (wool) from a sheep.

sheath rot   Inflammation of the prepuce in male sheep.

sheep bot   Any of a number of related flies whose larvae are parasitic in sheep. They usually are found in the sinuses.

shipping fever   A widespread respiratory disease of cattle and sheep.

shoat   A young pig of either sex.

shoe boil   Blemish of the horse caused by the horseshoe putting pressure on the elbow when the horse lies down.

shrink   Loss of weight—commonly used in the loss in live weight when animals are marketed or loss in weight from grease wool to clean wool.

sib   A brother or sister.

sickle hocks   Hocks that have too much set, causing the hind feet to be too far forward and too far under the animal.

side bones   Ossification of the lateral cartilages of the foot of a horse.

sigmoid flexure   The S-curve in the penis of boars, rams, bucks, and bulls.

silage   Forage, corn fodder, or sorghum preserved by fermentation that produces acids similar to the acids that are used to make pickled foods for people.

sire   Male parent.

Sire Index   A dairy bull test record obtained by comparing a bull's daughters with their contemporary herd mates.

skins   Skins come from smaller animals such as pigs, sheep, goats, and wild animals. A beef hide weighing less than 30 lb is called a skin.

sleeping sickness   An infectious disease common in tropical Africa and transmitted by the bite of a tsetse fly.

slotted floor   Floor having any kind of openings through which excreta may fall.

SNF   See *solids-not-fat.*

snood   The relatively long, fleshy extension at the base of the turkey's beak.

software   Program instructions to make computer hardware function.

soilage   Green forage that is cut and brought to animals as food.

solids-non-fat   Total milk solids minus fat. It includes protein, lactose, and minerals.

somatotropin   The growth hormone from the anterior pituitary that stimulates nitrogen retention and growth.

soremouth   A virus-caused disease affecting primarily lambs.

sow   A female swine that has farrowed one litter or has reached 12 months of age.

spawn   Act of fish laying eggs.

spay   To remove the ovaries.

specific combining ability   The ability of a line or population to exhibit superiority or inferiority when combined with other lines or populations.

spermatid   The haploid germ cell prior to spermiogenesis.

spermatogenesis   The process by which spermatozoa are formed.

spermiogenesis   The process by which the spermatid loses most of its cytoplasm and develops a tail to become a mature sperm.

spider syndrome   A recessive genetic abnormality common to black-faced sheep. The front legs are usually bent out from the knees and the hind legs typically show some deformities.

spinning count   The number of hanks of yarn that can be spun from a pound of clean wool. One method of evaluating fineness of wool.

splayfooted   See *toeing-out.*

spool joint   The joint where the foot and pastern are removed from the front leg. Used to identify a mutton carcass.

spur   A sharp projection on the back of a male bird's shank.

stags   Castrated male sheep, cattle, goats, or swine that have reached sexual maturity prior to castration.

stallion   A sexually mature male horse.

staple length   Length of wool fibers.

steer   A castrated bovine male that was castrated early in life before puberty.

sterility   Inability to produce offspring.

steroid   Artificially produced drug similar to the natural hormone that controls inflammation and regulates water balance.

stifle   Joint of the hind leg between the femur and tibia.

stifled   Injury of the stifle joint.

stillborn   Offspring born dead.

stocker (cattle)   Weaned cattle that are fed high–roughage diets (including grazing) before going into the feedlot.

stocker (fish)   Usually 6–12 inches in length and less than .75 pounds.

stomach worms   *Haemonchus contortus,* or worms of the stomach of cattle, swine, sheep, and goats.

strangles   An infectious disease of horses, characterized by inflammation of the mucous membranes of the respiratory tract.

streptococcus   Sperical, gram-positive bacteria that divide in only one plane and occur in chains. Some species cause serious disease.

stress   An unusual or abnormal influence causing a change in an animal's function, structure, or behavior.

stringhalt   A sudden and extreme flexion of the back of a horse, producing a jerking motion of the hindleg in walking.

strongyles   Any of various roundworms living as parasites, especially in domestic animals.

stud   Usually the same as stallion. Also a place where male animals are maintained (i.e., bull stud).

suckling gain   The gain that a young animal makes from birth until it is weaned.

subcutaneous   Situated beneath, or occurring beneath, the skin. A subcutaneous injection is an injection made under the skin.

sulfonamides   A sulfa drug capable of killing bacteria.

superovulation   The hormonally induced ovulation of a greater than normal number of eggs.

supplement   A feed used with another to improve the nutritive balance of performance of the total and intended to be (1) fed undiluted as a supplement to other feeds, (2) offered free-choice with other parts of the ration separately available, or (3) further diluted and mixed to produce a complete feed.

sweeney   Atrophy of muscle (typically shoulder) in horses.

sweetbread   An edible by-product also known as the pancreas.

switch   The tuft of long hair at the end of tail (cattle and horses).

syndactyly   Union of two or more digits (e.g., in cattle, the two toes would be a solid hoof).

synthetic breeds   See *composite breed.*

tagging   Clipping wool from the dock, udder, and vulva regions of the ewe prior to breeding and lambing.

tags   (1) Wool covered with manure. (2) Abbreviated form of ear tags, used for identification.

tallow   The fat of cattle and sheep.

tandem selection   Selection for one trait for a given period of time followed by selection for a second trait and continuing in this way until all important traits are selected.

TDN   Total digestible nutrients; includes the total amounts of digestible protein, nitrogen-free extract, fiber, and fat (multiplied by 2.25), all summed together.

teaser ram   A ram made incapable of impregnating a ewe by vasectomy or by use of an apron to prevent copulation, which is used to find ewes in heat.

teasing   The stallion in the presence of the mare to see if she will mate.

teat   The protuberance of the udder through which milk is drawn.

tendon   Tough, fibrous connective tissue at ends of muscle bundles that attach muscle to bones or cartilage structures.

terminal sire   The sire used in a terminal crossbreeding program. It is intended that all offspring from a terminal sire be sold as market animals.

testicle   The male sex gland that produces sperm and testosterone.

testosterone   The male sex hormone that stimulates the accessory sex glands, causes the male sex drive, and causes the development of masculine characteristics.

tetanus   Rigid paralytic disease caused by *Clostridium tetani*, an anaerobic bacterium that lives in soil and feces.

tetrad   A group of four similar chromotids formed by the splitting longitudinally of a pair of homologous chromosomes during meiotic prophase.

thermoneutral zone (TNZ)   Range in temperature where rate and efficiency of gain is maximized. Comfort zone.

thoroughpin   A hard swelling that is located between the Achilles tendon and the bone of the hock joint.

thrush   Foot disease characterized by degeneration of the frog and a thick, foul-smelling discharge.

thyroid gland   Two-lobed endocrine gland in the neck that controls the rate at which basic body functions proceed.

toeing-in   Toes of front feet turn in. Also called *pigeon-toed*.

toeing-out   Toes of front feet turn out. Also called *splayfooted*.

tom   A male turkey.

TPI   Total prediction index used in dairy cattle breeding. It includes the predicted differences for milk production, fat percentage, and type into one figure in a ratio of milk production × 3:fat percentage × 1:type × 1.

transcription   The synthesis of RNA from DNA in the nucleus by matching the sequences of the bases.

transgenic animals   Animals that contain genes transferred from other animals, usually from a different species.

transmissible gastroenteritis (TGE)   A serious, contagious diarrhea disease in baby pigs.

tripe   Edible product from walls of ruminant stomach.

trot   A diagonal two-beat gait in which the right front and left rear feet strike the ground in unison, and the left front and right rear feet strike the ground in unison.

twist   Vertical measurement from top of the rump to point where hindlegs separate.

twitch   To squeeze tightly the upper lip of a horse by means of a small rope that is twisted.

type   (1) The physical conformation of an animal. (2) All those physical attributes that contribute to the value of an animal for a specific purpose.

udder   The encased groups travel from the fetus to and from the placenta, respectively. This cord is broken when the young are born.

undershot jaw   Lower jaw is longer than upper jaw.

unsoundness   Any defect or injury that interferes with the usefulness of an animal.

urinary calculi   Disease where mineral deposits crystallize in the urinary tract. The deposits may block the tract, causing difficulty in urination.

uterus   That portion of the female reproductive tract where the young develop during pregnancy.

vaccination   The act of administering a vaccine or antigens.

vaccine   Suspension of attenuated or killed microbes or toxins administered to induce active immunity.

vagina   The copulatory portion of the female's reproductive tract. The vestibule portion of the vagina also serves for passage of urine during urination. The vagina also serves as a canal through which young pass when born.

variety meats   Edible organ by-products (e.g., liver, heart, tongue, tripe).

vas deferens   Ducts that carry sperm from the epididymis to the urethra.

vasectomy   The removal of a portion of the vas deferens. As a result, sperm are prevented from traveling from the testicles to become part of the semen.

veal   The meat from very young cattle, under 3 months of age.

vein   Vessel through which blood passes from various organs or parts back to the heart.

vermifuge   A chemical substance given to the animals to kill internal parasitic worms.

VFA   See *volatile fatty acids.*

villi   Projections of the inner lining of the small intestine.

virus   Ultramicroscopic bundle of genetic material capable of multiplying only in living cells. Viruses cause a wide range of disease in plants, animals, and humans, such as rabies and measles.

viscera   Internal organs and glands contained in the thoracic and abdominal cavities.

vitamin   An organic catalyst, or component thereof, that facilitates specific and necessary functions.

volatile fatty acids (VFA)   A group of fatty acids produced from microbial action in the rumen; examples are acetic, propionic, and butyric acids.

vulva   The external genitalia of a female mammal.

walk   A four-beat gait of a horse in which each foot strikes the ground at a time different from each of the other three feet.

warble The larval stage of the heel fly that burrows out through the hide of cattle in springtime.

wattle Method of identification in cattle where strips of skin (3–6 inches) long are usually cut on the nose, jaw, throat, or brisket.

weaner An animal that has been weaned or is nearing weaning age.

weaning Separating young animals from their dams so that the offspring can no longer suckle.

weaning weight EPD A genetic estimate of the weaning weight of a beef bull's calves when compared to other bulls in the sire summary.

weanling An animal of weaning age.

wet Used to describe a milking female (e.g., wet cow or wet ewe).

wether A male sheep castrated before reaching puberty.

white cells (leukocytes, white blood cells) Colorless blood cells active in the body's defense against infection or other assult. There are five types: neutrophils, lymphocytes, eosinoiphils, monocytes and basophils.

white muscle disease A muscular disease caused by a deficiency of selenium or vitamin E.

winking Indication of estrus in the mare where the vulva opens and closes.

withdrawal time The time before slaughter that a drug should not be given to an animal.

withers Top of the shoulders.

wool The fibers that grow from the skin of sheep.

wool blindness Sheep cannot see, owing to wool covering their eyes.

wool top A continuous untwisted strand of combed wool in which the fibers lie parallel and the short fibers have been combed out.

woolens Cloth made from short wool fibers that are intermingled in the making of the cloth by carding.

worsteds Cloth made from wool that is long enough to comb and spin into yarn. The finish of worsteds is harder than woolens, and worsted clothes hold a press better.

yearling Animals that are approximately 1 year old.

yearling weight Expected Progeny Difference (EPD) A breeding value that measures genetic differences in yearling weight in beef cattle.

yield Used interchangeably with *dressing percentage.*

yield grades The grouping of animals according to the estimated trimmed lean meat that their carcass would provide; cutability.

yolk (1) The yellow part of the egg. (2) The natural grease (lanolin) of wool.

yolk sac Layer of tissue encompassing the yolk of an egg.

zone of thermoneutrality The environmental temperature (about 65°F) at which heat production and heat elimination are approximately equal for most farm animals.

zygote (1) The cell formed by the union of two gametes. (2) An individual from the time of fertilization until death.

Index

Table A Weights and Measures

Conversion Factors

Length

12 inches	= 1 foot (ft)
3 feet	= 1 yard (yd)
16 1/2 ft	= 1 rod
40 rods	= 1 furlong
220 yds	= 1 furlong
5,280 ft	= 1 mile
1,760 yds	= 1 mile
8 furlongs	= 1 mile
1 hand (for measuring horses)	= 4 inches

Area

144 sq in.	= 1 sq ft
9 sq ft	= 1 sq yd
43,560 sq ft	= 1 acre
4,840 sq yds	= 1 sq mile
160 sq rods	= 1 acre
640 acres	= 1 sq mile

Dry Measure

2 pints	= 1 quart
2 pints	= 67.2 cu in.
8 quarts	= 1 peck
4 pecks	= 1 bushel

Liquid Measure

60 drops	= 1 teaspoon or 1/6 oz
3 teaspoons	= 1 tablespoon
16 tablespoons	= 1 cup
2 cups	= 1 pint or 16 fl oz
2 pints	= 1 quart
4 quarts	= 1 gallon
7.48 gal	= 1 cu ft
1 gal (milk)	= 8.6 lb
1 gal (water)	= 8.32 lb
1 acre ft	= 325,900 gal

Length

1 millimeter (mm)	= 0.04 in.
1 centimeter (cm)	= 0.39 in.
1 meter (m)	= 3.28 ft
1 kilometer (km)	= 0.62 mile
1 inch	= 25.40 cm
1 inch	= 2.54 cm
1 foot	= 0.30 m
1 yard	= 0.90 m
1 mile	= 0.60 km

Area

1 sq cm (cm^2)	= 0.16 sq in.
1 sq meter (m^2)	= 10.76 sq ft
1 hectare	= 2.47 acres
1 sq in.	= 6.50 cm^2
1 sq ft	= 0.09 m^2
1 acre	= 0.40 hectare

Mass

1 gram (g)	= 0.035 oz
1 gram (g)	= 0.0022 lb
1 kilogram (kg)	= 2.205 lb
1 metric ton	= 1.102 ton
1 oz (dry)	= 28.30 g
1 pound	= 453.60 g
1 pound	= 0.45 kg
1 ton	= 907.20 kg

Volume

1 teaspoon	= 5 milliliters (ml)
1 fl oz	= 29.60 ml
1 qt (liq)	= 0.95 liter (l)
1 gallon	= 3.78 liters
1 ml	= 0.2 teaspoon
1 liter	= 0.91 qt (dry)
1 liter	= 1.06 qt (liquid)
1 liter	= 0.264 gal